T0201551

FOUNDATIONS OF APPLIED ELECTRODYNAMICS

FOUNDATIONS OF APPLIED ELECTRODYNAMICS

Wen Geyi

Waterloo, Canada

A John Wiley and Sons, Ltd., Publication

This edition first published 2010
© 2010 John Wiley & Sons, Ltd

Registered office
John Wiley & Sons Ltd, The Atrium, Southern Gate, Chichester, West Sussex, PO19 8SQ, United Kingdom

For details of our global editorial offices, for customer services and for information about how to apply for permission to reuse the copyright material in this book please see our website at www.wiley.com.

The right of the author to be identified as the author of this work has been asserted in accordance with the Copyright, Designs and Patents Act 1988.

Library of Congress Cataloging-in-Publication Data

Wen, Geyi.
  Foundations of applied electrodynamics / Geyi Wen.
     p. cm.
  Includes bibliographical references and index.
  ISBN 978-0-470-68862-5 (cloth)
1. Electrodynamics–Mathematics.   I. Title.
  QC631.3.W46   2010
  537.601′51–dc22

A catalogue record for this book is available from the British Library.

ISBN 978-0-470-68862-5 (H/B)

Typeset in 10/12pt Times by Aptara Inc., New Delhi, India
Printed and Bound in Singapore by Markono Print Media Pte Ltd

To my parents
To Jun and Lan

Contents

Preface

Electrodynamics is an important course in both physics and electrical engineering curricula. The graduate students majoring in applied electromagnetics are often confronted with a large number of new concepts and mathematical techniques found in a number of courses, such as *Advanced Electromagnetic Theory*, *Field Theory of Guided Waves*, *Advanced Antenna Theory*, *Electromagnetic Wave Propagation*, *Network Theory and Microwave Circuits*, *Computational Electromagnetics*, *Relativistic Electronics*, and *Quantum Electrodynamics*. Frequently, students have to consult a large variety of books and journals in order to understand and digest the materials in these courses, and this turns out to be a time-consuming process. For this reason, it would be helpful for the students to have a book that gathers the essential parts of these courses together and treats them according to the similarity of mathematical techniques.

Engineers, applied mathematicians and physicists who have been doing research for many years often find it necessary to renew their knowledge and want a book that contains the fundamental results of these courses with a fresh and advanced approach. With this goal in mind, inevitably this is beyond the conventional treatment in these courses. For example, the completeness of eigenfunctions is a key result in mathematical physics but is often mentioned without rigorous proof in most books due to the involvement of generalized function theory. As a result, many engineers lack confidence in applying the theory of eigenfunction expansions to solve practical problems. In order to fully understand the theory of eigenfunction expansions, it is imperative to go beyond the classical solutions of partial differential equations and introduce the concept of generalized solutions.

The contents of this book have been selected according to the above considerations, and many topics are approached in contemporary ways. The book intends to provide a whole picture of the fundamental theory of electrodynamics in most active areas of engineering applications. It is self-contained and is adapted to the needs of graduate students, engineers, applied physicists and mathematicians, and is aimed at those readers who wish to acquire more advanced analytical techniques in studying applied electrodynamics. It is hoped that the book will be a useful tool for readers saving them time and effort consulting a wide range of books and technical journals. After reading this book, the readers should be able to pursue further studies in applied electrodynamics without too much difficulty.

The book consists of ten chapters and four appendices. Chapter 1 begins with experimental laws and reviews Maxwell equations, constitutive relations, as well as the important properties derived from them. In addition, the basic electromagnetic theorems are summarized. Since most practical electromagnetic signals can be approximated by a temporal or a spatial wavepacket, the theory of wavepackets and various propagation velocities of wavepackets are also examined.

In applications, the solution of a partial differential equation is usually understood to be a classical solution that satisfies the smooth condition required by the highest derivative in the equation. This requirement may be too stringent in some situations. A rectangular pulse is not smooth in the classical sense yet it is widely used in digital communication systems. The first derivative of the Green's function of a wave equation is not continuous, but is broadly accepted by physicists and engineers. Chapter 2 studies the solutions of Maxwell equations. Three main analytical methods for solving partial differential equations are discussed: (1) the separation of variables; (2) the Green's function; and (3) the variational method. In order to be free of the constraint of classical solutions, the theory of generalized solutions of differential equation is introduced. The Lagrangian and Hamiltonian formulations of Maxwell equations are the foundations of quantization of electromagnetic fields, and they are studied through the use of the generalized calculus of variations. The integral representations of the solutions of Maxwell equations and potential theory are also included.

Eigenvalue problems frequently appear in physics, and have their roots in the method of separation of variables. An eigenmode of a system is a possible state when the system is free of excitation, and the corresponding eigenvalue often represents an important quantity of the system, such as the total energy and the natural oscillation frequency. The theory of eigenvalue problems is of fundamental importance in physics. One of the important tasks in studying the eigenvalue problems is to prove the completeness of the eigenmodes, in terms of which an arbitrary state of the system can be expressed as a linear combination of the eigenmodes. To rigorously investigate the completeness of the eigenmodes, one has to use the concept of generalized solutions of partial differential equations. Chapter 3 discusses the eigenvalue problems from a unified perspective. The theory of symmetric operators is introduced and is then used to study the interior eigenvalue problems in electromagnetic theory, which involves metal waveguides and cavity resonators. This chapter also treats the mode theory of spherical waveguides and the method of singular function expansion for scattering problems, which are useful in solving exterior boundary value problems.

An antenna is a device for radiating or receiving radio waves. It is an overpass connecting a feeding line in a wireless system to free space. The antenna is characterized by a number of parameters such as gain, bandwidth, and radiation pattern. The free space may be viewed as a spherical waveguide, and the spherical wave modes excited by the antenna depend on the antenna size. The bigger the antenna size, the more the propagating modes are excited. For a small antenna, most spherical modes turn out to be evanescent, making the stored energy around the antenna very large and the gain of the antenna very low. For this reason, most of the antenna parameters are subject to certain limitations. From time to time, there arises a question of how to achieve better antenna performance than previously obtained. Chapter 4 attempts to answer this question and deals with the fundamentals of radiation theory. The most important antenna parameters are reviewed and summarized. A complete theory of spherical vector wave functions is introduced, and is then used to study the upper bounds of the product of gain and bandwidth for an arbitrary antenna. In this chapter, the Foster reactance theorem for an ideal antenna without Ohmic loss, and the relationship between antenna bandwidth and antenna quality factor are investigated. In addition, the methods for evaluating antenna quality factor are also developed.

Electromagnetic boundary value problems can be characterized either by a differential equation or an integral equation. The integral equation is most appropriate for radiation and scattering problems, where the radiation condition at infinity is automatically incorporated in the formulation. The integral equation formulation has certain unique features that a differential equation formulation does not have. For example, the smooth requirement for the solution of integral equation is weaker than the corresponding differential equation. Another feature is that the discretization error of the integral equation is limited on the boundary of the solution region, which leads to more accurate numerical results. Chapter 5 summarizes integral equations for various electromagnetic field problems encountered in microwave and antenna engineering, including waveguides, metal cavities, radiation, and scattering problems by conducting and dielectric objects. The spurious solutions of integral equations are examined. Numerical methods generally applicable to both differential equations and integral equations are introduced.

Field theory and circuit theory are complementary to each other in electromagnetic engineering, and the former is the theoretical foundation of the latter while the latter is much easier to master. The circuit formulation has removed unnecessary details in the field problem and has preserved most useful overall information, such as the terminal voltages and currents. Chapter 6 studies the network representation of electromagnetic field systems and shows how the network parameters of multi-port microwave systems can be calculated by the field theory through the use of reciprocity theorem, which provides a deterministic approach to wireless channel modeling. Also discussed in this chapter is the optimization of power transfer between antennas, a foundation for wireless power transfer.

The wave propagation in an inhomogeneous medium is a very complicated process, and it is characterized by a partial differential equation with variable coefficients. The inhomogeneous waveguides are widely used in microwave engineering. If the waveguides are bounded by a perfect conductor, only a number of discrete modes called guided modes can exist in the waveguides. If the waveguides are open, an additional continuum of radiating modes will appear. In order to obtain a complete picture of the modes in the inhomogeneous waveguides, one has to master a sophisticated tool called spectral analysis in operator theory. Chapter 7 investigates the wave propagation problems in inhomogeneous media and contains an introduction to spectral analysis. It covers the propagation of plane waves in inhomogeneous media, inhomogeneous metal waveguides, optical fibers and inhomogeneous metal cavity resonators.

Time-domain analysis has become a vital research area in recent years due to the rapid progress made in ultra-wideband technology. The traditional time-harmonic field theory is based on an assumption that a monotonic electromagnetic source turns on at $t = -\infty$ so that the initial conditions or causality are ignored. This assumption does not cause any problems if the system has dissipation or radiation loss. When the system is lossless, the assumption may lead to physically unacceptable solutions. In this case, one must resort to time-domain analysis. Chapter 8 discusses the time-domain theory of electromagnetic fields, including the transient fields in waveguides and cavity resonators, spherical wave expansion in time domain, and time-domain theory for radiation and scattering.

Modern physics has its origins deeply rooted in electrodynamics. A cornerstone of modern physics is relativity, which is composed of both special relativity and general relativity. The special theory of relativity studies the physical phenomena perceived by different observers traveling at a constant speed relative to each other, and it is a theory about the structure of space–time. The general theory studies the phenomena perceived by different observers traveling at an arbitrary relative speed and is a theory of gravitation. The relativity, especially the special relativity, is usually considered as an integral part of electrodynamics. Relativity has many practical applications. For example, in the design of the global positioning system

(GPS), the relativistic effects predicted by the special and general theories of relativity must be taken into account to enhance the positioning precision. Chapter 9 deals with both special relativity and general relativity. The tensor algebra and tensor analysis on manifolds are used throughout the chapter.

Another cornerstone of modern physics is quantum mechanics. Quantum electrodynamics is a quantum field theory of electromagnetics, which describes the interaction between light and matter or between two charged particles through the exchange of photons. It is remarkable for its extremely accurate predictions of some physical quantities. Quantum electrodynamics is especially needed in today's research and education activities in order to understand the interactions of new electromagnetic materials with the fields. Chapter 10 provides a short introduction to quantum electrodynamics and a review of the fundamental concepts of quantum mechanics. The interactions of fields with charged particles are investigated by use of the perturbation method, in terms of which the dielectric constant for atom media is derived. Furthermore, the Klein–Gordon equation and the Dirac equation in relativistic mechanics are briefly discussed.

The book features a wide coverage of the fundamental topics in applied electrodynamics, including microwave theory, antenna theory, wave propagation, relativistic and quantum electrodynamics, as well as the advanced mathematical techniques that often appear in the study of theoretical electrodynamics. For the convenience of readers, four appendices are also included to present the fundamentals of set theory, vector analysis, special functions, and the SI unit system. The prerequisite for reading the book is advanced calculus. The SI units are used throughout the book. A $e^{j\omega t}$ time variation is assumed for time-harmonic fields. A special symbol □ is used to indicate the end of an example or a remark.

During the writing and preparation of this book, the author had the pleasure of discussing the book with many colleagues and cannot list them all here. In particular, the author would like to thank Prof. Robert E. Collin of Case Western Reserve University for his comments and input on many topics discussed in the book, and Prof. Thomas T. Y. Wong of Illinois Institute of Technology for his useful suggestions on the selection of the contents of the book.

Finally, the author is grateful to his family. Without their constant support, the author could not have made this book a reality.

Wen Geyi
Waterloo, Ontario, Canada

1

Maxwell Equations

Ten thousand years from now, there can be little doubt that the most significant event of the 19th century will be judged as Maxwell's discovery of the laws of electrodynamics.
—Richard Feynman (American physicist, 1918–1988)

To master the theory of electromagnetics, we must first understand its history, and find out how the notions of electric charge and field arose and how electromagnetics is related to other branches of physical science. Electricity and magnetism were considered to be two separate branches in the physical sciences until Oersted, Ampère and Faraday established a connection between the two subjects. In 1820, Hans Christian Oersted (1777–1851), a Danish professor of physics at the University of Copenhagen, found that a wire carrying an electric current would change the direction of a nearby compass needle and thus disclosed that electricity can generate a magnetic field. Later the French physicist André Marie Ampère (1775–1836) extended Oersted's work to two parallel current-carrying wires and found that the interaction between the two wires obeys an inverse square law. These experimental results were then formulated by Ampère into a mathematical expression, which is now called Ampère's law. In 1831, the English scientist Michael Faraday (1791–1867) began a series of experiments and discovered that magnetism can also produce electricity, that is, electromagnetic induction. He developed the concept of a magnetic field and was the first to use lines of force to represent a magnetic field. Faraday's experimental results were then extended and reformulated by James Clerk Maxwell (1831–1879), a Scottish mathematician and physicist. Between 1856 and 1873, Maxwell published a series of important papers, such as 'On Faraday's line of force' (1856), 'On physical lines of force' (1861), and 'On a dynamical theory of the electromagnetic field' (1865). In 1873, Maxwell published 'A Treatise on Electricity and Magnetism' on a unified theory of electricity and magnetism and a new formulation of electromagnetic equations since known as Maxwell equations. This is one of the great achievements of nineteenth-century physics. Maxwell predicted the existence of electromagnetic waves traveling at the speed of light and he also proposed that light is an electromagnetic phenomenon. In 1888, the German physicist Heinrich Rudolph Hertz (1857–1894) proved that an electric signal can travel through the air and confirmed the existence of electromagnetic waves, as Maxwell had predicted.

Foundations of Applied Electrodynamics   Geyi Wen
© 2010 John Wiley & Sons, Ltd

Maxwell's theory is the foundation for many future developments in physics, such as special relativity and general relativity. Today the words 'electromagnetism', 'electromagnetics' and 'electrodynamics' are synonyms and all represent the merging of electricity and magnetism. Electromagnetic theory has greatly developed to reach its present state through the work of many scientists, engineers and mathematicians. This is due to the close interplay of physical concepts, mathematical analysis, experimental investigations and engineering applications. Electromagnetic field theory is now an important branch of physics, and has expanded into many other fields of science and technology.

1.1   Experimental Laws

It is known that nature has four fundamental forces: (1) the strong force, which holds a nucleus together against the enormous forces of repulsion of the protons, and does not obey the inverse square law and has a very short range; (2) the weak force, which changes one flavor of quark into another and regulates radioactivity; (3) gravity, the weakest of the four fundamental forces, which exists between any two masses and obeys the inverse square law and is always attractive; and (4) electromagnetic force, which is the force between two charges. Most of the forces in our daily lives, such as tension forces, friction and pressure forces are of electromagnetic origin.

1.1.1   Coulomb's Law

Charge is a basic property of matter. Experiments indicate that certain objects exert repulsive or attractive forces on each other that are not proportional to the mass, therefore are not gravitational. The source of these forces is defined as the charge of the objects. There are two kinds of charges, called positive and negative charge respectively. Charges are quantitized and come in integer multiples of an **elementary charge**, which is defined as the magnitude of the charge on the electron or proton. An arrangement of one or more charges in space forms a charge distribution. The **volume charge density**, the **surface charge density** and the **line charge density** describe the amount of charge per unit volume, per unit area and per unit length respectively. A net motion of electric charge constitutes an electric current. An electric current may consist of only one sign of charge in motion or it may contain both positive and negative charge. In the latter case, the current is defined as the net charge motion, the algebraic sum of the currents associated with both kinds of charges.

In the late 1700s, the French physicist Charles-Augustin de Coulomb (1736–1806) discovered that the force between two charges acts along the line joining them, with a magnitude proportional to the product of the charges and inversely proportional to the square of the distance between them. Mathematically the force $\mathbf{F}$ that the charge q_1 exerts on q_2 in vacuum is given by **Coulomb's law**

$$\mathbf{F} = \frac{q_1 q_2}{4\pi \varepsilon_0 R^2} \mathbf{u}_R \tag{1.1}$$

where $R = |\mathbf{r} - \mathbf{r}'|$ is the distance between the two charges with $\mathbf{r}'$ and $\mathbf{r}$ being the position vectors of q_1 and q_2 respectively; $\mathbf{u}_R = (\mathbf{r} - \mathbf{r}')/|\mathbf{r} - \mathbf{r}'|$ is the unit vector pointing from q_1 to q_2, and $\varepsilon_0 = 8.85 \times 10^{-12}$ is the permittivity of the medium in vacuum. In order that the distance between the two charges can be clearly defined, strictly speaking, Coulomb's law applies only to the point charges, the charged objects of zero size. Dividing (1.1) by q_2 gives a force exerting on a unit charge, which is defined as the **electric field intensity E** produced by the charge q_1. Thus the electric field produced by an arbitrary charge q is

$$\mathbf{E}(\mathbf{r}) = \frac{q}{4\pi\varepsilon_0 R^2}\mathbf{u}_R = -\nabla\phi(\mathbf{r}) \tag{1.2}$$

where $\phi(\mathbf{r}) = q/4\pi\varepsilon_0 R$ is called the **Coulomb potential**. Here $R = |\mathbf{r} - \mathbf{r}'|$, $\mathbf{r}'$ is the position vector of the point charge q and $\mathbf{r}$ is the observation point. For a continuous charge distribution in a finite volume V with charge density $\rho(\mathbf{r})$, the electric field produced by the charge distribution is obtained by superposition

$$\mathbf{E}(\mathbf{r}) = \int_V \frac{\rho(\mathbf{r}')}{4\pi\varepsilon_0 R^2}\mathbf{u}_R dV(\mathbf{r}') = -\nabla\phi(\mathbf{r}) \tag{1.3}$$

where

$$\phi(\mathbf{r}) = \int_V \frac{\rho(\mathbf{r}')}{4\pi\varepsilon_0 R} dV(\mathbf{r}')$$

is the potential. Taking the divergence of (1.3) and making use of $\nabla^2(1/R) = -4\pi\delta(R)$ leads to

$$\nabla \cdot \mathbf{E}(\mathbf{r}) = \frac{\rho(\mathbf{r})}{\varepsilon_0}. \tag{1.4}$$

This is called **Gauss's law**, named after the German scientist Johann Carl Friedrich Gauss (1777–1855). Taking the rotation of (1.3) gives

$$\nabla \times \mathbf{E}(\mathbf{r}) = 0. \tag{1.5}$$

The above results are valid in a vacuum. Consider a dielectric placed in an external electric field. If the dielectric is ideal, there are no free charges inside the dielectric but it does contain bound charges which are caused by slight displacements of the positive and negative charges of the dielectric's atoms or molecules induced by the external electric field. These slight displacements are very small compared to atomic dimensions and form small electric dipoles. The **electric dipole moment** of an induced dipole is defined by $\mathbf{p} = ql\mathbf{u}_l$, where l is the separation of the two charges and $\mathbf{u}_l$ is the unit vector directed from the negative charge to the positive charge (Figure 1.1).

Example 1.1: Consider the dipole shown in Figure 1.1. The distances from the charges to a field point P are denoted by R_+ and R_- respectively, and the distance from the center of the

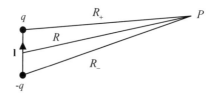

Figure 1.1   Induced dipole

dipole to the field point P is denoted by R. The potential at P is

$$\phi = \frac{q}{4\pi \varepsilon_0} \left(\frac{1}{R_+} - \frac{1}{R_-} \right).$$

If $l \ll R$, we have

$$\frac{1}{R_+} = \frac{1}{\sqrt{(l/2)^2 + R^2 - lR\mathbf{u}_l \cdot \mathbf{u}_R}} \approx \frac{1}{R}\left(1 + \frac{1}{2}\frac{l}{R}\mathbf{u}_l \cdot \mathbf{u}_R\right),$$

$$\frac{1}{R_-} = \frac{1}{\sqrt{(l/2)^2 + R^2 + lR\mathbf{u}_l \cdot \mathbf{u}_R}} \approx \frac{1}{R}\left(1 - \frac{1}{2}\frac{l}{R}\mathbf{u}_l \cdot \mathbf{u}_R\right),$$

where $\mathbf{u}_R$ is the unit vector directed from the center of the dipole to the field point P. Thus the potential can be written as

$$\phi \approx \frac{1}{4\pi \varepsilon_0 R^2}\mathbf{p} \cdot \mathbf{u}_R. \tag{1.6}$$

$\square$

The dielectric is said to be polarized when the induced dipoles occur inside the dielectric. To describe the macroscopic effect of the induced dipoles, we define the **polarization vector P** as

$$\mathbf{P} = \lim_{\Delta V \to 0} \frac{1}{\Delta V} \sum_i \mathbf{p}_i \tag{1.7}$$

where ΔV is a small volume and $\sum_i \mathbf{p}_i$ denotes the vector sum of all dipole moments induced inside ΔV. The polarization vector is the volume density of the induced dipole moments. The dipole moment of an infinitesimal volume dV is given by $\mathbf{P}dV$, which produces the potential (see (1.6))

$$d\phi \approx \frac{dV}{4\pi \varepsilon_0 R^2}\mathbf{P} \cdot \mathbf{u}_R.$$

The total potential due to a polarized dielectric in a region V bounded by S may be expressed as

$$\phi(\mathbf{r}) \approx \int_V \frac{\mathbf{P} \cdot \mathbf{u}_R}{4\pi \varepsilon_0 R^2} dV(\mathbf{r}') = \frac{1}{4\pi \varepsilon_0} \int_V \mathbf{P} \cdot \nabla' \frac{1}{R} dV(\mathbf{r}')$$

$$= \frac{1}{4\pi \varepsilon_0} \int_V \nabla' \cdot \left(\frac{\mathbf{P}}{R}\right) dV(\mathbf{r}') + \frac{1}{4\pi \varepsilon_0} \int_V \frac{-\nabla' \cdot \mathbf{P}}{R} dV(\mathbf{r}') \qquad (1.8)$$

$$= \frac{1}{4\pi \varepsilon_0} \int_S \frac{\mathbf{P} \cdot \mathbf{u}_n(\mathbf{r}')}{R} dV(\mathbf{r}') + \frac{1}{4\pi \varepsilon_0} \int_V \frac{-\nabla' \cdot \mathbf{P}}{R} dV(\mathbf{r}')$$

where the divergence theorem has been used. In the above, $\mathbf{u}_n$ is the outward unit normal to the surface. The first term of (1.8) can be considered as the potential produced by a surface charge density $\rho_{ps} = \mathbf{P} \cdot \mathbf{u}_n$, and the second term as the potential produced by a volume charge density $\rho_p = -\nabla \cdot \mathbf{P}$. Both ρ_{ps} and ρ_p are the bound charge densities. The total electric field inside the dielectric is the sum of the fields produced by the free charges and bound charges. Gauss's law (1.4) must be modified to incorporate the effect of dielectric as follows

$$\nabla \cdot \varepsilon_0 \mathbf{E} = \rho + \rho_p.$$

This can be written as

$$\nabla \cdot \mathbf{D} = \rho \qquad (1.9)$$

where $\mathbf{D} = \varepsilon_0 \mathbf{E} + \mathbf{P}$ is defined as the **electric induction intensity**. When the dielectric is linear and isotropic, the polarization vector is proportional to the electric field intensity so that $\mathbf{P} = \varepsilon_0 \chi_e \mathbf{E}$, where χ_e is a dimensionless number, called **electric susceptibility**. In this case we have

$$\mathbf{D} = \varepsilon_0 (1 + \chi_e) \mathbf{E} = \varepsilon_r \varepsilon_0 \mathbf{E} = \varepsilon \mathbf{E}$$

where $\varepsilon_r = 1 + \chi_e = \varepsilon/\varepsilon_0$ is a dimensionless number, called relative permittivity. Note that (1.5) holds in the dielectric.

1.1.2 Ampère's Law

There is no evidence that magnetic charges or magnetic monopoles exist. The source of the magnetic field is the moving charge or current. **Ampère's law** asserts that the force that a current element $\mathbf{J}_2 dV_2$ exerts on a current element $\mathbf{J}_1 dV_1$ in vacuum is

$$d\mathbf{F}_1 = \frac{\mu_0}{4\pi} \frac{\mathbf{J}_1 dV_1 \times (\mathbf{J}_2 dV_2 \times \mathbf{u}_R)}{R^2} \qquad (1.10)$$

where R is the distance between the two current elements, $\mathbf{u}_R$ is the unit vector pointing from current element $\mathbf{J}_2 dV_2$ to current element $\mathbf{J}_1 dV_1$, and $\mu_0 = 4\pi \times 10^{-7}$ is the permeability in vacuum. Equation (1.10) can be written as

$$d\mathbf{F}_1 = \mathbf{J}_1 dV_1 \times d\mathbf{B}$$

where $d\mathbf{B}$ is defined as the **magnetic induction intensity** produced by the current element $\mathbf{J}_2 dV_2$

$$d\mathbf{B} = \frac{\mu_0}{4\pi} \frac{\mathbf{J}_2 dV_2 \times \mathbf{u}_R}{R^2}.$$

By superposition, the magnetic induction intensity generated by an arbitrary current distribution $\mathbf{J}$ is

$$\mathbf{B}(\mathbf{r}) = \frac{\mu_0}{4\pi} \int_V \frac{\mathbf{J}(\mathbf{r}') \times \mathbf{u}_R}{R^2} dV(\mathbf{r}'). \tag{1.11}$$

This is called the **Biot-Savart law**, named after the French physicists Jean-Baptiste Biot (1774–1862) and Félix Savart (1791–1841). Equation (1.11) may be written as

$$\mathbf{B} = \nabla \times \mathbf{A}$$

where $\mathbf{A}$ is known as the **vector potential** defined by

$$\mathbf{A}(\mathbf{r}) = \frac{\mu_0}{4\pi} \int_V \frac{\mathbf{J}(\mathbf{r}')}{R} dV(\mathbf{r}').$$

Thus

$$\nabla \cdot \mathbf{B} = 0. \tag{1.12}$$

This is called Gauss's law for magnetism, which says that the magnetic flux through any closed surface S is zero

$$\int_S \mathbf{B} \cdot \mathbf{u}_n dS = 0.$$

Taking the rotation of magnetic induction intensity and using $\nabla^2(1/R) = -4\pi\delta(R)$ and $\nabla \cdot \mathbf{J} = 0$ yields

$$\nabla \times \mathbf{B} = \mu_0 \mathbf{J}(\mathbf{r}). \tag{1.13}$$

This is the differential form of Ampère's law.

Example 1.2: Consider a small circular loop of radius a that carries current I. The center of the loop is chosen as the origin of the spherical coordinate system as shown in Figure 1.2. The

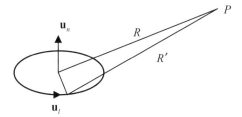

Figure 1.2   Small circular loop

vector potential is given by

$$\mathbf{A}(\mathbf{r}) = \frac{\mu_0 I}{4\pi} \int_l \frac{1}{R'} \mathbf{u}_l dl(\mathbf{r}')$$

where $\mathbf{u}_l$ is the unit vector along current flow and l stands for the loop. Due to the symmetry, the vector potential is independent of the angle φ of the field point P. Making use of the following identity

$$\int_l \phi \mathbf{u}_l dl = \int_S \mathbf{u}_n \times \nabla\phi dS$$

where S is the area bounded by the loop l, the vector potential can be written as

$$\mathbf{A}(\mathbf{r}) = \frac{\mu_0 I}{4\pi} \int_S \mathbf{u}_n \times \nabla' \frac{1}{R'} dS(\mathbf{r}')$$

$$= -\frac{\mu_0 I}{4\pi} \int_S \mathbf{u}_n \times \nabla \frac{1}{R'} dS(\mathbf{r}') = \frac{\mu_0 I}{4\pi} \nabla \times \int_S \mathbf{u}_n \frac{1}{R'} dS(\mathbf{r}').$$

If the loop is very small, we can let $R' \approx R$. Thus

$$\mathbf{A}(\mathbf{r}) = \frac{\mu_0 I}{4\pi} \nabla \times \int_S \mathbf{u}_n \frac{1}{R'} dS(\mathbf{r}')$$

$$\approx \frac{\mu_0}{4\pi} \nabla \times \frac{\mathbf{m}}{R} = \frac{\mu_0}{4\pi R^2} \mathbf{m} \times \mathbf{u}_R$$

$$(1.14)$$

where $\mathbf{u}_R$ is the unit vector from the center of the loop to the field point P and

$$\mathbf{m} = I \int_S \mathbf{u}_n(\mathbf{r}') dS(\mathbf{r}') = I \mathbf{u}_n \pi a^2$$

is defined as the **magnetic dipole moment** of the loop.   □

The above results are valid in a vacuum. All materials consist of atoms. An orbiting electron around the nucleus of an atom is equivalent to a tiny current loop or a magnetic dipole. In the absence of external magnetic field, these tiny magnetic dipoles have random orientations for most materials so that the atoms show no net magnetic moment. The application of an external magnetic field causes all these tiny current loops to be aligned with the applied magnetic field, and the material is said to be magnetized and the **magnetization current** occurs. To describe the macroscopic effect of magnetization, we define a **magnetization vector M** as

$$\mathbf{M} = \lim_{\Delta V \to 0} \frac{1}{\Delta V} \sum_i \mathbf{m}_i \tag{1.15}$$

where ΔV is a small volume and $\sum_i \mathbf{m}_i$ denotes the vector sum of all magnetic dipole moments induced inside ΔV. The magnetization vector is the volume density of the induced magnetic dipole moments. The magnetic dipole moments of an infinitesimal volume dV is given by $\mathbf{M}dV$, which produces a vector potential (see (1.14))

$$d\mathbf{A} = \frac{\mu_0}{4\pi R^2} \mathbf{M} \times \mathbf{u}_R dV(\mathbf{r}') = \frac{\mu_0}{4\pi} \mathbf{M} \times \nabla' \frac{1}{R} dV(\mathbf{r}').$$

The total vector potential due to a magnetized material in a region V bounded by S is then given by

$$\begin{aligned}
\mathbf{A} &= \frac{\mu_0}{4\pi} \int_V \mathbf{M} \times \nabla' \frac{1}{R} dV(\mathbf{r}') \\
&= \frac{\mu_0}{4\pi} \int_V \frac{\nabla' \times \mathbf{M}}{R} dV(\mathbf{r}') - \frac{\mu_0}{4\pi} \int_V \nabla' \times \frac{\mathbf{M}}{R} dV(\mathbf{r}') \\
&= \frac{\mu_0}{4\pi} \int_V \frac{\nabla' \times \mathbf{M}}{R} dV(\mathbf{r}') + \frac{\mu_0}{4\pi} \int_S \frac{\mathbf{M} \times \mathbf{u}_n(\mathbf{r}')}{R} dS(\mathbf{r}')
\end{aligned} \tag{1.16}$$

where $\mathbf{u}_n$ is the unit outward normal of S. The first term of (1.16) can be considered as the vector potential produced by a volume current density $\mathbf{J}_M = \nabla \times \mathbf{M}$, and the second term as the vector potential produced by a surface current density $\mathbf{J}_{Ms} = \mathbf{M} \times \mathbf{u}_n$. Both $\mathbf{J}_M$ and $\mathbf{J}_{Ms}$ are magnetization current densities. The total magnetic field inside the magnetized material is the sum of the fields produced by the conduction current and the magnetized current and Ampère's law (1.13) must be modified as

$$\nabla \times \mathbf{B} = \mu_0(\mathbf{J} + \mathbf{J}_M).$$

This can be rewritten as

$$\nabla \times \mathbf{H} = \mathbf{J} \tag{1.17}$$

where $\mathbf{H} = \mathbf{B}/\mu_0 - \mathbf{M}$ is called **magnetic field intensity**. When the material is linear and isotropic, the magnetization vector is proportional to the magnetic field intensity so that $\mathbf{M} = \chi_m \mathbf{H}$, where χ_m is a dimensionless number, called **magnetic susceptibility**. In this case we have

$$\mathbf{B} = \mu_0(1 + \chi_m)\mathbf{H} = \mu_r \mu_0 \mathbf{H} = \mu \mathbf{E}$$

where $\mu_r = 1 + \chi_m = \mu/\mu_0$ is a dimensionless number, called relative permeability. Notice that (1.12) holds in a magnetized material.

1.1.3 Faraday's Law

Faraday's law asserts that the induced electromotive force in a closed circuit is proportional to the rate of change of magnetic flux through any surface bounded by that circuit. The direction of the induced current is such as to oppose the change giving rise to it. Mathematically, this can be expressed as

$$\int_\Gamma \mathbf{E} \cdot \mathbf{u}_t d\Gamma = -\frac{\partial}{\partial t} \int_S \mathbf{B} \cdot \mathbf{u}_n dS$$

where Γ is a closed contour and S is the surface spanning the contour as shown in Figure 1.3; $\mathbf{u}_n$ and $\mathbf{u}_t$ are the unit normal to S and unit tangent vector along Γ respectively, and they satisfy the right-hand rule.

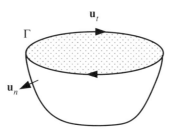

Figure 1.3   A two-sided surface

Loosely speaking, Faraday's law says that a changing magnetic field produces an electric field. The differential form of Faraday's law is

$$\nabla \times \mathbf{E} = -\frac{\partial \mathbf{B}}{\partial t}. \tag{1.18}$$

1.1.4 Law of Conservation of Charge

The law of conservation of charge states that the net charge of an isolated system remains constant. Mathematically, the amount of the charge flowing out of the surface S per second is equal to the decrease of the charge per second in the region V bounded by S

$$\int_S \mathbf{J} \cdot \mathbf{u}_n dS = -\frac{\partial}{\partial t} \int_V \rho dV.$$

The law of charge conservation is also known as the **continuity equation**. The differential form of the continuity equation is

$$\nabla \cdot \mathbf{J} = -\frac{\partial \rho}{\partial t}. \tag{1.19}$$

1.2   Maxwell Equations, Constitutive Relation, and Dispersion

From (1.18) and (1.17), one can find that a changing magnetic field produces an electric field by magnetic induction, but a changing electric field would not produce a magnetic field. In addition, equation (1.17) implies $\nabla \cdot \mathbf{J} = 0$, which contradicts the continuity equation for a time-dependent field. To solve these problems, Maxwell added an extra term $\mathbf{J}_d$ to Equation (1.17)

$$\nabla \times \mathbf{H} = \mathbf{J} + \mathbf{J}_d.$$

It then follows that

$$\nabla \cdot \mathbf{J} + \nabla \cdot \mathbf{J}_d = 0.$$

Introducing the continuity equation yields

$$\nabla \cdot \mathbf{J}_d = \frac{\partial \rho}{\partial t}.$$

Substituting Gauss's law (1.4) into the above equation, one may obtain $\mathbf{J}_d = \partial \mathbf{D}/\partial t$. Thus (1.17) must be modified to

$$\nabla \times \mathbf{H} = \frac{\partial \mathbf{D}}{\partial t} + \mathbf{J}. \tag{1.20}$$

The term $\partial \mathbf{D}/\partial t$ is called the **displacement current**. Equation (1.20) implies that a changing electric field generates a magnetic field by electric induction. It is this new electric induction postulate that makes it possible for Maxwell to predict the existence of electromagnetic waves. The mutual electric and magnetic induction produces a self-sustaining electromagnetic vibration moving through space.

1.2.1 Maxwell Equations and Boundary Conditions

It follows from (1.4), (1.12), (1.18) and (1.20) that

$$\nabla \times \mathbf{H}(\mathbf{r}, t) = \frac{\partial \mathbf{D}(\mathbf{r}, t)}{\partial t} + \mathbf{J}(\mathbf{r}, t),$$

$$\nabla \times \mathbf{E}(\mathbf{r}, t) = -\frac{\partial \mathbf{B}(\mathbf{r}, t)}{\partial t}, \tag{1.21}$$

$$\nabla \cdot \mathbf{D}(\mathbf{r}, t) = \rho(\mathbf{r}, t),$$

$$\nabla \cdot \mathbf{B}(\mathbf{r}, t) = 0.$$

The above equations are called **Maxwell equations**, and they describe the behavior of electric and magnetic fields, as well as their interactions with matter. It must be mentioned that the above vectorial form of Maxwell equations is due to the English engineer Oliver Heaviside (1850–1925), and is presented with neatness and clarity compared to the large set of scalar equations proposed by Maxwell. Maxwell equations are the starting point for the investigation of all macroscopic electromagnetic phenomena. In (1.21), $\mathbf{r}$ is the observation point of the fields in meters and t is the time in seconds; $\mathbf{H}$ is the magnetic field intensity measured in amperes per meter (A/m); $\mathbf{B}$ is the magnetic induction intensity measured in tesla (N/A·m); $\mathbf{E}$ is electric field intensity measured in volts per meter (V/m); $\mathbf{D}$ is the electric induction intensity measured in coulombs per square meter (C/m^2); $\mathbf{J}$ is electric current density measured in amperes per square meter (A/m^2); ρ is the electric charge density measured in coulombs per cubic meter (C/m^3). The first equation is Ampère's law, and it describes how the electric field changes according to the current density and magnetic field. The second equation is Faraday's law, and it characterizes how the magnetic field varies according to the electric field. The minus sign is required by Lenz's law, that is, when an electromotive force is generated by a change of magnetic flux, the polarity of the induced electromotive force is such that it produces a current whose magnetic field opposes the change, which produces it. The third equation is Coulomb's law, and it says that the electric field depends on the charge distribution and obeys the inverse square law. The final equation shows that there are no free magnetic monopoles and that the magnetic field also obeys the inverse square law. It should be understood that none of the experiments had anything to do with waves at the time when Maxwell derived his equations. Maxwell equations imply more than the experimental facts. The continuity equation can be derived from (1.21) as

$$\nabla \cdot \mathbf{J}(\mathbf{r}, t) = -\frac{\partial \rho(\mathbf{r}, t)}{\partial t}. \tag{1.22}$$

Remark 1.1: The charge density ρ and the current density $\mathbf{J}$ in Maxwell equations are free charge density and currents and they exclude charges and currents forming part of the structure of atoms and molecules. The bound charges and currents are regarded as material, which are not included in ρ and $\mathbf{J}$. The current density normally consists of two parts: $\mathbf{J} = \mathbf{J}_{con} + \mathbf{J}_{imp}$. Here $\mathbf{J}_{imp}$ is referred to as external or impressed current source, which is independent of the field and delivers energy to electric charges in a system. The impressed current source can be of electric and magnetic type as well as of non-electric or non-magnetic origin. $\mathbf{J}_{con} = \sigma \mathbf{E}$, where σ is the conductivity of the medium in mhos per meter, denotes the conduction current induced by the impressed source $\mathbf{J}_{imp}$. Sometimes it is convenient to introduce an external or impressed electric field $\mathbf{E}_{imp}$ defined by $\mathbf{J}_{imp} = \sigma \mathbf{E}_{imp}$. In a more general situation, one can write $\mathbf{J} = \mathbf{J}_{ind}(\mathbf{E}, \mathbf{B}) + \mathbf{J}_{imp}$, where $\mathbf{J}_{ind}(\mathbf{E}, \mathbf{B})$ is the induced current by the impressed current $\mathbf{J}_{imp}$. $\square$

Remark 1.2 (Duality): Sometimes it is convenient to introduce, magnetic current $\mathbf{J}_m$ and magnetic charges ρ_m, which are related by

$$\nabla \cdot \mathbf{J}_m(\mathbf{r}, t) = -\frac{\partial \rho_m(\mathbf{r}, t)}{\partial t} \tag{1.23}$$

and the Maxwell equations must be modified as

$$\nabla \times \mathbf{H}(\mathbf{r}, t) = \frac{\partial \mathbf{D}(\mathbf{r}, t)}{\partial t} + \mathbf{J}(\mathbf{r}, t),$$
$$\nabla \times \mathbf{E}(\mathbf{r}, t) = -\frac{\partial \mathbf{B}(\mathbf{r}, t)}{\partial t} - \mathbf{J}_m(\mathbf{r}, t),$$
$$\nabla \cdot \mathbf{D}(\mathbf{r}, t) = \rho(\mathbf{r}, t),$$
$$\nabla \cdot \mathbf{B}(\mathbf{r}, t) = \rho_m(\mathbf{r}, t). \tag{1.24}$$

The inclusion of $\mathbf{J}_m$ and ρ_m makes Maxwell equations more symmetric. However, there has been no evidence that the magnetic current and charge are physically present. The validity of introducing such concepts in Maxwell equations is justified by the equivalence principle, that is, they are introduced as a mathematical equivalent to electromagnetic fields. Equations (1.24) will be called the **generalized Maxwell equations**.

If all the sources are of magnetic type, Equations (1.24) reduce to

$$\nabla \times \mathbf{H}(\mathbf{r}, t) = \frac{\partial \mathbf{D}(\mathbf{r}, t)}{\partial t},$$
$$\nabla \times \mathbf{E}(\mathbf{r}, t) = -\frac{\partial \mathbf{B}(\mathbf{r}, t)}{\partial t} - \mathbf{J}_m(\mathbf{r}, t),$$
$$\nabla \cdot \mathbf{D}(\mathbf{r}, t) = 0,$$
$$\nabla \cdot \mathbf{B}(\mathbf{r}, t) = \rho_m(\mathbf{r}, t). \tag{1.25}$$

Mathematically (1.21) and (1.25) are similar. One can obtain one of them by simply interchanging symbols as shown in Table 1.1. This property is called **duality**. The importance of

Table 1.1 Duality.

Electric source	Magnetic source
E	H
H	−E
J	$\mathbf{J}_m$
ρ	ρ_m
μ	ε
ε	μ

duality is that one can obtain the solution of magnetic type from the solution of electric type by interchanging symbols and vice versa.   □

Remark 1.3: For the time-harmonic (sinusoidal) fields, Equations (1.21) and (1.22) can be expressed as

$$\nabla \times \mathbf{H}(\mathbf{r}) = j\omega \mathbf{D}(\mathbf{r}) + \mathbf{J}(\mathbf{r}),$$
$$\nabla \times \mathbf{E}(\mathbf{r}) = -j\omega \mathbf{B}(\mathbf{r}),$$
$$\nabla \cdot \mathbf{D}(\mathbf{r}) = \rho(\mathbf{r}),$$
$$\nabla \cdot \mathbf{B}(\mathbf{r}, \omega) = 0,$$
$$\nabla \cdot \mathbf{J}(\mathbf{r}) = -j\omega\rho(\mathbf{r}),$$

$$(1.26)$$

where the field quantities denote the complex amplitudes (phasors) defined by

$$\mathbf{E}(\mathbf{r}, t) = \mathrm{Re}[\mathbf{E}(\mathbf{r})e^{j\omega t}], \text{ etc.} \qquad \square$$

We use the same notations for both time-domain and frequency-domain quantities.

Remark 1.4: Maxwell equations summarized in (1.21) hold for macroscopic fields. For microscopic fields, the assumption that the charges and currents are continuously distributed is no longer valid. Instead, the charge density and current density are represented by

$$\rho(\mathbf{r}) = \sum_i q_i \delta(\mathbf{r} - \mathbf{r}_i), \ \mathbf{J}(\mathbf{r}) = \sum_i q_i \dot{\mathbf{r}}_i \delta(\mathbf{r} - \mathbf{r}_i) \qquad (1.27)$$

where q_i denotes the charge of i th particle and $\dot{\mathbf{r}}_i$ (the dot denotes the time derivative) its velocity. Correspondingly, Maxwell equations become

$$\nabla \times \mathbf{H}(\mathbf{r}, t) = \varepsilon_0 \frac{\partial \mathbf{E}(\mathbf{r}, t)}{\partial t} + \mathbf{J}(\mathbf{r}, t),$$
$$\nabla \times \mathbf{E}(\mathbf{r}, t) = -\mu_0 \frac{\partial \mathbf{H}(\mathbf{r}, t)}{\partial t},$$
$$\nabla \cdot \mathbf{E}(\mathbf{r}, t) = \frac{\rho(\mathbf{r}, t)}{\varepsilon_0},$$
$$\nabla \cdot \mathbf{H}(\mathbf{r}, t) = 0.$$

$$(1.28)$$

All charged particles have been included in (1.27). The macroscopic field equations (1.21) can be obtained from the microscopic field equations (1.28) by the method of averaging.   □

Remark 1.5: Ampère's law and Coulomb's law can be derived from the continuity equation. If we take electric charge Q as a primitive smoothly distributed over a volume V, we can define a charge density $\rho(\mathbf{r}, t)$ such that $Q = \int_V \rho(\mathbf{r}', t) dV(\mathbf{r}')$. Now the assumption that the electric charges are always conserved may be applied, which implies that if the charges within a region V have changed, the only possibility is that some charges have left or entered the region. Based on this assumption, it can be shown that there exists a vector $\mathbf{J}$, called current density, such that the continuity equation (1.22) holds (Duvaut and Lions, 1976; Kovetz, 2000). We can define a vector $\mathbf{D}$, called electric induction intensity, so that Coulomb's law holds

$$\nabla \cdot \mathbf{D}(\mathbf{r}, t) = \rho(\mathbf{r}, t).$$

Then the continuity equation (1.22) implies that the divergence of vector $\partial \mathbf{D}/\partial t + \mathbf{J}$ is zero. As a result, there exists at least one vector $\mathbf{H}$, called the magnetic field intensity, so that Ampère's law holds

$$\nabla \times \mathbf{H}(\mathbf{r}, t) = \frac{\partial \mathbf{D}(\mathbf{r}, t)}{\partial t} + \mathbf{J}(\mathbf{r}, t). \qquad \square$$

Remark 1.6: Maxwell equations might be derived from the laws of electrostatics (Elliott, 1993; Schwinger *et al.*, 1998) or from quantum mechanics (Dyson, 1990). $\qquad \square$

Remark 1.7: The force acting on a point charge q, moving with a velocity $\mathbf{v}$ with respect to an observer, by the electromagnetic field is given by

$$\mathbf{F}(\mathbf{r}, t) = q[\mathbf{E}(\mathbf{r}, t) + \mathbf{v}(\mathbf{r}, t) \times \mathbf{B}(\mathbf{r}, t)] \qquad (1.29)$$

where $\mathbf{E}$ and $\mathbf{B}$ are the total fields, including the field generated by the moving charge q. Equation (1.29) is referred to as **Lorentz force equation**, named after Dutch physicist Hendrik Antoon Lorentz (1853–1928). It is known that there are two different formalisms in classical physics. One is mechanics that deals with particles, and the other is electromagnetic field theory that deals with radiated waves. The particles and waves are coupled through the Lorentz force equation, which usually appears as an assumption separate from Maxwell equations. The Lorentz force is the only way to detect electromagnetic fields. For a continuous charge distribution, the Lorentz force equation becomes

$$\mathbf{f}(\mathbf{r}, t) = \rho \mathbf{E}(\mathbf{r}, t) + \mathbf{J}(\mathbf{r}, t) \times \mathbf{B}(\mathbf{r}, t) \qquad (1.30)$$

where $\mathbf{f}$ is the force density acting on the charge distribution ρ, that is, the force acting on the charge distribution per unit volume. Maxwell equations, Lorentz force equation and continuity equation constitute the fundamental equations in electrodynamics. To completely determine the interaction between charged particles and electromagnetic fields, we must introduce Newton's second law. An exact solution to the interaction problem is very difficult. Usually the fields are first determined by the known source without considering the influence of the moving charged particles. Then the dynamics of the charged particles can be studied by Newton's second law. The electromagnetic force causes like-charged things to repel and oppositely charged things to attract. Notice that the force that holds the atoms together to form molecules is essentially an electromagnetic force, called residual electromagnetic force. $\qquad \square$

Remark 1.8: Maxwell equations (1.21) are differential equations, which apply locally at each point in a continuous medium. At the interfaces of two different media, the charge and current and the corresponding fields are discontinuous and the differential (local) form of Maxwell equations becomes meaningless. Thus we must resort to the integral (global) form of Maxwell equations in this case. Let Γ be a closed contour and S be a regular two-sided surface spanning the contour as shown in Figure 1.3. Applying Stokes's theorem to the two curl equations in (1.21) yields

$$\int_{\Gamma} \mathbf{H} \cdot \mathbf{u}_t d\Gamma = \int_{S} \left(\mathbf{J} + \frac{\partial \mathbf{D}}{\partial t} \right) \cdot \mathbf{u}_n dS, \int_{\Gamma} \mathbf{E} \cdot \mathbf{u}_t d\Gamma = -\int_{S} \frac{\partial \mathbf{B}}{\partial t} \cdot \mathbf{u}_n dS \qquad (1.31)$$

If S is a closed surface, applying Gauss's theorem to the two divergence equations in (1.21) gives

$$\int_{S} \mathbf{D} \cdot \mathbf{u}_n dS = \int_{V} \rho dV, \int_{S} \mathbf{B} \cdot \mathbf{u}_n dS = 0. \qquad (1.32)$$

□

Remark 1.9: The boundary conditions on the surface between two different media can be easily obtained from (1.31) and (1.32), and they are

$$\mathbf{u}_n \times (\mathbf{H}_1 - \mathbf{H}_2) = \mathbf{J}_s,$$
$$\mathbf{u}_n \times (\mathbf{E}_1 - \mathbf{E}_2) = 0,$$
$$\mathbf{u}_n \cdot (\mathbf{D}_1 - \mathbf{D}_2) = \rho_s, \qquad (1.33)$$
$$\mathbf{u}_n \cdot (\mathbf{B}_1 - \mathbf{B}_2) = 0,$$

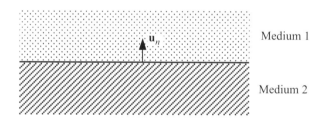

Figure 1.4   Interface between two different media

where $\mathbf{u}_n$ is the unit normal of the boundary directed from medium 2 to medium 1 as shown in Figure 1.4; $\mathbf{J}_s$ and ρ_s are the surface current density and surface charge density respectively. These boundary conditions can also be obtained from the differential form of Maxwell equations in the sense of generalized functions (see Chapter 2). □

1.2.2   Constitutive Relations

Maxwell equations are a set of seven equations involving 16 unknowns (that is five vector functions $\mathbf{E}, \mathbf{H}, \mathbf{B}, \mathbf{D}, \mathbf{J}$ and one scalar function ρ and the last equation of (1.21) is not independent). To determine the fields, nine more equations are needed, and they are given by the **generalized constitutive relations**:

$$\mathbf{D} = f_1(\mathbf{E}, \mathbf{H}), \mathbf{B} = f_2(\mathbf{E}, \mathbf{H})$$

together with the **generalized Ohm's law**:

$$\mathbf{J} = f_3(\mathbf{E}, \mathbf{H})$$

if the medium is conducting. The constitutive relations establish the connections between field quantities and reflect the properties of the medium, and they are totally independent of the Maxwell equations. In most cases, the constitutive relations can be expressed as

$$D_i(\mathbf{r}, t) = \sum_{j=x,y,z} [a_i^j(\mathbf{r})E_j(\mathbf{r}, t) + b_i^j(\mathbf{r})H_j(\mathbf{r}, t)]$$

$$\sum_{j=x,y,z} [(G_i^j * E_j)(\mathbf{r}, t) + (K_i^j * H_j)(\mathbf{r}, t)],$$

$$B_i(\mathbf{r}, t) = \sum_{j=x,y,z} [c_i^j(\mathbf{r})E_j(\mathbf{r}, t) + d_i^j(\mathbf{r})H_j(\mathbf{r}, t)]$$

$$= \sum_{j=x,y,z} [(L_i^j * E_j)(\mathbf{r}, t) + (F_i^j * H_j)(\mathbf{r}, t)],$$

where $i = x, y, z$; $*$ denotes the convolution with respect to time; a_i^j, b_i^j, c_i^j, d_i^j are independent of time; and G_i^j, K_i^j, L_i^j, F_i^j are functions of $(\mathbf{r}, t)$. The medium defined by the above equations is called **bianisotropic**. An **anisotropic medium** is defined by

$$D_i(\mathbf{r}, t) = \sum_{j=x,y,z} [a_i^j(\mathbf{r})E_j(\mathbf{r}, t) + (G_i^j * E_j)(\mathbf{r}, t)],$$
$$B_i(\mathbf{r}, t) = \sum_{j=x,y,z} [d_i^j(\mathbf{r})H_j(\mathbf{r}, t) + (F_i^j * H_j)(\mathbf{r}, t)].$$

A **biisotropic medium** is defined by

$$\mathbf{D}(\mathbf{r}, t) = a(\mathbf{r})\mathbf{E}(\mathbf{r}, t) + b(\mathbf{r})\mathbf{H}(\mathbf{r}, t)$$
$$+ (G * \mathbf{E})(\mathbf{r}, t) + (K * \mathbf{H})(\mathbf{r}, t)$$
$$\mathbf{B}(\mathbf{r}, t) = c(\mathbf{r})\mathbf{E}(\mathbf{r}, t) + d(\mathbf{r})\mathbf{H}(\mathbf{r}, t)$$
$$+ (L * \mathbf{E})(\mathbf{r}, t) + (F * \mathbf{H})(\mathbf{r}, t)$$

where a, b, c, d are independent of time and G, K, L, F are functions of $(\mathbf{r}, t)$. An **isotropic medium** is defined by

$$\mathbf{D}(\mathbf{r}, t) = a(\mathbf{r})\mathbf{E}(\mathbf{r}, t) + (G * \mathbf{E})(\mathbf{r}, t),$$
$$\mathbf{B}(\mathbf{r}, t) = d(\mathbf{r})\mathbf{H}(\mathbf{r}, t) + (F * \mathbf{H})(\mathbf{r}, t).$$

For monochromatic fields, the constitutive relations for a bianisotropic medium are usually expressed by

$$\mathbf{D} = \overset{\leftrightarrow}{\varepsilon} \cdot \mathbf{E} + \overset{\leftrightarrow}{\xi} \cdot \mathbf{H}, \mathbf{B} = \overset{\leftrightarrow}{\varsigma} \cdot \mathbf{E} + \overset{\leftrightarrow}{\mu} \cdot \mathbf{H}.$$

For an anisotropic medium, both $\overset{\leftrightarrow}{\xi}$ and $\overset{\leftrightarrow}{\varsigma}$ vanish.

Remark 1.10: The effects of the current $\mathbf{J} = \mathbf{J}_{imp} + \mathbf{J}_{ind}$ can be included in the constitutive relations by introducing a new vector $\mathbf{D}''$ such that

$$\mathbf{D}''(\mathbf{r}, t) = \int_{-\infty}^{t} \mathbf{J}(\mathbf{r}, t')dt' + \mathbf{D}(\mathbf{r}, t).$$

Thus (1.21) can be written as

$$\nabla \times \mathbf{H}(\mathbf{r}, t) = \frac{\partial \mathbf{D}''(\mathbf{r}, t)}{\partial t},$$

$$\nabla \times \mathbf{E}(\mathbf{r}, t) = -\frac{\partial \mathbf{B}(\mathbf{r}, t)}{\partial t},$$

$$\nabla \cdot \mathbf{D}''(\mathbf{r}, t) = 0,$$

$$\nabla \cdot \mathbf{B}(\mathbf{r}, t) = 0.$$

So the current source has been absorbed in the displacement current $\partial \mathbf{D}''(\mathbf{r}, t)/\partial t$, and the Maxwell equations are defined in a lossless and source-free region.   □

The constitutive relations are often written as

$$\begin{aligned} \mathbf{D}(\mathbf{r}, t) &= \varepsilon_0 \mathbf{E}(\mathbf{r}, t) + \mathbf{P}(\mathbf{r}, t) + \cdots, \\ \mathbf{B}(\mathbf{r}, t) &= \mu_0[\mathbf{H}(\mathbf{r}, t) + \mathbf{M}(\mathbf{r}, t) + \cdots], \end{aligned} \tag{1.34}$$

where $\mathbf{M}$ is the magnetization vector and $\mathbf{P}$ is the polarization vector. Equations (1.34) may contain higher order terms, which have been omitted since in most cases only the magnetization and polarization vectors are significant. The vectors $\mathbf{M}$ and $\mathbf{P}$ reflect the effects of the Lorentz force on elemental particles in the medium and therefore they depend on both $\mathbf{E}$ and $\mathbf{B}$ in general. Since the elemental particles in the medium have finite masses and are mutually interacting, $\mathbf{M}$ and $\mathbf{P}$ are also functions of time derivatives of $\mathbf{E}$ and $\mathbf{B}$ as well as their magnitudes. The same applies for the current density $\mathbf{J}_{ind}$.

A detailed study of magnetization and polarization process belongs to the subject of quantum mechanics. However, a macroscopic description of electromagnetic properties of the medium is simple as compared to the microscopic description. When the field quantities are replaced by their respective volume averages, the effects of the complicated array of atoms and electrons constituting the medium may be represented by a few parameters. The macroscopic description is satisfactory only when the large-scale effects of the presence of the medium are considered, and the details of the physical phenomena occurring on an atomic scale can be ignored. Since the averaging process is linear, any linear relation between the microscopic fields remains valid for the macroscopic fields.

In most cases, **M** is only dependent on the magnetic field **B** and its time derivatives while **P** and **J** depend only on the electric field **E** and its time derivatives. If these dependences are linear, the medium is said to be **linear**. These linear dependences are usually expressed as

$$\mathbf{D} = \tilde{\varepsilon}\mathbf{E} + \tilde{\varepsilon}_1 \frac{\partial \mathbf{E}}{\partial t} + \tilde{\varepsilon}_2 \frac{\partial^2 \mathbf{E}}{\partial t^2} + \cdots,$$

$$\mathbf{B} = \tilde{\mu}\mathbf{H} + \tilde{\mu}_1 \frac{\partial \mathbf{H}}{\partial t} + \tilde{\mu}_2 \frac{\partial^2 \mathbf{H}}{\partial t^2} + \cdots, \qquad (1.35)$$

$$\mathbf{J}_{ind} = \tilde{\sigma}\mathbf{E} + \tilde{\sigma}_1 \frac{\partial \mathbf{E}}{\partial t} + \tilde{\sigma}_2 \frac{\partial^2 \mathbf{E}}{\partial t^2} + \cdots,$$

where all the scalar coefficients are constants. For the monochromatic fields, the first two expressions of (1.35) reduce to

$$\mathbf{D} = \varepsilon\mathbf{E}, \mathbf{B} = \mu\mathbf{H}$$

where

$$\begin{aligned}
\varepsilon &= \varepsilon' - j\varepsilon'', & \mu &= \mu' - j\mu'', \\
\varepsilon' &= \tilde{\varepsilon} - \omega^2\tilde{\varepsilon}_2 + \cdots, & \mu' &= \tilde{\mu} - \omega^2\tilde{\mu}_2 + \cdots, & (1.36) \\
\varepsilon'' &= -\omega\tilde{\varepsilon}_1 + \omega^3\tilde{\varepsilon}_3 - \cdots, & \mu'' &= -\omega\tilde{\mu}_1 + \omega^3\tilde{\mu}_3 - \cdots.
\end{aligned}$$

The parameters ε' and ε'' are real and are called **capacitivity** and **dielectric loss factor** respectively. The parameters μ' and μ'' are real and are called **inductivity** and **magnetic loss factor** respectively.

Remark 1.11: According to the transformation of electromagnetic fields under the Lorentz transform (see Chapter 9), the constitutive relations depend on the reference systems.   □

1.2.3   Wave Equations

The electromagnetic wave equations are second-order partial differential equations that describe the propagation of electromagnetic waves through a medium. If the medium is homogeneous and isotropic and non-dispersive, we have $\mathbf{B} = \mu\mathbf{H}$ and $\mathbf{D} = \varepsilon\mathbf{E}$, where μ and ε are constants. On elimination of **E** or **H** in the generalized Maxwell equations, we obtain

$$\nabla \times \nabla \times \mathbf{E} + \mu\varepsilon \frac{\partial^2 \mathbf{E}}{\partial t^2} = -\nabla \times \mathbf{J}_m - \mu \frac{\partial \mathbf{J}}{\partial t},$$

$$\nabla \times \nabla \times \mathbf{H} + \mu\varepsilon \frac{\partial^2 \mathbf{H}}{\partial t^2} = \nabla \times \mathbf{J} - \varepsilon \frac{\partial \mathbf{J}_m}{\partial t}. \qquad (1.37)$$

These are known as the **wave equations**. Making use of $\nabla \cdot \mathbf{E} = -\rho/\varepsilon$ and $\nabla \cdot \mathbf{H} = -\rho_m/\mu$, the equations become

$$\left(\nabla^2 - \mu\varepsilon\frac{\partial^2}{\partial t^2}\right)\mathbf{E} = \nabla \times \mathbf{J}_m + \mu\frac{\partial \mathbf{J}}{\partial t} + \nabla\left(\frac{\rho}{\varepsilon}\right),$$

$$\left(\nabla^2 - \mu\varepsilon\frac{\partial^2}{\partial t^2}\right)\mathbf{H} = -\nabla \times \mathbf{J} + \varepsilon\frac{\partial \mathbf{J}_m}{\partial t} + \nabla\left(\frac{\rho_m}{\mu}\right).$$

$$(1.38)$$

In a source-free region, Equations (1.38) reduce to homogeneous equations, which have non-trivial solutions. The existence of the non-trivial solutions in a source-free region indicates the possibility of a self-sustaining electromagnetic field outside the source region. For the time-harmonic fields, Equations (1.37) and (1.38) respectively reduce to

$$\nabla \times \nabla \times \mathbf{E} - k^2\mathbf{E} = -\nabla \times \mathbf{J}_m - j\omega\mu\mathbf{J},$$

$$\nabla \times \nabla \times \mathbf{H} - k^2\mathbf{H} = \nabla \times \mathbf{J} - j\omega\varepsilon\mathbf{J}_m,$$

$$(1.39)$$

and

$$(\nabla^2 + k^2)\mathbf{E} = \nabla \times \mathbf{J}_m + j\omega\mu\mathbf{J} - \frac{\nabla(\nabla \cdot \mathbf{J})}{j\omega\varepsilon},$$

$$(\nabla^2 + k^2)\mathbf{H} = -\nabla \times \mathbf{J} + j\omega\varepsilon\mathbf{J}_m - \frac{\nabla(\nabla \cdot \mathbf{J}_m)}{j\omega\mu},$$

$$(1.40)$$

where $k = \omega\sqrt{\mu\varepsilon}$. It can be seen that the source terms on the right-hand side of (1.37) and (1.40) are very complicated. To simplify the analysis, the electromagnetic potential functions may be introduced (see Section 2.6.1). The wave equations may be used to solve the following three different field problems:

1. Electromagnetic fields in source-free region: wave propagations in space and waveguides, wave oscillation in cavity resonators, etc.
2. Electromagnetic fields generated by known source distributions: antenna radiations, excitations in waveguides and cavity resonators, etc.
3. Interaction of field and sources: wave propagation in plasma, coupling between electron beams and propagation mechanism, etc.

In a source-free region, Equations (1.39) and (1.40) become

$$\nabla \times \nabla \times \mathbf{E} - k^2\mathbf{E} = 0, \nabla \times \nabla \times \mathbf{H} - k^2\mathbf{H} = 0, \qquad (1.41)$$

and

$$(\nabla^2 + k^2)\mathbf{E} = 0, (\nabla^2 + k^2)\mathbf{H} = 0, \qquad (1.42)$$

respectively. It should be noted that Equation (1.41) is not equivalent to Equation (1.42). The former implies

$$\nabla \cdot \mathbf{E} = 0, \nabla \cdot \mathbf{H} = 0 \tag{1.43}$$

but the latter does not. Therefore the solutions of (1.41) satisfy Maxwell equations while those of (1.42) may not. For example, $\mathbf{E} = \mathbf{u}_z e^{-jkz}$ is a solution of (1.42) but it does not satisfy $\nabla \cdot \mathbf{E} = 0$. So it is not a solution of Maxwell equations. For this reason, it is imperative that one must incorporate (1.42) with (1.43). This can be accomplished by solving one of the equations in (1.42) to get one field quantity, say $\mathbf{E}$, and then using Maxwell equations to get the other field quantity $\mathbf{H}$. Such an approach guarantees that the fields satisfy (1.43).

If the medium is inhomogeneous and anisotropic so that $\mathbf{D} = \overleftrightarrow{\varepsilon} \cdot \mathbf{E}$ and $\mathbf{B} = \overleftrightarrow{\mu} \cdot \mathbf{H}$, the wave equations for the time-harmonic fields are

$$\begin{aligned} \nabla \times \overleftrightarrow{\mu}^{-1} \cdot \nabla \times \mathbf{E}(\mathbf{r}) - \omega^2 \overleftrightarrow{\varepsilon} \cdot \mathbf{E}(\mathbf{r}) = -j\omega\mathbf{J}(\mathbf{r}) - \nabla \times \overleftrightarrow{\mu}^{-1} \cdot \mathbf{J}_m, \\ \nabla \times \overleftrightarrow{\varepsilon}^{-1} \cdot \nabla \times \mathbf{H}(\mathbf{r}) - \omega^2 \overleftrightarrow{\mu} \cdot \mathbf{H}(\mathbf{r}) = -j\omega\mathbf{J}_m(\mathbf{r}) + \nabla \times \overleftrightarrow{\varepsilon}^{-1} \cdot \mathbf{J}. \end{aligned} \tag{1.44}$$

1.2.4   Dispersion

If the speed of the wave propagation and the wave attenuation in a medium depend on the frequency, the medium is said to be dispersive. Dispersion arises from the fact that the polarization and magnetization and the current density cannot follow the rapid changes of the electromagnetic fields, which implies that the electromagnetic energy can be absorbed by the medium. Thus, dissipation or absorption always occurs whenever the medium shows the dispersive effects. In reality, all media show some dispersive effects. The medium can be divided into normal dispersive and anomalous dispersive. A **normal dispersive medium** refers to the situation where the refractive index increases as the frequency increases. Most naturally occurring transparent media exhibit normal dispersion in the visible range of electromagnetic spectrum. In an **anomalous dispersive medium**, the refractive index decreases as frequency increases. The dispersive effects are usually recognized by the existence of elementary solutions (plane wave solution) of Maxwell equations in a source-free region

$$A(\mathbf{k})e^{j(\omega t - \mathbf{k} \cdot \mathbf{r})} \tag{1.45}$$

where $A(\mathbf{k})$ is the amplitude, $\mathbf{k}$ is the wave vector and ω is the frequency. When the elementary solutions are introduced into Maxwell equations, it will be found that $\mathbf{k}$ and ω must be related by an equation

$$f(\omega, \mathbf{k}) = 0. \tag{1.46}$$

This is called the **dispersion equation**. The plane wave $e^{j\omega t - j\mathbf{k} \cdot \mathbf{r}}$ has four-dimensional space-time orthorgonality properties, and is a solution of Maxwell equations in a source-free region when it satisfies the dispersion relation. It can be assumed that the frequency can be expressed in terms of the wave vector by solving the above dispersion equation

$$\omega = W(\mathbf{k}). \tag{1.47}$$

In general, a number of such solutions exist, which give different functions $W(\mathbf{k})$. Each solution is called a mode. To ensure that the solution $e^{j\omega t - j\mathbf{k}\cdot\mathbf{r}}$ is a plane wave, some restrictions must be put on the solution of dispersion equation, which are (Whitham, 1974)

$$\det\left|\frac{\partial^2 W}{\partial k_i \partial k_j}\right| \neq 0,\ W(\mathbf{k})\ \text{is real}. \tag{1.48}$$

These conditions have excluded all non-dispersive waves. A medium is called dispersive if there are solutions of (1.45) and (1.47) that satisfy (1.48). This definition applies to uniform medium. For a non-uniform medium, the definition of dispersive waves can be generalized to allow more general separable solutions of Maxwell equations, such as $A(\mathbf{k}, \mathbf{r})e^{j\omega t}$, where $A(\mathbf{k}, \mathbf{r})$ is an oscillatory function (for example, a Bessel function). It is hard to give a general definition of dispersion of waves. Roughly speaking, the dispersive effects may be expected whenever oscillations in space are coupled with oscillations in time.

If (1.45) is an elementary solution for a linear equation, then formally

$$\varphi(\mathbf{r}, t) = \int_{-\infty}^{\infty} A(\mathbf{k})e^{j(\omega t - \mathbf{k}\cdot\mathbf{r})}d\mathbf{k} \tag{1.49}$$

is also a solution of the linear equation. The arbitrary function $A(\mathbf{k})$ may be chosen to satisfy the initial or boundary condition. If there are n modes with n different choices of $W(\mathbf{k})$, there will be n terms like (1.49) with n arbitrary functions $A(\mathbf{k})$. For a single linear differential equation with constant coefficients, there is a one-to-one correspondence between the equation and the dispersion relation. We only need to consider the following correspondences:

$$\frac{\partial}{\partial t} \leftrightarrow j\omega, \nabla \leftrightarrow -j\mathbf{k},$$

which yield a polynomial dispersion relation. More complicated dispersion relation may be obtained for other different type of differential equations.

Example 1.3: To find the dispersion relation of the medium, the plane wave solutions may be assumed for Maxwell equations as follows

$$\mathbf{E}(\mathbf{r}, t) = \text{Re}[\mathbf{E}(\mathbf{r}, \omega)e^{j\omega t - j\mathbf{k}\cdot\mathbf{r}}],\ \text{etc.} \tag{1.50}$$

Similar expressions hold for other quantities. In the following, the wave vector $\mathbf{k}$ is allowed to be a complex vector and there is no impressed source inside the medium. Introducing (1.50) into (1.26) and using the calculation $\nabla e^{-j\mathbf{k}\cdot\mathbf{r}} = -j\mathbf{k}e^{-j\mathbf{k}\cdot\mathbf{r}}$, we obtain

$$-j\mathbf{k} \times \mathbf{H}(\mathbf{r}, \omega) + \nabla \times \mathbf{H}(\mathbf{r}, \omega) = j\omega\mathbf{D}(\mathbf{r}, \omega) + \mathbf{J}_{con}(\mathbf{r}, \omega),$$
$$-j\mathbf{k} \times \mathbf{E}(\mathbf{r}, \omega) + \nabla \times \mathbf{E}(\mathbf{r}, \omega) = -j\omega\mathbf{B}(\mathbf{r}, \omega).$$

In most situations, the complex amplitudes of the fields are slowly varying functions of space coordinates. The above equations may reduce to

$$\mathbf{k} \times \mathbf{H}(\mathbf{r}, \omega) = -\omega \mathbf{D}(\mathbf{r}, \omega) + j\mathbf{J}_{con}(\mathbf{r}, \omega),$$
$$\mathbf{k} \times \mathbf{E}(\mathbf{r}, \omega) = \omega \mathbf{B}(\mathbf{r}, \omega). \tag{1.51}$$

If the medium is isotropic, dispersive and lossy, we may write

$$\mathbf{J}_{con} = \sigma \mathbf{E}, \mathbf{D} = (\varepsilon' - j\varepsilon'')\mathbf{E}, \mathbf{B} = (\mu' - j\mu'')\mathbf{H}.$$

Substituting these equations into (1.51) yields

$$\mathbf{k} \cdot \mathbf{k} = \omega^2(\mu' - j\mu'')[\varepsilon' - j(\varepsilon'' + \sigma/\omega)].$$

Assuming $\mathbf{k} = \mathbf{u}_k(\beta - j\alpha)$, then

$$\beta - j\alpha = \omega\sqrt{(\mu' - j\mu'')[\varepsilon' - j(\varepsilon'' + \sigma/\omega)]}$$

from which we may find that

$$\beta = \frac{\omega}{\sqrt{2}}\sqrt{(A^2 + B^2)^{1/2} + A}, \alpha = \frac{\omega}{\sqrt{2}}\sqrt{(A^2 + B^2)^{1/2} - A}$$

where $A = \mu'\varepsilon' - \mu''(\varepsilon'' + \sigma/\omega), B = \mu''\varepsilon' + \mu'(\varepsilon'' + \sigma/\omega)$.   □

1.3   Theorems for Electromagnetic Fields

A number of theorems can be derived from Maxwell equations, and they usually bring deep physical insight into the problems. When applied properly, these theorems can simplify the problems dramatically.

1.3.1   Superposition Theorem

Superposition theorem applies to all linear systems. Suppose that the impressed current source $\mathbf{J}_{imp}$ can be expressed as a linear combination of independent impressed current sources $\mathbf{J}_{imp}^k$ $(k = 1, 2, \cdots, n)$

$$\mathbf{J}_{imp} = \sum_{k=1}^{n} a_k \mathbf{J}_{imp}^k,$$

where a_k $(k = 1, 2, \cdots, n)$ are arbitrary constants. If $\mathbf{E}^k$ and $\mathbf{H}^k$ are fields produced by the source $\mathbf{J}_{imp}^k$, the **superposition theorem** for electromagnetic fields asserts that the fields

$\mathbf{E} = \sum_{k=1}^{n} a_k \mathbf{E}^k$ and $\mathbf{H} = \sum_{k=1}^{n} a_k \mathbf{H}^k$ are a solution of Maxwell equations produced by the source $\mathbf{J}_{imp}$.

1.3.2 Compensation Theorem

The compensation theorem in network theory is well known, which says that any component in the network can be substituted by an ideal current generator with the same current intensity as in the element. Similarly the **compensation theorem** for electromagnetic fields states that the influences of the medium on the electromagnetic fields can be substituted by the equivalent impressed sources. Let $\mathbf{E}$, $\mathbf{H}$, $\mathbf{M}$, $\mathbf{P}$ and $\mathbf{J}_{ind}$ be the field quantities induced by the impressed current $\mathbf{J}_{imp}$, which satisfy the Maxwell equations (1.21) and the constitutive relations (1.34) with $\mathbf{J} = \mathbf{J}_{ind} + \mathbf{J}_{imp}$. Suppose that the medium is arbitrary and we can write

$$\mathbf{M} = \mathbf{M}_1 + \mathbf{M}_2, \mathbf{P} = \mathbf{P}_1 + \mathbf{P}_2, \mathbf{J} = \mathbf{J}_1 + \mathbf{J}_2, \tag{1.52}$$

Then Equation (1.34) becomes

$$\mathbf{D} = (\varepsilon_0 \mathbf{E} + \mathbf{P}_1) + \mathbf{P}_2 = \mathbf{D}_1 + \mathbf{P}_2,$$
$$\mathbf{B} = \mu_0(\mathbf{H} + \mathbf{M}_1) + \mu_0\mathbf{M}_2 = \mathbf{B}_1 + \mu_0\mathbf{M}_2,$$

with $\mathbf{B}_1 = \mu_0(\mathbf{H} + \mathbf{M}_1)$ and $\mathbf{D}_1 = (\varepsilon_0 \mathbf{E} + \mathbf{P}_1)$. Accordingly, the Maxwell equations (1.21) can be written as

$$\nabla \times \mathbf{H} = \frac{\partial \mathbf{D}_1}{\partial t} + \mathbf{J}_1 + (\mathbf{J}_{imp} + \mathbf{J}'_{imp}),$$
$$\nabla \times \mathbf{E} = -\frac{\partial \mathbf{B}_1(\mathbf{r}, t)}{\partial t} - \mathbf{J}'_{m,imp},$$
$$\nabla \cdot \mathbf{D}_1 = \rho + \rho', \rho' = -\nabla \cdot \mathbf{P}_2, \tag{1.53}$$
$$\nabla \cdot \mathbf{B}_1 = \rho_m, \rho_m = -\mu_0\nabla \cdot \mathbf{M}_2,$$

where the new impressed electric current $\mathbf{J}'_{imp} = \mathbf{J}_2 + \partial \mathbf{P}_2/\partial t$ and magnetic current $\mathbf{J}'_{m,imp} = \mu_0\partial \mathbf{M}_2/\partial t$ have been introduced to represent the influences of the medium partly or completely, depending on how the division is made in (1.52). Equations (1.53) are the mathematical formulation of compensation theorem. Note that both impressed electric and magnetic current source are needed to replace the medium, and the magnetic current density and the magnetic charge density satisfy the continuity equation $\nabla \cdot \mathbf{J}'_{m,imp} = -\partial \rho_m/\partial t$.

1.3.3 Conservation of Electromagnetic Energy

The law of **conservation of electromagnetic energy** is known as the **Poynting theorem**, named after the English physicist John Henry Poynting (1852–1914). It can be found from

(1.21) that

$$-\mathbf{J}_{imp} \cdot \mathbf{E} - \mathbf{J}_{ind} \cdot \mathbf{E} = \nabla \cdot \mathbf{S} + \mathbf{E} \cdot \frac{\partial \mathbf{D}}{\partial t} + \mathbf{H} \cdot \frac{\partial \mathbf{B}}{\partial t}. \tag{1.54}$$

In a region V bounded by S, the integral form of (1.54) is

$$-\int_V \mathbf{J}_{imp} \cdot \mathbf{E} dV = \int_V \mathbf{J}_{ind} \cdot \mathbf{E} dV + \int_S \mathbf{S} \cdot \mathbf{u}_n dS + \int_V \left(\mathbf{E} \cdot \frac{\partial \mathbf{D}}{\partial t} + \mathbf{H} \cdot \frac{\partial \mathbf{B}}{\partial t} \right) dV, \tag{1.55}$$

where $\mathbf{u}_n$ is the unit outward normal of S, and $\mathbf{S} = \mathbf{E} \times \mathbf{H}$ is the **Poynting vector** representing the electromagnetic power-flow density measured in watts per square meter (W/m^2). It will be assumed that this explanation holds for all media. Thus, the left-hand side of Equation (1.55) stands for the power supplied by the impressed current source. The first term on the right-hand side is the work done per second by the electric field to maintain the current in the conducting part of the system. The second term denotes the electromagnetic power flowing out of S. The last term can be interpreted as the work done per second by the impressed source to establish the fields. The energy density w required to establish the electromagnetic fields may be defined as follows

$$dw = \left(\mathbf{E} \cdot \frac{\partial \mathbf{D}}{\partial t} + \mathbf{H} \cdot \frac{\partial \mathbf{B}}{\partial t} \right) dt. \tag{1.56}$$

Assuming all the sources and fields are zero at $t = -\infty$, we have

$$w = w_e + w_m, \tag{1.57}$$

where w_e and w_m are the **electric field energy density** and **magnetic field energy density** respectively

$$w_e = \frac{1}{2} \mathbf{E} \cdot \mathbf{D} + \int_{-\infty}^{t} \frac{1}{2} \left(\mathbf{E} \cdot \frac{\partial \mathbf{D}}{\partial t} - \mathbf{D} \cdot \frac{\partial \mathbf{E}}{\partial t} \right) dt,$$

$$w_m = \frac{1}{2} \mathbf{H} \cdot \mathbf{B} + \int_{-\infty}^{t} \frac{1}{2} \left(\mathbf{H} \cdot \frac{\partial \mathbf{B}}{\partial t} - \mathbf{B} \cdot \frac{\partial \mathbf{H}}{\partial t} \right) dt.$$

Equation (1.55) can be written as

$$-\int_V \mathbf{J}_{imp} \cdot \mathbf{E} dV = \int_V \mathbf{J}_{ind} \cdot \mathbf{E} dV + \int_S \mathbf{S} \cdot \mathbf{u}_n dS + \frac{\partial}{\partial t} \int_V (w_e + w_m) dV. \tag{1.58}$$

In general, the energy density w does not represent the stored energy density in the fields: the energy temporarily located in the fields and completely recoverable when the fields are reduced to zero. The energy density w given by (1.57) can be considered as the stored energy density only if the medium is lossless (that is, $\nabla \cdot \mathbf{S} = 0$). If medium is isotropic and time-invariant, we have

$$w_e = \frac{1}{2}\mathbf{E} \cdot \mathbf{D}, \ w_m = \frac{1}{2}\mathbf{H} \cdot \mathbf{B}.$$

If the fields are time-harmonic, the Poynting theorem takes the following form

$$-\frac{1}{2}\int_V \mathbf{E} \cdot \bar{\mathbf{J}}_{imp}dV = \frac{1}{2}\int_V \mathbf{E} \cdot \bar{\mathbf{J}}_{ind}dV + \int_S \frac{1}{2}(\mathbf{E} \times \bar{\mathbf{H}}) \cdot \mathbf{u}_n dS$$

$$+ j2\omega \int_V \left(\frac{1}{4}\mathbf{B} \cdot \bar{\mathbf{H}} - \frac{1}{4}\mathbf{E} \cdot \bar{\mathbf{D}}\right) dV, \tag{1.59}$$

where the bar denotes complex conjugate. The time averages of the Poynting vector, energy densities over one period of the sinusoidal wave $e^{j\omega t}$, denoted T, are

$$\bar{\bar{\mathbf{S}}} = \frac{1}{T}\int_0^T \mathbf{E} \times \bar{\mathbf{H}}dt = \frac{1}{2}\mathrm{Re}(\mathbf{E} \times \bar{\mathbf{H}}),$$

$$\frac{1}{T}\int_0^T \frac{1}{2}\mathbf{E} \cdot \mathbf{D}dt = \frac{1}{4}\mathrm{Re}(\mathbf{E} \cdot \bar{\mathbf{D}}),$$

$$\frac{1}{T}\int_0^T \frac{1}{2}\mathbf{H} \cdot \mathbf{B}dt = \frac{1}{4}\mathrm{Re}(\mathbf{H} \cdot \bar{\mathbf{B}}),$$

where the double line indicates the time average.

1.3.4 Conservation of Electromagnetic Momentum

The force acting on a charged particle by electromagnetic fields is given by the Lorentz force equation

$$\mathbf{F}(\mathbf{r}, t) = q[\mathbf{E}(\mathbf{r}, t) + \mathbf{v}(\mathbf{r}, t) \times \mathbf{B}(\mathbf{r}, t)],$$

where $\mathbf{v}$ is the velocity of the particle. Let m be the mass of the particle and $\mathbf{G}_p = m\mathbf{v}$ its momentum. By Newton's law, we have

$$\frac{d\mathbf{G}_p(\mathbf{r}, t)}{dt} = q[\mathbf{E}(\mathbf{r}, t) + \mathbf{v}(\mathbf{r}, t) \times \mathbf{B}(\mathbf{r}, t)]. \tag{1.60}$$

Let $W_p = m\mathbf{v} \cdot \mathbf{v}/2$ denote the kinetic energy of the particle. It follows from (1.60) that

$$\frac{dW_p(\mathbf{r}, t)}{dt} = q\mathbf{v}(\mathbf{r}, t) \cdot \mathbf{E}(\mathbf{r}, t). \tag{1.61}$$

For a continuous charge distribution ρ, Equations (1.60) and (1.61) should be changed to

$$\frac{d\mathbf{g}_p}{dt} = \rho\mathbf{E} + \mathbf{J} \times \mathbf{B} = \mathbf{f}, \tag{1.62}$$

$$\frac{dw_p}{dt} = \mathbf{J} \cdot \mathbf{E}, \tag{1.63}$$

where $\mathbf{J} = \rho\mathbf{v}$; $\mathbf{g}_p = \rho_m\mathbf{v}$ and $w_p = \rho_m\mathbf{v} \cdot \mathbf{v}/2$ are the density of momentum and density of kinetic energy of the charge distribution respectively, and ρ_m is the mass density. Equations (1.62) and (1.63) indicate that the charged system gains energy and momentum from the electromagnetic fields if $dw_p/dt > 0$ and $d\mathbf{g}_p/dt > 0$ or releases energy and momentum to the electromagnetic fields if $dw_p/dt < 0$ and $d\mathbf{g}_p/dt < 0$. From the conservation laws of energy and momentum, it may be concluded that electromagnetic fields have energy and momentum. From the Maxwell equations and Lorentz force equation in free space, we obtain

$$\begin{aligned}\mathbf{f} = \rho\mathbf{E} + \mathbf{J} \times \mathbf{B} &= \mathbf{E}\nabla \cdot \mathbf{D} + \left(\nabla \times \mathbf{H} - \frac{\partial \mathbf{D}}{\partial t}\right) \times \mathbf{B} \\ &= -\frac{\partial}{\partial t}\left(\frac{1}{c^2}\mathbf{E} \times \mathbf{H}\right) + \nabla \cdot \left[\varepsilon_0\mathbf{E}\mathbf{E} + \mu_0\mathbf{H}\mathbf{H} - \frac{1}{2}(\varepsilon_0\mathbf{E} \cdot \mathbf{E} + \mu_0\mathbf{H} \cdot \mathbf{H})\overset{\leftrightarrow}{\mathbf{I}}\right],\end{aligned}$$

where $c = 1/\sqrt{\mu_0\varepsilon_0}$, $\mathbf{E}\mathbf{E}$ and $\mathbf{H}\mathbf{H}$ are dyads. By means of (1.62), the above equation can be written as

$$\nabla \cdot \overset{\leftrightarrow}{\mathbf{T}} - \frac{\partial}{\partial t}(\mathbf{g} + \mathbf{g}_p) = 0. \tag{1.64}$$

where $\overset{\leftrightarrow}{\mathbf{T}} = \varepsilon_0\mathbf{E}\mathbf{E} + \mu_0\mathbf{H}\mathbf{H} - \frac{1}{2}(\varepsilon_0\mathbf{E} \cdot \mathbf{E} + \mu_0\mathbf{H} \cdot \mathbf{H})\overset{\leftrightarrow}{\mathbf{I}}$ is referred to as the **Maxwell stress tensor** and $\mathbf{g} = \mathbf{E} \times \mathbf{H}/c^2$ is known as the **electromagnetic momentum density**. The integral form of (1.64) over a region V bounded by S is

$$\frac{\partial}{\partial t}\int_V (\mathbf{g} + \mathbf{g}_p)dV = \int_S \mathbf{u}_n \cdot \overset{\leftrightarrow}{\mathbf{T}}dS, \tag{1.65}$$

Equation (1.65) indicates that the increase of total momentum (the electromagnetic momentum plus the momentum of the charged system) inside V per unit time is equal to the force acting on the fields inside V through the boundary S by the fields outside S. For this reason, $\mathbf{u}_n \cdot \overset{\leftrightarrow}{\mathbf{T}}$ may be interpreted as the force per unit area acting on the surface. We can also interpret $-\mathbf{u}_n \cdot \overset{\leftrightarrow}{\mathbf{T}}$ as the momentum flow density into S and call $-\overset{\leftrightarrow}{\mathbf{T}}$ the **electromagnetic momentum flow density tensor** or the **electromagnetic energy-momentum tensor**.

1.3.5   Conservation of Electromagnetic Angular Momentum

It follows from (1.64) that

$$\nabla \cdot (\mathbf{r} \times \overset{\leftrightarrow}{\mathbf{T}}) + \frac{\partial}{\partial t}(\mathbf{r} \times \mathbf{g} + \mathbf{r} \times \mathbf{g}_p) = 0.$$

The integral form of the above equation over a region V bounded by S is

$$\frac{\partial}{\partial t} \int_V (\mathbf{r} \times \mathbf{g} + \mathbf{r} \times \mathbf{g}_p) dV = - \int_S \mathbf{u}_n \cdot (\mathbf{r} \times \overset{\leftrightarrow}{\mathbf{T}}) dS.$$

Here $\mathbf{r} \times \mathbf{g}$ may be interpreted as the **electromagnetic angular momentum density** and $\mathbf{r} \times \overset{\leftrightarrow}{\mathbf{T}}$ as the **electromagnetic angular momentum flow density tensor**.

Remark 1.12: The quantities of a dynamic system that do not change with time play an important role in theoretical physics. These conserved quantities can be the energy, momentum, and angular momentum. Noether's theorem, named after the German mathematician Amalie Emmy Noether (1882–1935), states that the conservation laws are the consequences of continuous symmetry transformations under which the action integral of the system is left invariant. For example, time translation symmetry gives conservation of energy; space translation symmetry gives conservation of momentum; rotation symmetry gives conservation of angular momentum.   □

1.3.6   Uniqueness Theorems

It is important to know the conditions under which the solution of Maxwell equations is unique. Let us consider a multiple-connected region V bounded by $S = \sum_{i=0}^{N} S_i$, as shown in Figure 1.5. Assume that the medium inside V is linear, isotropic and time-invariant, and it may contain some impressed source $\mathbf{J}_{imp}$. So we have $\mathbf{D} = \varepsilon\mathbf{E}, \mathbf{B} = \mu\mathbf{H}$, and $\mathbf{J}_{ind} = \sigma\mathbf{E}$. Let $\mathbf{E}_1, \mathbf{H}_1$ and $\mathbf{E}_2, \mathbf{H}_2$ be two solutions of Maxwell equations. Then the difference fields $\mathbf{E} = \mathbf{E}_1 - \mathbf{E}_2$ and $\mathbf{H} = \mathbf{H}_1 - \mathbf{H}_2$ are a solution of the Maxwell equations free of impressed source. The requirements that the difference fields must be identically zero are the conditions

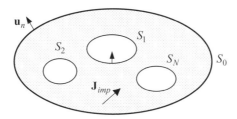

Figure 1.5   Multiple-connected region

for uniqueness that we seek. According to the Poynting theorem in the time domain, we write

$$\int_V \left(\mathbf{E} \cdot \frac{\partial \mathbf{D}}{\partial t} + \mathbf{H} \cdot \frac{\partial \mathbf{B}}{\partial t} \right) dV + \int_V \sigma \, |\mathbf{E}|^2 \, dV = - \int_S (\mathbf{E} \times \mathbf{H}) \cdot \mathbf{u}_n dS, \qquad (1.66)$$

where $\mathbf{u}_n$ is the unit outward normal of S. If $\mathbf{E}_1 = \mathbf{E}_2$ or $\mathbf{H}_1 = \mathbf{H}_2$ holds on the boundary S for $t > 0$, the above equation reduces to

$$\frac{\partial}{\partial t} \int_V \left(\frac{1}{2} \mathbf{E} \cdot \mathbf{D} + \frac{1}{2} \mathbf{H} \cdot \mathbf{B} \right) dV = - \int_V \sigma \, |\mathbf{E}|^2 \, dV.$$

Suppose that the source is turned on at $t = 0$. Taking the integration with respect to time over $[0, t]$ yields

$$\int_V \left(\frac{1}{2} \varepsilon \, |\mathbf{E}(\mathbf{r}, t)|^2 + \frac{1}{2} \mu \, |\mathbf{H}(\mathbf{r}, t)|^2 \right) dV - \int_V \left(\frac{1}{2} \varepsilon \, |\mathbf{E}(\mathbf{r}, 0)|^2 + \frac{1}{2} \mu \, |\mathbf{H}(\mathbf{r}, 0)|^2 \right) dV$$

$$= - \int_0^t dt \int_V \sigma \, |\mathbf{E}|^2 \, dV.$$

If $\mathbf{E}_1(\mathbf{r}, 0) = \mathbf{E}_2(\mathbf{r}, 0)$ and $\mathbf{H}_1(\mathbf{r}, 0) = \mathbf{H}_2(\mathbf{r}, 0)$ hold in V, the second term on the left-hand side vanishes. Since the right-hand side is a negative number while the left-hand side is a positive number, this is possible only when $\mathbf{E}_1(\mathbf{r}, t) = \mathbf{E}_2(\mathbf{r}, t)$ and $\mathbf{H}_1(\mathbf{r}, t) = \mathbf{H}_2(\mathbf{r}, t)$ for all $t > 0$.

If the region extends to infinity ($S_0 \to \infty$), we can assume that $\mathbf{E}_1 = \mathbf{E}_2$ or $\mathbf{H}_1 = \mathbf{H}_2$ on the boundary $\sum_{i=1}^{N} S_i$ for $t > 0$, and $\mathbf{E}_1(\mathbf{r}, 0) = \mathbf{E}_2(\mathbf{r}, 0)$ and $\mathbf{H}_1(\mathbf{r}, 0) = \mathbf{H}_2(\mathbf{r}, 0)$ in V. It follows from (1.66) that

$$\int_V \left(\frac{1}{2} \varepsilon \, |\mathbf{E}(\mathbf{r}, t)|^2 + \frac{1}{2} \mu \, |\mathbf{H}(\mathbf{r}, t)|^2 \right) dV$$

$$= - \int_0^t dt \int_V \sigma \, |\mathbf{E}|^2 \, dV - \int_0^t dt \int_{S_0} \frac{1}{\eta_0} |\mathbf{E}|^2 \, dS. \qquad (1.67)$$

Here $\eta_0 = \sqrt{\mu_0 / \varepsilon_0}$ is the wave impedance in free space. Equation (1.67) implies $\mathbf{E}_1(\mathbf{r}, t) = \mathbf{E}_2(\mathbf{r}, t)$ and $\mathbf{H}_1(\mathbf{r}, t) = \mathbf{H}_2(\mathbf{r}, t)$ for all $t > 0$. Note that the preceding discussions are valid even if σ is zero. Thus the following uniqueness theorem for electromagnetic fields in time domain has been proved.

Theorem 1.1 **Uniqueness theorem for time-domain fields:** *Suppose that the electromagnetic sources are turned on at at $t = 0$. The electromagnetic fields in a region are uniquely determined by the sources within the region, the initial electric field and the initial magnetic field at $t = 0$ inside the region, together with the tangential electric field (or the tangential magnetic field) on the boundary for $t > 0$, or together with the tangential electric field on part of the boundary and the tangential magnetic field on the rest of the boundary for $t > 0$.*   □

We now derive the uniqueness theorem in the frequency domain. Let $\mathbf{E}_1$, $\mathbf{H}_1$ and $\mathbf{E}_2$, $\mathbf{H}_2$ be two solutions of the time-harmonic Maxwell equations. For the difference fields $\mathbf{E} = \mathbf{E}_1 - \mathbf{E}_2$ and $\mathbf{H} = \mathbf{H}_1 - \mathbf{H}_2$, we may use the Poynting theorem in the frequency domain to write

$$\int_S \frac{1}{2}(\mathbf{E} \times \bar{\mathbf{H}}) \cdot \mathbf{u}_n dS + j2\omega \int_V \left(\frac{1}{4}\mathbf{B} \cdot \bar{\mathbf{H}} - \frac{1}{4}\mathbf{E} \cdot \bar{\mathbf{D}} \right) dV = -\frac{1}{2} \int_V \sigma \, |\mathbf{E}|^2 \, dV. \qquad (1.68)$$

If $\mathbf{E}_1 = \mathbf{E}_2$ or $\mathbf{H}_1 = \mathbf{H}_2$ holds on S, the first term on the left-hand side vanishes and we have

$$j\omega \int_V \frac{1}{2}\mu \, |\mathbf{H}|^2 \, dV - j\omega \int_V \frac{1}{2}\bar{\varepsilon} \, |\mathbf{E}|^2 \, dV + \frac{1}{2} \int_V \sigma \, |\mathbf{E}|^2 \, dV = 0.$$

This implies

$$\omega \int_V \frac{1}{2}\mathrm{Re}\,\varepsilon \, |\mathbf{E}|^2 \, dV - \omega \int_V \frac{1}{2}\mathrm{Re}\,\mu \, |\mathbf{H}|^2 \, dV = 0,$$

$$\omega \int_V \frac{1}{2}\mathrm{Im}\,\varepsilon \, |\mathbf{E}|^2 \, dV + \omega \int_V \frac{1}{2}\mathrm{Im}\,\mu \, |\mathbf{H}|^2 \, dV = \frac{1}{2} \int_V \sigma \, |\mathbf{E}|^2 \, dV.$$

For a dissipative medium, we have $\mathrm{Im}\,\varepsilon < 0$ and $\mathrm{Im}\,\mu < 0$. It is easy to see that if one of the following two conditions is met

$$\mathrm{Im}\,\varepsilon < 0, \, \mathrm{Im}\,\mu < 0, \qquad (1.69)$$

$$\sigma > 0, \qquad (1.70)$$

then the difference fields $\mathbf{E}$ and $\mathbf{H}$ vanish in V, which implies that the fields in V can be uniquely determined. Therefore, a loss must be assumed for time-harmonic fields in order to obtain the uniqueness.

In an unbounded region where $S_0 \to \infty$, we may assume that $\mu \to \mu_0$, $\varepsilon \to \varepsilon_0$ on S_0. Thus (1.68) may be written as

$$\sum_{i=1}^{N} \int_{S_i} \frac{1}{2}(\mathbf{E} \times \bar{\mathbf{H}}) \cdot \mathbf{u}_n dS + j\omega \int_V \frac{1}{2}\mu \, |\mathbf{H}|^2 \, dV - j\omega \int_V \frac{1}{2}\bar{\varepsilon} \, |\mathbf{E}|^2 \, dV$$

$$= -\int_{S_0} \frac{1}{2\eta_0} \, |\mathbf{E}|^2 \, dS - \frac{1}{2} \int_V \sigma \, |\mathbf{E}|^2 \, dV. \qquad (1.71)$$

If $\mathbf{E}_1 = \mathbf{E}_2$ or $\mathbf{H}_1 = \mathbf{H}_2$ holds on $\sum_{i=1}^{N} S_i$, the first term on the left-hand side of (1.71) vanishes and (1.71) reduces to

$$j\omega \int_V \frac{1}{2} \mu \, |\mathbf{H}|^2 \, dV - j\omega \int_V \frac{1}{2} \bar{\varepsilon} \, |\mathbf{E}|^2 \, dV = - \int_{S_0} \frac{1}{2\eta_0} |\mathbf{E}|^2 \, dS - \frac{1}{2} \int_V \sigma \, |\mathbf{E}|^2 \, dV. \qquad (1.72)$$

This leads to

$$\omega \int_V \frac{1}{2} \mathrm{Re}\, \varepsilon \, |\mathbf{E}|^2 \, dV - \omega \int_V \frac{1}{2} \mathrm{Re}\, \mu \, |\mathbf{H}|^2 \, dV = 0,$$

$$\omega \int_V \frac{1}{2} \mathrm{Im}\, \varepsilon \, |\mathbf{E}|^2 \, dV + \omega \int_V \frac{1}{2} \mathrm{Im}\, \mu \, |\mathbf{H}|^2 \, dV = \int_{S_0} \frac{1}{2\eta_0} |\mathbf{E}|^2 \, dS + \frac{1}{2} \int_V \sigma \, |\mathbf{E}|^2 \, dV. \qquad (1.73)$$

The difference fields vanish in the infinite region if either condition (1.69) or (1.70) is satisfied. We can further show that the difference fields vanish in the infinite region where radiation exists, even if the medium is lossless. Assuming that the medium is lossless, the second equation of (1.73) implies

$$\int_{S_0} \frac{1}{2\eta_0} |\mathbf{E}|^2 \, dS = 0, \; S_0 \to \infty.$$

It follows that

$$|\mathbf{E}|^2 = 0, \; S_0 \to \infty. \qquad (1.74)$$

This relation implies $\mathbf{E} = \mathbf{H} = 0$ in the region V, which can be proved as follows. Consider a sufficiently large sphere that contains all the impressed sources and inhomogeneities. The fields on the sphere may be expanded in terms of the spherical vector wavefunctions as follows (see Section 4.3)

$$\mathbf{E} = - \sum_{n,m,l} \left(\alpha_{nml}^{(2)} \mathbf{M}_{nml}^{(2)} + \beta_{nml}^{(2)} \mathbf{N}_{nml}^{(2)} \right),$$

$$\mathbf{H} = \frac{1}{j\eta_0} \sum_{n,m,l} \left(\alpha_{nml}^{(2)} \mathbf{N}_{nml}^{(2)} + \beta_{nml}^{(2)} \mathbf{M}_{nml}^{(2)} \right).$$

A simple calculation gives

$$|\mathbf{E}|^2 = \frac{1}{k_0^2} \sum_{n,m,l} N_{nm}^2 \left(\left| \alpha_{nml}^{(2)} \right|^2 + \left| \beta_{nml}^{(2)} \right|^2 \right), \qquad (1.75)$$

where $k_0 = \omega\sqrt{\mu_0\varepsilon_0}$ and N_{nm} is a constant. Combining (1.74) and (1.75), we obtain $\alpha_{nml}^{(2)} = \beta_{nml}^{(2)} = 0$. As a result, the fields outside a sufficiently large sphere are identically zero. By the analyticity of the electromagnetic fields, one must have $\mathbf{E} = \mathbf{H} = 0$ in the region V. Consequently the uniqueness theorem for time-harmonic field may be stated as follows.

Theorem 1.2 *Uniqueness theorem for time-harmonic fields: For a region that contains the dissipation loss or radiation loss, the electromagnetic fields are uniquely determined by the sources within the region, together with the tangential electric field (or the tangential magnetic field) on the boundary, or together with the tangential electric field on part of the boundary and the tangential magnetic field on the rest of the boundary.* □

The uniqueness for time-harmonic fields is guaranteed if the system has radiation loss, regardless whether the medium is lossy or not. This property has been widely validated by the study of antenna radiation problems, in which the surrounding medium is often assumed to be lossless.

Remark 1.13: The uniqueness for time-harmonic fields fails for a system that contains no dissipation loss and radiation loss. The uniqueness in a lossless medium is usually obtained by considering the fields in a lossless medium to be the limit of the corresponding fields in a lossy medium as the loss goes to zero, which is based on an assumption that the limit of a unique solution is also unique. However, this limiting process may lead to physically unacceptable solutions (see Section 3.3.2 and Section 8.2.1). Note that there is no need to introduce loss for a unique solution in the time-domain analysis. □

Example 1.4 (Image principle): To solve the boundary value problem with a perfect electric conductor, one can use the image principle that is based on the uniqueness theorem. The perfect electric conductor may be removed by introducing an 'image' of the original field source. The image is constructed in such a way so that the tangential component of the total electric field produced by the original source and its image vanishes on the perfect electric conductor. For example, an electric current element parallel to an infinitely large conducting plane has an image that is positioned symmetrically relative to the conducting plane and has a reverse orientation. The images for electric and magnetic current elements placed near the conducting plane are shown in Figure 1.6. □

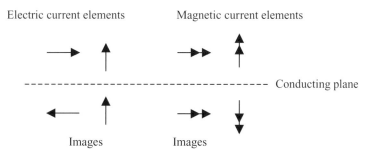

Figure 1.6   Image principle

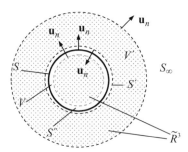

Figure 1.7 Equivalence theorem

1.3.7 Equivalence Theorems

It is known that there is no answer to the question of whether field or source is primary.
The equivalence principles indicate that the distinction between the field and source is kind
of blurred. Let V be an arbitrary region bounded by S; let S' be a closed surface pressed
tightly over S from outside; let S'' be a closed surface pressed tightly to S from inside; let
V' be the domain outside S'. A large closed surface S_∞ encloses S' as shown in Figure 1.7.
Two sources that produce the same fields inside a region are said to be equivalent within that
region. Similarly, two electromagnetic fields $\{E_1, D_1, H_1, B_1\}$ and $\{E_2, D_2, H_2, B_2\}$ are said
to be equivalent inside a region if they both satisfy the Maxwell equations and are equal in
that region.

The main application of the equivalence theorem is to find equivalent sources to replace the
influences of substance (the medium is homogenized), so that the formulae for retarding potentials can be used. The equivalent sources may be located inside S (equivalent volume sources)
or on S (equivalent surface sources). The most general form of the equivalent principles is as
follows.

General equivalence principle: Let us consider two electromagnetic field problems in two
different media:

$$
\text{Problem 1}: \begin{cases}
\nabla \times \mathbf{H}_1(\mathbf{r}, t) = \partial \mathbf{D}_1(\mathbf{r}, t)/\partial t + \mathbf{J}_1(\mathbf{r}, t), \\
\nabla \times \mathbf{E}_1(\mathbf{r}, t) = -\partial \mathbf{B}_1(\mathbf{r}, t)/\partial t - \mathbf{J}_{m1}(\mathbf{r}, t), \\
\nabla \cdot \mathbf{D}_1(\mathbf{r}, t) = \rho_1(\mathbf{r}, t), \ \nabla \cdot \mathbf{B}_1(\mathbf{r}, t) = \rho_{m1}(\mathbf{r}, t), \\
\mathbf{D}_1(\mathbf{r}, t) = \varepsilon_1(\mathbf{r})\mathbf{E}_1(\mathbf{r}, t), \ \mathbf{B}_1(\mathbf{r}, t) = \mu_1(\mathbf{r})\mathbf{H}_1(\mathbf{r}, t)
\end{cases}
$$

$$
\text{Problem 2}: \begin{cases}
\nabla \times \mathbf{H}_2(\mathbf{r}, t) = \partial \mathbf{D}_2(\mathbf{r}, t)/\partial t + \mathbf{J}_2(\mathbf{r}, t), \\
\nabla \times \mathbf{E}_2(\mathbf{r}, t) = -\partial \mathbf{B}_2(\mathbf{r}, t)/\partial t - \mathbf{J}_{m2}(\mathbf{r}, t), \\
\nabla \cdot \mathbf{D}_2(\mathbf{r}, t) = \rho_2(\mathbf{r}, t), \ \nabla \cdot \mathbf{B}_2(\mathbf{r}, t) = \rho_{m2}(\mathbf{r}, t), \\
\mathbf{D}_2(\mathbf{r}, t) = \varepsilon_2(\mathbf{r})\mathbf{E}_2(\mathbf{r}, t), \ \mathbf{B}_2(\mathbf{r}, t) = \mu_2(\mathbf{r})\mathbf{H}_2(\mathbf{r}, t).
\end{cases}
$$

If a new set of electromagnetic fields $\{\mathbf{E}, \mathbf{D}, \mathbf{H}, \mathbf{B}\}$ satisfying

$$
\begin{cases}
\nabla \times \mathbf{H}(\mathbf{r}, t) = \partial \mathbf{D}(\mathbf{r}, t)/\partial t + \mathbf{J}(\mathbf{r}, t), \\
\nabla \times \mathbf{E}(\mathbf{r}, t) = -\partial \mathbf{B}(\mathbf{r}, t)/\partial t - \mathbf{J}_m(\mathbf{r}, t), \\
\nabla \cdot \mathbf{D}(\mathbf{r}, t) = \rho(\mathbf{r}, t), \nabla \cdot \mathbf{B}(\mathbf{r}, t) = \rho_m(\mathbf{r}, t), \\
\mathbf{D}(\mathbf{r}, t) = \varepsilon(\mathbf{r})\mathbf{E}(\mathbf{r}, t), \mathbf{B}(\mathbf{r}, t) = \mu(\mathbf{r})\mathbf{H}(\mathbf{r}, t),
\end{cases}
\tag{1.76}
$$

is constructed in such a way that the sources of the fields $\{\mathbf{E}, \mathbf{D}, \mathbf{H}, \mathbf{B}\}$ and the parameters of the medium satisfy

$$
\begin{cases}
\mathbf{J} = \mathbf{J}_1, \mathbf{J}_m = \mathbf{J}_{m1} \\
\rho = \rho_1, \rho_m = \rho_{m1}, \mathbf{r} \in V \\
\mu = \mu_1, \varepsilon = \varepsilon_2
\end{cases}
;
\begin{cases}
\mathbf{J} = \mathbf{J}_2, \mathbf{J}_m = \mathbf{J}_{m2} \\
\rho = \rho_2, \rho_m = \rho_{m2}, \mathbf{r} \in R^3 - V \\
\mu = \mu_2, \varepsilon = \varepsilon_2
\end{cases}
$$

and

$$
\begin{cases}
\mathbf{J} = \mathbf{u}_n \times (\mathbf{H}_{2+} - \mathbf{H}_{1-}) \\
\mathbf{J}_m = -\mathbf{u}_n \times (\mathbf{E}_{2|} - \mathbf{E}_{1-}) \\
\rho = \mathbf{u}_n \cdot (\mathbf{D}_{2+} - \mathbf{D}_{1-}) \\
\rho_m = \mathbf{u}_n \cdot (\mathbf{B}_{2+} - \mathbf{B}_{1-})
\end{cases}
, \mathbf{r} \in S
$$

where $\mathbf{u}_n$ is the unit outward normal to S, and the subscripts $+$ and $-$ signify the values obtained as S is approached from outside S and inside S respectively, then we have

$$
\{\mathbf{E}, \mathbf{D}, \mathbf{H}, \mathbf{B}\} = \{\mathbf{E}_1, \mathbf{D}_1, \mathbf{H}_1, \mathbf{B}_1\}, \mathbf{r} \in V
$$
$$
\{\mathbf{E}, \mathbf{D}, \mathbf{H}, \mathbf{B}\} = \{\mathbf{E}_2, \mathbf{D}_2, \mathbf{H}_2, \mathbf{B}_2\}, \mathbf{r} \in R^3 - V
$$

To prove this theorem, we only need to show that the difference fields

$$
\begin{cases}
\delta\mathbf{E} = \mathbf{E} - \mathbf{E}_1 \\
\delta\mathbf{H} = \mathbf{H} - \mathbf{H}_1 \\
\delta\mathbf{D} = \mathbf{D} - \mathbf{D}_1 \\
\delta\mathbf{B} = \mathbf{B} - \mathbf{B}_1
\end{cases}
, \mathbf{r} \in V;
\begin{cases}
\delta\mathbf{E} = \mathbf{E} - \mathbf{E}_2 \\
\delta\mathbf{H} = \mathbf{H} - \mathbf{H}_2 \\
\delta\mathbf{D} = \mathbf{D} - \mathbf{D}_2 \\
\delta\mathbf{B} = \mathbf{B} - \mathbf{B}_2
\end{cases}
, \mathbf{r} \in R^3 - V
$$

in the shadowed region bounded by $S' + S'' + S_\infty$, denoted by $\tilde{R}^3$, are identically zero. The difference fields satisfy

$$
\begin{aligned}
&\nabla \times \delta\mathbf{H}(\mathbf{r}, t) = \partial \delta\mathbf{D}(\mathbf{r}, t)/\partial t, \\
&\nabla \times \delta\mathbf{E}(\mathbf{r}, t) = -\partial \delta\mathbf{B}(\mathbf{r}, t)/\partial t, \\
&\nabla \cdot \delta\mathbf{D}(\mathbf{r}, t) = 0, \nabla \cdot \delta\mathbf{B}(\mathbf{r}, t) = 0, \\
&\delta\mathbf{D}(\mathbf{r}, t) = \varepsilon_\delta(\mathbf{r})\mathbf{E}(\mathbf{r}, t), \\
&\delta\mathbf{B}(\mathbf{r}, t) = \mu_\delta(\mathbf{r})\delta\mathbf{H}(\mathbf{r}, t),
\end{aligned}
\tag{1.77}
$$

where $\varepsilon_\delta = \varepsilon_1, \mu_\delta = \mu_1$ for $\mathbf{r} \in V$ and $\varepsilon_\delta = \varepsilon_2, \mu_\delta = \mu_2$ for $\mathbf{r} \in R^3 - V$. From

$$\mathbf{u}_n \times (\mathbf{H}_{2+} - \mathbf{H}_{1-}) = \mathbf{u}_n \times (\mathbf{H}_+ - \mathbf{H}_-),$$
$$\mathbf{u}_n \times (\mathbf{E}_{2+} - \mathbf{E}_{1-}) = \mathbf{u}_n \times (\mathbf{E}_+ - \mathbf{E}_-),$$

we can find $\mathbf{u}_n \times \delta\mathbf{E}_+ = \mathbf{u}_n \times \delta\mathbf{E}_-$ and $\mathbf{u}_n \times \delta\mathbf{H}_+ = \mathbf{u}_n \times \delta\mathbf{H}_-$, which imply that the tangential components of $\delta\mathbf{E}$ and $\delta\mathbf{H}$ are continuous on S. It follows from (1.77) that

$$-\nabla \cdot (\delta\mathbf{E} \times \delta\mathbf{H}) = \frac{1}{2} \frac{\partial}{\partial t} (\varepsilon_\delta |\delta\mathbf{E}|^2 + \mu_\delta |\delta\mathbf{H}|^2).$$

Taking the integration over the shadowed region $\tilde{R}^3$ yields

$$-\int_{S'+S''+S_\infty} (\delta\mathbf{E} \times \delta\mathbf{H}) \cdot \mathbf{u}_n dS = \frac{1}{2} \frac{\partial}{\partial t} \int_{\tilde{R}^3} (\varepsilon_\delta |\delta\mathbf{E}|^2 + \mu_\delta |\delta\mathbf{H}|^2) dV.$$

If all the fields are produced after a finite moment $t_0 > -\infty$, one may take the integration with respect to time from $-\infty$ to t

$$-\int_{-\infty}^{t} dt \int_{S'+S''+S_\infty} [\delta\mathbf{E}(\mathbf{r}, t) \times \delta\mathbf{H}(\mathbf{r}, t)] \cdot \mathbf{u}_n dS$$

$$= \frac{1}{2} \int_{\tilde{R}^3} (\varepsilon_\delta |\delta\mathbf{E}(\mathbf{r}, t)|^2 + \mu_\delta |\delta\mathbf{H}(\mathbf{r}, t)|^2) dV. \qquad (1.78)$$

When S' and S'' approach S, the values of $\delta\mathbf{E}(\mathbf{r}, t) \times \delta\mathbf{H}(\mathbf{r}, t)$ on S' and S'' tend to be the same since $\delta\mathbf{E}(\mathbf{r}, t) \times \delta\mathbf{H}(\mathbf{r}, t)$ is continuous on S. Thus

$$\int_{S'+S''} [\delta\mathbf{E}(\mathbf{r}, t) \times \delta\mathbf{H}(\mathbf{r}, t)] \cdot \mathbf{u}_n dS = 0.$$

The electromagnetic wave travels at finite speed. It is thus possible to choose S_∞ to be large enough so that

$$\int_{S_\infty} [\delta\mathbf{E}(\mathbf{r}, t) \times \delta\mathbf{H}(\mathbf{r}, t)] \cdot \mathbf{u}_n dS = 0.$$

Consequently, Equation (1.78) reduces to

$$\int_{\tilde{R}^3} (\varepsilon_\delta |\delta\mathbf{E}(\mathbf{r}, t)|^2 + \mu_\delta |\delta\mathbf{H}(\mathbf{r}, t)|^2) dV = 0,$$

which implies $\delta\mathbf{E}(\mathbf{r}, t) = 0$ and $\delta\mathbf{H}(\mathbf{r}, t) = 0$. The proof is completed.

By the equivalence principle, the magnetic current $\mathbf{J}_m$ and magnetic charge ρ_m, introduced in the generalized Maxwell equations, are justified in the sense of equivalence. The difference between the compensation theorem and equivalence theorem is that the compensation implies replacement of induced sources or part of them by the imaginary impressed sources at the same locations. Equivalence implies replacement of any sources (impressed and/or induced) by another set of impressed sources, usually distributed in a different location.

If $\mathbf{E}_1 = \mathbf{D}_1 = \mathbf{H}_1 = \mathbf{B}_1 = \mathbf{J}_1 = \mathbf{J}_{m1} = 0$ in the general equivalence theorem, we can choose $\mu = \mu_2$, $\varepsilon = \varepsilon_2$ in (1.76) inside S. If all the sources for Problem 2 are contained inside S, the following sources

$$\begin{cases} \mathbf{J}_s = \mathbf{u}_n \times \mathbf{H}_{2+}, \mathbf{J}_{ms} = -\mathbf{u}_n \times \mathbf{E}_{2+} \\ \rho_s = \mathbf{u}_n \cdot \mathbf{D}_{2+}, \rho_{ms} = \mathbf{u}_n \cdot \mathbf{B}_{2+} \end{cases}, \quad \mathbf{r} \in S$$

produce the electromagnetic fields $\{\mathbf{E}, \mathbf{D}, \mathbf{H}, \mathbf{B}\}$ in (1.76). In other words, the above sources generate the fields $\{\mathbf{E}_2, \mathbf{D}_2, \mathbf{H}_2, \mathbf{B}_2\}$ in $R^3 - V$ and a zero field in V. Thus we have:

Theorem 1.3 *Schelkunoff–Love equivalence: (named after the American mathematician Sergei Alexander Schelkunoff, 1897–1992; and the English mathematician Augustus Edward Hough Love, 1863–1940):* Let $\{\mathbf{E}, \mathbf{D}, \mathbf{H}, \mathbf{B}\}$ be the electromagnetic fields with source confined in S. The following surface sources

$$\begin{cases} \mathbf{J}_s = \mathbf{u}_n \times \mathbf{H}, \mathbf{J}_{ms} = -\mathbf{u}_n \times \mathbf{E} \\ \rho_s = \mathbf{u}_n \cdot \mathbf{D}, \rho_{ms} = \mathbf{u}_n \cdot \mathbf{B} \end{cases}, \quad \mathbf{r} \in S \tag{1.79}$$

produce the same fields $\{\mathbf{E}, \mathbf{D}, \mathbf{H}, \mathbf{B}\}$ outside S and a zero field inside S. $\qquad\square$

It must be mentioned that the electromagnetic fields generated by a single electric source or a single magnetic source will never be zero within a finite region if the medium is homogeneous. The fields can be made to vanish inside a region only if both electric source and magnetic source exist so that the fields generated by both sources cancel each other in the region. In other words, only the solution of the generalized Maxwell equations can be zero within a finite region of homogeneous space. However, the solution of Maxwell equations can be zero within a finite region if the medium is inhomogeneous. Since the sources in (1.79) produce a zero field inside S, the interior of S may be filled with a perfect electric conductor. By use of the Lorentz reciprocity theorem (see Example 1.6), it can be shown that the surface electric current pressed tightly on the perfect conductor does not produce fields. As a result, only the surface magnetic current is needed in (1.79). Similarly, the interior of S may be filled with a perfect magnetic conductor, and in this case the surface magnetic current does not produce fields and only the surface electric current is needed in (1.79). In both cases, one cannot directly apply the vector potential formula even if the medium outside S is homogeneous.

Example 1.5 (An aperture problem): A general aperture coupling problem between two regions a and b is shown in Figure 1.8 (a). The impressed electric current $\mathbf{J}_{imp}$ and magnetic current $\mathbf{J}_{m,imp}$ are assumed to be located in region a only and there is no source in region b. The conductors in region b are assumed to be extended to infinity. By equivalent principle, the original problem can be separated into two equivalent problems as shown in Figure 1.8 (b). In region a, the fields are produced by the impressed sources $\mathbf{J}_{imp}$ and $\mathbf{J}_{m,imp}$, and the equivalent magnetic current $\mathbf{J}_{ms} = -\mathbf{u}_n \times \mathbf{E}$ over the aperture region S_a, with the aperture covered by an

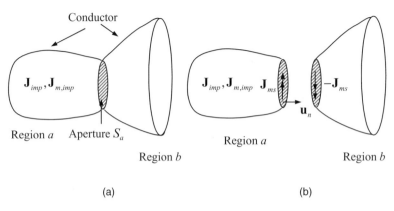

Figure 1.8 An aperture problem

electric conductor. In region b, the field is produced by the equivalent magnetic current $-\mathbf{J}_{ms}$ (the minus sign ensures that the tangential electrical field is continuous across the aperture). The tangential magnetic field in region a over the aperture, denoted $\mathbf{H}_t^a$, can be decomposed into two parts (Harrington and Mautz, 1976)

$$\mathbf{H}_t^a = \mathbf{H}_t^i + \mathbf{H}_t^m(\mathbf{J}_{ms}),$$

where $\mathbf{H}_t^i$ is due to the impressed source and $\mathbf{H}_t^m(\mathbf{J}_{ms})$ due to the equivalent source $\mathbf{J}_{ms}$, both being calculated with the aperture covered by an electric conductor. If $\mathbf{H}_t^b(-\mathbf{J}_{ms})$ denotes the tangential magnetic field in region b over the aperture, then the condition that the tangential magnetic field must be continuous across the aperture yields

$$\mathbf{H}_t^b(-\mathbf{J}_{ms}) = \mathbf{H}_t^i + \mathbf{H}_t^m(\mathbf{J}_{ms}).$$

This can be used to determine the magnetic current $\mathbf{J}_{ms}$.   □

1.3.8   Reciprocity

A linear system is said to be reciprocal if the response of the system with a particular load and a source is the same as the response when the source and the load are interchanged. The earliest study of reciprocity can be traced back to the work done by the English physicist Lord Rayleigh (1842–1919) in 1894 and the work by Lorentz in 1895. The reciprocity theorems are the most important analytical tools in the simplification and solution of various practical problems (Rumsey, 1961; Monteath, 1973; Richmond, 1961).

A linear system is characterized in an abstract way by a known source f, a response u and a system operator $\hat{L}$

$$\hat{L}(u) = f. \tag{1.80}$$

The system operator $\hat{L}$ is not unique for a given linear system and it depends on how the source and the response are defined. In what follows, it is assumed that $\hat{L}$ is a linear partial differential operator, and both f and u are defined in a region V with boundary S. For arbitrary functions u_1 and u_2, the following identity can be easily derived using integration by parts (Courant and Hilbert, 1953)

$$\int_V u_2 \hat{L}(u_1) dV = \int_V u_1 \hat{L}^*(u_2) dV + T(u_1, u_2; S) \tag{1.81}$$

for a time-independent system or

$$\int_{T_1}^{T_2} dt \int_V u_2 \hat{L}(u_1) dV = \int_{T_1}^{T_2} dt \int_V u_1 \hat{L}^*(u_2) dV + T(u_1, u_2; S, T_1, T_2) \tag{1.82}$$

for a time-dependent system. In (1.81) and (1.82), $\hat{L}^*$ is known as **formal adjoint** of $\hat{L}$; $T(u_1, u_2; S)$ and $T(u_1, u_2; S, T_1, T_2)$ are bilinear forms (boundary terms); and $[T_1, T_2]$ is an arbitrary time interval.

Equations (1.81) and (1.82) may be interpreted as **Huygens' principle**, named after the Dutch physicist Christiaan Huygens (1629–1695). For a time-independent system, Huygens' principle states that, given a source inside a hypothetical surface S, there is a certain source spreading over S, which gives the same field outside S as the original source inside S. For a time-dependent system, it states that the position of a wavefront and the magnitude of the wave at each point of the wavefront may be determined by the wavefront at any earlier time. Huygens' principle can be traced back to 1690 when Huygens published his classical work *Treatise on Light* (Huygens, 1690). Huygens was not able to formulate his principle precisely at that time. A number of famous scientists have worked in this area and elaborated this principle since then. It should be mentioned that different authors use the term 'Huygens' principle' with different meanings. The best-known representation of Huygens' principle is to express the field at some observation point in terms of a surface integral over a closed surface separating the observation point from the source. Such an expression can easily be obtained from (1.81) or (1.82), which in general gives a relationship between some volume integrals defined in the region V and some surface integrals defined on the boundary S. The idea behind Huygens' principle could apply not only to electromagnetics but also to any branch of physics, such as gravitation, elasticity, acoustics and many more (Rumsey, 1959).

One can consider three situations: (1) $\hat{L}^* = \hat{L}$ ($\hat{L}$ is formally self adjoint); (2) $\hat{L}^* \neq \hat{L}$; and (3) $\hat{L}^* = -\hat{L}$ ($\hat{L}$ is skew adjoint). For the first two situations, one can choose u_2 as a solution of the adjoint system $\hat{L}^*(u_2) = f_2$, where f_2 is a known source function. If the boundary term in (1.81) or (1.82) can be made to vanish, we have

$$\int_V u_2 f_1 dV = \int_V u_1 f_2 dV, \int_{T_1}^{T_2} dt \int_V u_2 f_1 dV = \int_{T_1}^{T_2} dt \int_V u_1 f_2 dV, \tag{1.83}$$

$$T(u_1, u_2; S) = 0, T(u_1, u_2; S, T_1, T_2) = 0. \tag{1.84}$$

Equations (1.83) are called reciprocity theorems of **Rayleigh-Carson form**, and Equations (1.84) are called reciprocity theorems of **Lorentz form**. These two forms are equivalent. For the situation $\hat{L}^* = -\hat{L}$, we may choose u_2 as a solution of the original system $\hat{L}(u_2) = f_2$. If the boundary term in (1.81) or (1.82) can be made to vanish, then

$$\int_V u_2 f_1 dV = -\int_V u_1 f_2 dV, \quad \int_{T_1}^{T_2} dt \int_V u_2 f_1 dV = -\int_{T_1}^{T_2} dt \int_V u_1 f_2 dV. \qquad (1.85)$$

The above relations may be called skew-reciprocity theorems of Rayleigh-Carson form.

The quantity $\int_V u_2 f_1 dV$ is called the **reaction** of field u_2 on source f_1 (Rumsey, 1954). Equations (1.83) (or (1.85)) simply state that the reaction of field u_2 on source f_1 is equal to the reaction (or the negative reaction) of field u_1 on source f_2. Apparently this kind of relations exists in various fields of physics and engineering. The concept of reaction is very useful and it can be used to answer some difficult questions with simplicity. If f_1 is a testing source of unit strength, the reaction $\int_V u_2 f_1 dV$ gives the numerical value of response u_2 at the point of the testing source. Thus, a method can be established to solve various boundary value problems based on the reaction, which is basically a theory of measurement rather than a field theory (Rumsey, 1963).

A number of reciprocity theorems for electromagnetic fields in both time domain and frequency domain can be derived by choosing different forms of the operator $\hat{L}$. But most of them are useless. Only one of the reciprocity theorems in frequency domain will be discussed here, which states that all possible time-harmonic fields of the same frequency are to some extent interrelated. Suppose that the sources $\mathbf{J}_1(\mathbf{r})$ and $\mathbf{J}_{m1}(\mathbf{r})$ give rise to the fields $\mathbf{E}_1(\mathbf{r})$ and $\mathbf{H}_1(\mathbf{r})$. Then the Maxwell equations in an isotropic medium can be rewritten in the operator form $\hat{L}(u_1) = f_1$ with

$$\hat{L} = \begin{bmatrix} -j\omega\varepsilon\cdot & \nabla\times \\ \nabla\times & j\omega\mu\cdot \end{bmatrix}, u_1 = \begin{bmatrix} \mathbf{E}_1 \\ \mathbf{H}_1 \end{bmatrix}, f_1 = \begin{bmatrix} \mathbf{J}_1 \\ -\mathbf{J}_{m1} \end{bmatrix}.$$

For an arbitrary $u_2 = [\mathbf{E}_2, \mathbf{H}_2]^T$ (the superscript T stands for the transpose operation), the formal adjoint of $\hat{L}$ and the boundary term may be found through integration by parts as

$$\hat{L}^* = \hat{L},$$

$$T(u_1, u_2; S) = \int_S (\mathbf{E}_1 \times \mathbf{H}_2 - \mathbf{E}_2 \times \mathbf{H}_1) \cdot \mathbf{u}_n dS,$$

where $\mathbf{u}_n$ is the outward unit normal to S. If $u_2 = [\mathbf{E}_2, \mathbf{H}_2]^T$ is a solution of the transposed system $\hat{L}^*(u_2) = f_2$ with $f_2 = [\mathbf{J}_2, -\mathbf{J}_{m2}]^T$, Equation (1.81) becomes

$$\int_V (\mathbf{E}_2 \cdot \mathbf{J}_1 - \mathbf{H}_2 \cdot \mathbf{J}_{m1}) dV = \int_V (\mathbf{E}_1 \cdot \mathbf{J}_2 - \mathbf{H}_1 \cdot \mathbf{J}_{m2}) dV$$

$$+ \int_S (\mathbf{E}_1 \times \mathbf{H}_2 - \mathbf{E}_2 \times \mathbf{H}_1) \cdot \mathbf{u}_n dS. \qquad (1.86)$$

If both sources are outside S, the surface integral in (1.86) is zero. If both sources are inside S, it can be shown that the surface integral is also zero by using the radiation condition. Therefore we obtain the Lorentz form of reciprocity

$$\int_S (\mathbf{E}_1 \times \mathbf{H}_2 - \mathbf{E}_2 \times \mathbf{H}_1) \cdot \mathbf{u}_n dS = 0,$$

and the Rayleigh-Carson form of reciprocity

$$\int_V (\mathbf{E}_2 \cdot \mathbf{J}_1 - \mathbf{H}_2 \cdot \mathbf{J}_{m1}) dV = \int_V (-\mathbf{H}_1 \cdot \mathbf{J}_{m2} + \mathbf{E}_1 \cdot \mathbf{J}_2) dV \qquad (1.87)$$

If the surface S only contains the sources $\mathbf{J}_1(\mathbf{r})$ and $\mathbf{J}_{m1}(\mathbf{r})$, Equation (1.86) becomes

$$\int_V (\mathbf{E}_2 \cdot \mathbf{J}_1 - \mathbf{H}_2 \cdot \mathbf{J}_{m1}) dV = \int_S (\mathbf{E}_2 \cdot \mathbf{u}_n \times \mathbf{H}_1 - \mathbf{H}_2 \cdot \mathbf{E}_1 \times \mathbf{u}_n) dS.$$

This is the familiar form of Huygens' principle. The electromagnetic reciprocity theorem can also be generalized to an anisotropic medium (Kong, 1990; Tai, 1961; Harrington, 1958).

Example 1.6: An interesting application of the reciprocity theorem is to prove that a surface electric (or magnetic) current pressed tightly on a perfect electric (or magnetic conductor) does not radiate. Let $\mathbf{J}_{s1}$ be a surface electric current pressed tightly on a perfect electric conductor, which generates electromagnetic fields $\mathbf{E}_1$ and $\mathbf{H}_1$. Now remove the surface electric current $\mathbf{J}_{s1}$ and place an arbitrary current source $\mathbf{J}_2$ in space that produces electromagnetic fields $\mathbf{E}_2$ and $\mathbf{H}_2$. According to (1.87), we have $\int_V \mathbf{E}_2 \cdot \mathbf{J}_{s1} dV = \int_V \mathbf{E}_1 \cdot \mathbf{J}_2 dV$, where V denotes the region outside the conductor. Since $\mathbf{E}_2$ only has a normal component on the surface of the conductor while $\mathbf{J}_{s1}$ is a tangential vector, the left side of the above equation must be zero. Thus, we have $\int_V \mathbf{E}_1 \cdot \mathbf{J}_2 dV = 0$. For $\mathbf{J}_2$ is arbitrary, we obtain $\mathbf{E}_1 = 0$. □

1.4   Wavepackets

A time-domain field can be expressed as the superposition of individual plane waves of the form $e^{j\omega t - j\mathbf{k} \cdot \mathbf{r}}$. Each plane wave travels with a **phase velocity** defined by $\mathbf{v}_p = \mathbf{u}_k \omega / |\mathbf{k}|$, where $\mathbf{u}_k = \mathbf{k}/|\mathbf{k}|$ and $\mathbf{k}$ is the wave vector. The phase velocity is the velocity at which the points of constant phase move in the medium. It is well known that, to transmit energy or a signal, the waves must come in a range of frequencies to form a wavepacket. The **wavepacket** was first introduced by Schrödinger and is used to represent a small group of plane waves. There are two different ways of building wavepackets. A waveform is called a **spatial wavepacket** (or **paraxial approximation**) if it is monochromatic and is confined to a narrow region of space along the path of propagation. A spatial wavepacket is basically a beam of wave. A waveform is called a **temporal wavepacket** (or **narrow-band approximation**) if it propagates in only one direction and its frequency spectrum is confined to a narrow band around a central frequency.

The propagation of wavepackets in an absorbing medium was studied for the first time by the German physicist, Arnold Johannes Wilhelm Sommerfeld (1868–1951), and the French physicist, Léon Nicolas Brillouin (1889–1969) (Brillouin, 1960). The group velocity, signal velocity and energy velocity are important quantities for characterizing the propagation of wavepackets and there are certain relationships among them. The speed of each frequency component is the phase velocity while the speed of the envelope of the wavepacket is called **group velocity**. The velocity at which the main part of the wavepacket propagates is called **signal velocity**. When a signal propagates in a dispersive medium, it does not retain its original form. At certain depth of the medium, very weak signal components appear at first and are called **forerunners** or **fronts** whose speed is always equal to the light speed in vacuum. The **energy velocity** of the wavepacket is defined as the ratio of the Poynting vector to the energy density.

1.4.1   Spatial Wavepacket and Temporal Wavepacket

By definition, a spatial wavepacket can be represented by

$$
\begin{aligned}
\mathbf{F}(\mathbf{r}, t) &= \frac{1}{(2\pi)^3} \int_{-\infty}^{\infty} \tilde{\mathbf{F}}(\mathbf{k}) e^{j\omega(\mathbf{k})t - j\mathbf{k}\cdot\mathbf{r}} d\mathbf{k} \\
&= \frac{1}{(2\pi)^3} \int_{-\infty}^{\infty} \int_{-\infty}^{\infty} \mathbf{F}(\xi, 0) e^{j\omega(\mathbf{k})t - j\mathbf{k}\cdot(\mathbf{r}-\xi)} d\xi d\mathbf{k},
\end{aligned}
\tag{1.88}
$$

where $\tilde{\mathbf{F}}(\mathbf{k})$ is given by $\tilde{\mathbf{F}}(\mathbf{k}) = \int_{-\infty}^{\infty} \mathbf{F}(\mathbf{r}, 0) e^{j\mathbf{k}\cdot\mathbf{r}} d\mathbf{r}$ and ω is the angular frequency of the wavepacket. Since $\mathbf{F}(\mathbf{r}, 0)$ is narrow-band in $\mathbf{k}$-space, a rapid phase variation $e^{-j\mathbf{k}_c\cdot\mathbf{r}}$ may be factored out so that one may write $\mathbf{F}(\mathbf{r}, 0) = \mathbf{A}(\mathbf{r}, 0) e^{-j\mathbf{k}_c\cdot\mathbf{r}}$, where $\mathbf{k}_c$ is the central wave vector and $\mathbf{A}(\mathbf{r}, 0)$ is the complex envelope that describes the slowly varying transverse beam profile or the spatial modulation as the wave propagates. If the dispersion of the medium is not strong, we may use the first-order approximation for the dispersion relation

$$
\omega(\mathbf{k}) \approx \omega_c + \delta\mathbf{k} \cdot \nabla\omega(\mathbf{k}_c),
$$

where $\omega_c = \omega(\mathbf{k}_c)$ and $\delta\mathbf{k} = \mathbf{k} - \mathbf{k}_c$. As a result, Equation (1.88) can be approximated by

$$
\begin{aligned}
\mathbf{F}(\mathbf{r}, t) &= \int_{-\infty}^{\infty} \mathbf{F}(\xi, 0) e^{j[\omega_c t - \mathbf{k}_c\cdot(\mathbf{r}-\xi)]} \delta\left[\nabla\omega(\mathbf{k}_c)t - (\mathbf{r} - \xi)\right] d\xi \\
&= \mathbf{F}(\mathbf{r} - \mathbf{v}_g t, 0) e^{j(\omega_c t - \mathbf{k}_c\cdot\mathbf{v}_g t)},
\end{aligned}
$$

where $\mathbf{v}_g = \nabla\omega(\mathbf{k}_c)$ is defined as the group velocity. Making use of $\mathbf{F}(\mathbf{r}, 0) = \mathbf{A}(\mathbf{r}, 0) e^{-j\mathbf{k}_c\cdot\mathbf{r}}$ we have

$$
\mathbf{F}(\mathbf{r}, t) = \mathbf{A}(\mathbf{r} - \mathbf{v}_g t, 0) e^{j(\omega_c t - \mathbf{k}_c\cdot\mathbf{r})}.
\tag{1.89}
$$

Hence the group velocity represents the speed of the envelope of the wavepacket. In a medium where the dispersion is not strong, the shape of the envelope of the wavepacket does not change very much as it propagates. When the wavepacket propagates in a highly dispersive medium, the shape of the envelope of the wavepacket will not remain the same. The phase of the wavepacket will change as the propagation distance and time increase. As a result, the concept of group velocity is no longer valid in a highly dispersive medium.

By definition, an arbitrary temporal wavepacket may be expressed as

$$\mathbf{F}(\mathbf{r}, t) = \frac{1}{2\pi} \int\limits_{-\infty}^{\infty} \tilde{\mathbf{F}}(\omega) e^{j[\omega t - \mathbf{k}(\omega) \cdot \mathbf{r}]} d\omega$$

$$= \frac{1}{2\pi} \int\limits_{-\infty}^{\infty} \int\limits_{-\infty}^{\infty} \mathbf{F}(0, t') e^{j[\omega(t-t') - \mathbf{k}(\omega) \cdot \mathbf{r}]} dt' d\omega,$$

(1.90)

where $\tilde{\mathbf{F}}(\omega) = \int\limits_{-\infty}^{\infty} \mathbf{F}(0, t) e^{-j\omega t} dt$. The narrow-band approximation assumes that the frequency spectrum of the time variation is confined to a narrow band around a carrier. Therefore $\mathbf{F}(0, t)$ is a bandpass signal and can be written as $\mathbf{F}(0, t) = \mathbf{A}(0, t) e^{j\omega_c t}$, where ω_c is the carrier frequency and $\mathbf{A}(0, t)$ is a slowly varying function of time. If the dispersion of the medium is not very strong we may make the first order approximation

$$\mathbf{k}(\omega) \approx \mathbf{k}_c + \delta\omega \frac{d\mathbf{k}(\omega_c)}{d\omega},$$

where $\delta\omega = \omega - \omega_c$ and $\mathbf{k}_c$ is the wave vector at the carrier frequency ω_c. Hence (1.90) can be expressed as

$$\mathbf{F}(\mathbf{r}, t) = \frac{1}{2\pi} \int\limits_{-\infty}^{\infty} \int\limits_{-\infty}^{\infty} \mathbf{F}(0, t') e^{j\omega_c(t-t') - j\mathbf{k}_c \cdot \mathbf{r} + j\delta\omega\left[(t-t') - \frac{d\mathbf{k}(\omega_c)}{d\omega} \cdot \mathbf{r}\right]} dt' d\delta\omega$$

$$= \mathbf{F}\left(0, t - \frac{d\mathbf{k}(\omega_c)}{d\omega} \cdot \mathbf{r}\right) e^{j\omega_c \frac{d\mathbf{k}(\omega_c)}{d\omega} \cdot \mathbf{r} - j\mathbf{k}_c \cdot \mathbf{r}}$$

(1.91)

$$= \mathbf{A}\left(0, t - \frac{d\mathbf{k}(\omega_c)}{d\omega} \cdot \mathbf{r}\right) e^{j\omega_c t - j\mathbf{k}_c \cdot \mathbf{r}}.$$

If the wavepacket propagates mainly in $\mathbf{k}_c$ direction, that is, $\mathbf{k} \approx \mathbf{k}_c$, we have $d\mathbf{k}(\omega_c)/d\omega = \mathbf{u}_{k_c}/v_g$, where $\mathbf{u}_{k_c}$ is the unit vector along the direction of $\mathbf{k}_c$, and $v_g = d\omega(\omega_c)/dk$ is the group velocity. Equation (1.91) indicates that the propagation velocity of the envelope of a temporal wavepacket is equal to group velocity. Again, the shape of the envelope as well as the phase of the wavepacket will change as it propagates in a strongly dispersive medium, and the concept of group velocity becomes invalid.

An analogy exists between the spatial diffraction of beams and temporal dispersion of pulses. It can be shown that the equation that describes how a wavepacket spreads in time due to dispersion is equivalent to the equation for the transverse spreading due to diffraction. The temporal imaging technique is based on this space–time analogy (Kolner, 1994).

1.4.2   Signal Velocity and Group Velocity

According to Sommerfeld and Brillouin, the signal velocity represents the velocity of the main part of the signal. Thus we may define signal velocity as $\mathbf{v}_s = d\mathbf{r}_p/dt$, where $\mathbf{r}_p$ is the position of the main part of the wavepacket $\mathbf{F}$, defined by (Vichnevetsky, 1988)

$$\mathbf{r}_p(t) = \frac{\int\limits_{-\infty}^{\infty} \mathbf{r}\,|\mathbf{F}(\mathbf{r},t)|^2\,d\mathbf{r}}{\int\limits_{-\infty}^{\infty} |\mathbf{F}(\mathbf{r},t)|^2\,d\mathbf{r}}. \tag{1.92}$$

It should be understood that the concept of signal velocity of a wavepacket is useful only when the dispersion of the medium is not very strong (so that the first-order approximation for the dispersion relation is valid). Substituting (1.89) into (1.92) and using the transformation $\mathbf{r} = \mathbf{u} + \mathbf{v}_g t$, we obtain

$$\mathbf{r}_p(t) = \frac{\int\limits_{-\infty}^{\infty} \mathbf{r}\,\left|\mathbf{A}(\mathbf{r}-\mathbf{v}_g t,0)\right|^2\,d\mathbf{r}}{\int\limits_{-\infty}^{\infty} \left|\mathbf{A}(\mathbf{r}-\mathbf{v}_g t,0)\right|^2\,d\mathbf{r}} = \frac{\int\limits_{-\infty}^{\infty} (\mathbf{u}+\mathbf{v}_g t)\,|\mathbf{A}(\mathbf{u},0)|^2\,d\mathbf{u}}{\int\limits_{-\infty}^{\infty} |\mathbf{A}(\mathbf{u},0)|^2\,d\mathbf{u}}.$$

By taking the time derivative of the above equation, we obtain $\mathbf{v}_s = \mathbf{v}_g$, and the signal velocity of a spatial wavepacket is equal to the group velocity. Similarly substituting (1.91) into (1.92) and making use of the transformation $\mathbf{r} = \mathbf{u} + v_g t \mathbf{u}_{k_c}$ yields

$$\mathbf{r}_p(t) = \frac{\int\limits_{-\infty}^{\infty} \mathbf{r}\,\left|\mathbf{A}(0,t-\mathbf{u}_{k_c}\cdot\mathbf{r}/v_g)\right|^2\,d\mathbf{r}}{\int\limits_{-\infty}^{\infty} \left|\mathbf{A}(0,t-\mathbf{u}_{k_c}\cdot\mathbf{r}/v_g)\right|^2\,d\mathbf{r}} = \frac{\int\limits_{-\infty}^{\infty} (\mathbf{u}+v_g t \mathbf{u}_{k_c})\,\left|\mathbf{A}(0,-\mathbf{u}_{k_c}\cdot\mathbf{u}/v_g)\right|^2\,d\mathbf{u}}{\int\limits_{-\infty}^{\infty} \left|\mathbf{A}(0,-\mathbf{u}_{k_c}\cdot\mathbf{u}/v_g)\right|^2\,d\mathbf{u}}.$$

Hence $\mathbf{v}_s = v_g \mathbf{u}_{k_c} = \mathbf{v}_g$, and the signal velocity of a temporal wavepacket is equal to the group velocity.

1.4.3   Energy Density for Wavepackets

An expression for the electromagnetic energy density that does not involve any medium properties (such as isotropic, anisotropic) is useful. Such an expression exists for a monochromatic wave (Tonning, 1960), and can be generalized to the wavepackets, which are more realistic in applications. In order to find the general expression of the energy density for a wavepacket, we have to assume that the dispersion of the medium is not very strong so that the first-order approximation of the dispersion is valid. From (1.89) and (1.91), the fields for both spatial and temporal wavepackets can be expressed as

$$\mathbf{E}(\mathbf{r},t) = \mathrm{Re}[\mathbf{E}_{en}(\mathbf{r},t;\omega_c)e^{j\omega_c t}], \text{ etc.} \tag{1.93}$$

where $\mathbf{E}_{en}$ etc. are the envelopes and they are slowly varying functions of time compared to $e^{j\omega_c t}$, and ω_c is the angular frequency of the monochromatic paraxial wave or the carrier wave frequency for the narrow-band signal. We cannot apply (1.57) to a wavepacket directly because the fields might not be zero at $t = -\infty$. To make use of (1.57), the standard way is to introduce a damping mechanism for the fields first so that all the fields are zero at $t = -\infty$, and then let the damping tend to zero after the calculation is finished. To this end, we can introduce a complex frequency $\tilde{\omega} = -j\alpha + \omega_c$ to replace the real frequency ω_c, where α is a small positive number (the damping factor). Equation (1.93) may be rewritten as

$$\tilde{\mathbf{E}}(\mathbf{r}, t) = \mathrm{Re}[\tilde{\mathbf{E}}_{en}(\mathbf{r}, t; \tilde{\omega})e^{j\tilde{\omega}t}], \text{ etc.} \tag{1.94}$$

which approach zero when $t \to -\infty$ and approach the corresponding real fields as $\alpha \to 0$. Assuming that the fields are analytic functions of frequency, the following first-order expansion can be made

$$\tilde{\mathbf{E}}_{en} \approx \mathbf{E}_{en} - j\alpha \frac{\partial \mathbf{E}_{en}}{\partial \omega_c}, \text{ etc.}$$

for α is assumed to be small. A simple calculation shows that

$$\tilde{\mathbf{E}} \cdot \frac{\partial \tilde{\mathbf{D}}}{\partial t} - \tilde{\mathbf{D}} \cdot \frac{\partial \tilde{\mathbf{E}}}{\partial t} + \tilde{\mathbf{H}} \cdot \frac{\partial \tilde{\mathbf{B}}}{\partial t} - \tilde{\mathbf{B}} \cdot \frac{\partial \tilde{\mathbf{H}}}{\partial t}$$

$$= -\omega_c e^{2\alpha t} \mathrm{Im}(\bar{\mathbf{E}}_{en} \cdot \mathbf{D}_{en} + \bar{\mathbf{H}}_{en} \cdot \mathbf{B}_{en})$$

$$+ \alpha \omega_c e^{2\alpha t} \mathrm{Re} \left(\bar{\mathbf{E}}_{en} \cdot \frac{\partial \mathbf{D}_{en}}{\partial \omega_c} - \mathbf{D}_{en} \cdot \frac{\partial \bar{\mathbf{E}}_{en}}{\partial \omega_c} + \bar{\mathbf{H}}_{en} \cdot \frac{\partial \mathbf{B}_{en}}{\partial \omega_c} - \mathbf{B}_{en} \cdot \frac{\partial \bar{\mathbf{H}}_{en}}{\partial \omega_c} \right).$$

From the lossless condition $\overline{\nabla \cdot (\mathbf{E} \times \mathbf{H})} = 0$ and Maxwell equations, we obtain

$$\mathrm{Im}(\mathbf{E}_{en} \cdot \mathbf{D}_{en} - \bar{\mathbf{H}}_{en} \cdot \mathbf{B}_{en}) = -\mathrm{Im}(\bar{\mathbf{E}}_{en} \cdot \mathbf{D}_{en} + \bar{\mathbf{H}}_{en} \cdot \mathbf{B}_{en}) = 0. \tag{1.95}$$

Thus the quantity $\bar{\mathbf{E}}_{en} \cdot \mathbf{D}_{en} + \bar{\mathbf{H}}_{en} \cdot \mathbf{B}_{en}$ is real. Consequently the integral of (1.57) can be expressed as

$$\int_{-\infty}^{t} \frac{1}{2} \left(\mathbf{E} \cdot \frac{\partial \mathbf{D}}{\partial t} - \mathbf{D} \cdot \frac{\partial \mathbf{E}}{\partial t} + \mathbf{H} \cdot \frac{\partial \mathbf{B}}{\partial t} - \mathbf{B} \cdot \frac{\partial \mathbf{H}}{\partial t} \right) dt$$

$$= \lim_{\alpha \to 0} \frac{1}{2} \int_{-\infty}^{t} \tilde{\mathbf{E}} \cdot \frac{\partial \tilde{\mathbf{D}}}{\partial t} - \tilde{\mathbf{D}} \cdot \frac{\partial \tilde{\mathbf{E}}}{\partial t} + \tilde{\mathbf{H}} \cdot \frac{\partial \tilde{\mathbf{B}}}{\partial t} - \tilde{\mathbf{B}} \cdot \frac{\partial \tilde{\mathbf{H}}}{\partial t} dt \tag{1.96}$$

$$= \frac{\omega_c}{4} \mathrm{Re} \left(\bar{\mathbf{E}}_{en} \cdot \frac{\partial \mathbf{D}_{en}}{\partial \omega_c} - \bar{\mathbf{D}}_{en} \cdot \frac{\partial \mathbf{E}_{en}}{\partial \omega_c} + \bar{\mathbf{H}}_{en} \cdot \frac{\partial \mathbf{B}_{en}}{\partial \omega_c} - \bar{\mathbf{B}}_{en} \cdot \frac{\partial \mathbf{H}_{en}}{\partial \omega_c} \right).$$

The quantity in the bracket is real. Actually taking the derivative of (1.95) with respect to the frequency gives

$$\text{Im}\left(\bar{\mathbf{E}}_{en} \cdot \frac{\partial \mathbf{D}_{en}}{\partial \omega_c} + \mathbf{D}_{en} \cdot \frac{\partial \bar{\mathbf{E}}_{en}}{\partial \omega_c} + \bar{\mathbf{H}}_{en} \cdot \frac{\partial \mathbf{B}_{en}}{\partial \omega_c} + \mathbf{B}_{en} \cdot \frac{\partial \bar{\mathbf{H}}_{en}}{\partial \omega_c}\right) = 0.$$

This relation still holds if we take the complex conjugate of the second and the fourth term and change their sign simultaneously. Since the envelopes can be considered as constants over one period of the carrier wave $e^{j\omega_c t}$, the time average of (1.96) over one period of the carrier wave is

$$\frac{1}{2}\overline{\int_{-\infty}^{t}\left(\mathbf{H} \cdot \frac{\partial \mathbf{B}}{\partial t} - \mathbf{B} \cdot \frac{\partial \mathbf{H}}{\partial t} + \mathbf{E} \cdot \frac{\partial \mathbf{D}}{\partial t} - \mathbf{D} \cdot \frac{\partial \mathbf{E}}{\partial t}\right) dt}$$

$$= \frac{\omega_c}{4}\left(\bar{\mathbf{E}}_{en} \cdot \frac{\partial \mathbf{D}_{en}}{\partial \omega_c} - \bar{\mathbf{D}}_{en} \cdot \frac{\partial \mathbf{E}_{en}}{\partial \omega_c} + \bar{\mathbf{H}}_{en} \cdot \frac{\partial \mathbf{B}_{en}}{\partial \omega_c} - \bar{\mathbf{B}}_{en} \cdot \frac{\partial \mathbf{H}_{en}}{\partial \omega_c}\right). \tag{1.97}$$

The above expression can be interpreted as the energy density related to dispersion and will be denoted $\bar{w}_d$. Similarly the time average of the rest of (1.57) over one period of the carrier wave is

$$\frac{1}{2}\overline{(\mathbf{E} \cdot \mathbf{D} + \mathbf{H} \cdot \mathbf{B})} = \frac{1}{4}\text{Re}(\mathbf{E}_{en} \cdot \bar{\mathbf{D}}_{en} + \mathbf{H}_{en} \cdot \bar{\mathbf{B}}_{en})$$

$$= \frac{1}{4}(\mathbf{E}_{en} \cdot \bar{\mathbf{D}}_{en} + \mathbf{H}_{en} \cdot \bar{\mathbf{B}}_{en}). \tag{1.98}$$

It follows from (1.57), (1.97) and (1.98) that the time average of the energy density over one period of the carrier wave $e^{j\omega_c t}$ can be expressed as

$$\bar{\bar{w}} = \frac{1}{4}(\bar{\mathbf{E}}_{en} \cdot \mathbf{D}_{en} + \bar{\mathbf{H}}_{en} \cdot \mathbf{B})$$

$$+ \frac{\omega_c}{4}\left(\bar{\mathbf{E}}_{en} \cdot \frac{\partial \mathbf{D}_{en}}{\partial \omega_c} - \bar{\mathbf{D}}_{en} \cdot \frac{\partial \mathbf{E}_{en}}{\partial \omega_c} + \bar{\mathbf{H}}_{en} \cdot \frac{\partial \mathbf{B}_{en}}{\partial \omega_c} - \bar{\mathbf{B}}_{en} \cdot \frac{\partial \mathbf{H}_{en}}{\partial \omega_c}\right) \tag{1.99}$$

$$= \frac{1}{4}\left[\bar{\mathbf{E}}_{en} \cdot \frac{\partial(\omega_c \mathbf{D}_{en})}{\partial \omega_c} - \omega_c \bar{\mathbf{D}}_{en} \cdot \frac{\partial \mathbf{E}_{en}}{\partial \omega_c} + \bar{\mathbf{H}}_{en} \cdot \frac{\partial(\omega_c \mathbf{B}_{en})}{\partial \omega_c} - \omega_c \bar{\mathbf{B}}_{en} \cdot \frac{\partial \mathbf{H}_{en}}{\partial \omega_c}\right].$$

Taking the time average of (1.55) over one period of the carrier wave $e^{j\omega_c t}$, the Poynting theorem in a lossless medium without impressed sources becomes

$$\int_S \bar{\bar{\mathbf{S}}} \cdot \mathbf{u}_n dS + \int_V \frac{\partial \bar{\bar{w}}}{\partial t} dV = 0,$$

where $\bar{\bar{\mathbf{S}}} = \text{Re}(\mathbf{E}_{en} \times \bar{\mathbf{H}}_{en})/2$ is the time average of the Poynting vector over one period of the wave $e^{j\omega_c t}$ and the calculation $\overline{\partial w/\partial t} = \partial \bar{\bar{w}}/\partial t$ has been used. Note that $\partial \bar{\bar{w}}/\partial t = 0$ for a monochromatic wave. As a special case, let us consider an isotropic medium defined by $\mathbf{D}_{en} = \varepsilon\mathbf{E}_{en}$, $\mathbf{B}_{en} = \mu\mathbf{H}_{en}$. In this case, Equation (1.99) reduces to the well-known expression

$$\overline{\overline{w}} = \frac{1}{4}\left[\frac{\partial(\omega_c\varepsilon)}{\partial\omega_c}|\mathbf{E}_{en}|^2 + \frac{\partial(\omega_c\mu)}{\partial\omega_c}|\mathbf{H}_{en}|^2\right].$$

1.4.4   Energy Velocity and Group Velocity

The energy velocity is defined as the ratio of the Poynting vector to the energy density, that is, $\mathbf{v}_e = \overline{\overline{\mathbf{S}}}/\overline{\overline{w}}$. If the dispersion of the medium is not very strong, a spatial or temporal wavepacket in its first-order approximation can be expressed as

$$\mathbf{E}(\mathbf{r}, t) = \mathrm{Re}[\mathbf{E}_0(\mathbf{r}, t; \omega_c)e^{j\omega_c t - j\mathbf{k}_c\cdot\mathbf{r}}], \text{ etc.} \tag{1.100}$$

where the fast phase variation $e^{-j\mathbf{k}_c\cdot\mathbf{r}}$ of the fields has been factored out. The new envelopes $\mathbf{E}_0$, etc. are slowly varying functions of both spatial coordinates and time. Introducing (1.100) into Maxwell equations in a source-free and lossless region and using the calculation $\nabla e^{-j\mathbf{k}_c\cdot\mathbf{r}} = -j\mathbf{k}_c e^{-j\mathbf{k}_c\cdot\mathbf{r}}$, we obtain

$$\mathbf{k}_c \times \mathbf{H}_0 + j\nabla \times \mathbf{H}_0 = -\omega_c\mathbf{D}_0,$$
$$\mathbf{k}_c \times \mathbf{E}_0 + j\nabla \times \mathbf{E}_0 = \omega_c\mathbf{B}_0.$$

Since $\mathbf{E}_0$ and $\mathbf{H}_0$ are slowly varying function of space, we can let $\nabla \times \mathbf{E}_0 \approx 0$ and $\nabla \times \mathbf{H}_0 \approx 0$. Thus the above equation may be rewritten as

$$\mathbf{k}_c \times \mathbf{H}_{en} + \omega_c\mathbf{D}_{en} \approx 0,$$
$$\mathbf{k}_c \times \mathbf{E}_{en} - \omega_c\mathbf{B}_{en} \approx 0. \tag{1.101}$$

By letting $\mathbf{k}_c = k_{cx}\mathbf{u}_x + k_{cy}\mathbf{u}_y + k_{cz}\mathbf{u}_z$ and taking the derivative of (1.101) with respect to k_{cx}, we obtain

$$\mathbf{u}_x \times \mathbf{H}_{en} + \left[\mathbf{k}_c \times \frac{\partial\mathbf{H}_{en}}{\partial\omega_c} + \frac{\partial(\omega_c\mathbf{D}_{en})}{\partial\omega_c}\right]\frac{\partial\omega_c}{\partial k_{cx}} \approx 0,$$
$$\mathbf{u}_x \times \mathbf{E}_{en} + \left[\mathbf{k}_c \times \frac{\partial\mathbf{E}_{en}}{\partial\omega_c} - \frac{\partial(\omega_c\mathbf{B}_{en})}{\partial\omega_c}\right]\frac{\partial\omega_c}{\partial k_{cx}} \approx 0.$$

Multiplying the first equation by $-\bar{\mathbf{E}}_{en}$ and second by $\bar{\mathbf{H}}_{en}$ and adding the resultant equations and using (1.99), we get

$$\mathbf{u}_x \cdot \overline{\overline{\mathbf{S}}} = \frac{\partial\omega_c}{\partial k_{cx}} \cdot \frac{1}{4}\left[\bar{\mathbf{E}}_{en} \cdot \frac{\partial(\omega\mathbf{D}_{en})}{\partial\omega_c} - \omega_c\bar{\mathbf{D}}_{en} \cdot \frac{\partial\mathbf{E}_{en}}{\partial\omega_c} + \bar{\mathbf{H}}_{en} \cdot \frac{\partial(\omega_c\mathbf{B}_{en})}{\partial\omega_c} - \omega_c\bar{\mathbf{B}}_{en} \cdot \frac{\partial\mathbf{H}_{en}}{\partial\omega_c}\right]$$
$$= \frac{\partial\omega_c}{\partial k_{cx}}\bar{w}.$$

Similarly we have $\mathbf{u}_y \cdot \overline{\overline{\mathbf{S}}} = \overline{\overline{w}} \partial \omega_c / \partial k_{cy}$ and $\mathbf{u}_z \cdot \overline{\overline{\mathbf{S}}} = \overline{\overline{w}} \partial \omega_c / \partial k_{cz}$. Therefore

$$\mathbf{v}_g = \frac{\overline{\overline{\mathbf{S}}}}{\overline{\overline{w}}} = \nabla \omega_c(\mathbf{k}_c) = \mathbf{v}_e.$$

This indicates that the group velocity is always equal to the energy velocity for a spatial wavepacket.

Taking the derivative of (1.101) with respect to the frequency, we obtain

$$\frac{d\mathbf{k}_c}{d\omega_c} \times \mathbf{H}_{en} + \mathbf{k}_c \times \frac{\partial \mathbf{H}_{en}}{\partial \omega_c} + \frac{\partial (\omega_c \mathbf{D}_{en})}{\partial \omega_c} \approx 0,$$

$$\frac{d\mathbf{k}_c}{d\omega_c} \times \mathbf{E}_{en} + \mathbf{k}_c \times \frac{\partial \mathbf{E}_{en}}{\partial \omega_c} - \frac{\partial (\omega_c \mathbf{B}_{en})}{\partial \omega_c} \approx 0.$$

Multiplying the first equation by $-\bar{\mathbf{E}}_{en}$ and second by $\bar{\mathbf{H}}_{en}$ and adding the resultant equations and using (1.99) yields

$$\frac{d\mathbf{k}_c}{d\omega_c} \cdot \overline{\overline{\mathbf{S}}} = \frac{1}{4} \left[\bar{\mathbf{E}}_{en} \cdot \frac{\partial (\omega \mathbf{D}_{en})}{\partial \omega_c} - \omega_c \bar{\mathbf{D}}_{en} \cdot \frac{\partial \mathbf{E}_{en}}{\partial \omega_c} + \bar{\mathbf{H}}_{en} \cdot \frac{\partial (\omega_c \mathbf{B}_{en})}{\partial \omega_c} - \omega_c \bar{\mathbf{B}}_{en} \cdot \frac{\partial \mathbf{H}_{en}}{\partial \omega_c} \right] = \overline{\overline{w}}.$$

It follows that

$$\mathbf{v}_e \cdot \frac{d\mathbf{k}_c}{d\omega_c} \approx 1 \text{ or } \mathbf{v}_e \cdot \mathbf{u}_{k_c} \approx v_g.$$

The above equation shows that the projection of the energy velocity in the direction of wave propagation is always equal to the group velocity for a temporal wavepacket.

Remark 1.14: In deriving the electromagnetic energy density for a wavepacket in a general lossless medium, a damping mechanism (that is, the small parameter α) has been introduced. This process appears to be a bit contrived. Nonetheless it is required by the uniqueness theorem for solutions of Maxwell equations. In a steady state, the information about the initial condition of the field has been lost and many possible solutions may exist. Introducing the loss is equivalent to introducing causality. □

Remark 1.15: One of the essential assumptions in special relativity is that the light speed is the greatest speed at which energy, information and signals can be transmitted. This is also the requirement of causality. Sommerfeld and Brillouin were the first to note that group velocity could be faster than light in the regions of anomalous dispersion. Some experiments in recent years have shown that the group velocity can exceed the light speed c or even become negative (for example, Wong, 2000). In all these experiments, the wavepackets experience a very strong dispersion when they travel in the medium, and the concept of group velocity that relies on the first-order approximation of dispersion relation is actually invalid. □

Giving an exact definition for the propagation velocity of wavepackets in a highly dispersive medium is essentially difficult. Several definitions have been proposed for various specific situations (Fushchych, 1998; Diener, 1998).

1.4.5 Narrow-band Stationary Stochastic Vector Field

As a linear modulation technique, an easy way to translate the spectrum of low-pass or baseband signal to a higher frequency is to multiply or heterodyne the baseband signal with a carrier wave. A narrowband bandpass stochastic vector field $\mathbf{F}$ (**modulated signal**) in the time domain can be expressed as

$$\mathbf{F}(\mathbf{r}, t) = \begin{cases} \mathbf{a}(\mathbf{r}, t) \cos[\omega_c t + \varphi(\mathbf{r}, t)], \\ \mathbf{x}(\mathbf{r}, t) \cos \omega_c t - \mathbf{y}(\mathbf{r}, t) \sin \omega_c t, \\ \text{Re}\, \mathbf{F}_{en}(\mathbf{r}, t) e^{j\omega_c t}, \end{cases}$$

where $\omega_c = 2\pi f_c$, $\mathbf{a}(\mathbf{r}, t)$ and $\varphi(\mathbf{r}, t)$ are the carrier frequency, envelope and phase of the modulated signal respectively, and

$$\mathbf{F}_{en}(\mathbf{r}, t) = \mathbf{x}(\mathbf{r}, t) + j\mathbf{y}(\mathbf{r}, t),$$
$$\mathbf{x}(\mathbf{r}, t) = \mathbf{a}(\mathbf{r}, t) \cos \varphi(\mathbf{r}, t),$$
$$\mathbf{y}(\mathbf{r}, t) = \mathbf{a}(\mathbf{r}, t) \sin \varphi(\mathbf{r}, t).$$

Here $\mathbf{F}_{en}(\mathbf{r}, t)$, $\mathbf{x}(\mathbf{r}, t)$ and $\mathbf{y}(\mathbf{r}, t)$ are the complex envelope, in-phase component, and quadrature component of the modulated signal respectively. The complex envelope $\mathbf{F}_{en}(\mathbf{r}, t)$ is a slowly varying function of time compared to $e^{j\omega_c t}$. It is easy to show that the complex envelopes of the electromagnetic fields satisfy the time-harmonic Maxwell equations

$$\nabla \times \mathbf{H}_{en}(\mathbf{r}, t) = j\omega_c \varepsilon \mathbf{E}_{en}(\mathbf{r}, t) + \mathbf{J}_{en}(\mathbf{r}, t),$$
$$\nabla \times \mathbf{E}_{en}(\mathbf{r}, t) = -j\omega_c \mu \mathbf{H}_{en}(\mathbf{r}, t). \tag{1.102}$$

Therefore the theoretical results about the time-harmonic fields can be applied to the complex envelopes of the fields. Let $\langle \mathbf{F} \rangle$ denote the ensemble average of $\mathbf{F}$. For a stationary and ergodic vector field $\mathbf{F}$, the ensemble average equals the time average, that is, $\langle \mathbf{F} \rangle = \overline{\overline{\mathbf{F}}} = \lim_{T \to \infty} \frac{1}{T} \int_{-T/2}^{T/2} \mathbf{F}(t) dt$. For a stationary and ergodic electromagnetic field, we may take the ensemble average of (1.102) to get

$$\nabla \times \overline{\overline{\mathbf{H}_{en}(\mathbf{r})}} = j\omega_c \varepsilon \overline{\overline{\mathbf{E}_{en}(\mathbf{r})}} + \overline{\overline{\mathbf{J}_{en}(\mathbf{r})}},$$
$$\nabla \times \overline{\overline{\mathbf{E}_{en}(\mathbf{r})}} = -j\omega_c \mu \overline{\overline{\mathbf{H}_{en}(\mathbf{r})}}.$$

Hence the theoretical results about the time-harmonic fields can also be applied to the ensemble averages of the complex envelopes of the fields.

> All the mathematical sciences are founded on relations between physical laws and laws of numbers, so that the aim of exact science is to reduce the problems of nature to the determination of quantities by operations with numbers.
>
> —James Maxwell

2

Solutions of Maxwell Equations

At the very beginning the given elliptic differential operator is only defined on a space of C^2-functions. The operator is extended to an abstractly defined operator using a formal closure. The main task is to show that the extended operator is self-adjoint. In this case it is possible to apply the methods of von Neumann.

—Kurt Otto Friedrichs

Maxwell equations are a set of partial differential equations (PDEs) relating to the field quantities and their partial derivatives with respect to space and time. The study of partial differential equations goes back to eighteenth century when they were first used to describe the problems in physical sciences. Various methods for the solution of partial differential equation have been proposed since then. Linear PDEs are generally solved by means of variational method, the method of separation of variables, and the method of Green's function, named after the British mathematician George Green (1793–1841). There are no generally applicable methods to solve non-linear PDEs. In most situations, numerical methods must be adopted for both linear and non-linear PDEs.

The solutions of partial differential equations had been understood to be classical solutions that satisfy the smooth conditions required by the differential operators until the 1920s when people began to realize that the smoothness requirements of classical solutions might be too stringent in some applications. This led to the notion of the generalized solution and the weak solution. The theory of generalized solutions is founded on the theory of generalized functions, which had been used by physicist before the French mathematician Laurent-Moïse Schwartz (1915–2002) set up a rigorous mathematical theory around 1950.

The physical laws are typically characterized by differential equations, and they can also be expressed as a variational principle. Newton's second law, described by a differential equation of second order, expresses the vector relationship between the forces acting on an object and the motion of the object. When the system becomes complicated, for example a multiple particle system with mutual interaction, applying Newton's second law would become tedious. For this reason, more efficient formulations are needed so that the complicated problems can be addressed. The Lagrangian formulation, named after the French mathematician Joseph-Louis Lagrange (1736–1813), and the Hamiltonian formulation, named after the Irish physicist

Foundations of Applied Electrodynamics   Geyi Wen
© 2010 John Wiley & Sons, Ltd

William Rowan Hamilton (1805–1865), provide such alternatives. Newton's law is focused on the vector nature of motion while the Lagrangian and Hamiltonian formulations are focused on the scalar nature of motion. In order to understand the Lagrangian formulation, we have to accept the least action principle, first proposed by French mathematician Pierre-Louis Moreau de Maupertuis (1698–1759) in 1746 and later developed by Leonhard Paul Euler (Swiss mathematician, 1707–1783), Lagrange and Hamilton. The least action principle is regarded as a universal principle, the origin of all laws. It allows us to replace the problem of integrating the differential equation under the specified boundary condition by the problem of seeking a function that minimizes the action.

2.1 Linear Space and Linear Operator

Mathematically an operator is defined as a rule, denoted by $\hat{A}$ (we shall use a small caret over a letter to designate an operator), which associates to each element x in a set M an element y in another set F, denoted by $y = \hat{A}(x)$. The set M is called the **domain of definition** of $\hat{A}$, denoted by $D(\hat{A})$. The set $\hat{A}(M) = \{y \in F \,|\, y = \hat{A}(x) \,, x \in M\}$ is called the **range** of $\hat{A}$, denoted by $R(\hat{A})$. The operator $\hat{A} : M \to F$ is called **surjective** if $\hat{A}(M) = F$. The operator $\hat{A} : M \to F$ is called **injective** if $\hat{A}(x_1) = \hat{A}(x_2)$ implies $x_1 = x_2$. The operator $\hat{A} : M \to F$ is called **bijective** if $\hat{A}$ is both surjective and injective. Operators are also called **maps** and **functions**. In particular, if F consists of real or complex numbers, $\hat{A}$ is called a **functional**. In order to show that the domain of definition of the operator $\hat{A} : M \to F$ is contained in the set E, we write $\hat{A} : M \subset E \to F$. In mathematical physics, most common operators encountered are **differential operators**, which contain the operation of differentiation, and **integral operators**, which contain the operation of integration.

2.1.1 Linear Space, Normed Space and Inner Product Space

Let R be the set of real numbers and C the set of complex numbers. A **linear space** E is a set of elements for which rules called **addition**, denoted by '+' and **scalar multiplication**, denoted by '·' are specified, satisfying the following axioms for all elements $x, y, z \cdots \in E$ and $\alpha, \beta \in R$ or C

1. Commutativity: $x + y = y + x$.
2. Associativity: $(x + y) + z = x + (y + z)$.
3. Existence of a zero element: There is a zero element, denoted 0, such that $x + 0 = x$.
4. Existence of a negative element: For each x there is negative element y such that $x + y = 0$.
5. Distributivity: $\alpha \cdot (x + y) = \alpha \cdot x + \alpha \cdot y$.
6. Distributivity: $(\alpha + \beta) \cdot x = \alpha \cdot x + \beta \cdot x$
7. $1 \cdot x = x$.
8. $\alpha \cdot (\beta \cdot x) = (\alpha \cdot \beta) \cdot x$.

If $\alpha, \beta \in R$ (resp. C), the space E is said to be a **real** (resp. **complex**) **linear space**. The scalar multiplication $\alpha \cdot x$ is usually abbreviated as αx. A subset M of a linear space E is called a **linear subspace** if $\alpha x + \beta y \in M$, for all $x, y \in M$ and $\alpha, \beta \in R$ or C. If $\{x_1, x_2, \ldots, x_n\} \subset E$, the set of elements $\sum_{i=1}^{n} \alpha_i x_i$ generated by all possible $\alpha_i \in R$ or C is called the space spanned by $\{x_1, x_2, \ldots, x_n\}$. A set of elements $\{x_1, x_2, \ldots, x_n\} \subset E$ is said to be **linearly independent** if there exist constants $\alpha_i (i = 1, 2 \ldots, n)$ such that $\sum_{i=1}^{n} \alpha_i x_i = 0$ implies $\alpha_i = 0$ $(i = 1, 2, \ldots, n)$.

Otherwise, it is said to be **linearly dependent**. A linearly independent subset of E, which spans E, is called a **basis** of E. Thus every element of E can be expressed uniquely as a linear combination of the basis elements. If the basis is finite, E is said to be **finite dimension**. Otherwise, it is said to be **infinite dimension**.

A linear space E is a **normed space** if there exists a real-valued function $\|x\|$, called the **norm** of x, such that

1. Positive definiteness: $\|x\| \geq 0$ and $\|x\| = 0$ if and only if $x = 0$.
2. Homogeneity: $\|\alpha x\| = |\alpha| \cdot \|x\|$.
3. Triangle inequality: $\|x + y\| \leq \|x\| + \|y\|$.

A sequence $\{x_n\}$ is said to be **convergent** to x if $\|x_n - x\| \xrightarrow[n \to \infty]{} 0$. A sequence $\{x_n\}$ is said to be a **fundamental** or **Cauchy sequence** (named after the French mathematician Augustin-Louis Cauchy, 1789–1857) if $\|x_m - x_n\| \xrightarrow[m,n \to \infty]{} 0$. A normed space E is said to be **complete** if every Cauchy sequence in E converges to an element in E. A complete normed space is called **Banach space**.

The set $B(x_0, r) = \{x \in E \mid \|x - x_0\| < r\}$ is called an open ball at x_0 with radius r. A subset $U(x_0)$ is called a neighborhood of the point x_0 if $U(x_0)$ contains x_0 and an open ball at x_0. A point $x_0 \in M \subset E$ is called an **interior point** of the subset M if M contains a neighborhood of the point x_0. The subset M is called **open** if all its points are interior points. The subset M is called **closed** if $E - M$ is open. A point x_0 is called a **limit point** of the subset $M \subset E$ if for every neighborhood $U(x_0)$, we have $U(x_0) \cap M \neq 0$. A limit point is called an **accumulation point** if the intersection $U(x_0) \cap M$ always contains a point different from x_0. Otherwise it is called an **isolated point**. The intersection of all closed sets that contain the subset M is called the **closure** of M, denoted by $\bar{M}$. A subset $M \subset E$ is called **dense** in E if $\bar{M} = E$. A space is called **separable** if it has a countable dense set. A subset $M \subset E$ is called **compact** if every sequence of elements from M has a convergent subsequence whose limit lies in M.

An **inner product** on a linear space E is a mapping $E \times E \to R$ or C, denoted by $(\cdot, \cdot)$, which satisfies

1. Positive definiteness: $(x, x) \geq 0$ and $(x, x) = 0$ if and only if $x = 0$.
2. Hermitian property: $(x, y) = \overline{(y, x)}$.
3. Homogeneity: $(\alpha x, y) = \alpha(x, y)$.
4. Additivity: $(x + y, z) = (x, z) + (y, z)$.

Here $x, y, z \in H, \alpha \in R$ or C. An **inner product space** is a linear space with an inner product. An inner product space has a natural norm $\|x\| = (x, x)^{1/2}$, called the norm induced by the inner product. A complete inner product space in the induced norm is called a **Hilbert space**, named after the German mathematician David Hilbert (1862–1943). In an inner product space, the **Cauchy–Schwartz inequality** holds

$$|(x, y)| \leq \sqrt{(x, x)}\sqrt{(y, y)} = \|x\| \cdot \|y\| .\tag{2.1}$$

2.1.2 Linear and Multilinear Maps

An operator $\hat{B}$ is said to be an extension of $\hat{A}$, denoted by $\hat{A} \subset \hat{B}$, if $D(\hat{A}) \subset D(\hat{B})$ and $\hat{A}(x) = \hat{B}(x)$ for all $x \in D(\hat{A})$. By a **linear operator** $\hat{A}$, we mean it satisfies the following linearity

$$\hat{A}(\alpha x_1 + \beta x_2) = \alpha \hat{A}(x_1) + \beta \hat{A}(x_2),$$

where α, β are scalars. The sum of two linear operators $\hat{A}$ and $\hat{B}$ is defined by

$$(\hat{A} + \hat{B})(x) = \hat{A}(x) + \hat{B}(x), x \in D(\hat{A}) \cap D(\hat{B}).$$

If α is a scalar, the product $\alpha \hat{A}$ is defined by

$$(\alpha \hat{A})(x) = \alpha \hat{A}(x), x \in D(\hat{A}).$$

The **product** of two linear operators $\hat{A}$ and $\hat{B}$ is defined by

$$(\hat{A}\hat{B})(x) = \hat{A}[\hat{B}(x)], x \in \left\{ x \,\middle|\, x \in D(\hat{B}), \hat{B}(x) \in D(\hat{A}) \right\} .$$

The sum and product of linear operators are linear operators. The **power** $\hat{A}^n (n = 0, 1, 2, \ldots)$ is defined inductively by $\hat{A}^0 = \hat{I}$ ($\hat{I}$ is a unit operator: $\hat{I}x = x$) and $\hat{A}^{n+1} = \hat{A}\hat{A}^n$. It is easy to show that $\hat{A}^n \hat{A}^m = \hat{A}^{n+m}, n, m \geq 0$. The **graph** of an operator $\hat{A}$, denoted by $\Gamma(\hat{A})$, is defined as the set of all pairs $\{(u, \hat{A}u) \,|\, u \in D(\hat{A})\}$. If $\hat{A} \subset \hat{B}$, then $\Gamma(\hat{A}) \subset \Gamma(\hat{B})$. A linear operator $\hat{A} : D(\hat{A}) \subset E \rightarrow R(\hat{A}) \subset F$ is said to be **invertible** when it is a bijection from $D(\hat{A})$ onto $R(\hat{A})$. The inverse map is also a linear map and is called the **inverse operator**, denoted by $\hat{A}^{-1}$. An operator $\hat{A}$ is invertible if and only if the equation $\hat{A}(x) = 0$ has a unique solution $x = 0$.

Remark 2.1 (Isomorphism): An isomorphism is a bijection $\hat{A}$ such that both $\hat{A}$ and $\hat{A}^{-1}$ are homomorphism (a homomorphism is a structure-preserving map between two algebraic structures such as vector spaces). □

Example 2.1: Let $\hat{A}u = -d^2u/dx^2$. The domain of definition of $\hat{A}$ consists of all possible functions having first and second derivatives in the open interval $(0, 1)$ and satisfying $u(0) = u(1) = 0$. To find the inverse of $\hat{A}$, let us consider the equation

$$\hat{A}u = -\frac{d^2u}{dx^2} = f(x), u(0) = u(1) = 0,$$

where $f(x)$ is an arbitrary continuous function. Integrating the above equation twice leads to

$$u(x) = -\int_0^x ds \int_0^s f(t)dt + c_1 x + c_2.$$

The boundary condition $u(0) = 0$ gives $c_2 = 0$. Applying integration by parts yields

$$u(x) = -\int_0^x (x - t)f(t)dt + c_1 x.$$

From the boundary condition $u(1) = 0$, we obtain

$$c_1 = \int_0^1 (1 - t)f(t)dt = \int_0^x (1 - t)f(t)dt + \int_x^1 (1 - t)f(t)dt.$$

It follows that

$$u(x) = \hat{A}^{-1} f = \int_0^x t(1 - x)f(t)dt + \int_x^1 x(1 - t)f(t)dt = \int_0^1 K(x, t)f(t)dt,$$

where $K(x, t) = \begin{cases} x(1 - t), & x \leq t \\ t(1 - x), & x \geq t \end{cases}$ is the kernel.   □

Let E and F be normed spaces. A linear operator $\hat{A} : E \rightarrow F$ is **continuous** at $x_0 \in E$ if, for any $\varepsilon > 0$, there exists a positive number $\delta(x_0, \varepsilon)$ such that $\|\hat{A}(x) - \hat{A}(x_0)\| < \varepsilon$ whenever $\|x - x_0\| < \delta(x_0, \varepsilon)$. If $\hat{A}$ is continuous at every point in its domain, we say $\hat{A}$ is continuous. A linear operator $\hat{A} : E \rightarrow F$ is **bounded** if there is a finite constant $c_1 > 0$ such that $\|\hat{A}(x)\| \leq c_1 \|x\|$ for all $u \in D(\hat{A})$. The least constant c_1 such that $\|\hat{A}(x)\| \leq c_1 \|x\|$ for all $x \in D(\hat{A})$ is called the **norm** of $\hat{A}$ and is denoted by $\|\hat{A}\|$. If $\hat{A} \subset \hat{B}$ and $\hat{B}$ is bounded, $\hat{A}$ is also bounded and $\|\hat{A}\| \leq \|\hat{B}\|$. Here are some important properties of bounded operators:

1. If $\hat{A}$ is a bounded operator, then

$$\|\hat{A}\| = \sup_{\|x\| \leq 1} \|\hat{A}(x)\| = \sup_{\|x\| = 1} \|\hat{A}(x)\|.$$

2. If $\hat{A}$ and $\hat{B}$ are bounded and defined on the whole space, then

$$\|\hat{A} + \hat{B}\| \leq \|\hat{A}\| + \|\hat{B}\|, \|\hat{A}\hat{B}\| \leq \|\hat{A}\| \|\hat{B}\|.$$

Let $E_i(i = 1, 2, \ldots, k)$ and F be linear spaces. A map $\hat{A} : E_1 \times E_2 \times \ldots \times E_k \to F$ is called k-**multilinear** if $\hat{A}(x_1, x_2, \ldots, x_k)$ is linear in each argument separately. The space of all continuous k-multilinear maps of $E_1 \times E_2 \times \ldots \times E_k$ to F is denoted by $L(E_1, E_2, \ldots, E_k; F)$. If $E_i = E(i = 1, 2, \ldots, k)$, the space is denoted by $L^k(E, F)$. A k-multilinear map $\hat{A}$ is continuous if and only if there exists a $c_1 > 0$, such that

$$\left\| \hat{A}(x_1, x_2, \ldots, x_k) \right\| \leq c_1 \|x_1\| \cdot \|x_2\| \ldots \|x_k\|$$

for all $x_i \in E_i(i = 1, 2, \ldots, k)$. Here $E_i(i = 1, 2, \ldots, k)$ and F are assumed to be normed spaces. Similarly, the norm of the k-multilinear map is defined by

$$\left\| \hat{A} \right\| = \sup_{\|x_i\|=1(i=1,2,\ldots,k)} \left\| \hat{A}(x_1, x_2, \ldots, x_k) \right\|.$$

This norm makes $L(E_1, E_2, \ldots, E_k; F)$ into a normed space, which is complete if F is. Especially $L(E, R)$ or $L(E, C)$ is called real or complex **dual space** of E, and all denoted by E^*.

2.2 Classification of Partial Differential Equations

Given any type of partial differential equations (PDEs), one must investigate the following three properties that any physical phenomena should have: (1) existence of solution; (2) uniqueness of solution; and (3) stability of solution, that is the solution continuously depends on the boundary values and initial values. A problem which has these three properties is said to be **well posed**. Here are some usual trinities for PDEs (Gustafson, 1987):

1. Three types of PDEs: elliptical, hyperbolic and parabolic.
2. Three types of problems: boundary value problems, initial value problems, and eigenvalue problems.
3. Three types of boundary conditions: Dirichlet boundary condition, named after the German mathematician Johann Peter Gustav Lejeune Dirichlet (1805–1859); Neumann boundary condition, named after the German mathematician Carl Gottfried Neumann (1832–1925); and Robin boundary condition, named after the French mathematician Victor Gustave Robin (1855–1897).
4. Three analytical solution methods: separation of variables, Green's function method, and variational method.
5. Three important mathematical tools: divergence theorem, inequalities and convergence theorems.

Let us consider the general second-order equation in two variables

$$A\frac{\partial^2 u}{\partial x^2} + 2B\frac{\partial^2 u}{\partial x \partial y} + C\frac{\partial^2 u}{\partial y^2} + D\frac{\partial u}{\partial x} + E\frac{\partial u}{\partial y} + Fu + G = 0, \qquad (2.2)$$

which occurs most frequently in practice. Introducing the transformation of coordinates

$$\xi = \varphi(x, y), \eta = \psi(x, y) \tag{2.3}$$

we can rewrite (2.2) as

$$\alpha \frac{\partial^2 u}{\partial \xi^2} + 2\beta \frac{\partial^2 u}{\partial \xi \partial \eta} + \gamma \frac{\partial^2 u}{\partial \eta^2} + \Phi \left(u, \frac{\partial u}{\partial \xi}, \frac{\partial u}{\partial \eta}, \xi, \eta \right) = 0, \tag{2.4}$$

where

$$\alpha = A \left(\frac{\partial \varphi}{\partial x} \right)^2 + 2B \frac{\partial \varphi}{\partial x} \frac{\partial \varphi}{\partial y} + C \left(\frac{\partial \varphi}{\partial y} \right)^2,$$

$$\beta = A \frac{\partial \varphi}{\partial x} \frac{\partial \psi}{\partial x} + B \left(\frac{\partial \varphi}{\partial x} \frac{\partial \psi}{\partial y} + \frac{\partial \varphi}{\partial y} \frac{\partial \psi}{\partial x} \right) + C \frac{\partial \varphi}{\partial y} \frac{\partial \psi}{\partial y},$$

$$\gamma = A \left(\frac{\partial \psi}{\partial x} \right)^2 + 2B \frac{\partial \psi}{\partial x} \frac{\partial \psi}{\partial y} + C \left(\frac{\partial \psi}{\partial y} \right)^2,$$

and Φ is linear function with respect to u, $\partial u / \partial \xi$ and $\partial u / \partial \eta$.

Theorem 2.1: *Assume $A \neq 0$ and consider the first-order PDE*

$$A \left(\frac{\partial v}{\partial x} \right)^2 + 2B \frac{\partial v}{\partial x} \frac{\partial v}{\partial y} + C \left(\frac{\partial v}{\partial y} \right)^2 = 0. \tag{2.5}$$

If $v = v(x, y)$ is a solution of (2.5), then $v(x, y) = c_1$ (c_1 is an arbitrary constant) is the general solution of the following ordinary differential equation

$$A \left(\frac{dy}{dx} \right)^2 - 2B \frac{dy}{dx} + C = 0. \tag{2.6}$$

The converse is also true.   □

The theorem can be proved as follows. Since $v = v(x, y)$ is a solution of (2.5), $\partial v / \partial y$ is not zero (otherwise v is a constant). It follows from $v(x, y) = c_1$ that

$$\frac{dy}{dx} = -\frac{\partial v}{\partial x} \bigg/ \frac{\partial v}{\partial y}.$$

From (2.5), we obtain

$$A \left(\frac{dy}{dx} \right)^2 - 2B \frac{dy}{dx} + C = A \left(-\frac{\partial v}{\partial x} \bigg/ \frac{\partial v}{\partial y} \right)^2 - 2B \left(-\frac{\partial v}{\partial x} \bigg/ \frac{\partial v}{\partial y} \right) + C = 0.$$

Therefore $v(x, y) = c_1$ is the general solution of (2.6). We now prove that the converse is also true. Let $\varphi(x, y) = c_2$ (c_2 is an arbitrary constant) be the general solution of (2.6). From the implicit function theorem and $\varphi(x, y) = c_2$, y can be expressed in terms of x as follows

$$y = f(x, c_2) \tag{2.7}$$

if $\partial\varphi/\partial y \neq 0$. Along the curve (2.7), we have

$$\frac{dy}{dx} = -\frac{\partial\varphi}{\partial x} \Big/ \frac{\partial\varphi}{\partial y}.$$

Substituting this into (2.6) yields

$$A\left(-\frac{\partial\varphi}{\partial x} \Big/ \frac{\partial\varphi}{\partial y}\right)^2 - 2B\left(-\frac{\partial\varphi}{\partial x} \Big/ \frac{\partial\varphi}{\partial y}\right) + C\,\Bigg|_{y=f(x,c_2)} = 0,$$

which simply states that $v = \varphi(x, y)$ satisfies (2.5). The proof is complete.

We can choose the transformation (2.3) to make α and γ vanish. This can be achieved by solving (2.5) or (2.6) as indicated by the above theorem. Equation (2.6) is known as the **characteristic equation** of (2.2) and the solutions of (2.6) are referred to as **characteristics** or **characteristic curves**. The characteristic equation (2.6) may be written as

$$\frac{dy}{dx} = \frac{B \pm \sqrt{d}}{A}, \tag{2.8}$$

where $d = B^2 - AC$ is the discriminant. Equation (2.2) is respectively called **elliptical, hyperbolic** or **parabolic** if $d < 0$, $d > 0$ or $d = 0$.

2.2.1   Canonical Form of Elliptical Equations

Since $d < 0$ for elliptical equation, Equation (2.8) has two solutions $\varphi(x, y) = c_1$ and $\bar{\varphi}(x, y) = c_2$. Making use of the following transformation

$$\xi = \varphi(x, y), \eta = \bar{\varphi}(x, y),$$

Equation (2.4) may be reduced to

$$\frac{\partial^2 u}{\partial\xi\partial\eta} + \frac{1}{2\beta}\Phi\left(u, \frac{\partial u}{\partial\xi}, \frac{\partial u}{\partial\eta}, \xi, \eta\right) = 0. \tag{2.9}$$

Introducing another transformation

$$\rho = \frac{\xi + \eta}{2}, \sigma = \frac{\xi - \eta}{2j},$$

Equation (2.9) may be written as

$$\frac{\partial^2 u}{\partial \rho^2} + \frac{\partial^2 u}{\partial \sigma^2} + \Phi_1 = 0, \; \Phi_1 = \frac{2\Phi}{\beta}. \tag{2.10}$$

This is the canonical form of elliptical equation.

2.2.2   Canonical Form of Hyperbolic Equations

Since $d > 0$ for hyperbolic equation, Equation (2.8) has two real solutions $\varphi(x, y) = c_1$ and $\psi(x, y) = c_2$. By use of the transformation (2.3), Equation (2.4) may be reduced to (2.9). Similarly, we may introduce another transform

$$\rho = \frac{\xi + \eta}{2}, \sigma = \frac{\xi - \eta}{2}$$

to obtain

$$\frac{\partial^2 u}{\partial \rho^2} - \frac{\partial^2 u}{\partial \sigma^2} + \Phi_2 = 0, \; \Phi_2 = \frac{2\Phi}{\beta}. \tag{2.11}$$

This is the canonical form of hyperbolic equation.

2.2.3   Canonical Form of Parabolic Equations

For parabolic equations, we have $d = 0$, and Equation (2.8) has only one real solution $\varphi(x, y) = c_1$. We may choose an arbitrary function $\psi(x, y)$ linearly independent of $\varphi(x, y)$ such that $\gamma \neq 0$. Using (2.3), we have

$$\alpha = A \left(\frac{\partial \varphi}{\partial x} \right)^2 + 2B \frac{\partial \varphi}{\partial x} \frac{\partial \varphi}{\partial y} + C \left(\frac{\partial \varphi}{\partial y} \right)^2 = 0$$

from which we obtain

$$\sqrt{A} \frac{\partial \varphi}{\partial x} + \sqrt{C} \frac{\partial \varphi}{\partial y} = 0.$$

Thus

$$\beta = A \frac{\partial \varphi}{\partial x} \frac{\partial \psi}{\partial x} + \sqrt{AC} \left(\frac{\partial \varphi}{\partial x} \frac{\partial \psi}{\partial y} + \frac{\partial \varphi}{\partial y} \frac{\partial \psi}{\partial x} \right) + C \frac{\partial \varphi}{\partial y} \frac{\partial \psi}{\partial y}$$

$$= \left(\sqrt{A} \frac{\partial \varphi}{\partial x} + \sqrt{C} \frac{\partial \varphi}{\partial y} \right) \left(\sqrt{A} \frac{\partial \psi}{\partial x} + \sqrt{C} \frac{\partial \psi}{\partial y} \right) = 0$$

and (2.4) reduces to

$$\frac{\partial^2 u}{\partial \eta^2} + \Phi_3 = 0, \Phi_3 = \frac{1}{\gamma}\Phi. \tag{2.12}$$

This is the canonical form of parabolic equation.

The classification of partial differential equations with more than two independent variables can be carried out in a similar manner (for example, Rubinsten and Rubinsten, 1998).

2.3 Modern Theory of Partial Differential Equations

When different methods (for example, separation of variables) are used to find a solution of a partial differential equation, a formal solution is often obtained. If the formal solution is smooth and satisfies the equation as well as the boundary conditions or initial values, it is called a **classical solution** in the usual sense. However, a formal solution is not necessarily a classical solution although it may be practically meaningful. For this reason, we have to extend the concept of the classical solutions.

2.3.1 Limitation of Classical Solutions

We now use several examples to demonstrate that the smoothness requirement of the classical solutions may be too stringent in practical situations.

Example 2.2 (the Cauchy problem): Let us consider the Cauchy problem of one-dimensional wave equation

$$\frac{\partial^2 u(x,t)}{\partial t^2} - a^2 \frac{\partial^2 u(x,t)}{\partial x^2} = 0, t > 0, -\infty < x < \infty,$$

$$u(x,0) = \varphi(x), \frac{\partial u(x,0)}{\partial t} = \psi(x), -\infty < x < \infty. \tag{2.13}$$

The formal solution of the above equation is given by the D'Alembert formula:

$$u(x,t) = \frac{1}{2}[\varphi(x-at) + \varphi(x+at)] + \frac{1}{2a}\int_{x-at}^{x+at} \psi(\xi)d\xi,$$

named after the French mathematician Jean le Rond d'Alembert (1717–1783). This formal solution becomes a classical solution if

$$u(x,t) \in C^2(\Omega) \cap C^1(\bar{\Omega}), \Omega = \{(x,t)\,|t > 0, -\infty < x < \infty\}.$$

Here $C^m(\Omega)$ $(m \geq 1)$ stands for the set of all functions, which have continuous partial derivatives to the order m, and $C^m(\bar{\Omega})$ is the set of all $u \in C^m(\Omega)$ for which all partial derivatives can be extended to the closure $\bar{\Omega}$ of Ω. It can be shown that if $\varphi(x) \in C^2(R)$ and $\psi(x) \in C^1(R)$, the formal solution given by the D'Alembert formula is a classical solution. However, the requirement that the initial data $\varphi(x)$ and $\psi(x)$ must be smooth is impractical in some situations. For instance, if $\psi(x)$ is a rectangular pulse

$$\psi(x) = \begin{cases} 1, & -\Delta < x < \Delta \\ 0, & |x| \geq \Delta, \Delta > 0 \end{cases}$$

for small Δ, $\psi(x)$ is not continuous. Nonetheless, it is known that the rectangular pulse is often used in engineering analysis. Hence we must generalize the concept of solutions. One of the strategies is to construct smooth sequences $\{\varphi_n\} \in C^2(R)$ and $\{\psi_n\} \in C^1(R)$ such that

$$\lim_{n \to \infty} \varphi_n = \varphi, \ \lim_{n \to \infty} \psi_n = \psi \tag{2.14}$$

hold, and consider a series of Cauchy problems

$$\frac{\partial^2 u_n(x,t)}{\partial t^2} - a^2 \frac{\partial^2 u_n(x,t)}{\partial x^2} = 0, t > 0, -\infty < x < \infty,$$

$$u_n(x,0) = \varphi_n(x), \ \frac{\partial u_n(x,0)}{\partial t} = \psi_n(x), -\infty < x < \infty$$

for $n = 1, 2, \ldots$ The classical solutions of these Cauchy problems are then given by the D'Alembert formula

$$u_n(x,t) = \frac{1}{2}[\varphi_n(x-at) + \varphi_n(x+at)] + \frac{1}{2a}\int_{x-at}^{x+at} \psi_n(\xi)d\xi.$$

As $n \to \infty$, u_n may approach a limit u

$$\lim_{n \to \infty} u_n = u. \tag{2.15}$$

In this case, the limit function u can be considered as a generalized solution. This generalized solution may not have continuous derivatives and thus cannot be considered as a classical solution. To set the stage for the generalized solution, we first need to clarify the convergence implied by (2.14) and (2.15). Second, if the limit function u in (2.15) is considered as a generalized solution of (2.13), we must specify the meaning of the partial derivatives in (2.13). In other words, we must generalize the concept of derivatives in the usual sense.   □

Example 2.3 (the Dirichlet problem): Let Ω be a non-empty open set in R^N, $N \geq 1$ and $\mathbf{x} = (x_1, x_2, \ldots, x_N) \in R^N$. Consider the **variational problem**

$$F(u) = \frac{1}{2}\int_\Omega \sum_{j=1}^N (\partial u/\partial x_j)^2 d\mathbf{x} - \int_\Omega f u d\mathbf{x} = \min,$$

$$u = g, \mathbf{x} \in \partial\Omega, \tag{2.16}$$

where f and g are continuous functions and $d\mathbf{x} = dx_1 dx_2 \ldots dx_n$. If we assume that $u \in C^2(\bar{\Omega})$, it is easy to show that u is a solution of the following boundary value problem

$$-\nabla^2 u = f, \mathbf{x} \in \Omega; u = g, \mathbf{x} \in \partial\Omega. \tag{2.17}$$

If the solution of (2.16) is not sufficiently smooth, for example, $u \notin C^2(\bar{\Omega})$, u may not be a solution of (2.17) in the usual sense. There exist reasonable situations where (2.16) lacks smooth conditions. In order to investigate these situations, we must add 'ideal elements' to the class of smooth solutions to ensure that the space of solutions is closed (Courant and Hilbert, 1989; Zeidler, 1995), and we also need to understand how the partial derivatives are applied to these 'ideal elements'.   □

2.3.2   Theory of Generalized Functions

The theory of generalized functions was introduced by Schwartz around 1950 to get rid of the restrictions in classical analysis. He invented a new calculus and extended the concept of ordinary functions while preserving many of the basic operations of analysis, including addition, multiplication by C^∞ functions, differentiation as well as convolution and Fourier transforms. The generalized functions have now been widely used in theoretical physics, engineering science, differential equations, group theory and functional analysis.

Let Ω be a non-empty open set in R^N, $N \geq 1$ and $\mathbf{x} = (x_1, x_2, \ldots, x_N) \in R^N$. A **multi-index** is denoted by $\alpha = (\alpha_1, \alpha_2, \ldots, \alpha_N)$ with $\alpha_i \geq 0$, $i = 1, 2, \ldots, N$, and $|\alpha| = \alpha_1 + \alpha_2 + \ldots + \alpha_N$ is the length of α. We use the following shorthand notations

$$\partial^\alpha u(\mathbf{x}) = \frac{\partial^{|\alpha|} u(\mathbf{x})}{\partial x_1^{\alpha_1} \partial x_2^{\alpha_2} \ldots \partial x_N^{\alpha_N}}, \mathbf{x}^\alpha = x_1^{\alpha_1} \cdot x_2^{\alpha_2} \cdots x_N^{\alpha_N}.$$

By convention, we let $\partial^\alpha u = u$ if $\alpha = (0, 0, \ldots, 0)$. Let $C^m(\Omega) m = 1, 2, \ldots)$ stand for the set of all functions that have continuous partial derivatives of order $k = 0, 1, 2, \ldots m$, and $C^m(\bar{\Omega})$ for the set of all $u \in C^m(\Omega)$ for which all partial derivatives can be extended to the closure $\bar{\Omega}$ of Ω. If $u \in C^k(\Omega)$ (resp. $u \in C^k(\bar{\Omega})$) for all $k(k = 0, 1, 2 \ldots)$, we write $u \in C^\infty(\Omega)$ (resp. $u \in C^\infty(\bar{\Omega})$). Let $C_0^\infty(\Omega)$ denote the set of all functions $u \in C^\infty(\Omega)$ that vanish outside a compact set K of Ω (Different u may have different K), and Let $L^2(\Omega)$ denote the set of all functions such that $\int_\Omega |u|^2 \, d\mathbf{x} < \infty$. It can be shown that $L^2(\Omega)$ is a Hilbert space with the inner product $(u, v) = \int_\Omega u\bar{v} d\mathbf{x}$. In addition, $L^2(\Omega)$ is separable, and we have (Adams, 1975):

1. $C_0^\infty(\Omega)$ is dense in $L^2(\Omega)$.
2. $C(\bar{\Omega})$ is dense in $L^2(\Omega)$.

The **support** of a function u is defined by supp $u(\mathbf{x}) = \overline{\{\mathbf{x} \,|\, \mathbf{x} \in \Omega, u(\mathbf{x}) \neq 0\}}$. The limit concept in $C_0^\infty(\Omega)$ can be defined as follows. Let $\varphi_n, \varphi \in C_0^\infty(\Omega)$, we say that φ_n approaches φ in $C_0^\infty(\Omega)$ if

1. There exists a compact subset K of open set Ω such that all $\varphi_n(\mathbf{x})$ vanish in $\Omega - K$.
2. For all multi-index α, the sequence $\{\partial^\alpha \varphi_n\}$ uniformly converges to $\partial^\alpha \varphi$ on K

$$\max_{\mathbf{x} \in K} \left| \partial^\alpha \varphi_n - \partial^\alpha \varphi \right| \to 0, n \to \infty.$$

The space $C_0^\infty(\Omega)$ equipped with the above convergence is denoted by $D(\Omega)$.

Let $D'(\Omega)$ be the dual space of $D(\Omega)$. The element in $D'(\Omega)$ is called the **generalized function**. $D(\Omega)$ is called the **fundamental space** of $D'(\Omega)$. We define the **addition** and **scalar multiplication** in $D'(\Omega)$ as follows:

1. Addition: $(T_1 + T_2)\varphi = T_1\varphi + T_2\varphi$, $T_1, T_2 \in D'(\Omega)$, $\varphi \in D(\Omega)$.
2. Scalar multiplication: $(aT)\varphi = aT\varphi$, $T \in D'(\Omega)$, $\varphi \in D(\Omega)$.

where a is a real or complex number. Thus $D'(\Omega)$ becomes a linear space. Let $\{T_n\}$ be a sequence in $D'(\Omega)$ and $T \in D'(\Omega)$. We say that the sequence $\{T_n\}$ converges to T, denoted by $\lim_{n \to \infty} T_n = T$, if $\lim_{n \to \infty} T_n(\varphi) = T(\varphi)$ for all $\varphi \in D(\Omega)$.

The variational lemma plays a fundamental role in the calculus of variations.

Variational lemma: *Let Ω be a nonempty set in R^N and $u \in L^2(\Omega)$. If $\int_\Omega u(\mathbf{x})\varphi(\mathbf{x})d\mathbf{x} = 0$ holds for all $\varphi \in C_0^\infty(\Omega)$, then $u(\mathbf{x}) = 0$ for almost all $\mathbf{x} \in \Omega$. If, in addition, $u \in C(\Omega)$, then $u(\mathbf{x}) = 0$ for all $\mathbf{x} \in \Omega$.*

Example 2.4: The space $L^2(\Omega)$ can be identified as a linear subspace of $D'(\Omega)$. For $u, v \in L^2(\Omega)$, we define

$$U(\varphi) = \int_\Omega u(\mathbf{x})\varphi(\mathbf{x})d\mathbf{x}, \ V(\varphi) = \int_\Omega v(\mathbf{x})\varphi(\mathbf{x})d\mathbf{x}, \ \varphi \in D(\Omega).$$

Thus, U and V are generalized functions. It is obvious that $u = v$ implies $U = V$. Moreover, the map from $u \in L^2(\Omega)$ to $U \in D'(\Omega)$ is injective. In other words, $U = V$ implies $u = v$. In fact, suppose $u, v \in L^2(\Omega)$ and

$$U(\varphi) = \int_\Omega u(\mathbf{x})\varphi(\mathbf{x})d\mathbf{x} = \int_\Omega v(\mathbf{x})\varphi(\mathbf{x})d\mathbf{x} = V(\varphi).$$

Then $\int_\Omega [u(\mathbf{x}) - v(\mathbf{x})] \varphi(\mathbf{x})d\mathbf{x} = 0$ holds for all $\varphi \in D(\Omega)$. By the variational lemma, we have $u = v$. $\qquad \square$

The function $u \in L^2(\Omega)$ is called a representation of U, or simply called a generalized function (we identify u with U). For $T \in D'(\Omega)$, we formally write

$$T(\varphi) \equiv \langle T, \varphi \rangle \equiv \int_\Omega T(\mathbf{x})\varphi(\mathbf{x})d\mathbf{x}, \ \varphi \in D(\Omega),$$

where $\langle \cdot, \cdot \rangle$ is used to denote a bilinear form (pairing). In engineering, we often use $T(\mathbf{x})$ instead of T to denote the generalized function to explicitly indicate its dependence on $\mathbf{x}$.

Remark 2.2 (Physical explanation of generalized function): Let $T(\mathbf{r})$ be a scalar field distribution in space. If we attempt to measure the field at a specific point $\mathbf{r}$ using an instrument, what is actually obtained is the weighted average of the field distribution in a small area centered at $\mathbf{r}$. In other words, the measured field is not exactly $T(\mathbf{r})$ but a number given by $\langle T, \varphi \rangle = \int_{R^3} T(\mathbf{r})\varphi(\mathbf{r})d\mathbf{r}$, where $\varphi(\mathbf{r})$ denotes the weight. Different measurements correspond to different weights in the function space $D(\Omega)$. If we let the weight $\varphi(\mathbf{r})$ go through all the elements in $D(\Omega)$, the field $T(\mathbf{r})$ can then be determined by $\langle T, \varphi \rangle$. Therefore, the functional $\varphi \mapsto \langle T, \varphi \rangle$ can be used to determine the field distribution $T(\mathbf{r})$. □

Example 2.5: Let $\mathbf{y} \in \Omega$. Define $\langle \delta(\mathbf{x} - \mathbf{y}), \varphi(\mathbf{x}) \rangle = \varphi(\mathbf{y})$, $\varphi \in D(\Omega)$. Obviously $\delta(\mathbf{x} - \mathbf{y})$ is a generalized function and is called **δ-function**. Formally we have

$$\int_\Omega \delta(\mathbf{x} - \mathbf{y})\varphi(\mathbf{x})d\mathbf{x} = \varphi(\mathbf{y}), \mathbf{x}, \mathbf{y} \in \Omega, \varphi \in D(\Omega).$$

□

The definition of derivative of generalized functions is based on the integration by parts

$$\int_\Omega u\partial^\alpha v d\mathbf{x} = (-1)^{|\alpha|} \int_\Omega v\partial^\alpha u d\mathbf{x}, u, v \in D(\Omega).$$

For $T \in D'(\Omega)$ and all multi-index α, the **generalized derivative** $\partial^\alpha T$ is defined by

$$(\partial^\alpha T)\varphi = (-1)^{|\alpha|}T(\partial^\alpha \varphi), \varphi \in D(\Omega).$$

It is readily found that $\partial^\alpha T \in D'(\Omega)$ for all multi-index α.

Example 2.6: The generalized derivative of **unit step function**

$$H(x) = \begin{cases} 0, x < 0 \\ 1, x \geq 0 \end{cases} \tag{2.18}$$

is $H'(x) = \delta(x)$. □

Let $T \in D'(\Omega)$, $a(\mathbf{x}) \in C^\infty(\Omega)$. The **scalar multiplication** aT is defined by

$$(aT)\varphi = T(a\varphi), \varphi \in D(\Omega).$$

The **Leibniz rule**, named after the German mathematician Gottfried Wilhelm Leibniz (1646–1716), holds for the scalar multiplication

$$\partial^\alpha(aT) = \sum_{\beta \leq \alpha} C_\alpha^\beta(\partial^\beta a) \cdot (\partial^{\alpha - \beta} T),$$

where α and β are multi-index; $\beta \le \alpha$ means $|\beta| \le |\alpha|$; $C_\alpha^\beta = C_{|\alpha|}^{|\beta|} = |\alpha|!/|\beta|!(|\alpha| - |\beta|)!$ is the combinatorial number. For a vector function $\mathbf{A}(\mathbf{x}) = \sum_{i=1}^{N} A_i \mathbf{u}_{x_i}$ ($\mathbf{u}_{x_i}$ is the unit vector along x_i direction), we define the generalized divergence and rotation as follows

$$\langle \nabla \cdot \mathbf{A}, \varphi \rangle = \sum_{i=1}^{N} (A_i, -\partial\varphi/\partial x_i),$$

$$\langle \nabla \times \mathbf{A}, \varphi \rangle = \sum_{i=1}^{N} \sum_{j=1}^{N} \sum_{k=1}^{N} \varepsilon^{ijk}(A_k, -\partial\varphi/\partial x_j)\mathbf{u}_{x_i},$$

where $\varepsilon^{ijk} = \mathbf{u}_{x_i} \cdot (\mathbf{u}_{x_j} \times \mathbf{u}_{x_k})$.

Example 2.7: Consider a function $f(x)$ which is differentiable (in the usual sense) except for some finite jump discontinuities at point set $\{x^i\}$. The generalized derivative of $f(x)$ is

$$\frac{df}{dx} = \left\{ \frac{df}{dx} \right\} + \sum_{i} \Delta_i f \delta(x - x^i),$$

where $\{df/dx\}$ denotes the derivative in the usual sense and $\Delta_i f = f(x^i + 0) - f(x^i - 0)$. The above relation can be generalized to multivariable functions. Let $f(\mathbf{r})$ be differentiable except on the surface $S \subset R^3$. The generalized derivative of $f(\mathbf{r})$ is

$$\frac{\partial f}{\partial x_i} = \left\{ \frac{\partial f}{\partial x_i} \right\} + n_i \Delta f \delta(S),$$

where n_i is the projection of the unit outward normal $\mathbf{u}_n$ of S along the x_i-axis; Δf is the jump of f when crossing S along the unit outward normal; $\delta(S)$ is a generalized function defined by

$$\langle \delta(S), \varphi(\mathbf{r}) \rangle = \int_{R^3} \delta(S)\varphi(\mathbf{r})d\mathbf{r} = \int_{S} \varphi(\mathbf{r})dS(\mathbf{r}).$$

In R^3, we have

$$\nabla f = \{\nabla f\} + \mathbf{u}_n \Delta f \delta(S),$$
$$\nabla \cdot \mathbf{e} = \{\nabla \cdot \mathbf{e}\} + \Delta(\mathbf{u}_n \cdot \mathbf{e})\delta(S), \tag{2.19}$$
$$\nabla \times \mathbf{e} = \{\nabla \times \mathbf{e}\} + \Delta(\mathbf{u}_n \times \mathbf{e})\delta(S).$$

These relations are useful when dealing with current or charge distribution on a surface. If the sources contain current and charge on the surface S, we write

$$\mathbf{J} = \{\mathbf{J}\} + \mathbf{J}_s\delta(S), \quad \rho = \{\rho\} + \rho_s\delta(S). \tag{2.20}$$

Making use of (2.19) and (2.20), Maxwell equations can be written as

$$\{\nabla \times \mathbf{H}\} + \Delta(\mathbf{u}_n \times \mathbf{H})\delta(S) = \{\mathbf{J}\} + \mathbf{J}_s\delta(S) + \partial \mathbf{D}/\partial t,$$

$$\{\nabla \times \mathbf{E}\} + \Delta(\mathbf{u}_n \times \mathbf{E})\delta(S) = -\partial \mathbf{B}/\partial t,$$

$$\{\nabla \cdot \mathbf{D}\} + \Delta(\mathbf{u}_n \cdot \mathbf{D})\delta(S) = \{\rho\} + \rho_s\delta(S),$$

$$\{\nabla \cdot \mathbf{B}\} + \Delta(\mathbf{u}_n \cdot \mathbf{B})\delta(S) = 0.$$

Thus, we have

$$\mathbf{J}_s = \Delta(\mathbf{u}_n \times \mathbf{H}),\ \Delta(\mathbf{u}_n \times \mathbf{E}) = 0,$$

$$\Delta(\mathbf{u}_n \cdot \mathbf{D}) = \rho_s,\ \Delta(\mathbf{u}_n \cdot \mathbf{B}) = 0.$$

These are the boundary conditions on the surface between two different media.   □

Let $\varphi \in C^\infty(R^N)$ and $\boldsymbol{\alpha}, \mathbf{p}$ be multi-index. The function φ is called **rapidly decreasing** at infinity if

$$\lim_{|\mathbf{x}| \to \infty} \mathbf{x}^{\boldsymbol{\alpha}} \partial^{\mathbf{p}} \varphi(\mathbf{x}) = 0$$

holds for all $\boldsymbol{\alpha}, \mathbf{p}$. The set of all rapidly decreasing functions is denoted by $S(R^N)$, which forms a linear space under usual addition and scalar multiplication. We now introduce the convergence in $S(R^N)$. Let $\varphi_n, \varphi \in S(R^N)$, we say φ_n approaches φ in $S(R^N)$ if the sequence $\{\partial^{\boldsymbol{\alpha}}\varphi_n\}$ for all multi-index $\boldsymbol{\alpha}$ uniformly converges to $\partial^{\boldsymbol{\alpha}}\varphi$ in R^N

$$\max_{\mathbf{x} \in R^N} \left| \partial^{\boldsymbol{\alpha}} \varphi_n - \partial^{\boldsymbol{\alpha}} \varphi \right| \to 0, n \to \infty.$$

The set $S(R^N)$ equipped with the above convergence is called the **rapid decreasing function space**. The dual space of $S(R^N)$ is denoted by $S'(R^N)$, whose elements are called generalized functions. $S(R^N)$ is called the **fundamental space** of $S'(R^N)$. Evidently we have $C_0^\infty(R^N) \subset S(R^N) \subset C^\infty(R^N)$; $C_0^\infty(R^N)$ is dense in $S(R^N)$; and $S(R^N)$ is dense in $C^\infty(R^N)$. All the operations (such as limit, derivative, etc.) in $S'(R^N)$ can be defined in the same way as in $D'(\Omega)$. It can be shown that $D'(R^N) \supset S'(R^N)$.

In classical analysis, the convolution of two functions $f(\mathbf{x})$ and $g(\mathbf{x})$ is defined by

$$f * g(\mathbf{x}) = \int_{R^N} f(\mathbf{y})g(\mathbf{x} - \mathbf{y})d\mathbf{y}.$$

If $f * g(\mathbf{x})$ are considered as the generalized function in $D'(R^N)$ or $S'(R^N)$, then for $\varphi \in D(R^N)$ or $S(R^N)$ we have

$$\langle f * g(\mathbf{x}), \varphi(\mathbf{x}) \rangle = \int_{R^N} d\mathbf{x} \varphi(\mathbf{x}) \int_{R^N} f(\mathbf{y}) g(\mathbf{x} - \mathbf{y}) d\mathbf{y}$$

$$= \int_{R^N} d\mathbf{x} f(\mathbf{x}) \int_{R^N} g(\mathbf{y}) \varphi(\mathbf{x} + \mathbf{y}) d\mathbf{y}$$

$$= \langle f(\mathbf{x}), \langle g(\mathbf{y}), \varphi(\mathbf{x} + \mathbf{y}) \rangle \rangle.$$

It is readily found that

$$\langle f(\mathbf{x}), \langle g(\mathbf{y}), \varphi(\mathbf{x} + \mathbf{y}) \rangle \rangle = \langle g(\mathbf{x}), \langle f(\mathbf{y}), \varphi(\mathbf{x} + \mathbf{y}) \rangle \rangle.$$

Let T_1 and T_2 be two generalized functions. If the operation $T_1 * T_2$ defined by

$$\langle T_1 * T_2(\mathbf{x}), \varphi(\mathbf{x}) \rangle = \langle T_1(\mathbf{x}), \langle T_2(\mathbf{y}), \varphi(\mathbf{x} + \mathbf{y}) \rangle \rangle, \quad \varphi \in D(R^N) \text{ or } S(R^N)$$

exists, it is called the **convolution** of generalized function T_1 and T_2. It is easy to see that

1. $T_1 * T_2 = T_2 * T_1$, $(T_1 * T_2) * T_3 = T_1 * (T_2 * T_3)$.
2. $\delta * T = T * \delta = T$, $\frac{\partial(\delta * T)}{\partial x_i} = T * \frac{\partial \delta}{\partial x_i} = \frac{\partial T}{\partial x_i}$.
3. $\frac{\partial(T_1 * T_2)}{\partial x_i} = \frac{\partial T_1}{\partial x_i} * T_2 = T_1 * \frac{\partial T_2}{\partial x_i}$.

We recall that the classical Fourier transform (named after the French mathematician Jean Baptiste Joseph Fourier, 1768–1830) and the inverse Fourier transform are defined respectively by

$$\mathcal{F}(f)(\xi) = \int_{R^N} f(\mathbf{x}) e^{-j\mathbf{x} \cdot \xi} d\mathbf{x}, \quad \mathcal{F}^{-1}(g)(\mathbf{x}) = \frac{1}{(2\pi)^N} \int_{R^N} g(\xi) e^{j\mathbf{x} \cdot \xi} d\xi,$$

where $\mathbf{x} = (x_1, x_2, \ldots, x_N)$ and $\xi = (\xi_1, \xi_2, \ldots, \xi_N)$. Some important properties of Fourier transform can be easily found and are summarized below:

1. Differential properties: $\mathcal{F}(\partial^\alpha f) = j^{|\alpha|} \xi^\alpha \mathcal{F}(f)$, $\mathcal{F}(-j^{|\alpha|} \mathbf{x}^\alpha f) = \partial^\alpha \mathcal{F}(f)$;
2. Convolution properties: $\mathcal{F}(f * g) = \mathcal{F}(f) \cdot \mathcal{F}(g)$, $\mathcal{F}(f \cdot g) = (2\pi)^{-N} \mathcal{F}(f) * \mathcal{F}(g)$;
3. Parseval equality: $\int_{R^N} f \cdot \bar{g} d\mathbf{x} = (2\pi)^{-N} \int_{R^N} \mathcal{F}(f) \cdot \overline{\mathcal{F}(g)} d\xi$.

The Fourier transform can be extended to the generalized functions. For $T \in S'(R^N)$, its **Fourier transform**, denoted $\mathcal{F}(T)$, is defined by

$$\langle \mathcal{F}(T), \varphi \rangle = \langle T, \mathcal{F}(\varphi) \rangle, \varphi \in S(R^N).$$

The inverse Fourier transform of the generalized function T, denoted $\mathcal{F}^{-1}(T)$, is defined by

$$\langle \mathcal{F}^{-1}(T), \varphi \rangle = \langle T, \mathcal{F}^{-1}(\varphi) \rangle, \varphi \in S(R^N).$$

The differential properties and convolution properties still hold for the generalized Fourier transform.

2.3.3 Sobolev Spaces

A differential operator defined on $C[a, b]$ is unbounded in the usual norm $\|x\| = \max_{a \le x \le b} |u(x)|$ on $C[a, b]$, but it may be bounded if we adopt a different norm. For example, if we introduce the norm

$$\|u\|_1 = \left[\int_a^b (|u|^2 + |u'|^2) dx \right]^{1/2}$$

on the space $C[a, b]$, the differentiation is a continuous operator on the space $C[a, b]$. The space $C[a, b]$ equipped with the above norm is not complete. However, if the function u is considered as a generalized function and the derivative is understood in the generalized sense, the space of the generalized functions with the norm $\|\cdot\|_1$ is complete. Such spaces are called **Sobolev spaces**, named after the Russian mathematician Sergei Lvovich Sobolev (1908–1989), and they play a fundamental role in modern theory of partial differential equations. In terms of the generalized function theory, the Sobolev spaces $H^m(\Omega)$ are defined by

$$H^m(\Omega) = \left\{ u \mid u \in L^2(\Omega), \partial^\alpha u \in L^2(\Omega), |\alpha| \le m \right\}$$

with an inner product

$$(u, v)_m = \sum_{|\alpha| \le m} (\partial^\alpha u, \partial^\alpha v), u, v \in H^m(\Omega) \tag{2.21}$$

where ∂^α denotes the generalized derivative and $(\cdot, \cdot)$ is the inner product of $L^2(\Omega)$. The norm corresponding to (2.21) is denoted by $\|\cdot\|_m$. It is easy to show that $H^m(\Omega)$ is a Hilbert space, and we have $H^{m_1} \subset H^{m_2}$ for $m_1 \ge m_2 \ge 0$. Evidently, $C_0^\infty(\Omega) \subset H^m(\Omega)$. The completion of $C_0^\infty(\Omega)$ in $H^m(\Omega)$ in the norm on $H^m(\Omega)$ is denoted by $H_0^m(\Omega)$, that is $H_0^m(\Omega) = \overline{C_0^\infty(\Omega)}$. Thus, $H_0^m(\Omega)$ is also a Hilbert space and $C_0^\infty(\Omega)$ is dense in $H_0^m(\Omega)$. For $m = 1$, we have

$$H^1(\Omega) = \left\{ u \mid u \in L^2(\Omega), \partial u \in L^2(\Omega) \right\},$$

$$(u, v)_1 = \int_\Omega \left(\sum_{i=1}^N \frac{\partial u}{\partial x_i} \cdot \frac{\partial \bar{v}}{\partial x_i} + u\bar{v} \right) dx,$$

$$\|u\|_1 = \left\{ \int_\Omega \left[\sum_{i=1}^N |\partial u/\partial x_i|^2 + |u|^2 \right] dx \right\}^{1/2}.$$

We can also define the Sobolev spaces $H^s(\Omega)$, where s can be factional and negative values. For $0 < s = [s] + \lambda, 0 < \lambda < 1$, where $[s]$ denotes the integral part of s, we define $H^s(\Omega)$ to be the closure of $C^\infty(\Omega)$ with respect to the norm

$$\|u\|_s^2 = \|u\|_{[s]}^2 + \sum_{|\alpha|=[s]} \int_\Omega \int_\Omega \frac{|\partial^\alpha u(\mathbf{x}) - \partial^\alpha u(\mathbf{y})|^2}{|\mathbf{x} - \mathbf{y}|^{N+2\lambda}} d\mathbf{x} d\mathbf{y}, \tag{2.22}$$

where N is the dimension of Ω. Similarly, $H_0^s(\Omega)$ is the closure of $C_0^\infty(\Omega)$ with respect to the norm (2.22).

To solve the partial differential equations, we often need to know the boundary values of the unknown functions. If $u(\mathbf{x}) \in C(\bar{\Omega})$, the restriction of the function $u(\mathbf{x})$ to the boundary Γ denoted $u(\mathbf{x})|_\Gamma$, is called the **trace** of the function $u(\mathbf{x})$ on Γ. Obviously $u(\mathbf{x})|_\Gamma \in L^2(\Gamma)$. If $u(\mathbf{x}) \in L^2(\Omega)$, $u(\mathbf{x})$ may not be continuous on Γ and the above definition of trace is not valid. However, we can make use of a sequence of smooth functions $\{u_n(\mathbf{x})\}$ defined in $\bar{\Omega}$ to approximate the function $u(\mathbf{x})$. The trace of the function $u(\mathbf{x})$ is defined as the limit of the traces $\{u_n(\mathbf{x})|_\Gamma\}$. Let Ω be a bounded region in R^N with boundary Γ. For $u(\mathbf{x}) \in C^m(\bar{\Omega})$, we define the **trace operator** as follows

$$\gamma_0 u = u|_\Gamma, \gamma_j u = \left.\frac{\partial^j u}{\partial n^j}\right|_\Gamma = \gamma_0 \frac{\partial^j u}{\partial n^j}, j = 1, 2, \ldots, m - 1, \tag{2.23}$$

where $\partial^j u/\partial n^j$ denotes the jth derivative of u in the direction of $\mathbf{u}_n$ (the unit outward normal of Γ). The trace operators γ_j can be extended to bounded linear operators mapping from $H^m(\Omega)$ onto $H^{m-j-1/2}(\Gamma)$. The kernel of the trace operator $\gamma_j(u)$ is $H_0^m(\Omega)$, $\gamma_j[H_0^m(\Omega)] = 0$, $j = 0, 1, 2, \ldots, m - 1$ (Adams, 1975).

2.3.4   Generalized Solutions of Partial Differential Equations

Many practical situations do not admit sufficiently smooth solutions. For this reason, we must generalize the concept of solutions. Mathematically, the generalized solution and weak solution come naturally from the direct method of the calculus of variations as well as the problems where the solution is constructed as the limit of an approximate procedure. Let us consider the partial differential equation

$$\hat{A}u(\mathbf{x}) = f(\mathbf{x}), \mathbf{x} \in \Omega, \tag{2.24}$$

where $\hat{A} = \sum_{|\alpha| \leq m} a_\alpha(\mathbf{x})\partial^\alpha$ is a linear partial differential operator of order m. The coefficients $a_\alpha(\mathbf{x})$ and the source term f are assumed to be smooth enough. The classical solution of (2.24) requires that $u \in C^m(\Omega), f \in C(\Omega)$. Thus, $\hat{A}u \in C(\Omega)$. It follows from the integration by parts that

$$(\hat{A}u, \varphi) = (u, \hat{A}^*\varphi) = (f, \varphi), \varphi \in C_0^\infty(\Omega), u \in C^m(\Omega), \tag{2.25}$$

where $(\cdot, \cdot)$ is the inner product of $L^2(\Omega)$ and $\hat{A}^*$ is the formal adjoint of $\hat{A}$:

$$\hat{A}^*\varphi = \sum_{|\alpha|\leq m} (-1)^{|\alpha|}\partial^\alpha [\bar{a}_\alpha(\mathbf{x})\varphi].$$

If u is a classical solution of (2.24) with $u \in C^m(\Omega)$ and $\hat{A}u \in C(\Omega)$, we have

$$(u, \hat{A}^*\varphi) = (f, \varphi), \varphi \in C_0^\infty(\Omega), u \in C^m(\Omega). \tag{2.26}$$

If $u \notin C^m(\Omega)$ or $f \notin C(\Omega)$, u is not a classical solution of (2.24), and it does not satisfy (2.24) in the usual sense. However, it may satisfy (2.26). For $f \in L^2(\Omega)$, if there exists $u \in L^2(\Omega)$ such that

$$(u, \hat{A}^*\varphi) = (f, \varphi) \tag{2.27}$$

holds for all $\varphi \in C_0^\infty(\Omega)$, u is called a **weak solution** of (2.24), and (2.27) is called the **weak formulation**. Weak solutions are important because many differential equations encountered in practice do not have smooth solutions. The only way of solving such equations is through the weak formulation. Notice that the above approach does not involve the concept of generalized derivatives. In the case of linear problems, especially for elliptic and parabolic equations, it is often possible to show that solutions in the weakest sense (2.27) are classical solutions. Let $f \in D'(\Omega)$. If there exists $u \in D'(\Omega)$ such that $\hat{A}u = f$ in $D'(\Omega)$ (all the derivatives in $\hat{A}$ are viewed as generalized derivatives), u is called a **generalized solution**. If $u \in L^2(\Omega)$ is a weak solution of (2.24), it can be considered as a generalized function in $D'(\Omega)$

$$\langle u, \varphi \rangle = \int_\Omega u(\mathbf{x})\bar{\varphi}(\mathbf{x})d\mathbf{x} = (u, \varphi), \varphi \in D(\Omega).$$

Similarly, $f \in L^2(\Omega)$ can also be viewed as a generalized function

$$\langle f, \varphi \rangle = \int_\Omega f(\mathbf{x})\bar{\varphi}(\mathbf{x})d\mathbf{x} = (f, \varphi), \varphi \in D(\Omega).$$

If the coefficients $a_\alpha(\mathbf{x})$ are sufficiently smooth, then

$$\langle u, \hat{A}^*\varphi \rangle = \left\langle u, \sum_{|\alpha|\leq m} (-1)^{|\alpha|}\partial^\alpha [\bar{a}_\alpha(\mathbf{x})\varphi] \right\rangle$$

$$= \sum_{|\alpha|\leq m} (-1)^{|\alpha|}\langle u, \partial^\alpha [\bar{a}_\alpha(\mathbf{x})\varphi] \rangle = \sum_{|\alpha|\leq m} \langle a_\alpha(\mathbf{x})\partial^\alpha u, \varphi \rangle$$

$$= \left\langle \sum_{|\alpha|\leq m} a_\alpha(\mathbf{x})\partial^\alpha u, \varphi \right\rangle = \langle \hat{A}u, \varphi \rangle, \varphi \in D(\Omega).$$

From (2.27), it follows that

$$\langle \hat{A}u, \varphi \rangle = \langle f, \varphi \rangle, \varphi \in D(\Omega),$$

which implies $\hat{A}u = f$ in $D'(\Omega)$. Therefore, a weak solution is also a generalized solution. The inverse is not necessarily true. The modern theory of partial differential equations is based on the notion of generalized solutions. The strategy is to break up the investigation of partial differential equations into two steps:

1. Prove the existence of generalized solutions by the method of functional analysis.
2. Prove the regularity of generalized solutions, that is prove that the generalized solutions are also classical solutions by assuming that the conditions (such as the boundary data and the source function) are sufficiently regular.

2.4 Method of Separation of Variables

The **method of separation of variables** is also called the **method of eigenfunction expansion**. The basic idea of separation of variables is to seek a solution of the form of a product of functions, each of which depends on one variable only, so that the solution of original partial differential equations may reduce to the solution of ordinary differential equations. We use the Helmholtz equation to illustrate the procedure. The **Helmholtz equation**, named after the German physicist Hermann Ludwig Ferdinand von Helmholtz (1821–1894), is the time-independent form of wave equation, and is given by

$$(\nabla^2 + k^2)u = 0, \tag{2.28}$$

where k is a constant. When k is zero, the Helmholtz equation reduces to the Laplace equation, named after the French mathematician Pierre-Simon Marquis de Laplace (1749–1827). The Helmholtz equation is separable in eleven orthogonal coordinate systems (Eisenhart, 1934).

2.4.1 Rectangular Coordinate System

In the rectangular coordinate system, Equation (2.28) becomes

$$\frac{\partial^2 u}{\partial x^2} + \frac{\partial^2 u}{\partial y^2} + \frac{\partial^2 u}{\partial z^2} + k^2 u = 0. \tag{2.29}$$

We seek a solution in the form

$$u = X(x)Y(y)Z(z). \tag{2.30}$$

Substituting (2.30) into (2.29) gives

$$\frac{1}{X}\frac{d^2 X}{dx^2} + \frac{1}{Y}\frac{d^2 Y}{dy^2} + \frac{1}{Z}\frac{d^2 Z}{dz^2} + k^2 = 0. \tag{2.31}$$

Since k is a constant and each term depends on one variable only and can change independently, the left-hand side of (2.31) can sum to zero for all coordinate values only if each term is a constant. Thus, we have

$$\frac{d^2 X}{dx^2} + k_x^2 X = 0,$$

$$\frac{d^2 Y}{dy^2} + k_y^2 Y = 0, \qquad (2.32)$$

$$\frac{d^2 Z}{dz^2} + k_z^2 Z = 0,$$

where k_x, k_y and k_z are separation constants and satisfy

$$k_x^2 + k_y^2 + k_z^2 = k^2. \qquad (2.33)$$

The solutions of (2.32) are harmonic functions, denoted by $X(k_x x)$, $Y(k_y y)$ and $Z(k_z z)$, and they are any linear combination of the following independent **harmonic functions**:

$$e^{ik_\alpha \alpha}, e^{-ik_\alpha \alpha}, \cos k_\alpha \alpha, \sin k_\alpha \alpha (\alpha = x, y, z). \qquad (2.34)$$

Consequently (2.30) may be expressed as

$$u = X(k_x x) Y(k_y y) Z(k_z z). \qquad (2.35)$$

The separation constants k_x, k_y and k_z are also called eigenvalues, and they are determined by the boundary conditions. The corresponding solutions (2.35) are called eignefunctions or elementary wavefunctions. The general solution of (2.29) can be expressed as a linear combination of the elementary wavefunctions.

2.4.2 Cylindrical Coordinate System

In cylindrical coordinate system, Equation (2.28) can be written as

$$\frac{1}{\rho} \frac{\partial}{\partial \rho} \left(\rho \frac{\partial u}{\partial \rho} \right) + \frac{1}{\rho^2} \frac{\partial^2 u}{\partial \varphi^2} + \frac{\partial^2 u}{\partial z^2} + k^2 u = 0. \qquad (2.36)$$

By the method of separation of variables, the solutions may be assumed to be

$$u = R(\rho) \Phi(\varphi) Z(z). \qquad (2.37)$$

Introducing (2.37) into (2.36) yields

$$\frac{d^2 R}{d\rho^2} + \frac{1}{\rho}\frac{dR}{d\rho} + \left(\mu^2 - \frac{p^2}{\rho^2}\right)R = 0,$$

$$\frac{d^2 \Phi}{d\varphi^2} + p^2 \Phi = 0, \qquad\qquad (2.38)$$

$$\frac{d^2 Z}{dz^2} + \beta^2 Z = 0,$$

where μ, p and β are separation constants and satisfy

$$\beta^2 + \mu^2 = k^2. \qquad\qquad (2.39)$$

The first equation of (2.38) is the **Bessel equation**, named after the German mathematician Friedrich Wilhelm Bessel (1784–1846), whose solutions are **Bessel functions**:

$$J_p(\mu\rho),\; N_p(\mu\rho),\; H_p^{(1)}(\mu\rho),\; H_p^{(2)}(\mu\rho),$$

where $J_p(\mu\rho)$ and $N_p(\mu\rho)$ are the Bessel functions of the first and second kind, $H_p^{(1)}(\mu\rho)$ and $H_p^{(2)}(\mu\rho)$ are the Bessel functions of the third and fourth kind, also called **Hankel functions** of the first and second kind respectively, named after the German mathematician Hermann Hankel (1839–1873). The solutions of second and third equation of (2.38) are harmonic functions. Note that only $J_p(\mu\rho)$ is finite at $\rho = 0$. The separation constants μ and p are determined by the boundary conditions. For example, if the field u is finite and satisfies the homogeneous Dirichlet boundary condition $u = 0$ at $\rho = a$, the separation constant μ is determined by $J_p(\mu\rho) = 0$. If the cylindrical region contains all φ from 0 to 2π, the separation constant p is usually determined by the requirement that the field is single-valued, that is $\Phi(0) = \Phi(2\pi)$. In this case p must be integers. If the cylindrical region only contains a circular sector, p will be fractional numbers.

2.4.3 Spherical Coordinate System

In spherical coordinate system, Equation (2.28) can be expressed as

$$\frac{1}{r^2}\frac{\partial}{\partial r}\left(r^2 \frac{\partial u}{\partial r}\right) + \frac{1}{r^2 \sin\theta}\frac{\partial}{\partial\theta}\left(\sin\theta \frac{\partial u}{\partial\theta}\right) + \frac{1}{r^2 \sin\theta}\frac{\partial^2 u}{\partial\varphi^2} + k^2 u = 0. \qquad (2.40)$$

By means of the separation of variables, we let

$$u = R(r)\Theta(\theta)\Phi(\varphi). \qquad\qquad (2.41)$$

Substitution of (2.41) into (2.40) leads to

$$\frac{1}{R}\frac{d}{dr}\left(r^2\frac{dR}{dr}\right) + k^2 r^2 = \beta^2,$$

$$\frac{1}{\Theta \sin\theta}\frac{d}{d\theta}\left(\sin\theta\frac{d\Theta}{d\theta}\right) - \frac{m^2}{\sin^2\theta} = -\beta^2, \tag{2.42}$$

$$\frac{d^2\Phi}{d\varphi^2} + m^2\Phi = 0.$$

Let $x = \cos\theta$ and $P(x) = \Theta(\theta)$, the second equation of (2.42) becomes

$$(1-x^2)\frac{d^2 P}{dx^2} - 2x\frac{dP}{dx} + \left(\beta^2 - \frac{m^2}{1-x^2}\right)P = 0. \tag{2.43}$$

This is called the **Legendre equation**, named after the French mathematician Adrien-Marie Legendre (1752–1833). The points $x = \pm 1$ are singular. Equation (2.43) has two linearly independent solutions and can be expressed as a power series at $x = 0$. In general, the series solution diverges at $x = \pm 1$. But if we let $\beta^2 = n(n+1)$, $n = 0, 1, 2\ldots$, the series will be finite at $x = \pm 1$ and have finite terms. Thus the separation constant β is determined naturally and (2.42) can be written as

$$\frac{d}{dr}\left(r^2\frac{dR}{dr}\right) + \left[k^2 r^2 - n(n+1)\right]R = 0,$$

$$(1-x^2)\frac{d^2 P}{dx^2} - 2x\frac{dP}{dx} + \left[n(n+1) - \frac{m^2}{1-x^2}\right]P = 0, \tag{2.44}$$

$$\frac{d^2\Phi}{d\varphi^2} + m^2\Phi = 0.$$

The solutions of the first equation of (2.44) are **spherical Bessel functions**

$$j_n(kr) = \sqrt{\frac{\pi}{2kr}}J_{n+1/2}(kr), \quad n_n(kr) = \sqrt{\frac{\pi}{2kr}}N_{n+1/2}(kr),$$

$$h_n^{(1)}(kr) = \sqrt{\frac{\pi}{2kr}}H_{n+1/2}^{(1)}(kr), \quad h_n^{(2)}(kr) = \sqrt{\frac{\pi}{2kr}}H_{n+1/2}^{(2)}(kr).$$

The solutions of the second equation of (2.44) are **associated Legendre functions**

$$P_n^m(\cos\theta), \quad Q_n^m(\cos\theta).$$

The solutions of the third equation of (2.44) are harmonic functions. Note that in spherical coordinate system the separation coefficients are not related.

2.5   Method of Green's Function

Physically, the Green's function represents the field produced by a point source. Through the use of the Green's function, the solution of Maxwell equations can be represented by an integral defined over the source region or on a closed surface enclosing the source. Mathematically, the solution of a partial differential equation $\hat{L}u = f$ can be expressed as $u = \hat{L}^{-1}f$, where $\hat{L}^{-1}$ stands for the inverse of $\hat{L}$ and is often represented by an integral operator whose kernel is the Green's function.

2.5.1   Fundamental Solutions of Partial Differential Equations

Consider a partial differential operator with constant coefficients: $\hat{A} = \sum\limits_{|\alpha| \leq m} a_\alpha \partial^\alpha$. If there exists a generalized function $G \in D'(R^N)$ such that $\hat{A}(G) = -\delta$, G is called the **fundamental solution** or **Green's function** of the equation $\hat{A}(G) = f$. If f is a function such that $u = -G * f$ exists as a generalized function, then u is a generalized solution of $\hat{A}(u) = f$. Let $\rho = (x, y)$, $\mathbf{r} = (x, y, z)$ and v be a constant. The fundamental solutions of wave equations are summarized below:

1. Two-dimensional Laplace equation

$$\nabla^2 G(\rho, \rho') = -\delta(\rho - \rho'), G(\rho, \rho') = -\frac{1}{2\pi} \ln|\rho - \rho'|.$$

2. Three-dimensional Laplace equation

$$\nabla^2 G(\mathbf{r}, \mathbf{r}') = -\delta(\mathbf{r} - \mathbf{r}'), G(\mathbf{r}, \mathbf{r}') = \frac{1}{4\pi |\mathbf{r} - \mathbf{r}'|}.$$

3. Two-dimensional Helmholtz equation

$$(\nabla^2 + k^2)G(\rho, \rho') = -\delta(\rho - \rho'), G(\rho, \rho') = \frac{1}{4j} H_0^{(2)}(k|\rho - \rho'|).$$

4. Three-dimensional Helmholtz equation

$$(\nabla^2 + k^2)G(\mathbf{r}, \mathbf{r}') = -\delta(\mathbf{r} - \mathbf{r}'), G(\mathbf{r}, \mathbf{r}') = \frac{e^{-jk|\mathbf{r}-\mathbf{r}'|}}{4\pi |\mathbf{r} - \mathbf{r}'|}.$$

5. One-dimensional wave equation

$$\begin{cases} \left(\dfrac{\partial^2}{\partial z^2} - \dfrac{1}{v^2}\dfrac{\partial^2}{\partial t^2}\right) G(z, z'; t, t') = -\delta(z - z')\delta(t - t') \\ G(z, z'; t, t') = 0, t < t' \end{cases},$$

$$G(z, z'; t, t') = \frac{v}{2} H(t - t' - |z - z'|/v).$$

6. Two-dimensional wave equation

$$
\begin{cases}
\left(\nabla^2 - \dfrac{1}{v^2}\dfrac{\partial^2}{\partial t^2}\right) G(\boldsymbol{\rho}, \boldsymbol{\rho}'; t, t') = -\delta(\boldsymbol{\rho} - \boldsymbol{\rho}')\delta(t - t') \\[2mm]
G(\boldsymbol{\rho}, \boldsymbol{\rho}'; t, t') = 0, \, t < t'
\end{cases},
$$

$$
G(\boldsymbol{\rho}, \boldsymbol{\rho}'; t, t') = \frac{H(t - t' - |\boldsymbol{\rho} - \boldsymbol{\rho}'|/v)}{2\pi\sqrt{(t - t')^2 - |\boldsymbol{\rho} - \boldsymbol{\rho}'|/v}}.
$$

7. Three-dimensional wave equation

$$
\begin{cases}
\left(\nabla^2 - \dfrac{1}{v^2}\dfrac{\partial^2}{\partial t^2}\right) G(\mathbf{r}, \mathbf{r}'; t, t') = -\delta(\mathbf{r} - \mathbf{r}')\delta(t - t') \\[2mm]
G(\mathbf{r}, \mathbf{r}'; t, t') = 0, \, t < t'
\end{cases},
$$

$$
G(\mathbf{r}, \mathbf{r}'; t, t') = \frac{\delta(t - t' - |\mathbf{r} - \mathbf{r}'|/v)}{4\pi|\mathbf{r} - \mathbf{r}'|}.
$$

In the above, $H(x)$ is the unit step function. More examples of Green's functions and their derivations can be found in Chapter 8.

2.5.2 Integral Representations of Arbitrary Fields

The integral expressions of the fields are important for the study of boundary value problems, especially for the radiation and scattering problems. The results obtained in this section are applicable to any time-domain fields, not necessarily of electromagnetic origin.

Representation theorem for time-domain fields: *Let V be a bounded region with boundary S, and* $\mathbf{u}_n$ *be the unit outward normal to S. For any time-dependent smooth vector fields* $\mathbf{e}$ *and* $\mathbf{h}$, *we have*

$$
-\nabla \times \int_{-\infty}^{\infty} dt' \int_{S} G(\mathbf{r}, \mathbf{r}'; t, t')\mathbf{u}_n(\mathbf{r}') \times \mathbf{e}(\mathbf{r}', t') dS(\mathbf{r}')
$$

$$
+\nabla \int_{-\infty}^{\infty} dt' \int_{S} G(\mathbf{r}, \mathbf{r}'; t, t')\mathbf{u}_n(\mathbf{r}') \cdot \mathbf{e}(\mathbf{r}', t') dS(\mathbf{r}')
$$

$$
\pm\frac{\partial}{v\partial t} \int_{-\infty}^{\infty} dt' \int_{S} G(\mathbf{r}, \mathbf{r}'; t, t')\mathbf{u}_n(\mathbf{r}') \times \mathbf{h}(\mathbf{r}', t') dS(\mathbf{r}')
$$

$$
+\nabla \times \int_{-\infty}^{\infty} dt' \int_{V} \left\{ G(\mathbf{r}, \mathbf{r}'; t, t')\left[\nabla' \times \mathbf{e}(\mathbf{r}', t') \pm \frac{\partial\mathbf{h}(\mathbf{r}', t')}{v\partial t'}\right] \right\} dV(\mathbf{r}')
$$

$$-\nabla \int_{-\infty}^{\infty} dt' \int_V \left[G(\mathbf{r}, \mathbf{r}') \nabla' \cdot \mathbf{e}(\mathbf{r}', t') \right] dV(\mathbf{r}')$$

$$\mp \frac{\partial}{v \partial t} \int_{-\infty}^{\infty} dt' \int_V G(\mathbf{r}, \mathbf{r}'; t, t') \left[\nabla' \times \mathbf{h}(\mathbf{r}', t') \mp \frac{\partial \mathbf{e}(\mathbf{r}', t')}{v \partial t'} \right] dV(\mathbf{r}')$$

$$= \begin{cases} \mathbf{e}(\mathbf{r}, t), \mathbf{r} \in V \\ 0, \mathbf{r} \in R^3 - V \end{cases} \tag{2.45}$$

where $G(\mathbf{r}, \mathbf{r}'; t, t') = \delta(t - t' - |\mathbf{r} - \mathbf{r}'|/v)/4\pi |\mathbf{r} - \mathbf{r}'|$ is Green's function of the wave equation in three-dimensional space.

Equation (2.45) is called the **Stratton–Chu formula**, named after the American scientist Julius Adams Stratton (1901–1994) and the Chinese scientist Lan Jen Chu (1913–1973), and can be derived as follows. For an arbitrary field point $\mathbf{r} \in V$ shown in Figure 2.1, the following vector identities can easily be established

$$\nabla' \times \left[G(\mathbf{r}, \mathbf{r}'; t, t') \mathbf{e}(\mathbf{r}', t') \right] = G(\mathbf{r}, \mathbf{r}'; t, t') \nabla' \times \mathbf{e}(\mathbf{r}', t')$$
$$- \nabla \times \left[G(\mathbf{r}, \mathbf{r}'; t, t') \mathbf{e}(\mathbf{r}', t') \right],$$

$$\nabla' \cdot \left[G(\mathbf{r}, \mathbf{r}'; t, t') \mathbf{e}(\mathbf{r}', t') \right] = G(\mathbf{r}, \mathbf{r}'; t, t') \nabla' \cdot \mathbf{e}(\mathbf{r}', t')$$
$$- \nabla \cdot \left[G(\mathbf{r}, \mathbf{r}'; t, t') \mathbf{e}(\mathbf{r}', t') \right],$$

$$\mp \nabla' \times \left[G(\mathbf{r}, \mathbf{r}'; t, t') \frac{\partial \mathbf{h}(\mathbf{r}', t')}{v \partial t'} \right] = \mp G(\mathbf{r}, \mathbf{r}'; t, t') \nabla' \times \frac{\partial \mathbf{h}(\mathbf{r}', t')}{v \partial t'}$$

$$\pm \nabla \times \left[G(\mathbf{r}, \mathbf{r}'; t, t') \frac{\partial \mathbf{h}(\mathbf{r}', t')}{v \partial t'} \right], \tag{2.46}$$

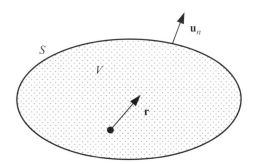

Figure 2.1 An arbitrary region

where the property $\nabla G = -\nabla' G$ has been used. Applying $\nabla \times$ to the first equation, $-\nabla$ to the second, and using the relation $\nabla \times \nabla \times = -\nabla^2 + \nabla\nabla\cdot$, we obtain

$$\nabla \times \nabla' \times \left[G(\mathbf{r}, \mathbf{r}'; t, t')\mathbf{e}(\mathbf{r}', t') \right]$$

$$= \nabla \times \left[G(\mathbf{r}, \mathbf{r}'; t, t')\nabla' \times \mathbf{e}(\mathbf{r}', t') \right] - \nabla\nabla \cdot \left[G(\mathbf{r}, \mathbf{r}'; t, t')\mathbf{e}(\mathbf{r}', t') \right]$$

$$+ \left[\frac{\partial^2 G(\mathbf{r}, \mathbf{r}'; t, t')}{v^2 \partial t'^2} - \delta(\mathbf{r} - \mathbf{r}')\delta(t - t') \right] \mathbf{e}(\mathbf{r}', t'),$$

and

$$-\nabla\nabla' \cdot \left[G(\mathbf{r}, \mathbf{r}'; t, t')\mathbf{e}(\mathbf{r}', t') \right] = -\nabla \left[G(\mathbf{r}, \mathbf{r}'; t, t')\nabla' \cdot \mathbf{e}(\mathbf{r}', t') \right]$$

$$+\nabla\nabla \cdot \left[G(\mathbf{r}, \mathbf{r}'; t, t')\mathbf{e}(\mathbf{r}', t') \right].$$

Adding the above two equations as well as the last equation of (2.46) yields

$$\nabla \times \nabla' \times \left[G(\mathbf{r}, \mathbf{r}'; t, t')\mathbf{e}(\mathbf{r}', t') \right] - \nabla\nabla' \cdot \left[G(\mathbf{r}, \mathbf{r}'; t, t')\mathbf{e}(\mathbf{r}', t') \right]$$

$$\mp\nabla' \times \left[G(\mathbf{r}, \mathbf{r}'; t, t')\frac{\partial \mathbf{h}(\mathbf{r}', t')}{v\partial t'} \right]$$

$$= \nabla \times \left\{ G(\mathbf{r}, \mathbf{r}'; t, t') \left[\nabla' \times \mathbf{e}(\mathbf{r}', t') \pm \frac{\partial \mathbf{h}(\mathbf{r}', t')}{v\partial t'} \right] \right\} - \nabla \left[G(\mathbf{r}, \mathbf{r}'; t, t')\nabla' \cdot \mathbf{e}(\mathbf{r}', t') \right]$$

$$\mp G(\mathbf{r}, \mathbf{r}'; t, t')\nabla' \times \frac{\partial \mathbf{h}(\mathbf{r}', t')}{v\partial t'} + \frac{\partial^2 G(\mathbf{r}, \mathbf{r}'; t, t')}{v^2 \partial t'^2}\mathbf{e}(\mathbf{r}', t') - \delta(\mathbf{r} - \mathbf{r}')\delta(t - t')\mathbf{e}(\mathbf{r}', t').$$

Taking the integration over V and interchanging differentiation and integration, we obtain

$$\nabla \times \int_{-\infty}^{\infty} dt' \int_S \mathbf{u}_n(\mathbf{r}') \times \mathbf{e}(\mathbf{r}')G(\mathbf{r}, \mathbf{r}'; t, t')dS(\mathbf{r}')$$

$$-\nabla \int_{-\infty}^{\infty} dt' \int_S \mathbf{u}_n(\mathbf{r}') \cdot \mathbf{e}(\mathbf{r}', t')G(\mathbf{r}, \mathbf{r}'; t, t')dS(\mathbf{r}')$$

$$\mp \int_{-\infty}^{\infty} dt' \int_S \mathbf{u}_n(\mathbf{r}') \times \left[G(\mathbf{r}, \mathbf{r}'; t, t')\frac{\partial \mathbf{h}(\mathbf{r}', t')}{v\partial t'} \right] dS(\mathbf{r}')$$

$$= \nabla \times \int_{-\infty}^{\infty} dt' \int_V \left\{ G(\mathbf{r}, \mathbf{r}'; t, t') \left[\nabla' \times \mathbf{e}(\mathbf{r}', t') \pm \frac{\partial \mathbf{h}(\mathbf{r}', t')}{v\partial t'} \right] \right\} dV(\mathbf{r}')$$

$$-\nabla \int_{-\infty}^{\infty} dt' \int_V G(\mathbf{r}, \mathbf{r}')\nabla' \cdot \mathbf{e}(\mathbf{r}')dV(\mathbf{r}')$$

$$
\mp \int\limits_{-\infty}^{\infty} dt' \int\limits_{V} \left[G(\mathbf{r}, \mathbf{r}'; t, t') \nabla' \times \frac{\partial \mathbf{h}(\mathbf{r}', t')}{v \partial t'} \mp \frac{\partial^2 G(\mathbf{r}, \mathbf{r}'; t, t')}{v^2 \partial t'^2} \mathbf{e}(\mathbf{r}', t') \right] dV(\mathbf{r}')
$$

$$
= \begin{cases} \mathbf{e}(\mathbf{r}, t), \ \mathbf{r} \in V \\ 0, \ \mathbf{r} \in R^3 - V \end{cases}.
$$

Equation (2.45) can be obtained by making use of $\partial G / \partial t = -\partial G / \partial t'$ and the following calculations

$$
\int\limits_{-\infty}^{\infty} dt' \int\limits_{V} \left[G(\mathbf{r}, \mathbf{r}'; t, t') \nabla' \times \frac{\partial \mathbf{h}(\mathbf{r}', t')}{v \partial t'} \mp \frac{\partial^2 G(\mathbf{r}, \mathbf{r}'; t, t')}{v^2 \partial t'^2} \mathbf{e}(\mathbf{r}', t') \right] dV(\mathbf{r}')
$$

$$
= \frac{\partial}{v \partial t} \int\limits_{-\infty}^{\infty} dt' \int\limits_{V} G(\mathbf{r}, \mathbf{r}'; t, t') \left[\nabla' \times \mathbf{h}(\mathbf{r}', t') \mp \frac{\partial \mathbf{e}(\mathbf{r}', t')}{v \partial t'} \right] dV(\mathbf{r}'),
$$

$$
\int\limits_{-\infty}^{\infty} dt' \int\limits_{S} \mathbf{u}_n(\mathbf{r}') \times \left[G(\mathbf{r}, \mathbf{r}'; t, t') \frac{\partial \mathbf{h}(\mathbf{r}', t')}{v \partial t'} \right] dS(\mathbf{r}')
$$

$$
= \frac{\partial}{v \partial t} \int\limits_{-\infty}^{\infty} dt' \int\limits_{S} G(\mathbf{r}, \mathbf{r}'; t, t') \mathbf{u}_n(\mathbf{r}') \times \mathbf{h}(\mathbf{r}', t') dS(\mathbf{r}').
$$

In a similar manner, we can prove the Stratton–Chu formula for time-harmonic fields.

Representation theorem for time-harmonic fields: *Let V be a bounded region with boundary S, and $\mathbf{u}_n$ be the outward unit normal to S. For any time-independent smooth vector fields* **e** *and* **h**, *we have the Stratton–Chu formula*

$$
-\nabla \times \int\limits_{S} G(\mathbf{r}, \mathbf{r}') \mathbf{u}_n(\mathbf{r}') \times \mathbf{e}(\mathbf{r}') dS(\mathbf{r}') + \nabla \int\limits_{S} G(\mathbf{r}, \mathbf{r}') \mathbf{u}_n(\mathbf{r}') \cdot \mathbf{e}(\mathbf{r}') dS(\mathbf{r}')
$$

$$
\pm jk \int\limits_{S} G(\mathbf{r}, \mathbf{r}') \mathbf{u}_n(\mathbf{r}') \times \mathbf{h}(\mathbf{r}') dS(\mathbf{r}')
$$

$$
+\nabla \times \int\limits_{V} G(\mathbf{r}, \mathbf{r}') \left[\nabla' \times \mathbf{e}(\mathbf{r}') \pm jk\mathbf{h}(\mathbf{r}') \right] dV(\mathbf{r}')
$$

$$
-\nabla \int\limits_{V} G(\mathbf{r}, \mathbf{r}') \nabla \cdot \mathbf{e}(\mathbf{r}') dv(\mathbf{r}') \mp jk \int\limits_{V} G(\mathbf{r}, \mathbf{r}') \left[\nabla' \times \mathbf{h}(\mathbf{r}') \mp jk\mathbf{e}(\mathbf{r}') \right] dV(\mathbf{r}')
$$

$$
= \begin{cases} \mathbf{e}(\mathbf{r}, t), \ \mathbf{r} \in V \\ 0, \ \mathbf{r} \in R^3 - V \end{cases} \tag{2.47}
$$

where $G(\mathbf{r}, \mathbf{r}') = e^{-jk|\mathbf{r}-\mathbf{r}'|}/4\pi \, |\mathbf{r} - \mathbf{r}'|$ *is Green's function of the Helmholtz equation in three-dimensional space.*

2.5.3   Integral Representations of Electromagnetic Fields

In our preceding discussions, the field quantities $\mathbf{e}$ and $\mathbf{h}$ do not have to satisfy the Maxwell equations. Let V be a region of finite extension, which contains the sources $\mathbf{J}$ and $\mathbf{J}_m$. For the fields satisfying the time-domain Maxwell equations

$$\nabla \times \mathbf{H}(\mathbf{r}, t) - \frac{1}{v\eta} \frac{\partial \mathbf{E}(\mathbf{r}, t)}{\partial t} = \mathbf{J}(\mathbf{r}, t),$$

$$\nabla \times \mathbf{E}(\mathbf{r}, t) + \frac{\eta}{v} \frac{\partial \mathbf{H}(\mathbf{r}, t)}{\partial t} = -\mathbf{J}_m(\mathbf{r}, t),$$

$$\nabla \cdot \mathbf{E}(\mathbf{r}, t) = \eta v \rho(\mathbf{r}, t), \ \nabla \cdot \mathbf{H}(\mathbf{r}, t) = \frac{v}{\eta} \rho_m(\mathbf{r}, t),$$

where $v = 1/\sqrt{\mu\varepsilon}$ and $\eta = \sqrt{\mu/\varepsilon}$, we let

$$\mathbf{e} = \sqrt{\varepsilon}\mathbf{E}, \mathbf{h} = \sqrt{\mu}\mathbf{H} \tag{2.48}$$

in (2.45) to obtain

$$\nabla \times \int_{-\infty}^{\infty} dt' \int_S G(\mathbf{r}, \mathbf{r}'; t, t') \mathbf{J}_{ms}(\mathbf{r}', t') dS(\mathbf{r}') + \eta v \nabla \int_{-\infty}^{\infty} dt' \int_S G(\mathbf{r}, \mathbf{r}'; t, t') \rho_s(\mathbf{r}', t') dS(\mathbf{r}')$$

$$+ \frac{\eta}{v} \frac{\partial}{\partial t} \int_{-\infty}^{\infty} dt' \int_S G(\mathbf{r}, \mathbf{r}'; t, t') \mathbf{J}_s(\mathbf{r}', t') dS(\mathbf{r}')$$

$$- \nabla \times \int_{-\infty}^{\infty} dt' \int_V G(\mathbf{r}, \mathbf{r}'; t, t') \mathbf{J}_m(\mathbf{r}', t') dV(\mathbf{r}') - \eta v \nabla \int_{-\infty}^{\infty} dt' \int_V \left[G(\mathbf{r}, \mathbf{r}'; t, t') \rho(\mathbf{r}', t')\right] dV(\mathbf{r}')$$

$$- \frac{\eta}{v} \frac{\partial}{\partial t} \int_{-\infty}^{\infty} dt' \int_V G(\mathbf{r}, \mathbf{r}'; t, t') \mathbf{J}(\mathbf{r}', t') dV(\mathbf{r}') = \begin{cases} \mathbf{E}(\mathbf{r}, t), \mathbf{r} \in V \\ 0, \mathbf{r} \in R^3 - V \end{cases} \tag{2.49}$$

where $\mathbf{J}_s = \mathbf{u}_n \times \mathbf{H}$, $\mathbf{J}_{ms} = -\mathbf{u}_n \times \mathbf{E}$, $\rho_s = \varepsilon\mathbf{u}_n \cdot \mathbf{E}$, $\rho_{ms} = \mu\mathbf{u}_n \cdot \mathbf{H}$. Interchanging $\mathbf{e}$ and $\mathbf{h}$ in (2.45) and using (2.48) yields

$$- \nabla \times \int_{-\infty}^{\infty} dt' \int_S G(\mathbf{r}, \mathbf{r}'; t, t') \mathbf{J}_s(\mathbf{r}', t') dS(\mathbf{r}') + \frac{v}{\eta} \nabla \int_{-\infty}^{\infty} dt' \int_S \rho_{ms}(\mathbf{r}', t') G(\mathbf{r}, \mathbf{r}'; t, t') dS(\mathbf{r}')$$

$$+ \frac{1}{\eta v} \frac{\partial}{\partial t} \int_{-\infty}^{\infty} dt' \int_S G(\mathbf{r}, \mathbf{r}'; t, t') \mathbf{J}_{ms}(\mathbf{r}', t') dS(\mathbf{r}')$$

$$+\nabla \times \int_{-\infty}^{\infty} dt' \int_{V} G(\mathbf{r}, \mathbf{r}'; t, t') \mathbf{J}(\mathbf{r}', t') dV(\mathbf{r}') - \frac{v}{\eta} \nabla \int_{-\infty}^{\infty} dt' \int_{V} G(\mathbf{r}, \mathbf{r}') \rho_m(\mathbf{r}', t') dV(\mathbf{r}')$$

$$-\frac{1}{\eta v} \frac{\partial}{\partial t} \int_{-\infty}^{\infty} dt' \int_{V} G(\mathbf{r}, \mathbf{r}'; t, t') \mathbf{J}_m(\mathbf{r}', t') dV(\mathbf{r}') = \begin{cases} \mathbf{H}(\mathbf{r}, t), \mathbf{r} \in V \\ 0, \mathbf{r} \in R^3 - V \end{cases}. \qquad (2.50)$$

Similarly, for the time-harmonic fields satisfying the Maxwell equations,

$$\nabla \times \mathbf{H}(\mathbf{r}) - j\omega \frac{1}{v\eta} \mathbf{E}(\mathbf{r}) = \mathbf{J}(\mathbf{r}),$$

$$\nabla \times \mathbf{E}(\mathbf{r}) + j\omega \frac{\eta}{v} \mathbf{H}(\mathbf{r}) = -\mathbf{J}_m(\mathbf{r}),$$

$$\nabla \cdot \mathbf{E}(\mathbf{r}) = \eta v \rho(\mathbf{r}), \nabla \cdot \mathbf{H}(\mathbf{r}) = \frac{v}{\eta} \rho_m(\mathbf{r}),$$

we may use (2.47) to obtain

$$jk\eta \int_{S} G(\mathbf{r}, \mathbf{r}') \mathbf{J}_s(\mathbf{r}') dS(\mathbf{r}') + \nabla \times \int_{S} G(\mathbf{r}, \mathbf{r}') \mathbf{J}_{ms}(\mathbf{r}') dS(\mathbf{r}')$$

$$+\eta v \nabla \int_{S} G(\mathbf{r}, \mathbf{r}') \rho_s(\mathbf{r}') dS(\mathbf{r}') - jk\eta \int_{V} G(\mathbf{r}, \mathbf{r}') \mathbf{J}(\mathbf{r}') dV(\mathbf{r}')$$

$$-\nabla \times \int_{V} G(\mathbf{r}, \mathbf{r}') \mathbf{J}_m(\mathbf{r}') dV(\mathbf{r}') - \eta v \nabla \int_{V} G(\mathbf{r}, \mathbf{r}') \rho(\mathbf{r}') dV(\mathbf{r}') \qquad (2.51)$$

$$= \begin{cases} \mathbf{E}(\mathbf{r}), \mathbf{r} \in V \\ 0, \mathbf{r} \in R^3 - V \end{cases},$$

$$j\frac{k}{\eta} \int_{S} G(\mathbf{r}, \mathbf{r}') \mathbf{J}_{ms}(\mathbf{r}') dS(\mathbf{r}') - \nabla \times \int_{S} G(\mathbf{r}, \mathbf{r}') \mathbf{J}_s(\mathbf{r}') dS(\mathbf{r}')$$

$$+\frac{v}{\eta} \nabla \int_{S} \rho_{ms}(\mathbf{r}') G(\mathbf{r}, \mathbf{r}') dS(\mathbf{r}') - j\frac{k}{\eta} \int_{V} G(\mathbf{r}, \mathbf{r}') \mathbf{J}_m(\mathbf{r}') dV(\mathbf{r}')$$

$$+\nabla \times \int_{V} G(\mathbf{r}, \mathbf{r}') \mathbf{J}(\mathbf{r}') dV(\mathbf{r}') - \frac{v}{\eta} \nabla \int_{V} G(\mathbf{r}, \mathbf{r}') \rho_m(\mathbf{r}') dV(\mathbf{r}') \qquad (2.52)$$

$$= \begin{cases} \mathbf{H}(\mathbf{r}), \mathbf{r} \in V \\ 0, \mathbf{r} \in R^3 - V \end{cases}.$$

The electromagnetic fields are called radiating if they satisfy

$$\lim_{\mathbf{r} \to \infty} r(\mathbf{u}_r \times \mathbf{E} - \eta \mathbf{H}) = 0, \tag{2.53}$$

where $r = |\mathbf{r}|$. Equation (2.53) is called the **Silver-Müller radiation condition**, named after the American scientist Samuel Silver (1915–1976) and the German mathematician Claus Müller (1920–2008). Let all the sources be confined in V_0, and S be any closed surface that encloses V_0. Using the Silver-Müller radiation condition, the time-domain radiating fields in $R^3 - V_0$ can be expressed as

$$\mathbf{E}(\mathbf{r}, t) = -\nabla \times \int_S \frac{\mathbf{J}_{ms}(\mathbf{r}', t - R/v)}{4\pi R} dS(\mathbf{r}') - \eta v \nabla \int_S \frac{\rho_s(\mathbf{r}', t - R/v)}{4\pi R} dS(\mathbf{r}')$$
$$-\frac{\eta}{v} \frac{\partial}{\partial t} \int_S \frac{\mathbf{J}_s(\mathbf{r}', t - R/v)}{4\pi R} dS(\mathbf{r}'), \tag{2.54}$$

$$\mathbf{H}(\mathbf{r}, t) = \nabla \times \int_S \frac{\mathbf{J}_s(\mathbf{r}', t - R/v)}{4\pi R} dS(\mathbf{r}') - \frac{v}{\eta} \nabla \int_S \frac{\rho_{ms}(\mathbf{r}', t - R/v)}{4\pi R} dS(\mathbf{r}')$$
$$-\frac{1}{\eta v} \frac{\partial}{\partial t} \int_S \frac{\mathbf{J}_{ms}(\mathbf{r}', t - R/v)}{4\pi R} dS(\mathbf{r}'), \tag{2.55}$$

$$\mathbf{E}(\mathbf{r}, t) = -\nabla \times \int_{V_0} \frac{\mathbf{J}_m(\mathbf{r}', t - R/v)}{4\pi R} dV(\mathbf{r}') - \eta v \nabla \int_{V_0} \frac{\rho(\mathbf{r}', t - R/v)}{4\pi R} dV(\mathbf{r}')$$
$$-\frac{\eta}{v} \frac{\partial}{\partial t} \int_{V_0} \frac{\mathbf{J}(\mathbf{r}', t - R/v)}{4\pi R} dV(\mathbf{r}'), \tag{2.56}$$

$$\mathbf{H}(\mathbf{r}, t) = \nabla \times \int_{V_0} \frac{\mathbf{J}(\mathbf{r}', t - R/v)}{4\pi R} dV(\mathbf{r}') - \frac{v}{\eta} \nabla \int_{V_0} \frac{\rho_m(\mathbf{r}', t - R/v)}{4\pi R} dV(\mathbf{r}')$$
$$-\frac{1}{\eta v} \frac{\partial}{\partial t} \int_{V_0} \frac{\mathbf{J}_m(\mathbf{r}', t - R/v)}{4\pi R} dV(\mathbf{r}'), \tag{2.57}$$

where $R = |\mathbf{r} - \mathbf{r}'|$. Similarly the time-harmonic radiating fields in $R^3 - V_0$ may be represented by

$$
\mathbf{E}(\mathbf{r}) = -jk\eta \int_S G(\mathbf{r}, \mathbf{r}')\mathbf{J}_s(\mathbf{r}')dS(\mathbf{r}') - \nabla \times \int_S \mathbf{J}_{ms}(\mathbf{r}')G(\mathbf{r}, \mathbf{r}')dS(\mathbf{r}')
$$
$$
-\eta v \nabla \int_S \rho_s(\mathbf{r}')G(\mathbf{r}, \mathbf{r}')dS(\mathbf{r}'),
\tag{2.58}
$$

$$
\mathbf{H}(\mathbf{r}) = -j\frac{k}{\eta} \int_S G(\mathbf{r}, \mathbf{r}')\mathbf{J}_{ms}(\mathbf{r}')dS(\mathbf{r}') + \nabla \times \int_S \mathbf{J}_s(\mathbf{r}')G(\mathbf{r}, \mathbf{r}')dS(\mathbf{r}')
$$
$$
-\frac{v}{\eta}\nabla \int_S \rho_{ms}(\mathbf{r}')G(\mathbf{r}, \mathbf{r}')dS(\mathbf{r}'),
\tag{2.59}
$$

$$
\mathbf{E}(\mathbf{r}) = -jk\eta \int_{V_0} G(\mathbf{r}, \mathbf{r}')\mathbf{J}(\mathbf{r}')dV(\mathbf{r}') - \nabla \times \int_{V_0} \mathbf{J}_m(\mathbf{r}')G(\mathbf{r}, \mathbf{r}')dV(\mathbf{r}')
$$
$$
-\eta v \nabla \int_{V_0} \rho(\mathbf{r}')G(\mathbf{r}, \mathbf{r}')dV(\mathbf{r}'),
\tag{2.60}
$$

$$
\mathbf{H}(\mathbf{r}) = -j\frac{k}{\eta} \int_{V_0} G(\mathbf{r}, \mathbf{r}')\mathbf{J}_m(\mathbf{r}')dV(\mathbf{r}') + \nabla \times \int_{V_0} \mathbf{J}(\mathbf{r}')G(\mathbf{r}, \mathbf{r}')dV(\mathbf{r}')
$$
$$
-\frac{v}{\eta}\nabla \int_{V_0} \rho_m(\mathbf{r}')G(\mathbf{r}, \mathbf{r}')dV(\mathbf{r}'),
\tag{2.61}
$$

where $G(\mathbf{r}, \mathbf{r}') = e^{-jkR}/4\pi R$. From these integral representations, it is readily found that the radiating fields satisfy the **finiteness condition**

$$
\mathbf{E}(\mathbf{r}) = o(1/r), \quad \mathbf{H}(\mathbf{r}) = o(1/r)
\tag{2.62}
$$

for sufficiently large r. It must be mentioned that the integral expressions (2.54), (2.55), (2.58), and (2.59) are only valid for a closed surface S. If the surface S is open, these integral expressions no longer satisfy the Maxwell equations. In this case, special treatment is needed in order for the integral expressions to satisfy the Maxwell equations. Consider an open surface S with boundary Γ. Let $\mathbf{u}_n$ be the outward normal on S, $\mathbf{u}_t$ be the tangent along the boundary Γ, and $\mathbf{u}_b$ be a unit vector perpendicular to both $\mathbf{u}_n$ and $\mathbf{u}_t$, as shown in Figure 2.2. To maintain the continuity equation along the boundary Γ, electric and magnetic charges along Γ must be

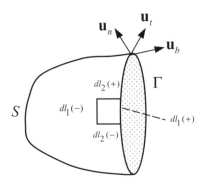

Figure 2.2   An open surface

introduced. To see this, we can make a small rectangle of length dl_1 and width dl_2. According to the continuity equation, we have

$$\mathbf{J}_s|_{dl_2(+)} \cdot \mathbf{u}_t dl_2 - \mathbf{J}_s|_{dl_2(-)} \cdot \mathbf{u}_t dl_2 + \mathbf{J}_s|_{dl_1(+)} \cdot \mathbf{u}_b dl_1 - \mathbf{J}_s|_{dl_1(-)} \cdot \mathbf{u}_b dl_1 = -dl_1 dl_2 j\omega\rho.$$

Letting dl_2 approach zero yields

$$\lim_{dl_2 \to 0} dl_2 j\omega\rho = j\omega\rho_l = \mathbf{J}_s|_{dl_1(-)} \cdot \mathbf{u}_b = -\mathbf{u}_t \cdot \mathbf{H}.$$

Similarly, for the magnetic current we may find that

$$\lim_{dl_2 \to 0} dl_2 j\omega\rho_m = j\omega\rho_{ml} = \mathbf{J}_{ms}|_{dl_1(-)} \cdot \mathbf{u}_b = \mathbf{u}_t \cdot \mathbf{E}.$$

The radiation fields in $R^3 - V_0$ are the superposition of contributions from surface sources and line sources along Γ, and can be expressed by

$$\mathbf{E}(\mathbf{r}) = -\int_S \left[jk\eta G(\mathbf{r}, \mathbf{r}')\mathbf{J}_s(\mathbf{r}') + \mathbf{J}_{ms}(\mathbf{r}') \times \nabla' G(\mathbf{r}, \mathbf{r}') - \eta c\rho_s(\mathbf{r}')\nabla' G(\mathbf{r}, \mathbf{r}') \right] dS(\mathbf{r}')$$

$$+ \int_\Gamma \eta c\rho_l(\mathbf{r}')\nabla' G(\mathbf{r}, \mathbf{r}') d\Gamma(\mathbf{r}'),$$

$$\mathbf{H}(\mathbf{r}) = -\int_S \left[j\frac{k}{\eta} G(\mathbf{r}, \mathbf{r}')\mathbf{J}_{ms}(\mathbf{r}') - \mathbf{J}_s(\mathbf{r}') \times \nabla' G(\mathbf{r}, \mathbf{r}') - \frac{c}{\eta}\rho_{ms}(\mathbf{r}')\nabla' G(\mathbf{r}, \mathbf{r}') \right] dS(\mathbf{r}')$$

$$+ \int_\Gamma \frac{c}{\eta}\rho_{ml}(\mathbf{r}')\nabla' G(\mathbf{r}, \mathbf{r}') d\Gamma(\mathbf{r}').$$

$$(2.63)$$

It can be verified that Equations (2.63) satisfy the Maxwell equations.

2.6   Potential Theory

Potential theory is the mathematical treatment of the potential-energy functions used in physics to study gravitation and electromagnetism, and has developed into a major field of mathematical research (Kellog, 1953; MacMillan, 1958). In the nineteenth century, it was believed that all forces in nature could be derived from a potential which satisfies the Laplace equation. These days, the term 'potential' is used in a broad sense, and the potential is not necessarily a solution of Laplace equation. As long as the solution of a partial differential equation can be expressed as the first derivative of a new function, this new function can be considered as a potential. Usually the solution and its potential function satisfy the same type of equation, while the equation for the latter has a simpler source term.

2.6.1   Vector Potential, Scalar Potential, and Gauge Conditions

From the equations $\nabla \cdot \mathbf{B} = 0$ and $\nabla \times \mathbf{E} = -\partial \mathbf{B}/\partial t$, a **vector potential A** and a **scalar potential** ϕ can be introduced such that

$$\mathbf{E} = -\nabla\phi - \frac{\partial \mathbf{A}}{\partial t}, \mathbf{B} = \nabla \times \mathbf{A}. \tag{2.64}$$

If the medium is isotropic and homogeneous, we can substitute (2.64) into $\nabla \times \mathbf{H} = \mathbf{J} + \partial \mathbf{D}/\partial t$, and insert the first of (2.64) into $\nabla \cdot \mathbf{D} = \rho$ to obtain

$$\left(\nabla^2 - \frac{1}{v^2}\frac{\partial^2}{\partial t^2}\right)\mathbf{A} = -\mu\mathbf{J} + \nabla\left(\nabla \cdot \mathbf{A} + \frac{1}{v^2}\frac{\partial\phi}{\partial t}\right),$$
$$\left(\nabla^2 - \frac{1}{v^2}\frac{\partial^2}{\partial t^2}\right)\phi = -\frac{\rho}{\varepsilon} - \frac{\partial}{\partial t}\left(\nabla \cdot \mathbf{A} + \frac{1}{v^2}\frac{\partial\phi}{\partial t}\right), \tag{2.65}$$

where $v = 1/\sqrt{\mu\varepsilon}$. The term $\nabla \cdot \mathbf{A} + \partial\phi/v^2\partial t$ on the right-hand sides can be set to zero by means of the gauge transform. In fact, we can define a new vector potential $\mathbf{A}'$ and a new scalar potential ϕ' through

$$\mathbf{A}' = \mathbf{A} + \nabla\psi, \tag{2.66}$$

$$\phi' = \phi - \frac{\partial\psi}{\partial t}, \tag{2.67}$$

where ψ is called the **gauge function**. The transformation from $(\mathbf{A}, \phi)$ to $(\mathbf{A}', \phi')$ defined by (2.66) and (2.67) is called the **gauge transformation**. The electromagnetic fields remain unchanged under the gauge transformation. The new vector potential $\mathbf{A}'$ and scalar potential ϕ' satisfy

$$\nabla \cdot \mathbf{A}' + \frac{1}{v^2}\frac{\partial\phi'}{\partial t} = \nabla \cdot \mathbf{A} + \frac{1}{v^2}\frac{\partial\phi}{\partial t} + \nabla^2\psi - \frac{1}{v^2}\frac{\partial^2\psi}{\partial t^2}.$$

If the term $\nabla \cdot \mathbf{A} + \partial \phi / v^2 \partial t$ is not zero, the left-hand side can be sent to zero by forcing the gauge function ψ to satisfy

$$\nabla^2 \psi - \frac{1}{v^2}\frac{\partial^2 \psi}{\partial t^2} = -\left(\nabla \cdot \mathbf{A} + \frac{1}{v^2}\frac{\partial \phi}{\partial t}\right).$$

Thus, the equation

$$\nabla \cdot \mathbf{A} + \frac{1}{v^2}\frac{\partial \phi}{\partial t} = 0 \tag{2.68}$$

may be assumed and is called the **Lorenz gauge condition**, named after the Danish physicist Ludvig Valentin Lorenz (1829–1891). If $\mathbf{A}$ and ϕ satisfy the Lorenz gauge condition, Equations (2.65) reduce to

$$\left(\nabla^2 - \frac{1}{v^2}\frac{\partial^2}{\partial t^2}\right)\mathbf{A} = -\mu \mathbf{J}, \quad \left(\nabla^2 - \frac{1}{v^2}\frac{\partial^2}{\partial t^2}\right)\phi = -\frac{\rho}{\varepsilon}, \tag{2.69}$$

and they become uncoupled. The retarded solutions of (2.69) are given by

$$\mathbf{A}(\mathbf{r}, t) = \int_{V_0} \frac{\mu \mathbf{J}(\mathbf{r}', t - R/v)}{4\pi R}dV(\mathbf{r}'), \ \phi(\mathbf{r}, t) = \int_{V_0} \frac{\rho(\mathbf{r}', t - R/v)}{4\pi \varepsilon R}dV(\mathbf{r}'),$$

where V_0 denotes the source region. Note that the Lorenz gauge condition implies the continuity equation of the current.

Another important gauge condition is the **Coulomb gauge**:

$$\nabla \cdot \mathbf{A} = 0. \tag{2.70}$$

The existence of such a gauge condition can be justified by the following argument. From (2.66) we obtain $\nabla \cdot \mathbf{A}' = \nabla \cdot \mathbf{A} + \nabla^2 \psi$. Therefore, if $\nabla \cdot \mathbf{A}$ is not zero, we may set $\nabla \cdot \mathbf{A}'$ to zero by letting $\nabla^2 \psi = -\nabla \cdot \mathbf{A}$. If $\mathbf{A}$ and ϕ satisfy the Coulomb gauge condition, Equations (2.65) become

$$\left(\nabla^2 - \frac{1}{v^2}\frac{\partial^2}{\partial t^2}\right)\mathbf{A} = -\mu \mathbf{J} + \frac{1}{v^2}\nabla\frac{\partial \phi}{\partial t}, \ \nabla^2 \phi = -\frac{\rho}{\varepsilon}. \tag{2.71}$$

By means of $\nabla^2(1/4\pi R) = -\delta(\mathbf{r} - \mathbf{r}')$, the current source $\mathbf{J}$ can be divided into the sum of two components

$$\mathbf{J}(\mathbf{r}, t) = \int_{V_0} \mathbf{J}(\mathbf{r}', t)\delta(\mathbf{r} - \mathbf{r}')dV(\mathbf{r}') = -\nabla^2 \int_{V_0} \frac{\mathbf{J}(\mathbf{r}', t)}{4\pi R}dV(\mathbf{r}') = \mathbf{J}^{\|} + \mathbf{J}^{\perp} \tag{2.72}$$

where

$$\mathbf{J}^{\|}(\mathbf{r}, t) = -\nabla\nabla \cdot \int\limits_{V_0} \frac{\mathbf{J}(\mathbf{r}', t)}{4\pi R} dV(\mathbf{r}'), \mathbf{J}^{\perp}(\mathbf{r}, t) = \nabla \times \nabla \times \int\limits_{V_0} \frac{\mathbf{J}(\mathbf{r}', t)}{4\pi R} dV(\mathbf{r}') \qquad (2.73)$$

are referred to as the **irrotational component** and the **solenoidal component** of $\mathbf{J}$ respectively. By using Gauss's theorem, the irrotational component can further be written as

$$
\begin{aligned}
\mathbf{J}^{\|}(\mathbf{r}, t) &= -\nabla \int\limits_{V_0} \nabla' \cdot \left[\frac{\mathbf{J}(\mathbf{r}', t)}{4\pi R} \right] dV(\mathbf{r}') - \nabla \int\limits_{V_0} \frac{\nabla' \cdot \mathbf{J}(\mathbf{r}', t)}{4\pi R} dV(\mathbf{r}') \\
&= \nabla \int\limits_{\partial V_0} \frac{\mathbf{J}(\mathbf{r}', t)}{4\pi R} \cdot \mathbf{u}_n(\mathbf{r}') dV(\mathbf{r}') - \nabla \int\limits_{V_0} \frac{\nabla' \cdot \mathbf{J}(\mathbf{r}', t)}{4\pi R} dV(\mathbf{r}') \qquad (2.74) \\
&= -\nabla \int\limits_{V_0} \frac{\nabla' \cdot \mathbf{J}(\mathbf{r}', t)}{4\pi R} dV(\mathbf{r}') = \nabla \frac{\partial}{\partial t} \int\limits_{V_0} \frac{\rho(\mathbf{r}', t)}{4\pi R} dV(\mathbf{r}') = \varepsilon \nabla \frac{\partial \phi}{\partial t},
\end{aligned}
$$

where ∂V_0 denotes the boundary of V_0. The surface integral in (2.74) is zero for we have assumed that the source is confined inside V_0. Then Equations (2.71) become

$$\left(\nabla^2 - \frac{1}{v^2} \frac{\partial^2}{\partial t^2} \right) \mathbf{A} = -\mu \mathbf{J}^{\perp}, \nabla^2 \phi = -\frac{\rho}{\varepsilon}. \qquad (2.75)$$

These equations indicate that the vector potential is determined by the solenoidal component of the current distribution while the scalar potential is determined by the instantaneous distribution of charges. This would not violate the fact that the field travels at finite speed. In fact, the finite propagation time effects have been included in the vector potential (Heras, 2007).

The electric field can also be decomposed into the sum of the irrotational component and the solenoidal component $\mathbf{E} = \mathbf{E}^{\|} + \mathbf{E}^{\perp}$. From $\mathbf{E} = -\nabla\phi - \partial\mathbf{A}/\partial t$ and the Coulomb gauge condition, we obtain

$$\mathbf{E}^{\|} = -\nabla\phi, \mathbf{E}^{\perp} = -\frac{\partial\mathbf{A}}{\partial t}. \qquad (2.76)$$

So the Coulomb gauge condition allows us to separate the field into two parts, one part being described solely by the vector potential $\mathbf{A}$, and the other part solely by the scalar potential ϕ. The Coulomb gauge is often applied to source-free region. In this case, we have $\phi = 0$, and $\mathbf{A}$ satisfies the homogeneous wave equation. Considering (2.70), we conclude that the electromagnetic fields in a source-free region can be represented by two scalar potential functions.

Remark 2.3: As shown in (2.72), any vector $\mathbf{F}$ can be expressed as the sum of a solenoidal component $\mathbf{F}^{\perp}$ and an irrotational component $\mathbf{F}^{\|}$ with

$$\nabla \times \mathbf{F}^{\|} = 0, \nabla \cdot \mathbf{F}^{\perp} = 0. \qquad (2.77)$$

This is called the **Helmholtz theorem**. Introducing the Fourier transform pair

$$\tilde{\mathbf{F}}(\mathbf{k}) = \mathcal{F}(\mathbf{F})(\mathbf{k}) = \int \mathbf{F}(\mathbf{r}) e^{-j\mathbf{k}\cdot\mathbf{r}} d\mathbf{r},$$

$$\mathbf{F}(\mathbf{r}) = \mathcal{F}^{-1}(\tilde{\mathbf{F}})(\mathbf{r}) = \frac{1}{(2\pi)^3} \int \tilde{\mathbf{F}}(\mathbf{k}) e^{j\mathbf{k}\cdot\mathbf{r}} d\mathbf{r},$$

and using (2.77) we have

$$\mathbf{k} \cdot \tilde{\mathbf{F}}^{\perp} = \mathbf{k} \cdot \mathcal{F}(\mathbf{F}^{\perp}) = 0, \ \mathbf{k} \times \tilde{\mathbf{F}}^{\parallel} = \mathbf{k} \times \mathcal{F}(\mathbf{F}^{\parallel}) = 0. \tag{2.78}$$

These equations imply that the solenoidal field is transverse in $\mathbf{k}$ space while the irrotational field is longitudinal in $\mathbf{k}$ space. For this reason, the solenoidal field and irrotational field are also known as the transverse field and the longitudinal field respectively. Apparently we have

$$\tilde{\mathbf{F}}^{\parallel} = \frac{1}{k^2}(\tilde{\mathbf{F}} \cdot \mathbf{k})\mathbf{k}, \ \tilde{\mathbf{F}}^{\perp} = \tilde{\mathbf{F}} - \tilde{\mathbf{F}}^{\parallel}$$

where $k = |\mathbf{k}|$. Hence

$$\mathbf{F}^{\parallel}(\mathbf{r}) = \mathcal{F}^{-1}(\tilde{\mathbf{F}}^{\parallel})(\mathbf{r}) = \frac{1}{(2\pi)^3} \int \frac{1}{k^2} \left[\tilde{\mathbf{F}}(\mathbf{k}) \cdot \mathbf{k} \right] \mathbf{k} e^{j\mathbf{k}\cdot\mathbf{r}} d\mathbf{k}$$

$$= \frac{1}{(2\pi)^3} \int \int \frac{1}{k^2} \left[\mathbf{F}(\mathbf{r}') \cdot \mathbf{k} \right] \mathbf{k} e^{j\mathbf{k}\cdot(\mathbf{r}-\mathbf{r}')} d\mathbf{k} d\mathbf{r}'$$

$$= \int \mathbf{F}(\mathbf{r}') \cdot \overset{\leftrightarrow}{\delta}{}^{\parallel}(\mathbf{r} - \mathbf{r}') d\mathbf{r}',$$

$$\mathbf{F}^{\perp}(\mathbf{r}) = \frac{1}{(2\pi)^3} \int \int \left\{ \mathbf{F}(\mathbf{r}') - \frac{1}{k^2} \left[\mathbf{F}(\mathbf{r}') \cdot \mathbf{k} \right] \mathbf{k} \right\} e^{j\mathbf{k}\cdot(\mathbf{r}-\mathbf{r}')} d\mathbf{k} d\mathbf{r}'$$

$$= \frac{1}{(2\pi)^3} \int \int \mathbf{F}(\mathbf{r}') \cdot \left(\overset{\leftrightarrow}{\mathbf{I}} - \frac{1}{k^2}\mathbf{k}\mathbf{k} \right) e^{j\mathbf{k}\cdot(\mathbf{r}-\mathbf{r}')} d\mathbf{k} d\mathbf{r}'$$

$$= \int \mathbf{F}(\mathbf{r}') \cdot \overset{\leftrightarrow}{\delta}{}^{\perp}(\mathbf{r} - \mathbf{r}') d\mathbf{r}',$$

where $\overset{\leftrightarrow}{\mathbf{I}}$ is the identity dyadic; $\overset{\leftrightarrow}{\delta}{}^{\parallel}$ and $\overset{\leftrightarrow}{\delta}{}^{\perp}$ are the **longitudinal and transverse δ-dyadics**, defined respectively by

$$\overset{\leftrightarrow}{\delta}{}^{\parallel}(\mathbf{r}) = \frac{1}{(2\pi)^3} \int \frac{1}{k^2}\mathbf{k}\mathbf{k} e^{j\mathbf{k}\cdot\mathbf{r}} d\mathbf{k}, \ \overset{\leftrightarrow}{\delta}{}^{\perp}(\mathbf{r}) = \frac{1}{(2\pi)^3} \int \left(\overset{\leftrightarrow}{\mathbf{I}} - \frac{1}{k^2}\mathbf{k}\mathbf{k} \right) e^{j\mathbf{k}\cdot\mathbf{r}} d\mathbf{k}.$$

Note that $\overset{\leftrightarrow}{\delta}(\mathbf{r}) \equiv \overset{\leftrightarrow}{\mathbf{I}}\delta(\mathbf{r}) = \overset{\leftrightarrow}{\delta}{}^{\parallel}(\mathbf{r}) + \overset{\leftrightarrow}{\delta}{}^{\perp}(\mathbf{r})$. From $\nabla^2(1/4\pi R) = -\delta(\mathbf{r} - \mathbf{r}')$, we obtain

$$\overset{\leftrightarrow}{\delta}{}^{\parallel}(\mathbf{r}) = \frac{1}{3}\overset{\leftrightarrow}{\delta}(\mathbf{r}) + \frac{1}{4\pi r^3} \left(\overset{\leftrightarrow}{\mathbf{I}} - 3\frac{\mathbf{r}\mathbf{r}}{r^2} \right), \ \overset{\leftrightarrow}{\delta}{}^{\perp}(\mathbf{r}) = \frac{2}{3}\overset{\leftrightarrow}{\delta}(\mathbf{r}) - \frac{1}{4\pi r^3} \left(\overset{\leftrightarrow}{\mathbf{I}} - 3\frac{\mathbf{r}\mathbf{r}}{r^2} \right),$$

where $r = |\mathbf{r}|$ $\qquad\qquad\qquad\qquad\qquad\qquad\qquad\qquad\qquad\qquad\qquad\qquad\qquad$ $\square$

We mention in passing that the **temporal gauge** or the **Hamiltonian gauge** is defined by $\phi = 0$. The **velocity gauge** is defined by $\nabla \cdot \mathbf{A} + u^{-2}\partial\phi/\partial t = 0$, where u is a complex constant. The velocity gauge includes the Lorenz gauge ($u = 1/\sqrt{\mu\varepsilon}$) and the Coulomb gauge ($u = \infty$).

2.6.2 Hertz Vectors and Debye Potentials

In addition to vector potential $\mathbf{A}$ and scalar potential ϕ, other potential functions can be introduced to simplify the problems. If the current source $\mathbf{J}$ is irrotational, it only has a longitudinal component and can be written as

$$\mathbf{J} = \frac{\partial \mathbf{P}}{\partial t} \tag{2.79}$$

by (2.74). Here

$$\mathbf{P}(\mathbf{r}, t) = \nabla \int_{V_0} \frac{\rho(\mathbf{r}', t)}{4\pi R} dV(\mathbf{r}')$$

is the equivalent polarization vector. From the continuity equation, the corresponding polarization charge density is given by $\rho = -\nabla \cdot \mathbf{P}$. Substituting (2.79) into the first equation of (2.69), we have

$$\left(\nabla^2 - \frac{1}{v^2} \frac{\partial^2}{\partial t^2} \right) \mathbf{A} = -\mu \frac{\partial \mathbf{P}}{\partial t}.$$

To get rid of the differential operation on the source term, we may introduce a new potential function $\mathbf{\Pi}_e$ such that

$$\mathbf{A} = \frac{1}{v^2} \frac{\partial \mathbf{\Pi}_e}{\partial t}. \tag{2.80}$$

The new potential function $\mathbf{\Pi}_e$ is called the **electric Hertz vector** and satisfies

$$\left(\nabla^2 - \frac{1}{v^2} \frac{\partial^2}{\partial t^2} \right) \mathbf{\Pi}_e = -\frac{\mathbf{P}}{\varepsilon}. \tag{2.81}$$

From (2.68) and (2.80), we obtain

$$\phi = -\nabla \cdot \mathbf{\Pi}_e. \tag{2.82}$$

In terms of the electric Hertz vector, the electromagnetic fields may be represented by

$$\mathbf{B} = \frac{1}{v^2} \nabla \times \frac{\partial \mathbf{\Pi}_e}{\partial t}, \quad \mathbf{E} = \nabla(\nabla \cdot \mathbf{\Pi}_e) - \frac{1}{v^2} \frac{\partial^2 \mathbf{\Pi}_e}{\partial t^2}. \tag{2.83}$$

If the current source $\mathbf{J}$ is solenoidal, it only has a transverse component and may be written as

$$\mathbf{J} = \nabla \times \mathbf{M} \tag{2.84}$$

where $\mathbf{M}$ is the equivalent magnetization vector

$$\mathbf{M}(\mathbf{r}, t) = \nabla \times \int_{V_0} \frac{\mathbf{J}(\mathbf{r}', t)}{4\pi R} dV(\mathbf{r}')$$

by (2.73). Introducing (2.84) into the first equation of (2.69) gives

$$\left(\nabla^2 - \frac{1}{v^2} \frac{\partial^2}{\partial t^2} \right) \mathbf{A} = -\mu \nabla \times \mathbf{M}.$$

To get rid of the differential operation on the source term, we can introduce a new potential function $\mathbf{\Pi}_m$, called the **magnetic Hertz vector** such that $\mathbf{A} = -\mu \nabla \times \mathbf{\Pi}_m$. The magnetic Hertz vector satisfies

$$\left(\nabla^2 - \frac{1}{v^2} \frac{\partial^2}{\partial t^2} \right) \mathbf{\Pi}_m = -\mathbf{M}. \tag{2.85}$$

Since $\nabla \cdot \mathbf{A} = 0$ implies $\phi = 0$, the electromagnetic fields can be expressed as

$$\mathbf{B} = \mu \nabla \times \nabla \times \mathbf{\Pi}_m, \mathbf{E} = -\mu \nabla \times \frac{\partial \mathbf{\Pi}_m}{\partial t}. \tag{2.86}$$

In general, the current source $\mathbf{J}$ is of the form $\mathbf{J} = \partial \mathbf{P}/\partial t + \nabla \times \mathbf{M}$ from (2.73). For a linear medium, the superposition theorem applies and the electromagnetic fields for a general current source can be expressed as the sum of (2.83) and (2.86):

$$\mathbf{E} = \nabla(\nabla \cdot \mathbf{\Pi}_e) - \frac{1}{v^2} \frac{\partial^2 \mathbf{\Pi}_e}{\partial t^2} - \mu \nabla \times \frac{\partial \mathbf{\Pi}_m}{\partial t}, \mathbf{H} = \varepsilon \nabla \times \frac{\partial \mathbf{\Pi}_e}{\partial t} + \nabla \times \nabla \times \mathbf{\Pi}_m.$$

In source-free region, these equations may be written as

$$\mathbf{E} = \nabla \times \nabla \times \mathbf{\Pi}_e - \mu \nabla \times \frac{\partial \mathbf{\Pi}_m}{\partial t}, \mathbf{H} = \varepsilon \nabla \times \frac{\partial \mathbf{\Pi}_e}{\partial t} + \nabla \times \nabla \times \mathbf{\Pi}_m,$$

by use of (2.81) and (2.85).

In source-free region, the electromagnetic fields can be represented by two scalar potential functions. Hence we may use the spherical coordinate system (r, θ, φ) and choose $\mathbf{\Pi}_e = \mathbf{r}u_e$ and $\mathbf{\Pi}_m = \mathbf{r}u_m$ to represent the electromagnetic fields. Here u_e and u_m satisfy the homogeneous wave equations:

$$\left(\nabla^2 - \frac{1}{v^2} \frac{\partial^2}{\partial t^2} \right) u_e = 0, \left(\nabla^2 - \frac{1}{v^2} \frac{\partial^2}{\partial t^2} \right) u_m = 0,$$

and they are called **Debye potentials**, named after the Dutch physicist Peter Joseph William Debye (1884–1966). Let $\mathbf{u}_r$, $\mathbf{u}_\theta$ and $\mathbf{u}_\varphi$ denote the unit vectors in the direction of increasing r, θ and φ respectively. A simple calculation gives

$$\nabla \times \nabla \times (r u_e) = \left(-\frac{1}{r}\nabla^2_{\theta\varphi} u_e\right)\mathbf{u}_r + \nabla_{\theta\varphi}\left[\frac{1}{r}\frac{\partial(r u_e)}{\partial r}\right],$$

$$\nabla \times (r u_m) = \nabla_{\theta\varphi} u_m \times \mathbf{u}_r,$$

where

$$\nabla_{\theta\varphi} = \mathbf{u}_\theta\frac{\partial}{\partial\theta} + \mathbf{u}_\varphi\frac{1}{\sin\theta}\frac{\partial}{\partial\varphi},$$

$$\nabla^2_{\theta\varphi} = \frac{1}{\sin\theta}\frac{\partial}{\partial\theta}\left(\sin\theta\frac{\partial}{\partial\theta}\right) + \frac{1}{\sin^2\theta}\frac{\partial^2}{\partial\varphi^2}.$$

Thus the electromagnetic fields in a source-free region can be expressed as

$$\mathbf{E} = -\left(\frac{1}{r}\nabla^2_{\theta\varphi} u_e\right)\mathbf{u}_r + \nabla_{\theta\varphi}\left[\frac{1}{r}\frac{\partial(r u_e)}{\partial r}\right] + \mu\mathbf{u}_r \times \frac{\partial}{\partial t}\nabla_{\theta\varphi} u_m,$$

$$\mathbf{H} = -\left(\frac{1}{r}\nabla^2_{\theta\varphi} u_m\right)\mathbf{u}_r + \nabla_{\theta\varphi}\left[\frac{1}{r}\frac{\partial(r u_m)}{\partial r}\right] - \varepsilon\mathbf{u}_r \times \frac{\partial}{\partial t}\nabla_{\theta\varphi} u_e. \tag{2.87}$$

The derivation of (2.87) relies on an assumption that the electromagnetic fields can be represented by Debye potentials. It can be verified that this assumption is valid in a region between two concentric spheres (Wilcox, 1957).

2.6.3 Jump Relations in Potential Theory

In electromagnetic theory, we are often faced with the following potential integrals

$$\mathbf{A}(\mathbf{r}) = \int_S \mathbf{a}(\mathbf{r}')G(\mathbf{r}, \mathbf{r}')dS(\mathbf{r}'), \quad \varphi(\mathbf{r}) = \int_S f(\mathbf{r}')G(\mathbf{r}, \mathbf{r}')dS(\mathbf{r}'),$$

where $G(\mathbf{r}, \mathbf{r}') = e^{-jkR}/4\pi R$. When the field point $\mathbf{r}$ is on the surface S, these potential integrals are defined as the improper but convergent integrals as follows

$$\mathbf{A}(\mathbf{r}) = \lim_{\delta\to 0}\int_{S-S_\delta} \mathbf{a}(\mathbf{r}')G(\mathbf{r}, \mathbf{r}')dS(\mathbf{r}'), \quad \varphi(\mathbf{r}) = \lim_{\delta\to 0}\int_{S-S_\delta} f(\mathbf{r}')G(\mathbf{r}, \mathbf{r}')dS(\mathbf{r}'),$$

where $\mathbf{r} \in S$, S_δ is a small area of arbitrary shape containing $\mathbf{r}$ and δ is the maximum chord of S_δ. For $\mathbf{r} \in S$, the following **jump relations** can be established

$$\nabla \cdot \mathbf{A}_\pm(\mathbf{r}) = \int_S \nabla G(\mathbf{r}, \mathbf{r}') \cdot \mathbf{a}(\mathbf{r}') dS(\mathbf{r}') \mp \frac{1}{2} \mathbf{u}_n(\mathbf{r}) \cdot \mathbf{a}(\mathbf{r}),$$

$$\nabla \times \mathbf{A}_\pm(\mathbf{r}) = \int_S \nabla G(\mathbf{r}, \mathbf{r}') \times \mathbf{a}(\mathbf{r}') dS(\mathbf{r}') \mp \frac{1}{2} \mathbf{u}_n(\mathbf{r}) \times \mathbf{a}(\mathbf{r}), \qquad (2.88)$$

$$\nabla \varphi_\pm(\mathbf{r}) = \int_S f(\mathbf{r}') \nabla G(\mathbf{r}, \mathbf{r}') dS(\mathbf{r}') \mp \frac{1}{2} \mathbf{u}_n(\mathbf{r}) f(\mathbf{r}),$$

where $\mathbf{u}_n(\mathbf{r})$ is the unit outward normal of S at $\mathbf{r}$ and

$$\nabla \cdot \mathbf{A}_\pm(\mathbf{r}) \equiv \lim_{h \to +0} \nabla \cdot \mathbf{A}[\mathbf{r} \pm h\mathbf{u}_n(\mathbf{r})],$$

$$\nabla \times \mathbf{A}_\pm(\mathbf{r}) \equiv \lim_{h \to +0} \nabla \times \mathbf{A}[\mathbf{r} \pm h\mathbf{u}_n(\mathbf{r})],$$

$$\nabla \varphi_\pm(\mathbf{r}) \equiv \lim_{h \to +0} \nabla \varphi[\mathbf{r} \pm h\mathbf{u}_n(\mathbf{r})].$$

The subscripts $+$ and $-$ respectively indicate the limit values as $\mathbf{r}$ approaches S from the exterior and interior of S. All the integrals in (2.88) stand for the Cauchy principal values. Moreover, we have

$$\lim_{h \to +0} \mathbf{u}_n(\mathbf{r}) \times \{\nabla \times \nabla \times \mathbf{A}[\mathbf{r} + h\mathbf{u}_n(\mathbf{r})] - \nabla \times \nabla \times \mathbf{A}[\mathbf{r} - h\mathbf{u}_n(\mathbf{r})]\} = 0, \mathbf{r} \in S.$$

We only show the derivation of the last relation in (2.88). Let the closed surface S be split into two parts, S' and S_δ, of which S_δ is a small region surrounding $\mathbf{r}$, and S' the remainder of S. If S is smooth around $\mathbf{r}$, S_δ may be considered as a circular disk of radius δ centered at $\mathbf{r}$, as shown in Figure 2.3. Thus

$$\nabla \varphi_\pm(\mathbf{r}) = \lim_{h \to 0} \nabla \int_S f(\mathbf{r}') G(\mathbf{r} \pm h\mathbf{u}_n(\mathbf{r}), \mathbf{r}') dS(\mathbf{r}')$$

$$= \lim_{h \to 0} \nabla \int_{S'} f(\mathbf{r}') G(\mathbf{r} \pm h\mathbf{u}_n(\mathbf{r}), \mathbf{r}') dS(\mathbf{r}')$$

$$+ \lim_{h \to 0} \nabla \int_{S_\delta} f(\mathbf{r}') G(\mathbf{r} \pm h\mathbf{u}_n(\mathbf{r}), \mathbf{r}') dS(\mathbf{r}').$$

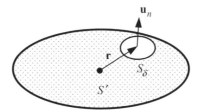

Figure 2.3   Cauchy principal value

The first integral on the right-hand side approaches the principal value as $\delta \to 0$. The integral over S_δ can be calculated through the approximation

$$
\lim_{h \to 0} \nabla \int_{S_\delta} f(\mathbf{r}')G(\mathbf{r} \pm h\mathbf{u}_n(\mathbf{r}), \mathbf{r}')dS(\mathbf{r}')
$$

$$
= f(\mathbf{r}) \lim_{h \to 0} \left(\nabla_t \pm \mathbf{u}_n \frac{\partial}{\partial h} \right) \int_{S_\delta} \frac{dS(\mathbf{r}')}{|\mathbf{r} \pm h\mathbf{u}_n(\mathbf{r}) - \mathbf{r}'|}
$$

$$
\approx f(\mathbf{r}) \lim_{h \to 0} \frac{1}{4\pi} \left(\nabla_t \pm \mathbf{u}_n \frac{\partial}{\partial h} \right) \int_0^{2\pi} \int_0^{\delta} \frac{\rho\,d\rho\,d\phi}{\sqrt{\rho^2 + h^2}}
$$

$$
= \frac{1}{2} f(\mathbf{r}) \lim_{h \to 0} \left(\nabla_t \pm \mathbf{u}_n \frac{\partial}{\partial h} \right) \left(\delta - h + \frac{h^2}{2\delta} \right) = \mp \frac{1}{2} \mathbf{u}_n f(\mathbf{r}),
$$

which gives the last expression of (2.88).
  The function

$$
\varphi(\mathbf{r}) = \int_S f(\mathbf{r}')G(\mathbf{r}, \mathbf{r}')dS(\mathbf{r}'), \, \mathbf{r} \in R^3 - S
$$

is called a **single-layer potential** with density f and the function

$$
\psi(\mathbf{r}) = \int_S f(\mathbf{r}')\frac{\partial G(\mathbf{r}, \mathbf{r}')}{\partial n(\mathbf{r}')}dS(\mathbf{r}'), \, \mathbf{r} \in R^3 - S
$$

is called a **double-layer potential** with density f.

Example 2.8: The potential generated by a charge distribution ρ on a surface S is

$$
\varphi(\mathbf{r}) = \int_S \frac{\rho(\mathbf{r}')}{|\mathbf{r} - \mathbf{r}'|}dS(\mathbf{r}'), \, \mathbf{r} \in R^3 - S.
$$

The potential produced by a layer of electric dipoles on a surface S is

$$\psi(\mathbf{r}) = \int_S \tau(\mathbf{r}') \frac{\partial}{\partial n(\mathbf{r}')} \left(\frac{1}{\mathbf{r} - \mathbf{r}'} \right) dS(\mathbf{r}'), \mathbf{r} \in R^3 - S,$$

where τ is dipole moment density. Physically, the double layer potential is very much like a charged battery.   □

The potential theory can be used to establish integral equations on the surface of a scattering object. It can also be used to show the existence of a unique solution to the scattering problems (Colton and Kress, 1983, 1998; Jones, 1979).

For a perfect conducting surface S, we may introduce the following boundary value problems for electromagnetic fields

1. Interior boundary value problem: $\mathbf{u}_n \times \mathbf{E}^{sc}(\mathbf{r}) = -\mathbf{u}_n \times \mathbf{E}^{in}(\mathbf{r}), \mathbf{r} \in S$ and $\mathbf{E}^{in}(\mathbf{r})$ is a known continuous function.
2. Exterior boundary value problem: $\mathbf{u}_n \times \mathbf{E}^{sc}(\mathbf{r}) = -\mathbf{u}_n \times \mathbf{E}^{in}(\mathbf{r})$, $\mathbf{r} \in S$ and $\mathbf{E}^{in}(\mathbf{r})$ is a known continuous function; $\mathbf{E}^{sc}(\mathbf{r})$ and $\mathbf{H}^{sc}$ satisfies the Silver–Müller radiation condition at infinity.

Example 2.9: From the jump relations, we may find that the electromagnetic fields

$$\mathbf{E}^{sc}(\mathbf{r}) = \int_S \mathbf{J}_m(\mathbf{r}) \times \nabla' G(\mathbf{r}, \mathbf{r}') dS(\mathbf{r}'),$$

$$\mathbf{H}^{sc}(\mathbf{r}) = -\frac{1}{jk\eta} \nabla \times \mathbf{E}^{sc}(\mathbf{r}),$$

generated by a magnetic current distribution $\mathbf{J}_m$ on S, are a solution of the interior boundary value problem if $\mathbf{J}_m$ is a solution of the integral equation

$$(\hat{I} - \hat{G}) \mathbf{J}_m(\mathbf{r}) = 2\mathbf{u}_n \times \mathbf{E}^{in}(\mathbf{r}), \mathbf{r} \in S \tag{2.89}$$

where

$$\hat{G}(\mathbf{J})(\mathbf{r}) = 2\mathbf{u}_n(\mathbf{r}) \times \int_S \mathbf{J}(\mathbf{r}') \times \nabla' G(\mathbf{r}, \mathbf{r}') dS(\mathbf{r}'), \mathbf{r} \in S.$$

Similarly, the electromagnetic fields

$$\mathbf{E}^{sc}(\mathbf{r}) = -\int_S \mathbf{J}_m(\mathbf{r}) \times \nabla' G(\mathbf{r}, \mathbf{r}') dS(\mathbf{r}')$$

$$\mathbf{H}^{sc}(\mathbf{r}) = -\frac{1}{jk\eta} \nabla \times \mathbf{E}^{sc}(\mathbf{r})$$

generated by a magnetic current distribution $\mathbf{J}_m$ on S, are a solution of the exterior boundary value problem if $\mathbf{J}_m$ is a solution of the integral equation

$$(\hat{I} + \hat{G})\mathbf{J}_m(\mathbf{r}) = 2\mathbf{u}_n \times \mathbf{E}^{in}(\mathbf{r}), \mathbf{r} \in S \tag{2.90}$$

where the surface S is assumed to be smooth.   □

2.7   Variational Principles

The **principle of least action** leads to the development of the Lagrangian and Hamiltonian formulations of classical mechanics. Although these formulations seem difficult to grasp at first, they have some merits that Newton's formulation does not have. For example, they can easily be transferred to the frameworks of relativistic and quantum mechanical physics. The principle of least action is considered the core strategy of modern physics. In terms of the principle of least action, the differential equations of a given physical system can be derived by minimizing the action of the system. The original problem, governed by the differential equations, is thus replaced by an equivalent variational problem. Such a procedure is also called the **energy method**. It is commonly believed that the theoretical formulation of a physical law is not complete until the law can be reformulated as a variational problem.

2.7.1   Generalized Calculus of Variation

To study the extreme value problem involving operators, it is necessary to generalize the usual concepts of derivative in classical calculus to the operators. Let E and F be normed spaces and $\hat{f}, \hat{g} : U \subset E \to F$ be two maps where U is open in E. The two maps are said to be **tangent** at $u_0 \in U$ if (for example, Abraham *et al.*, 1988)

$$\lim_{u \to u_0} \frac{\|\hat{f}(u) - \hat{g}(u)\|}{\|u - u_0\|} = 0.$$

For $\hat{f} : U \subset E \to F$ and $u_0 \in U$ it can be shown that there is at most one $\hat{L} \in L(E, F)$ such that the map $\hat{g} : U \subset E \to F$ defined by $\hat{g}(u) = \hat{f}(u_0) + \hat{L}(u - u_0)$ is tangent to $\hat{f}$ at u_0. If such $\hat{L}$ exists, we say $\hat{f}$ is **differentiable** at u_0, and we define the derivative of $\hat{f}$ at u_0 to be $D\hat{f}(u_0) = \hat{L}$. The map $D\hat{f} : U \to L(E, F)$ is called the **derivative** of $\hat{f}$. Moreover, if $D\hat{f}$ is continuous, we say $\hat{f}$ is of **class** C^1 or **continuously differentiable**.

Example 2.10: Let F be a normed space and $\hat{g} : U \subset R \to F$ be differentiable. Then $D\hat{g}(t) \in L(R, F)$, and

$$\hat{g}'(t) = \frac{d\hat{g}}{dt} = \lim_{h \to 0} \frac{\hat{g}(t+h) - \hat{g}(t)}{h} = D\hat{g}(t) \cdot 1, 1 \in R.$$

                                                                                            □

Similar to the classical calculus, we may introduce the concept of directional derivative. Let us consider an operator $\hat{f}$ from $U \subset E$ to F. If, for a given $u \in U$ and a non-zero $v \in E$, there is an operator $D\hat{f} : U \to L(E, F)$ such that

$$\lim_{t \to 0} \left\| \frac{\hat{f}(u + tv) - \hat{f}(u)}{t} - D\hat{f}(u) \cdot v \right\| = 0,$$

then $D\hat{f}(u) \in F$ is called the **directional derivative** in the direction of v. If $\hat{f}$ is differentiable at u, the directional derivatives of $\hat{f}$ exists at u and is given by

$$D\hat{f}(u) \cdot v = \frac{d}{dt} \hat{f}(u + tv)\Big|_{t=0}. \tag{2.91}$$

A general approach of variational method may be formulated by the functional derivative. Let $\langle \cdot, \cdot \rangle$ be a continuous bilinear form $E \times F$ to R. The bilinear form $\langle \cdot, \cdot \rangle$ is called E-**nondegenerate** if $\langle x, y \rangle = 0$ for all $y \in F$ implies $x = 0$. Similarly the bilinear form $\langle \cdot, \cdot \rangle$ is called F-**nondegenerate** if $\langle x, y \rangle = 0$ for all $x \in E$ implies $y = 0$. Let E and F be normed spaces and $\langle \cdot, \cdot \rangle : E \times F \to R$ be E-nondegenerate. Let $\hat{f} : F \to R$ be differentiable at the point $y \in F$. The **functional derivative** $\delta \hat{f}/\delta y$ of $\hat{f}$ with respect to y is the unique element in E, if it exists, such that for all $y' \in F$ we have

$$D\hat{f}(y) \cdot y' = \langle \delta \hat{f}/\delta y, y' \rangle. \tag{2.92}$$

Similarly if $\hat{g} : E \to R$ is differentiable at the point $x \in E$ and $\langle \cdot, \cdot \rangle : E \times F \to R$ is F-nondegenerate, the functional derivative $\delta \hat{g}/\delta x$ is defined, if it exists, by

$$D\hat{g}(x) \cdot x' = \langle x', \delta \hat{g}/\delta x \rangle \tag{2.93}$$

for all $x' \in E$.

Example 2.11: Let E be a normed space and $\hat{f} : U \subset E \to R$ be differentiable. Then $D\hat{f}(u) \in L(E, R) = E^*$. If E is a Hilbert space, the **gradient** of $\hat{f}$ is the map $\mathrm{grad}\,\hat{f} = \nabla \hat{f} : U \subset E \to E$ defined by

$$D\hat{f}(u) \cdot v = (\nabla \hat{f}(u), v). \tag{2.94}$$

Thus $\nabla \hat{f}(u) = \delta \hat{f}/\delta u$.   □

Example 2.12: Let $E = F$ be a Banach space of functions defined on a region $\Omega \subset R^N$. The L^2-bilinear form on $E \times E$ is defined by

$$\langle \varphi, \psi \rangle = \int_\Omega \varphi(\mathbf{x}) \psi(\mathbf{x}) d\mathbf{x}.$$

For $\hat{f} : E \to R$, the functional derivative $\delta \hat{f}/\delta \varphi$ is defined by

$$D\hat{f}(\varphi) \cdot \psi = \langle \delta \hat{f}/\delta \varphi, \psi \rangle = \int_{\Omega} \frac{\delta \hat{f}}{\delta \varphi}(\mathbf{x})\psi(\mathbf{x})d\mathbf{x}$$

for all $\psi \in E$. From the above equation we can immediately write

$$\int_{\Omega} \frac{\delta \hat{f}}{\delta \varphi}(\mathbf{x})\psi(\mathbf{x})d\mathbf{x} = \frac{d}{d\varepsilon} \hat{f}(\varphi + \varepsilon \psi)\bigg|_{\varepsilon=0}. \tag{2.95}$$

This relation can be used to find the functional derivative $\delta \hat{f}/\delta \varphi$.   □

If $\hat{f}$ is a differential function of n variables: $\hat{f} : F_1 \times F_2 \times \ldots F_n \to R$, we have n bilinear forms $\langle \cdot, \cdot \rangle_i : E_i \times F_i \to R$ $(i = 1, 2, \ldots, n)$. The ith **partial functional derivative** $\delta \hat{f}/\delta y_i$ of $\hat{f}$ with respect to $y_i \in F_i$ is defined by

$$\langle \delta \hat{f}/\delta y_i, y_i' \rangle_i = \frac{d}{d\varepsilon} \hat{f}(y_1, y_2, \ldots, y_i + \varepsilon y_i', \ldots, y_n)\bigg|_{\varepsilon=0}. \tag{2.96}$$

The **total functional derivative** can be expressed as

$$D\hat{f}(y_1, y_2, \ldots, y_n) \cdot (y_1', y_2', \ldots, y_n') = \sum_{i=1}^{n} \langle \delta \hat{f}/\delta y_i, y_i' \rangle. \tag{2.97}$$

Extremum theorem: *Let $F_i(i = 1, 2, \ldots, n)$ be spaces of functions. A necessary condition for a differentiable function $\hat{f} : F_1 \times F_2 \times \ldots F_n \to R$ to have an extremum at $(y_1, y_2, \ldots, y_n)$ is*

$$\frac{\delta \hat{f}}{\delta y_i} = 0, i = 1, 2, \ldots, n. \tag{2.98}$$

These equations are referred to as the **Lagrangian equations**.

2.7.2   Lagrangian Formulation

The **action** of a system is an integral over time of a function called the **Lagrangian function**. The Lagrangian function is usually expressed as the difference between kinetic energy and potential energy, and depends on the scalar properties of the system. The least action principle requires that the action of the system must be a minimum. This leads to the Lagrangian equations, which are the differential equations describing the system.

2.7.2.1 Lagrangian Equation for Charged Particle

Classical physics deals with two different phenomena: the particles governed by Newtonian mechanics, and the fields governed by Maxwell equations. These two phenomena are coupled through the Lorentz force equation. For a particle of mass m, Newton's second law states that

$$\frac{d}{dt}[m\dot{\mathbf{q}}(t)] = \mathbf{F}[t, \mathbf{q}(t), \dot{\mathbf{q}}(t)] \tag{2.99}$$

where $\mathbf{q} = (q_1, q_2, q_3)$ is its position; $\mathbf{F}$ is the force acting on the particle; and the dot denotes the derivative with respect to time. The above equation is the well-known differential equation formulation of classical mechanics. From theoretical point of view, a more useful formulation for the classical mechanics is based on the principle of least action, by which (2.99) is derived by minimizing certain functional, called the action of the mechanical system. The action of the system is an integral over time of a function called Lagrangian function L

$$S(\mathbf{q}) = \int_{t_1}^{t_2} L\,[t, \mathbf{q}(t), \dot{\mathbf{q}}(t)]\,dt,$$

where $\mathbf{q}(t)$ are known as the generalized coordinates. Equation (2.99) is found by demanding the action S to be at its minimum so that the functional derivative vanishes

$$\frac{\delta S}{\delta q_i} = 0, i = 1, 2, 3.$$

Suppose $\mathbf{q}(t)$ is the path that renders S to be a minimum. Let the path $\mathbf{q}(t)$ be changed to $\mathbf{q}(t) + \varepsilon\Delta\mathbf{q}(t)$, where $\Delta\mathbf{q} = (\Delta q_1, \Delta q_2, \Delta q_3)$ is small everywhere in the time interval $[t_1, t_2]$ and the endpoints of the path are supposed to be fixed: $\Delta\mathbf{q}(t_1) = \Delta\mathbf{q}(t_2) = 0$. Then the partial functional derivative with respect to q_1 can be found by (2.95) as

$$\int_{t_1}^{t_2} \frac{\delta S}{\delta q_1}\Delta q_1 dt = \frac{d}{d\varepsilon} \int_{t_1}^{t_2} L(t, q_1 + \varepsilon\Delta q_1, q_2, q_3, \dot{q}_1 + \varepsilon\Delta\dot{q}_1, \dot{q}_2, \dot{q}_3)dt \bigg|_{\varepsilon=0}$$

$$= \int_{t_1}^{t_2} \left(\Delta q_1 \frac{\partial L}{\partial q_1} - \Delta q_1 \frac{d}{dt}\frac{\partial L}{\partial \dot{q}_1} \right) dt + \Delta q_1 \frac{\partial L}{\partial q_1}\bigg|_{t_1}^{t_2}$$

$$= \int_{t_1}^{t_2} \left(\frac{\partial L}{\partial q_1} - \frac{d}{dt}\frac{\partial L}{\partial \dot{q}_1} \right) \Delta q_1 dt.$$

Similar expressions can be obtained for the partial functional derivative with respect to q_2 and q_3. So if S has a local extremum at $\mathbf{q}(t)$, the path $\mathbf{q}(t)$ must satisfy

$$\frac{\delta S}{\delta q_i} = \frac{d}{dt}\frac{\partial L}{\partial \dot{q}_i} - \frac{\partial L}{\partial q_i} = 0, i = 1, 2, 3. \tag{2.100}$$

These are the Lagrangian equations. In order to transform (2.99) into the form of (2.100), we need to find an appropriate Lagrangian function L, which constitutes the so-called inverse variational problem. For a conservative mechanical system, the Lagrangian function is

$$L[t, \mathbf{q}(t), \dot{\mathbf{q}}(t)] = T - V = \frac{1}{2}m\, |\dot{\mathbf{q}}|^2 - V(\mathbf{q})$$

with $T = m\, |\dot{\mathbf{q}}|^2 /2$ being the kinetic energy and V being the potential energy of the mechanical system. It can be shown that (2.100) reduces to (2.99) with the above Lagrangian function.

Remark 2.4: In the Lagrangian function, only the first derivative of $\mathbf{q}(t)$ is involved. As a result, the smoothness requirement for the solution of original Newton's second law is reduced in the least action principle, which implies that we seek a solution in an expanded solution space.   □

Remark 2.5: Let $f(t, \mathbf{q}(t))$ be an arbitrary function, the new Lagrangian function formed by

$$L'[t, \mathbf{q}(t), \dot{\mathbf{q}}(t)] = L[t, \mathbf{q}(t), \dot{\mathbf{q}}(t)] + \frac{d}{dt} f[t, \mathbf{q}(t)]$$

is equivalent to L in the sense that they both give the same Lagrangian equations (2.100).   □

A charged particle q of velocity $\dot{\mathbf{q}}$ in an electromagnetic field is subject to a force given by Lorentz force equation

$$\mathbf{F} = q(\mathbf{E} + \dot{\mathbf{q}} \times \mathbf{B}).$$

The equation of motion for the charged particle of mass m is

$$\frac{d(m\dot{\mathbf{q}})}{dt} = q(\mathbf{E} + \dot{\mathbf{q}} \times \mathbf{B}).$$

To transform the above equation into the Lagrangian equation, the Lagrangian function must be chosen as

$$L[t, \mathbf{q}(t), \dot{\mathbf{q}}(t)] = -m_0 c \sqrt{c^2 - |\dot{\mathbf{q}}|^2} - q(\phi - \dot{\mathbf{q}} \cdot \mathbf{A}),$$

where m_0 is the rest mass of the particle; ϕ and $\mathbf{A}$ are scalar potential and vector potential function respectively. The generalized momentum is (see (2.106))

$$\mathbf{p} = m\dot{\mathbf{q}} + q\mathbf{A},$$

where $m = m_0/\sqrt{1 - |\dot{\mathbf{q}}|^2 /c^2}$. The generalized momentum for a charged particle in an electromagnetic field is not equal to the mechanical momentum $m\dot{\mathbf{q}}$ and an additional term $q\mathbf{A}$ must be added.

2.7.2.2 Lagrangian Equation for Electromagnetic Fields

Let $\mathbf{r} = (x_1, x_2, x_3)$ and consider a dynamic system $\boldsymbol{\eta}(\mathbf{r}, t) = (\eta_1(\mathbf{r}, t), \eta_2(\mathbf{r}, t), \ldots, \eta_n(\mathbf{r}, t))$, whose components $\eta_i(\mathbf{r}, t)$ $(i = 1, 2, \ldots, n)$ are called **generalized coordinates**. The Lagrangian function $L(t, \boldsymbol{\eta})$ is the integral of **Lagrangian density function** $\mathcal{L}$ over a region Ω in R^3

$$L(t, \boldsymbol{\eta}) = \int_{\Omega \subset R^3} \mathcal{L}\left(t, x_j, \eta_i, \frac{\partial \eta_i}{\partial x_j}, \frac{\partial \eta_i}{\partial t}\right) dx_1 dx_2 dx_3.$$

The action of the system is then defined by

$$S(\boldsymbol{\eta}) = \int_{t_1}^{t_2} L(t, \boldsymbol{\eta}) dt = \int_{t_1}^{t_2} dt \int_{\Omega \subset R^3} \mathcal{L}\left(t, x_j, \eta_i, \frac{\partial \eta_i}{\partial x_j}, \frac{\partial \eta_i}{\partial t}\right) dx_1 dx_2 dx_3.$$

Suppose η is the function that makes $S(\boldsymbol{\eta})$ reach a local minimum. Let $\boldsymbol{\eta}$ change to $\boldsymbol{\eta} + \varepsilon \Delta \boldsymbol{\eta}$, where $\Delta \boldsymbol{\eta} = (\Delta \eta_1, \Delta \eta_2, \ldots, \Delta \eta_n)$ is small everywhere and is zero on the boundaries of Ω and at the endpoints of the time interval $[t_1, t_2]$: $\Delta \boldsymbol{\eta}|_{\partial\Omega} = \Delta \boldsymbol{\eta}|_{t_1} = \Delta \boldsymbol{\eta}|_{t_2} = 0$. Thus

$$\int_{t_1}^{t_2} \int_{\Omega \subset R^3} \frac{\delta S(\boldsymbol{\eta})}{\delta \eta_i} \Delta \eta_i dt dx_1 dx_2 dx_3 = \frac{d}{d\varepsilon} S(\eta_1, \eta_2, \ldots, \eta_i + \varepsilon \Delta \eta_i, \ldots, \eta_n)\bigg|_{\varepsilon=0}.$$

$$= \int_{t_1}^{t_2} dt \int_{\Omega \subset R^3} \left(\frac{\partial \mathcal{L}}{\partial \eta_i} - \sum_{j=1}^{3} \frac{d}{dx_j} \frac{\partial \mathcal{L}}{\partial \eta_i / \partial x_j} - \frac{d}{dt} \frac{\partial \mathcal{L}}{\partial \eta_i / \partial t}\right) \Delta \eta_i dx_1 dx_2 dx_3.$$

Hence the Lagrangian equations for the dynamic system are given by

$$\frac{\delta S}{\delta \eta_i} = \frac{d}{dt} \frac{\partial \mathcal{L}}{\partial(\partial \eta_i / \partial t)} + \sum_{j=1}^{3} \frac{d}{dx_j} \frac{\partial \mathcal{L}}{\partial(\partial \eta_i / \partial x_j)} - \frac{\partial \mathcal{L}}{\partial \eta_i} = 0, i = 1, 2, \ldots, n. \qquad (2.101)$$

Comparing (2.100) with (2.101), the second term on the right-hand side of (2.101) emerges since the generalized coordinates η_i are functions of position. The Maxwell equations in free space can be expressed as

$$\nabla \times \mathbf{E} + \frac{\partial \mathbf{B}}{\partial t} = 0, \nabla \cdot \mathbf{B} = 0, \qquad (2.102)$$

$$\nabla \times \mathbf{B} - \mu_0 \varepsilon_0 \frac{\partial \mathbf{E}}{\partial t} = \mu_0 \mathbf{J}, \nabla \cdot \mathbf{E} = \frac{1}{\varepsilon_0} \rho. \qquad (2.103)$$

The fields $\mathbf{E}$ and $\mathbf{B}$ have six components that are not independent, and therefore they cannot be used as the generalized coordinates. We have to resort to the vector potential $\mathbf{A}$ and the scalar potential ϕ. The electromagnetic fields can then be represented by

$$\mathbf{E} = -\nabla\phi - \frac{\partial\mathbf{A}}{\partial t}, \quad \mathbf{B} = \nabla \times \mathbf{A}. \tag{2.104}$$

These are equivalent to (2.102). For this reason, only (2.103) should be considered as the equation of motion if we use $\mathbf{A}$ and ϕ as the generalized coordinates, and (2.102) simply gives the definition of $\mathbf{A}$ and ϕ. The Lagrangian density function for the electromagnetic fields is given by

$$\mathcal{L}(t, x_j, \phi, \mathbf{A}) = \frac{1}{2}(\varepsilon_0 |\mathbf{E}|^2 - \mu_0^{-1} |\mathbf{B}|^2) - \rho\phi + \mathbf{J} \cdot \mathbf{A}. \tag{2.105}$$

Here $\mathbf{E}$ and $\mathbf{B}$ are related to the generalized coordinates $\mathbf{A}$ and ϕ through (2.104). Considering

$$\frac{\partial\mathcal{L}}{\partial\phi} = -\rho,$$

$$\frac{\partial\mathcal{L}}{\partial(\partial\phi/\partial x_j)} = \varepsilon_0 E_j \frac{\partial E_j}{\partial(\partial\phi/\partial x_j)} = -\varepsilon_0 E_j,$$

$$\frac{\partial\mathcal{L}}{\partial(\partial\phi/\partial t)} = 0,$$

and substituting these relations into (2.101), we obtain the Lagrangian equation corresponding to the generalized coordinate ϕ as follows

$$-\varepsilon_0 \sum_{j=1}^{3} \frac{\partial E}{\partial x_j} + \rho = 0,$$

which is the second equation of (2.103). Taking the following calculations

$$\frac{\partial\mathcal{L}}{\partial A_1} = J_1,$$

$$\frac{\partial\mathcal{L}}{\partial(\partial A_1/\partial t)} = \varepsilon_0 E_1 \frac{\partial E_1}{\partial(\partial A_1/\partial t)} = -\varepsilon_0 E_1,$$

$$\frac{\partial\mathcal{L}}{\partial(\partial A_1/\partial x_2)} = -\frac{1}{\mu_0} B_3 \frac{\partial B_3}{\partial(\partial A_1/\partial x_2)} = \frac{1}{\mu_0} B_3,$$

$$\frac{\partial\mathcal{L}}{\partial(\partial A_1/\partial x_3)} = -\frac{1}{\mu_0} B_2,$$

into account, we have

$$\frac{1}{\mu_0}\left(\frac{\partial B_3}{\partial x_2} - \frac{\partial B_2}{\partial x_3}\right) - \varepsilon_0 \frac{\partial E_1}{\partial t} - J_1 = 0,$$

and similarly

$$\frac{1}{\mu_0}\left(\frac{\partial B_1}{\partial x_3} - \frac{\partial B_3}{\partial x_1}\right) - \varepsilon_0 \frac{\partial E_2}{\partial t} - J_2 = 0,$$

$$\frac{1}{\mu_0}\left(\frac{\partial B_2}{\partial x_1} - \frac{\partial B_1}{\partial x_2}\right) - \varepsilon_0 \frac{\partial E_3}{\partial t} - J_3 = 0.$$

These are the first equation of (2.103).

2.7.3 Hamiltonian Formulation

The **Hamiltonian function** is often defined as the sum of kinetic energy and potential energy, that is, the total energy. As a result, the Hamiltonian function must remain constant in a closed system. The Hamiltonian mechanics is the starting point of Schrödinger's development for his wave mechanics.

2.7.3.1 Hamiltonian Equation for Charged Particle

The second-order equations in (2.100) can be further reduced to first-order equations in terms of the Hamiltonian formulation. The Hamiltonian function is defined by

$$H(t, \mathbf{q}, \mathbf{p}) = \sum_{i=1}^{3} p_i \dot{q}_i - L[t, \mathbf{q}(t), \dot{\mathbf{q}}(t)],$$

where $\mathbf{p} = (p_1, p_2, p_3)$ is called the **generalized momentum**, whose components are

$$p_i = \frac{\partial L}{\partial \dot{q}_i}, i = 1, 2, 3. \tag{2.106}$$

From (2.100), we obtain

$$\dot{p}_i = \frac{\partial L}{\partial q_i}, i = 1, 2, 3. \tag{2.107}$$

The first-order equations of motion can be obtained by considering the total differential of the Hamiltonian function

$$dH(t, \mathbf{q}, \mathbf{p}) = \sum_{i=1}^{3}(\dot{q}_i dp_i + p_i d\dot{q}_i) - \sum_{i=1}^{3}\left(\frac{\partial L}{\partial q_i} dq_i + \frac{\partial L}{\partial \dot{q}_i} d\dot{q}_i\right) - \frac{\partial L}{\partial t} dt$$

$$= \sum_{i=1}^{3}\left(\dot{q}_i dp_i - \frac{\partial L}{\partial q_i} dq_i\right) - \frac{\partial L}{\partial t} dt = \sum_{i=1}^{3}(\dot{q}_i dp_i - \dot{p}_i dq_i) - \frac{\partial L}{\partial t} dt,$$

where Equations (2.107) have been used. Comparing the above equation with the following calculation

$$dH(t, \mathbf{q}, \mathbf{p}) = \sum_{i=1}^{3} \left(\frac{\partial H}{\partial q_i} dq_i + \frac{\partial H}{\partial p_i} dp_i \right) + \frac{\partial H}{\partial t} dt$$

we obtain $\partial H/\partial t = -\partial L/\partial t$ and

$$\dot{q}_i = \frac{\partial H}{\partial p_i}, \quad \dot{p}_i = -\frac{\partial H}{\partial q_i}, \quad i = 1, 2, 3. \tag{2.108}$$

These are the first-order equations we seek. Equations (2.108) are called **Hamiltonian equations** and they are equivalent to the second-order Lagrangian equations. The preceding discussion can easily be generalized to a system consisting of multiple particles. The Hamiltonian function for the charged particle in an electromagnetic field is given by

$$H(t, \mathbf{q}, \mathbf{p}) = \sum_{i=1}^{3} p_i \dot{q}_i - L = mc^2 + q\phi = c\sqrt{(\mathbf{p} - q\mathbf{A})^2 + m_0^2 c^2} + q\phi.$$

If $|\dot{\mathbf{q}}| \ll c$ this reduces to

$$H(t, \mathbf{q}, \mathbf{p}) = \frac{1}{2m_0}(\mathbf{p} - q\mathbf{A})^2 + m_0 c^2 + q\phi,$$

which is the non-relativistic Hamiltonian function.

2.7.3.2 Hamiltonian Equation for Electromagnetic Fields

Let $\boldsymbol{\eta}(\mathbf{r}, t) = (\eta_1(\mathbf{r}, t), \eta_2(\mathbf{r}, t), \ldots, \eta_n(\mathbf{r}, t))$ be the generalized coordinates with $\mathbf{r} = (x_1, x_2, x_3)$. We may introduce the **Hamiltonian density function**

$$\mathcal{H}(t, x_j, \eta_i, \pi_i) = \sum_{i=1}^{n} \pi_i \frac{\partial \eta_i}{\partial t} - \mathcal{L}\left(t, x_j, \eta_i, \frac{\partial \eta_i}{\partial x_j}, \frac{\partial \eta_i}{\partial t} \right), \tag{2.109}$$

where $\pi_i = \partial \mathcal{L}/\partial(\partial \eta_i/\partial t)(i = 1, 2, \ldots, n)$ are the **generalized momentum density functions**. The Hamiltonian function is the integral of the Hamiltonian density function

$$H(t, \eta_i, \pi_i) = \int_{\Omega \subset R^3} \mathcal{H}(t, x_j, \eta_i, \pi_i) dx_1 dx_2 dx_3$$

$$= \int_{\Omega \subset R^3} \left[\sum_{i=1}^{n} \pi_i (\partial \eta_i/\partial t) - \mathcal{L} \right] dx_1 dx_2 dx_3.$$

The total differential of the Hamiltonian function is

$$DH(t, \eta_i, \pi_i) \cdot (\Delta t, \Delta \eta_i, \Delta \pi_i)$$

$$= \sum_{i=1}^{n} \langle \delta H/\delta \eta_i, \Delta \eta_i \rangle + \sum_{i=1}^{n} \langle \delta H/\delta \pi_i, \Delta \pi_i \rangle + \langle \delta H/\delta t, \Delta t \rangle$$

$$= \sum_{i=1}^{n} \int_{\Omega \subset R^3} \frac{\delta H}{\delta \eta_i} \Delta \eta_i dx_1 dx_2 dx_3 + \sum_{i=1}^{n} \int_{\Omega \subset R^3} \frac{\delta H}{\delta \pi_i} \Delta \pi_i dx_1 dx_2 dx_3$$

$$+ \int_{\Omega \subset R^3} \frac{\delta H}{\delta t} \Delta t dx_1 dx_2 dx_3,$$

(2.110)

which can be written as

$$DH(t, \eta_i, \pi_i) \cdot (\Delta t, \Delta \eta_i, \Delta \pi_i)$$

$$= \sum_{i=1}^{n} \int_{\Omega \subset R^3} \Delta \eta_i \left[-\frac{\partial \mathcal{L}}{\partial \eta_i} + \sum_{j=1}^{3} \frac{d}{dx_j} \frac{\partial \mathcal{L}}{\partial (\partial \eta_i / \partial x_j)} \right] dx_1 dx_2 dx_3$$

$$+ \sum_{i=1}^{n} \int_{\Omega \subset R^3} \Delta \pi_i \frac{\partial \eta_i}{\partial t} dx_1 dx_2 dx_3 - \int_{\Omega \subset R^3} \frac{\partial \mathcal{L}}{\partial t} \Delta t dx_1 dx_2 dx_3.$$

(2.111)

Comparing (2.110) with (2.111), we obtain $\delta H/\delta t = -\partial \mathcal{L}/\partial t$ and

$$\frac{\delta H}{\delta \eta_i} = -\frac{\partial \mathcal{L}}{\partial \eta_i} + \sum_{j=1}^{3} \frac{d}{dx_j} \frac{\partial \mathcal{L}}{\partial (\partial \eta_i / \partial x_j)}$$

$$\frac{\delta H}{\delta \pi_i} = \frac{\partial \eta_i}{\partial t}, i = 1, 2, \ldots, n.$$

Making use of (2.101), the above equations become

$$\dot{\eta}_i = \frac{\delta H}{\delta \pi_i}, \dot{\pi}_i = -\frac{\delta H}{\delta \eta_i}, i = 1, 2, \ldots, n.$$

(2.112)

The Hamiltonian density function for the electromagnetic field can be constructed through (2.105) and (2.109). Let $\eta_1 = \phi$, $\eta_i = A_{i-1}$, $i = 2, 3, 4$. Taking the following calculations

$$\pi_1 = \frac{\partial \mathcal{L}}{\partial (\partial \phi / \partial t)} = 0, \pi_i = \frac{\partial \mathcal{L}}{\partial (\partial A_{i-1} / \partial t)} = -\varepsilon_0 E_{i-1}, i = 2, 3, 4,$$

into account, the Hamiltonian density function can be expressed as

$$\mathcal{H}(t, x_j, \phi, \mathbf{A}, \pi_i) = -\varepsilon_0 \mathbf{E} \cdot \frac{\partial \mathbf{A}}{\partial t} - \frac{1}{2}\left(\varepsilon_0 |\mathbf{E}|^2 - \mu_0^{-1} |\mathbf{B}|^2\right) + \rho\phi - \mathbf{J} \cdot \mathbf{A}$$

$$= \frac{1}{2}\left[\varepsilon_0^{-1}\left(\pi_2^2 + \pi_3^2 + \pi_4^2\right) + \mu_0^{-1} |\mathbf{B}|^2\right] - \left(\pi_2 \mathbf{u}_{x_1} + \pi_3 \mathbf{u}_{x_2} + \pi_4 \mathbf{u}_{x_3}\right) \cdot \nabla\phi + \rho\phi - \mathbf{J} \cdot \mathbf{A}$$

where $\mathbf{u}_{x_i} (i = 1, 2, 3)$ are the unit vectors along x_i. The Hamiltonian function is thus of the form

$$H(t, \phi, \mathbf{A}, \pi_i) = \int_{\Omega \subset R^3} \mathcal{H}(t, x_j, \phi, \mathbf{A}, \pi_i) dx_1 dx_2 dx_3$$

$$= \int_{\Omega \subset R^3} \left\{ \frac{1}{2}[\varepsilon_0^{-1}(\pi_2^2 + \pi_3^2 + \pi_4^2) + \mu_0^{-1} |\mathbf{B}|^2] \right. \tag{2.113}$$

$$\left. - (\pi_2 \mathbf{u}_{x_1} + \pi_3 \mathbf{u}_{x_2} + \pi_4 \mathbf{u}_{x_3}) \cdot \nabla\phi + \rho\phi - \mathbf{J} \cdot \mathbf{A} \right\} dx_1 dx_2 dx_3.$$

It is readily found that

$$\langle \delta H/\delta\phi, \Delta\phi \rangle = \int_{\Omega \subset R^3} (-\varepsilon_0 \nabla \cdot \mathbf{E} + \rho)\Delta\phi dx_1 dx_2 dx_3,$$

$$\langle \delta H/\delta A_i, \Delta A_i \rangle = \int_{\Omega \subset R^3} \left[-\mu_0^{-1}\nabla \cdot (\mathbf{u}_{x_i} \times \mathbf{B}) - J_i\right] \Delta A_i dx_1 dx_2 dx_3, i = 1, 2, 3.$$

Therefore

$$\frac{\delta H}{\delta\phi} = -\varepsilon_0 \nabla \cdot \mathbf{E} + \rho, \quad \frac{\delta H}{\delta A_i} = -\frac{1}{\mu_0}\nabla \cdot (\mathbf{u}_{x_i} \times \mathbf{B}) - J_i,$$

and the equations $\dot{\pi}_i = -\delta H/\delta\eta_i$ imply $\mu_0^{-1}\nabla \times \mathbf{B} = \mathbf{J} + \varepsilon_0 \partial \mathbf{E}/\partial t$ and $\nabla \cdot \mathbf{E} = \rho/\varepsilon_0$. Furthermore, it can be shown that

$$\frac{\delta H}{\delta\pi_1} = 0,$$

$$\frac{\delta H}{\delta\pi_i} = \frac{1}{\varepsilon_0}\pi_i - \mathbf{u}_{x_{i-1}} \cdot \nabla\phi = \mathbf{u}_{x_{i-1}} \cdot (\mathbf{E}_{i-1} - \nabla\phi), i = 2, 3, 4.$$

So the equations $\dot{\eta}_i = \delta H/\delta\pi_i, i = 2, 3, 4$ imply $\mathbf{E} = -\nabla\phi - \partial \mathbf{A}/\partial t$ and the equation $\dot{\eta}_1 = \delta H/\delta\pi_1$ implies $\partial\phi/\partial t = 0$ or $\phi = 0$. Thus, the temporal gauge or Hamiltonian gauge comes into the picture naturally, and (2.113) may be written as

$$H(t, \mathbf{A}, \mathbf{\Pi}) = \int\limits_{\Omega \subset R^3} \left\{ \frac{1}{2} \left[\varepsilon_0^{-1}(\Pi_1^2 + \Pi_2^2 + \Pi_3^2) + \mu_0^{-1} |\mathbf{B}|^2 \right] - \mathbf{J} \cdot \mathbf{A} \right\} dx_1 dx_2 dx_3,$$

where the components of the vector potential $\mathbf{A}$ are considered as the generalized coordinates and the generalized momentum is given by the vector $\mathbf{\Pi} = (\Pi_1, \Pi_2, \Pi_3) = -\varepsilon_0 \mathbf{E}$.

> The partial differential equation entered theoretical physics as a handmaid, but has gradually become mistress.
>
> —Albert Einstein

3

Eigenvalue Problems

The validity of theorems on eigenfunctions can be made plausible by the following observation made by Daniel Bernoulli (1700–1782). A mechanical system of n degrees of freedom possesses exactly n eigensolutions. A membrane is, however, a system with an infinite number of degrees of freedom. This system will, therefore, have an infinite number of eigenoscillations.

—Arnold Sommerfeld

In 1894, the French mathematician Jules Henri Poincaré (1854–1912) established the existence of an infinite sequence of eigenvalues and the corresponding eigenfunctions for the Laplace operator under Dirichlet boundary condition. This key result signifies the beginning of spectral theory, which extends the eigenvector and eigenvalue theory of a square matrix and has played an important role in mathematics and physics. The study of eigenvalue problems has its roots in the method of separation of variables. An eigenmode of a system is a possible state when the system is free of excitation, which might exist in the system on its own under certain conditions, and is also called an **eigenstate** of the system. The corresponding eigenvalue often represents an important quantity of the system, for example the total energy of the system and the natural oscillation frequency. The eigenmode analysis has been extensively used in physics and engineering science as an arbitrary state of the system can be expressed as a linear combination of the eigenmodes. When the eigenvalue problem is solved, what remains is to determine the expansion coefficients in the linear combination by using the source conditions or the initial values of the system. In most situations, only one or a few eigenmodes dominate in the linear combination.

The electromagnetic eigenvalue problems are often expressed by differential equations. The corresponding differential operators are typically positive-bounded-below and symmetric if the medium is isotropic and homogeneous. The variational method may be used to solve the eigenvalue problem for a positive-bounded-below symmetric operator, and to prove the existence and completeness of eigenmodes. The electromagnetic eigenvalue problems can also be formulated by integral equations, and the corresponding integral operators are frequently compact.

Eigenvalue problems result from the boundary value problems defined in a finite region. When the defining region is unbounded, the discrete eigenvalues may become continuous and the Fourier integrals enter the picture. In some cases, the eigenvalue theory can be applied to study the boundary value problems whose defining region is infinite. For example, the method of singular function expansion in electromagnetics is rooted in the integral equation for the scattering problem with a compact operator, and the singular function is the eigenfunction of the product of the compact operator and its adjoint.

3.1 Introduction to Linear Operator Theory

The basic concepts of linear operators are introduced in Chapter 2. More topics on the linear operators will be discussed in this section for later use. The domain of definition of a linear operator $\hat{A}$ is denoted by $D(\hat{A})$, and it is a subset of some functional space. An integral operator is often defined on the whole functional space while a differential operator is typically defined in a proper subset of the functional space.

3.1.1 Compact Operators and Embeddings

A set S in a normed space is called **relatively compact** if and only if each sequence in S has a convergent subsequence. If, in addition, the limits of these convergent subsequences belong to S, then S is compact. Each relatively compact set is bounded. In R^N, a closed and bounded set is compact. Let E and F be two normed spaces. An operator $\hat{A} : D(\hat{A}) \subset E \to F$ is called **compact** if and only if $\hat{A}$ is continuous and $\hat{A}$ transforms bounded sets into relatively compact sets. The compact operators are always bounded. If there exists a compact operator $\hat{A}_\varepsilon$ depending on a parameter ε such that $\|\hat{A}u - \hat{A}_\varepsilon u\| \leq \varepsilon \|u\|$ holds for all $\varepsilon > 0$ and all $u \in E$, $\hat{A}$ is compact. This property can be used to study the compactness of the integral operators. The following Arzelà–Ascoli theorem, named after the Italian mathematicians Cesare Arzelà (1847–1912) and Giulio Ascoli (1843–1896), gives a sufficient condition to decide if a set of continuous functions forms a relative compact set.

Theorem 3.1 *Arzelà–Ascoli theorem: Let $H := C[a, b]$ with $\|u\| := \max_{a \leq x \leq b} |u(x)|$. Suppose that we are given a set M in H such that*

1. *M is bounded: $\|u\| \leq c_1$ for all $u \in M$ and for fixed $c_1 \geq 0$.*
2. *M is equicontinuous: For each $\varepsilon > 0$, there is a $\delta > 0$ such that $|x - y| < \delta$ implies $|u(x) - u(y)| < \varepsilon$ for all $u \in M$.*

Then M is a relatively compact subset of H. □

Example 3.1: Consider the following integral equation

$$\int_a^b K(x, y)v(y) = \lambda v(x), \quad -\infty < a \leq x \leq b < \infty \tag{3.1}$$

where the function $K : [a, b] \times [a, b] \to R$ is continuous (in this case we call $K(x, y)$ a continuous kernel) and is symmetric: $K(x, y) = K(y, x)$. Set $H = L^2(a, b)$ with the usual inner product, and define the integral operator $\hat{A}$ as follows

$$\hat{A}v(x) = \int_a^b K(x, y)v(y)dy, \; v \in H. \tag{3.2}$$

Since the set $[a, b] \times [a, b]$ is compact, the function $K : [a, b] \times [a, b] \to R$ is uniformly continuous. As a result, for each $\varepsilon > 0$, there is a $\delta > 0$ such that $|x - z| < \delta$ implies

$$\alpha = \max_{a \leq y \leq b} |K(x, y) - K(z, y)| < \varepsilon.$$

Let $u = \hat{A}v$. For $|x - z| < \delta$ we thus have

$$|u(x) - u(z)| \leq \alpha \int_a^b |v(y)| \, dy \leq \varepsilon(b - a)^{1/2} \|v\|, \, x, z \in [a, b] \tag{3.3}$$

where use is made of Cauchy-Schwartz inequality:

$$\int_a^b 1 \cdot |v(y)| \, dy \leq \left(\int_a^b 1 \cdot dy \right)^{1/2} \left(\int_a^b |v(y)|^2 \, dy \right)^{1/2} = (b - a)^{1/2} \|v\|.$$

Consequently $u(x)$ is continuous function on $[a, b]$. Also we have

$$\max_{a \leq y \leq b} |u(x)| \leq \max_{a \leq x, y \leq b} |K(x, y)| \int_a^b |v(y)dy| \leq c_1 \|v\| \tag{3.4}$$

for some constant c_1. The above expression implies

$$\|\hat{A}v\| = \left(\int_a^b |u(y)|^2 \, dy \right)^{1/2} \leq c_2 \|v\|$$

for some constant c_2. This indicates that the operator $\hat{A}$ is bounded.

Let M be bounded set in H. It follows from (3.3), (3.4) and the Arzelà–Ascoli theorem that the set $\hat{A}(M)$ is relatively compact in $C[a, b]$. We may further show that $\hat{A}(M)$ is relatively compact in $L^2(a, b)$. In fact, if $u_n \to u$ in $C[a, b]$ as $n \to \infty$, then

$$\|u_n - u\| = \left\{ \int_a^b |u_n(x) - u(x)|^2 \, dx \right\}^{1/2} \leq \max_{a \leq x \leq b} |u_n(x) - u(x)| (b - a)^{1/2} \to 0.$$

Therefore $\hat{A}$ is compact. $\qquad\qquad\qquad\qquad\qquad\qquad\qquad\qquad\qquad\qquad\square$

Example 3.2: Consider (3.1) again. If $K(x, y)$ is square integrable

$$\int_a^b \int_a^b |K(x, y)|^2 \, dx dy < \infty$$

(in this case $K(x, y)$ is called L^2 kernel), the operator $\hat{A}$ defined by (3.2) is a compact operator from $L^2(a, b)$ to $L^2(a, b)$. This can be proved as follows. Since $C\{[a, b] \times [a, b]\}$ is dense in $L^2[(a, b) \times (a, b)]$, there is a sequence of continuous kernel $\{K_n(x, y)\}$ such that

$$\lim_{n \to \infty} \int_a^b \int_a^b |K(x, y) - K_n(x, y)|^2 \, dx dy = 0.$$

The operator $\hat{A}_n$ defined by $\hat{A}_n v(x) = \int_a^b K_n(x, y) v(y) dy$, $v \in H$ is compact by the previous theorem. Making use of Schwarz inequality, we have

$$\left\| (\hat{A} - \hat{A}_n) v \right\| = \left\| \int_a^b [K(x, y) - K_n(x, y)] v(y) dy \right\|$$

$$\leq \left[\int_a^b dx \int_a^b |K(x, y) - K_n(x, y)|^2 \, dy \int_a^b |v(y)|^2 \, dy \right]^{1/2}$$

$$= \left[\int_a^b dx \int_a^b |K(x, y) - K_n(x, y)|^2 \, dy \right]^{1/2} \|v\| .$$

Hence

$$\left\| \hat{A} - \hat{A}_n \right\| \leq \left[\int_a^b \int_a^b |K(x, y) - K_n(x, y)|^2 \, dx dy \right]^{1/2} .$$

This implies $\lim_{n \to \infty} \left\| \hat{A} - \hat{A}_n \right\| = 0$ and $\hat{A}$ is compact.   □

Example 3.3: Consider the integral operator

$$\hat{G}(\mathbf{J})(\mathbf{r}) = \mathbf{u}_n(\mathbf{r}) \times \int_S \mathbf{J}(\mathbf{r}') \times \nabla' G(\mathbf{r}, \mathbf{r}') dS(\mathbf{r}'), \mathbf{r} \in S, \mathbf{J} \in [L^2(S)]^3, \qquad (3.5)$$

where S is a closed surface, $\mathbf{u}_n(\mathbf{r})$ is the unit outward normal, $\mathbf{J}$ is a vector field tangent to S, and $G(\mathbf{r}, \mathbf{r}') = e^{-jk|\mathbf{r}-\mathbf{r}'|}/4\pi |\mathbf{r} - \mathbf{r}'|$. For $\delta > 0$, we may define

$$\mathbf{G}_\delta = \begin{cases} \nabla' G(\mathbf{r}, \mathbf{r}'), \, |\mathbf{r} - \mathbf{r}'| \geq \delta \\ 0, \, |\mathbf{r} - \mathbf{r}'| < \delta \end{cases}, \mathbf{G}' = \begin{cases} \nabla' G(\mathbf{r}, \mathbf{r}'), \, |\mathbf{r} - \mathbf{r}'| < \delta \\ 0, \, |\mathbf{r} - \mathbf{r}'| \geq \delta \end{cases}.$$

The integral operator $\hat{G}$ can be split into the sum of two integral operators $\hat{G} = \hat{G}_\delta + \hat{G}'$ with

$$\hat{G}_\delta(\mathbf{J})(\mathbf{r}) = \mathbf{u}_n(\mathbf{r}) \times \int_S \mathbf{J}(\mathbf{r}') \times \mathbf{G}_\delta dS(\mathbf{r}'),$$

$$\hat{G}'(\mathbf{J})(\mathbf{r}) = \mathbf{u}_n(\mathbf{r}) \times \int_S \mathbf{J}(\mathbf{r}') \times \mathbf{G}' dS(\mathbf{r}').$$

The kernel of $\hat{G}_\delta$ has no singularity and is square integrable. Hence $\hat{G}_\delta$ is compact. It can be shown that the norm of $\hat{G}'$ can be arbitrarily small (Jones, 1979). Therefore, $\hat{G}$ is a compact operator. □

Let E and F be normed spaces. We say that the **embedding** $'E \subset F'$ is continuous if and only if there exists a linear injective operator $\hat{J} : E \to F$, which is continuous. The embedding $'E \subset F'$ is called compact if and only if there exists a linear injective operator $\hat{J} : E \to F$, which is compact. Since the operator $\hat{J} : E \to F$ is injective, we may identify $u \in E$ with $\hat{J}(u) \in F$. The space E may thus be considered as a subspace of F and we may write $E \subset F$ instead of $'E \subset F'$.

Example 3.4: Let Ω be a nonempty bounded open set in R^n, $n > 1$. Then we have:

1. The embedding $C(\bar{\Omega}) \subset L^2(\Omega)$ is continuous.
2. The embedding $H_0^1(\Omega) \subset L^2(\Omega)$ is compact.
3. The embedding $H^1(\Omega) \subset L^2(\Omega)$ is compact.

Note that the norm of $u \in C(\bar{\Omega})$ is defined by $\|u\| = \max_{\mathbf{x} \in \bar{\Omega}} |u(\mathbf{x})|$. □

If E is a subset of F, we can define $\hat{J}(u) := u$ for all $u \in E$. In this case, we have

1. The embedding $E \subset F$ is continuous if and only if $u_n \xrightarrow{n \to \infty} u$ in E implies $u_n \xrightarrow{n \to \infty} u$ in F.
2. The embedding $E \subset F$ is compact if and only if it is continuous and each bounded sequence $\{u_n\} \subset E$ has a subsequence that converges in F.

3.1.2   Closed Operators

We assume that all the operators are defined in a Hilbert space H. Let $\hat{A}$ be an operator with domain $D(\hat{A})$. To extend $\hat{A}$ to a larger domain, we can use the following strategy. Suppose that u is not in $D(\hat{A})$, but that there exists a sequence $\{u_n\} \subset D(\hat{A})$ such that $u_n \to u$ and $\hat{A}(u_n) \to v$.

Then we can define $\hat{A}(u) = v$. This definition is reasonable only if v does not depend on the sequence $\{u_n\}$. In other words, if $\{u'_n\} \subset D(\hat{A})$ is such that $u'_n \to u$ and $\hat{A}(u'_n) \to v'$, we must have $v = v'$. An operator having this property is called **closable**, and the extended operator is called its **closure**, denoted by $\underline{\hat{A}}$. Equivalently an operator $\hat{A}$ is closable if the closure of its graph $\Gamma(\hat{A})$, denoted $\overline{\Gamma(\hat{A})}$, is the graph of some linear operator. This latter linear operator is the closure of $\hat{A}$. Thus $\Gamma(\underline{\hat{A}}) = \overline{\Gamma(\hat{A})}$. Since $\Gamma(\hat{A}) \subset \overline{\Gamma(\hat{A})}$, if $\underline{\hat{A}}$ exists, it extends $\hat{A}$. An operator $\hat{A}$ is **closed** if it is closable and $\hat{A} = \underline{\hat{A}}$. The closure $\underline{\hat{A}}$, whenever it exists, is a closed operator. If $\underline{\hat{A}} \subset \underline{\hat{B}}$ and $\hat{B}$ is closable, then $\overline{\Gamma(\hat{A})} \subset \Gamma(\bar{B})$; hence $\hat{A}$ is closable and $\underline{\hat{A}} \subset \hat{B}$. Thus the closure of a closable operator $\hat{A}$ is equal to the least closed operator which extends $\hat{A}$. Closedness is very important to the convergence of the solutions of separation of variables. An operator is not usually continuous but it may have the property of being closed.

Remark 3.1: If E and F are normed spaces, the map $\|\cdot\|_{E\times F} : E \times F \to R$ defined by

$$\|(u, v)\|_{E\times F} = \left(\|u\|_E^2 + \|v\|_F^2\right)^{1/2}, u \in E, v \in F$$

is a norm on $E \times F$. Equivalent norms on $E \times F$ are $(u, v) \mapsto \max\{\|u\|_E, \|v\|_F\}$ and $(u, v) \mapsto \|u\|_E + \|v\|_F$. The normed space $E \times F$ is denoted by $E \oplus F$ and called the **direct sum** of E and F.   □

Remark 3.2: If $\hat{A}$ is a closed operator, the map $u \mapsto (u, \hat{A}u)$ is an isomorphism between $D(\hat{A})$ and the closed subspace $\Gamma(\hat{A})$. If we set $\|u\|_\Gamma = (\|u\| + \|\hat{A}u\|)^{1/2}$, $D(\hat{A})$ becomes a Banach space. The norm $\|\cdot\|_\Gamma$ on $D(\hat{A})$ is called **graph norm**.   □

The closure $\underline{\hat{A}}$ may not be linear even if $\hat{A}$ is linear. If an operator is only densely defined, 'closed' is weaker than 'bounded' or 'continuous'. To find the closure of an operator is, in general, not an easy task. To determine the closure of an operator $\hat{A}$, we must define the domain $D(\bar{A})$ and the map $\underline{\hat{A}}$, which requires the extension of the notion of derivatives and the introduction of generalized derivatives. Almost all differential operators are either closed or closable.

3.1.3 Spectrum and Resolvent of Linear Operators

For any scalar λ and a linear operator $\hat{A} : D(\hat{A}) \subset H \to H$, where H is a Hilbert space, we define $\hat{A}_\lambda = \lambda \hat{I} - \hat{A}$. The set of those λ for which $\hat{A}_\lambda^{-1}$ does not exist is called the **point spectrum** (or **discrete spectrum**) of the operator $\hat{A}$ and is denoted by $\sigma_p(\hat{A})$. Thus $\lambda \in \sigma_p(\hat{A})$ if and only if the equation

$$\hat{A}u - \lambda u = 0 \tag{3.6}$$

has a solution $u \neq 0$. The elements of the point spectrum are called the **eigenvalues** of $\hat{A}$ and the corresponding non-zero solutions of the above equation are called **eigenfunctions** of $\hat{A}$. Now suppose $\hat{A}_\lambda^{-1}$ exists. The set of those λ for which $R(\hat{A}_\lambda)$ is not dense in H is called the **residual spectrum** of $\hat{A}$ and is denoted by $\sigma_r(\hat{A})$. The set of those λ for which $R(\hat{A}_\lambda)$ is dense in H and $\hat{A}_\lambda^{-1}$ is bounded is called **resolvent set** of $\hat{A}$ and is denoted by $\rho(\hat{A})$ (in this case, we can resolve the problem completely). The set of those λ for which $R(\hat{A}_\lambda)$ is dense in H and $\hat{A}_\lambda^{-1}$ is not bounded is called **continuous spectrum** of $\hat{A}$ and is denoted by $\sigma_c(\hat{A})$ (the term continuous derives from the fact that in most cases such λ form a continuum in the plane). These definitions are summarized below:

1. $\lambda \in \sigma_p(\hat{A})$ if and only if $\hat{A}_\lambda^{-1}$ does not exist. Uniqueness of (3.6) fails but existence holds.
2. $\lambda \in \sigma_r(\hat{A})$ if and only if $\hat{A}_\lambda^{-1}$ exists and $\overline{R(\hat{A}_\lambda)} \neq H$.
3. $\lambda \in \sigma_c(\hat{A})$ if and only if $\hat{A}_\lambda^{-1}$ exists and is unbounded and $\overline{R(\hat{A}_\lambda)} = H$. Uniqueness of (3.6) holds but existence fails.
4. $\lambda \in \rho(\hat{A})$ if and only if $\hat{A}_\lambda^{-1}$ exists and is bounded and $\overline{R(\hat{A}_\lambda)} = H$.

The union $\sigma_p(\hat{A}) \cup \sigma_r(\hat{A}) \cup \sigma_c(\hat{A})$ is called the **spectrum** of $\hat{A}$ and is denoted by $\sigma(\hat{A})$. We have

$$\sigma_p(\hat{A}) \cup \sigma_r(\hat{A}) \cup \sigma_c(\hat{A}) \cup \rho(\hat{A}) = C,$$

where C is the set of all complex numbers. Note that the residual spectrum does not occur for a large class of operators, such as self-adjoint operators. For $\lambda \in \rho(\hat{A})$, the operator $\hat{A}_\lambda^{-1}$ is called the **resolvent** of $\hat{A}$. Here are some important properties:

1. $\hat{A}_\lambda$ has a bounded inverse if and only if $\|\hat{A}_\lambda u\| \geq c_1 \|u\|$ holds for all $u \in D(\hat{A})$ and for some constant $c_1 > 0$.
2. If $\hat{A}$ is a closed operator, its resolvent $\hat{A}_\lambda^{-1}$ is bounded and defined on the whole space.
3. If $\hat{A}$ is a closable operator, then $\sigma(\hat{A}) = \sigma(\underline{\hat{A}})$.

Example 3.5: Consider the heat-conduction equation defined in a finite interval $[0, l]$

$$\frac{\partial^2 v(x, t)}{\partial x^2} - \frac{\partial v(x, t)}{\partial t} = 0,$$

$$v(0, t) = v(l, t) = 0, t > 0; v(x, 0) = f(x). \tag{3.7}$$

Let $H = L^2[0, 1]$, and $\hat{A} : u(x) \rightarrow -d^2u/dx^2$, with $D(\hat{A}) = \{u \in C^2[0, l] | u(0) = u(l) = 0\}$. We first solve the eigenvalue problem

$$-\frac{d^2u}{dx^2} = \lambda^2 u, u(0) = u(l) = 0.$$

The normalized eigenfunctions of the eigenvalue problem are easily found to be

$$u_n(x) = \sqrt{2/l} \sin \lambda_n x \tag{3.8}$$

where $\lambda_n = n\pi/l$. The spectrum of $\hat{A}$ is $\sigma(\hat{A}) = \sigma_p(\hat{A}) = \{\lambda_n^2 \mid n = 1, 2, \cdots\}$. The eigenfunctions (3.8) can be used to expand the solution of (3.7), and we have

$$v(x, t) = \frac{2}{l} \sum_{n=1}^{\infty} a_n \exp\left(-\lambda_n^2 t\right) \sin \lambda_n x, \qquad (3.9)$$

where $a_n = \int_0^l f(x) \sin \lambda_n x dx$. As $l \to \infty$, the difference $\Delta\lambda = \lambda_{n+1} - \lambda_n = \pi/l$ between any two neighboring eigenvalues approaches zero. As a result, the point spectrum becomes a continuous spectrum and the series (3.9) becomes an integral

$$v(x, t) = \frac{2}{\pi} \int_0^{\infty} a(\lambda) \exp(-\lambda^2 t) \sin(\lambda t) d\lambda,$$

where $a(\lambda) = \int_0^{\infty} f(x) \sin \lambda x dx$. This is how the Fourier integral enters the picture. □

3.1.4 Adjoint Operators and Symmetric Operators

Let H be a Hilbert space and $(\cdot, \cdot)$ stand for its inner product and $\|\cdot\|$ its corresponding norm. Let $\hat{A}: D(\hat{A}) \subset H \to R(\hat{A}) \subset H$ be a linear operator, where $D(\hat{A})$ denotes the domain of definition of $\hat{A}$ and $R(\hat{A})$ the range of $\hat{A}$. We assume that $D(\hat{A})$ is dense in Hilbert space H. Thus there is a unique linear operator, denoted by $\hat{A}^*$ and defined in $D(\hat{A}^*)$, such that

$$(\hat{A}u, v) = (u, \hat{A}^*v), u \in D(\hat{A}), v \in D(\hat{A}^*). \qquad (3.10)$$

The operator $\hat{A}^*$ is called the **adjoint operator** of $\hat{A}$. An operator $\hat{A}$ is called **self-adjoint** (resp. **skew adjoint**) if $\hat{A}^*$ exists and $\hat{A} = \hat{A}^*$ (resp. $\hat{A} = -\hat{A}^*$). Some important properties of adjoint operators are listed below (Yosida, 1988):

1. If $D(\hat{A}) = H$, $\hat{A}^*$ is bounded.
2. $(\alpha\hat{A})^* = \bar{\alpha}\hat{A}^*$, $\hat{A}^* + \hat{B}^* \subset (\hat{A} + \hat{B})^*$, $\hat{A}^*\hat{B}^* \subset (\hat{B}\hat{A})^*$ ($\bar{\alpha}$ is the complex conjugate of α). If $\hat{A}$ is bounded operator defined in the whole space H, we have $\hat{A}^* + \hat{B}^* = (\hat{A} + \hat{B})^*$ and $\hat{A}^*\hat{B}^* = (\hat{B}\hat{A})^*$.
3. If $\hat{A}$ is closable and $\hat{A}^*$ exists, then $(\underline{\hat{A}})^* = \hat{A}^*$. $\hat{A}^*$ is a closed operator. Every self-adjoint operator is closed.

Example 3.6: Let S be a closed surface in R^3 and $L^2(S)$ be the set of square-integrable complex functions defined on S. Define the inner product by $(u, v) = \int_S u(\mathbf{r})\bar{v}(\mathbf{r})dS(\mathbf{r})$ and consider the operator $\hat{A} : L^2(S) \to L^2(S)$

$$\hat{A}u(\mathbf{r}) = \int_S \frac{\partial G(\mathbf{r}, \mathbf{r}')}{\partial n(\mathbf{r}')} u(\mathbf{r}') dS(\mathbf{r}'), \tag{3.11}$$

where $G(\mathbf{r}, \mathbf{r}') = e^{-jk|\mathbf{r}-\mathbf{r}'|}/4\pi |\mathbf{r} - \mathbf{r}'|$. Then the adjoint of $\hat{A}$ is

$$\hat{A}^* v(\mathbf{r}) = \int_S \frac{\partial \bar{G}(\mathbf{r}, \mathbf{r}')}{\partial n(\mathbf{r})} v(\mathbf{r}') dS(\mathbf{r}'). \tag{3.12}$$

In fact, we have

$$(\hat{A}u, v) = \int_S \bar{v}(\mathbf{r}) \left[\int_S \frac{\partial G(\mathbf{r}, \mathbf{r}')}{\partial n(\mathbf{r}')} u(\mathbf{r}') dS(\mathbf{r}') \right] dS(\mathbf{r})$$

$$= \int_S u(\mathbf{r}) \left[\int_S \frac{\partial \bar{G}(\mathbf{r}', \mathbf{r})}{\partial n(\mathbf{r})} v(\mathbf{r}') dS(\mathbf{r}') \right] dS(\mathbf{r}) = (u, \hat{A}^* v).$$

The integral operators defined by (3.11) and (3.12) are also compact.   □

An operator $\hat{A}$ is said to be **symmetric** if and only if $\hat{A} \subset \hat{A}^*$ and

$$(\hat{A}u, v) = (u, \hat{A}v) \tag{3.13}$$

for all $u, v \in D(\hat{A})$. Since $\hat{A}^*$ is closed, a symmetric operator $\hat{A}$ is closable. Moreover, the closure of a symmetric operator is a symmetric operator since $(\underline{\hat{A}})^* = \hat{A}^*$. If $\hat{A}$ is a symmetric operator, its closure $\underline{\hat{A}}$ exists.

For technical reasons, it is the notion of self-adjoint rather than symmetric operator that is important in applications. However verifying self-adjointness is very difficult while verifying symmetry is usually trivial. In addition, self-adjointness is just a special situation of symmetry.

Example 3.7: Let $\hat{A}u = -d^2u/dx^2$, $D(\hat{A})$ consists of all possible functions having first and second derivatives in the open interval $(0, 1)$ and satisfying $u(0) = u(1) = 0$. Then the operator $\hat{A}$ is symmetric since

$$(\hat{A}u, v) - (u, \hat{A}v) = -\int_0^1 \frac{d}{dx} \left(v \frac{du}{dx} - u \frac{dv}{dx} \right) dx = -\left(v \frac{du}{dx} - u \frac{dv}{dx} \right)_0^1 = 0.$$

Notice that the symmetry of a differential operator depends on the boundary conditions.   □

Let E and F be two inner product spaces and $\hat{A} : D(\hat{A}) \subset E \to F$ be a linear differential operator. If there exists an operator $\hat{A}^* : D(\hat{A}^*) \subset F \to E$ such that

$$(\hat{A}u, v) = (u, \hat{A}^*v) + b(u, v), u \in D(\hat{A}), v \in D(\hat{A}^*),$$

$\hat{A}^*$ is called the **adjoint differential operator** of $\hat{A}$. Here $b(u, v)$ is the boundary term. If $\hat{A} = \hat{A}^*$, the operator $\hat{A}$ is called the **self-adjoint differential operator**.

Example 3.8: Let $u(x, y) \in C^2(\Omega) \cap C^1(\bar{\Omega}) \subset L^2(\Omega)$, $p(x, y) \in C^1(\bar{\Omega})$, $q(x, y) \in C(\bar{\Omega})$ and

$$\hat{A}u = \frac{\partial}{\partial x} p \frac{\partial u}{\partial x} + \frac{\partial}{\partial y} p \frac{\partial u}{\partial y} + qu.$$

From the integration by parts, we have

$$\int_\Omega v \hat{A}u \, dxdy = \int_\Omega u \hat{A}^* v \, dxdy + b(u, v),$$

where

$$\hat{A}^*v = \frac{\partial}{\partial x} p \frac{\partial v}{\partial x} + \frac{\partial}{\partial y} p \frac{\partial v}{\partial y} + qv,$$

$$b(u, v) = \int_\Gamma p \left(v \frac{\partial u}{\partial n} - u \frac{\partial v}{\partial n} \right) d\Gamma,$$

and Γ is the boundary of Ω. Hence $\hat{A} = \hat{A}^*$ and $\hat{A}$ is self-adjoint.   □

3.1.5   Energy Space

Let H be a real Hilbert space. A symmetric operator $\hat{B}$ is said to be **positive definite** if $(\hat{B}u, u) \geq 0$ for all $u \in D(\hat{B}) \subset H$ and the equality holds only for $u = 0$. A symmetric operator $\hat{B}$ is said to be **positive-bounded-below** if there is a positive constant c_1 such that

$$(\hat{B}u, u) > c_1 \|u\|^2 \tag{3.14}$$

for all $u \in D(\hat{B}) \subset H$. If $\hat{B}$ is positive definite, the quantity $(\hat{B}u, u)$ is called energy of u. We introduce the **energy inner product** of the operator $\hat{B}$

$$(u, v)_{\hat{B}} = (\hat{B}u, v)$$

and let $\|\cdot\|_{\hat{B}}$ stand for the corresponding **energy norm** defined by $\|u\|_{\hat{B}} = (\hat{B}u, u)^{1/2}$. We say that the sequence u_n **converges in energy** to u if $\|u_n - u\|_{\hat{B}} \xrightarrow[n \to \infty]{} 0$. Evidently if $\hat{B}$ is positive-bounded-below and $\|u_n - u\|_{\hat{B}} \xrightarrow[n \to \infty]{} 0$, then $\|u_n - u\| \xrightarrow[n \to \infty]{} 0$.

Remark 3.3: It is interesting to examine the difference between the positive definite and the positive-bounded-below physically. For a system involving a positive definite operator, an arbitrarily small input may give rise to an arbitrarily large response. For a system involving a positive-bounded-below operator, a large response needs a sufficiently large input. □

Remark 3.4: For most eigenvalue problems in mathematical physics, the operator $\hat{B}$ is a differential operator. To check if the operator $\hat{B}$ is positive-bounded-below, we may use **Poincaré inequality**

$$\int_{\Omega} u^2 d\Omega \leq c_1 \int_{\Omega} \sum_{j=1}^{N} (\partial_j u)^2 d\Omega + c_2 \left(\int_{\Omega} u d\Omega \right)^2, \tag{3.15}$$

where Ω is a bounded open region in R^N, c_1 and c_2 are positive constants. We can also use the inequality

$$c_3 \int_{\Omega} u^2 d\Omega \leq \int_{\Omega} \sum_{j=1}^{N} (\partial_j u)^2 d\Omega, \tag{3.16}$$

where c_3 is a positive constant and u is assumed to be equal to zero on the boundary of Ω. The expression (3.16) is called the **Friedrichs inequality**, named after the German mathematician Kurt Otto Friedrichs (1901–1982). □

Suppose that $\hat{B}$ is positive-bounded-below. The **energy space** $H_{\hat{B}}$ consists of all $u \in H$ that have the following two properties:

1. There exists a sequence $\{u_n\} \subset D(\hat{B})$ such that $\|u_n - u\| \xrightarrow[n \to \infty]{} 0$, $u \in H$.
2. The sequence $\{u_n\} \subset D(\hat{B})$ is a Cauchy sequence with respect to energy norm $\|\cdot\|_{\hat{B}}$.

The sequence $\{u_n\} \subset D(\hat{B})$ satisfying the above properties is called the **admissible sequence** for u. For all $u, v \in H_{\hat{B}}$ we define $(u, v)_{\hat{B}} = \lim_{n \to \infty} (u_n, v_n)_{\hat{B}}$, where $\{u_n\}$ and $\{v_n\}$ are the admissible sequences for u and v respectively. To show that the limit $\lim_{n \to \infty} (u_n, v_n)_{\hat{B}}$ exists, we note that

$$(u_n, v_n)_{\hat{B}} - (u_m, v_m)_{\hat{B}} = (u_n, v_n - v_m)_{\hat{B}} + (u_n - u_m, v_m)_{\hat{B}}$$
$$\leq \|u_n\|_{\hat{B}} \|v_n - v_m\|_{\hat{B}} + \|u_n - u_m\|_{\hat{B}} \|v_m\|_{\hat{B}} \xrightarrow[n,m \to \infty]{} 0.$$

Let $\hat{B}$ be positive-bounded-below. It is readily found that

1. $H_{\hat{B}}$ is a real Hilbert space with respect to the energy product $(\cdot, \cdot)_{\hat{B}}$, and $D(\hat{B})$ is dense in $H_{\hat{B}}$.
2. The embedding $H_{\hat{B}} \subset H$ is continuous, that is, $\|u\| \leq c_1 \|u\|_{\hat{B}}$ for all $u \in H_{\hat{B}}$.

3. The operator $\hat{J} : H \to H_{\hat{B}}^*$ defined by $\hat{J}(f)(v) = (f, v)$, $v \in H_{\hat{B}}$ is a continuous embedding.

We can identify f with $\hat{J}(f)$ so that H becomes a subset of $H_{\hat{B}}^*$. Thus we have $H_{\hat{B}} \subset H \subset H_{\hat{B}}^*$ and

$$\langle f, v \rangle_{\hat{B}} = (f, v), \ f \in H, v \in H_{\hat{B}}, \tag{3.17}$$

where the note $\langle g, v \rangle_{\hat{B}}$ is defined by $\langle g, v \rangle_{\hat{B}} = g(v)$ for $g \in H_{\hat{B}}^*$ and all $v \in H_{\hat{B}}$.

3.1.6 Energy Extension, Friedrichs Extension and Generalized Solution

Let $\hat{B} : D(\hat{B}) \subset H \to H$ be a linear, symmetric, positive-bounded-below operator on the real Hilbert space H. The **duality map** $\hat{B}_E : H_{\hat{B}} \to H_{\hat{B}}^*$ is defined by

$$\langle \hat{B}_E u, v \rangle_{\hat{B}} = (u, v)_{\hat{B}}, u, v \in H_{\hat{B}}. \tag{3.18}$$

For $u \in D(\hat{B})$, we have

$$\langle \hat{B}_E u, v \rangle_{\hat{B}} = (u, v)_{\hat{B}} = (\hat{B}u, v) = \langle \hat{B}u, v \rangle_{\hat{B}}, u, v \in H_{\hat{B}}.$$

This implies $\hat{B}_E u = \hat{B}u$ and $\hat{B}_E$ is the extension of $\hat{B}$. The duality map $\hat{B}_E$ is called the **energy extension** of $\hat{B}$. We can also introduce a new operator $\hat{B}_F : D(\hat{B}_F) \subset H \to H$ defined by

$$\hat{B}_F u = \hat{B}_E u$$

with $D(\hat{B}_F) = \{u \in H_{\hat{B}} \mid \hat{B}_E u \in H\}$. Then it is easy to show that $\hat{B} \subset \hat{B}_F \subset \hat{B}_E$. The new operator $\hat{B}_F$ is called the **Friedrichs extension** of $\hat{B}$, which has the following properties (Zeidler, 1995):

1. The operator $\hat{B}_F$ is self-adjoint and bijective, and $(\hat{B}_F u, u) \geq c_1 \|u\|^2$ for all $u \in D(\hat{B}_F)$.
2. The inverse operator $\hat{B}_F^{-1} : H \to H$ is linear, continuous and self-adjoint. If the embedding $H_{\hat{B}} \subset H$ is compact, the inverse operator $\hat{B}_F^{-1} : H \to H$ is compact.

Let us consider the operator equation

$$\hat{B}u = f, u \in D(\hat{B}), f \in H. \tag{3.19}$$

This is considered the original classical problem and its solution is the classical solution, which is confined in $D(\hat{B})$ so that $\hat{B}u$ exists. We now introduce three new equations

$$\hat{B}_F u = f, u \in D(\hat{B}_F), \tag{3.20}$$

$$(u, \hat{B}v) = (f, v), u \in H_{\hat{B}}, \text{ for all } v \in D(\hat{B}), \tag{3.21}$$

$$\frac{1}{2}(u, u)_{\hat{B}} - (f, u) = \min, u \in H_{\hat{B}}. \tag{3.22}$$

These are the generalized versions of the original classical problem (3.19), and their solutions are sought in a broader domain of definition than $D(\hat{B})$. The three generalized versions are equivalent and they have a unique solution $u_0 = \hat{B}_F^{-1} f$. Actually, by the properties of $\hat{B}_F$, Equation (3.20) has a unique solution $u_0 = \hat{B}_F^{-1} f$. For all $u \in H_{\hat{B}}$, we have

$$(f, u) = \langle f, u \rangle_{\hat{B}} = \langle \hat{B}_F u_0, u \rangle_{\hat{B}} = \langle \hat{B}_E u_0, u \rangle_{\hat{B}} = (u_0, u)_{\hat{B}},$$

which allows us to write

$$\frac{1}{2}(u - u_0, u - u_0)_{\hat{B}} - \frac{1}{2}(u_0, u_0)_{\hat{B}} = \frac{1}{2}(u, u)_{\hat{B}} - (f, u).$$

Equation (3.22) is thus equivalent to

$$\frac{1}{2}(u - u_0, u - u_0)_{\hat{B}} - \frac{1}{2}(u_0, u_0)_{\hat{B}} = \min, u \in H_{\hat{B}},$$

which has a unique solution $u = u_0$. Since H is a real Hilbert space, Equation (3.21) is equivalent to

$$(\hat{B}_F v, u) = (f, v), u \in H_{\hat{B}}$$

for all $v \in D(\hat{B})$. Considering

$$(\hat{B}_F v, u) = \langle \hat{B}_F v, u \rangle_{\hat{B}} = \langle \hat{B}_E v, u \rangle_{\hat{B}} = (v, u)_{\hat{B}}, u \in H_{\hat{B}}, v \in D(\hat{B})$$

and $(v, u)_{\hat{B}} = (u, v)_{\hat{B}}$, Equation (3.21) is equivalent to

$$\langle \hat{B}_E u, v \rangle_{\hat{B}} = (f, v) = \langle f, v \rangle_{\hat{B}}, u \in H_{\hat{B}}$$

for all $v \in D(\hat{B})$. Since $D(\hat{B})$ is dense in $H_{\hat{B}}$, the above equation is equivalent to

$$\hat{B}_E u = f, u \in H_{\hat{B}}.$$

This is equivalent to (3.20) since $f \in H$. Thus we have proved the equivalence of (3.20), (3.21) and (3.22). The solution $u_0 = \hat{B}_F^{-1} f$ is called the **generalized solution** of the original classical problem (3.19).

Example 3.9: Consider the classical problem

$$\hat{B}u = u'' = -\frac{d^2 u}{dx^2} = f, u \in D(\hat{B}) = C_0^\infty(0, 1), f \in H = L^2(0, 1). \tag{3.23}$$

We now show that $H_{\hat{B}} = H_0^1(0, 1)$ with the inner product

$$(u, v)_{\hat{B}} = (\hat{B}u, u) = \int_0^1 u'v'dx, u, v \in H_{\hat{B}}. \tag{3.24}$$

Here $u' = du/dx$ denotes the generalized derivative. First we demonstrate that $\hat{B}$ is positive-bounded-below. The inner product in H is defined by $(u, v) = \int_0^1 u(x)v(x)dx$. Using integration by parts and the boundary conditions, we have

$$(u, v)_{\hat{B}} = (\hat{B}u, v) = -\int_0^1 u''vdx = \int_0^1 u'v'dx, u, v \in D(\hat{B}). \tag{3.25}$$

Moreover, $(\hat{B}u, u) = 0$ implies $\int_0^1 u'^2dx = 0$, which leads to $u' = 0$. Hence $u(x)$ is a constant, which must be zero according to the boundary conditions. Consequently, the operator $\hat{B}$ is positive definite. Since $u(0) = 0$, one may write $u(x) = \int_0^x u'(t)dt$. Applying Cauchy's inequality, we obtain

$$u^2(x) \leq \int_0^x 1^2dt \int_0^x [u'(t)]^2dt = x \int_0^x [u'(t)]^2dt \leq x \int_0^1 [u']^2dt.$$

Taking the integration over the interval $[0, 1]$ yields

$$\|u\|^2 = \int_0^1 u^2(x)dx \leq \frac{1}{2} \int_0^1 u'^2dt = \frac{1}{2}(\hat{B}u, u).$$

This shows that the operator $\hat{B}$ is positive-bounded-below. Next we prove $H_{\hat{B}} = H_0^1(0, 1)$. For $u \in H_{\hat{B}}$, there exists an admissible sequence $\{u_n\}$ such that

$$\|u_n - u\| \xrightarrow[n \to \infty]{} 0, \tag{3.26}$$

and the sequence $\{u_n\} \subset D(\hat{B}) = C_0^\infty(0, 1)$ is a Cauchy sequence with respect to the energy norm $\|\cdot\|_{\hat{B}}$. So we have

$$\|u_n - u_m\|_{\hat{B}}^2 = \int_0^1 (u_n' - u_m')^2dx \to 0,$$

and hence $\{u'_n\}$ is a Cauchy sequence in H. As a result, there exists a function $v \in H$ such that

$$\|u'_n - v\| \xrightarrow[n \to \infty]{} 0. \tag{3.27}$$

Making use of $\int_0^1 u_n \varphi' dx = -\int_0^1 u'_n \varphi dx$ for all $\varphi \in C_0^\infty(0, 1)$, and letting $n \to \infty$, we obtain

$$\int_0^1 u \varphi' dx = -\int_0^1 v \varphi dx.$$

Thus v is the generalized derivative of u and $u \in H^1(0, 1)$. By (3.26) and (3.27), we may conclude that $u \in H_0^1(0, 1)$. On the other hand, if we assume that $u \in H_0^1(0, 1)$, there exists a sequence $\{u_n\} \subset D(\hat{B})$ such that (3.26) and (3.27) hold. Consequently, $\{u_n\}$ is an admissible sequence for u and $u \in H_{\hat{B}}$. For any $u, v \in H_{\hat{B}}$, the inner product $(u, v)_{\hat{B}}$ is defined by

$$(u, v)_{\hat{B}} = \lim_{n \to \infty} (u_n, v_n)_{\hat{B}} = \lim_{n \to \infty} \int_0^1 u'_n v'_n dx = \int_0^1 u' v' dx,$$

where $\{u_n\}$ and $\{v_n\}$ are the admissible sequences for u and v respectively, and the derivatives in the last integral are understood in the generalized sense.

The weak formulation of the classical problem (3.23) is given by (3.21)

$$\int_0^1 u(-v'')dx = \int_0^1 f v dx, u \in H_{\hat{B}} \tag{3.28}$$

for all $v \in D(\hat{B})$. By definition, the energy extension $\hat{B}_E$ of $\hat{B}$ is defined as

$$\langle \hat{B}_E u, v \rangle_{\hat{B}} = (u, v)_{\hat{B}}, u, v \in H_{\hat{B}}.$$

This implies

$$\langle \hat{B}_E u, v \rangle_{\hat{B}} = \int_0^1 u' v' dx = \int_0^1 -u'' v dx, u, v \in H_{\hat{B}},$$

where all the derivatives are understood in the generalized sense. From the above equation, one may find that $\hat{B}_E u = -u'', u \in H_{\hat{B}}$. The Friedrichs extension $\hat{B}_F$ is

$$\hat{B}_F u = \hat{B}_E u, u \in D(\hat{B}_F),$$

where $D(\hat{B}_F)$ consists of all $u \in H_{\hat{B}}$ such that $\hat{B}_E u \in H = L^2(0, 1)$. Explicitly

$$D(\hat{B}_F) = \left\{ u \in H_0^1(0, 1) \middle| u'' \in L^2(0, 1) \right\},$$

where the double prime denotes the generalized derivative. Note that (3.20) is

$$-u'' = f, u \in D(\hat{B}_F), f \in L^2(0, 1). \qquad \square$$

Remark 3.5: In general, if $\hat{B}$ is a $2m$ th order positive-bounded-below differential operator defined over a finite region Ω, we take $H = L^2(\Omega)$. The definition of domain $D(\hat{B})$ consists of the functions u for which $\hat{B}u$ exists, with u and its derivatives satisfying the homogeneous boundary conditions. The energy space $H_{\hat{B}}$ contains the functions that have the generalized derivatives up to m (including m). These generalized derivatives are square-integrable and satisfy the homogeneous boundary conditions. $\qquad \square$

3.2 Eigenvalue Problems for Symmetric Operators

The concepts of eigenvalues and eigenfunctions play a major role in mathematics and physics, and many mathematical equations lead to eigenvalue problems after separation of variables.

3.2.1 Positive-Bounded-Below Symmetric Operators

Consider the eigenvalue problem defined by

$$\hat{B}v - \lambda v = 0, v \in D(\hat{B}) \subset H, v \neq 0, \qquad (3.29)$$

where H is a real Hilbert space, and $D(\hat{B})$ is dense in H. The operators $\hat{B}$ is assumed to be symmetric and positive-bounded-below, that is, there is a positive constant c_1 such that

$$\|v\|_{\hat{B}}^2 = (v, v)_{\hat{B}} \geq c_1 \|v\|^2 \qquad (3.30)$$

for all $v \in D(\hat{B})$. This assumption ensures that $(\hat{B}v, v)$ indeed gives an inner product, so that the Cauchy-Schwarz's inequality holds

$$\left| (v_1, v_2)_{\hat{B}} \right|^2 \leq (v_1, v_1)_{\hat{B}} (v_2, v_2)_{\hat{B}}. \qquad (3.31)$$

For the classical problem (3.29), we may introduce the generalized eigenvalue problem

$$\hat{B}_F u - \lambda u = 0, u \in D(\hat{B}_F), u \neq 0. \qquad (3.32)$$

The variational method may be used to solve the eigenvalue problems (Mikhlin, 1964; Weinberger, 1974). We have the following properties:

1. The eigenvalues are real and all the corresponding eigenfunctions can be assumed to be real. Furthermore all the eigenvalues are positive.
2. The eigenfunctions of different eigenvalues are orthogonal: If λ_1 and λ_2 are distinct eigenvalues and v_1 and v_2 are the corresponding eigenfunctions, then $(v_1, v_2) = 0$.
3. Let λ_1 be the lower bound of the functional $(\hat{B}v, v)/(v, v)$. If there exists a function v_0 such that $\lambda_1 = (\hat{B}v_0, v_0)/(v_0, v_0)$, then $\lambda_1 = (\hat{B}v_0, v_0)/(v_0, v_0)$ is the lowest eigenvalue of (3.29) and v_0 is the corresponding eigenfunction.
4. Let $\lambda_1, \lambda_2, \cdots, \lambda_n$ be (in the order of increasing value) the first n eigenvalues and let $v_1, v_2, \cdots, v_n$ be the corresponding eigenfunctions. If there exists a function v_{n+1}, which minimizes the functional $(\hat{B}v, v)/(v, v)$ under the supplementary conditions $(v, v_1) = (v, v_2) = \cdots = (v, v_n) = 0$, then v_{n+1} is the eigenfunction corresponding to the eigenvalue $\lambda_{n+1} = (\hat{B}v_{n+1}, v_{n+1})/(v_{n+1}, v_{n+1})$. This eigenvalue is the next following after λ_n. This procedure reduces the eigenvalue problem to the variational problem of finding the minimum of the functional $(\hat{B}v, v)/(v, v)$.
5. Let H be a real separable Hilbert space with dim $H = \infty$. If the embedding $H_{\hat{B}} \subset H$ is compact, then the generalized eigenvalue problem (3.32) has an infinite set of eigenvalues $c_1 \leq \lambda_1 \leq \lambda_2 \leq \cdots \leq \lambda_n \cdots$, and $\lambda_n \to \infty$ as $n \to \infty$. In addition, the corresponding eigenfunctions $\{v_n\}$ constitute a complete system in H, and $v_n \in H_{\hat{B}}$ for all n.

The first four properties are straightforward, and we only prove the last one. Let

$$\varphi(v) = \frac{(\hat{B}v, v)}{(v, v)} \geq c_1, v \in D(\hat{B}) \tag{3.33}$$

and $\lambda_1 > 0$ be the exact lower bound of the functional $\varphi(v)$

$$\lambda_1 = \inf_{v \in D(\hat{B})} \varphi(v).$$

By the definition of exact lower bound, for any positive integer n, it is possible to find a function $u_n \in D(\hat{B})$ such that $c_1 \leq \lambda_1 \leq \varphi(u_n) \leq \lambda_1 + 1/n$. Obviously we have $\lim_{n \to \infty} \varphi(u_n) = \lambda_1$. The functional $\varphi(u_n)$ does not change if u_n is multiplied by a constant. Thus we may choose $\|u_n\| = 1$, then $\varphi(u_n) = (\hat{B}u_n, u_n)$ and

$$(\hat{B}u_n, u_n) \to \lambda_1. \tag{3.34}$$

We now prove that there is a function $v_1 \in H_{\hat{B}}$ such that $(\hat{B}_F v_1, v_1) = \lambda_1$. Let η be an arbitrary function from $D(\hat{B})$ and t an arbitrary real number. Then

$$\frac{(\hat{B}(u_n + t\eta), u_n + t\eta)}{\|u_n + t\eta\|^2} \geq \lambda_1.$$

Since $\|u_n\| = 1$, we have

$$t^2 \left[(\hat{B}\eta, \eta) - \lambda_1 \|\eta\|^2 \right] + 2t \left[(\hat{B}u_n, \eta) - \lambda_1(u_n, \eta) \right] + (\hat{B}u_n, u_n) - \lambda_1 \geq 0.$$

The discriminant of the above expression must be negative as it is a quadratic expression and does not change sign. Hence

$$\left[(\hat{B}u_n, \eta) - \lambda_1(u_n, \eta)\right]^2 \leq \left[(\hat{B}\eta, \eta) - \lambda_1 \|\eta\|^2\right]\left[(\hat{B}u_n, u_n) - \lambda_1\right]. \tag{3.35}$$

From (3.34) and (3.35) it follows that

$$(\hat{B}u_n, \eta) - \lambda_1(u_n, \eta) \xrightarrow[n\to\infty]{} 0 \tag{3.36}$$

for all $\eta \in D(\hat{B})$. Note that the sequence $\{u_n\}$ is contained in $D(\hat{B}) \subset H_{\hat{B}}$. By assumption that the embedding $H_{\hat{B}} \subset H$ is compact, we may therefrom select a subsequence, still denoted $\{u_n\}$, such that $u_n \xrightarrow[n\to\infty]{} v_1 \in H$. We now show that $v_1 \in H_{\hat{B}}$. In fact, let $\eta = u_n - u_m$ in (3.36), we obtain

$$(\hat{B}u_n, u_n - u_m) - \lambda_1(u_n, u_n - u_m) \xrightarrow[n\to\infty]{} 0. \tag{3.37}$$

Interchanging m and n gives

$$(\hat{B}u_m, u_m - u_n) - \lambda_1(u_m, u_m - u_n) \xrightarrow[m\to\infty]{} 0. \tag{3.38}$$

Adding (3.37) and (3.38) yields

$$(\hat{B}(u_n - u_m), u_n - u_m) - \lambda_1(u_n - u_m, u_n - u_m) \xrightarrow[m,n\to\infty]{} 0.$$

Since $u_n \to v_1$ in H, the above equation implies

$$\|u_n - u_m\|_{\hat{B}} \xrightarrow[m,n\to\infty]{} 0.$$

Thus $\{u_n\}$ is an admissible sequence and $v_1 \in H_{\hat{B}}$. By symmetric property of $\hat{B}$, Equation (3.36) can be written as

$$(u_n, \hat{B}\eta) - \lambda_1(u_n, \eta) \xrightarrow[n\to\infty]{} 0 \tag{3.39}$$

for all $\eta \in D(\hat{B})$. Proceeding to the limit as $n \to \infty$ in (3.39) gives

$$(v_1, \hat{B}\eta) - \lambda_1(v_1, \eta) = 0, \ v_1 \in H_{\hat{B}}$$

for all $\eta \in D(\hat{B})$. This is equivalent to $\hat{B}_F v_1 - \lambda_1 v_1 = 0$, $v_1 \in D(\hat{B}_F)$, and $\lambda_1 = (\hat{B}_F v_1, v_1)$.

We denote by λ_2 the exact lower bound of functional $\varphi(v)$ under the supplementary condition $(v, v_1) = 0$. By repeating the foregoing arguments, we may find that λ_2 is the second eigenvalue of operator $\hat{B}_F$ and that it corresponds to a normalized eigenfunction v_2, orthogonal to v_1. Continuing this process, we obtain an increasing sequence of eigenvalues of $\lambda_1 \leq \lambda_2 \leq \cdots \leq \lambda_n \cdots$ and the corresponding orthonormal sequence of eigenfunctions $v_1, v_2, \cdots, v_n, \cdots$.

In order to prove that λ_n approaches infinity as n increases, we may use the method of contradiction. Suppose that the converse is true. Then there exists a constant c_1 such that $\lambda_n \le c_1$ for all n. Since $(\hat{B}v_n, v_n) = \lambda_n \le c_1$, the sequence $\{v_n\}$ of eigenfunctions is bounded. By the hypothesis of the theorem that the embedding $H_{\hat{B}} \subset H$ is compact, it is possible to choose from the sequence $\{v_n\}$ a subsequence $\{v_{nk}\}$, which converges as k increases: $\lim_{k,l \to \infty} \|v_{nk} - v_{nl}\|^2 = 0$. This contradicts the fact that $\|v_{nk} - v_{nl}\|^2 = 2$.

To prove the completeness of the system of eigenfunctions $\{v_n\}$, we first demonstrate the completeness of $\{v_n\}$ in $H_{\hat{B}}$. If the system of the eigenfunctions were incomplete in $H_{\hat{B}}$, there would exist nonzero functions in $H_{\hat{B}}$, which are orthogonal to all the eigenfunctions v_n. Denoting by $\tilde{\lambda}$ the exact lower bound of $\varphi(v)$ for these functions one would find that $\tilde{\lambda}$ is an eigenvalue of operator $\hat{B}_F$, greater than all λ_n and this is impossible since $|\lambda_n| \to \infty$. For an arbitrary $u \in H$, we may choose a $v \in H_{\hat{B}}$ such that $\|u - v\| < \varepsilon/2$ (since $H_{\hat{B}}$ is dense in H), where ε is an arbitrary small positive number. By the completeness of $\{v_n\}$ in $H_{\hat{B}}$, v can be approximated by a sum of the form $\sum_{i=1}^{N} a_i v_i$ such that $\left\| v - \sum_{i=1}^{N} a_i v_i \right\|_{\hat{B}} < \varepsilon c_1/2$, where c_1 is the constant in (3.30). Now we have

$$\left\| u - \sum_{i=1}^{N} a_i v_i \right\| < \|u - v\| + \left\| v - \sum_{i=1}^{N} a_i v_i \right\| < \varepsilon/2 + \frac{1}{c_1} \left\| v - \sum_{i=1}^{N} a_i v_i \right\|_{\hat{B}} < \varepsilon.$$

The proof is completed.

Therefore, any function $u \in H$ has the following expansion

$$u = \sum_{n} (u, v_n) v_n,$$

where the set of eigenfunctions $\{v_n\}$ is assumed to be orthonormal.

Remark 3.6: In practice, we only need to show that the operator $\hat{B}$ is positive definite. If $\hat{B}$ is positive definite, we may add ξ on both sides of (3.29), where ξ is a arbitrary positive number, to yield

$$(\hat{B} + \xi \hat{I})v = (\lambda + \xi)v.$$

Thus the new operator $\hat{B} + \xi \hat{I}$ is positive-bounded-below.   □

Example 3.10: The longitudinal component of a TM mode in a uniform metal waveguide is characterized by the Dirichlet problem

$$-\nabla^2 u(\mathbf{x}) = \lambda u(\mathbf{x}), \mathbf{x} \in \Omega,$$
$$u(\mathbf{x}) = 0, \mathbf{x} \in \Gamma. \tag{3.40}$$

By (3.16), the operator $\hat{B} = -\nabla^2$ is positive-bounded-below in $D(\hat{B}) = C_0^\infty(\Omega)$. It can be shown that $H_{\hat{B}} = H_0^1(\Omega)$ (Zeidler, 1995). Furthermore the embedding $H_0^1(\Omega) \subset H = L^2(\Omega)$

is compact. Thus the generalized eigenvalue problem $\hat{B}_F u = \lambda u, u \in D(\hat{B}_F)$ has a complete orthonormal set $\{u_n\}$ in $L^2(\Omega)$. The corresponding eigenvalues satisfy $\lambda_n \xrightarrow[n \to \infty]{} \infty$ with $0 < \lambda_1 \leq \lambda_2 \leq \cdots \leq \lambda_n \leq \cdots$. $\qquad \square$

Example 3.11: The longitudinal component of a TE mode in a uniform metal waveguide is characterized by the Neumann problem

$$-\nabla^2 u(\mathbf{x}) = \lambda u(\mathbf{x}), \mathbf{x} \in \Omega,$$
$$\partial u(\mathbf{x})/\partial n = 0, \mathbf{x} \in \Gamma. \qquad (3.41)$$

The domain of definition of the operator $\hat{B} = -\nabla^2$ is $D(\hat{B}) = \{u \in C^\infty(\bar{\Omega}) \mid \partial u/\partial n = 0\}$. Let $H = L^2(\Omega)$. For all $u, v \in D(\hat{B})$

$$(u, v)_{\hat{B}} = \int_\Omega -v\nabla^2 u d\mathbf{x} = \int_\Omega \sum_{j=1}^N \partial_j u \partial_j v d\mathbf{x} - \int_\Gamma \frac{\partial u}{\partial n} v d\Gamma$$

$$= \int_\Omega \sum_{j=1}^N \partial_j u \partial_j v d\mathbf{x}, u, v \in D(\hat{B}).$$

Therefore, the operator $\hat{B}$ is positive definite. By introducing a positive constant ξ, Equation (3.41) can be modified into the equivalent problem

$$-\nabla^2 u(\mathbf{x}) + \xi u(\mathbf{x}) = \hat{A}(u) = (\lambda + \xi)u(\mathbf{x}), \mathbf{x} \in \Omega,$$
$$\partial u(\mathbf{x})/\partial n = 0, \mathbf{x} \in \Gamma, \qquad (3.42)$$

where $\hat{A} = \hat{B} + \xi \hat{I}$. The new operator $\hat{A}$ is positive-bounded-below with $D(\hat{A}) = D(\hat{B})$. For all $u, v \in D(\hat{A})$

$$(u, v)_{\hat{A}} = \int_\Omega \sum_{j=1}^N \partial_j u \partial_j v d\mathbf{x} + \xi \int_\Omega uv d\mathbf{x}$$

$$= \int_\Omega \nabla u \cdot \nabla v d\mathbf{x} + \xi \int_\Omega uv d\mathbf{x}, u, v \in D(\hat{A}). \qquad (3.43)$$

The completion of $D(\hat{A})$ with respect to the norm $\|\cdot\|_{\hat{A}}$ is the energy space $H_{\hat{A}}$. For arbitrary $u, v \in H_{\hat{A}}$ there exist two admissible sequences $\{u_n\}$ and $\{v_n\}$ such that $\|u_n - u\| \to 0$ and $\|v_n - v\| \to 0$. We define

$$(u, v)_{\hat{A}} = \lim_{n \to \infty} (u_n, v_n)_{\hat{A}}. \qquad (3.44)$$

Let $u \in H_{\hat{A}}$, there exists an admissible sequence $\{u_n\} \subset D(\hat{A})$ such that

$$\|u_n - u\| \xrightarrow[n \to \infty]{} 0$$

and the sequence $\{u_n\} \subset D(\hat{A})$ is a Cauchy sequence with respect to energy norm $\|\cdot\|_{\hat{A}}$. From (3.43) we obtain

$$\|u_n - u_m\|_{\hat{A}}^2 = \sum_{j=1}^{N} \left\| \partial_j u_n - \partial_j u_m \right\|^2 + \xi \|u_n - u_m\|^2 \to 0.$$

Thus $\{\partial_j u_n\}$ is a Cauchy sequence in H. As a result, there exists a function $v_j \in H$ such that

$$\left\| \partial_j u_n - v_j \right\| \xrightarrow[n \to \infty]{} 0, \, v_j \in H, \, j = 1, 2, \cdots, N.$$

Making use of

$$\int_{\Omega} u_n \partial_j \varphi d\mathbf{x} = - \int_{\Omega} \partial_j u_n \varphi d\mathbf{x}, \, \varphi \in C_0^{\infty}(\Omega)$$

and letting $n \to \infty$ we obtain

$$\int_{\Omega} u \partial_j \varphi d\mathbf{x} = - \int_{\Omega} v_j \varphi d\mathbf{x}, \, \varphi \in C_0^{\infty}(\Omega).$$

Hence $v_j = \partial_j u$ holds in the generalized sense. As a result, $u \in H^1(\Omega)$ and $H_{\hat{A}}$ is a subspace of $H^1(\Omega)$. Since $\partial_j u_n \to \partial_j u$ as $n \to \infty$, we have $\partial u / \partial n = 0$ on Γ in the generalized sense. Therefore $H_{\hat{A}} \subset \{u \in H^1(\Omega) | \partial u / \partial n = 0 \text{ on } \Gamma\}$. Now we are able to write (3.44) as

$$(u, v)_{\hat{A}} = \int_{\Omega} \nabla u \cdot \nabla v d\mathbf{x} + \xi \int_{\Omega} u v d\mathbf{x}, \, u, v \in H_{\hat{A}}$$

in which the derivatives should be understood in the generalized sense.

To prove that the embedding $H_{\hat{A}}(\Omega) \subset L^2(\Omega)$ is compact, let $\hat{J}(u) = u, u \in H_{\hat{A}}(\Omega)$. The operator $\hat{J} : H_{\hat{A}}(\Omega) \to L^2(\Omega)$ is linear and continuous since

$$\left\| \hat{J}(u) \right\|^2 = \|u\|^2 \leq \xi^{-1}(\|\nabla u\|^2 + \xi \|u\|^2) = \xi^{-1} \|u\|_{\hat{A}}.$$

A bounded sequence $\{u_n\}$ in $H_{\hat{A}}$ implies $\|u_n\|_{\hat{A}} = \|\nabla u_n\|^2 + \xi \|u_n\|^2 \leq c_1$, where c_1 is a constant. The compactness of $\hat{J}$ can then be obtained by using the **Rellich theorem** (named after the Italian mathematician Franz Rellich, 1906–1955) stated below. Therefore the generalized eigenvalue problem $\hat{A}_F u = (\lambda + \xi)u, u \in D(\hat{A}_F)$ has a complete orthonormal set $\{u_n\}$ in $L^2(\Omega)$. The corresponding eigenvalues satisfy $0 \leq \lambda_1 \leq \lambda_2 \leq \cdots \leq \lambda_n \leq \cdots$, with $\lambda_n \xrightarrow[n \to \infty]{} \infty$. $\qquad \square$

Theorem 3.2 *Rellich's theorem: Any sequence* $\{f_n\}$ *that satisfies*

$$\|f_n\|^2 = \int_\Omega f_n^2 d\Omega \le c_1, \ \|\nabla f_n\|^2 = \int_\Omega (\nabla f_n)^2 d\Omega \le c_2,$$

where c_1 *and* c_2 *are constants, has a subsequence, still denoted by* $\{f_n\}$, *such that* $\lim_{m,n\to\infty} \int_\Omega (f_m - f_n)^2 d\Omega = 0$ *(Kurokawa, 1969).* $\qquad\qquad\square$

3.2.2 Compact Symmetric Operators

We now study the eigenvalue problem

$$\hat{A}v - \lambda v = 0, \tag{3.45}$$

where $\hat{A}$ is defined in the whole space H. The set of all the eigenfunctions corresponding to an eigenvalue λ is called the **eigenspace**. An eigenvalue λ is said to have **finite multiplicity** if the corresponding eigenspace has finite dimension. The following Hilbert-Schmidt theorem is a fundamental result concerning compact self-adjoint operators on Hilbert spaces.

Hilbert-Schmidt theorem: *Let* $\hat{A} : H \to H$ *be a linear compact symmetric operator on the Hilbert space* H *which is separable and* $H \ne \{0\}$. *Then*

1. *All the eigenvalues of* $\hat{A}$ *are real, and each eigenvalue* $\lambda \ne 0$ *has finite multiplicity.*
2. *Any two eigenfunctions of* $\hat{A}$ *that correspond to different eigenvalues are orthogonal.*
3. *The operator* $\hat{A}$ *has a complete orthonormal system of eigenfunctions.*
4. *If the operator* $\hat{A}$ *has a countable set of eigenvalues (for example* $\lambda = 0$ *is not an eigenvalue of* $\hat{A}$ *and* $\dim H = \infty$), *the eigenvalues of* $\hat{A}$ *form a sequence* $\{\lambda_n\}$ *such that* $\lambda_n \xrightarrow[n\to\infty]{} 0$.

   (This means the inverse of $\hat{A}$, *if it exists, is unbounded.)*

The first two properties are obvious and we only need to prove the last two. To this end, we first make the following assumption:

Assumption 1: $\hat{A}u = 0$ implies $u = 0$ (that is, $\lambda = 0$ is not an eigenvalue of $\hat{A}$) and $\dim H = \infty$.

Proof based on assumption 1. Since $H \ne \{0\}$, the above assumption implies $\hat{A} \ne 0$ and hence $\|\hat{A}\| \ne 0$. Let

$$\varphi(v) = \frac{(\hat{A}v, v)}{(v, v)}, \tag{3.46}$$

and consider the maximum problem max $|\varphi(v)|$. If l_1 is the exact upper bound of the functional $|\varphi(v)|$, that is

$$l_1 = \sup_{v\in H} |\varphi(v)|, \tag{3.47}$$

then, for any positive integer n, it is possible to find a function $u_n \in H$ such that $l_1 + 1/n \leq |\varphi(u_n)| \leq l_1$. Obviously $\lim_{n \to \infty} |\varphi(u_n)| = l_1$. Since the functional $\varphi(u_n)$ does not change if u_n is multiplied by a constant, we may choose $\|u_n\| = 1$. Then $\varphi(u_n) = (\hat{A}u_n, u_n)$ and

$$|\varphi(u_n)| = |(\hat{A}u_n, u_n)| \to l_1. \tag{3.48}$$

For the real bounded sequence $\{\varphi(u_n)\}$, there is a convergent subsequence of $\{\varphi(u_n)\}$, still denoted by $\{\varphi(u_n)\}$, and a real number λ_1 such that $\varphi(u_n) \xrightarrow[n \to \infty]{} \lambda_1$, with $|\lambda_1| = l_1$. We now prove that there exists a function v_1 such that $\varphi(v_1) = \lambda_1$. Let η be an arbitrary function from H and t an arbitrary real number. Then

$$|\varphi(u_n + t\eta)| = \frac{|(\hat{A}(u_n + t\eta), u_n + t\eta)|}{\|u_n + t\eta\|^2} \leq |\lambda_1|.$$

Using $\|u_n\| = 1$ we obtain

$$t^2 \left[(\hat{A}\eta, \eta)s - |\lambda_1|\|\eta\|^2 \right] + 2t \left[(\hat{A}u_n, \eta)s - |\lambda_1|(u_n, \eta) \right] + (\hat{A}u_n, u_n)s - |\lambda_1| \leq 0.$$

Here $s = \operatorname{sgn}(\hat{A}(u_n + t\eta), u_n + t\eta)$. The discriminant of the above expression must be negative as it is a quadratic expression and does not change sign. Hence we have

$$\begin{aligned}
\left| (\hat{A}u_n, \eta)s - |\lambda_1|(u_n, \eta) \right|^2 &\leq \left[(\hat{A}\eta, \eta)s - |\lambda_1|\|\eta\|^2 \right] \left[(\hat{A}u_n, u_n)s - |\lambda_1| \right] \\
&\leq \left[|(\hat{A}\eta, \eta)| + |\lambda_1|\|\eta\|^2 \right] \left[(\hat{A}u_n, u_n)s - |\lambda_1| \right].
\end{aligned} \tag{3.49}$$

If $t = 0$ is assumed, we have $s = \operatorname{sgn}(\hat{A}u_n, u_n)$ and $\lambda_1 = |\lambda_1|s$. From (3.48) and (3.49), it follows that

$$(\hat{A}u_n, \eta) - \lambda_1(u_n, \eta) \xrightarrow[n \to \infty]{} 0, \eta \in H. \tag{3.50}$$

Since $\hat{A}$ is compact, there exists a subsequence, still denoted $\{u_n\}$ such that $\{\hat{A}u_n\}$ converges. The above relation implies that there exists $v_1 \in H$, such that

$$u_n \xrightarrow[n \to \infty]{} v_1.$$

By symmetric property of $\hat{A}$, Equation (3.50) may be written as

$$(u_n, \hat{A}\eta) - \lambda_1(u_n, \eta) \xrightarrow[n \to \infty]{} 0, \eta \in H. \tag{3.51}$$

Proceeding to the limit as $n \to \infty$ in (3.51) and making use of the symmetric property again, we obtain

$$(\hat{A}v_1 - \lambda_1 v_1, \eta) = 0, \eta \in H,$$

which implies $\hat{A}v_1 - \lambda_1 v_1 = 0$, and v_1 is an eigenfunction. Now let $H_1 = \{u \in H \, |(u, v_1) = 0\}$ and $u \in H_1$. Then

$$(\hat{A}u, v_1) = (u, \hat{A}v_1) = \lambda_1(u, v_1) = 0.$$

Thus $\hat{A}u \bot v_1$. Since dim $H = \infty$, we have $H_1 \neq \{0\}$. Furthermore $\hat{A} \neq 0$ on H_1. Otherwise there would exist a non-zero element $u \in H_1$ such that $\hat{A}u = 0$, which contradicts our assumption 1. Therefore we may apply the previous procedure to the restricted operator $\hat{A} : H_1 \to H_1$ to find the second eigenfunction v_2:

$$\hat{A}v_2 - \lambda_2 v_2 = 0, v_2 \in H_1, \|v_1\| = 1.$$

Note that

$$|\lambda_2| = \sup_{v \in H_1} \frac{|(\hat{A}v, v)|}{(v, v)}.$$

Comparing this to (3.47) gives $|\lambda_1| \geq |\lambda_2| > 0$. If the above procedure is continued, we obtain a countable system of eigenfunctions

$$\hat{A}v_n - \lambda_n v_n = 0, n = 1, 2, \cdots, |\lambda_1| \geq |\lambda_2| \geq \cdots > 0,$$

where $\{v_n\}$ is orthonormal system. We now prove that $|\lambda_n| \xrightarrow[n \to \infty]{} 0$. Suppose that the converse is true. Then there exists a constant c_1 such that $|\lambda_n| \geq c_1 > 0$ for all n, where c_1 is a constant. As a result, the sequence $\{v_n/\lambda_n\}$ is bounded. Since $\hat{A}$ is compact and $\hat{A}(v_n/\lambda_n) = v_n$, $n = 1, 2, \cdots$, the sequence $\{v_n\}$ contains a convergent subsequence, still denoted $\{v_n\}$. But this is impossible since $(v_n, v_m) = 0$ for $n \neq m$. In fact, $\|v_n - v_m\|^2 = 2$. Hence $\{v_n\}$ is not a Cauchy sequence, contradicting the fact that $\{v_n\}$ is convergent.

We now prove the completeness of the system of eigenfunctions. If the system of the eigenfunctions were incomplete, there would exist nonzero functions, which are orthogonal to all the eigenfunctions v_n. Denoting by $|\tilde{\lambda}|$ the exact upper bound of $|\varphi(v)|$ for these functions, which must be greater than zero according to the assumption that $\lambda = 0$ is not an eigenvalue. Then $\tilde{\lambda}$ would be an eigenvalue of operator $\hat{A}$, smaller than all λ_n in absolute value and this is impossible since $|\lambda_n| \xrightarrow[n \to \infty]{} 0$.

In order to show that each eigenvalue of $\hat{A}$ has finite multiplicity, we may consider the eigenvalue λ_1 without loss of generality. Since $|\lambda_n| \neq 0$ and $|\lambda_n| \xrightarrow[n \to \infty]{} 0$, there is a number N such that $\lambda_1 = \cdots = \lambda_N$ and $\lambda_j \neq \lambda_1$ for all $j > N$. Assuming that u is an eigenfunction corresponding to λ_1, then u may be expanded as

$$u = \sum_{n=1}^{\infty} (u, v_n)v_n = \sum_{n=1}^{N} (u, v_n)v_n$$

for the eigenfunctions with different eigenvalues are orthogonal. Therefore $\{v_1, v_2, \cdots, v_N\}$ forms a basis of eigenspace of λ_1.

We now proceed to the proof based on the following assumption:

Assumption 2: $\hat{A}u = 0$ implies $u = 0$ and $\dim H = N$, where $N = 1, 2, \cdots$.

Remark 3.7: If M is a closed linear subspace of a Hilbert space H, then there exists a unique orthogonal decomposition for each $u \in H$

$$u = w + v, \ w \in M, \ v \in M^{\perp},$$

where $M^{\perp}$ is the orthogonal complement to M defined by

$$M^{\perp} = \{v \in H \,|\, (v, w) = 0, \text{ for all } w \in M\},$$

which is a closed linear subspace of H.   □

Proof based on assumption 2. Following the same procedure, there exists a system of eigenfunctions $\{v_n \,|\, n = 1, 2, \cdots, M\}$ such that

$$\hat{A}v_n - \lambda_n v_n = 0, n = 1, 2, \cdots, M, |\lambda_1| \geq |\lambda_2| \geq \cdots > |\lambda_M|.$$

Obviously we have $M = N$ by construction. Hence the system $\{v_n \,|\, n = 1, 2, \cdots, N\}$ is complete.

Proof of general situation: We only need to consider the case when $\lambda = 0$ is an eigenvalue of $\hat{A}$. Let $N(\hat{A}) = \{u \in H \,|\, \hat{A}u = 0\}$ denote the **kernel** (or null space) of $\hat{A}$. Then $N(\hat{A}) \neq \{0\}$. Since $\hat{A}$ is continuous, $N(\hat{A})$ is a closed linear subspace of H. There exists a complete orthonormal system in a separable Hilbert space H with $H \neq 0$. Therefore, there exists a complete orthonormal system $\{w_m\}$ in $N(\hat{A})$, which satisfies $\hat{A}w_m = 0, m = 1, 2, \cdots$.

For each $u \in H$ there exists a unique orthogonal decomposition

$$u = w + v, \ w \in N(\hat{A}), \ v \in N(\hat{A})^{\perp}.$$

If $v \in N(\hat{A})^{\perp}$ then $(\hat{A}v, w) = (v, \hat{A}w) = 0$ for all $w \in N(\hat{A})$. Hence $\hat{A}v \in N(\hat{A})^{\perp}$ and the operator $\hat{A}$ maps $N(\hat{A})^{\perp}$ into $N(\hat{A})^{\perp}$.

If $\hat{A}v = 0$ with $v \in N(\hat{A})^{\perp}$, then $v = 0$. Actually if $\hat{A}v = 0$ with $v \in N(\hat{A})^{\perp}$, then $v \in N(\hat{A}) \cap N(\hat{A})^{\perp}$. By the uniqueness of orthogonal decomposition, we have $v = 0$. If the Hilbert-Schmidt theorem with assumption 1 is applied to the restricted operator $\hat{A} : N(\hat{A})^{\perp} \rightarrow N(\hat{A})^{\perp}$, we may obtain a complete orthonormal system $\{v_n\}$ on $N(\hat{A})^{\perp}$. Hence for each $u \in H$ we have

$$u = w + v = \sum_{m=1}^{\infty} (w, w_m)w_m + \sum_{m=1}^{\infty} (v, v_n)v_n$$

$$= \sum_{m=1}^{\infty} (v + w, w_m)w_m + \sum_{m=1}^{\infty} (v + w, v_n)v_n$$

$$= \sum_{m=1}^{\infty} (u, w_m)w_m + \sum_{m=1}^{\infty} (u, v_n)v_n,$$

where we have used $(v, w_m) = 0$ for all m and

$$(w, v_n) = \frac{(w, \hat{A}v_n)}{\lambda_n} = \frac{(\hat{A}w, v_n)}{\lambda_n} = 0$$

for all n. The proof is completed.

3.3 Interior Electromagnetic Problems

A metal waveguide is a conducting tube used to guide electromagnetic waves. A metal cavity resonator is a device consisting of a hollow space bounded by a conducting surface, and can be used to generate or select electromagnetic waves of specific frequencies. Both of them are typical examples of interior eigenvalue problems in electromagnetic theory.

3.3.1 Mode Theory for Waveguides

Let us consider the wave propagation in an arbitrary metal waveguide, which is uniform along z-axis. The medium filled in the waveguide is assumed to be isotropic and homogeneous with medium parameters μ, ε and σ. The cross-section of the waveguide is denoted by Ω and its boundary by Γ, as shown in Figure 3.1. The electric field in a source-free region in the waveguide satisfies the wave equation

$$\nabla \times \nabla \times \mathbf{E}(\mathbf{r}) - k^2 \mathbf{E}(\mathbf{r}) = 0, \mathbf{r} \in \Omega$$

$$\mathbf{u}_n(\mathbf{r}) \times \mathbf{E}(\mathbf{r}) = 0, \mathbf{r} \in \Gamma$$

(3.52)

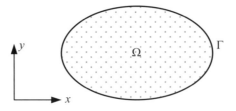

Figure 3.1   An arbitrary metal waveguide

where $k = \omega\sqrt{\mu\varepsilon_e}$, $\varepsilon_e = \varepsilon(1 - j\sigma/\omega\varepsilon)$, and $\mathbf{u}_n$ is the unit outward normal to the boundary Γ. The electric field can be decomposed into a transverse and a longitudinal component, both of which are assumed to be the separable functions of the transverse coordinates $\rho = (x, y)$ and the longitudinal component z

$$\mathbf{E}(\mathbf{r}) = [\mathbf{e}(\rho) + \mathbf{u}_z e_z(\rho)]u(z). \tag{3.53}$$

Substituting (3.53) into (3.52) and considering the boundary conditions yields

$$\nabla \times \nabla \times \mathbf{e} - \nabla(\nabla \cdot \mathbf{e}) = \hat{B}(\mathbf{e}) = k_c^2\mathbf{e}, \rho \in \Omega,$$

$$\mathbf{u}_n \times \mathbf{e} = 0, \nabla \cdot \mathbf{e} = 0, \rho \in \Gamma, \tag{3.54}$$

where k_c^2 is the separation constant, and $\hat{B} = \nabla \times \nabla \times (\cdot) - \nabla\nabla(\cdot)$ The function $u(z)$ satisfies

$$\frac{d^2u(z)}{dz^2} + (k^2 - k_c^2)u(z) = 0.$$

The domain of definition of operator $\hat{B}$ is

$$D(\hat{B}) = \left\{\mathbf{e} \,\middle|\, \mathbf{e} \in [C^\infty(\bar{\Omega})]^2, \mathbf{u}_n \times \mathbf{e} = 0, \nabla \cdot \mathbf{e} = 0 \text{ on } \Gamma\right\}.$$

Let $H = [L^2(\Omega)]^2 = L^2(\Omega) \times L^2(\Omega)$. For two transverse vector fields $\mathbf{e}_1$ and $\mathbf{e}_2$ in H, the inner product is defined by $(\mathbf{e}_1, \mathbf{e}_2) = \int_\Omega \mathbf{e}_1 \cdot \mathbf{e}_2 d\Omega$ and the corresponding norm is denoted by $\|\cdot\|$. For all $\mathbf{e}_1, \mathbf{e}_2 \in D(\hat{B})$, we have

$$(\hat{B}(\mathbf{e}_1), \mathbf{e}_2) = \int_\Omega [\nabla \times \nabla \times \mathbf{e}_1 - \nabla(\nabla \cdot \mathbf{e}_1)] \cdot \mathbf{e}_2 d\Omega$$

$$= \int_\Omega [\nabla \times \mathbf{e}_1 \cdot \nabla \times \mathbf{e}_2 + (\nabla \cdot \mathbf{e}_1)(\nabla \cdot \mathbf{e}_2)] d\Omega.$$

So $\hat{B}$ is symmetric and positive definite. To apply the previous eigenvalue theory, we may modify (3.54) into

$$\nabla \times \nabla \times \mathbf{e} - \nabla(\nabla \cdot \mathbf{e}) + \xi\mathbf{e} = \hat{A}(\mathbf{e}) = (k_c^2 + \xi)\mathbf{e}, \rho \in \Omega,$$

$$\mathbf{u}_n \times \mathbf{e} = 0, \nabla \cdot \mathbf{e} = 0, \rho \in \Gamma, \tag{3.55}$$

where ξ is an arbitrary positive constant and $\hat{A} = \hat{B} + \xi\hat{I}$. For all $\mathbf{e}_1, \mathbf{e}_2 \in D(\hat{A}) = D(\hat{B})$, we have

$$(\mathbf{e}_1, \mathbf{e}_2)_{\hat{A}} = (\hat{A}(\mathbf{e}_1), \mathbf{e}_2) = \int_\Omega [\nabla \times \nabla \times \mathbf{e}_1 - \nabla(\nabla \cdot \mathbf{e}_1) + \xi\mathbf{e}_1] \cdot \mathbf{e}_2 d\Omega$$

$$= \int_\Omega [\nabla \times \mathbf{e}_1 \cdot \nabla \times \mathbf{e}_2 + (\nabla \cdot \mathbf{e}_1)(\nabla \cdot \mathbf{e}_2) + \xi\mathbf{e}_1 \cdot \mathbf{e}_2] d\Omega. \tag{3.56}$$

Hence the new operator $\hat{A}$ is a symmetric, positive-bounded-below operator. Thus we can assume that $\mathbf{e}$ is real. The completion of $D(\hat{A})$ with respect to the norm $\|\cdot\|_{\hat{A}} = (\cdot, \cdot)_{\hat{A}}^{1/2}$ is denoted by $H_{\hat{A}}$. Assuming $\mathbf{e} \in H_{\hat{A}}$, by definition, there exists admissible sequence $\{\mathbf{e}_n \in D(\hat{A})\}$ for $\mathbf{e}$ such that

$$\|\mathbf{e}_n - \mathbf{e}\| \xrightarrow[n\to\infty]{} 0 \tag{3.57}$$

and $\{\mathbf{e}_n\}$ is a Cauchy sequence in $H_{\hat{A}}$. So we have

$$\|\mathbf{e}_n - \mathbf{e}_m\|_{\hat{A}}^2 = \|\nabla \times \mathbf{e}_n - \nabla \times \mathbf{e}_m\|^2 + \|\nabla \cdot \mathbf{e}_n - \nabla \cdot \mathbf{e}_m\|^2 + \xi \|\mathbf{e}_n - \mathbf{e}_m\|^2 \xrightarrow[n\to\infty]{} 0.$$

Consequently, $\{\nabla \times \mathbf{e}_n\}$ and $\{\nabla \cdot \mathbf{e}_n\}$ are Cauchy sequences in H and $L^2(\Omega)$ respectively. As a result, there exist $\mathbf{h} \in H$, and $\rho \in L^2(\Omega)$ such that

$$\|\nabla \times \mathbf{e}_n - \mathbf{h}\| \xrightarrow[n\to\infty]{} 0, \ \|\nabla \cdot \mathbf{e}_n - \rho\| \xrightarrow[n\to\infty]{} 0. \tag{3.58}$$

From (3.58) and

$$\int_\Omega \nabla \times \mathbf{e}_n \cdot \boldsymbol{\varphi} d\Omega = \int_\Omega \mathbf{e}_n \cdot \nabla \times \boldsymbol{\varphi} d\Omega, \ \boldsymbol{\varphi} \in [C_0^\infty(\Omega)]^2,$$

$$\int_\Omega (\nabla \cdot \mathbf{e}_n)\varphi d\Omega = -\int_\Omega \mathbf{e}_n \cdot \nabla\varphi d\Omega, \ \varphi \in C_0^\infty(\Omega),$$

we obtain

$$\int_\Omega \mathbf{h} \cdot \boldsymbol{\varphi} d\Omega = \int_\Omega \mathbf{e} \cdot \nabla \times \boldsymbol{\varphi} d\Omega, \ \boldsymbol{\varphi} \in [C_0^\infty(\Omega)]^2,$$

$$\int_\Omega \rho\varphi d\Omega = -\int_\Omega \mathbf{e} \cdot \nabla\varphi d\Omega, \ \varphi \in C_0^\infty(\Omega).$$

Therefore, $\nabla \times \mathbf{e} = \mathbf{h}$ and $\nabla \cdot \mathbf{e} = \rho$ hold in the generalized sense. For two smooth functions $\mathbf{e}$ and $\boldsymbol{\psi}$ defined in $\bar{\Omega}$, we have

$$\int_\Gamma (\mathbf{u}_n \times \mathbf{e}) \cdot \boldsymbol{\psi} d\Gamma = \int_\Omega (\boldsymbol{\psi} \cdot \nabla \times \mathbf{e} - \mathbf{e} \cdot \nabla \times \boldsymbol{\psi}) d\Omega,$$

$$\int_\Gamma (\nabla \cdot \mathbf{e})(\mathbf{u}_n \cdot \boldsymbol{\psi}) d\Gamma = \int_\Omega (\nabla \cdot \boldsymbol{\psi} \nabla \cdot \mathbf{e} + \boldsymbol{\psi} \cdot \nabla\nabla \cdot \mathbf{e}) d\Omega.$$

If we take $\mathbf{e}$ as the limit of an admissible sequence $\{\mathbf{e}_n \in D(\hat{A})\}$, we have

$$\int_\Gamma (\mathbf{u}_n \times \mathbf{e}) \cdot \boldsymbol{\psi} d\Gamma = 0, \ \int_\Gamma (\nabla \cdot \mathbf{e})(\mathbf{u}_n \cdot \boldsymbol{\psi}) d\Gamma = 0.$$

Hence $\mathbf{u}_n \times \mathbf{e} = 0, \nabla \cdot \mathbf{e} = 0$ hold in the generalized sense. Therefore

$$H_{\hat{A}} \subset \left\{ \mathbf{e} \in H | \nabla \times \mathbf{e} \in H, \nabla \cdot \mathbf{e} \in L^2(\Omega); \mathbf{u}_n \times \mathbf{e} = 0, \nabla \cdot \mathbf{e} = 0 \text{ on } \Gamma \right\}.$$

For arbitrary $\mathbf{e}_1, \mathbf{e}_2 \in H_{\hat{A}}$, there are two admissible functions $\{\mathbf{e}_{1n}\}$ and $\{\mathbf{e}_{2n}\}$ such that $\|\mathbf{e}_{1n} - \mathbf{e}_1\| \xrightarrow[n \to \infty]{} 0$ and $\|\mathbf{e}_{2n} - \mathbf{e}_2\| \xrightarrow[n \to \infty]{} 0$. Then

$$(\mathbf{e}_1, \mathbf{e}_2)_{\hat{A}} = \lim_{n \to \infty} (\mathbf{e}_{1n}, \mathbf{e}_{2n})_{\hat{A}} = \int_\Omega [\nabla \times \mathbf{e}_1 \cdot \nabla \times \mathbf{e}_2 + (\nabla \cdot \mathbf{e}_1)(\nabla \cdot \mathbf{e}_2) + \xi \mathbf{e}_1 \cdot \mathbf{e}_2] d\Omega,$$

where the derivatives are understood in the generalized sense.

We now show that the embedding $H_{\hat{A}} \subset H$ is compact. Let $\hat{J}(\mathbf{e}) = \mathbf{e}, \mathbf{e} \in H_{\hat{A}}$. Then the linear operator $\hat{J} : H_{\hat{A}} \to H$ is continuous since

$$\left\| \hat{J}(\mathbf{e}) \right\|^2 = \|\mathbf{e}\|^2 \le \xi^{-1}(\xi \|\mathbf{e}\|^2 + \|\nabla \times \mathbf{e}\|^2 + \|\nabla \cdot \mathbf{e}\|^2) = \xi^{-1} \|\mathbf{e}\|_{\hat{A}}^2.$$

A bounded sequence $\{\mathbf{e}_n\} \subset H_{\hat{A}}$ implies

$$\|\mathbf{e}_n\|_{H_A}^2 = \xi \|\mathbf{e}_n\|^2 + \|\nabla \times \mathbf{e}_n\|^2 + \|\nabla \cdot \mathbf{e}_n\|$$

$$= \int_\Omega \left[\xi(e_{nx})^2 + \xi(e_{ny})^2 + (\nabla e_{nx})^2 + (\nabla e_{ny})^2 \right] d\Omega \le c_1,$$

where c_1 is a constant. The compactness of the operator $\hat{J}$ follows from Rellich's theorem. Thus, according to the eigenvalue theory of symmetric operators, Equation (3.54) has an infinite set of eigenvalues $0 \le k_{c1}^2 \le k_{c2}^2 \le \cdots \le k_{cn}^2 \le \cdots$, and $k_{cn}^2 \to \infty$ as $n \to \infty$. The corresponding set of eigenfunctions $\{\mathbf{e}_n\}$ constitutes a complete system in H. Multiplying the first of (3.55) by $\mathbf{e}$ and taking the integration across Ω yield

$$k_c^2 = \frac{(\hat{A}(\mathbf{e}), \mathbf{e})}{(\mathbf{e}, \mathbf{e})} = \frac{\int_\Omega (|\nabla \times \mathbf{e}|^2 + |\nabla \cdot \mathbf{e}|^2) d\Omega}{\int_\Omega |\mathbf{e}|^2 d\Omega}. \tag{3.59}$$

The eigenvalue k_{cn} is called the **cut-off wavenumber** of the n th mode in the waveguide and the corresponding eigenfunction $\mathbf{e}_n$ is called the n th **waveguide vector modal function**. It should be noted that the waveguide vector modal functions do not depend on the operating frequency of the waveguide and they are dependent merely on the geometry of the cross-section of the waveguide. Apparently the waveguide vector modal functions $\mathbf{e}_n$ belong to one of the following four categories:

$$1. \quad \nabla \times \mathbf{e}_n = 0, \ \nabla \cdot \mathbf{e}_n = 0.$$
$$2. \quad \nabla \times \mathbf{e}_n \neq 0, \ \nabla \cdot \mathbf{e}_n = 0.$$
$$3. \quad \nabla \times \mathbf{e}_n = 0, \ \nabla \cdot \mathbf{e}_n \neq 0.$$
$$4. \quad \nabla \times \mathbf{e}_n \neq 0, \ \nabla \cdot \mathbf{e}_n \neq 0.$$

Since the vector modal functions belonging to category 4 can be expressed as a linear combination of the vector modal functions of the second and third categories due to (3.54), a complete set of eigenfunctions can be constructed from the waveguide modal functions of the first three categories only.

The waveguide vector modal functions that belong to the first category are called transverse electromagnetic (TEM) modes, which satisfy $\nabla \times \mathbf{e}_n = 0$ and $\nabla \cdot \mathbf{e}_n = 0$. We may introduce a potential function $\varphi(\boldsymbol{\rho})$ such that $\mathbf{e}_n = \nabla \varphi$ and

$$\nabla \cdot \nabla \varphi = 0, \ \boldsymbol{\rho} \in \Omega,$$
$$\mathbf{u}_n \times \nabla \varphi = 0, \ \boldsymbol{\rho} \in \Gamma.$$

The second equation implies that φ is a constant along the boundary Γ. If Ω is a simply connected region, the above equations imply $\mathbf{e}_n = 0$, and no TEM mode exists in the waveguide. Actually, from

$$\int_\Omega \nabla \cdot (\varphi \nabla \varphi) d\Omega = \int_\Gamma \varphi \nabla \varphi \cdot \mathbf{u}_n d\Gamma = \varphi \int_\Gamma \nabla \varphi \cdot \mathbf{u}_n d\Gamma$$

$$= \varphi \int_\Omega \nabla \cdot \nabla \varphi d\Omega = 0$$

and

$$\int_\Omega \nabla \cdot (\varphi \nabla \varphi) d\Omega = \int_\Omega (\nabla \varphi \cdot \nabla \varphi + \varphi \nabla \cdot \nabla \varphi) d\Omega = \int_\Omega \nabla \varphi \cdot \nabla \varphi d\Omega,$$

we can obtain $\mathbf{e}_n = \nabla \varphi = 0$. If Ω is a multiply connected region (for example, a coaxial cable), φ can take different values on different conductors, which can support a TEM mode. If $\mathbf{e}_n$ is a TEM mode, we have $k_{cn}^2 = 0$ from (3.59).

The waveguide vector modal functions that belong to the second category are called transverse electric (TE) modes, which satisfy $\nabla \times \mathbf{e}_n \neq 0$ and $\nabla \cdot \mathbf{e}_n = 0$. Since

$$\mathbf{u}_z \times \nabla \times \mathbf{e}_n = \nabla(\mathbf{u}_z \cdot \mathbf{e}_n) - (\mathbf{u}_z \cdot \nabla)\mathbf{e}_n = 0,$$

we may introduce a new function h_{zn} such that

$$\nabla \times \mathbf{e}_n = -\mathbf{u}_z k_{cn} h_{zn}. \tag{3.60}$$

The new function h_{zn} is proportional to the longitudinal magnetic field. It follows from (3.54) and (3.60) that

$$\nabla^2 h_{zn} + k_{cn}^2 h_{zn} = 0, \, \rho \in \Omega, \tag{3.61}$$
$$\mathbf{u}_n \cdot \nabla h_{zn} = 0, \, \rho \in \Gamma.$$

The eigenfunctions of (3.61) form a complete set (see Example 3.11). By definition, we have

$$\int_\Omega h_{zm} h_{zn} d\Omega = \frac{1}{k_{cm} k_{cn}} \int_\Omega \nabla \times \mathbf{e}_m \cdot \nabla \times \mathbf{e}_n d\Omega = \frac{k_{cm}}{k_{cn}} \int_\Omega \mathbf{e}_m \cdot \mathbf{e}_n d\Omega.$$

Therefore, if the set $\{\mathbf{e}_n\}$ is orthonormal, the set $\{h_{zn}\}$ is also orthonormal. On the contrary, if h_{zn} is an eigenfunction of (3.61), we may find

$$\mathbf{e}_n = \frac{1}{k_{cn}} \mathbf{u}_z \times \nabla h_{zn}. \tag{3.62}$$

It is easy to show that $\mathbf{e}_n$ satisfies (3.54) and $\nabla \cdot \mathbf{e}_n = 0$. Thus $\mathbf{e}_n$ is a TE mode. In addition we have

$$\int_\Omega \mathbf{e}_m \cdot \mathbf{e}_n d\Omega = \frac{1}{k_{cm} k_{cn}} \int_\Omega \nabla h_{zm} \cdot \nabla h_{zn} d\Omega = \frac{k_{cm}}{k_{cn}} \int_\Omega h_{zm} \cdot h_{zn} d\Omega.$$

So the set $\{\mathbf{e}_n\}$ is orthonormal if the set $\{h_{zn}\}$ is. The foregoing analysis indicates that there is a one-to-one correspondence between the set of TE modes and the set of eigenfunctions $\{h_{zn}\}$.

The waveguide vector modal functions that belong to the third category are called transverse magnetic (TM) modes, which satisfy $\nabla \cdot \mathbf{e}_n \neq 0$. We may introduce a new function e_{zn} such that

$$\nabla \cdot \mathbf{e}_n = k_{cn} e_{zn}. \tag{3.63}$$

The new function e_{zn} is proportional to the longitudinal electric field. It follows from (3.54) and (3.63) that

$$\nabla^2 e_{zn} + k_{cn}^2 e_{zn} = 0, \, \rho \in \Omega, \tag{3.64}$$
$$e_{zn} = 0, \, \rho \in \Gamma.$$

The eigenfunctions of (3.64) form a complete set (see Example 3.10). Moreover, we have

$$\int_\Omega e_{zm} e_{zn} d\Omega = \frac{1}{k_{cm} k_{cn}} \int_\Omega \nabla \cdot \mathbf{e}_m \cdot \nabla \cdot \mathbf{e}_n d\Omega = \frac{k_{cm}}{k_{cn}} \int_\Omega \mathbf{e}_m \cdot \mathbf{e}_n d\Omega.$$

Thus if the set $\{\mathbf{e}_n\}$ is orthonormal, so is the set $\{e_{zn}\}$. Conversely, the TM modes can be derived from the eigenfunctions e_{zn} of (3.64) through

$$\mathbf{e}_n = \frac{1}{k_{cn}} \nabla e_{zn}. \tag{3.65}$$

It is easy to show that $\mathbf{e}_n$ defined by (3.65) satisfies (3.54) and is a TM mode. Furthermore, we have

$$\int_\Omega \mathbf{e}_m \cdot \mathbf{e}_n d\Omega = \frac{1}{k_{cm} k_{cn}} \int_\Omega \nabla e_{zm} \cdot \nabla e_{zn} d\Omega = \frac{k_{cm}}{k_{cn}} \int_\Omega e_{zm} \cdot e_{zn} d\Omega,$$

and if the set $\{e_{zn}\}$ is orthonormal so is the set $\{\mathbf{e}_n\}$. Therefore, a one-to-one correspondence between the set of TM modes and the set of eigenfunctions $\{e_{zn}\}$ is established.

Example 3.12: Consider a rectangular waveguide shown in Figure 3.2. The longitudinal component of the TE modes can be obtained from (3.61) by means of the method of separation of variables, and they are

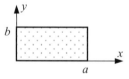

Figure 3.2   Rectangular waveguide

$$h_{zn} = \sqrt{\frac{\varepsilon_p \varepsilon_q}{ab}} \cos \frac{p\pi}{a} x \cos \frac{q\pi}{b} y, \varepsilon_p = \begin{cases} 1, p = 0 \\ 2, p \neq 0 \end{cases},$$

$$k_{cn} = \sqrt{(p\pi/a)^2 + (q\pi/b)^2}, p, q = 0, 1, 2, \cdots.$$

Here the subscript n is used to designate the multi-index (p, q). The transverse TE vector modal functions $\mathbf{e}_n$ can be obtained from (3.62) as

$$\mathbf{e}_n = \mathbf{u}_x \sqrt{\frac{\varepsilon_p \varepsilon_q}{ab}} \frac{q\pi}{k_{cn} b} \cos \frac{p\pi}{a} x \sin \frac{q\pi}{b} y - \mathbf{u}_y \sqrt{\frac{\varepsilon_p \varepsilon_q}{ab}} \frac{p\pi}{k_{cn} a} \sin \frac{p\pi}{a} x \cos \frac{q\pi}{b} y.$$

Similarly the longitudinal component of the TM modes can be derived from (3.64)

$$e_{zn} = \sqrt{\frac{4}{ab}} \sin \frac{p\pi}{a} x \sin \frac{q\pi}{b} y,$$

$$k_{cn} = \sqrt{(p\pi/a)^2 + (q\pi/b)^2}, \, p, q = 1, 2, \cdots.$$

The transverse TM vector modal functions $\mathbf{e}_n$ are then determined by (3.65)

$$\mathbf{e}_n = \mathbf{u}_x \sqrt{\frac{4}{ab}} \frac{p\pi}{k_{cn}a} \cos \frac{p\pi}{a} x \sin \frac{q\pi}{b} y + \mathbf{u}_y \sqrt{\frac{4}{ab}} \frac{q\pi}{k_{cn}b} \sin \frac{p\pi}{a} x \cos \frac{q\pi}{b} y.$$

$\square$

The set of waveguide vector modal functions $\mathbf{e}_n$ provides a complete system for the transverse fields in the waveguide. Other complete sets of functions derived from $\mathbf{e}_n$ are listed below

$$\{\mathbf{u}_z \times \mathbf{e}_n | \mathbf{u}_n \cdot \mathbf{u}_z \times \mathbf{e}_n = 0, \boldsymbol{\rho} \in \Gamma\},$$

$$\{e_{zn} = \nabla \cdot \mathbf{e}_n / k_{cn} | e_{zn} = 0, \boldsymbol{\rho} \in \Gamma\},$$

$$\{h_{zn} = \mathbf{u}_z \cdot \nabla \times \mathbf{e}_n / k_{cn}, \tilde{c} | \mathbf{u}_n \cdot \nabla h_{zn} = 0, \boldsymbol{\rho} \in \Gamma\}.$$

Here $\tilde{c}$ is a constant. According to the boundary conditions that the eigenfunctions satisfy, $\{\mathbf{e}_n\}$ are electric field-like and most appropriate for the expansion of the transverse electric field; $\{\mathbf{u}_z \times \mathbf{e}_n\}$ are magnetic field-like and most appropriate for the expansion of the transverse magnetic field; $\{e_{zn}\}$ are electric field-like and most appropriate for the expansion of the longitudinal electric field; $\{h_{zn}\}$ are magnetic field-like and most appropriate for the expansion of the longitudinal magnetic field. Notice that $\nabla \times \mathbf{E}$ is magnetic field-like while $\nabla \times \mathbf{H}$ is electric field-like. Hereafter, the waveguide modal functions are assumed to satisfy the orthonormal condition:

$$\int_\Omega \mathbf{e}_m \cdot \mathbf{e}_n d\Omega = \delta_{mn}. \tag{3.66}$$

Introducing the **modal voltage** and the **modal current**

$$V_n = \int_\Omega \mathbf{E} \cdot \mathbf{e}_n d\Omega, \, I_n = \int_\Omega \mathbf{H} \cdot \mathbf{u}_z \times \mathbf{e}_n d\Omega, \tag{3.67}$$

the electromagnetic fields in the waveguide have the following expansions

$$\mathbf{E} = \sum_{n=1}^\infty \mathbf{e}_n V_n + \mathbf{u}_z \sum_{n=1}^\infty \frac{\nabla \cdot \mathbf{e}_n}{k_{cn}} \int_\Omega \mathbf{u}_z \cdot \mathbf{E} \frac{\nabla \cdot \mathbf{e}_n}{k_{cn}} d\Omega,$$

$$\mathbf{H} = \sum_{n=1}^\infty I_n \mathbf{u}_z \times \mathbf{e}_n + \mathbf{u}_z \frac{1}{\Omega^{1/2}} \int_\Omega \frac{\mathbf{u}_z \cdot \mathbf{H}}{\Omega^{1/2}} d\Omega + \sum_{n=1}^\infty \frac{\nabla \times \mathbf{e}_n}{k_{cn}} \int_\Omega \mathbf{H} \cdot \frac{\nabla \times \mathbf{e}_n}{k_{cn}} d\Omega.$$

Similarly we have

$$\nabla \times \mathbf{E} = \sum_{n=1}^{\infty} \mathbf{u}_z \times \mathbf{e}_n \int_{\Omega} \nabla \times \mathbf{E} \cdot \mathbf{u}_z \times \mathbf{e}_n d\Omega$$

$$+ \mathbf{u}_z \frac{1}{\Omega^{1/2}} \int_{\Omega} \frac{\mathbf{u}_z \cdot \nabla \times \mathbf{E}}{\Omega^{1/2}} d\Omega + \sum_{n=1}^{\infty} \frac{\nabla \times \mathbf{e}_n}{k_{cn}} \int_{\Omega} \nabla \times \mathbf{E} \cdot \frac{\nabla \times \mathbf{e}_n}{k_{cn}} d\Omega,$$

$$\nabla \times \mathbf{H} = \sum_{n=1}^{\infty} \mathbf{e}_n \int_{\Omega} \nabla \times \mathbf{H} \cdot \mathbf{e}_n d\Omega + \mathbf{u}_z \sum_{n=1}^{\infty} \frac{\nabla \cdot \mathbf{e}_n}{k_{cn}} \int_{\Omega} \mathbf{u}_z \cdot \nabla \times \mathbf{H} \frac{\nabla \cdot \mathbf{e}_n}{k_{cn}} d\Omega.$$

For a perfectly conducting waveguide, the expansion coefficients in the above expansions can be calculated as follows

$$\int_{\Omega} \nabla \times \mathbf{E} \cdot \mathbf{u}_z \times \mathbf{e}_n d\Omega = \frac{dV_n}{dz} + \int_{\Omega} \mathbf{u}_z \cdot \mathbf{E} \nabla \cdot \mathbf{e}_n d\Omega,$$

$$\int_{\Omega} \nabla \times \mathbf{E} \cdot \frac{\nabla \times \mathbf{e}_n}{k_{cn}} d\Omega = k_{cn} V_n,$$

$$\int_{\Omega} \nabla \times \mathbf{H} \cdot \mathbf{e}_n d\Omega = -\frac{dI_n}{dz} + \int_{\Omega} \mathbf{H} \cdot \nabla \times \mathbf{e}_n d\Omega,$$

$$\int_{\Omega} \nabla \times \mathbf{H} \cdot \mathbf{u}_z \frac{\nabla \cdot \mathbf{e}_n}{k_{cn}} d\Omega = k_{cn} I_n.$$

Substituting these equations into the Maxwell equations in a lossy medium:

$$\nabla \times \mathbf{H} = j\omega\varepsilon \mathbf{E} + \sigma \mathbf{E},$$
$$\nabla \times \mathbf{E} = -j\omega\mu \mathbf{H},$$

we obtain

$$\sum_{n=1}^{\infty} \left\{ -\frac{dI_n}{dz} + \int_{\Omega} \mathbf{H} \cdot \nabla \times \mathbf{e}_t d\Omega \right\} = j\omega\varepsilon_e \sum_{n=1}^{\infty} \mathbf{e}_n V_n,$$

$$\sum_{n=1}^{\infty} \frac{\nabla \cdot \mathbf{e}_n}{k_{cn}} k_{cn} I_n = j\omega\varepsilon_e \sum_{n=1}^{\infty} \frac{\nabla \cdot \mathbf{e}_n}{k_{cn}} \int_{\Omega} \mathbf{u}_z \cdot \mathbf{E} \frac{\nabla \cdot \mathbf{e}_n}{k_{cn}} d\Omega,$$

$$\sum_{n=1}^{\infty} \mathbf{u}_z \times \mathbf{e}_{tn} \left\{ \frac{dV_n}{dz} + \int_{\Omega} \mathbf{u}_z \cdot \mathbf{E} \nabla \cdot \mathbf{e}_n d\Omega \right\} = -j\omega\mu \sum_{n=1}^{\infty} \mathbf{u}_z \times \mathbf{e}_n I_n,$$

$$\sum_{n=1}^{\infty} \frac{\nabla \times \mathbf{e}_n}{k_{cn}} k_{cn} V_n = -j\omega\mu \left\{ \sum_{n=1}^{\infty} \frac{\nabla \times \mathbf{e}_n}{k_{cn}} \int_{\Omega} \mathbf{H} \cdot \frac{\nabla \times \mathbf{e}_n}{k_{cn}} d\Omega + \frac{\mathbf{u}_z}{\Omega^{1/2}} \int_{\Omega} \mathbf{H} \cdot \frac{\mathbf{u}_z}{\Omega^{1/2}} d\Omega \right\}.$$

Comparing the coefficients of the preceding equations gives

$$-\frac{dI_n}{dz} + \int_{\Omega} \mathbf{H} \cdot \nabla \times \mathbf{e}_n d\Omega = j\omega\varepsilon_e V_n,$$

$$k_{cn}^2 I_n = j\omega\varepsilon_e \int_{\Omega} \mathbf{u}_z \cdot \mathbf{E} \nabla \cdot \mathbf{e}_n d\Omega,$$

$$\frac{dV_n}{dz} + \int_{\Omega} \mathbf{u}_z \cdot \mathbf{E} \nabla \cdot \mathbf{e}_n d\Omega = -j\omega\mu I_n,$$

$$k_{cn}^2 V_n = -j\omega\mu \int_{\Omega} \mathbf{H} \cdot \nabla \times \mathbf{e}_n d\Omega,$$

$$\int_{\Omega} \mathbf{H} \cdot \frac{\mathbf{u}_z}{\Omega^{1/2}} d\Omega = 0.$$

It follows that the modal voltage and current satisfy the **transmission line equations**

$$\frac{dV_n}{dz} = -j\beta_n Z_{wn} I_n(z), \quad \frac{dI_n}{dz} = -j\beta_n Y_{wn} V_n(z), \tag{3.68}$$

where β_n and Z_{wn} are called the **propagation constant** and **wave impedance** for the nth mode respectively

$$\beta_n = \begin{cases} k, \text{ TEM} \\ \sqrt{k^2 - k_{cn}^2}, \text{ TE or TM} \end{cases},$$

$$Z_{wn} = \begin{cases} \eta, \text{ TEM} \\ \eta k/\beta_n, \text{ TE} \\ \eta\beta_n/k, \text{ TM} \end{cases}, \tag{3.69}$$

$$Y_{wn} = 1/Z_{wn}.$$

Here $k = \omega\sqrt{\mu\varepsilon_e}$, $\eta = \sqrt{\mu/\varepsilon_e}$. If $\beta_n \neq 0$, the solution of (3.68) can be expressed as

$$V_n(z) = A_n e^{-j\beta_n z} + B_n e^{j\beta_n z}, \quad I_n(z) = (A_n e^{-j\beta_n z} - B_n e^{j\beta_n z}) Z_{wn}^{-1}. \tag{3.70}$$

Other expansion coefficients can be represented as

$$\int_{\Omega} \mathbf{H} \cdot \frac{\nabla \times \mathbf{e}_n}{k_{cn}} d\Omega = -\frac{k_{cn}}{j\beta_n} \frac{V_n(z)}{Z_{wn}}, \quad \int_{\Omega} \mathbf{u}_z \cdot \mathbf{E} \frac{\nabla \cdot \mathbf{e}_n}{k_{cn}} d\Omega = \frac{k_{cn}}{j\beta_n} I_n(z) Z_{wn}.$$

Accordingly the fields in the waveguide can be written as

$$\mathbf{E} = \sum_{n=1}^{\infty} \left[\mathbf{e}_n V_n(z) + \mathbf{u}_z \frac{\nabla \cdot \mathbf{e}_n}{j\beta_n} I_n(z) Z_{wn} \right],$$

$$\mathbf{H} = \sum_{n=1}^{\infty} \left[\mathbf{u}_z \times \mathbf{e}_n I_n(z) - \mathbf{u}_z \frac{\nabla \times \mathbf{e}_n}{j\beta_n} \frac{V_n(z)}{Z_{wn}} \right].$$

(3.71)

The above expansions are fundamental in the analysis of waveguide discontinuities.

3.3.2   Mode Theory for Cavity Resonators

A cavity resonator is an important passive device at microwave frequency, whose counterpart is the LC resonant circuit at low frequency. We assume that the cavity resonator has a perfectly conducting wall and the medium in the cavity is homogeneous and isotropic with medium parameters σ, μ and ε. The volume occupied by the cavity is denoted by V and its boundary by S (Figure 3.3). Since the metallic wall is a perfect conductor, the fields in the cavity satisfy the Maxwell equations

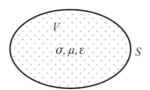

Figure 3.3   An arbitrary metal cavity

$$\nabla \times \mathbf{H}(\mathbf{r}, t) = \varepsilon \frac{\partial}{\partial t} \mathbf{E}(\mathbf{r}, t) + \sigma \mathbf{E},$$

$$\nabla \times \mathbf{E}(\mathbf{r}, t) = -\mu \frac{\partial}{\partial t} \mathbf{H}(\mathbf{r}, t),$$

(3.72)

$$\nabla \cdot \mathbf{E}(\mathbf{r}, t) = 0, \nabla \cdot \mathbf{H}(\mathbf{r}, t) = 0,$$

with boundary conditions $\mathbf{u}_n \times \mathbf{E} = 0$ and $\mathbf{u}_n \cdot \mathbf{H} = 0$, where $\mathbf{u}_n$ is the unit outward normal to the boundary S. From (3.72), the following wave equations can easily be obtained

$$\nabla \times \nabla \times \mathbf{E}(\mathbf{r}, t) + \mu\varepsilon \frac{\partial^2 \mathbf{E}(\mathbf{r}, t)}{\partial t^2} + \mu\sigma \frac{\partial \mathbf{E}(\mathbf{r}, t)}{\partial t} = 0, \mathbf{r} \in V,$$

$$\mathbf{u}_n \times \mathbf{E}(\mathbf{r}, t) = 0, \mathbf{r} \in S,$$

(3.73)

$$\nabla \times \nabla \times \mathbf{H}(\mathbf{r}, t) + \mu\varepsilon \frac{\partial^2 \mathbf{H}(\mathbf{r}, t)}{\partial t^2} + \mu\sigma \frac{\partial \mathbf{H}(\mathbf{r}, t)}{\partial t} = 0, \mathbf{r} \in V,$$

$$\mathbf{u}_n \cdot \mathbf{H}(\mathbf{r}, t) = 0, \mathbf{u}_n \times \nabla \times \mathbf{H}(\mathbf{r}, t) = 0, \mathbf{r} \in S.$$

(3.74)

Assuming that the solutions of (3.73) and (3.74) can be expressed as a separable function of space and time

$$\mathbf{E}(\mathbf{r}, t) = \mathbf{e}(\mathbf{r})u(t), \mathbf{H}(\mathbf{r}, t) = \mathbf{h}(\mathbf{r})v(t),$$

and introducing these into (3.73) and (3.74), we obtain

$$\nabla \times \nabla \times \mathbf{e} - k_e^2 \mathbf{e} = 0, \nabla \cdot \mathbf{e} = 0, \mathbf{r} \in V,$$
$$\mathbf{u}_n \times \mathbf{e} = 0, \mathbf{r} \in S, \tag{3.75}$$

$$\nabla \times \nabla \times \mathbf{h} - k_h^2 \mathbf{h} = 0, \nabla \cdot \mathbf{h} = 0, \mathbf{r} \in V,$$
$$\mathbf{u}_n \cdot \mathbf{h} = 0, \mathbf{u}_n \times \nabla \times \mathbf{h} = 0, \mathbf{r} \in S, \tag{3.76}$$

$$\frac{1}{v^2} \frac{\partial^2 u}{\partial t^2} + \sigma \frac{\eta}{v} \frac{\partial u}{\partial t} + k_e^2 u = 0, \tag{3.77}$$

$$\frac{1}{v^2} \frac{\partial^2 v}{\partial t^2} + \sigma \frac{\eta}{v} \frac{\partial v}{\partial t} + k_h^2 v = 0, \tag{3.78}$$

where k_e^2 and k_h^2 are separation constants, $\eta = \sqrt{\mu/\varepsilon}$, and $v = 1/\sqrt{\mu\varepsilon}$. Both (3.75) and (3.76) form an eigenvalue problem. However, their eigenfunctions do not form a complete set. Considering $\nabla \cdot \mathbf{e} = 0$ and $\nabla \cdot \mathbf{h} = 0$, Equations (3.75) and (3.76) may be modified as

$$\nabla \times \nabla \times \mathbf{e} - \nabla\nabla \cdot \mathbf{e} - k_e^2 \mathbf{e} = 0, \mathbf{r} \in V,$$
$$\mathbf{u}_n \times \mathbf{e} = 0, \nabla \cdot \mathbf{e} = 0, \mathbf{r} \in S, \tag{3.79}$$

$$\nabla \times \nabla \times \mathbf{h} - \nabla\nabla \cdot \mathbf{h} - k_h^2 \mathbf{h} = 0, \mathbf{r} \in V,$$
$$\mathbf{u}_n \cdot \mathbf{h} = 0, \mathbf{u}_n \times \nabla \times \mathbf{h} = 0, \mathbf{r} \in S. \tag{3.80}$$

These are the eigenvalue equations for the metal cavity system. Equations (3.79) and (3.54) are similar, and they can be approached in an exact manner. Consequently, Equation (3.79) has an infinite set of eigenvalues $0 \le k_{e1}^2 \le k_{e2}^2 \le \cdots \le k_{en}^2 \le \cdots$, and $k_{en}^2 \to \infty$ as $n \to \infty$. The corresponding eigenfunctions $\{\mathbf{e}_n\}$, called **cavity vector modal functions**, constitute a complete orthonormal system in $[L^2(V)]^3$, and they can be chosen from the following three categories

1. $\nabla \times \mathbf{e}_n = 0, \nabla \cdot \mathbf{e}_n = 0.$
2. $\nabla \times \mathbf{e}_n \ne 0, \nabla \cdot \mathbf{e}_n = 0.$
3. $\nabla \times \mathbf{e}_n = 0, \nabla \cdot \mathbf{e}_n \ne 0.$

In the same way, we can show that the vector modal functions $\mathbf{h}_n$ of (3.80) constitute a complete orthonormal system, and they can be chosen from the following three categories

1. $\nabla \times \mathbf{h}_n = 0, \nabla \cdot \mathbf{h}_n = 0.$
2. $\nabla \times \mathbf{h}_n \ne 0, \nabla \cdot \mathbf{h}_n = 0.$
3. $\nabla \times \mathbf{h}_n = 0, \nabla \cdot \mathbf{h}_n \ne 0.$

The vector modal functions belonging to category 2 in the two sets of vector modal functions $\{\mathbf{e}_n\}$ and $\{\mathbf{h}_n\}$ are related to each other. In fact, if $\mathbf{e}_n$ belong to category 2, we can define a function $\mathbf{h}_n$ through

$$\nabla \times \mathbf{e}_n = k_{en} \mathbf{h}_n. \tag{3.81}$$

and $\mathbf{h}_n$ belongs to category 2. Furthermore, we have

$$\nabla \times \nabla \times \mathbf{h}_n - k_{en}^2 \mathbf{h}_n = k_{en}^{-1}\nabla \times \left(\nabla \times \nabla \times \mathbf{e}_n - k_{en}^2 \mathbf{e}_n\right) = 0, \mathbf{r} \in V,$$
$$\mathbf{u}_n \times \nabla \times \mathbf{h}_n = k_{en}^{-1}\mathbf{u}_n \times \nabla \times \nabla \times \mathbf{e}_n = k_{en}^{-1}\mathbf{u}_n \times k_{en}^2 \mathbf{e}_n = 0, \mathbf{r} \in S.$$

Consider the integration of $\mathbf{u}_n \cdot \mathbf{h}_n$ over an arbitrary part of S, denoted ΔS

$$\int_{\Delta S} \mathbf{u}_n \cdot \mathbf{h}_n dS = k_{en}^{-1}\int_{\Delta S} \mathbf{u}_n \cdot \nabla \times \mathbf{e}_n dS = k_{en}^{-1}\int_{\Delta\Gamma} \mathbf{e}_n \cdot \mathbf{u}_\Gamma d\Gamma, \tag{3.82}$$

where $\Delta\Gamma$ is the closed contour around ΔS and $\mathbf{u}_\Gamma$ is the unit tangent vector along the contour. The right-hand side of (3.82) is zero. So we have $\mathbf{u}_n \cdot \mathbf{h}_n = 0$ for ΔS is arbitrary. Therefore $\mathbf{h}_n$ satisfies (3.80) and the corresponding eigenvalue is k_{en}^2. If $\mathbf{h}_m$ is another vector modal function corresponding to $\mathbf{e}_m$ belonging to category 2, then

$$\int_V \mathbf{h}_m \cdot \mathbf{h}_n dV = (k_{em}k_{en})^{-1}\int_V \nabla \times \mathbf{e}_m \cdot \nabla \times \mathbf{e}_n dV$$
$$= (k_{em}k_{en})^{-1}\int_S \mathbf{u}_n \times \mathbf{e}_m \cdot \nabla \times \mathbf{e}_n dS + (k_{en}/k_{em})\int_V \mathbf{e}_m \cdot \mathbf{e}_n dV = \delta_{mn}.$$

Consequently the vector modal functions $\mathbf{h}_n$ in category 2 can be derived from the vector modal functions $\mathbf{e}_n$ in category 2 and they are orthonormal. Conversely if $\mathbf{h}_n$ is in category 2, one can define $\mathbf{e}_n$ through

$$\nabla \times \mathbf{h}_n = k_{hn}\mathbf{e}_n \tag{3.83}$$

and a similar discussion shows that $\mathbf{e}_n$ is an eigenfunction of (3.75) with k_{hn} being the eigenvalue. So the completeness of the two sets is still guaranteed if the vector modal functions belonging to category 2 in $\{\mathbf{e}_n\}$ and $\{\mathbf{h}_n\}$ are related through either (3.81) or (3.83). From now on, Equations (3.81) and (3.83) will be assumed and $k_{en} = k_{hn}$ will be denoted by k_n. Note that the complete set $\{\mathbf{e}_n\}$ is most appropriate for the expansion of electric field, and $\{\mathbf{h}_n\}$ is most appropriate for the expansion of the magnetic field.

Example 3.13: Assume that the cavity contains a time-harmonic impressed electric current source with frequency ω

$$\mathbf{J}(\mathbf{r}, t) = \mathbf{J}(\mathbf{r}) \sin \omega t = \mathrm{Re} e^{j\omega t} \mathbf{J}(\mathbf{r}) e^{-j\pi/2}. \tag{3.84}$$

The electromagnetic fields (phasors) excited by the source satisfy

$$\nabla \times \mathbf{H}(\mathbf{r}) = j\omega\varepsilon\mathbf{E} + \sigma\mathbf{E} + \mathbf{J}(\mathbf{r})e^{-j\pi/2},$$
$$\nabla \times \mathbf{E}(\mathbf{r}) = -j\omega\mu\mathbf{H}(\mathbf{r}). \tag{3.85}$$

The fields can then be expanded in terms of the vector modal functions as follows

$$
\mathbf{E}(\mathbf{r}) = \sum_n V_n \mathbf{e}_n(\mathbf{r}) + \sum_\nu V_\nu \mathbf{e}_\nu(\mathbf{r}),
$$
$$
\mathbf{H}(\mathbf{r}) = \sum_n I_n \mathbf{h}_n(\mathbf{r}) + \sum_\tau I_\tau \mathbf{h}_\tau(\mathbf{r}),
$$
$$(3.86)$$

$$
\nabla \times \mathbf{E}(\mathbf{r}) = \sum_n \mathbf{h}_n(\mathbf{r}) \int_V \nabla \times \mathbf{E}(\mathbf{r}) \cdot \mathbf{h}_n(\mathbf{r}) dV + \sum_\tau \mathbf{h}_\tau(\mathbf{r}) \int_V \nabla \times \mathbf{E}(\mathbf{r}) \cdot \mathbf{h}_\tau(\mathbf{r}) dV,
$$
$$(3.87)$$
$$
\nabla \times \mathbf{H}(\mathbf{r}) = \sum_n \mathbf{e}_n(\mathbf{r}) \int_V \nabla \times \mathbf{H}(\mathbf{r}) \cdot \mathbf{e}_n(\mathbf{r}) dV + \sum_\nu \mathbf{e}_\nu(\mathbf{r}) \int_V \nabla \times \mathbf{H}(\mathbf{r}) \cdot \mathbf{e}_\nu(\mathbf{r}) dV,
$$

where the subscript n denotes the modes belonging to category 2, and the Greek subscript ν and τ for the modes belonging to categories 1 and 3 respectively, and

$$
V_{n(\nu)} = \int_V \mathbf{E}(\mathbf{r}) \cdot \mathbf{e}_{n(\nu)}(\mathbf{r}) dV, \; I_{n(\tau)} = \int_V \mathbf{H}(\mathbf{r}) \cdot \mathbf{h}_{n(\tau)}(\mathbf{r}) dV.
$$
$$(3.88)$$

Taking the following calculations

$$
\int_V \nabla \times \mathbf{E} \cdot \mathbf{h}_n dV = \int_V \mathbf{E} \cdot \nabla \times \mathbf{h}_n dV + \int_S (\mathbf{E} \times \mathbf{h}_n) \cdot \mathbf{u}_n dS = k_n V_n,
$$
$$
\int_V \nabla \times \mathbf{E} \cdot \mathbf{h}_\tau dV = \int_V \mathbf{E} \cdot \nabla \times \mathbf{h}_\tau dV + \int_S (\mathbf{E} \times \mathbf{h}_\tau) \cdot \mathbf{u}_n dS = 0,
$$
$$
\int_V \nabla \times \mathbf{H} \cdot \mathbf{e}_n dS = \int_V \mathbf{H} \cdot \nabla \times \mathbf{e}_n dV + \int_S (\mathbf{H} \times \mathbf{e}_n) \cdot \mathbf{u}_n dS = k_n I_n,
$$
$$
\int_V \nabla \times \mathbf{H} \cdot \mathbf{e}_\nu dS = \int_V \mathbf{H} \cdot \nabla \times \mathbf{e}_\nu dV + \int_S (\mathbf{H} \times \mathbf{e}_\nu) \cdot \mathbf{u}_n dS = 0,
$$

into account, Equation (3.87) can be written as

$$
\nabla \times \mathbf{E} = \sum_n k_n V_n \mathbf{h}_n, \; \nabla \times \mathbf{H} = \sum_n k_n I_n \mathbf{e}_n.
$$

Substituting the above expansions into time-harmonic equations (3.85) and comparing the expansion coefficients, we may see that

$$j\omega V_n + \frac{\sigma}{\varepsilon} V_n - \frac{k_n}{\varepsilon} I_n = \frac{j}{\varepsilon} \int_V \mathbf{J} \cdot \mathbf{e}_n dV,$$

$$j\omega V_\nu + \frac{\sigma}{\varepsilon} V_\nu = \frac{j}{\varepsilon} \int_V \mathbf{J} \cdot \mathbf{e}_\nu dV,$$

$$j\omega I_n + \frac{\sigma_m}{\mu} I_n + \frac{k_n}{\mu} V_n = 0, \qquad (3.89)$$

$$I_\tau = 0.$$

From (3.89), we obtain

$$V_n = -\frac{1}{k_n} \frac{\eta \omega \omega_n}{\omega_n^2 - \omega^2 + 2j\omega\gamma} \int_V \mathbf{J}(\mathbf{r}) \cdot \mathbf{e}_n(\mathbf{r}) dV,$$

$$I_n = -\frac{j\omega_n \nu}{\omega_n^2 - \omega^2 + 2j\omega\gamma} \int_V \mathbf{J}(\mathbf{r}) \cdot \mathbf{e}_n(\mathbf{r}) dV, \qquad (3.90)$$

$$V_\nu = \frac{j}{j\omega\varepsilon + \sigma} \int_V \mathbf{J} \cdot \mathbf{e}_\nu dV,$$

$$I_\tau = 0,$$

where $\omega_n = k_n/\sqrt{\mu\varepsilon}$ stand for the resonant frequencies of the metal cavity resonator; $\gamma = \sigma/2\varepsilon$ is the attenuation constant; $\nu = 1/\sqrt{\mu\varepsilon}$; and $\eta = \sqrt{\mu/\varepsilon}$ is the wave impedance. Introducing (3.90) into (3.86) gives

$$\mathbf{E}(\mathbf{r}) = \sum_n \frac{1}{k_n} \frac{-\eta \omega \omega_n}{\omega_n^2 - \omega^2 + 2j\omega\gamma} \mathbf{e}_n(\mathbf{r}) \int_V \mathbf{J}(\mathbf{r}) \cdot \mathbf{e}_n(\mathbf{r}) dV + \sum_\nu \frac{j\mathbf{e}_\nu(\mathbf{r})}{j\omega\varepsilon + \sigma} \int_V \mathbf{J}(\mathbf{r}) \cdot \mathbf{e}_\nu(\mathbf{r}) dV,$$

$$\mathbf{H}(\mathbf{r}) = \sum_n \frac{-j\omega_n c}{\omega_n^2 - \omega^2 + 2j\omega\gamma} \mathbf{h}_n(\mathbf{r}) \int_V \mathbf{J}(\mathbf{r}) \cdot \mathbf{e}_n(\mathbf{r}) dV.$$

The fields in time domain are then given by

$$\mathbf{E}(\mathbf{r}, t) = \mathrm{Re}\mathbf{E}(\mathbf{r})e^{j\omega t} = \sum_n \frac{-\eta \omega \omega_n}{k_n} \frac{(\omega_n^2 - \omega^2)\cos\omega t + 2\omega\gamma \sin\omega t}{(\omega_n^2 - \omega^2)^2 + 4\omega^2\gamma^2} \mathbf{e}_n(\mathbf{r}) \int_V \mathbf{J}(\mathbf{r}) \cdot \mathbf{e}_n(\mathbf{r}) dV$$

$$+ \frac{1}{\varepsilon} \sum_\nu \frac{\omega \cos\omega t - 2\gamma \sin\omega t}{\omega^2 + 4\gamma^2} \mathbf{e}_\nu(\mathbf{r}) \int_V \mathbf{J} \cdot \mathbf{e}_\nu dV, \qquad (3.91)$$

$$\mathbf{H}(\mathbf{r}, t) = \mathrm{Re}\mathbf{H}(\mathbf{r})e^{j\omega t} = \sum_n \omega_n \nu \frac{(\omega_n^2 - \omega^2)\sin\omega t - 2\omega\gamma \cos\omega t}{(\omega_n^2 - \omega^2)^2 + 4\omega^2\gamma^2} \mathbf{h}_n(\mathbf{r}) \int_V \mathbf{J}(\mathbf{r}) \cdot \mathbf{e}_n(\mathbf{r}) dV.$$

$$(3.92)$$

If the loss σ is sent to zero, these equations reduce to

$$\mathbf{E}(\mathbf{r}, t) = \sum_n \frac{-\eta \omega \omega_n}{k_n} \frac{\cos \omega t}{\omega_n^2 - \omega^2} \mathbf{e}_n(\mathbf{r}) \int_V \mathbf{J}(\mathbf{r}) \cdot \mathbf{e}_n(\mathbf{r}) dV + \frac{1}{\varepsilon} \sum_v \frac{\cos \omega t}{\omega} \mathbf{e}_v(\mathbf{r}) \int_V \mathbf{J} \cdot \mathbf{e}_v dV,$$

$$(3.93)$$

$$\mathbf{H}(\mathbf{r}, t) = \sum_n \omega_n v \frac{\sin \omega t}{\omega_n^2 - \omega^2} \mathbf{h}_n(\mathbf{r}) \int_V \mathbf{J}(\mathbf{r}) \cdot \mathbf{e}_n(\mathbf{r}) dV. \qquad (3.94)$$

Note that both (3.93) and (3.94) have singularities whenever the frequency of the excitation source coincides with one of the resonant frequencies. As a result, the field distributions are infinite everywhere inside the cavity, leading to a physically unacceptable solution. This result indicates that the time-harmonic problem is not stable under the perturbation of system loss and a lossless time-harmonic system cannot be considered to be the limit of the corresponding lossy system as the loss goes to zero. The lossy condition is necessary in solving time-harmonic problems, which is implied by the uniqueness theorem in frequency domain.   □

3.4   Exterior Electromagnetic Problems

The eigenvalue theory can also be applied to the exterior problems. The free space can be considered as a spherical waveguide, and the spherical wave functions, derived from an eigenvalue problem of Laplace operator in spherical coordinates, are widely used in radiation and scattering problems. The eigenmode expansion method (EEM) has been used to solve exterior boundary value problems (Ramm, 1982; Marks, 1989). The method is based on the eigenfunctions of integral equations and lacks a solid mathematical foundation. The integral operator involved in the EEM is not symmetric, and it is thus hard to prove the existence and the completeness of its eigenfunctions. A more useful method for the study of scattering problem is the singular function expansion, which was first introduced by the German mathematician Erhard Schmidt (1876-1959) in 1907 (Cochran, 1972), and has been applied to study various scattering problems (Inagaki, 1982; Pozar, 1984).

3.4.1   Mode Theory for Spherical Waveguides

The free space can be considered as a spherical waveguide. In a spherical coordinate system, the fields can be decomposed into a transverse component and a radial component

$$\mathbf{E} = \mathbf{E}_t + \mathbf{u}_r E_r, \mathbf{H} = \mathbf{H}_t + \mathbf{u}_r H_r,$$

where $\mathbf{u}_r$ is the unit vector in the direction of increasing r. Taking the vector and scalar product of the Maxwell equations

$$\nabla \times \mathbf{H} = j \omega \varepsilon \mathbf{E} + \mathbf{J}, \nabla \times \mathbf{E} = -j \omega \mu \mathbf{H} - \mathbf{J}_m$$

with the vector $\mathbf{r}$, we obtain

$$j\omega\varepsilon(\mathbf{r} \times \mathbf{E}) + \mathbf{r} \times \mathbf{J} = \nabla(\mathbf{r} \cdot \mathbf{H}) - (\mathbf{r} \cdot \nabla)\mathbf{H} - \mathbf{H},$$

$$-j\omega\mu(\mathbf{r} \times \mathbf{H}) - \mathbf{r} \times \mathbf{J}_m = \nabla(\mathbf{r} \cdot \mathbf{E}) - (\mathbf{r} \cdot \nabla)\mathbf{E} - \mathbf{E},$$

$$(3.95)$$

$$-\nabla \cdot (\mathbf{r} \times \mathbf{H}_t) = j\omega\varepsilon(\mathbf{r} \cdot \mathbf{E}) + \mathbf{r} \cdot \mathbf{J},$$

$$\nabla \cdot (\mathbf{r} \times \mathbf{E}_t) = j\omega\mu(\mathbf{r} \cdot \mathbf{H}) + \mathbf{r} \cdot \mathbf{J}_m.$$

$$(3.96)$$

Making use of

$$(\mathbf{r} \cdot \nabla)\mathbf{F} = r\frac{\partial\mathbf{F}}{\partial r} = r\frac{\partial}{\partial r}(F_r\mathbf{u}_r + F_\theta\mathbf{u}_\theta + F_\varphi\mathbf{u}_\varphi)$$

$$= r\mathbf{u}_r\frac{\partial}{\partial r}F_r + r\frac{\partial}{\partial r}(F_\theta\mathbf{u}_\theta + F_\varphi\mathbf{u}_\varphi) = r\mathbf{u}_r\frac{\partial}{\partial r}F_r + r\frac{\partial}{\partial r}\mathbf{F}_t,$$

and comparing the transverse components of (3.95) yields

$$\frac{1}{r}\nabla_{\theta\varphi}(\mathbf{r} \cdot \mathbf{H}) - r\frac{\partial\mathbf{H}_t}{\partial r} - \mathbf{H}_t = j\omega\varepsilon(\mathbf{r} \times \mathbf{E}_t) + \mathbf{r} \times \mathbf{J}_t,$$

$$\frac{1}{r}\nabla_{\theta\varphi}(\mathbf{r} \cdot \mathbf{E}) - r\frac{\partial\mathbf{E}_t}{\partial r} - \mathbf{E}_t = -j\omega\mu(\mathbf{r} \times \mathbf{H}_t) - \mathbf{r} \times \mathbf{J}_{mt},$$

$$(3.97)$$

where $\nabla_{\theta\varphi} = \mathbf{u}_\theta\partial/\partial\theta + \mathbf{u}_\varphi(\sin\theta)^{-1}\partial/\partial\varphi$. The radial components in (3.97) can be eliminated by using (3.96), to get the equations for the transverse fields

$$-j\omega\mu\frac{\partial}{\partial r}(r\mathbf{H}_t) + \frac{1}{r^2}\nabla_{\theta\varphi}\nabla_{\theta\varphi} \cdot (\mathbf{r} \times \mathbf{E}_t) + k^2(\mathbf{r} \times \mathbf{E}_t) = j\omega\mu(\mathbf{r} \times \mathbf{J}_t) + \frac{1}{r}\nabla_{\theta\varphi}(\mathbf{r} \cdot \mathbf{J}_m),$$

$$-j\omega\varepsilon\frac{\partial}{\partial r}(r\mathbf{E}_t) - \frac{1}{r^2}\nabla_{\theta\varphi}\nabla_{\theta\varphi} \cdot (\mathbf{r} \times \mathbf{H}_t) - k^2(\mathbf{r} \times \mathbf{H}_t) = -j\omega\varepsilon(\mathbf{r} \times \mathbf{J}_{mt}) + \frac{1}{r}\nabla_{\theta\varphi}(\mathbf{r} \cdot \mathbf{J}).$$

$$(3.98)$$

From (2.87), we can write

$$r\mathbf{E}_t = \nabla_{\theta\varphi}u(r, \theta, \varphi) + \mathbf{u}_r \times \nabla_{\theta\varphi}u'(r, \theta, \varphi),$$

$$r\mathbf{H}_t = \nabla_{\theta\varphi}v(r, \theta, \varphi) + \mathbf{u}_r \times \nabla_{\theta\varphi}v'(r, \theta, \varphi),$$

$$(3.99)$$

where u, u', v and v' are scalar functions. Introducing the positive definite operator

$$-\nabla_{\theta\varphi}^2 = -\frac{1}{\sin\theta}\frac{\partial}{\partial\theta}\left(\sin\theta\frac{\partial}{\partial\theta}\right) - \frac{1}{\sin^2\theta}\frac{\partial^2}{\partial\varphi^2},$$

the eigenvalue problem

$$-\nabla_{\theta\varphi}^2[Y(\theta, \varphi)] = \lambda Y(\theta, \varphi),$$

together with the boundary conditions that $Y(\theta, \varphi)$ must be finite and $Y(\theta, 0) = Y(\theta, 2\pi)$ may be solved by using the method of separation of variables. The eigensolutions are called

spherical harmonics $Y_{nm}^l(\theta, \varphi)$. The spherical harmonics form a complete set in $L^2(\Omega)$ (Ω is the unit sphere), and satisfy

$$-\nabla_{\theta\varphi}^2 \left[Y_{nm}^l(\theta, \varphi) \right] = n(n+1) Y_{nm}^l(\theta, \varphi).$$

Here

$$Y_{nm}^l(\theta, \varphi) = P_n^m(\cos \theta) f_{ml}(\varphi),$$

$$f_{ml}(\varphi) = \begin{cases} \cos m\varphi, \, l = e \\ \sin m\varphi, \, l = o \end{cases},$$

$$n = 0, 1, 2, \cdots, m = 0, 1, 2, \cdots, n,$$

and $P_n^m(\cos \theta)$ are the associated Legendre functions. It is easy to show the following orthogonal relationships

$$\int_0^{2\pi} d\varphi \int_0^{\pi} \left\{ \sin \theta \left[\frac{\partial Y_{nm}^l}{\partial \theta} \frac{\partial Y_{n'm'}^{l'}}{\partial \theta} \right] + \frac{1}{\sin \theta} \left[\frac{\partial Y_{nm}^l}{\partial \varphi} \frac{\partial Y_{n'm'}^{l'}}{\partial \varphi} \right] \right\} d\theta$$

$$= \begin{cases} 0, [n, m, l] \neq [n', m', l'] \\ N_{nm}^2, [n, m, l] = [n', m', l'] \end{cases},$$

$$\int_0^{2\pi} d\varphi \int_0^{\pi} Y_{nm}^l Y_{n'm'}^{l'} \sin \theta d\theta = \begin{cases} 0, [n, m, l] \neq [n', m', l'] \\ (1 + \delta_{m0}) \dfrac{2\pi}{2n+1} \dfrac{(n+m)!}{(n-m)!}, [n, m, l] = [n', m', l'] \end{cases},$$

where

$$\delta_{m0} = \begin{cases} 1, m = 0 \\ 0, m \neq 0 \end{cases}, \quad N_{nm}^2 = (1 + \delta_{m0}) \frac{2\pi(n+m)!n(n+1)}{(n-m)!(2n+1)}.$$

We can introduce the vector basis functions

$$\mathbf{e}_{nml}(\theta, \varphi) = \frac{1}{N_{nm}} \nabla_{\theta\varphi} Y_{nm}^l(\theta, \varphi),$$

$$\mathbf{h}_{nml}(\theta, \varphi) = \mathbf{u}_r \times \mathbf{e}_{nml}(\theta, \varphi),$$

(3.100)

which satisfy the orthonormal relationships:

$$\int_{S'} \mathbf{e}_{nml} \cdot \mathbf{e}_{n'm'l'} d\Omega = \delta_{mm'} \delta_{nn'} \delta_{ll'},$$

$$\int_{S'} \mathbf{h}_{nml} \cdot \mathbf{h}_{n'm'l'} d\Omega = \delta_{mm'} \delta_{nn'} \delta_{ll'},$$

(3.101)

$$\int_{S'} \mathbf{e}_{nml} \cdot \mathbf{h}_{n'm'l'} d\Omega = 0,$$

where S' is a sphere enclosing the source and $d\Omega$ is the differential element of the solid angle. Furthermore,

$$\nabla_{\theta\varphi} \cdot \mathbf{e}_{nml}(\theta, \varphi) = \frac{1}{N_{nm}} \nabla_{\theta\varphi} \cdot \nabla_{\theta\varphi} Y_{nm}^l(\theta, \varphi) = -\frac{1}{N_{nm}} n(n+1) Y_{nm}^l(\theta, \varphi),$$

$$\nabla_{\theta\varphi} \cdot \mathbf{h}_{nml}(\theta, \varphi) = \nabla_{\theta\varphi} \cdot [\mathbf{u}_r \times \mathbf{e}_{nml}(\theta, \varphi)] = 0.$$

The radial components E_r and H_r can be expanded in terms of the spherical harmonics

$$E_r = \sum_{n,m,l} C_{nml}(r) Y_{nm}^l(\theta, \varphi), \quad H_r = \sum_{n,m,l} D_{nml}(r) Y_{nm}^l(\theta, \varphi).$$

Similar expansions exist for the scalar functions u, u', v and v' in (3.99). Considering these expansions and (3.99), the transverse electromagnetic fields can be expressed as

$$r\mathbf{E}_t = \sum_{n,m,l} \left[V_{nml}^{TM}(r)\mathbf{e}_{nml} + V_{nml}^{TE}(r)\mathbf{h}_{nml} \right],$$

$$r\mathbf{H}_t = \sum_{n,m,l} \left[I_{nml}^{TM}(r)\mathbf{h}_{nml} - I_{nml}^{TE}(r)\mathbf{e}_{nml} \right],$$

(3.102)

where V_{nml} and I_{nml} are modal voltages and modal currents respectively, and

$$r\mathbf{E}_t^{TM} = \sum_{n,m,l} V_{nml}^{TM}(r)\mathbf{e}_{nml}, \qquad r\mathbf{H}_t^{TM} = \sum_{n,m,l} I_{nml}^{TM}(r)\mathbf{h}_{nml},$$

$$r\mathbf{E}_t^{TE} = \sum_{n,m,l} V_{nml}^{TE}(r)\mathbf{h}_{nml}, \qquad r\mathbf{H}_t^{TE} = -\sum_{n,m,l} I_{nml}^{TE}(r)\mathbf{e}_{nml}.$$

For the field components $r\mathbf{E}_t^{TM}$ and $r\mathbf{H}_t^{TM}$, the corresponding radial field component H_r is zero in a source-free region. Hence they are a TM wave. Similarly the field components $r\mathbf{E}_t^{TE}$ and $r\mathbf{H}_t^{TE}$ are a TE wave, where the radial field component E_r is zero. We may substitute (3.102) into (3.98) to find that, in a source-free region, the modal voltages and currents satisfy the following **spherical transmission line equations**

$$\frac{dV_{nml}^{TM}}{dr} = -j\beta_n(r)Z_n^{TM}(r)I_{nml}^{TM},$$

(3.103)

$$\frac{dI_{nml}^{TM}}{dr} = -j\beta_n(r)Y_n^{TM}(r)V_{nml}^{TM},$$

$$\frac{dV_{nml}^{TE}}{dr} = -j\beta_n(r)Z_n^{TE}(r)I_{nml}^{TE},$$

(3.104)

$$\frac{dI_{nml}^{TE}}{dr} = -j\beta_n(r)Y_n^{TE}(r)V_{nml}^{TE},$$

where $\beta_n = \sqrt{k^2 - n(n+1)/r^2}$ and

$$Z_n^{TM}(r) = \frac{1}{Y_n^{TM}(r)} = \frac{\eta \beta_n(r)}{k} \xrightarrow[r \to \infty]{} \eta,$$

$$\tilde{Z}_n^{TE}(r) = \frac{1}{Y_n^{TE}(r)} = \frac{\eta k}{\beta_n(r)} \xrightarrow[r \to \infty]{} \eta$$

are the **characteristic impedances** for the corresponding modes. From the above relationships we obtain

$$\frac{d^2 I_{nml}^{TM}}{dr^2} + \beta_n^2(r) I_{nml}^{TM} = 0, \quad \frac{d^2 V_{nml}^{TE}}{dr^2} + \beta_n^2(r) V_{nml}^{TE} = 0. \tag{3.105}$$

Once V_{nml}^{TE} and I_{nml}^{TM} are known, V_{nml}^{TM} and I_{nml}^{TE} can then be determined from (3.103) and (3.104) respectively. It follows from (3.105) that the modal voltages V_{nml} and currents I_{nml} decrease as r increases if $kr < \sqrt{n(n+1)}$. The number $\sqrt{n(n+1)}/k$ is called **cut-off radius**.

3.4.2   Singular Functions and Singular Values

Given a compact operator $\hat{A}$, we can define a new operator by the composition $\hat{L} = \hat{A}\hat{A}^*$. This new operator is positive definite, symmetric and compact. Therefore, the eigenvalue theory for a compact symmetric operator can be applied to the eigenvalue problem

$$\hat{L}v_n = \lambda_n v_n. \tag{3.106}$$

We have the following properties:

1. All the eigenvalues of $\hat{L}$ are real and nonnegative, and each eigenvalue $\lambda \neq 0$ has finite multiplicity.
2. Two eigenfunctions of $\hat{L}$ that correspond to different eigenvalues are orthogonal.
3. The operator $\hat{L}$ has a complete orthonormal system of eigenfunctions.
4. If the operator $\hat{L}$ has a countable set of eigenvalues (for example, $\lambda = 0$ is not an eigenvalue of $\hat{L}$), the eigenvalues of $\hat{L}$ form a sequence $\{\lambda_n\}$ such that $\lambda_n \xrightarrow[n \to \infty]{} 0$. (This means the inverse of $\hat{L}$, if it exists, is unbounded.)

Since $\lambda_n \geq 0$, we can write $\lambda_n = \mu_n^2$. Then, the function

$$u_n = \mu_n^{-1} \hat{A}^* v_n, \ \mu_n > 0 \tag{3.107}$$

is an eigenfunction of $\hat{L}^*$ (the adjoint of $\hat{L}$). In fact,

$$\hat{L}^* u_n = \mu_n^{-1} \hat{A}^* \hat{A} \hat{A}^* v_n = \mu_n^{-1} \hat{A}^* \mu_n^2 v_n = \mu_n^2 u_n.$$

Furthermore, we have

$$(u_n, u_m) = (\mu_n^{-1}\hat{A}^* v_n, \mu_m^{-1}\hat{A}^* v_m) = \mu_n^{-1}\mu_m^{-1}(v_n, \hat{A}\hat{A}^* v_m) = \mu_n^{-1}\mu_m(v_n, v_m) = \delta_{nm}.$$

Thus, $\{u_n\}$ is an orthonormal set. The functions $\{u_n\}$ and $\{v_n\}$ are called **singular functions** of $\hat{A}$ and $\{\mu_n\}$ are the corresponding **singular values**. It must be mentioned that the singular functions do not have any singular behavior. The name arises in connection with the singular value decomposition of an operator. As a result, any function f has the following expansions

$$f = v + \sum_{n=1}^{\infty}(f, v_n)v_n, \; f = u + \sum_{n=1}^{\infty}(f, u_n)u_n,$$

where v and u are functions (depending on f) satisfying $\hat{A}^* v = 0$ and $\hat{A}u = 0$ respectively.

Example 3.14: For a perfectly conducting scatterer, the electric current $\mathbf{J}_s$ induced by an incident field $\mathbf{H}^{in}(\mathbf{r})$ satisfies the integral equation of second kind (see Section 5.5.1 for derivation)

$$(\hat{I} - \hat{G})\mathbf{J}_s(\mathbf{r}) = 2\mathbf{u}_n(\mathbf{r}) \times \mathbf{H}^{in}(\mathbf{r}), \mathbf{r} \in S,$$

where $\mathbf{u}_n$ is the unit outward normal of the surface S and

$$\hat{G}(\mathbf{J}_s)(\mathbf{r}) = 2\mathbf{u}_n(\mathbf{r}) \times \int_S \mathbf{J}_s(\mathbf{r}') \times \nabla' G_0(\mathbf{r}, \mathbf{r}')dS(\mathbf{r}'), \mathbf{r} \in S,$$

with $G_0(\mathbf{r}, \mathbf{r}') = e^{-jk|\mathbf{r}-\mathbf{r}'|}/4\pi|\mathbf{r} - \mathbf{r}'|$. Its adjoint is given by

$$\hat{G}^*(\mathbf{J}_s)(\mathbf{r}) = 2\int_S [\mathbf{u}_n(\mathbf{r}') \times \mathbf{J}_s(\mathbf{r}')] \times \nabla' \bar{G}_0(\mathbf{r}, \mathbf{r}')dS(\mathbf{r}').$$

Let $\hat{L} = (\hat{I} - \hat{G})^*(\hat{I} - \hat{G})$, and consider the eigenvalue problem

$$\hat{L}(\mathbf{J}_{sn}) = \lambda_n \mathbf{J}_{sn}, \mathbf{J}_{sn} \in [L^2(S)]^3.$$

Since $\hat{G}$ is a compact operator (Example 3.3), $\hat{L}$ is a compact symmetric operator and is also positive definite. Therefore the eigenvalue theory of compact symmetric operators applies for the operator $\hat{L}$.   □

3.5 Eigenfunctions of Curl Operator

The eigenvalue problem of the curl operator has important applications in electromagnetics (Good, 1957; Moses, 1959; Moses and Prosser, 1986). The eigenfunction ψ of a curl operator satisfies

$$\nabla \times \boldsymbol{\psi}(\mathbf{r}) = \lambda \boldsymbol{\psi}(\mathbf{r}), \, \mathbf{r} \in R^3. \tag{3.108}$$

One type of eigensolution of (3.108) may be assumed to be a plane wave solution

$$\boldsymbol{\psi}(\mathbf{r}) = \mathbf{V}e^{j\mathbf{k}\cdot\mathbf{r}} \tag{3.109}$$

where $\mathbf{k} = (k_x, k_y, k_z)$ and $\mathbf{V}$ is a constant vector. Introducing (3.109) into (3.108) yields

$$j\mathbf{k} \times \mathbf{V} = \lambda \mathbf{V}.$$

It is easy to show that the above equation has three eigenvalues $\lambda = \tau k (\tau = 0, \pm 1)$, $k = |\mathbf{k}|$, and the corresponding orthonormal eigensolutions are

$$\mathbf{V}_0(\mathbf{k}) = \frac{1}{k} \begin{bmatrix} k_x \\ k_y \\ k_z \end{bmatrix}, \, \mathbf{V}_\delta(\mathbf{k}) = \frac{1}{\sqrt{2}k(k_x^2 + k_y^2)^{1/2}} \begin{bmatrix} j\delta k k_y - k_x k_z \\ -j\delta k k_x - k_y k_z \\ k_x^2 + k_y^2 \end{bmatrix}, (\delta = \pm 1)$$

with $\mathbf{V}_\tau^* \cdot \mathbf{V}_\upsilon = \delta_{\tau\upsilon}(\tau, \upsilon = 0, \pm 1)$. We may introduce the following orthonormal vectors

$$\boldsymbol{\chi}_\tau(\mathbf{r}|\mathbf{k}) = [\chi_{\tau 1}(\mathbf{r}|\mathbf{k}), \chi_{\tau 2}(\mathbf{r}|\mathbf{k}), \chi_{\tau 3}(\mathbf{r}|\mathbf{k})]^T = (2\pi)^{-3/2}\mathbf{V}_\tau(\mathbf{k})e^{j\mathbf{k}\cdot\mathbf{r}},$$

which satisfy the orthonormal conditions:

$$\int_{R^3} \boldsymbol{\chi}_\tau^*(\mathbf{r}|\mathbf{k}) \cdot \boldsymbol{\chi}_\upsilon(\mathbf{r}|\mathbf{k}')d\mathbf{r} = \delta_{\tau\upsilon}\delta(\mathbf{k} - \mathbf{k}'), (\tau, \upsilon = 0, \pm 1),$$

$$\sum_\tau \int_{R^3} \chi_{\tau i}^*(\mathbf{r}|\mathbf{k}) \cdot \chi_{\tau j}(\mathbf{r}'|\mathbf{k})d\mathbf{k} = \delta_{ij}\delta(\mathbf{r} - \mathbf{r}'), (i, j = 1, 2, 3).$$

An arbitrary vector $\mathbf{Q}(\mathbf{r})$ can then be expanded as follows

$$\mathbf{Q}(\mathbf{r}) = \sum_\tau \int_{R^3} \boldsymbol{\chi}_\tau(\mathbf{r}|\mathbf{k}) \cdot \mathbf{q}_\tau(\mathbf{k})d\mathbf{k} = \sum_\tau \mathbf{Q}_\tau(\mathbf{r})$$

where

$$\mathbf{Q}_\tau(\mathbf{r}) = \int_{R^3} \boldsymbol{\chi}_\tau(\mathbf{r}|\mathbf{k}) \cdot \mathbf{q}_\tau(\mathbf{k})d\mathbf{k},$$

$$\mathbf{q}_\tau(\mathbf{k}) = \int_{R^3} \boldsymbol{\chi}_\tau^*(\mathbf{r}|\mathbf{k}) \cdot \mathbf{Q}_\tau(\mathbf{r})d\mathbf{r}.$$

Since $\nabla \cdot \mathbf{Q}_\tau(\mathbf{r}) = 0 (\tau = \pm 1)$ and $\nabla \times \mathbf{Q}_0(\mathbf{r}) = 0$, an arbitrary vector $\mathbf{Q}$ may be decomposed into three components: one is irrotational and the other two are solenoidal. This result may be regarded as the **generalized Helmholtz theorem**.

The Euler 'Calculus of Variations' from 1744 is one of the most beautiful mathematical works that has ever been written.

—Constantin Carathéodory (Greek mathematician, 1873–1950)

4

Antenna Theory

True optimization is the revolutionary contribution of modern research to decision processes.
—George Bernhard Dantzig (American mathematician, 1914–2005)

Electromagnetic radiation is caused by accelerated charge or time-changing current. When an external source is used to excite the current distribution on a nearby scatterer, the scatterer gives off electromagnetic waves. This constitutes a radiation problem. For the radiation problem, the scatterer must be carefully designed to control the electromagnetic energy distribution in free space. An antenna is a device for radiating or receiving radio waves. In most applications, the electromagnetic energy is fed to the scatterer by a waveguide, and the antenna is a bridge connecting the waveguide and free space. The antenna transmits the electromagnetic energy from the waveguide into free space in a transmitting mode, or receives electromagnetic energy from free space to the waveguide in a receiving mode. In other words, it transforms guided waves into free space waves and vice versa. A scattering problem typically refers to the situation where the external source is far away from the scatterer. A current distribution induced on the scatterer by the incident field from the external source produces a far field pattern. If the direction of the incident field changes diversely so that each portion of the scatterer faces a direct illumination at least once, the far field patterns obtained will include all the geometrical information of the scatterer, and the geometry of the scatterer can be recovered from them.

In 1886, Hertz invented the first wire antennas (a dipole and a loop) to confirm Maxwell's theory and the existence of electromagnetic waves. Modern antenna theory was started during the World War II and a number of classical antennas were introduced during that time (Silver, 1949). The sources of radiation fields are the current distributions, including both conduction current and displacement current. The antenna can thus be classified as conduction-current type and displacement-current type. For the conduction-current antenna, the source of radiation is conduction current on a metallic radiator surface. Linear antenna, loop, helix and spiral antenna are of the conduction-current type, and they are typically for lower frequency, lower gain, and wide beam width applications. For the displacement-current antenna, the source of the radiation is the electromagnetic fields at the antenna aperture or on the antenna surface. Horn antenna, slot antenna, aperture antenna, parabolic reflector, dielectric rod antenna belong to this type, and they are usually for higher frequency, higher gain, and narrow beamwidth

Foundations of Applied Electrodynamics   Geyi Wen
© 2010 John Wiley & Sons, Ltd

applications. The antennas can also be categorized into four basic types: electrically small antenna, resonant antenna, broadband antenna and aperture antenna. For the small antenna, its maximum extent is much less than a wavelength and it has low directivity, low radiation resistance, low radiation efficiency and high input reactance. Both the resonant antenna and broadband antenna have real input impedance but the bandwidth is narrow for the former and very wide for the latter. The aperture antenna has very high gain and moderate bandwidth. The radiation patterns of an antenna can be omni-directional or directional, depending on the antenna applications.

The most important parameters for characterizing antenna are gain, efficiency, input impedance, bandwidth, radiation pattern, beamwidth, sidelobes, front-to-back ratio, and polarization. There are trade-offs between these antenna parameters. To satisfy one parameter requirement, one may have to sacrifice one or more other parameter levels. Most of the antenna parameters are subject to certain limitations, which can be understood by spherical wavefunction expansion of the fields produced by antenna. The propagating modes supported by an antenna depend on the size of the smallest circumscribing sphere enclosing the antenna. The bigger the antenna size (the size of the sphere), the more propagating modes the antenna will generate. When the antenna is very small, no propagating modes can exist and all the spherical modes are rapidly cut off. As a result, the stored energy around the antenna becomes very large and the radiation power becomes very small, and the antenna has a high quality factor.

A more useful performance index for describing antenna is the product of antenna gain and bandwidth for they must be maximized simultaneously in most applications. It can be shown that antenna fractional bandwidth is reciprocal to antenna quality factor. Thus, the product of antenna gain and bandwidth can be expressed as the ratio of antenna gain over antenna quality factor. The maximum possible product of antenna gain and bandwidth is an upper bound of the antenna performance, which can be used to determine the antenna size required to achieve a specified antenna performance.

4.1 Antenna Parameters

An arbitrary transmitting antenna system and a receiving antenna system are shown in Figure 4.1. The power to the matching network is denoted by P^m; the power accepted by the antenna is denoted by P^a; and the power radiated by the antenna is denoted by P^{rad}. Due to the mismatch, portion of the power P^m is reflected back to the source by the matching network, which is denoted by P^{ref}. The power accepted by the antenna can be expressed as

$$P^a = \frac{1}{2} \mathrm{Re} V \bar{I} = P^m - P^{ref} - P^{match}_{loss},$$

where V and I are the modal voltage and modal current for the dominant mode in the feeding waveguide respectively and they are calculated at the reference plane, and P^{match}_{loss} is the power loss in the matching network. The radiated power of the antenna can be represented by

$$P^{rad} = \frac{1}{2} \int_S \mathrm{Re}(\mathbf{E} \times \bar{\mathbf{H}}) \cdot \mathbf{u}_n dS$$

where S is a surface, which encloses the antenna.

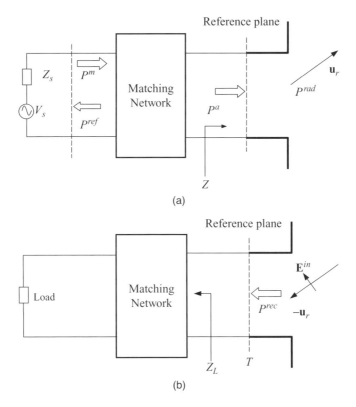

Figure 4.1   (a) Transmitting antenna (b) Receiving antenna

Antenna performances depend on the antenna geometry as well as how the antenna is used. In mobile devices, the antenna position keeps changing as the subscriber travels around, and reasonable antenna performances are expected for all different positions. The antenna design is based on those positions that are used most often. For example, the mobile phone has three typical positions:

1. Talking position: the phone is held by the user against the ear.
2. Dialing position: the phone is held in front of the user so the display can be seen.
3. Set-down position: the phone is lying down on a flat non-metal surface.

Blockage of the antenna from user's head and hand is also a major consideration for handset antenna design, which determines where the antenna should be located to minimize the influences of the hand and head.

4.1.1   Radiation Patterns and Radiation Intensity

The **radiation pattern** of antenna is a mathematical function or a graphical representation of the radiation properties of the antenna as a function of space coordinates. In most cases, the radiation pattern is determined in the far field region. Radiation properties can be power flux density, radiation density, field strength, phase or polarization. For a linearly polarized antenna, the radiation pattern is usually described by E-plane and H-plane patterns. The E-plane is defined as the plane containing the electric field vector and the direction of the maximum radiation and the H-plane is defined as the plane containing the magnetic field vector and the direction of maximum radiation. The antenna radiation pattern magnitude must be plotted relative to a recognized standard. The most common standard level is that of a perfect isotropic radiator (antenna), which would radiate energy equally in all directions.

Let $\mathbf{u}_r$ be the unit vector along a far field observation point $\mathbf{r} = r_\infty \mathbf{u}_r$. The **radiation intensity** of an antenna in the direction $\mathbf{u}_r$ is defined as the power radiated from the antenna per unit solid angle

$$U(\mathbf{u}_r) = \frac{r_\infty^2}{2} \mathrm{Re}\left[\mathbf{E}(\mathbf{r}) \times \bar{\mathbf{H}}(\mathbf{r})\right] = \frac{r_\infty^2}{2\eta} |\mathbf{E}(\mathbf{r})|^2 ,$$

where $\eta = \sqrt{\mu/\varepsilon}$. The radiation intensity for an isotropic radiator is $U(\mathbf{u}_r) = P^{rad}/4\pi$.

4.1.2 Radiation Efficiency, Antenna Efficiency and Matching Network Efficiency

Not all the input power to the antenna will be radiated to the free space. The power loss may come from the impedance mismatch that causes a portion of the input power to be reflected back to the transmitter, or from the imperfect conductors and dielectrics that cause a portion of the input power to be dissipated as heat. The **radiation efficiency** of the antenna is defined by

$$e_r = \frac{P^{rad}}{P^a}.$$

The radiation efficiency reflects the conduction and dielectric losses of the antenna. The **antenna efficiency** is defined by

$$e_t = \frac{P^{rad}}{P^m} = \frac{P^m - P^{ref}}{P^m} \cdot \frac{P^a}{P^m - P^{ref}} \cdot \frac{P^{rad}}{P^a} = (1 - |\Gamma|^2) e_s e_r,$$

where $e_s = P^a/(P^m - P^{ref})$ is the efficiency describing the loss in the matching network; Γ is the reflection coefficient at the input of the matching network; and

$$e_m = \frac{P^a}{P^m} = (1 - |\Gamma|^2) e_s$$

is the **matching network efficiency**. Better antenna efficiency means:

1. Better quality of communication.
2. Better wireless coverage.
3. Longer battery life for wireless terminals.

4.1.3 Directivity and Gain

The **directivity** of an antenna is defined as the ratio of the radiation intensity in a given direction from an antenna to the radiation intensity averaged over all directions

$$D(\mathbf{u}_r) = \frac{U(\mathbf{u}_r)}{P^{rad}/4\pi}.$$

Theoretically, there is no mathematical limit to the directivity that can be obtained from currents confined in an arbitrarily small volume. However, the high field intensities around a small antenna with a high directivity will produce high energy storage around the antenna, large power dissipation, low radiation efficiency and narrow bandwidth.

The (absolute) **gain** of an antenna is defined as the ratio of the radiation intensity in a given direction to the radiation intensity that would be obtained if the power accepted by the antenna were radiated isotropically.

$$G(\mathbf{u}_r) = \frac{U(\mathbf{u}_r)}{P^a/4\pi} = e_r D(\mathbf{u}_r).$$

The old definition of the gain is

$$G_{old}(\mathbf{u}_r) = \frac{U(\mathbf{u}_r)}{P^m/4\pi} = e_t D(\mathbf{u}_r).$$

This is also called absolute gain. The gain of an antenna often refers to the maximum gain and is usually given in decibels.

4.1.4 Input Impedance, Bandwidth and Antenna Quality Factor

The **input impedance** of antenna is defined as the ratio of the voltage to current at the input reference plane of the antenna. The **bandwidth** of an antenna is defined as the range of frequencies within which the performance of the antenna, with respect to some characteristics (such as the input impedance, return loss, gain, radiation efficiency, pattern, beamwidth, polarization, sidelobe level, and beam direction), conforms to a specified standard. Antenna bandwidth is an important quantity, which measures the quality of signal transmission such as signal distortion. For broadband antennas, the bandwidth is usually expressed as the ratio of the upper-to-lower frequencies of acceptable operation. For narrow band antennas, the bandwidth is expressed as a percentage of the frequency difference (upper minus lower) over the center frequency of the bandwidth (fractional bandwidth). The bandwidth can be enhanced by introducing losses, parasitic elements, loading or changing matching network.

If the impedance of an antenna is not perfectly matched to that of the source, some power will be reflected back and not transmitted. This reflected power relative to incident power is called **return loss**. A figure of merit for antenna is **return loss bandwidth**, which is defined as the frequency range where return loss is below an acceptable level. For example, a return loss of -10 dB indicates 90% of the power is transmitted. At -7 dB return loss, 80% of the power is transmitted. The return loss bandwidth is closely related to antenna physical volume. Increasing the return loss bandwidth is one of the challenges in small antenna design.

The **quality factor** of antenna is defined as 2π times the ratio of total energy stored around the antenna divided by the energy radiated per cycle:

$$Q = \frac{\omega \tilde{W}}{P_{rad}},$$

where $\tilde{W}$ is the time average energy stored in the antenna and ω is the frequency. In most applications of this definition, the antenna quality factor is evaluated at the resonant frequency, and the quality factor can be expressed as

$$Q = \frac{2\omega \tilde{W}_i}{P_{rad}},$$

where $\tilde{W}_i$ is either the average stored electric or magnetic energy. For a non-resonant antenna, it is tacitly assumed that the antenna system should be tuned to resonance by adding a capacitive or inductive energy storage element depending on whether the stored energy is predominantly magnetic or electric. For this reason, $\tilde{W}_i$ is chosen as the average stored magnetic energy $\tilde{W}_m$ or the average stored electric energy $\tilde{W}_e$ in the near field zone around the antenna, whichever is larger.

It will be shown later that, for a high quality factor antenna, the quality factor is reciprocal of antenna fractional bandwidth for input impedance. The antenna quality factor is a field quantity and is more convenient for theoretical study, while the antenna bandwidth requires more information on the frequency behavior of the input impedance. We will use Q_{real} to indicate that all the stored energy around an antenna has been included in the calculation of antenna quality factor, to distinguish it from another antenna quality factor, denoted by Q, to be introduced later, in which only the stored energy outside the circumscribing sphere of the antenna is included. Obviously we have $Q_{real} \gg Q$.

4.1.5   Vector Effective Length, Equivalent Area and Antenna Factor

Let $\mathbf{r} = r_\infty \mathbf{u}_r$ be a far field observation point. The far field of the antenna in a homogeneous and isotropic medium can be expressed as (see (4.8))

$$\mathbf{E}(\mathbf{r}) = -\frac{j\omega\mu I}{4\pi r_\infty} e^{-jkr_\infty} \mathbf{L}(\mathbf{u}_r).$$

Here I is the exciting current at the feeding plane, and $\mathbf{L}$ is called the **antenna vector effective length** defined by

$$\mathbf{L}(\mathbf{u}_r) = \frac{1}{I} \int_{V_0} \left\{ \mathbf{J}(\mathbf{r}') - \left[\mathbf{J}(\mathbf{r}') \cdot \mathbf{u}_r \right] \mathbf{u}_r \right\} e^{jk\mathbf{u}_r \cdot \mathbf{r}'} dV(\mathbf{r}'),$$

where V_0 is the source region and $\mathbf{J}$ is the current distribution inside the source region. The open circuit voltage at the antenna-feeding plane induced by an incident field $\mathbf{E}^{in}$ (Figure 4.1(b)) is given by (see Section 6.4.1)

$$V_{oc}(\mathbf{u}_r) = -\frac{1}{I} \int_{V_0} \mathbf{E}^{in}(\mathbf{r}') \cdot \mathbf{J}(\mathbf{r}') dV(\mathbf{r}')$$

which results from the reciprocity of transmitting and receiving antenna. Let the incident field be a plane wave from the direction $-\mathbf{u}_r$. The incident field may be written as

$$\mathbf{E}^{in}(\mathbf{r}) = \mathbf{E}^{in}(o)e^{jk\mathbf{u}_r \cdot \mathbf{r}},$$

where $\mathbf{E}^{in}(o)$ is the field strength at the origin (antenna position) and is perpendicular to $\mathbf{u}_r$. Thus

$$V_{oc}(\mathbf{u}_r) = -\frac{1}{I}\mathbf{E}^{in}(o) \cdot \int_{V_0} \mathbf{J}(\mathbf{r}')e^{jk\mathbf{u}_r \cdot \mathbf{r}'} dV(\mathbf{r}')$$

$$= -\frac{1}{I}\mathbf{E}^{in}(o) \cdot \int_{V_0} \left\{ \mathbf{J}(\mathbf{r}') - \left[\mathbf{J}(\mathbf{r}') \cdot \mathbf{u}_r \right] \mathbf{u}_r \right\} e^{jk\mathbf{u}_r \cdot \mathbf{r}'} dV(\mathbf{r}')$$

$$= -\mathbf{E}^{in}(o) \cdot \mathbf{L}(\mathbf{u}_r).$$

This relation has been used as the definition of the vector effective length in most literature. According to the equivalent circuit for the receiving antenna as shown in Figure 4.2, the received power by the load is

$$P^{rec}(\mathbf{u}_r) = \frac{1}{2}\left|\frac{V_{oc}(\mathbf{u}_r)}{Z + Z_L}\right|^2 \operatorname{Re}Z_L = \frac{1}{2}\left|\frac{\mathbf{E}^{in}(o) \cdot \mathbf{L}(\mathbf{u}_r)}{Z + Z_L}\right|^2 \operatorname{Re}Z_L,$$

where Z is the antenna input impedance.

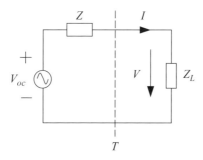

Figure 4.2   Equivalent circuit for receiving antenna

The **antenna equivalent area** is a transverse area defined as the ratio of received power to the power flux density of the incident plane wave

$$A_e(\mathbf{u}_r) = \frac{P^{rec}(\mathbf{u}_r)}{\left|E^{in}(o)\right|^2/2\eta} = \left|\frac{E^{in}(o)\cdot\mathbf{L}(\mathbf{u}_r)}{Z+Z_L}\right|^2 \frac{\eta\mathrm{Re}Z_L}{\left|E^{in}(o)\right|^2}.$$

If the receiving antenna is conjugately matched and there is no polarization loss, the antenna equivalent area can be simplified as

$$A_e(\mathbf{u}_r) = \frac{1}{4}\left(\frac{\eta}{\mathrm{Re}Z_L}\right)|\mathbf{L}(\mathbf{u}_r)|^2 .$$

The **antenna factor** is defined as the ratio of incident electric field strength to the induced terminal voltage

$$AF(\mathbf{u}_r) = \frac{\left|E^{in}(o)\right|}{|V(\mathbf{u}_r)|},$$

where $V(\mathbf{u}_r)$ stands for the induced terminal voltage at the reference plane of the receiving antenna due to the incident field. From the equivalent circuit of a receiving antenna, we obtain

$$V(\mathbf{u}_r) = \frac{Z_L}{Z_L+Z}V_{oc}(\mathbf{u}_r).$$

So the relationship between the antenna factor and vector effective length is

$$AF(\mathbf{u}_r) = \frac{\left|E^{in}(o)\right|}{\left|E^{in}(o)\cdot\mathbf{L}(\mathbf{u}_r)\right|}\left|1+\frac{Z}{Z_L}\right|.$$

If there is no polarization loss, this reduces to

$$AF(\mathbf{u}_r) = \left|1+\frac{Z}{Z_L}\right|\frac{1}{|\mathbf{L}(\mathbf{u}_r)|}.$$

Let S_∞ be a large closed surface, which encloses the antenna. The transmitting properties of antenna can be expressed as functions of the effective length and they are summarized below:

Poynting vector:

$$S(\mathbf{r}) = \frac{1}{2\eta}|\mathbf{E}(\mathbf{r})|^2 = \frac{\eta|I|^2}{8r_\infty^2}\left|\frac{\mathbf{L}(\mathbf{u}_r)}{\lambda}\right|^2.$$

Radiation intensity:

$$U(\mathbf{u}_r) = \frac{r_\infty^2}{2\eta}|\mathbf{E}(\mathbf{r})|^2 = \frac{\eta|I|^2}{8}\left|\frac{\mathbf{L}(\mathbf{u}_r)}{\lambda}\right|^2.$$

Radiated power:

$$P^{rad} = \frac{\eta \, |I|^2}{8} \int_{S_\infty} \left| \frac{\mathbf{L}(\mathbf{u}_r)}{\lambda} \right|^2 d\Omega.$$

Radiation resistance:

$$R^{rad} \equiv \frac{2 P^{rad}}{|I|^2} = \frac{\eta}{4} \int_{S_\infty} \left| \frac{\mathbf{L}(\mathbf{u}_r)}{\lambda} \right|^2 d\Omega.$$

Directivity:

$$D(\mathbf{u}_r) = \frac{4\pi \, U(\mathbf{u}_r)}{P^{rad}} = \frac{\pi \eta}{R^{rad}} \left| \frac{\mathbf{L}(\mathbf{u}_r)}{\lambda} \right|^2.$$

Gain:

$$G(\mathbf{u}_r) = e_r \, D(\mathbf{u}_r) = \frac{\pi \eta}{R^{rad} + R^{loss}} \left| \frac{\mathbf{L}(\mathbf{u}_r)}{\lambda} \right|^2.$$

4.1.6   Polarization and Coupling

The **polarization of a wave** is defined as the curve traced by the instantaneous electric field in a plane perpendicular to the propagation direction of the wave. If the direction of the electric field at a point of space is constant in time, we say the electric field at that point is **linearly polarized**. If the tip of the electric field is a circle (or ellipse) centered at the point in the course of time, we say the electric field is **circularly (elliptically) polarized** at that point. Elliptically polarized field is encountered in practice very often. The **polarization of an antenna** is defined as the curve traced by instantaneous electric field radiated by the antenna in a plane perpendicular to the radial direction. The radiation fields of all antennas aside from the dipoles are generally elliptically polarized, except in some preferred directions.

For perfect transmission of power between two antennas, their polarizations must match exactly. In practice, the polarization mismatch loss always exists. If two antennas have no coupling, their polarizations are said to be orthogonal. The polarization mismatch loss between a circularly polarized and linearly polarized antenna is 3 dB and half power is lost. Two linear polarized antennas orientated at an angle of 45 degrees will also have 3 dB polarization mismatch loss.

In a cellular environment, the degree of polarization match between the mobile and base station can vary considerably. In outdoor suburban environments, the polarization of the incident field would be mainly vertical while in indoors and in dense urban environments, scattering and multipath reflections can cause the incident polarization to change dramatically. Additionally, the degree of polarization match between the incident field and mobile antenna is impacted by the user. For example, how the device is held and placed changes the degree of polarization match.

Minimizing the antenna coupling is important if isolation between two signal paths is required. The coupling between two antennas can be measured using a network analyzer. The coupling is the amplitude of S_{21} over the frequency range of both bands when the network analyzer is connected to these two antennas. The coupling between two antennas in the far field is inversely related to the square of the distance between them (assume both antennas are in free space). For example, if the distance is doubled, the coupling is reduced by a factor of four (-6dB). Antenna coupling is strongly influenced by the out-of-band impedance of the antenna. For example, if one antenna is very poorly matched at the band of another antenna and vice versa, the coupling between the two antennas might be low even if they are placed in close proximity. Unbalanced antennas are fed against the ground, and they make the ground part of the antenna. Two unbalanced antennas fed against the same ground tend to have less isolation than balanced antennas. This problem can be improved if the antenna design can make the currents in the ground localized in the vicinity of the antenna.

4.2 Properties of Far Fields

If the antenna current distribution is known, all the antenna performances can be determined. Some performances of antenna are very sensitive to the antenna current distribution while some of them are relatively insensitive. Since the exact current distribution of antenna is very complicated and is not easily discovered, people usually use approximations to find a simplified current distribution in order to predict the antenna performances that are insensitive to the current distribution, such as gain, antenna pattern, and radiation resistance. In the feeding area, approximations have to be adopted on the basis of a good understanding of how the current distribution affects the various antenna performances. The factors that affect the antenna current distribution include antenna shape, size, excitation, and the environment of the antenna. Müller has systematically studied the properties of electromagnetic radiation patterns (Müller, 1956, 1969) and a summary has been given by Colton and Kress (Colton & Kress, 1983, 1998). From (2.60) and (2.61), the fields produced by a time-harmonic current source $\mathbf{J}$ can be expressed as

$$\mathbf{E}(\mathbf{r}) = -jk\eta \int_{V_0} G(\mathbf{r}, \mathbf{r}')\mathbf{J}(\mathbf{r}')dV(\mathbf{r}')$$

$$-\frac{\eta}{jk}\int_{V_0} \nabla' \cdot \mathbf{J}(\mathbf{r}')\nabla' G(\mathbf{r}, \mathbf{r}')dV(\mathbf{r}') - \int_{V_0} \mathbf{J}_m(\mathbf{r}') \times \nabla' G(\mathbf{r}, \mathbf{r}')dV(\mathbf{r}'), \quad (4.1)$$

$$\mathbf{H}(\mathbf{r}) = -j\frac{k}{\eta}\int_{V_0} G(\mathbf{r}, \mathbf{r}')\mathbf{J}_m(\mathbf{r}')dV(\mathbf{r}')$$

$$-\frac{1}{j\eta k}\int_{V_0} \nabla' \cdot \mathbf{J}_m(\mathbf{r}')\nabla' G(\mathbf{r}, \mathbf{r}')dV(\mathbf{r}') + \int_{V_0} \mathbf{J}(\mathbf{r}') \times \nabla' G(\mathbf{r}, \mathbf{r}')dV(\mathbf{r}'), \quad (4.2)$$

where $G(\mathbf{r}, \mathbf{r}') = e^{-jkR}/4\pi R$ with $R = |\mathbf{r} - \mathbf{r}'|$. Making use of the Gauss theorem, we have

$$\int_{V_0} \nabla' \cdot \mathbf{J}(\mathbf{r}')\nabla' G(\mathbf{r}, \mathbf{r}')dV(\mathbf{r}') = -\int_{V_0} \left[\mathbf{J}(\mathbf{r}') \cdot \nabla'\right]\nabla' G(\mathbf{r}, \mathbf{r}')dV(\mathbf{r}'),$$

and the electromagnetic fields can be rewritten as

$$\mathbf{E}(\mathbf{r}) = -jk\eta \int_{V_0} \left(\overset{\leftrightarrow}{\mathbf{I}} + \frac{1}{k^2}\nabla\nabla \right) G(\mathbf{r},\mathbf{r}') \cdot \mathbf{J}(\mathbf{r}')dV(\mathbf{r}') - \int_{V_0} \mathbf{J}_m(\mathbf{r}') \times \nabla'G(\mathbf{r},\mathbf{r}')dV(\mathbf{r}'), \quad (4.3)$$

$$\mathbf{H}(\mathbf{r}) = -j\frac{k}{\eta} \int_{V_0} \left(\overset{\leftrightarrow}{\mathbf{I}} + \frac{1}{k^2}\nabla\nabla \right) G(\mathbf{r},\mathbf{r}') \cdot \mathbf{J}_m(\mathbf{r}')dV(\mathbf{r}') + \int_{V_0} \mathbf{J}(\mathbf{r}') \times \nabla'G(\mathbf{r},\mathbf{r}'dV(\mathbf{r}'), \quad (4.4)$$

where $\overset{\leftrightarrow}{\mathbf{I}}$ is the identity dyadic tensor. Let $\mathbf{u}_R = (\mathbf{r}-\mathbf{r}')/|\mathbf{r}-\mathbf{r}'|$. Then

$$\left[\mathbf{J}(\mathbf{r}) \cdot \nabla' \right] \nabla'G(\mathbf{r},\mathbf{r}')$$

$$= G(\mathbf{r},\mathbf{r}')\left[-k^2 + \frac{3}{R}\left(jk + \frac{1}{R} \right) \right] \left[\mathbf{J}(\mathbf{r}') \cdot \mathbf{u}_R \right] \mathbf{u}_R - G(\mathbf{r},\mathbf{r}')\frac{\mathbf{J}(\mathbf{r}')}{R}\left(jk + \frac{1}{R} \right),$$

If R is sufficiently large, we can ignore the terms higher than R^{-1}. Thus,

$$\left[\mathbf{J}(\mathbf{r}') \cdot \nabla' \right] \nabla'G(\mathbf{r},\mathbf{r}') \approx -k^2 G(\mathbf{r},\mathbf{r}')\left[\mathbf{J}(\mathbf{r}') \cdot \mathbf{u}_R \right] \mathbf{u}_R. \quad (4.5)$$

In the far field region defined by $r \gg r'$, $kr \gg 1$, the following approximations can be made

$$R = |\mathbf{r}-\mathbf{r}'| \approx r - \mathbf{u}_r \cdot \mathbf{r}', \quad \frac{1}{|\mathbf{r}-\mathbf{r}'|} \approx \frac{1}{r},$$

$$\frac{e^{-jk|\mathbf{r}-\mathbf{r}'|}}{|\mathbf{r}-\mathbf{r}'|}\mathbf{u}_R \approx \frac{e^{-jkr}}{r}e^{jk\mathbf{u}_r\cdot\mathbf{r}'}\mathbf{u}_r, \quad (4.6)$$

where $\mathbf{u}_r$ is the unit vector along $\mathbf{r}$. It is readily found from (4.1), (4.2), (4.5) and (4.6) that the far fields have the following asymptotic forms

$$\mathbf{E}(\mathbf{r}) = \frac{e^{-jkr}}{r}\left[\mathbf{E}_\infty(\mathbf{u}_r) + O\left(\frac{1}{r} \right) \right], \quad \mathbf{H}(\mathbf{r}) = \frac{e^{-jkr}}{r}\left[\mathbf{H}_\infty(\mathbf{u}_r) + O\left(\frac{1}{r} \right) \right], \quad (4.7)$$

Here the vector fields $\mathbf{E}_\infty$ and $\mathbf{H}_\infty$ are defined on the unit sphere Ω, and are known as the **electric far field pattern** and **magnetic far field pattern** respectively. The far field patterns are independent of the distance r and are given by

$$\mathbf{E}_\infty(\mathbf{u}_r) = -\frac{jk\eta}{4\pi} \int_{V_0} \left[\mathbf{J} - (\mathbf{J}\cdot\mathbf{u}_r)\mathbf{u}_r + \frac{1}{\eta}\mathbf{J}_m \times \mathbf{u}_r \right] e^{jk\mathbf{u}_r\cdot\mathbf{r}'}dV(\mathbf{r}'),$$

$$\mathbf{H}_\infty(\mathbf{u}_r) = -\frac{jk}{4\pi\eta} \int_{V_0} \left[\mathbf{J}_m - (\mathbf{J}_m\cdot\mathbf{u}_r)\mathbf{u}_r - \eta\mathbf{J} \times \mathbf{u}_r \right] e^{jk\mathbf{u}_r\cdot\mathbf{r}'}dV(\mathbf{r}'), \quad (4.8)$$

and satisfy

$$\eta \mathbf{H}_\infty(\mathbf{u}_r) = \mathbf{u}_r \times \mathbf{E}_\infty(\mathbf{u}_r), \, \mathbf{u}_r \cdot \mathbf{E}_\infty(\mathbf{u}_r) = \mathbf{u}_r \cdot \mathbf{H}_\infty(\mathbf{u}_r) = 0. \tag{4.9}$$

It follows from (4.7) and (4.9) that the far fields satisfy the Silver–Müller radiation conditions

$$\lim_{r \to \infty} r(\mathbf{u}_r \times \mathbf{E} - \eta \mathbf{H}) = 0.$$

If there are no magnetic sources, the Poynting vector in the far field region can be expressed as

$$\mathbf{S} = \frac{1}{2} \text{Re}(\mathbf{E} \times \bar{\mathbf{H}}) = \mathbf{u}_r \frac{k^2 \eta}{32\pi^2 r^2} \left| \mathbf{u}_r \times \int_{V_0} \mathbf{J} e^{jk\mathbf{u}_r \cdot \mathbf{r}'} dV(\mathbf{r}') \right|^2. \tag{4.10}$$

The total radiated power by the current $\mathbf{J}$ is

$$P^{rad} = \int_{S_\infty} \mathbf{S} \cdot \mathbf{u}_n dS(\mathbf{r}') = \int_{S_\infty} \mathbf{S} \cdot \mathbf{u}_n r^2 d\Omega$$

$$= \frac{k^2 \eta}{32\pi^2} \int_{S_\infty} \left| \mathbf{u}_r \times \int_{V_0} \mathbf{J} e^{jk\mathbf{u}_r \cdot \mathbf{r}'} dV(\mathbf{r}') \right|^2 d\Omega(\mathbf{r}). \tag{4.11}$$

From the Poynting theorem and (4.3), the radiated power can also be calculated by

$$P^{rad} = -\frac{1}{2} \text{Re} \int_{V_0} \bar{\mathbf{J}}(\mathbf{r}') \cdot \mathbf{E}(\mathbf{r}') dV(\mathbf{r}')$$

$$= \frac{k\eta}{8\pi} \int_{V_0} \int_{V_0} \bar{\mathbf{J}}(\mathbf{r}) \cdot \left(\ddot{\mathbf{I}} + \frac{1}{k^2} \nabla \nabla \right) \frac{\sin(k |\mathbf{r} - \mathbf{r}'|)}{|\mathbf{r} - \mathbf{r}'|} \cdot \mathbf{J}(\mathbf{r}') dV(\mathbf{r}) dV(\mathbf{r}').$$

Remark 4.1: Let V_0 be a finite region. If the electric or magnetic far field pattern vanishes identically, the electromagnetic fields generated by the source confined in V_0 are identically zero in $R^3 - V_0$. This property can easily be proved by the analyticity of the fields. □

Remark 4.2: Many factors affect the propagation properties of radiated waves, such as the earth's atmosphere, the ground, mountains, buildings, and weather conditions. When a wave is incident upon an obstacle, it will be reflected, refracted and diffracted. **Reflection** is the change of direction of the wave at the surface of the obstacle so that the wave returns to the medium from which it is originated. **Refraction** is the change of direction of the wave, which happens when the wave passes into the obstacle and is accompanied by a change in speed and wavelength of the wave. The amount of reflection and refraction depends on the medium properties, wave polarization, the angle of incidence and the wave frequency. **Diffraction** is the change of direction of the wave to bend around the obstacle. A low frequency wave whose wavelength is longer than the maximum size of an obstacle can easily be propagated around the obstacle. When frequency increases, the obstacle causes more and more attenuation and a shadow zone on the opposite side of the incidence develops. The shadowed side consists of bright and dark regions, which are not expected from the analysis of geometrical optics. Therefore, diffraction often refers to the departure from the analysis of geometrical optics, which occurs whenever a portion of the wave front of the incident wave is obstructed. □

4.3   Spherical Vector Wavefunctions

If the antenna is very small compared to the wavelength, the radiated fields will be substantially spherical. Let the antenna be enclosed by a sphere. The radiated fields outside the sphere can be expanded as a linear combination of spherical vector wavefunctions (SVWF), which was first reported by the American physicist William Webster Hansen (1909–1949) (Hansen, 1935). For this reason, the spherical vector wavefunctions have found wide applications in antenna analysis.

4.3.1   Field Expansions in Terms of Spherical Vector Wavefunctions

Consider the vector Helmholtz equation

$$\nabla \times \nabla \times \mathbf{F}(\mathbf{r}) - \nabla\nabla \cdot \mathbf{F}(\mathbf{r}) - k^2 \mathbf{F}(\mathbf{r}) = 0. \tag{4.12}$$

To find its independent vector solutions, we may start with a scalar function ψ, which is a solution of the Helmholtz equation:

$$(\nabla^2 + k^2)\psi = 0. \tag{4.13}$$

It can be shown that (4.12) has three independent vector solutions

$$\mathbf{L} = \nabla\psi, \mathbf{M} = \nabla \times (\mathbf{r}\psi), \mathbf{N} = \frac{1}{k}\nabla \times \nabla \times (\mathbf{r}\psi).$$

If $\{\psi_n\}$ is a complete set, we may expect that the corresponding vector functions $\{\mathbf{L}_n, \mathbf{M}_n, \mathbf{N}_n\}$ will also form a complete set and can be used to represent an arbitrary vector wavefunction. In the spherical coordinate system, the solution of (4.13) is

$$\psi_{nml}^{(q)}(\mathbf{r}) = h_n^{(q)}(kr)Y_{nm}^l(\theta, \varphi).$$

Here $Y_{nm}^l(\theta, \varphi) = P_n^m(\cos\theta)f_{ml}(\varphi)$ $(n = 0, 1, 2, \cdots; m = 0, 1, 2, \cdots, n; l = e, o)$ are the spherical harmonics; $P_n^m(\cos\theta)$ are the associated Legendre functions; $h_n^{(q)}(kr)$ $(q = 1, 2)$

are the spherical Hankel functions; and

$$f_{ml}(\varphi) = \begin{cases} \cos m\varphi, l = e \\ \sin m\varphi, l = o \end{cases}.$$

The **spherical vector wavefunctions** are defined by

$$\begin{aligned}
\mathbf{L}_{nml}^{(q)}(\mathbf{r}) &= \nabla\left[\psi_{nml}^{(q)}(\mathbf{r})\right], \\
\mathbf{M}_{nml}^{(q)}(\mathbf{r}) &= \nabla \times \left[\mathbf{r}\psi_{nml}^{(q)}(\mathbf{r})\right] = \nabla\psi_{nml}^{(q)}(\mathbf{r}) \times \mathbf{r}, \\
\mathbf{N}_{nml}^{(q)}(\mathbf{r}) &= \frac{1}{k}\nabla \times \nabla \times \left[\mathbf{r}\psi_{nml}^{(q)}(\mathbf{r})\right] = \frac{1}{k}\nabla \times \mathbf{M}_{nml}^{(q)}(\mathbf{r}).
\end{aligned} \tag{4.14}$$

Explicitly,

$$\mathbf{M}_{nml}^{(q)}(\mathbf{r}) = \frac{h_n^{(q)}(kr)}{\sin\theta}\frac{\partial Y_{nm}^l(\theta,\varphi)}{\partial\varphi}\mathbf{u}_\theta - h_n^{(q)}(kr)\frac{\partial Y_{nm}^l(\theta,\varphi)}{\partial\theta}\mathbf{u}_\varphi,$$

$$\mathbf{N}_{nml}^{(q)}(\mathbf{r}) = \frac{n(n+1)}{kr}h_n^{(q)}(kr)Y_{nm}^l(\theta,\varphi)\mathbf{u}_r + \frac{1}{kr}\frac{d\left[rh_n^{(q)}(kr)\right]}{dr}\frac{\partial Y_{nm}^l(\theta,\varphi)}{\partial\theta}\mathbf{u}_\theta$$

$$+\frac{1}{kr\sin\theta}\frac{d\left[rh_n^{(q)}(kr)\right]}{dr}\frac{\partial Y_{nm}^l(\theta,\varphi)}{\partial\varphi}\mathbf{u}_\varphi.$$

It will be shown later that the spherical vector wavefunctions $\{\mathbf{L}_{nml}^{(q)}, \mathbf{M}_{nml}^{(q)}, \mathbf{N}_{nml}^{(q)}\}$ form a complete set. So the vector potential function can be expanded as follows

$$\mathbf{A} = \frac{1}{j\omega}\sum_{n,m,l,q}\left(\alpha_{nml}^{(q)}\mathbf{M}_{nml}^{(q)} + \beta_{nml}^{(q)}\mathbf{N}_{nml}^{(q)} + \gamma_{nml}^{(q)}\mathbf{L}_{nml}^{(q)}\right). \tag{4.15}$$

From Maxwell equations and $\mu\mathbf{H} = \nabla \times \mathbf{A}$, the electromagnetic fields can be expressed by

$$\mathbf{E} = -\sum_{n,m,l,q}\left(\alpha_{nml}^{(q)}\mathbf{M}_{nml}^{(q)} + \beta_{nml}^{(q)}\mathbf{N}_{nml}^{(q)}\right),$$

$$\mathbf{H} = \frac{1}{j\eta}\sum_{n,m,l,q}\left(\alpha_{nml}^{(q)}\mathbf{N}_{nml}^{(q)} + \beta_{nml}^{(q)}\mathbf{M}_{nml}^{(q)}\right), \tag{4.16}$$

where $\eta = \sqrt{\mu/\varepsilon}$, μ and ε are medium parameters.

Example 4.1 (Spherical waveguide): The free space may be considered as a spherical waveguide, and the transmission direction is along the radius r in a spherical coordinate system (r, θ, ϕ) while the waveguide cross-sections are spherical surfaces. The electromagnetic fields $\mathbf{E}$ and $\mathbf{H}$ in spherical coordinates (r, θ, ϕ) can be decomposed into transverse components $\mathbf{E}_t$, $\mathbf{H}_t$ and radial components $\mathbf{u}_r E_r$, $\mathbf{u}_r H_r$

$$\mathbf{E} = \mathbf{E}_t + \mathbf{u}_r E_r, \mathbf{H} = \mathbf{H}_t + \mathbf{u}_r H_r.$$

We can introduce the orthonormal vector basis functions

$$\mathbf{e}_{nml} = \frac{1}{N_{nm}} \nabla_{\theta\varphi} Y^l_{nm}(\theta, \varphi), \quad \mathbf{h}_{nml} = \mathbf{u}_r \times \mathbf{e}_{nml}.$$

Then the spherical vector wavefunctions can be expressed as

$$\mathbf{M}^{(q)}_{nml} = -\frac{N_{nm}}{kr} \tilde{h}^{(q)}_n(kr) \mathbf{h}_{nml},$$

$$\mathbf{N}^{(q)}_{nml} = \frac{N_{nm}}{kr} \dot{\tilde{h}}^{(q)}_n(kr) \mathbf{e}_{nml} + \mathbf{u}_r \gamma^{(q)}_{nml}, \tag{4.17}$$

where $\tilde{h}^{(q)}_n(kr) = krh^{(q)}_n(kr)$, $\gamma^{(q)}_{nml} = (kr)^{-1}n(n+1)h^{(q)}_n(kr)Y^l_{nm}(\theta, \varphi)$ and $\dot{\tilde{h}}^{(q)}_n(kr)$ is the derivative of $\tilde{h}^{(q)}_n(kr)$ with respect to its argument. Substituting (4.17) into (4.16) gives

$$r\mathbf{E}_t = \frac{1}{k} \sum_{m,n,l} \left[N_{nm}\tilde{h}^{(1)}_n(kr)\alpha^{(1)}_{nml} + N_{nm}\tilde{h}^{(2)}_n(kr)\alpha^{(2)}_{nml} \right] \mathbf{h}_{nml}$$

$$- \left[N_{nm}\dot{\tilde{h}}^{(1)}_n(kr)\beta^{(1)}_{nml} + N_{nm}\dot{\tilde{h}}^{(2)}_n(kr)\beta^{(2)}_{nml} \right] \mathbf{e}_{nml}, \tag{4.18}$$

$$r\mathbf{H}_t = \frac{1}{jk\eta} \sum_{m,n,l} \left[N_{nm}\dot{\tilde{h}}^{(1)}_n(kr)\alpha^{(1)}_{nml} + N_{nm}\dot{\tilde{h}}^{(2)}_n(kr)\alpha^{(2)}_{nml} \right] \mathbf{e}_{nml}$$

$$- \left[N_{nm}\tilde{h}^{(1)}_n(kr)\beta^{(1)}_{nml} + N_{nm}\tilde{h}^{(2)}_n(kr)\beta^{(2)}_{nml} \right] \mathbf{h}_{nml}.$$

These can be rewritten as

$$r\mathbf{E}_t = \sum_{n,m,l} \left[V^{TM}_{nml}(r)\mathbf{e}_{nml} + V^{TE}_{nml}(r)\mathbf{h}_{nml} \right],$$

$$r\mathbf{H}_t = \sum_{n,m,l} \left[I^{TM}_{nml}(r)\mathbf{h}_{nml} - I^{TE}_{nml}(r)\mathbf{e}_{nml} \right]. \tag{4.19}$$

Here

$$V^{TE}_{nml}(r) = V^{TE+}_{nml}(r) + V^{TE-}_{nml}(r),$$

$$I^{TE}_{nml}(r) = I^{TE+}_{nml}(r) + I^{TE-}_{nml}(r),$$

$$V^{TM}_{nml}(r) = V^{TM+}_{nml}(r) + V^{TM-}_{nml}(r),$$

$$I^{TM}_{nml}(r) = I^{TM+}_{nml}(r) + I^{TM-}_{nml}(r).$$

are the equivalent modal voltages and currents for TE and TM modes with

$$V_{nml}^{TM+}(r) = -\frac{N_{nm}\beta_{nml}^{(2)}}{k}\dot{h}_n^{(2)}(kr), \quad V_{nml}^{TM-}(r) = -\frac{N_{nm}\beta_{nml}^{(1)}}{k}\dot{h}_n^{(1)}(kr),$$

$$V_{nml}^{TE+}(r) = \frac{N_{nm}\alpha_{nml}^{(2)}}{k}\tilde{h}_n^{(2)}(kr), \quad V_{nml}^{TE-}(r) = \frac{N_{nm}\alpha_{nml}^{(1)}}{k}\tilde{h}_n^{(1)}(kr),$$

$$I_{nml}^{TE+}(r) = -\frac{N_{nm}\alpha_{nml}^{(2)}}{j\eta k}\dot{h}_n^{(2)}(kr), \quad I_{nml}^{TE-}(r) = -\frac{N_{nm}\alpha_{nml}^{(1)}}{j\eta k}\dot{h}_n^{(1)}(kr),$$

$$I_{nml}^{TM+}(r) = -\frac{N_{nm}\beta_{nml}^{(2)}}{j\eta k}\tilde{h}_n^{(2)}(kr), \quad I_{nml}^{TM-}(r) = -\frac{N_{nm}\beta_{nml}^{(1)}}{j\eta k}\tilde{h}_n^{(1)}(kr),$$

where the superscripts $+$ and $-$ denote outward-going and inward-going waves respectively. The radially directed wave impedances for TE modes and TM modes are defined by

$$Z_n^{TE}(r) = \frac{V_{nml}^{TE+}(r)}{I_{nml}^{TE+}(r)} = -j\eta\frac{\tilde{h}_n^{(2)}(kr)}{\dot{h}_n^{(2)}(kr)},$$

$$Z_n^{TM}(r) = \frac{V_{nml}^{TM+}(r)}{I_{nml}^{TM+}(r)} = j\eta\frac{\dot{h}_n^{(2)}(kr)}{\tilde{h}_n^{(2)}(kr)}.$$

As $r \to \infty$, the wave impedances approach η.   □

The expansion coefficients in (4.16) may be expressed in terms of the current distribution confined in the finite region V_0. The radial components of the fields are

$$\mathbf{r}\cdot\mathbf{E} = -\sum_{n,m,l,q}\beta_{nml}^{(q)}\mathbf{r}\cdot\mathbf{N}_{nml}^{(q)}, \quad \mathbf{r}\cdot\mathbf{H} = \frac{1}{j\eta}\sum_{n,m,l,q}\alpha_{nml}^{(q)}\mathbf{r}\cdot\mathbf{N}_{nml}^{(q)}.$$

Since the source is limited in a finite region V_0 and the observation point is outside the source region, the wave must be out-going and one can choose $q = 2$. The above equations become

$$\mathbf{r}\cdot\mathbf{E} = -\frac{1}{k}\sum_{n=0}^{\infty}\sum_{m=0}^{n}\sum_{l=e,o}\beta_{nml}^{(2)}n(n+1)h_n^{(2)}(kr)Y_{nm}^{l}(\theta,\varphi),$$

$$\mathbf{r}\cdot\mathbf{H} = \frac{1}{j\eta k}\sum_{n=0}^{\infty}\sum_{m=0}^{n}\sum_{l=e,o}\alpha_{nml}^{(2)}n(n+1)h_n^{(2)}(kr)Y_{nm}^{l}(\theta,\varphi).$$

(4.20)

From the wave equations

$$\nabla\times\nabla\times\mathbf{H} - k^2\mathbf{H} = \nabla\times\mathbf{J}, \quad \nabla\times\nabla\times\mathbf{E} - k^2\mathbf{E} = -j\omega\mu\mathbf{J}$$

we obtain

$$\mathbf{r} \cdot \nabla \times \nabla \times \mathbf{H} - k^2 \mathbf{r} \cdot \mathbf{H} = \mathbf{r} \cdot \nabla \times \mathbf{J},$$
$$\mathbf{r} \cdot \nabla \times \nabla \times \mathbf{E} - k^2 \mathbf{r} \cdot \mathbf{E} = -j\omega\mu \mathbf{r} \cdot \mathbf{J}.$$

By use of the following identity for an arbitrary vector function $\mathbf{F}$

$$(\nabla^2 + k^2)(\mathbf{r} \cdot \mathbf{F}) = 2\nabla \cdot \mathbf{F} + \mathbf{r} \cdot \nabla(\nabla \cdot \mathbf{F}) - \mathbf{r} \cdot \nabla \times \nabla \times \mathbf{F} + k^2 \mathbf{r} \cdot \mathbf{F},$$

one may find that the radial components satisfy

$$(\nabla^2 + k^2)\left(\mathbf{r} \cdot \mathbf{E} - \frac{j}{\omega\varepsilon}\mathbf{r} \cdot \mathbf{J}\right) = \frac{j}{\omega\varepsilon}\mathbf{r} \cdot \nabla \times \nabla \times \mathbf{J},$$
$$(\nabla^2 + k^2)(\mathbf{r} \cdot \mathbf{H}) = -\mathbf{r} \cdot \nabla \times \mathbf{J}.$$

The solutions of these equations are

$$\mathbf{r} \cdot \mathbf{E} = \frac{j}{\omega\varepsilon}\mathbf{r} \cdot \mathbf{J} + \frac{1}{j\omega\varepsilon}\int_{V_0} G(\mathbf{r}, \mathbf{r}')\mathbf{r}' \cdot \nabla' \times \nabla' \times \mathbf{J}(\mathbf{r}')dV(\mathbf{r}'),$$

$$\mathbf{r} \cdot \mathbf{H} = \int_{V_0} G(\mathbf{r}, \mathbf{r}')\mathbf{r}' \cdot \nabla' \times \mathbf{J}(\mathbf{r}')dV(\mathbf{r}'), \tag{4.21}$$

where $G(\mathbf{r}, \mathbf{r}') = e^{-jk|\mathbf{r}-\mathbf{r}'|}/4\pi|\mathbf{r} - \mathbf{r}'|$ is the Green's function. If the field point $\mathbf{r}$ is outside the source region V_0, we have $\mathbf{r} \cdot \mathbf{J} = 0$. The Green's function has the expansion

$$G(\mathbf{r}, \mathbf{r}') = -\frac{jk}{4\pi}\sum_{n=0}^{\infty}(2n + 1)j_n(kr_<)h_n^{(2)}(kr_>)P_n(\cos \gamma),$$

where $j_n(kr)$ are the spherical Bessel functions, γ is the angle between $\mathbf{r}$ and $\mathbf{r}'$, and $r_< = \min\{r, r'\}$, $r_> = \max\{r, r'\}$. Making use of the addition formula

$$P_n(\cos \gamma) = \sum_{m=0}^{n}\sum_{l=e,o}\frac{2(n - m)!}{(n + m)!(1 + \delta_{m0})}Y_{nm}^l(\theta, \varphi)Y_{nm}^l(\theta', \varphi'),$$

the Green's function may be represented by

$$G(\mathbf{r}, \mathbf{r}') = -jk\sum_{n=0}^{\infty}\sum_{m=0}^{n}\sum_{l=e,o}\frac{n(n + 1)}{N_{nm}^2}\chi_{nml}(r, \theta, \varphi)\tilde{\chi}_{nml}(r', \theta', \varphi'), \tag{4.22}$$

where $N_{nm}^2 = (1 + \delta_{m0}) \dfrac{2\pi(n+m)!n(n+1)}{(n-m)!(2n+1)}$ and

$$\chi_{nml}(r, \theta, \varphi) = \begin{cases} j_n(kr)Y_{nm}^l(\cos\theta), r < r', \\ h_n^{(2)}(kr)Y_{nm}^l(\cos\theta), r > r', \end{cases}$$

$$\tilde{\chi}_{nml}(r', \theta', \varphi') = \begin{cases} h_n^{(2)}(kr')Y_{nm}^l(\cos\theta'), r < r', \\ j_n(kr')Y_{nm}^l(\cos\theta'), r > r'. \end{cases}$$

Substituting (4.22) into (4.21) gives

$$\mathbf{r} \cdot \mathbf{E} = -\frac{k}{\omega\varepsilon} \sum_{n=0}^{\infty} \sum_{m=0}^{n} \sum_{l=e,o} \frac{n(n+1)}{N_{nm}^2} \chi_{nml}(r, \theta, \varphi)$$

$$\times \int_{V_0} \tilde{\chi}_{nml}(r', \theta', \varphi')\mathbf{r}' \cdot \nabla' \times \nabla' \times \mathbf{J}(\mathbf{r}')dV(\mathbf{r}'),$$

$$\mathbf{r} \cdot \mathbf{H} = -jk \sum_{n=0}^{\infty} \sum_{m=0}^{n} \sum_{l=e,o} \frac{n(n+1)}{N_{nm}^2} \chi_{nml}(r, \theta, \varphi)$$

$$\times \int_{V_0} \tilde{\chi}_{nml}(r', \theta', \varphi')\mathbf{r}' \cdot \nabla' \times \mathbf{J}(\mathbf{r}')dV(\mathbf{r}').$$

(4.23)

These equations can be rewritten as

$$\mathbf{r} \cdot \mathbf{E} = -\eta \sum_{n=0}^{\infty} \sum_{m=0}^{n} \sum_{l=e,o} \frac{n(n+1)}{N_{nm}^2} \chi_{nml}(r, \theta, \varphi)$$

$$\times \int_{V_0} \mathbf{J}(\mathbf{r}') \cdot \nabla' \times \nabla' \times \left[\mathbf{r}'\tilde{\chi}_{nml}(r', \theta', \varphi')\right]dV(\mathbf{r}'),$$

$$\mathbf{r} \cdot \mathbf{H} = -jk \sum_{n=0}^{\infty} \sum_{m=0}^{n} \sum_{l=e,o} \frac{n(n+1)}{N_{nm}^2} \chi_{nml}(r, \theta, \varphi)$$

$$\times \int_{V_0} \mathbf{J}(\mathbf{r}') \cdot \nabla' \times \left[\mathbf{r}'\tilde{\chi}_{nml}(r', \theta', \varphi')\right]dV(\mathbf{r}').$$

(4.24)

Comparing (4.20) and (4.24), we obtain the expansion coefficients

$$\alpha_{nml}^{(2)} = \frac{k^2\eta}{N_{nm}^2} \int_{V_0} \mathbf{J}(\mathbf{r}') \cdot \nabla' \times \left[\mathbf{r}'\tilde{\chi}_{nml}(r', \theta', \varphi')\right]dV(\mathbf{r}'),$$

$$\beta_{nml}^{(2)} = \frac{k\eta}{N_{nm}^2} \int_{V_0} \mathbf{J}(\mathbf{r}') \cdot \nabla' \times \nabla' \times \left[\mathbf{r}'\tilde{\chi}_{nml}(r', \theta', \varphi')\right]dV(\mathbf{r}').$$

(4.25)

4.3.2 Completeness of Spherical Vector Wavefunctions

The proof of the completeness of the spherical vector wavefunctions has been investigated by a number of authors (Vekua, 1953; Wilcox, 1957; Aydin & Hizal, 1986), and can also be carried out by the unified theory studied in Chapter 3. The **spherical vector wavefunctions of the second kind** are defined by letting $q = 2$ in (4.14)

$$\mathbf{L}_{nml}^{(2)}(\mathbf{r}) = \nabla\left[\psi_{nml}^{(2)}(\mathbf{r})\right],$$
$$\mathbf{M}_{nml}^{(2)}(\mathbf{r}) = \nabla \times \left[\mathbf{r}\psi_{nml}^{(2)}(\mathbf{r})\right] = \nabla\psi_{nml}^{(2)}(\mathbf{r}) \times \mathbf{r}, \qquad (4.26)$$
$$\mathbf{N}_{nml}^{(2)}(\mathbf{r}) = \frac{1}{k}\nabla \times \nabla \times \left[\mathbf{r}\psi_{nml}^{(2)}(\mathbf{r})\right] = \frac{1}{k}\nabla \times \mathbf{M}_{nml}^{(2)}(\mathbf{r}),$$

where $\psi_{nml}^{(2)}(\mathbf{r}) = h_n^{(2)}(kr)P_n^m(\cos\theta)f_{ml}(\varphi)$. The spherical vector wavefunctions of the second kind satisfy the radiation condition. Let S be a closed surface that contains the origin, the SVWF of the second kind may be used to expand an arbitrary function in $[L^2(S)]^3$.

Completeness of SVWF of second kind: *The set of spherical vector wavefunctions of the second kind is complete in $[L^2(S)]^3$.*

In order to prove the completeness, it suffices to prove that a vector $\mathbf{U}(\mathbf{r}) \in [L^2(S)]^3$ that is orthogonal to all the spherical vector wavefunctions of the second kind, that is,

$$\left(\mathbf{M}_{nml}^{(2)}, \mathbf{U}\right)_s = \int_S \mathbf{M}_{nml}^{(2)}(\mathbf{r}') \cdot \bar{\mathbf{U}}(\mathbf{r}')dS(\mathbf{r}') = 0,$$

$$\left(\mathbf{N}_{nml}^{(2)}, \mathbf{U}\right)_s = \int_S \mathbf{N}_{nml}^{(2)}(\mathbf{r}') \cdot \bar{\mathbf{U}}(\mathbf{r}')dS(\mathbf{r}') = 0, \qquad (4.27)$$

$$\left(\mathbf{L}_{nml}^{(2)}, \mathbf{U}\right)_s = \int_S \mathbf{L}_{nml}^{(2)}(\mathbf{r}') \cdot \bar{\mathbf{U}}(\mathbf{r}')dS(\mathbf{r}') = 0,$$

must be zero. Let $C_{nm} = -jk/N_{nm}^2$ and $\overset{\leftrightarrow}{\mathbf{I}}$ be the identity dyadic. For $r > r'$, the Green's function has the expansion (Morse and Feshback, 1953)

$$\frac{e^{-jk|\mathbf{r}-\mathbf{r}'|}}{4\pi|\mathbf{r}-\mathbf{r}'|}\overset{\leftrightarrow}{\mathbf{I}} = 2\sum_{n=0}^{\infty}\sum_{m=0}^{n}\sum_{l=e,o} C_{nm}$$
$$\times \left[\mathbf{M}_{nml}^{(1)}(\mathbf{r}')\mathbf{M}_{nml}^{(2)}(\mathbf{r}) + \mathbf{N}_{nml}^{(1)}(\mathbf{r}')\mathbf{N}_{nml}^{(2)}(\mathbf{r}) + \frac{n(n+1)\mathbf{L}_{nml}^{(1)}(\mathbf{r}')\mathbf{L}_{nml}^{(2)}(\mathbf{r})}{k^2}\right]. \qquad (4.28)$$

Let V_0 be a spherical region centered at the origin and contained in S. Multiplying the first, second and third equations of (4.27) by $C_{nm}\mathbf{M}_{nml}^{(1)}(\mathbf{r})$, $C_{nm}\mathbf{N}_{nml}^{(1)}(\mathbf{r})$, $C_{nm}n(n+1)\mathbf{L}_{nml}^{(1)}(\mathbf{r})/k^2$ respectively, with $\mathbf{r} \in V_0$, and summing the equations for all n, m and l, and using (4.28),

we obtain

$$
\mathbf{T}(\mathbf{r}) \equiv \int_S \bar{\mathbf{U}}(\mathbf{r}') \cdot \overleftrightarrow{\mathbf{I}} \frac{e^{-jk|\mathbf{r}-\mathbf{r}'|}}{2\pi \, |\mathbf{r}-\mathbf{r}'|} dS(\mathbf{r}') = 0, \mathbf{r} \in V_0.
$$

This is a solution of the vector Helmholtz equation inside S and is identically zero in V_0. Since the solutions of the Helmholtz equation are analytical, $\mathbf{T}(\mathbf{r})$ and all its derivatives must be zero inside S. Consequently

$$
\mathbf{T}(\mathbf{r})_- = 0, \quad \left. \frac{\partial \mathbf{T}(\mathbf{r})}{\partial n(\mathbf{r})} \right|_- = 0,
$$

where the subscript $-$ denotes the limit value as $\mathbf{r}$ approaches S from the interior of S. Since $\mathbf{T}(\mathbf{r})$ is continuous, this implies

$$
\mathbf{T}(\mathbf{r}) = 0, \mathbf{r} \in S. \tag{4.29}
$$

By the jump relation, we have

$$
\left. \frac{\partial \mathbf{T}(\mathbf{r})}{\partial n(\mathbf{r})} \right|_+ = -2\bar{\mathbf{U}}(\mathbf{r}), \mathbf{r} \in S. \tag{4.30}
$$

Notice that $\mathbf{T}(\mathbf{r})$ satisfies the radiation condition. From the uniqueness theorem of Helmholtz equation and (4.29), $\mathbf{T}(\mathbf{r})$ must be zero outside S, which implies $\partial \mathbf{T}(\mathbf{r})/ \partial n(\mathbf{r})|_+ = 0$. It follows from (4.30) that $\bar{\mathbf{U}} = 0$ in $[L^2(S)]^3$. The proof is completed.

Remark 4.3: The completeness of spherical vector wavefunctions can be generalized to the region between two concentric spheres. In this case, the complete set is

$$
\left\{ \mathbf{L}_{nml}^{(1)}, \mathbf{M}_{nml}^{(1)}, \mathbf{N}_{nml}^{(1)}, \mathbf{L}_{nml}^{(2)}, \mathbf{M}_{nml}^{(2)}, \mathbf{N}_{nml}^{(2)} \right\}. \qquad\qquad \square
$$

4.4 Foster Theorems and Relationship Between Quality Factor and Bandwidth

The **Foster theorems**, named after the American mathematician Ronald Martin Foster (1896–1998), state that the slope of the reactance curve or susceptance curve as a function of frequency is always positive for a lossless circuit. Although the Foster theorems are typically stated for a lossless network in text books, they can be generalized to a lossy network in numerous situations. For example, the Foster reactance theorem holds for a simple series RLC circuit or any lossy network that consists of a resistor serially connected to a lossless network. From the viewpoint of circuit theory, an **ideal antenna**, defined as an antenna without Ohmic loss, is a one-port lossy network with radiation loss. By using the complex frequency domain approach, the Foster theorems can be shown to hold for an ideal antenna (Geyi *et al.*, 2000; Geyi, 2007a).

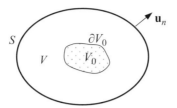

Figure 4.3   Poynting theorem

4.4.1   *Poynting Theorem and the Evaluation of Antenna Quality Factor*

The differential form of the complex Poynting theorem for time-harmonic fields in a homogeneous and isotropic medium is

$$\nabla \cdot \mathbf{S} = -\frac{1}{2}\bar{\mathbf{J}} \cdot \mathbf{E} - j2\omega(w_m - w_e), \tag{4.31}$$

where $\mathbf{S} = \mathbf{E} \times \bar{\mathbf{H}}/2$ is the complex Poynting vector, $w_m = \mu\mathbf{H} \cdot \bar{\mathbf{H}}/4$ and $w_e = \varepsilon\mathbf{E} \cdot \bar{\mathbf{E}}/4$ are the magnetic and electric field energy densities. Let V_0 be the volume occupied by the electric current source $\mathbf{J}$ and ∂V_0 be the surface surrounding V_0. Taking the integration of the imaginary part of (4.31) over a volume V containing V_0 as shown in Figure 4.3, we obtain

$$\text{Im} \int_S \mathbf{u}_n \cdot \mathbf{S}dS = -\text{Im} \int_{V_0} \frac{1}{2}\bar{\mathbf{J}} \cdot \mathbf{E}dV - 2\omega \int_V (w_m - w_e)dV, \tag{4.32}$$

where S is the boundary of V. Choosing $V = V_0$, Equation (4.32) becomes

$$\text{Im} \int_{\partial V_0} \mathbf{u}_n \cdot \mathbf{S}dS = -\text{Im} \int_{V_0} \frac{1}{2}\bar{\mathbf{J}} \cdot \mathbf{E}dV - 2\omega \int_{V_0} (w_m - w_e)dV. \tag{4.33}$$

If we choose $V = V_\infty$, where V_∞ is the region enclosed by a sphere S_∞ with radius r being sufficiently large so that it lies in the far field region of the antenna system, we get

$$\text{Im} \int_{S_\infty} \mathbf{u}_n \cdot \mathbf{S}dS = -\text{Im} \int_{V_0} \frac{1}{2}\bar{\mathbf{J}} \cdot \mathbf{E}dV - 2\omega \int_{V_\infty} (w_m - w_e)dV. \tag{4.34}$$

Since $\mathbf{S}$ is a real vector in the far field region, the above equation reduces to

$$-\text{Im} \int_{V_0} \frac{1}{2}\bar{\mathbf{J}} \cdot \mathbf{E}dV = 2\omega \int_{V_\infty} (w_m - w_e)dV. \tag{4.35}$$

It follows from (4.32), (4.33) and (4.34) that

$$\text{Im} \int_{\partial V_0} \mathbf{u}_n \cdot S dS = 2\omega \int_{V_\infty - V_0} (w_m - w_e) dV, \tag{4.36}$$

$$\text{Im} \int_{S} \mathbf{u}_n \cdot S dS = 2\omega \int_{V_\infty - V} (w_m - w_e) dV. \tag{4.37}$$

Taking the integration of the real part of (4.31) over the volume V containing the source region V_0, we obtain the radiated power

$$P^{rad} = \text{Re} \int_{S} \mathbf{u}_n \cdot S dS = -\text{Re} \int_{V_0} \frac{1}{2} \bar{\mathbf{J}} \cdot \mathbf{E} dV. \tag{4.38}$$

Equation (4.38) shows that the surface integral of the real part of the Poynting vector is independent of the surface S as long as it encloses the source region V_0. Equations (4.36) and (4.37) show that the surface integral of the imaginary part of the Poynting vector depends on the integration surface S in the near field region (in the far field region it becomes zero). Considering (4.32), (4.38) and (4.35) we may find that

$$\int_{S} \mathbf{u}_n \cdot S dS = -\int_{V_0} \frac{1}{2} \bar{\mathbf{J}} \cdot \mathbf{E} dV - j2\omega \int_{V} (w_m - w_e) dV$$

$$= P^{rad} - j\text{Im} \int_{V_0} \frac{1}{2} \bar{\mathbf{J}} \cdot \mathbf{E} dV - j2\omega \int_{V} (w_m - w_e) dV \tag{4.39}$$

$$= P^{rad} + j2\omega \int_{V_\infty - V} (w_m - w_e) dV.$$

The above relation indicates that the complex power flowing out of S is equal to the radiation power plus the reactive power outside S. This expression seems to be the most general form of the Poynting theorem for an open system. Let $\tilde{w}_e(w_e^{rad})$ and $\tilde{w}_m(w_m^{rad})$ denote the stored (radiated) electric field and magnetic field energy densities respectively. The **stored energies** are defined by (Collin and Rothschild, 1964)

$$\tilde{w}_m = w_m - w_m^{rad}, \; \tilde{w}_e = w_e - w_e^{rad}. \tag{4.40}$$

These calculations are physically appropriate since density is a summable quantity. It is readily seen from (4.37) that w_m is equal to w_e in the far field zone, since the complex Poynting vector becomes real as V approaches V_∞. This observation indicates that the electric field energy and the magnetic field energy for the radiated field are identical everywhere, that is,

$$w_e^{rad} = \frac{1}{4} \varepsilon \mathbf{E}^{rad} \cdot \bar{\mathbf{E}}^{rad} = \frac{1}{4} \mu \mathbf{H}^{rad} \cdot \bar{\mathbf{H}}^{rad} = w_m^{rad}. \tag{4.41}$$

The total energy of the radiated fields is simply twice the electric or magnetic energy density of the radiated fields. Mathematically, Equation (4.41) holds everywhere. Consequently, from (4.35), (4.36) and (4.39), we obtain

$$\tilde{W}_m - \tilde{W}_e = \int_{V_\infty - V_0} (\tilde{w}_m - \tilde{w}_e)dV = \int_{V_\infty - V_0} (w_m - w_e)dV = \frac{1}{2\omega}\mathrm{Im}\int_{\partial V_0} \mathbf{S}\cdot\mathbf{u}_n dS. \qquad (4.42)$$

Here $\tilde{W}_m$ and $\tilde{W}_e$ stand for the total stored magnetic and electric energy in the volume surrounding the radiator. Note that the total stored energy can be expressed as

$$\begin{aligned}
\tilde{W}_e + \tilde{W}_m &= \int_{V_\infty - V_0} \left(w_e - w_e^{rad} + w_m - w_m^{rad}\right)dV \\
&= \int_{V_\infty - V_0} (w_e + w_m)dV - \int_{V_\infty - V_0} \left(w_e^{rad} + w_m^{rad}\right)dV \qquad (4.43) \\
&= \int_{V_\infty - V_0} (w_e + w_m)dV - \frac{r}{c}\mathrm{Re}\int_{\partial V_0} \mathbf{S}\cdot\mathbf{u}_n dS,
\end{aligned}$$

where r is the radius of the sphere S_∞, and c is the wave velocity. Both terms on the right-hand side of (4.43) are divergent as $r \to \infty$, but it can be shown that their difference is finite. So the stored electric and magnetic field energies may be obtained from (4.42) and (4.43) as

$$\begin{aligned}
\tilde{W}_m &= \frac{1}{2}\left[\int_{V_\infty - V_0} (w_e + w_m)dV - \frac{r}{c}\mathrm{Re}\int_{\partial V_0} \mathbf{S}\cdot\mathbf{u}_n dS + \frac{1}{2\omega}\mathrm{Im}\int_{\partial V_0} \mathbf{S}\cdot\mathbf{u}_n dS\right], \\
&\qquad\qquad\qquad\qquad\qquad\qquad\qquad\qquad\qquad\qquad\qquad (4.44) \\
\tilde{W}_e &= \frac{1}{2}\left[\int_{V_\infty - V_0} (w_e + w_m)dV - \frac{r}{c}\mathrm{Re}\int_{\partial V_0} \mathbf{S}\cdot\mathbf{u}_n dS - \frac{1}{2\omega}\mathrm{Im}\int_{\partial V_0} \mathbf{S}\cdot\mathbf{u}_n dS\right].
\end{aligned}$$

The antenna Q_{real} is defined by

$$Q_{real} = \begin{cases} \dfrac{2\omega\tilde{W}_m}{P^{rad}}, & \tilde{W}_m > \tilde{W}_e, \\[2mm] \dfrac{2\omega\tilde{W}_e}{P^{rad}}, & \tilde{W}_e > \tilde{W}_m. \end{cases} \qquad (4.45)$$

Remark 4.4: The evaluation of the antenna quality factor can be traced back to the classical work of Chu, who derived the theoretical value of Q for an ideal antenna enclosed in a circumscribing sphere (Chu, 1948). Chu's analysis is based on the spherical mode expansions and is only valid for an omni-directional antenna that radiates either TE or TM modes. In order to avoid the difficulty that the total electric and magnetic field energies are infinite, Chu introduced the equivalent impedance for each mode and obtained an expression of antenna Q through the calculation of stored energies in the equivalent circuit for the impedance. The

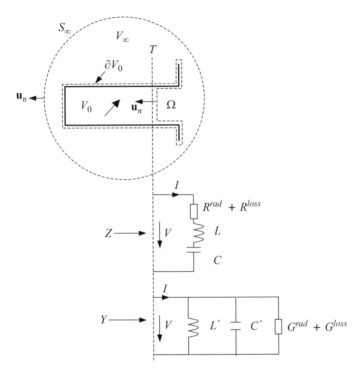

Figure 4.4   Equivalent circuit for transmitting antenna

major shortcomings of the Chu's method are that it is restricted to spherical modes and requires several approximations. Collin and Rothschild proposed a method for evaluating antenna Q (Collin and Rothschild, 1964). Their method is based on the idea that the total stored energy can be calculated by subtracting the radiated field energy from the total energy in the fields. Such method has been successfully used by Fante (Fante, 1969) and re-examined by McLean (McLean, 1996) to study the antenna Q. All these studies only utilize the stored energies outside the circumscribing sphere of the antenna in the calculation of antenna Q, which is much smaller than Q_{real} (for numerical examples, see Collardey et al., 2005, 2006).   □

4.4.2   Equivalent Circuit for Transmitting Antenna

We choose the source region V_0 in such a way that its surface ∂V_0 is coincident with the antenna surface (except at the antenna input terminal Ω where ∂V_0 crosses the antenna reference plane T), as shown in Figure 4.4. Applying Poynting theorem over the region $V_\infty - V_0$ yields

$$\frac{1}{2}\int_{S_\infty}(\mathbf{E}\times\bar{\mathbf{H}})\cdot\mathbf{u}_n dS + \frac{1}{2}\int_{\partial V_0}(\mathbf{E}\times\bar{\mathbf{H}})\cdot\mathbf{u}_n dS$$

$$= -j2\omega\int_{V_\infty-V_0}(w_m - w_e)dV - \frac{1}{2}\int_{V_\infty-V_0}\sigma\mathbf{E}\cdot\bar{\mathbf{E}}dV,$$

(4.46)

where σ is the conductivity of the medium in $V_\infty - V_0$. If we assume that the antenna surface is perfectly conducting, $\mathbf{E} \times \bar{\mathbf{H}}$ vanishes everywhere on ∂V_0 except over the input terminal Ω. We further assume that the antenna reference plane T is away from the antenna discontinuity so that the higher order modes excited by the discontinuity has negligible effects at the reference plane. For a single mode feeding waveguide, we have

$$\frac{1}{2} \int_{\partial V_0} (\mathbf{E} \times \bar{\mathbf{H}}) \cdot \mathbf{u}_n dS = \frac{1}{2} \int_\Omega (\mathbf{E} \times \bar{\mathbf{H}}) \cdot \mathbf{u}_n dS = -\frac{1}{2} V \bar{I}, \tag{4.47}$$

where V and I are equivalent modal voltage and current at the reference plane respectively. Introducing (4.47) into (4.46) and using the fact that $P^{rad} = \frac{1}{2} \int_{\partial V_\infty} (\mathbf{E} \times \bar{\mathbf{H}}) \cdot \mathbf{u}_n dS$, we may find that

$$\frac{1}{2} V \bar{I} = P^{rad} + P^{loss} + j2\omega \int_{V_\infty - V_0} (w_m - w_e) dV,$$

where $P^{loss} = \frac{1}{2} \int_{V_\infty - V_0} \sigma \mathbf{E} \cdot \bar{\mathbf{E}} dV$ stands for the power loss outside the source region V_0. The antenna impedance Z and admittance Y are defined by

$$Z = \frac{V}{I} = \frac{2P^{rad}}{|I|^2} + \frac{2P^{loss}}{|I|^2} + j\frac{4\omega(W_m - W_e)}{|I|^2},$$

$$Y = \frac{I}{V} = \frac{2P^{rad}}{|V|^2} + \frac{2P^{loss}}{|V|^2} + j\frac{4\omega(W_m - W_e)}{|V|^2},$$

where W_m and W_e are the total magnetic energy and electric energy produced by the antenna respectively, and both are infinite as integration region $V_\infty - V_0$ is infinite. By use of (4.42), we have $W_m - W_e = \tilde{W}_m - \tilde{W}_e$. Thus

$$Z = R^{rad} + R^{loss} + jX = R^{rad} + R^{loss} + j\left(\omega L - \frac{1}{\omega C}\right),$$

$$Y = G^{rad} + G^{loss} + jB = G^{rad} + G^{loss} + j\left(\omega C' - \frac{1}{\omega L'}\right),$$

where R, X, G, and B denote the resistance, reactance, conductance and susceptance respectively and their definitions are given below

$$R^{rad} = 2P^{rad}/|I|^2, G^{rad} = 2P^{rad}/|V|^2$$
$$R^{loss} = 2P^{loss}/|I|^2, G^{loss} = 2P^{loss}/|V|^2,$$
$$X = \omega L - 1/\omega C, B = \omega C' - 1/\omega L', \tag{4.48}$$
$$L = 4\tilde{W}_m/|I|^2, L' = |V|^2/4\omega^2 \tilde{W}_m$$
$$C = |I|^2/4\omega^2 \tilde{W}_e, C' = 4\tilde{W}_e/|V|^2.$$

The equivalent circuits for the antenna are shown in Figure 4.4. It should be notified that the values of all elements in the equivalent circuits depend on the frequency.

4.4.3   Foster Theorems for Ideal Antenna and Antenna Quality Factor

To prove that the Foster theorem holds for an ideal antenna, we introduce the complex frequency $s = \alpha + j\omega$ and all calculations are confined to the complex frequency plane. For clarity, all quantities in the complex frequency domain are denoted by the same symbols in real frequency domain but explicitly showing the dependence on s. Taking the Laplace transform of the time-domain Maxwell's equations in a lossless medium yields

$$\nabla \times \mathbf{E}(\mathbf{r}, s) = -s\mu\mathbf{H}(\mathbf{r}, s), \ \nabla \times \mathbf{H}(\mathbf{r}, s) = s\varepsilon\mathbf{E}(\mathbf{r}, s). \tag{4.49}$$

The frequency-domain quantities can be recovered by letting $\alpha = 0$ in (4.49). From (4.49) a relation similar to (4.31) can be obtained in the region outside V_0

$$\nabla \cdot \left[\frac{1}{2}\mathbf{E}(\mathbf{r}, s) \times \bar{\mathbf{H}}(\mathbf{r}, s) \right]$$

$$= -\frac{1}{2}\alpha \left[\mu \, |\mathbf{H}(\mathbf{r}, s)|^2 + \varepsilon \, |\mathbf{E}(\mathbf{r}, s)|^2 \right] - j\frac{1}{2}\omega \left[\mu \, |\mathbf{H}| \, (\mathbf{r}, s)|^2 - \varepsilon \, |\mathbf{E}(\mathbf{r}, s)|^2 \right]. \tag{4.50}$$

Taking the integration of (4.50) over the connected region $V_\infty - V_0$, as shown in Figure 4.4, gives

$$\int_{\partial V_0 + S_\infty} \frac{1}{2} \left[\mathbf{E}(\mathbf{r}, s) \times \bar{\mathbf{H}}(\mathbf{r}, s) \right] \cdot \mathbf{u}_n dS = -2\alpha \left[W_m(s) + W_e(s) \right] - 2j\omega \left[W_m(s) - W_e(s) \right], \tag{4.51}$$

where

$$W_m(s) = \frac{1}{4} \int_{V_\infty - V_0} \mu \, |\mathbf{H}(\mathbf{r}, s)|^2 \, dV(\mathbf{r}),$$

$$W_e(s) = \frac{1}{4} \int_{V_\infty - V_0} \varepsilon \, |\mathbf{E}(\mathbf{r}, s)|^2 \, dV(\mathbf{r}).$$

We assume again that the antenna reference plane is away from the antenna discontinuity so that the higher-order modes excited by the discontinuity have negligible effects at the reference plane. Thus, for a single-mode feeding waveguide, we can make the following approximation

$$\frac{1}{2} \int_{\partial V_0} \left[\mathbf{E}(\mathbf{r}, s) \times \bar{\mathbf{H}}(\mathbf{r}, s) \right] \cdot \mathbf{u}_n dS(\mathbf{r}) = -\frac{1}{2} V(s)\bar{I}(s). \tag{4.52}$$

Letting $P^{rad}(s) = \frac{1}{2} \int\limits_{S_\infty} \left[\mathbf{E}(\mathbf{r}, s) \times \bar{\mathbf{H}}(\mathbf{r}, s) \right] \cdot \mathbf{u}_n dS(\mathbf{r})$ and substituting it into (4.51), we get

$$\frac{1}{2} V(s) \bar{I}(s) = P^{rad}(s) + 2\alpha \left[W_m(s) + W_e(s) \right] + 2j\omega \left[W_m(s) - W_e(s) \right]. \tag{4.53}$$

If α is sufficiently small so that $\alpha \ll v/r$ with $v = 1/\sqrt{\mu\varepsilon}$, we can make a first-order approximation $e^{-\alpha r/c} \approx 1 - \alpha r/v$, and derive directly from the Maxwell equations, defined in the complex plane, the following:

$$\mathbf{E}^{rad}(\mathbf{r}, s) \approx -\left(1 - r\frac{\alpha}{v} \right) \frac{j\omega\mu}{4\pi r} e^{-jkr} \int\limits_{V_0} \left[\mathbf{J}(\mathbf{r}', s) - \mathbf{J}(\mathbf{r}', s) \cdot \mathbf{u}_r \right] e^{-jk\mathbf{u}_r \cdot \mathbf{r}'} dV(\mathbf{r}'),$$

$$\mathbf{H}^{rad}(\mathbf{r}, s) \approx -\eta \left(1 - r\frac{\alpha}{v} \right) \frac{j\omega\varepsilon}{4\pi r} e^{-jkr} \int\limits_{V_0} \mathbf{u}_r \times \mathbf{J}(\mathbf{r}', s) e^{-jk\mathbf{u}_r \cdot \mathbf{r}'} dV(\mathbf{r}').$$

Hence

$$P^{rad}(s) = P^{rad}(\omega)(1 - r\alpha/v)^2 \approx (1 - 2r\alpha/v) P^{rad}(\omega) \tag{4.54}$$

where $P^{rad}(\omega)$, previously defined in (4.38), is the radiated power in the frequency domain, which is independent of α. Substituting (4.54) into (4.53), we obtain

$$\frac{1}{2} V(s) \bar{I}(s) = P^{rad}(\omega) + 2\alpha \left[W_m(s) + W_e(s) - \frac{r}{c} P^{rad}(\omega) \right] + 2j\omega \left[W_m(s) - W_e(s) \right]. \tag{4.55}$$

The impedance and admittance in the complex frequency plane can then be expressed as

$$Z(s) = \frac{2 P^{rad}(\omega)}{|I(s)|^2} + \frac{4\alpha}{|I(s)|^2} \left[W_m(s) + W_e(s) - \frac{r}{c} P^{rad}(\omega) \right] + \frac{4j\omega}{|I(s)|^2} \left[W_m(s) - W_e(s) \right],$$

$$Y(s) = \frac{2 P^{rad}(\omega)}{|V(s)|^2} + \frac{4\alpha}{|V(s)|^2} \left[W_m(s) + W_e(s) - \frac{r}{c} P^{rad}(\omega) \right] + \frac{4j\omega}{|V(s)|^2} \left[W_m(s) - W_e(s) \right]. \tag{4.56}$$

Similarly, we can introduce the stored energies in the complex frequency domain

$$\tilde{W}_m(s) + \tilde{W}_e(s) = W_m(s) + W_e(s) - \frac{r}{c} P^{rad}(\omega),$$

$$\tilde{W}_m(s) - \tilde{W}_e(s) = W_m(s) - W_e(s),$$

and rewrite (4.56) as

$$Z(s) = \frac{2P^{rad}(\omega)}{|I(s)|^2} + \frac{4s\,\tilde{W}_m(s)}{|I(s)|^2} + \frac{4\bar{s}\,\tilde{W}_e(s)}{|I(s)|^2},$$

$$Y(s) = \frac{2P^{rad}(\omega)}{|V(s)|^2} + \frac{4s\,\tilde{W}_m(s)}{|V(s)|^2} + \frac{4\bar{s}\,\tilde{W}_e(s)}{|V(s)|^2}.$$

(4.57)

To get rid of the complex conjugation $\bar{s}$, we may introduce a new quantity $\tilde{W}'_e(s) = |s|^2\,\tilde{W}_s(s)$. Now (4.57) can be extended to an analytic function of s by replacing all complex conjugations $\bar{s}$ in $\tilde{W}_m(s)$ and $\tilde{W}'_e(s)$ with $-s$, and $j\omega$ (resp. $-j\omega$) by s (resp. $-s$) in $P^{rad}(\omega)$. Thus, (4.57) become analytic and can be written as

$$Z(s) = \frac{2P^{rad}(s)}{I(s)I(-s)} + \frac{4s\,\tilde{W}_m(s)}{I(s)I(-s)} + \frac{4s^{-1}\,\tilde{W}'_e(s)}{I(s)I(-s)},$$

$$Y(s) = \frac{2P^{rad}(s)}{V(s)V(-s)} + \frac{4s\,\tilde{W}_m(s)}{V(s)V(-s)} + \frac{4s^{-1}\,\tilde{W}'_e(s)}{V(s)V(-s)}.$$

(4.58)

Note that (4.57) and (4.58) are identical when $\alpha = 0$. If α is assumed to be small, a Taylor expansion may be assumed for an arbitrary analytic function $\mathbf{A}(s)$ so that

$$\mathbf{A}(s)\mathbf{A}(-s) = |\mathbf{A}(j\omega)|^2 + j\alpha T(\omega) + o(\alpha)$$

where $T(\omega)$ is a real function of ω. When this relation is used in (4.58) and use is made of the following decompositions

$$Z(s) = R(\alpha, \omega) + jX(\alpha, \omega),$$

$$Y(s) = G(\alpha, \omega) + jB(\alpha, \omega),$$

(4.59)

we may find that

$$R(\alpha, \omega) = \frac{2P^{rad}}{|I|^2} + \frac{4\alpha}{|I|^2}(\tilde{W}_m + \tilde{W}_e),$$

$$G(\alpha, \omega) = \frac{2P^{rad}}{|V|^2} + \frac{4\alpha}{|V|^2}(\tilde{W}_m + \tilde{W}_e),$$

(4.60)

where the power, energies, voltage and current are all calculated at $\alpha = 0$. Since $Z(s)$ and $Y(s)$ are analytic functions, their real and imaginary parts satisfy the Cauchy-Riemann conditions

$$\frac{\partial R(\alpha, \omega)}{\partial \alpha} = \frac{\partial X(\alpha, \omega)}{\partial \omega}, \quad \frac{\partial G(\alpha, \omega)}{\partial \alpha} = \frac{\partial B(\alpha, \omega)}{\partial \omega},$$

(4.61)

$$\frac{\partial R(\alpha, \omega)}{\partial \omega} = -\frac{\partial X(\alpha, \omega)}{\partial \alpha}, \quad \frac{\partial G(\alpha, \omega)}{\partial \omega} = -\frac{\partial B(\alpha, \omega)}{\partial \alpha}.$$

(4.62)

By direct calculation, we have

$$\frac{\partial R(\alpha, \omega)}{\partial \alpha}\bigg|_{\alpha=0} = \frac{4(\tilde{W}_m + \tilde{W}_e)}{|I|^2}, \quad \frac{\partial G(\alpha, \omega)}{\partial \alpha}\bigg|_{\alpha=0} = \frac{4(\tilde{W}_m + \tilde{W}_e)}{|V|^2}. \tag{4.63}$$

From (4.61) and (4.63), we obtain

$$\frac{\partial X}{\partial \omega}\bigg|_{\alpha=0} = \frac{4}{|I|^2}(\tilde{W}_m + \tilde{W}_e) > 0, \quad \frac{\partial B}{\partial \omega}\bigg|_{\alpha=0} = \frac{4}{|V|^2}(\tilde{W}_m + \tilde{W}_e) > 0. \tag{4.64}$$

These are the Foster theorems for a lossless antenna system, which indicate that the slope of the reactance curve or the susceptance curve as a function of the frequency for an ideal antenna is always positive. From (4.64), we obtain the stored magnetic and electric field energies

$$\tilde{W}_e = \frac{1}{8}|I|^2\left(\frac{\partial X}{\partial \omega} - \frac{X}{\omega}\right), \quad \tilde{W}_m = \frac{1}{8}|I|^2\left(\frac{\partial X}{\partial \omega} + \frac{X}{\omega}\right). \tag{4.65}$$

So the antenna quality factor may be written as

$$Q_{real} = \frac{1}{2R^{rad}}\left(\omega\frac{dX}{d\omega} \pm X\right) = \frac{1}{2G^{rad}}\left(\omega\frac{dB}{d\omega} \pm B\right), \tag{4.66}$$

where either $+$ or $-$ is chosen to give the higher Q_{real}.

Several remarks are needed to clarify some misunderstandings about the Foster theorems in antenna engineering.

Remark 4.5: Some antenna impedances or the admittances may have singularities depending on the position of the reference plane, where the reactance curve or susceptance curve abruptly changes from positive infinity to negative infinity so that the Foster theorems hold. It should be noted that all numerical methods cannot accurately model these singularities due to numerical errors, and thus a negative (but very steep) slope may occur. For the same reason, all antenna measurements cannot accurately handle singularities and in addition the Ohmic loss will be introduced in the measurements. Therefore the negative slope may appear in the measured reactance or susceptance curves, and the Foster theorems only hold approximately. □

Remark 4.6: The Foster theorems only hold for an ideal antenna with the feeding waveguide connected. The modeling of wire antennas is usually based on a number of approximations. The most dramatic approximation is that the feeding waveguide is replaced by a delta gap. This kind of approximation is questionable and cannot be checked experimentally since all practical antennas involve a feeding waveguide, and the feeding waveguide itself contributes significantly to the value of antenna impedance. When the delta gap is used to calculate the impedance of thick wires, one should limit the calculation to the low frequency range since the delta gap is only valid for thin wires. □

Remark 4.7: The Foster theorems hold only in the frequency range between the cut-off frequency of the dominant mode and the cut-off frequency of the first higher-order mode of the feeding waveguide as a single-mode assumption has been used at the antenna terminal in the proof of Foster theorems for an ideal antenna. When the operating frequency is higher than the cut-off frequency of the first higher-order mode, the feeding waveguide is equivalent to a multiple-transmission line system and the antenna becomes a multi-port device. In this case, the Foster theorems fail. ☐

Remark 4.8: All circuit parameters for a microwave network depend on the frequency. A common mistake made in the study of the Foster theorems for antenna is that either the terminal voltage or current of the antenna are assumed to be independent of frequency. This assumption will lead to additional terms on the right-hand side of (4.64), which cannot be explained physically. ☐

4.4.4   Relationship Between Antenna Quality Factor and Bandwidth

We now use the Foster theorems to prove a widely-held assumption that the antenna quality factor is the inversion of antenna fractional bandwidth. Consider a high quality factor system. Let ω_r denote one of the resonant frequencies of a single antenna system; then by definition, we have

$$X(\omega_r) = 0. \tag{4.67}$$

Since α is very small, we have $X(\alpha, \omega_r) \approx X(\omega_r) = 0$ at the resonant frequency ω_r. From (4.62), we obtain

$$\left.\frac{dR^{rad}}{d\omega}\right|_{\omega_r} = -\left.\frac{\partial X(\alpha, \omega_r)}{\partial \alpha}\right|_{\alpha=0} \approx 0.$$

Thus, as one moves off resonance, Z can be written as

$$Z \approx R^{rad} + j(\omega - \omega_r)\left.\frac{dX}{d\omega}\right|_{\omega_r} + \cdots.$$

The frequency at which the absolute value of the input impedance is equal to $\sqrt{2}$ times its value at resonance is the half-power point. The half-power points occur when

$$R^{rad} = \left|(\omega - \omega_r)\left.\frac{dX_A}{d\omega}\right|_{\omega_r}\right|, \tag{4.68}$$

so that the fractional bandwidth B_f can be written

$$B_f = \frac{2|\omega - \omega_r|}{\omega_r} \approx \frac{2R^{rad}}{\omega_r |dX/d\omega|_{\omega_r}} = \frac{4P^{rad}}{\omega_r |I|^2 |dX/d\omega|_{\omega_r}}. \tag{4.69}$$

From (4.65) we obtain

$$\frac{dX}{d\omega} = \frac{8\tilde{W}_e}{|I|^2} + \frac{X}{\omega}, \frac{dX}{d\omega} = \frac{8\tilde{W}_m}{|I|^2} - \frac{X}{\omega}.$$

Introducing the above into (4.69) and using (4.67), we find

$$B_f = \frac{4P^{rad}}{\left|8\omega_r \tilde{W} \pm |I|^2 X\right|_{\omega_r}} = \frac{4P^{rad}}{\left|8\omega_r \tilde{W}\right|_{\omega_r}} = \frac{1}{Q_{real}}. \tag{4.70}$$

Here $\tilde{W} = \tilde{W}_m$ or $\tilde{W}_e$, whichever is larger. Thus, we have proved that the antenna fractional bandwidth is the inversion of antenna Q_{real} when $Q_{real} \gg 1$.

4.5 Minimum Possible Antenna Quality Factor

The study of antenna quality factor was usually based on the spherical wavefunction expansion outside the circumscribing sphere of the antenna. The antenna quality factor resulting from the spherical wavefunction expansion is much lower than the real value Q_{real} as the stored energy inside the circumscribing sphere has been ignored.

4.5.1 Spherical Wavefunction Expansion for Antenna Quality Factor

Assume that the antenna is enclosed by the circumscribing sphere of radius a, denoted by V_a with bounding surface S_a. The stored energies outside the circumscribing sphere can be evaluated through (4.42) and (4.43)

$$\tilde{W}_m - \tilde{W}_e = \frac{1}{2\omega}\text{Im} \int_{S_a} \mathbf{S} \cdot \mathbf{u}_n dS = \frac{1}{4\omega}\text{Im} \int_0^{2\pi} d\phi \int_0^{\pi} (\mathbf{E} \times \bar{\mathbf{H}}) \cdot \mathbf{u}_r a^2 \sin\theta d\theta,$$

$$\tilde{W}_e + \tilde{W}_m = \int_a^{\infty} dr \left\{ \int_0^{2\pi} d\varphi \int_0^{\pi} r^2 \left(\frac{\varepsilon}{4}|\mathbf{E}|^2 + \frac{\mu}{4}|\mathbf{H}|^2\right) \sin\theta d\theta - \frac{1}{2c}\text{Re} \int_{\partial V_{\infty}} \mathbf{E} \times \bar{\mathbf{H}} \cdot \mathbf{u}_r dS \right\}.$$

From (4.16) and (4.18), we obtain

$$P^{rad} = \frac{1}{2}\text{Re} \int_{S_{\infty}} \mathbf{E} \times \bar{\mathbf{H}} \cdot \mathbf{u}_r dS = \frac{1}{2k^2\eta} \sum_{n,m,l} N_{nm}^2 \left(\left|\alpha_{nml}^{(2)}\right|^2 + \left|\beta_{nml}^{(2)}\right|^2\right), \tag{4.71}$$

$$\tilde{W}_e + \tilde{W}_m = \frac{\varepsilon a}{4k^2} \sum_{n,m,l} N_{nm}^2 \left(\left|\alpha_{nml}^{(2)}\right|^2 + \left|\beta_{nml}^{(2)}\right|^2\right)$$

$$\cdot \left\{ 2 - (ka)^2 \left[\left|h_n^{(2)}(ka)\right|^2 - j_{n-1}(ka)j_{n+1}(ka) - n_{n-1}(ka)n_{n+1}(ka)\right] \tag{4.72} \right.$$

$$\left. - \left|h_n^{(2)}(ka)\right|^2 - (ka)[j_n(ka)\dot{j}_n(ka) + n_n(kr)\dot{n}_n(ka)] \right\},$$

$$\tilde{W}_m - \tilde{W}_e = \frac{\varepsilon a}{4k^2} \sum_{n,m,l} N_{nm}^2 \left[\left|\beta_{nml}^{(2)}\right|^2 - \left|\alpha_{nml}^{(2)}\right|^2\right]$$

$$\cdot \left\{ \left|h_n^{(2)}(ka)\right|^2 + (ka)[j_n(ka)\dot{j}_n(ka) + n_n(ka)\dot{n}_n(ka)] \right\}.$$

It follows from (4.72) that

$$
\begin{aligned}
\tilde{W}_m &= \frac{\varepsilon}{4k^3} \sum_{n,m,l} N_{nm}^2 \left(\left| \alpha_{nml}^{(2)} \right|^2 Q_n + \left| \beta_{nml}^{(2)} \right|^2 Q_n' \right), \\
\tilde{W}_e &= \frac{\varepsilon}{4k^3} \sum_{n,m,l} N_{nm}^2 \left(\left| \alpha_{nml}^{(2)} \right|^2 Q_n' + \left| \beta_{nml}^{(2)} \right|^2 Q_n \right),
\end{aligned}
\tag{4.73}
$$

where

$$
\begin{aligned}
Q_n &= ka - \left| h_n^{(2)}(ka) \right|^2 \left[\frac{1}{2}(ka)^3 + ka(n+1) \right] - \frac{1}{2}(ka)^3 \left| h_{n+1}^{(2)}(ka) \right|^2 \\
&\quad + \frac{1}{2}(ka)^2(2n+3) \left[j_n(ka) j_{n+1}(ka) + n_n(ka) n_{n+1}(ka) \right], \\
Q_n' &= ka - \frac{1}{2}(ka)^3 \left[\left| h_n^{(2)}(ka) \right|^2 - j_{n-1}(ka) j_{n+1}(ka) - n_{n-1}(ka) n_{n+1}(ka) \right].
\end{aligned}
\tag{4.74}
$$

Plots of Q_n and Q_n' show that $Q_n > Q_n'$, $Q_{n+1} > Q_n$ and $Q_{n+1}' > Q_n'$ (Fante, 1969). Furthermore we have $Q_n \gg Q_n'$ for small ka. If ka has the order of n, Q_n and Q_n' are of the same order of magnitude. For the first three modes, we have

$$
\begin{aligned}
Q_1 &= \frac{1}{ka} + \frac{1}{(ka)^3}, & Q_1' &= \frac{1}{ka}, \\
Q_2 &= \frac{3}{ka} + \frac{6}{(ka)^3} + \frac{18}{(ka)^5}, & Q_2' &= \frac{3}{ka} + \frac{4}{(ka)^3}, \\
Q_3 &= \frac{6}{ka} + \frac{21}{(ka)^3} + \frac{135}{(ka)^5} + \frac{675}{(ka)^7}, & Q_3' &= \frac{6}{ka} + \frac{15}{(ka)^3} + \frac{45}{(ka)^5}.
\end{aligned}
$$

We now can define an antenna quality factor Q based on (4.71) and (4.73)

$$
Q = \begin{cases} \dfrac{2\omega \tilde{W}_m}{P_{rad}}, & \tilde{W}_m > \tilde{W}_e \\[2ex] \dfrac{2\omega \tilde{W}_e}{P_{rad}}, & \tilde{W}_e > \tilde{W}_m \end{cases} = \text{Larger of} \left\{ \frac{\sum\limits_{n=1}^{\infty} \left(a_n^2 Q_n + b_n^2 Q_n' \right)}{\sum\limits_{n=1}^{\infty} \left(a_n^2 + b_n^2 \right)}, \frac{\sum\limits_{n=1}^{\infty} \left(a_n^2 Q_n' + b_n^2 Q_n \right)}{\sum\limits_{n=1}^{\infty} \left(a_n^2 + b_n^2 \right)} \right\}
\tag{4.75}
$$

where

$$
a_n^2 = \sum_{m,l} N_{nm}^2 \left| \alpha_{nml}^{(2)} \right|^2, \quad b_n^2 = \sum_{m,l} N_{nm}^2 \left| \beta_{nml}^{(2)} \right|^2.
\tag{4.76}
$$

The antenna Q defined by (4.75) is much smaller than the real value Q_{real}.

4.5.2 Minimum Possible Antenna Quality Factor

To study the minimum possible antenna quality factor, we first assume that the antenna only radiates TE modes. Since $Q_n > Q'_n$, Equation (4.75) reduces to

$$Q^{TE} = \frac{\sum\limits_{n=1}^{\infty} a_n^2 Q_n}{\sum\limits_{n=1}^{\infty} a_n^2}.$$

In this case, the minimum possible antenna Q^{TE} will be Q_1. In fact, we have

$$Q^{TE} = \frac{\sum\limits_{n=1}^{\infty} a_n^2 Q_n}{\sum\limits_{n=1}^{\infty} a_n^2} \geq \frac{\sum\limits_{n=1}^{\infty} a_n^2 Q_1}{\sum\limits_{n=1}^{\infty} a_n^2} = Q_1 = \min Q^{TE}$$

since $Q_{n+1} > Q_n$, and $\min Q^{TE}$ can be achieved by setting $a_n = 0$ for $n \geq 2$. If the antenna only radiates TM modes, a similar discussion leads to

$$\min Q^{TE} = \min Q^{TM} = Q_1, \tag{4.77}$$

which can be reached by letting $b_n = 0$ for $n \geq 2$. In the general situation, both TE and TM modes exist. If the first expression of (4.75) is assumed to be the largest, the antenna Q is

$$Q = \frac{\sum\limits_{n=1}^{\infty} \left(a_n^2 Q_n + b_n^2 Q'_n \right)}{\sum\limits_{n=1}^{\infty} \left(a_n^2 + b_n^2 \right)}.$$

To insure that the first expression of (4.75) is always larger than the second during the optimizing process, a constraint on the coefficients a_n and b_n is needed. This can be achieved by assuming $a_n \geq b_n$. Under this condition, we have

$$Q = \frac{\sum\limits_{n=1}^{\infty} \left(a_n^2 Q_n + b_n^2 Q'_n \right)}{\sum\limits_{n=1}^{\infty} \left(a_n^2 + b_n^2 \right)} \geq \frac{\sum\limits_{n=1}^{\infty} \left(a_n^2 Q_1 + b_n^2 Q'_1 \right)}{\sum\limits_{n=1}^{\infty} \left(a_n^2 + b_n^2 \right)} = C(Q_1 - Q'_1) + Q'_1,$$

where $C = \sum\limits_{n=1}^{\infty} a_n^2 / \sum\limits_{n=1}^{\infty} \left(a_n^2 + b_n^2 \right) \geq \frac{1}{2}$ since $a_n \geq b_n$. The right-hand side of the above expression can be minimized by setting $C = 1/2$ or $a_n = b_n (n \geq 1)$. Therefore,

$$\min Q = \frac{Q_1 + Q'_1}{2} = \frac{1}{ka} + \frac{1}{2(ka)^3}, \tag{4.78}$$

which can be achieved by setting $a_1 = b_1$ and $a_n = b_n = 0$ for $n \geq 2$. If the second expression in (4.75) is the largest, exactly the same result can be obtained by interchanging a_n and b_n. Thus, the minimum Q problem has a unique lowest limit although the Q is defined conditionally. Therefore, the antenna will attain the lowest Q if only TE_{1m} and TM_{1m} modes are equally excited. In this case, the stored electric energy and magnetic energy outside the circumscribing sphere will be equal, and the antenna will be at resonance outside the sphere. The existence of a lower bound for antenna Q implies that the stored energy around antenna can never be made zero. Once the maximum antenna size is given, this lower bound is then determined. For a small antenna ($ka < 1$), Equations (4.77) and (4.78) can be approximated by

$$\min Q^{TE} = \min Q^{TM} \approx \frac{1}{(ka)^3}, \quad \min Q \approx \frac{1}{2(ka)^3}. \tag{4.79}$$

Since Q_{real} is always greater than the Q defined by (4.75), Equation (4.78) may be considered the minimum possible value for Q_{real}.

4.6   Maximum Possible Product of Gain and Bandwidth

In most applications, we need to maximize antenna gain and bandwidth simultaneously. For this reason, a reasonable quantity characterizing antenna would be the product of antenna gain and bandwidth, or the ratio of antenna gain to antenna Q_{real}. The optimization of the ratio of the gain to Q_{real} is more important from the practical point of view. The ratio of gain to Q_{real} is actually the ratio of radiation intensity over the averaged stored energy around the antenna. Since the antenna quality factor is defined conditionally as shown in (4.45), the optimization of the ratio of gain to Q_{real} is subject to certain constraints (Geyi, 2003a). In order to seek the maximum possible ratio of gain over antenna quality factor, we can use Q defined by (4.75) to replace Q_{real} in the optimization process.

4.6.1   Directive Antenna

We assume that the antenna is placed in a spherical coordinate system (r, θ, φ) and enclosed by the smallest circumscribing sphere of radius a, and the spherical coordinate system is oriented in such a way that the maximum radiation is in $(\theta, \varphi) = (0, 0)$ direction. Considering the basis functions in (4.18)

$$\mathbf{e}_{nml} = \frac{\mathbf{u}_\theta}{N_{nm} \sin \theta} \cdot f_{ml}(\varphi) \left[(n+1) \cos \theta \, P_n^m(\cos \theta) - (n-m+1) P_{n+1}^m(\cos \theta) \right]$$
$$- \frac{\mathbf{u}_\varphi}{N_{nm} \sin \theta} \cdot \left[P_n^m(\cos \theta) f'_{ml}(\varphi) \right],$$

we may find that only $m = 1$ contributes to the field in the direction of $\theta = 0$. Hence making use of the relationships

$$\lim_{\theta \to \pi} \frac{P_n^1(\cos \theta)}{\sin \theta} = \frac{1}{2}(-1)^n n(n+1), \quad \lim_{\theta \to 0} \frac{P_n^1(\cos \theta)}{\sin \theta} = \frac{1}{2} n(n+1),$$

we obtain

$$\lim_{\theta \to 0} \mathbf{e}_{n1l} = -\frac{1}{2N_{n1}}(n+1)n\left[\mathbf{u}_\theta f_{1l}(\varphi) + \mathbf{u}_\varphi f'_{1l}(\varphi)\right],$$

$$\lim_{\theta \to 0} \mathbf{h}_{n1l} = -\frac{1}{2N_{n1}}(n+1)n\left[\mathbf{u}_\varphi f_{1l}(\varphi) - \mathbf{u}_\theta f'_{1l}(\varphi)\right].$$

From (4.18), the far field components in the direction of $\varphi = 0$ are

$$E_\theta = -\frac{1}{2kr}\sum_n (n+1)n\left[\dot{\tilde{h}}_n^{(2)}(kr)\beta_{n1e}^{(2)} + \tilde{h}_n^{(2)}(kr)\alpha_{n1o}^{(2)}\right],$$

$$H_\varphi = \frac{1}{j2\eta kr}\sum_n (n+1)n\left[-\tilde{h}_n^{(2)}(kr)\beta_{n1e}^{(2)} + \dot{\tilde{h}}_n^{(2)}(kr)\alpha_{n1o}^{(2)}\right],$$

$$E_\varphi = -\frac{1}{2kr}\sum_n (n+1)n\left[\dot{\tilde{h}}_n^{(2)}(kr)\beta_{n1o}^{(2)} - \tilde{h}_n^{(2)}(kr)\alpha_{n1e}^{(2)}\right],$$

$$H_\theta = \frac{1}{j2\eta kr}\sum_n (n+1)n\left[\tilde{h}_n^{(2)}(kr)\beta_{n1o}^{(2)} + \dot{\tilde{h}}_n^{(2)}(kr)\alpha_{n1e}^{(2)}\right].$$

For sufficiently large r, the radiation intensity can be written as

$$\frac{1}{2}r^2\mathrm{Re}(\mathbf{E} \times \bar{\mathbf{H}}) \cdot \mathbf{u}_r = \frac{1}{2}r^2\mathrm{Re}(E_\theta \bar{H}_\varphi - E_\varphi \bar{H}_\theta)$$

$$= \frac{1}{8k^2\eta}\left|\sum_n (n+1)nj^n\left(\beta_{n1e}^{(2)} + j\alpha_{n1o}^{(2)}\right)\right|^2 + \frac{1}{8k^2\eta}\left|\sum_n (n+1)nj^n\left(\beta_{n1o}^{(2)} - j\alpha_{n1e}^{(2)}\right)\right|^2.$$

The directivity in the direction of $(\theta, \varphi) = (0, 0)$ is then given by

$$G = 4\pi r^2 \frac{\frac{1}{2}\mathrm{Re}(\mathbf{E} \times \bar{\mathbf{H}}) \cdot \mathbf{u}_r}{P_{rad}}$$

$$= \pi \frac{\left|\sum_n (n+1)nj^n\left(\beta_{n1e}^{(2)} + j\alpha_{n1o}^{(2)}\right)\right|^2 + \left|\sum_n (n+1)nj^n\left(\beta_{n1o}^{(2)} - j\alpha_{n1e}^{(2)}\right)\right|^2}{\sum_{n,m,l} N_{nm}^2\left(\left|\alpha_{nml}^{(2)}\right|^2 + \left|\beta_{nml}^{(2)}\right|^2\right)}. \tag{4.80}$$

From (4.75) and (4.80), we obtain

$$\left.\frac{G}{Q}\right|_{dir} = \frac{\pi}{d}\left[\left|\sum_{n=1}^\infty (n+1)nj^n\left(\beta_{n1e}^{(2)} + j\alpha_{n1o}^{(2)}\right)\right|^2 + \left|\sum_{n=1}^\infty (n+1)nj^n\left(\beta_{n1o}^{(2)} - j\alpha_{n1e}^{(2)}\right)\right|^2\right],$$

$$\tag{4.81}$$

where the subscript dir is used to indicate a directional pattern, and

$$d = \text{Larger of}$$

$$\left\{ \sum_{n,m,l} \left(N_{nm}^2 \left| \alpha_{nml}^{(2)} \right|^2 Q_n + N_{nm}^2 \left| \beta_{nml}^{(2)} \right|^2 Q_n' \right), \sum_{n,m,l} \left(N_{nm}^2 \left| \alpha_{nml}^{(2)} \right|^2 Q_n' + N_{nm}^2 \left| \beta_{nml}^{(2)} \right|^2 Q_n \right) \right\}.$$

$$(4.82)$$

Since only $\alpha_{n1l}^{(2)}$ and $\beta_{n1l}^{(2)}$ contribute to the numerator of (4.81), Equation (4.81) can be increased by setting $\alpha_{nml}^{(2)} = \beta_{nml}^{(2)} = 0(m \neq 1)$. Thus

$$d = \text{Larger of} \begin{cases} \displaystyle\sum_{n=1}^{\infty} N_{n1}^2 \left[\left(\left| \alpha_{n1o}^{(2)} \right|^2 + \left| \alpha_{n1e}^{(2)} \right|^2 \right) Q_n + \left(\left| \beta_{n1e}^{(2)} \right|^2 + \left| \beta_{n1o}^{(2)} \right|^2 \right) Q_n' \right], \\ \displaystyle\sum_{n=1}^{\infty} N_{n1}^2 \left[\left(\left| \alpha_{n1o}^{(2)} \right|^2 + \left| \alpha_{n1e}^{(2)} \right|^2 \right) Q_n' + \left(\left| \beta_{n1e}^{(2)} \right|^2 + \left| \beta_{n1o}^{(2)} \right|^2 \right) Q_n \right]. \end{cases}$$

Setting

$$A_{on} = j^{n+1} n(n+1) \alpha_{n1o}^{(2)}, \quad B_{en} = j^n n(n+1) \beta_{n1e}^{(2)},$$
$$A_{en} = -j^{n+1} n(n+1) \alpha_{n1o}^{(2)}, \quad B_{on} = j^n n(n+1) \beta_{n1o}^{(2)},$$

we have

$$\left. \frac{G}{Q} \right|_{dir} = \frac{1}{d} \left[\left| \sum_{n=1}^{\infty} (A_{on} + B_{en}) \right|^2 + \left| \sum_{n=1}^{\infty} (A_{en} + B_{on}) \right|^2 \right], \qquad (4.83)$$

where

$$d = \text{Larger of} \begin{cases} \displaystyle\sum_{n=1}^{\infty} \frac{2}{2n+1} \left[(|A_{on}|^2 + |A_{en}|^2)Q_n + (|B_{en}|^2 + |B_{on}|^2)Q_n' \right], \\ \displaystyle\sum_{n=1}^{\infty} \frac{2}{2n+1} \left[(|A_{on}|^2 + |A_{en}|^2)Q_n' + (|B_{en}|^2 + |B_{on}|^2)Q_n \right]. \end{cases} \qquad (4.84)$$

The denominator of (4.83) depends only on the magnitudes of A_n and B_n. If we adjust the phase of A_n and B_n so that they are in phase to maximize the numerator, the denominator will not change. Therefore, (4.83) can be rewritten as

$$\left. \frac{G}{Q} \right|_{dir} = \frac{1}{d} \left\{ \left[\sum_{n=1}^{\infty} (|A_{on}| + |B_{en}|) \right]^2 + \left[\sum_{n=1}^{\infty} (|A_{en}| + |B_{on}|) \right]^2 \right\}. \qquad (4.85)$$

Suppose that the upper expression of (4.84) is the largest. In order that the upper expression of (4.84) is always larger than the lower expression in the optimizing process, we must place a restriction on $|A_n|$ and $|B_n|$:

$$|A_{on}| \geq |B_{en}|, \ |A_{en}| \geq |B_{on}|. \tag{4.86}$$

Under this condition, the ratio in (4.85) can be maximized by letting $|A_{on}| = |B_{en}|$ and $|A_{en}| = |B_{on}|$. From Cauchy-Schwartz inequality, we obtain

$$\left.\frac{G}{Q}\right|_{dir} = 2 \frac{\left(\sum\limits_{n=1}^{\infty} |A_{on}|\right)^2 + \left(\sum\limits_{n=1}^{\infty} |A_{en}|\right)^2}{\sum\limits_{n=1}^{\infty} \frac{1}{2n+1}(|A_{on}|^2 + |A_{en}|^2)(Q_n + Q_n')}$$

$$= 2\frac{(\boldsymbol{\zeta}, \mathbf{C}_o)_E^2 + (\boldsymbol{\zeta}, \mathbf{C}_e)_E^2}{(\mathbf{C}_o, \mathbf{C}_o)_E + (\mathbf{C}_e, \mathbf{C}_e)_E} \leq 2\|\boldsymbol{\zeta}\|_E^2 . \tag{4.87}$$

where $\boldsymbol{\zeta} = (\zeta_1, \zeta_2, \cdots)$, $\mathbf{C}_{e(o)} = (C_{e(o)1}, C_{e(o)2}, \cdots)$ with

$$C_{e(o)n} = \sqrt{\frac{Q_n + Q_n'}{2n+1}} |A_{e(o)n}|, \ \zeta_n = \sqrt{\frac{2n+1}{Q_n + Q_n'}}.$$

Both $\boldsymbol{\zeta}$ and $\mathbf{C}_{e(o)}$ are vectors in the Euclidean space consisting of all vectors of infinite dimension with the inner product and norm defined by $(\boldsymbol{\zeta}, \mathbf{C})_E = \sum\limits_{n=1}^{\infty} \zeta_n C_n$ and $\|\boldsymbol{\zeta}\| = (\boldsymbol{\zeta}, \boldsymbol{\zeta})_E^{1/2}$, respectively. The ratio of gain to Q reaches maximum if $\mathbf{C}_e = \mathbf{C}_o = c_1 \boldsymbol{\zeta}$, that is,

$$|A_{en}| = |A_{on}| = |B_{en}| = |B_{on}| = c_1 \frac{2n+1}{Q_n + Q_n'}, \tag{4.88}$$

where c_1 is an arbitrary constant. The above conditions show that both the TE and TM modes must be equally excited to achieve the maximum possible ratio of gain to Q, which is in agreement with the condition for minimizing Q. From (4.87) the maximum possible ratio of gain to Q for a directional antenna will be $2\|\boldsymbol{\zeta}\|_E^2$. Thus, the upper limit of ratio of gain to Q for a directional antenna is

$$\max \left.\frac{G}{Q}\right|_{dir} = \sum_{n=1}^{\infty} \frac{2(2n+1)}{Q_n + Q_n'}. \tag{4.89}$$

4.6.2 Omni-Directional Antenna

We assume that the antenna has an omni-directional pattern and the field is independent of φ, and consider the maximum possible ratio of gain to Q in the direction of $\theta = \pi/2$. Since the field is independent of φ, the vector basis function in (4.18) can be chosen as

$$\mathbf{e}_{n0e} = -\mathbf{u}_\theta \frac{1}{N_{n0}} P_n^1(\cos\theta), \quad \mathbf{h}_{n0e} = -\mathbf{u}_\varphi \frac{1}{N_{n0}} P_n^1(\cos\theta).$$

The field components produced by the omni-directional antenna are then given by

$$E_\theta = -\frac{1}{kr} \sum_n \beta_{n0e}^{(2)} \dot{\tilde{h}}_n^{(2)}(kr) P_n^1(\cos\theta),$$

$$E_\varphi = \frac{1}{kr} \sum_n \alpha_{n0e}^{(2)} \tilde{h}_n^{(2)}(kr) P_n^1(\cos\theta),$$

$$H_\theta = \frac{1}{j\eta kr} \sum_n \alpha_{n0e}^{(2)} \dot{\tilde{h}}_n^{(2)}(kr) P_n^1(\cos\theta),$$

$$H_\varphi = -\frac{1}{j\eta kr} \sum_n \beta_{n0e}^{(2)} \tilde{h}_n^{(2)}(kr) P_n^1(\cos\theta).$$

For sufficiently large r, the radiation intensity can be written as

$$\frac{1}{2} r^2 \text{Re}(\mathbf{E}\times\bar{\mathbf{H}})\cdot\mathbf{u}_r = \frac{1}{2} r^2 \text{Re}(E_\theta \bar{H}_\varphi - E_\varphi \bar{H}_\theta)$$

$$= \frac{1}{2\eta k^2}\left(\left|\sum_n j^n \beta_{n0e}^{(2)} P_n^1(\cos\theta)\right|^2 + \left|\sum_n j^{n+1}\alpha_{n0e}^{(2)} P_n^1(\cos\theta)\right|^2 \right).$$

The directivity is

$$G = 4\pi r^2 \frac{\frac{1}{2}\text{Re}(\mathbf{E}\times\bar{\mathbf{H}})\cdot\mathbf{u}_r}{P^{rad}} = 4\pi \frac{\left|\sum_{n=1}^\infty j^n \beta_{n0e}^{(2)} P_n^1(\cos\theta)\right|^2 + \left|\sum_{n=1}^\infty j^{n+1}\alpha_{n0e}^{(2)} P_n^1(\cos\theta)\right|^2}{\sum_{n,m,l} N_{nm}^2 \left(\left|\alpha_{nml}^{(2)}\right|^2 + \left|\beta_{nml}^{(2)}\right|^2\right)}.$$

(4.90)

From (4.75) and (4.90), we obtain

$$\left.\frac{G}{Q}\right|_{omn} = \frac{4\pi}{d}\left[\left|\sum_n j^n \beta_{n0e}^{(2)} P_n^1(0)\right|^2 + \left|\sum_n j^{n+1}\alpha_{n0e}^{(2)} P_n^1(0)\right|^2 \right],$$

(4.91)

where

$$d = \text{Larger of}$$

$$\left\{ \sum_{n,m,l}\left(N_{nm}^2 \left|\alpha_{nml}^{(2)}\right|^2 Q_n + N_{nm}^2 \left|\beta_{nml}^{(2)}\right|^2 Q_n'\right), \sum_{n,m,l}\left(N_{nm}^2 \left|\alpha_{nml}^{(2)}\right|^2 Q_n' + N_{nm}^2 \left|\beta_{nml}^{(2)}\right|^2 Q_n\right)\right\}.$$

The first term in the numerator of (4.91) represents the contribution from the TM modes and the second term represents the contribution from the TE modes. Only $\alpha_{n0e}^{(2)}$ and $\beta_{n0e}^{(2)}$ contribute to the numerator of (4.91), so (4.91) can be increased by setting $\alpha_{nml}^{(2)} = \beta_{nml}^{(2)} = 0$ $(m \neq 0)$, $\alpha_{n0o}^{(2)} = \beta_{n0o}^{(2)} = 0$. Let $A_n = j^{n+1}\alpha_{n0e}^{(2)}$ and $B_n = j^n\beta_{n0e}^{(2)}$, we have

$$\left.\frac{G}{Q}\right|_{omn} = \frac{4\pi}{d}\left[\left|\sum_{n=1}^{\infty} A_n P_n^1(0)\right|^2 + \left|\sum_{n=1}^{\infty} B_n P_n^1(0)\right|^2\right], \tag{4.92}$$

where

$$d = \text{Larger of } \left\{\sum_{n=1}^{\infty} N_{n0}^2(|A_n|^2 Q_n + |B_n|^2 Q_n'), \sum_{n=1}^{\infty} N_{n0}^2(|A_n|^2 Q_n' + |B_n|^2 Q_n)\right\}. \tag{4.93}$$

Since the denominator of (4.92) depends only on the magnitude of A_n and B_n, the denominator is not changed if the phases of A_n and B_n are adjusted to maximize the ratio of gain to Q. If we choose the phases of A_n and B_n to be the negative of $P_n^1(0)$, the terms in the numerator will be added in phase. Thus

$$\left.\frac{G}{Q}\right|_{omn} = \frac{4\pi}{d}\left\{\left[\sum_{n=1}^{\infty} |A_n|\,|P_n^1(0)|\right]^2 + \left[\sum_{n=1}^{\infty} |B_n|\,|P_n^1(0)|\right]^2\right\}.$$

If the first expression of (4.93) is the largest we have

$$\left.\frac{G}{Q}\right|_{omn} = 4\pi \frac{\left[\sum_{n=1}^{\infty} |A_n|\,|P_n^1(0)|\right]^2 + \left[\sum_{n=1}^{\infty} |B_n|\,|P_n^1(0)|\right]^2}{\sum_{n=1}^{\infty} N_{n0}^2(|A_n|^2 Q_n + |B_n|^2 Q_n')}. \tag{4.94}$$

Similarly, we introduce the condition $|A_n| \geq |B_n|\ (n \geq 1)$ to guarantee that the first expression of (4.93) is always the largest during the optimizing process. Under this condition, the above ratio can be maximized by letting $|A_n| = |B_n|$. Therefore,

$$\left.\frac{G}{Q}\right|_{omn} = 8\pi \frac{\left[\sum_{n=1}^{\infty} |A_n|\,|P_n^1(0)|\right]^2}{\sum_{n=1}^{\infty} N_{n0}^2 |A_n|^2 (Q_n + Q_n')} = 8\pi \frac{(\boldsymbol{\xi}, \mathbf{C})_E}{(\mathbf{C}, \mathbf{C})_E} \leq 8\pi \, \|\boldsymbol{\xi}\|_E^2, \tag{4.95}$$

where $\boldsymbol{\xi} = (\xi_1, \xi_2, \cdots)$, $\mathbf{C} = (C_1, C_2, \cdots)$ with

$$C_n = \sqrt{N_{n0}^2(Q_n + Q_n')}\,|A_n|, \ \xi_n = |P_n^1(0)|\,/\sqrt{N_{n0}^2(Q_n + Q_n')}.$$

The ratio (4.95) reaches maximum if $\mathbf{C} = c_1\xi$, or

$$|A_n| = \frac{c_1 \left|P_n^1(\cos\theta)\right|}{N_{n0}^2(Q_n + Q_n')}.$$

The upper limit of the ratio of gain to Q for an omni-directional antenna is

$$\max \left.\frac{G}{Q}\right|_{omn} = \sum_{n=1}^{\infty} \frac{2(2n+1)\left|P_n^1(0)\right|^2}{n(n+1)(Q_n + Q_n')}. \tag{4.96}$$

Remark 4.9: Chu has shown that the maximum ratio of gain to Q for an omni-directional antenna is (Chu, 1948)

$$\max \left.\frac{G}{Q}\right|_{omn}^{Chu} = \sum_{n=1}^{\infty} \frac{(2n+1)\left|P_n^1(0)\right|^2}{n(n+1)Q_n^{Chu}}. \tag{4.97}$$

Here Q_n^{Chu} is the quality factor of nth TM modes and is a function of ka. Chu's theory is valid only for an omni-directional antenna that radiates either TE or TM modes, and is based on the equivalent ladder network representation of the wave impedance of each mode and the stored energies in some elements have been neglected. Hence Chu's limit just holds approximately. Also note that the new upper limit (4.96) can be twice as much as Chu's limit (4.97) if ka is small. □

4.6.3 Best Possible Antenna Performance

Since the antenna fractional bandwidth is reciprocal to antenna Q_{real} if Q_{real} is not very small, the product of antenna gain and bandwidth can be expressed as $GB_f \approx G/Q_{real}$. The antenna quality factor used in (4.89) and (4.96) does not include the stored energies inside the circumscribing sphere of the antenna, it is thus much smaller than the real antenna Q_{real}. It follows from (4.89) and (4.96) that the products of gain and bandwidth for an arbitrary antenna of dimension $2a$ are bounded by

$$GB_f\big|_{dir} \leq \max GB_f\big|_{dir} = \sum_{n=1}^{\infty} \frac{2(2n+1)}{Q_n(ka) + Q_n'(ka)},$$

$$GB_f\big|_{omn} \leq \max GB_f\big|_{omn} = \sum_{n=1}^{\infty} \frac{2(2n+1)\left|P_n^1(0)\right|^2}{n(n+1)\left[Q_n(ka) + Q_n'(ka)\right]}. \tag{4.98}$$

The first expression applies for the directional antennas, and the second one for the omni-directional antennas. It should be noted that the right-hand sides of (4.98) are finite numbers. From (4.78), the fractional bandwidth of an arbitrary antenna of dimension $2a$ has an upper limit too

$$B_f \leq \max B_f = \frac{2(ka)^3}{2(ka)^2 + 1}. \tag{4.99}$$

Equations (4.98) indicate that one can sacrifice the bandwidth to enhance the gain. If the bandwidth is rendered very small, a high gain antenna can be achieved. One can also sacrifice the gain to improve the bandwidth. But the improvement will be limited as the bandwidth itself is bounded by the right-hand side of (4.99).

The upper bounds $\max GB_f|_{dir}$, $\max GB_f|_{omn}$ and $\max B_f$ are all monotonically increasing functions of ka. It can be seen that $\max GB_f|_{dir}$ is always higher than $\max GB_f|_{omn}$. The rate of increase of these upper bounds for small ka is much higher than that for large ka, which implies that a little increase in the size of the small antennas will notably improve their performances. For the small antennas with $ka < 1$, only the first terms of the infinite series in (4.98) are significant. Thus, we may write

$$\max GB_f|_{dir} \approx \frac{6}{Q_1 + Q_1'} = \frac{6(ka)^3}{2(ka)^2 + 1},$$

$$\max GB_f|_{omn} \approx \frac{3}{Q_1 + Q_1'} = \frac{3(ka)^3}{2(ka)^2 + 1}.$$
(4.100)

The right-hand sides of (4.100) are the best possible antenna performances that a small antenna of maximum dimension $2a$ can achieve. They set up a target that can be approached by various methods and have been proven to be very useful for small antenna design for which trial and error method is often used.

4.7 Evaluation of Antenna Quality Factor

The minimum possible antenna quality factor for an arbitrary antenna provides a theoretical lower bound to the antenna quality factor when the maximum antenna size is given. This lower bound is usually far lower than the antenna Q_{real} since the stored energies inside the smallest circumscribing sphere are totally ignored. In this section, we introduce two methods for the evaluation of antenna Q_{real}.

4.7.1 Quality Factor for Arbitrary Antenna

The first method is based on the Foster reactance theorem for antennas. To find out the stored energies, we only need to know the energy difference $\tilde{W}_m - \tilde{W}_e$ (see (4.65)). It follows from (4.39) that

$$-\frac{1}{2} \int_{V_0} \bar{\mathbf{J}} \cdot \mathbf{E} dV(\mathbf{r}) = \int_S \mathbf{u}_n \cdot \mathbf{S} dS + j2\omega \int_V (w_m - w_e) dV.$$

Ignoring the energy difference in V_0, we may write

$$-\frac{1}{2} \int_{V_0} \bar{\mathbf{J}} \cdot \mathbf{E} dV(\mathbf{r}) = \int_S \mathbf{u}_n \cdot \mathbf{S} dS + j2\omega \int_{V-V_0} (\tilde{w}_m - \tilde{w}_e) dV.$$
(4.101)

The left-hand side can be expressed as

$$-\frac{1}{2}\int_{V_0}\bar{\mathbf{J}}\cdot\mathbf{E}dV(\mathbf{r})=-\frac{1}{2}\int_{V_0}\bar{\mathbf{J}}\cdot(-\nabla\phi-j\omega\mathbf{A})dV(\mathbf{r}),\tag{4.102}$$

where ϕ and $\mathbf{A}$ are the scalar and vector potential functions

$$\phi(\mathbf{r})=\frac{\eta v}{4\pi}\int_{V_0}\frac{\rho(\mathbf{r}')e^{-jkR}}{R}dV(\mathbf{r}'),\mathbf{A}(\mathbf{r})=\frac{\eta}{4\pi v}\int_{V_0}\frac{\mathbf{J}(\mathbf{r}')e^{-jkR}}{R}dV(\mathbf{r}'),$$

with $R=|\mathbf{r}-\mathbf{r}'|$, $\eta=\sqrt{\mu/\varepsilon}$ and $v=1/\sqrt{\mu\varepsilon}$. Inserting these equations into (4.102) yields

$$-\frac{1}{2}\int_{V_0}\bar{\mathbf{J}}\cdot\mathbf{E}dV(\mathbf{r})=\frac{\omega\eta v}{8\pi}\int_{V_0}\int_{V_0}\left[\frac{1}{c^2}\frac{\bar{\mathbf{J}}(\mathbf{r})\cdot\mathbf{J}(\mathbf{r}')}{R}-\frac{\bar{\rho}(\mathbf{r})\rho(\mathbf{r}')}{R}\right]\sin(kR)dV(\mathbf{r})dV(\mathbf{r}')$$

$$+j\frac{\omega\eta v}{8\pi}\int_{V_0}\int_{V_0}\left[\frac{1}{c^2}\frac{\bar{\mathbf{J}}(\mathbf{r})\cdot\mathbf{J}(\mathbf{r}')}{R}-\frac{\bar{\rho}(\mathbf{r})\rho(\mathbf{r}')}{R}\right]\cos(kR)dV(\mathbf{r})dV(\mathbf{r}').$$

From the above equation and (4.101), we obtain

$$P^{rad}=\frac{\omega\eta c}{8\pi}\int_{V_0}\int_{V_0}\left[\frac{1}{c^2}\frac{\bar{\mathbf{J}}(\mathbf{r})\cdot\mathbf{J}(\mathbf{r}')}{R}-\frac{\bar{\rho}(\mathbf{r})\rho(\mathbf{r}')}{R}\right]\sin(kR)dV(\mathbf{r})dV(\mathbf{r}'),$$

$$\tilde{W}_m-\tilde{W}_e=\frac{\eta c}{16\pi}\int_{V_0}\int_{V_0}\left[\frac{1}{c^2}\frac{\bar{\mathbf{J}}(\mathbf{r})\cdot\mathbf{J}(\mathbf{r}')}{R}-\frac{\bar{\rho}(\mathbf{r})\rho(\mathbf{r}')}{R}\right]\cos(kR)dV(\mathbf{r})dV(\mathbf{r}').$$

$$\tag{4.103}$$

Thus, once the current distribution is known, the calculation of the energy difference is simply an integration. The antenna Q_{real} can then be determined by (4.66), and the element values of the antenna equivalent circuit can be determined by (4.48).

4.7.2   Quality Factor for Small Antenna

We now introduce a method to calculate the quality factor Q_{real} for small antennas (Geyi, 2003(b)). The method is based on the understanding that, for a small antenna, the total energy in the Poynting theorem can easily be separated into the stored energy and radiated energy by using the low frequency expansions. The Poynting theorem in the frequency domain provides an equation on the stored electric and magnetic energy while the Poynting theorem in the time domain can be used as another independent equation for the stored electric and magnetic energy. By solving these equations, the stored electric and magnetic energy can be obtained, thus making the Q_{real} calculation possible.

If the maximum dimension of the source distribution is small compared to the wavelength λ so that $R = |\mathbf{r} - \mathbf{r}'| < \lambda/2\pi$, we may use the following expansion

$$e^{-jkR} = 1 - jkR - \frac{(kR)^2}{2} + \frac{j(kR)^3}{6} + \cdots.$$

Hence the potential functions can be approximated by

$$\phi(\mathbf{r}) \approx \frac{\eta v}{4\pi} \int_{V_0} \frac{\rho(\mathbf{r}')}{R} dV(\mathbf{r}') - jkq$$

$$-\frac{\eta v}{8\pi} k^2 \int_{V_0} R\rho(\mathbf{r}') dV(\mathbf{r}') + j\frac{\eta v}{24\pi} k^3 \int_{V_0} R^2 \rho(\mathbf{r}') dV(\mathbf{r}'),$$

$$\mathbf{A}(\mathbf{r}) \approx \frac{\eta}{4\pi v} \int_{V_0} \frac{\mathbf{J}(\mathbf{r}')}{R} dV(\mathbf{r}') + k^2 v\mathbf{p}$$

$$-\frac{\eta}{8\pi v} k^2 \int_{V_0} R\mathbf{J}(\mathbf{r}') dV(\mathbf{r}') + j\frac{\eta}{24\pi v} k^3 \int_{V_0} R^2 \mathbf{J}(\mathbf{r}') dV(\mathbf{r}'),$$

$$(4.104)$$

where $\mathbf{p} \equiv \int_{V_0} \mathbf{r}' \rho(\mathbf{r}') dV(\mathbf{r}')$ is the electric dipole moment of the source (using the continuity equation, one can easily show that $\int_{V_0} \mathbf{J}(\mathbf{r}') dV(\mathbf{r}') - j\omega\mathbf{p} = 0$), and q is the total charge of the source, which is zero if there is no current flowing out of ∂V_0 since

$$\int_{V_0} \nabla \cdot \mathbf{J} dV = \int_{\partial V} \mathbf{J} \cdot \mathbf{u}_n dS = 0 = -j\omega \int_{V_0} \rho dV = -j\omega q.$$

Substituting (4.104) into (4.102), we obtain

$$-\int_{V_0} \frac{1}{2} \bar{\mathbf{J}} \cdot \mathbf{E} dV(\mathbf{r}) = \frac{\eta k^4 v^2}{12\pi} |\mathbf{p}|^2 + \frac{\eta k^4}{12\pi} |\mathbf{m}|^2$$

$$-j\frac{\omega \eta v}{8\pi} \left\{ \int_{V_0} \int_{V_0} \left[\frac{\rho(\mathbf{r}')\bar{\rho}(\mathbf{r})}{R} - \frac{k^2 R}{2} \rho(\mathbf{r}')\bar{\rho}(\mathbf{r}) \right] dV(\mathbf{r}) dV(\mathbf{r}') \right.$$

$$(4.105)$$

$$\left. -\frac{1}{v^2} \int_{V_0} \int_{V_0} \left[\frac{\mathbf{J}(\mathbf{r}') \cdot \bar{\mathbf{J}}(\mathbf{r})}{R} - \frac{k^2 R}{2} \mathbf{J}(\mathbf{r}') \cdot \bar{\mathbf{J}}(\mathbf{r}) \right] dV(\mathbf{r}) dV(\mathbf{r}') \right\},$$

where $\mathbf{m} = \frac{1}{2} \int_{V_0} \mathbf{r} \times \mathbf{J}(\mathbf{r}) dV(\mathbf{r})$ is the magnetic dipole moment of the source. If S is large enough, the first term on the right-hand side of (4.101) becomes a real number. Thus, the following identifications can be made from (4.101) and (4.105)

$$P^{rad} \approx \frac{\eta k^4 v^2 |\mathbf{p}|^2}{12\pi} + \frac{\eta k^4 |\mathbf{m}|^2}{12\pi}, \tag{4.106}$$

$$\tilde{W}_e - \tilde{W}_m \approx \frac{\eta v}{16\pi} \left\{ \int_{V_0} \int_{V_0} \left[\frac{\rho(\mathbf{r}')\bar{\rho}(\mathbf{r})}{R} - \frac{k^2 R}{2} \rho(\mathbf{r}')\bar{\rho}(\mathbf{r}) \right] dV(\mathbf{r})dV(\mathbf{r}') \right.$$
$$\left. - \frac{1}{v^2} \int_{V_0} \int_{V_0} \frac{\mathbf{J}(\mathbf{r}') \cdot \bar{\mathbf{J}}(\mathbf{r})}{R} dV(\mathbf{r})dV(\mathbf{r}') \right\}, \tag{4.107}$$

where the terms higher than the order of $1/v^2$ in the stored energy have been neglected. The radiated power in (4.106) consists of the contributions from the electric dipole moment and the magnetic dipole moment of the source current distribution. Equation (4.107) gives the difference between the stored magnetic energy and the stored electric energy around the antenna. In order to determine $\tilde{W}_e$ and $\tilde{W}_m$, another equation is needed, which can be obtained by the time-domain Poynting theorem. We will use the same notations as in the frequency domain for all electromagnetic quantities in the time domain, with the time t explicitly appearing as an independent variable. In the time domain, the Poynting theorem can be expressed as

$$-\int_{V_0} \mathbf{J}(\mathbf{r}, t) \cdot \mathbf{E}(\mathbf{r}, t)dV(\mathbf{r}) = P^{rad}(t) + \frac{d}{dt}[W_e(t) + W_m(t)], \tag{4.108}$$

with

$$P^{rad}(t) = \int_S [\mathbf{E}(\mathbf{r}, t) \times \mathbf{H}(\mathbf{r}, t)] \cdot \mathbf{u}_n dS(\mathbf{r}),$$

$$W_e(t) = \frac{1}{2} \int_V \mathbf{E}(\mathbf{r}, t) \cdot \mathbf{D}(\mathbf{r}, t)dV,$$

$$W_m(t) = \frac{1}{2} \int_V \mathbf{H}(\mathbf{r}, t) \cdot \mathbf{B}(\mathbf{r}, t)dV.$$

Here V is a large volume enclosing the source region V_0, and S is the boundary of V. The physical implication of time-domain Poynting theorem is different from its counterpart in frequency domain, and hence provides another independent equation. The rate of energy transfer from the charged particles to the electromagnetic fields can be calculated as follows

$$-\int_{V_0} \mathbf{J}(\mathbf{r}, t) \cdot \mathbf{E}(\mathbf{r}, t)dV(\mathbf{r}) = -\int_{V_0} \mathbf{J}(\mathbf{r}, t) \cdot \left[-\nabla\phi(\mathbf{r}, t) - \frac{\partial \mathbf{A}(\mathbf{r}, t)}{\partial t} \right] dV(\mathbf{r}), \tag{4.109}$$

where

$$\mathbf{A}(\mathbf{r}, t) = \int_{V_0} \frac{\mu \mathbf{J}(\mathbf{r}', T)}{4\pi R} dV(\mathbf{r}'), \phi(\mathbf{r}, t) = \int_{V_0} \frac{\rho(\mathbf{r}', T)}{4\pi \varepsilon R} dV(\mathbf{r}'), \tag{4.110}$$

with $R = |\mathbf{r} - \mathbf{r}'|$, $T = t - R/v$. If the sources are confined in a small region, we can make the following approximations for the sources in (4.110)

$$\rho(\mathbf{r}', T) = \rho(\mathbf{r}', t) - \frac{\partial \rho(\mathbf{r}', t)}{\partial t} \frac{R}{v}$$
$$+ \frac{1}{2} \frac{\partial^2 \rho(\mathbf{r}', t)}{\partial t^2} \left(\frac{R}{v}\right)^2 - \frac{1}{6} \frac{\partial^3 \rho(\mathbf{r}', t)}{\partial t^3} \left(\frac{R}{v}\right)^3 + \cdots,$$
$$\mathbf{J}(\mathbf{r}', T) = \mathbf{J}(\mathbf{r}', t) - \frac{\partial \mathbf{J}(\mathbf{r}', t)}{\partial t} \frac{R}{v} \qquad (4.111)$$
$$+ \frac{1}{2} \frac{\partial^2 \mathbf{J}(\mathbf{r}', t)}{\partial t^2} \left(\frac{R}{v}\right)^2 - \frac{1}{6} \frac{\partial^3 \mathbf{J}(\mathbf{r}', t)}{\partial t^3} \left(\frac{R}{v}\right)^3 + \cdots.$$

It follows from (4.109), (4.110) and (4.111) that

$$-\int_{V_0} \mathbf{J}(\mathbf{r}, t) \cdot \mathbf{E}(\mathbf{r}, t) dv = \frac{\eta}{6\pi v^2} |\ddot{\mathbf{p}}(t)|^2 + \frac{\eta}{6\pi v^4} |\ddot{\mathbf{m}}(t)|$$

$$+ \frac{\eta v}{8\pi} \frac{d}{dt} \left\{ \int_{V_0} \int_{V_0} \frac{1}{R} \rho(\mathbf{r}, t)\rho(\mathbf{r}', t) dV(\mathbf{r}) dV(\mathbf{r}') \right.$$

$$+ \frac{1}{2v^2} \int_{V_0} \int_{V_0} R \frac{\partial \rho(\mathbf{r}, t)}{\partial t} \frac{\partial \rho(\mathbf{r}', t)}{\partial t} dV(\mathbf{r}) dV(\mathbf{r}') \qquad (4.112)$$

$$\left. + \frac{1}{v^2} \int_{V_0} \int_{V_0} \frac{\mathbf{J}(\mathbf{r}, t) \cdot \mathbf{J}(\mathbf{r}', t)}{R} dV(\mathbf{r}) dV(\mathbf{r}') \right\},$$

where $\mathbf{p}(t) = \int_{V_0} \mathbf{r}\rho(\mathbf{r}, t) dV(\mathbf{r})$ and $\mathbf{m}(t) = \frac{1}{2} \int_{V_0} \mathbf{r} \times \mathbf{J}(\mathbf{r}, t) dV(\mathbf{r})$ are the electric dipole moment and the magnetic dipole moment of the source respectively. Comparing the above expression to (4.108), we can identify the radiated power in the time domain and the total energy around the antenna as

$$P^{rad}(t) = \frac{\eta}{6\pi v^2} |\ddot{\mathbf{p}}(t)|^2 + \frac{\eta}{6\pi v^4} |\ddot{\mathbf{m}}(t)|^2, \qquad (4.113)$$

$$\tilde{W}_e(t) + \tilde{W}_m(t) \approx \frac{\eta v}{8\pi} \left\{ \int_{V_0} \int_{V_0} \frac{1}{R} \rho(\mathbf{r}, t)\rho(\mathbf{r}', t) dV(\mathbf{r}) dV(\mathbf{r}') \right.$$

$$+ \frac{1}{2v^2} \int_{V_0} \int_{V_0} R \frac{\partial \rho(\mathbf{r}, t)}{\partial t} \frac{\partial \rho(\mathbf{r}', t)}{\partial t} dV(\mathbf{r}) dV(\mathbf{r}') \qquad (4.114)$$

$$\left. + \frac{1}{v^2} \int_{V_0} \int_{V_0} \frac{\mathbf{J}(\mathbf{r}, t) \cdot \mathbf{J}(\mathbf{r}', t)}{R} dV(\mathbf{r}) dV(\mathbf{r}') \right\},$$

where we have assumed that $W_e(t) \approx \tilde{W}_e(t)$, $W_m(t) \approx \tilde{W}_m(t)$ since the total radiated energy for a small antenna is of the order $1/v^2$ and is very small compared to the total energy. For a time-harmonic field, we may write $\rho(\mathbf{r}, t) = \text{Re}\rho(\mathbf{r})e^{j\omega t}$, $\mathbf{J}(\mathbf{r}, t) = \text{Re}\mathbf{J}(\mathbf{r})e^{j\omega t}$, and the time average of the total stored energy is

$$
\tilde{W}_m + \tilde{W}_e = \overline{\tilde{W}_m(t) + \tilde{W}_e(t)} = \frac{1}{T} \int_0^T \left[\tilde{W}_m(t) + \tilde{W}_e(t) \right] dt
$$

$$
= \frac{\eta v}{16\pi} \frac{1}{v^2} \int_{V_0} \int_{V_0} \frac{\mathbf{J}(\mathbf{r}) \cdot \bar{\mathbf{J}}(\mathbf{r}')}{R} dV(\mathbf{r}) dV(\mathbf{r}') \qquad (4.115)
$$

$$
+ \frac{\eta v}{16\pi} \int_{V_0} \int_{V_0} \left(\frac{k^2 R}{2} + \frac{1}{R} \right) \rho(\mathbf{r}) \bar{\rho}(\mathbf{r}') dV(\mathbf{r}) dV(\mathbf{r}').
$$

Combining (4.107) and (4.115) gives the stored energies around the antenna:

$$
\tilde{W}_e = \frac{\eta v}{16\pi} \int_{V_0} \int_{V_0} \frac{1}{R} \rho(\mathbf{r}) \bar{\rho}(\mathbf{r}') dV(\mathbf{r}) dV(\mathbf{r}'),
$$

$$
\tilde{W}_m = \frac{\eta v}{16\pi} \frac{1}{v^2} \int_{V_0} \int_{V_0} \frac{\mathbf{J}(\mathbf{r}) \cdot \bar{\mathbf{J}}(\mathbf{r}')}{R} dV(\mathbf{r}) dV(\mathbf{r}') \qquad (4.116)
$$

$$
+ \frac{\eta v}{16\pi} \frac{k^2}{2} \int_{V_0} \int_{V_0} R\rho(\mathbf{r}) \bar{\rho}(\mathbf{r}') dV(\mathbf{r}) dV(\mathbf{r}').
$$

The time average of (4.113) is exactly the same as (4.106). In (4.116), the stored energies are directly related to the source distributions and numerical integrations can be easily carried out once the current distributions are known. The antenna Q_{real} can be calculated by (4.45).

Remark 4.10: To show that the energies in Equations (4.116) are positive, let us consider the integral

$$
I = \int_{V_0} \int_{V_0} \frac{\rho(\mathbf{r}) \bar{\rho}(\mathbf{r}')}{|\mathbf{r} - \mathbf{r}'|} dV(\mathbf{r}) dV(\mathbf{r}').
$$

Making use of the relation (Byron and Fuller, 1969)

$$
\frac{1}{4\pi |\mathbf{r} - \mathbf{r}'|} = \frac{1}{(2\pi)^3} \int_{R^3} \frac{e^{j\mathbf{k}\cdot(\mathbf{r}-\mathbf{r}')}}{|\mathbf{k}|^2} d^3\mathbf{k},
$$

we have

$$I = \frac{1}{2\pi^2} \int_{R^3} \frac{1}{|\mathbf{k}|^2} \left[\int_{V_0} \int_{V_0} \rho(\mathbf{r})\bar{\rho}(\mathbf{r}')e^{j\mathbf{k}\cdot(\mathbf{r}-\mathbf{r}')}dV(\mathbf{r})dV(\mathbf{r}') \right] d^3\mathbf{k}$$

$$= \frac{1}{2\pi^2} \int_{R^3} \frac{1}{|\mathbf{k}|^2} \tilde{\rho}(\mathbf{k})\bar{\tilde{\rho}}(\mathbf{k})d^3\mathbf{k} > 0,$$

where $\tilde{\rho}(\mathbf{k}) = \int_{V_0} \rho(\mathbf{r})e^{j\mathbf{k}\cdot\mathbf{r}}d^3\mathbf{k}$.   □

Example 4.2: A dipole is a one-dimensional structure that only radiates TM modes. Since the dipole only uses a small space within the circumscribing sphere, the real antenna Q_{real} should be much larger than the theoretical limit (4.78). We assume that the dipole antenna has radius a_0 and length $2a$. It is readily found that, for $a_0 \ll 2a$, we have

$$P^{rad} \approx 10a^2k^2 |I_0|^2 , \quad \tilde{W}_e \approx \frac{\eta c |I_0|^2}{4\pi\omega^2 a} \left(\ln \frac{a}{a_0} - 1 \right).$$

So the antenna Q_{real} is given by

$$Q_{real} = \frac{2\omega\tilde{W}_e}{P^{rad}} \approx \frac{6}{(ka)^3} \left(\ln \frac{a}{a_0} - 1 \right). \tag{4.117}$$

Equation (4.66) can be used to demonstrate the validity of the above expression. In fact, the input impedance of a small dipole antenna has been given by Schelkunoff as follows (Schelkunoff, 1952)

$$R^{rad} = 20 (ka)^2 , \quad X = -\frac{\eta}{ka\pi} \left(\ln \frac{a}{a_0} - 1 \right).$$

Substituting the above equations into (4.66), we can find the same result as (4.117).

Now consider a small circular loop antenna. The radius of the loop is a and the the radius of the wire is a_0. The radiated power and the stored magnetic energy can be found as

$$P^{rad} \approx \frac{1}{12}k^4\pi\eta |I_0|^2 a^4, \quad \tilde{W}_m \approx \frac{\eta a}{4c} |I_0|^2 \ln \frac{a}{a_0}.$$

Therefore, the quality factor Q_{real} of the small loop antenna is

$$Q_{real} = \frac{2\omega\tilde{W}_m}{P^{rad}} \approx \frac{6}{\pi(ka)^3} \ln \frac{a}{a_0}. \tag{4.118}$$

It can be seen that the quality factor of a small loop antenna is about three times lower than that of a small dipole of the same maximum dimension $2a$. Again, the input impedance of a small loop antenna has been given by Schelkunoff as follows

$$R^{rad} = \frac{\eta\pi}{6}(ka)^4, \quad X = k\eta a \ln\frac{a}{a_0}.$$

Substituting the above equations into (4.66), we obtain the same result as (4.118). $\qquad\qquad\square$

4.7.3 Some Remarks on Electromagnetic Stored Energy

The calculation of stored energies is a key step in calculating antenna Q_{real} and antenna input impedance. The expressions for stored energies tend to infinity as some parameters of the geometry (for example, the radius of the wire) approach zero. This infinity problem has haunted physicist for years, and even quantum electrodynamics does not satisfactorily resolve this problem. As an example, let us consider the energy of a stationary charge distribution. The interaction energy W_{inter} of two point charges q_1 and q_2, located at $\mathbf{r}_1$ and $\mathbf{r}_2$ respectively, is equal to the work to move one of the charges into place from infinity while the other charge is fixed, that is, $W_{inter} = q_1 q_2/4\pi\varepsilon_0\,|\mathbf{r}_1 - \mathbf{r}_2|$. For a distribution of n point charges, the interaction energy is

$$W_{inter} = \frac{1}{2}\sum_{i=1}^{n} q_i \sum_{j=1,j\neq i}^{n} \frac{q_j}{4\pi\varepsilon_0\,|\mathbf{r}_i - \mathbf{r}_j|} = \frac{1}{2}\sum_{i=1}^{n} q_i\phi(\mathbf{r}_i) \qquad (4.119)$$

where $\phi(\mathbf{r}_i) = \sum_{j=1,j\neq i}^{n} \dfrac{q_j}{4\pi\varepsilon_0\,|\mathbf{r}_i - \mathbf{r}_j|}$ is the potential at $\mathbf{r}_i$ produced by all charges except q_i.
In deriving (4.119), the contribution of q_i to the potential $\phi(\mathbf{r}_i)$ has been excluded, which corresponds to the self-energy of the point charge and is infinite. The total energy of the system is the sum of self-energy and interaction energy, and is also infinite. For a continuous charge distribution, the total energy of the system can be calculated by the integral

$$W_{total} = \frac{1}{2}\int_{V_0} \rho(\mathbf{r})\phi(\mathbf{r})dV(\mathbf{r}) \qquad (4.120)$$

where $\phi(\mathbf{r}) = \frac{1}{2}\int_{V_0} \frac{\rho(\mathbf{r}')}{4\pi\varepsilon_0\,|\mathbf{r} - \mathbf{r}'|}dV(\mathbf{r}')$. It should be noted that (4.119) and (4.120) are essentially different and the latter includes the self-energy. For a continuous charge distribution, the contribution of $\rho(\mathbf{r}')$ to $\phi(\mathbf{r})$ becomes infinitely small as $dV(\mathbf{r})$ approaches zero. Equation (4.120) can be rewritten as

$$\begin{aligned}
W_{total} &= \frac{\varepsilon_0}{2}\int_{V_0} |\mathbf{E}(\mathbf{r})|^2\,dV(\mathbf{r}) + \frac{\varepsilon_0}{2}\int_{\partial V_0} \phi(\mathbf{r})\mathbf{E}(\mathbf{r})\cdot\mathbf{u}_n(\mathbf{r})dS(\mathbf{r}) \\
&= \frac{\varepsilon_0}{2}\int_{\text{All space}} |\mathbf{E}(\mathbf{r})|^2\,dV(\mathbf{r}).
\end{aligned} \qquad (4.121)$$

The surface integral in (4.121) approaches zero as the surface becomes infinitely large. Equation (4.120) can be interpreted as the total energy of the charge system and the charge is considered to be the carrier of the energy, while (4.121) represents the total energy of the static field and the electric field is considered to be the carrier of the energy. These two different interpretations are all appropriate for a static field. When the field is time-dependent, the electromagnetic field exists in free space and has energy, and in this case the field must be considered the carrier of the energy.

To overcome the infinity problem for a point charge, one may assume that the charge is distributed on a sphere of radius a and mass m_{bare}, carrying total charge q. The total (potential) energy of this single charge system is

$$W_{total} = \frac{1}{2} \int_{V_0} \rho(\mathbf{r})\phi(\mathbf{r})dV(\mathbf{r}) = \frac{1}{2} \int_0^{4\pi} \frac{q}{4\pi a^2} \frac{q}{4\pi\varepsilon_0 a} a^2 d\Omega(\mathbf{r}) = \frac{q^2}{8\pi\varepsilon_0 a},$$

which is finite. The above calculation, however, raises two questions:

1. If the charge particle, say, the electron, is of finite size what is its internal structure?
2. It would be difficult to take into account that forces acting on the particle must be transmitted by a speed less than light speed.

Therefore, the finite size particle cannot serve as a good basis for a theory of elementary objects and the radius a of the sphere representing an electron has to be sent to zero. According to Einstein's special theory of relativity, $q^2/8\pi\varepsilon_0 a = m_{extra}c^2$, where m_{extra} is the extra mass due to energy. Thus, the total mass of the particle of mass m_{bare} is $m_{phys} = m_{bare} + m_{extra}$. It is the mass m_{phys}, called **physical mass**, that an experimenter would measure if the particle were subject to Newton's law (Hooft, 2000). Again, the physical mass tends to infinity as the radius a is sent to zero.

> Every existence above a certain rank has its singular points; the higher the rank, the more of them. At these points, influences whose physical magnitude is too small to be taken account of by a finite being may produce results of the greatest importance.
>
> —James Maxwell

5

Integral Equation Formulations

Integral equation had now become a new mathematical tool not confined to symmetrical kernels. It was developed during several decades and was seen as a universal tool with which it was possible to solve the majority of boundary value problems of physics.
—Lars Gårding (Swedish mathematician, 1919–)

An equation that contains an unknown function under one or more signs of integration is called an **integral equation**. The integral equation formulation of boundary value problems can be traced back to the early work of Helmholtz, Gustav Robert Kirchhoff (German physicist, 1824–1887) and Rayleigh. The establishment of the general theory of integral equations is attributed to a number of famous scientists, including Ivar Erik Fredholm (Swedish mathematician, 1866–1927), Vito Volterra (Italian mathematician, 1860–1940), Hilbert and Schmidt, to name just a few. The derivation of integral equations is either based on the **direct method** or the **indirect method**. The former uses the representation theorem of fields, and the latter utilizes layer ansatz. The integral equations hold a wide range of engineering applications and offer some unique features that the differential equations do not have. The integral equation method is most appropriate for solving a field problem whose domain extends to infinity, such as the radiation and scattering problems in electromagnetics. The boundary condition at infinity is automatically incorporated into the integral equation formulation, and the unbounded-domain problem is transformed into a bounded-domain problem. Since the unknowns are restricted on the boundary of the physical problem, the dimension of the problem is decreased by one. As a result, the number of unknowns is reduced and the numerical accuracy is improved once the integral equation is discretized into a matrix equation.

The requirements of smoothness of the unknown functions are relaxed when a differential equation is transformed into an integral equation. This raises the question of whether the integral equation obtained is equivalent to the original differential equation. In fact, the spurious solutions may occur in the integral equation formulation. Therefore, a challenge in the integral equation formulation is to find various methods to remove or distinguish the spurious solutions.

5.1   Integral Equations

Let R^N be the N-dimensional Euclidean space, and $\Omega \subset R^N$ be a bounded region and Γ its boundary. Let $\hat{A}$ be a linear differential operator and $\hat{A}^*$ its formal adjoint. Consider a differential equation

$$\hat{A}u(\mathbf{x}) = f(\mathbf{x}), \mathbf{x} \in \Omega,$$

where u is the unknown function and f is a known function. We may use the direct method to establish the integral equation. From integration by parts, we obtain

$$\int_{\Omega} \left[\varphi(\mathbf{x})\hat{A}\psi(\mathbf{x}) - \psi(\mathbf{x})\hat{A}^*\varphi(\mathbf{x}) \right] d\mathbf{x} = \int_{\Gamma} b\left[\varphi(\mathbf{x}), \psi(\mathbf{x}) \right] d\Gamma(\mathbf{x}), \tag{5.1}$$

where φ and ψ are two arbitrary smooth functions and $b(\cdot, \cdot)$ is a bilinear form. If $G(\mathbf{x}, \mathbf{x}')$ is the Green's function of $\hat{A}^*$

$$\hat{A}^*G(\mathbf{x}, \mathbf{x}') = -\delta(\mathbf{x} - \mathbf{x}'),$$

we can let $\varphi(\mathbf{x}) = G(\mathbf{x}, \mathbf{x}')$, $\psi(\mathbf{x}) = u(\mathbf{x})$ in (5.1), yielding

$$\int_{\Omega} u(\mathbf{x})\delta(\mathbf{x} - \mathbf{x}')d\mathbf{x} - \int_{\Gamma} b\left[G(\mathbf{x}, \mathbf{x}'), u(\mathbf{x}) \right] d\Gamma(\mathbf{x}) = -\int_{\Omega} G(\mathbf{x}, \mathbf{x}')f(\mathbf{x})d\mathbf{x}.$$

If Γ is smooth, we may let $\mathbf{x}' \to \Gamma$ to obtain

$$\frac{1}{2}u(\mathbf{x}') - \int_{\Gamma} b\left[G(\mathbf{x}, \mathbf{x}'), u(\mathbf{x}) \right] d\Gamma(\mathbf{x}) = -\int_{\Omega} G(\mathbf{x}, \mathbf{x}')f(\mathbf{x})d\mathbf{x}.$$

By use of the symmetric property of the Green's function $G(\mathbf{x}, \mathbf{x}') = G(\mathbf{x}', \mathbf{x})$, the above equation can be written as

$$\frac{1}{2}u(\mathbf{x}) - \int_{\Gamma} b\left[G(\mathbf{x}, \mathbf{x}'), u(\mathbf{x}') \right] d\Gamma(\mathbf{x}') = -\int_{\Omega} G(\mathbf{x}, \mathbf{x}')f(\mathbf{x}')d\mathbf{x}'.$$

This is the integral equation defined on the boundary Γ.

Remark 5.1: A Fredholm equation of the first type is defined by

$$\int_{\Gamma} K(\mathbf{x}, \mathbf{x}')u(\mathbf{x}')d\Gamma(\mathbf{x}') = g(\mathbf{x}),$$

where u is the unknown function and appears inside of the integral; $g(\mathbf{x})$ is a known function; $K(\mathbf{x}, \mathbf{x}')$ is also known and is called the **kernel function**. If the unknown function occurs both inside and outside the integral, we have a **Fredholm equation of the second type** defined by

$$u(\mathbf{x}) - \lambda \int_{\Gamma} K(\mathbf{x}, \mathbf{x}')u(\mathbf{x}')d\Gamma(\mathbf{x}') = g(\mathbf{x}).$$

Here λ is a parameter. A Fredholm equation features that the domain of integral is constant. The domain of integral can also be variable. In this case, the integral equations

$$\int_{\Gamma(\mathbf{x})} K(\mathbf{x}, \mathbf{x}')u(\mathbf{x}')d\Gamma(\mathbf{x}') = g(\mathbf{x}),$$

$$u(\mathbf{x}) - \lambda \int_{\Gamma(\mathbf{x})} K(\mathbf{x}, \mathbf{x}')u(\mathbf{x}')d\Gamma(\mathbf{x}') = g(\mathbf{x}),$$

are called **Volterra equations** of the first and second type respectively.   □

Many integral equations are characterized by compact operators. For this reason, many properties of compact operators are applicable to integral equations. Especially we have:

The Fredholm alternative theorem: *Let H be a Hilbert space with an inner product* $(\cdot, \cdot)$. *If the operator* $\hat{A} : H \rightarrow H$ *is compact, EITHER the equations* $(\hat{I} - \lambda \hat{A})u = 0$ *and* $(\hat{I} - \lambda \hat{A}^*)v = 0$ *have only the trivial solutions* $u = 0$ *and* $v = 0$, *and the equation* $(\hat{I} - \lambda \hat{A})u = f$ *has a unique solution OR the equations* $(\hat{I} - \lambda \hat{A})u = 0$ *and* $(\hat{I} - \lambda \hat{A}^*)v = 0$ *have the same number of nontrivial solutions, and* $(\hat{I} - \lambda \hat{A})u = f$ *has a solution if and only if* $(f, v) = 0$ *for all solutions v of the equation* $(\hat{I} - \lambda \hat{A}^*)v = 0$.

5.2   TEM Transmission Lines

A transverse electromagnetic (TEM) transmission line filled with homogeneous medium is shown in Figure 5.1, where the cross-section Ω of the transmission line is assumed to be a multiple connected region enclosed by a perfectly conducting boundary $\Gamma = \sum_{i=0}^{p} \Gamma_i$. The TEM

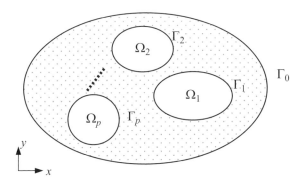

Figure 5.1   A TEM transmission line

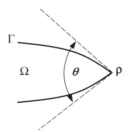

Figure 5.2   An arbitrary boundary point

mode in the transmission line may be characterized by the electric potential function φ that satisfies Laplace equation:

$$\left(\frac{\partial^2}{\partial x^2} + \frac{\partial^2}{\partial y^2}\right)\varphi(x, y) = 0, (x, y) \in \Omega,$$

$$\varphi(x, y) = V(x, y), (x, y) \in \Gamma,$$

(5.2)

where $V(x, y)$ is a known potential distribution on the boundary Γ. By use of Green's identity, the following integral equation for (5.2) can easily be established

$$c(\boldsymbol{\rho})\varphi(\boldsymbol{\rho}) = \int_{\Gamma}\left[G(\boldsymbol{\rho}, \boldsymbol{\rho}')\frac{\partial\varphi(\boldsymbol{\rho}')}{\partial n(\boldsymbol{\rho}')} - \varphi(\boldsymbol{\rho}')\frac{\partial G(\boldsymbol{\rho}, \boldsymbol{\rho}')}{\partial n(\boldsymbol{\rho}')}\right]d\Gamma(\boldsymbol{\rho}'),$$

(5.3)

where $\boldsymbol{\rho} = (x, y)$; $c(\boldsymbol{\rho}) = \theta/2\pi$ and θ is the angle formed by the two half tangents at the boundary point $\boldsymbol{\rho}$, as shown in Figure 5.2; $G(\boldsymbol{\rho}, \boldsymbol{\rho}') = -(2\pi)^{-1}\ln|\boldsymbol{\rho} - \boldsymbol{\rho}'|$ is the Green's function of the Laplace equation:

$$\nabla^2 G(\boldsymbol{\rho}, \boldsymbol{\rho}') = -\delta\left(\boldsymbol{\rho} - \boldsymbol{\rho}'\right).$$

(5.4)

The boundary value problem (5.2) can also be solved by using the complex variable integral equation formulation. Let z_i be the fixed points inside Ω_i in the complex plane, and ψ_i be the total flux through Γ_i ($i = 1, 2, \cdots, p$). We can introduce an analytic function (Geyi et al., 1989)

$$W(z) = \varphi(x, y) + j\psi(x, y) - \sum_{i=1}^{p}\frac{\psi_i}{2\pi}\ln(z - z_i),$$

where $\psi(x, y)$ is the flux function. Applying the Cauchy integral formula results in a complex variable integral equation

$$j\theta W(z) = \int_{\Gamma}\frac{W(z')}{z' - z}dz'.$$

(5.5)

5.3   Waveguide Eigenvalue Problems

Consider an arbitrary metal waveguide filled with homogeneous medium. Let the cross-section of the waveguide be denoted by Ω and its boundary by Γ. For convenience, we introduce the following integral operators

$$\hat{G}_{k_c} = 2 \int_{\Gamma} d\Gamma(\rho') G(\rho, \rho'),$$

$$\hat{G}_{k_c}^{n'} = 2 \int_{\Gamma} d\Gamma(\rho') \frac{\partial G(\rho, \rho')}{\partial n(\rho')},$$

$$\hat{G}_{k_c}^{n} = 2 \int_{\Gamma} d\Gamma(\rho') \frac{\partial G(\rho, \rho')}{\partial n(\rho)},$$

$$\hat{N}_{k_c} = -\frac{1}{2} \int_{\Gamma} d\Gamma(\rho') N_0(k_c |\rho - \rho'|),$$

$$\hat{N}_{k_c}^{n'} = -\frac{1}{2} \int_{\Gamma} d\Gamma(\rho') \frac{\partial N_0(k_c |\rho - \rho'|)}{\partial n(\rho')},$$

$$\hat{N}_{k_c}^{n} = -\frac{1}{2} \int_{\Gamma} d\Gamma(\rho') \frac{\partial N_0(k_c |\rho - \rho'|)}{\partial n(\rho)},$$

$$\hat{H}_{k_c} = \frac{1}{2j} \int_{\Gamma} d\Gamma(\rho') H_0^{(2)}(k_c |\rho - \rho'|),$$

$$\hat{H}_{k_c}^{n'} = \frac{1}{2j} \int_{\Gamma} d\Gamma(\rho') \frac{\partial H_0^{(2)}(k_c |\rho - \rho'|)}{\partial n(\rho')},$$

$$\hat{H}_{k_c}^{n} = \frac{1}{2j} \int_{\Gamma} d\Gamma(\rho') \frac{\partial H_0^{(2)}(k_c |\rho - \rho'|)}{\partial n(\rho)},$$

where $\rho = (x, y)$; $G(\rho, \rho')$ is the Green's function of two-dimensional Helmholtz equation

$$(\nabla^2 + k_c^2)G(\rho, \rho') = -\delta(\rho - \rho'), \tag{5.6}$$

and k_c is the cut-off wavenumber; $N_0(k_c |\rho - \rho'|)$ is the Neumann function; and $H_0^{(2)}(k_c |\rho - \rho'|)$ is the Hankel function of the second. The longitudinal component of magnetic field of a TE mode in the waveguide, denoted by h, constitutes the Neumann eigenvalue problem:

$$(\nabla^2 + k_c^2)h = 0, \ \rho \in \Omega,$$
$$\partial h / \partial n = 0, \ \rho \in \Gamma. \tag{5.7}$$

The corresponding integral equation can be easily obtained by Green's identity as follows

$$(\hat{I} + \hat{G}_{k_c}^{n'})h = 0, \tag{5.8}$$

where $\hat{I}$ is the identity operator. The longitudinal component of electric field of a TM mode in the waveguide, denoted e, satisfies the Dirichlet eigenvalue problem:

$$(\nabla^2 + k_c^2)e = 0, \rho \in \Omega,$$
$$e = 0, \rho \in \Gamma, \tag{5.9}$$

and the corresponding integral equation is

$$\hat{G}_{k_c}(q) = 0, \tag{5.10}$$

where $q = \partial e/\partial n$.

The cut-off wavenumbers for the TE or TM modes are obtained by requiring that the integral equation has a nontrivial solution. The eigenfunctions and corresponding cut-off wavenumbers satisfying the differential equations (5.7) and (5.9) also satisfy the integral equations (5.8) and (5.10) respectively, but the converse is not necessarily true. For example, if $G = -N_0/4$ is exploited, the spurious wavenumbers do occur in (5.8) and (5.10) (Harrington, 1993). For this reason, it is necessary to investigate the properties of the spurious solutions and find a method to distinguish or remove them.

5.3.1   Spurious Solutions and their Discrimination

If $G = -N_0/4$ is used, Equation (5.8) becomes

$$(\hat{I} + \hat{N}_{k_c}^{n'})h = 0. \tag{5.11}$$

Let k_s be a spurious wavenumber and h_s the corresponding spurious eigenfunction. Then $(\hat{I} + \hat{N}_{k_s}^{n'})h_s = 0$. According to the Fredholm alternative theorem, the transpose of the above equation has a nontrivial solution h_s^t

$$(\hat{I} + \hat{N}_{k_s}^{n})h_s^t = 0. \tag{5.12}$$

We define a single-layer potential function with h_s^t as the density

$$V(\rho) = \hat{N}_{k_s}(h_s^t), \rho \in R^2, \tag{5.13}$$

which is a solution of the Helmholtz equation. It follows from the jump relation that

$$\left(\frac{\partial V}{\partial n}\right)_- = (\hat{I} + \hat{N}_{k_s}^{n})h_s^t, \rho \in \Gamma, \tag{5.14}$$

where the subscript $-$ stands for the limit value of $(\partial V/\partial n)$ as ρ approaches Γ from the interior of Ω. It follows from (5.12) and (5.14) that $(\partial V/\partial n)_- = 0$. Therefore, V satisfies (5.7). Since k_s and the corresponding eigenfunction h'_s are assumed to be a spurious solution of (5.7), this is possible only when V is zero everywhere inside Ω. By continuity of V, we have

$$V(\rho) = 0, \rho \in \Gamma. \tag{5.15}$$

Furthermore, we have the jump relation

$$\left(\frac{\partial V}{\partial n}\right)_+ = (-\hat{I} + \hat{N}^n_{k_s})h'_s, \rho \in \Gamma,$$

where the subscript $+$ represents the limit value as ρ approaches Γ from the exterior of Ω. On account of the above equation and (5.14) we have $(\partial V/\partial n)_+ = -2h'_s$. Therefore, $V(\rho)$ is not zero in $R^2 - \Omega$. Thus the spurious wavenumbers of the interior Neumann eigenvalue equation (5.11) are eigenvalues of the exterior Dirichlet eigenvalue problem.

A similar discussion can be carried out for TM modes. If $G = -N_0/4$, Equation (5.10) becomes

$$\hat{N}_{k_c}(q) = 0. \tag{5.16}$$

Assume that k_s is a spurious wavenumber and h_s is the corresponding spurious eigenfunction. Then $\hat{N}_{k_s}(q_s) = 0$ has a nontrivial solution and the single-layer potential function $V = \hat{N}_{k_s}(q_s)$ vanishes on Γ. As a result, the potential V must be zero inside Ω. Otherwise it would be a solution of (5.9), contradicting the assumption that k_s is a spurious wavenumber. Since V is zero inside Ω, $(\partial V/\partial n)_-$ vanishes. The jump relations yield

$$\left(\frac{\partial V}{\partial n}\right)_- = (\hat{I} + \hat{N}^n_{k_s})q_s, \rho \in \Gamma,$$

$$\left(\frac{\partial V}{\partial n}\right)_+ = (-\hat{I} + \hat{N}^n_{k_s})q_s, \rho \in \Gamma. \tag{5.17}$$

Subtracting the second equation from the first gives

$$\left(\frac{\partial V}{\partial n}\right)_+ = -2q_s. \tag{5.18}$$

This implies that the potential V is not zero in $R^2 - \Omega$. Thus, the spurious wavenumbers of the interior Dirichlet eigenvalue equation (5.16) are eigenvalues of the exterior Dirichlet eigenvalue problem.

Consequently, the spurious wavenumbers of both interior Dirichlet and Neumann eigenvalue problems are eigenvalues of the exterior Dirichlet eigenvalue problem. This property can be used to discriminate the spurious solutions for a waveguide with an edge as shown in Figure 5.3. In the vicinity of the edge, the longitudinal electric field e for a TM mode and the longitudinal

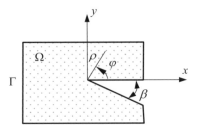

Figure 5.3   A waveguide with an edge

magnetic field h for a TE mode inside the waveguide have the following expansions

$$e = \sum_{l=1}^{\infty} a_l J_\nu(k_c \rho) \sin \nu\varphi, \quad h = \sum_{l=1}^{\infty} b_l J_\nu(k_c \rho) \cos \nu\varphi,$$

where $\nu = l\pi/(2\pi - \beta)$. For small ρ and $\varphi \in (0, 2\pi - \beta)$, it is easy to show that

$$e \propto \rho^{\pi/(2\pi-\beta)}, \quad \frac{\partial e}{\partial n} \propto \rho^{(\beta-\pi)/(2\pi-\beta)},$$

$$h \propto c_0 + c_1 \rho^{\pi/(2\pi-\beta)}, \quad \frac{\partial h}{\partial n} \propto \rho^{(\beta-\pi)/(2\pi-\beta)},$$

where c_0 and c_1 are constants. Thus, e and h are always finite while their normal derivatives are finite only when $\beta > \pi$. The field in the vicinity of the edge outside the waveguide for an exterior Dirichlet eigenvalue problem has a similar asymptotic expression, but the normal derivatives of the field are infinite for $\beta > \pi$. Therefore, the following criterion for discriminating spurious wavenumbers is obtained (Geyi, 1990).

Criterion for discriminating spurious wavenumbers: Assume that the waveguide has an edge with $\beta > \pi$. If the normal derivative of the eigenfunction corresponding to an eigenvalue approaches a large number at the vortex of the edge, the eigenvalue is a spurious wavenumber.

5.3.2   Integral Equations Without Spurious Solutions

An integral equation without spurious solutions can be established by using the Green's function that satisfies the radiation condition. Let us consider the Neumann problem first. Without loss of generality, we use $G(\rho, \rho') = H_0^{(2)}(k_c |\rho - \rho'|)/4j$. For the Neumann problem, Equation (5.8) becomes

$$(\hat{I} + \hat{H}_{k_c}^{n'})h = 0. \tag{5.19}$$

If k_s is a spurious wavenumber such that (5.19) has a nontrivial solution

$$(\hat{I} + \hat{H}_{k_s}^{n'})h_s = 0,$$

then the transpose of the above equation has a nontrivial solution h_s^t

$$(\hat{I} + \hat{H}_{k_s}^n)h_s^t = 0.$$

From the solution h_s^t, we may construct a single-layer potential function $V = \hat{H}_{k_s}(h_s^t)$. On account of the jump relation

$$\left(\frac{\partial V}{\partial n}\right)_- = (\hat{I} + \hat{H}_{k_s}^n)h_s^t = 0, \, \rho \in \Gamma,$$

and that k_s is not an eigenvalue of the Neumann problem, the potential V must vanish inside Ω. By continuity, the potential V vanishes on Γ. Since V satisfies the radiation condition, it must vanish in $R^2 - \Omega$ by the uniqueness theorem for the solution of Helmholtz equation. This implies $h_s^t = 0$, contradicting the assumption that h_s^t is a nontrivial solution. Thus, the Neumann problem (5.19) has no spurious solutions.

For Dirichlet problem, Equation (5.10) becomes $\hat{H}_{k_c}(q) = 0$. If k_s is a spurious solution so that the following equation

$$\hat{H}_{k_s}(q_s) = 0 \tag{5.20}$$

has a nontrivial solution q_s, we can construct a single-layer potential function $V = \hat{H}_{k_s}(q_s)$. Since the potential V vanishes on Γ from (5.20), it must vanish in Ω. By the uniqueness theorem for the solution of the Helmholtz equation, the potential V must vanish in $R^2 - \Omega$ since V satisfies the radiation condition. This implies $q_s = 0$, which is against our assumption that q_s is a nontrivial solution. Thus, the Dirichlet eigenvalue problem (5.10) has no spurious solutions if the Green's function G satisfies the radiation condition.

5.4 Metal Cavity Resonators

An idealized model for a resonant cavity consists of a finite space filled with a homogeneous medium and completely enclosed by a perfect conductor. Its resonant frequencies and corresponding resonant modes satisfy the Maxwell equations

$$\left.\begin{array}{l} \nabla \times \mathbf{E}(\mathbf{r}) = -j\omega\mu\mathbf{H}(\mathbf{r}) \\ \nabla \times \mathbf{H}(\mathbf{r}) = j\omega\varepsilon\mathbf{E}(\mathbf{r}) \end{array}\right\}, \mathbf{r} \in V \tag{5.21}$$
$$\mathbf{u}_n(\mathbf{r}) \times \mathbf{E}(\mathbf{r}) = 0, \mathbf{r} \in S,$$

where V is the region enclosed by the metal boundary S and $\mathbf{u}_n$ its outward unit vector. To derive an integral equation for the resonant cavity problem, we can use the integral representation of the magnetic field inside V

$$\mathbf{H}(\mathbf{r}) = -\int_S \left[\mathbf{u}_n(\mathbf{r}') \times \mathbf{H}(\mathbf{r}')\right] \times \nabla' G(\mathbf{r}, \mathbf{r}')dS(\mathbf{r}'), \mathbf{r} \in V, \tag{5.22}$$

where $G(\mathbf{r}, \mathbf{r}') = e^{-jk|\mathbf{r}-\mathbf{r}'|}/4\pi |\mathbf{r} - \mathbf{r}'|$ and $k = \omega\sqrt{\mu\varepsilon}$. In deriving (5.22), the boundary conditions in (5.21) have been used. Letting $\mathbf{r}$ approach S from inside V and using the jump relation yields

$$\frac{1}{2}\mathbf{H}_-(\mathbf{r}) = -\int_S \left[\mathbf{u}_n(\mathbf{r}') \times \mathbf{H}_-(\mathbf{r}')\right] \times \nabla' G(\mathbf{r}, \mathbf{r}')dS(\mathbf{r}'),$$

where $\mathbf{H}_-(\mathbf{r})$ denotes the limit value of $\mathbf{H}(\mathbf{r})$ when $\mathbf{r}$ approaches the boundary S from inside V. Introducing the surface current density $\mathbf{J}_s(\mathbf{r}) = -\mathbf{u}_n(\mathbf{r}) \times \mathbf{H}_-(\mathbf{r})$, the above equation can be written as

$$\frac{1}{2}\mathbf{J}_s(\mathbf{r}) + \mathbf{u}_n(\mathbf{r}) \times \int_S \mathbf{J}_s(\mathbf{r}') \times \nabla' G(\mathbf{r}, \mathbf{r}')dS(\mathbf{r}') = 0. \qquad (5.23)$$

The condition that (5.23) has a nontrivial solution determines the resonant frequencies. Numerical discretization of (5.23) is straightforward (Geyi and Hongshi, 1988(a)). Evidently any resonant frequencies of (5.21) satisfy the integral equation (5.23). It can be shown that the converse is also true. In fact, if $\mathbf{J}_s(\mathbf{r})$ is a nontrivial solution corresponding to a frequency ω obtained from (5.23), one can construct the fields

$$\mathbf{E}(\mathbf{r}) = \int_S \left[j\omega\mu\mathbf{J}_s(\mathbf{r}')G(\mathbf{r}, \mathbf{r}') - \frac{\rho_s(\mathbf{r}')}{\varepsilon}\nabla' G(\mathbf{r}, \mathbf{r}') \right] dS(\mathbf{r}'),$$

$$\mathbf{H}(\mathbf{r}) = -\int_S \mathbf{J}_s(\mathbf{r}') \times \nabla' G(\mathbf{r}, \mathbf{r}')dS(\mathbf{r}'),$$

$$(5.24)$$

where $\mathbf{r} \in R^3$, $\rho_s = \nabla_s \cdot \mathbf{J}_s/j\omega$ and ∇_s is the surface divergence. From the jump relations, we obtain

$$\mathbf{E}_+(\mathbf{r}) = -\frac{\rho_s(\mathbf{r})}{2\varepsilon}\mathbf{u}_n(\mathbf{r}) + \int_S \left[j\omega\mu\mathbf{J}_s(\mathbf{r}')G(\mathbf{r}, \mathbf{r}') - \frac{\rho_s(\mathbf{r}')}{\varepsilon}\nabla' G(\mathbf{r}, \mathbf{r}') \right] dS(\mathbf{r}'),$$

$$\mathbf{E}_-(\mathbf{r}) = \frac{\rho_s(\mathbf{r})}{2\varepsilon}\mathbf{u}_n(\mathbf{r}) + \int_S \left[j\omega\mu\mathbf{J}_s(\mathbf{r}')G(\mathbf{r}, \mathbf{r}') - \frac{\rho_s(\mathbf{r}')}{\varepsilon}\nabla' G(\mathbf{r}, \mathbf{r}') \right] dS(\mathbf{r}'),$$

$$\mathbf{H}_+(\mathbf{r}) = -\frac{1}{2}\mathbf{J}_s(\mathbf{r}) \times \mathbf{u}_n(\mathbf{r}) - \int_S \mathbf{J}_s(\mathbf{r}') \times \nabla' G(\mathbf{r}, \mathbf{r}')dS(\mathbf{r}'),$$

$$\mathbf{H}_-(\mathbf{r}) = \frac{1}{2}\mathbf{J}_s(\mathbf{r}) \times \mathbf{u}_n(\mathbf{r}) - \int_S \mathbf{J}_s(\mathbf{r}') \times \nabla' G(\mathbf{r}, \mathbf{r}')dS(\mathbf{r}'),$$

where $+$ and $-$ denote the limit value from inside V and outside V respectively. These equations imply

$$\mathbf{E}_+(\mathbf{r}) - \mathbf{E}_-(\mathbf{r}) = -\frac{\rho_s(\mathbf{r})}{\varepsilon}\mathbf{u}_n(\mathbf{r}), \tag{5.25}$$

$$\mathbf{u}_n(\mathbf{r}) \times \mathbf{H}_+(\mathbf{r}) = 0, \mathbf{r} \in S. \tag{5.26}$$

It is easy to show that the fields defined by (5.24) satisfy the Maxwell equations in whole space and the radiation condition at infinity. From (5.26) and the uniqueness theorem of Maxwell equations, the electromagnetic fields defined by (5.24) are zero outside Ω. Therefore, we have $\mathbf{E}_+(\mathbf{r}) = 0$, $\mathbf{r} \in S$ and by (5.25) we obtain $\mathbf{u}_n(\mathbf{r}) \times \mathbf{E}_-(\mathbf{r}) = 0$, $\mathbf{r} \in S$, which shows that ω and the fields defined by (5.24) satisfy (5.21). Hence ω is a resonant frequency of the cavity resonator.

5.5 Scattering Problems

The methods used to study the scattering problem depend on the electrical length of the scatterer. When the electrical length of the scatterer is very small, the low frequency analysis, such as Stevenson's approach (Stevenson, 1953) and Neumann's method (Ahner and Kleimann, 1973; Colton & Kleimann, 1980; Geyi, 1995) may be used. When the wavelength and the size of the scatterer are comparable, one may adopt numerical methods, such as the finite element method (FEM) and the moment method. If the electrical length of the scatterer is very large, the high frequency methods, such as the geometric theory of diffraction, may be applied to solve the scattering problem.

5.5.1 Three-Dimensional Scatterers

5.5.1.1 Conducting scatterer

An arbitrary conducting scatterer is shown in Figure 5.4. It is assumed that the scatterer is a perfect conductor and occupies a finite region V bounded by S. A current distribution $\mathbf{J}$ is located in the source region V_0. Let S_∞ be a closed surface large enough to enclose both V_0 and the scatterer V. From the representation theorems, the total fields inside the region bounded

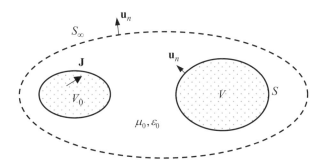

Figure 5.4   A three dimensional scatterer

by S and S_∞ can be expressed by

$$
\begin{aligned}
\mathbf{E}(\mathbf{r}) = &- \int_S j\omega\mu_0 G_0(\mathbf{r},\mathbf{r}')\mathbf{u}_n(\mathbf{r}') \times \mathbf{H}(\mathbf{r}')dS(\mathbf{r}') + \int_S \left[\mathbf{u}_n(\mathbf{r}') \times \mathbf{E}(\mathbf{r}')\right] \times \nabla'G_0(\mathbf{r},\mathbf{r}')dS(\mathbf{r}') \\
&+ \int_S \left[\mathbf{u}_n(\mathbf{r}') \cdot \mathbf{E}(\mathbf{r}')\right] \nabla'G_0(\mathbf{r},\mathbf{r}')dS(\mathbf{r}') + \int_{S_\infty} j\omega\mu_0 G_0(\mathbf{r},\mathbf{r}')\mathbf{u}_n(\mathbf{r}') \times \mathbf{H}(\mathbf{r}')dS(\mathbf{r}') \\
&- \int_{S_\infty} \left[\mathbf{u}_n(\mathbf{r}') \times \mathbf{E}(\mathbf{r}')\right] \times \nabla'G_0(\mathbf{r},\mathbf{r}')dS(\mathbf{r}') - \int_{S_\infty} \left[\mathbf{u}_n(\mathbf{r}') \cdot \mathbf{E}(\mathbf{r}')\right] \nabla'G_0(\mathbf{r},\mathbf{r}')dS(\mathbf{r}') \\
&- \int_{V_0} \left[j\omega\mu_0 G_0(\mathbf{r},\mathbf{r}')\mathbf{J}(\mathbf{r}') - \frac{\rho(\mathbf{r}')}{\varepsilon_0}\nabla'G_0(\mathbf{r},\mathbf{r}')\right]dV(\mathbf{r}'),
\end{aligned}
$$

$$
\begin{aligned}
\mathbf{H}(\mathbf{r}) = &\int_S j\omega\varepsilon_0 G_0(\mathbf{r},\mathbf{r}')\mathbf{u}_n(\mathbf{r}') \times \mathbf{E}(\mathbf{r}')dS(\mathbf{r}') + \int_S \left[\mathbf{u}_n(\mathbf{r}') \times \mathbf{H}(\mathbf{r}')\right] \times \nabla'G_0(\mathbf{r},\mathbf{r}')dS(\mathbf{r}') \\
&+ \int_S \left[\mathbf{u}_n(\mathbf{r}') \cdot \mathbf{H}(\mathbf{r}')\right] \nabla'G_0(\mathbf{r},\mathbf{r}')dS(\mathbf{r}') - \int_{S_\infty} j\omega\varepsilon_0 G_0(\mathbf{r},\mathbf{r}')\mathbf{u}_n(\mathbf{r}') \times \mathbf{E}(\mathbf{r}')dS(\mathbf{r}') \\
&- \int_{S_\infty} \left[\mathbf{u}_n(\mathbf{r}') \times \mathbf{H}(\mathbf{r}')\right] \times \nabla'G_0(\mathbf{r},\mathbf{r}')dS(\mathbf{r}') - \int_{S_\infty} \left[\mathbf{u}_n(\mathbf{r}') \cdot \mathbf{H}(\mathbf{r}')\right] \nabla'G_0(\mathbf{r},\mathbf{r}')dS(\mathbf{r}') \\
&+ \int_{V_0} \mathbf{J}(\mathbf{r}') \times \nabla'G_0(\mathbf{r},\mathbf{r}')dV(\mathbf{r}'),
\end{aligned}
$$

where $G_0(\mathbf{r},\mathbf{r}') = e^{-jk_0|\mathbf{r}-\mathbf{r}'|}/4\pi|\mathbf{r}-\mathbf{r}'|$ is the Green's function in free space and $k_0 = \omega\sqrt{\mu_0\varepsilon_0}$. Taking the radiation conditions into account, the integral over S_∞ must be zero when S_∞ becomes infinite. Therefore,

$$
\begin{aligned}
\mathbf{E}(\mathbf{r}) = &- \int_S j\omega\mu_0 G_0(\mathbf{r},\mathbf{r}')\mathbf{u}_n(\mathbf{r}') \times \mathbf{H}(\mathbf{r}')dS(\mathbf{r}') \\
&+ \int_S \left[\mathbf{u}_n(\mathbf{r}') \times \mathbf{E}(\mathbf{r}')\right] \times \nabla'G_0(\mathbf{r},\mathbf{r}')dS(\mathbf{r}') \\
&+ \int_S \left[\mathbf{u}_n(\mathbf{r}') \cdot \mathbf{E}(\mathbf{r}')\right] \nabla'G_0(\mathbf{r},\mathbf{r}')dS(\mathbf{r}') \\
&- \int_{V_0} \left[j\omega\mu_0 G_0(\mathbf{r},\mathbf{r}')\mathbf{J}(\mathbf{r}') - \frac{\rho(\mathbf{r}')}{\varepsilon_0}\nabla'G_0(\mathbf{r},\mathbf{r}')\right]dV(\mathbf{r}'), \\
\mathbf{H}(\mathbf{r}) = &\int_S j\omega\varepsilon_0 G_0(\mathbf{r},\mathbf{r}')\mathbf{u}_n(\mathbf{r}') \times \mathbf{E}(\mathbf{r}')dS(\mathbf{r}')
\end{aligned}
$$

$$+ \int_S \left[\mathbf{u}_n(\mathbf{r}') \times \mathbf{H}(\mathbf{r}')\right] \times \nabla' G_0(\mathbf{r}, \mathbf{r}') dS(\mathbf{r}')$$

$$+ \int_S \left[\mathbf{u}_n(\mathbf{r}') \cdot \mathbf{H}(\mathbf{r}')\right] \nabla' G_0(\mathbf{r}, \mathbf{r}') dS(\mathbf{r}')$$

$$+ \int_{V_0} \mathbf{J}(\mathbf{r}') \times \nabla' G_0(\mathbf{r}, \mathbf{r}') dV(\mathbf{r}').$$

If the **incident field** is defined as the portion of the total field when the scatterer is not present and the **scattered field** as the portion of the total field with the incident field subtracted, the above equations can be written as

$$\mathbf{E}(\mathbf{r}) = \mathbf{E}^s(\mathbf{r}) + \mathbf{E}^{in}(\mathbf{r})$$
$$= - \int_S j\omega\mu_0 G_0(\mathbf{r}, \mathbf{r}')\mathbf{u}_n(\mathbf{r}') \times \mathbf{H}(\mathbf{r}') dS(\mathbf{r}')$$
$$+ \int_S \left[\mathbf{u}_n(\mathbf{r}') \times \mathbf{E}(\mathbf{r}')\right] \times \nabla' G_0(\mathbf{r}, \mathbf{r}') dS(\mathbf{r}')$$
$$- \frac{1}{j\omega\varepsilon_0} \int_S \nabla'_s \cdot \left[\mathbf{u}_n(\mathbf{r}') \times \mathbf{H}(\mathbf{r}')\right] \nabla' G_0(\mathbf{r}, \mathbf{r}') dS(\mathbf{r}') + \mathbf{E}^{in}(\mathbf{r}), \qquad (5.27)$$
$$\mathbf{H}(\mathbf{r}) = \mathbf{H}^s(\mathbf{r}) + \mathbf{H}^{in}(\mathbf{r})$$
$$= \int_S j\omega\varepsilon_0 G_0(\mathbf{r}, \mathbf{r}')\mathbf{u}_n(\mathbf{r}') \times \mathbf{E}(\mathbf{r}') dS(\mathbf{r}')$$
$$+ \int_S \left[\mathbf{u}_n(\mathbf{r}') \times \mathbf{H}(\mathbf{r}')\right] \times \nabla' G_0(\mathbf{r}, \mathbf{r}') dS(\mathbf{r}')$$
$$+ \frac{1}{j\omega\mu_0} \int_S \nabla'_s \cdot \left[\mathbf{u}_n(\mathbf{r}') \times \mathbf{E}(\mathbf{r}')\right] \nabla' G_0(\mathbf{r}, \mathbf{r}') dS(\mathbf{r}') + \mathbf{H}^{in}(\mathbf{r}), \qquad (5.28)$$

where $\mathbf{E}^s(\mathbf{r})$ and $\mathbf{H}^s(\mathbf{r})$ denote the scattered fields, and $\mathbf{E}^{in}(\mathbf{r})$ and $\mathbf{H}^{in}(\mathbf{r})$ are the incident fields:

$$\mathbf{E}^{in}(\mathbf{r}) = - \int_{V_0} \left[j\omega\mu_0 G_0(\mathbf{r}, \mathbf{r}')\mathbf{J}(\mathbf{r}') - \frac{\rho(\mathbf{r}')}{\varepsilon_0}\nabla' G_0(\mathbf{r}, \mathbf{r}')\right] dV(\mathbf{r}'),$$
$$\mathbf{H}^{in}(\mathbf{r}) = \int_{V_0} \mathbf{J}(\mathbf{r}') \times \nabla' G_0(\mathbf{r}, \mathbf{r}') dV(\mathbf{r}').$$

$$(5.29)$$

Here we have used the relations

$$\nabla_s \cdot (\mathbf{u}_n \times \mathbf{H}) = -j\omega\varepsilon_0 \mathbf{u}_n \cdot \mathbf{E},$$
$$\nabla_s \cdot (\mathbf{u}_n \times \mathbf{E}) = j\omega\mu_0 \mathbf{u}_n \cdot \mathbf{H},$$

where $\nabla_s\cdot$ denotes the surface divergence. From the jump relations, it follows that

$$
\begin{aligned}
\mathbf{E}(\mathbf{r}) = &-\frac{1}{2}\mathbf{u}_n(\mathbf{r}) \times [\mathbf{u}_n(\mathbf{r}) \times \mathbf{E}(\mathbf{r})] - \frac{1}{2}\frac{1}{j\omega\varepsilon_0}\mathbf{u}_n(\mathbf{r})\nabla_s \cdot [\mathbf{u}_n(\mathbf{r}) \times \mathbf{H}(\mathbf{r})] \\
&- \int_S j\omega\mu_0 G_0(\mathbf{r},\mathbf{r}')\mathbf{u}_n(\mathbf{r}') \times \mathbf{H}(\mathbf{r}')dS(\mathbf{r}') + \int_S [\mathbf{u}_n(\mathbf{r}') \times \mathbf{E}(\mathbf{r}')] \times \nabla'G_0(\mathbf{r},\mathbf{r}')dS(\mathbf{r}') \\
&- \frac{1}{j\omega\varepsilon_0}\int_S \nabla'_s \cdot [\mathbf{u}_n(\mathbf{r}') \times \mathbf{H}(\mathbf{r}')]\nabla'G_0(\mathbf{r},\mathbf{r}')dS(\mathbf{r}') + \mathbf{E}^{in}(\mathbf{r}),
\end{aligned}
$$

$$
\begin{aligned}
\mathbf{H}(\mathbf{r}) = &-\frac{1}{2}\mathbf{u}_n(\mathbf{r}) \times [\mathbf{u}_n(\mathbf{r}) \times \mathbf{H}(\mathbf{r})] + \frac{1}{2}\mathbf{u}_n(\mathbf{r})\nabla_s \cdot [\mathbf{u}_n(\mathbf{r}) \times \mathbf{E}(\mathbf{r})] \\
&+ \int_S j\omega\varepsilon_0 G_0(\mathbf{r},\mathbf{r}')\mathbf{u}_n(\mathbf{r}') \times \mathbf{E}(\mathbf{r}')dS(\mathbf{r}') + \int_S [\mathbf{u}_n(\mathbf{r}') \times \mathbf{H}(\mathbf{r}')] \times \nabla'G_0(\mathbf{r},\mathbf{r}')dS(\mathbf{r}') \\
&+ \frac{1}{j\omega\mu_0}\int_S \nabla'_s \cdot [\mathbf{u}_n(\mathbf{r}') \times \mathbf{E}(\mathbf{r}')]\nabla'G_0(\mathbf{r},\mathbf{r}')dS(\mathbf{r}') + \mathbf{H}^{in}(\mathbf{r}).
\end{aligned}
$$

Applying the boundary conditions on a perfect conductor $\mathbf{u}_n \times \mathbf{E} = 0$, $\mathbf{u}_n \cdot \mathbf{H} = 0$ leads to

$$
\begin{aligned}
\mathbf{u}_n(\mathbf{r}) \times &\int_S j\omega\mu_0 G_0(\mathbf{r},\mathbf{r}')\mathbf{J}_s(\mathbf{r}')dS(\mathbf{r}') \\
&+ \frac{1}{j\omega\varepsilon_0}\mathbf{u}_n(\mathbf{r}) \times \int_S \nabla'_s \cdot \mathbf{J}_s(\mathbf{r}')\nabla'G_0(\mathbf{r},\mathbf{r}')dS(\mathbf{r}') = \mathbf{u}_n(\mathbf{r}) \times \mathbf{E}^{in}(\mathbf{r}),
\end{aligned}
\tag{5.30}
$$

$$
\frac{1}{2}\mathbf{J}_s(\mathbf{r}) - \mathbf{u}_n(\mathbf{r}) \times \int_S \mathbf{J}_s(\mathbf{r}') \times \nabla'G_0(\mathbf{r},\mathbf{r}')dS(\mathbf{r}') = \mathbf{u}_n(\mathbf{r}) \times \mathbf{H}^{in}(\mathbf{r}),
\tag{5.31}
$$

where $\mathbf{J}_s = \mathbf{u}_n \times \mathbf{H}$. Equation (5.30) is an integral equation of first kind and is called an **electric field integral equation** (EFIE) while (5.31) is an integral equation of second kind and is called a **magnetic field integral equation** (MFIE). Both integral equations have singular kernels. The singularities of the kernel are weak in the sense that they are integrable.

Remark 5.2: If both the source and the scatterer are two-dimensional (see Figure 5.5) so that all the field quantities are independent of one coordinate, say, z, the integral equations (5.30) and (5.31) may be simplified to scalar integral equations. To find the simplified scalar integral equations, we may use the following relation

$$
G_0(\boldsymbol{\rho},\boldsymbol{\rho}') = \frac{1}{4j}H_0^{(2)}(k_0|\boldsymbol{\rho}-\boldsymbol{\rho}'|) = \int_{-\infty}^{\infty} \frac{e^{-jk_0|\mathbf{r}-\mathbf{r}'|}}{4\pi|\mathbf{r}-\mathbf{r}'|}dz',
\tag{5.32}
$$

where $|\mathbf{r}-\mathbf{r}'| = [|\boldsymbol{\rho}-\boldsymbol{\rho}'|^2 + (z-z')^2]^{1/2}$, and $\boldsymbol{\rho} = (x,y)$ is the position vector in the (x,y)-plane. If the incident field is a TM wave (that is, the magnetic field has no z-component), the induced current on the scatterer has a z-component only

$$
\mathbf{J}_s = \mathbf{u}_n \times (H_n\mathbf{u}_n + H_l\mathbf{u}_l) = \mathbf{u}_z J_\Gamma,
$$

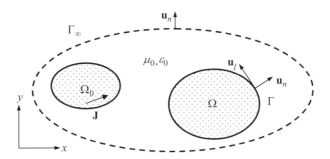

Figure 5.5   A two dimensional scatterer

where $J_\Gamma = H_l$, H_n and H_l are the normal and tangential components on the scatterer boundary Γ. Thus, (5.30) may be written as

$$\int_\Gamma j\omega\mu_0 G_0(\boldsymbol{\rho}, \boldsymbol{\rho}')J_\Gamma(\boldsymbol{\rho}')d\Gamma(\boldsymbol{\rho}') = E_z^{in}(\boldsymbol{\rho}). \qquad (5.33)$$

If the incident field is a TE wave (that is, the electric field has no z-component), the surface current has a $\mathbf{u}_l$ component only

$$\mathbf{J}_s = \mathbf{u}_n \times H_z\mathbf{u}_z = -H_z\mathbf{u}_l = \mathbf{u}_l J_\Gamma,$$

where $J_\Gamma = -H_z$. Then (5.31) reduces to

$$\frac{1}{2}J_\Gamma(\boldsymbol{\rho}) - \int_\Gamma J_\Gamma(\boldsymbol{\rho}')\frac{\partial G_0(\boldsymbol{\rho}, \boldsymbol{\rho}')}{\partial n(\boldsymbol{\rho}')}d\Gamma(\boldsymbol{\rho}') = -H_z^{in}(\boldsymbol{\rho}). \qquad (5.34)$$

Equations (5.33) and (5.34) can also be derived from Green's identity or layer ansatz.   □

Both the EFIE and the MFIE do not have uniqueness at those values of k corresponding to the interior resonance (Jones, 1979). The uniqueness of the EFIE can be determined by examining if there exists $\mathbf{J}_p \neq 0$ satisfying the homogeneous equation

$$\mathbf{u}_n(\mathbf{r}) \times \int_S j\omega\mu_0 G_0(\mathbf{r}, \mathbf{r}')\mathbf{J}_p(\mathbf{r}')dS(\mathbf{r}')$$

$$+\frac{1}{j\omega\varepsilon_0}\mathbf{u}_n(\mathbf{r}) \times \int_S \nabla_s' \cdot \mathbf{J}_p(\mathbf{r}')\nabla'G_0(\mathbf{r}, \mathbf{r}')dS(\mathbf{r}') = 0. \qquad (5.35)$$

If this is true, we may construct the following electromagnetic fields

$$\mathbf{E}_p(\mathbf{r}) = \int_S j\omega\mu_0 G_0(\mathbf{r}, \mathbf{r}')\mathbf{J}_p(\mathbf{r}')dS(\mathbf{r}') + \frac{1}{j\omega\varepsilon_0}\int_S \nabla_s' \cdot \mathbf{J}_p(\mathbf{r}')\nabla'G_0(\mathbf{r}, \mathbf{r}')dS(\mathbf{r}'),$$

$$\mathbf{H}_p(\mathbf{r}) = -\int_S \mathbf{J}_p(\mathbf{r}') \times \nabla'G_0(\mathbf{r}, \mathbf{r}')dS(\mathbf{r}').$$

From the jump relation and (5.35) it follows that $(\mathbf{u}_n \times \mathbf{E}_p)_+ = 0$. It is easy to demonstrate that $\mathbf{E}_p$ and $\mathbf{H}_p$ satisfy the Maxwell equations and radiation conditions. By the uniqueness theorem, the fields $\mathbf{E}_p$ and $\mathbf{H}_p$ must be identically zero outside S. Similarly, by the jump relation we have

$$(\mathbf{u}_n \times \mathbf{E}_p)_- = 0. \tag{5.36}$$

Hence it may be concluded that the fields $\mathbf{E}_p$ and $\mathbf{H}_p$, which satisfy the Maxwell equations inside S and (5.36), are not zero inside S. Otherwise from the jump relations, we obtain

$$(\mathbf{E}_p)_+ - (\mathbf{E}_p)_- = \frac{\mathbf{u}_n}{j\omega\varepsilon_0}\nabla_s \cdot \mathbf{J}_p, (\mathbf{H}_p)_+ - (\mathbf{H}_p)_- = -\mathbf{J}_p \times \mathbf{u}_n,$$

which would lead to $\mathbf{J}_p = 0$, contradicting our previous assumption that $\mathbf{J}_p \neq 0$.

In a similar manner, we can discuss the uniqueness of the MFIE. Let us assume that there exists a non-trivial solution $\mathbf{J}_p \neq 0$ satisfying the homogeneous equation

$$\frac{1}{2}\mathbf{J}_p(\mathbf{r}) - \mathbf{u}_n(\mathbf{r}) \times \int_S \mathbf{J}_p(\mathbf{r}') \times \nabla' G_0(\mathbf{r}, \mathbf{r}')dS(\mathbf{r}') = 0. \tag{5.37}$$

If we introduce the inner product $(\mathbf{J}_{p1}, \mathbf{J}_{p2}) = \int_S \mathbf{J}_{p1} \cdot \bar{\mathbf{J}}_{p2} dS$ for two tangential vectors $\mathbf{J}_{p1}$ and $\mathbf{J}_{p2}$, the adjoint of (5.37) is

$$\frac{1}{2}\mathbf{J}_p^a(\mathbf{r}) + \int_S \left[\mathbf{u}_n(\mathbf{r}') \times \mathbf{J}_p^a(\mathbf{r}')\right] \times \nabla' \bar{G}_0(\mathbf{r}, \mathbf{r}')dS(\mathbf{r}') = 0. \tag{5.38}$$

By the Fredholm alternative, (5.37) has a non-trivial solution if (5.38) does. Taking the complex conjugate, the above equation can be written as

$$\frac{1}{2}\mathbf{J}'_p(\mathbf{r}) + \mathbf{u}_n(\mathbf{r}) \times \int_S \mathbf{J}'_p(\mathbf{r}) \times \nabla' G_0(\mathbf{r}, \mathbf{r}')dS(\mathbf{r}') = 0, \tag{5.39}$$

where $\mathbf{J}'_p = \mathbf{u}_n \times \bar{\mathbf{J}}_p^a(\mathbf{r})$. Thus, (5.37) possesses a non-trivial solution if (5.39) does. We can construct the fields

$$\mathbf{E}'_p(\mathbf{r}) = \int_S \mathbf{J}'_p(\mathbf{r}') \times \nabla' G_0(\mathbf{r}, \mathbf{r}')dS(\mathbf{r}'),$$

$$\mathbf{H}'_p(\mathbf{r}) = \int_S j\omega\varepsilon_0 G_0(\mathbf{r}, \mathbf{r}')\mathbf{J}'_p(\mathbf{r}')dS(\mathbf{r}') + \frac{1}{j\omega\mu_0}\int_S \nabla'_s \cdot \mathbf{J}'_p(\mathbf{r}')\nabla' G_0(\mathbf{r}, \mathbf{r}')dS(\mathbf{r}').$$

It is easy to show that $\mathbf{E}'_p$ and $\mathbf{H}'_p$ satisfy Maxwell equations, the radiation condition and $(\mathbf{u}_n \times \mathbf{E}'_p)_+ = 0$ due to (5.39). Thus, the fields $\mathbf{E}'_p$ and $\mathbf{H}'_p$ are identically zero outside S. By virtue of the jump relations

$$(\mathbf{E}'_p)_+ - (\mathbf{E}'_p)_- = \mathbf{J}'_p \times \mathbf{u}_n,$$

$$(\mathbf{H}'_p)_+ - (\mathbf{H}'_p)_- = -\frac{\mathbf{u}_n}{j\omega\mu_0}\nabla_s \cdot \mathbf{J}'_p, \tag{5.40}$$

we obtain $(\mathbf{u}_n \times \mathbf{H}'_p)_- = 0$. As a result, the fields $\mathbf{E}'_p$ and $\mathbf{H}'_p$, which satisfy Maxwell equations inside S, are not zero inside S. Otherwise the first equation of (5.40) will lead to a contradiction that $\mathbf{J}'_p = 0$.

In conclusion, the uniqueness of the EFIE fails at those values of k_0, which are the interior modes of electric resonance and satisfy

$$\nabla \times \mathbf{H}_p = j\frac{k_0}{\eta_0}\mathbf{E}_p, \nabla \times \mathbf{E}_p = -jk_0\eta_0\mathbf{E}_p,$$

$$\mathbf{u}_n \times \mathbf{E}_p = 0, \mathbf{r} \in S,$$

where $\eta_0 = \sqrt{\mu_0/\varepsilon_0}$. The uniqueness of the MFIE fails at those values of k_0, which are the interior modes of magnetic resonance and satisfy

$$\nabla \times \mathbf{H}'_p = j\frac{k_0}{\eta_0}\mathbf{E}'_p, \nabla \times \mathbf{E}'_p = -jk_0\eta_0\mathbf{E}'_p,$$

$$\mathbf{u}_n \times \mathbf{H}' = 0, \mathbf{r} \in S.$$

5.5.1.2 Dielectric scatterer

A dielectric scatterer involves the penetration of the fields inside the dielectric body, and its analysis is more complicated. It will be assumed that the dielectric body with medium parameters μ and ε is finite and homogeneous, which occupies the region V bounded by S, as shown in Figure 5.4. When the field, generated by a nearby source, is incident upon the dielectric body, it will generate a scattered field outside S and a transmitted field inside S. The tangential components of the total fields must be continuous across S

$$(\mathbf{u}_n \times \mathbf{E})_+ = (\mathbf{u}_n \times \mathbf{E}_d)_-, (\mathbf{u}_n \times \mathbf{H})_+ = (\mathbf{u}_n \times \mathbf{H}_d)_-, \tag{5.41}$$

where the subscript d is used to represent the total field inside S. The total fields in the region bounded by S and S_∞ are given by (5.27) and (5.28). Similarly, the total field inside S can be expressed as

$$\mathbf{E}_d(\mathbf{r}) = \int_S j\omega\mu\mathbf{u}_n(\mathbf{r}') \times \mathbf{H}_d(\mathbf{r}')G(\mathbf{r}, \mathbf{r}')dS(\mathbf{r}') - \int_S \left[\mathbf{u}_n(\mathbf{r}') \times \mathbf{E}_d(\mathbf{r}')\right] \times \nabla'G(\mathbf{r}, \mathbf{r}')dS(\mathbf{r}')$$

$$+ \frac{1}{j\omega\varepsilon} \int_S \nabla'_s \cdot \left[\mathbf{u}_n(\mathbf{r}') \times \mathbf{H}(\mathbf{r}')\right] \nabla'G(\mathbf{r}, \mathbf{r}')dS(\mathbf{r}'),$$

$$\mathbf{H}_d(\mathbf{r}) = -\int_S j\omega\varepsilon\mathbf{u}_n(\mathbf{r}') \times \mathbf{E}_d(\mathbf{r}')G(\mathbf{r}, \mathbf{r}')dS(\mathbf{r}') - \int_S \left[\mathbf{u}_n(\mathbf{r}') \times \mathbf{H}_d(\mathbf{r}')\right] \times \nabla'G(\mathbf{r}, \mathbf{r}')dS(\mathbf{r}')$$

$$- \frac{1}{j\omega\mu} \int_S \nabla'_s \cdot \left[\mathbf{u}_n(\mathbf{r}') \times \mathbf{E}(\mathbf{r}')\right] \nabla'G(\mathbf{r}, \mathbf{r}')dS(\mathbf{r}'),$$

where $G(\mathbf{r}, \mathbf{r}') = e^{-jk|\mathbf{r}-\mathbf{r}'|}/4\pi |\mathbf{r} - \mathbf{r}'|$ and $k = \omega\sqrt{\mu\varepsilon}$. Introducing the equivalent surface electric current and surface magnetic current on the dielectric body

$$\mathbf{J}_s = \mathbf{u}_n \times \mathbf{H} = \mathbf{u}_n \times \mathbf{H}_d, \mathbf{J}_{ms} = -\mathbf{u}_n \times \mathbf{E} = -\mathbf{u}_n \times \mathbf{E}_d,$$

the fields outside S and inside S can be represented by

$$\mathbf{E}(\mathbf{r}) = -\int_S j\omega\mu_0 G_0(\mathbf{r}, \mathbf{r}')\mathbf{J}_s(\mathbf{r}')dS(\mathbf{r}') - \int_S \mathbf{J}_{ms}(\mathbf{r}') \times \nabla' G_0(\mathbf{r}, \mathbf{r}')dS(\mathbf{r}')$$
$$- \frac{1}{j\omega\varepsilon_0} \int_S \nabla'_s \cdot \mathbf{J}_s(\mathbf{r}')\nabla' G_0(\mathbf{r}, \mathbf{r}')dS(\mathbf{r}') + \mathbf{E}^{in}(\mathbf{r}),$$

$$\mathbf{H}(\mathbf{r}) = -\int_S j\omega\varepsilon_0 G_0(\mathbf{r}, \mathbf{r}')\mathbf{J}_{ms}(\mathbf{r}')dS(\mathbf{r}') + \int_S \mathbf{J}_s(r') \times \nabla' G_0(\mathbf{r}, \mathbf{r}')dS(\mathbf{r}')$$
$$- \frac{1}{j\omega\mu_0} \int_S \nabla'_s \cdot \mathbf{J}_{ms}(\mathbf{r}')\nabla' G_0(\mathbf{r}, \mathbf{r}')dS(\mathbf{r}') + \mathbf{H}^{in}(\mathbf{r}),$$

and

$$\mathbf{E}_d(\mathbf{r}) = \int_S j\omega\mu\mathbf{J}_s(\mathbf{r}')G(\mathbf{r}, \mathbf{r}')dS(\mathbf{r}') + \int_S \mathbf{J}_{ms}(\mathbf{r}') \times \nabla' G(\mathbf{r}, \mathbf{r}')dS(\mathbf{r}')$$
$$+ \frac{1}{j\omega\varepsilon} \int_S \nabla'_s \cdot \mathbf{J}_s(\mathbf{r}')\nabla' G(\mathbf{r}, \mathbf{r}')dS(\mathbf{r}'),$$

$$\mathbf{H}_d(\mathbf{r}) = \int_S j\omega\varepsilon\mathbf{J}_{ms}(\mathbf{r}')G(\mathbf{r}, \mathbf{r}')dS(\mathbf{r}') - \int_S \mathbf{J}_s(\mathbf{r}') \times \nabla' G(\mathbf{r}, \mathbf{r}')dS(\mathbf{r}')$$
$$+ \frac{1}{j\omega\mu} \int_S \nabla'_s \cdot \mathbf{J}_{ms}(\mathbf{r}')\nabla' G(\mathbf{r}, \mathbf{r}')dS(\mathbf{r}'),$$

respectively. If the observation point $\mathbf{r}$ approaches a point of S and the jump relations are used, we obtain

$$\mathbf{E}(\mathbf{r}) = \frac{1}{2}\mathbf{u}_n(\mathbf{r}) \times \mathbf{J}_{ms}(\mathbf{r}) - \frac{1}{j2\omega\varepsilon_0}\mathbf{u}_n(\mathbf{r})\nabla_s \cdot \mathbf{J}_s(\mathbf{r})$$
$$- \int_S j\omega\mu_0 G_0(\mathbf{r}, \mathbf{r}')\mathbf{J}_s(\mathbf{r}')dS(\mathbf{r}') - \int_S \mathbf{J}_{ms}(\mathbf{r}') \times \nabla' G_0(\mathbf{r}, \mathbf{r}')dS(\mathbf{r}')$$
$$- \frac{1}{j\omega\varepsilon_0} \int_S \nabla'_s \cdot \mathbf{J}_s(\mathbf{r}')\nabla' G_0(\mathbf{r}, \mathbf{r}')dS(\mathbf{r}') + \mathbf{E}^{in}(\mathbf{r}),$$

$$\mathbf{H}(\mathbf{r}) = -\frac{1}{2}\mathbf{u}_n(\mathbf{r}) \times \mathbf{J}_s(\mathbf{r}) - \frac{1}{j2\omega\mu_0}\mathbf{u}_n(\mathbf{r})\nabla_s \cdot \mathbf{J}_{ms}(\mathbf{r})$$

$$-\int_S j\omega\varepsilon_0 G_0(\mathbf{r}, \mathbf{r}')\mathbf{J}_{ms}(\mathbf{r}')dS(\mathbf{r}') + \int_S \mathbf{J}_s(\mathbf{r}') \times \nabla'G_0(\mathbf{r}, \mathbf{r}')dS(\mathbf{r}')$$

$$-\frac{1}{j\omega\mu_0}\int_S \nabla'_s \cdot \mathbf{J}_{ms}(\mathbf{r}')\nabla'G_0(\mathbf{r}, \mathbf{r}')dS(\mathbf{r}') + \mathbf{H}^{in}(\mathbf{r}),$$

$$\mathbf{E}_d(\mathbf{r}) = \frac{1}{2}\mathbf{u}_n(\mathbf{r}) \times \mathbf{J}_{ms}(\mathbf{r}) - \frac{1}{j2\omega\varepsilon}\mathbf{u}_n(\mathbf{r})\nabla_s \cdot \mathbf{J}_s(\mathbf{r})$$

$$+\int_S j\omega\mu\mathbf{J}_s(\mathbf{r}')G(\mathbf{r}, \mathbf{r}')dS(\mathbf{r}') + \int_S \mathbf{J}_{ms}(\mathbf{r}') \times \nabla'G(\mathbf{r}, \mathbf{r}')dS(\mathbf{r}')$$

$$+\frac{1}{j\omega\varepsilon}\int_S \nabla'_s \cdot \mathbf{J}_s(\mathbf{r}')\nabla'G(\mathbf{r}, \mathbf{r}')dS(\mathbf{r}'),$$

$$\mathbf{H}_d(\mathbf{r}) = -\frac{1}{2}\mathbf{u}_n(\mathbf{r}) \times \mathbf{J}_s(\mathbf{r}) - \frac{1}{j2\omega\mu}\mathbf{u}_n(\mathbf{r})\nabla_s \cdot \mathbf{J}_{ms}(\mathbf{r})$$

$$+\int_S j\omega\varepsilon\mathbf{J}_{ms}(\mathbf{r}')G(\mathbf{r}, \mathbf{r}')dS(\mathbf{r}') - \int_S \mathbf{J}_s(\mathbf{r}') \times \nabla'G(\mathbf{r}, \mathbf{r}')dS(\mathbf{r}')$$

$$+\frac{1}{j\omega\mu}\int_S \nabla'_s \cdot \mathbf{J}_{ms}(\mathbf{r}')\nabla'G(\mathbf{r}, \mathbf{r}')dS(\mathbf{r}').$$

Multiplying these equations vectorially by $\mathbf{u}_n$ yields

$$-\frac{1}{2}\varepsilon_0\mathbf{J}_{ms}(\mathbf{r}) = -\mathbf{u}_n(\mathbf{r}) \times \int_S j\omega\mu_0\varepsilon_0 G_0(\mathbf{r}, \mathbf{r}')\mathbf{J}_s(\mathbf{r}')dS(\mathbf{r}')$$

$$-\mathbf{u}_n(\mathbf{r}) \times \int_S \mathbf{J}_{ms}(\mathbf{r}') \times \varepsilon_0\nabla'G_0(\mathbf{r}, \mathbf{r}')dS(\mathbf{r}') \qquad (5.42)$$

$$-\frac{1}{j\omega}\mathbf{u}_n(\mathbf{r}) \times \int_S \nabla'_s \cdot \mathbf{J}_s(\mathbf{r}')\nabla'G_0(\mathbf{r}, \mathbf{r}')dS(\mathbf{r}')$$

$$+\mathbf{u}_n(\mathbf{r}) \times \varepsilon_0\mathbf{E}^{in}(\mathbf{r}),$$

$$\frac{1}{2}\mu_0\mathbf{J}_s(\mathbf{r})(\mathbf{r}) = -\mathbf{u}_n(\mathbf{r}) \times \int_S j\omega\mu_0\varepsilon_0 G_0(\mathbf{r}, \mathbf{r}')\mathbf{J}_{ms}(\mathbf{r}')dS(\mathbf{r}')$$

$$+\mathbf{u}_n(\mathbf{r}) \times \int_S \mathbf{J}_s(\mathbf{r}') \times \mu_0\nabla'G_0(\mathbf{r}, \mathbf{r}')dS(\mathbf{r}') \qquad (5.43)$$

$$-\frac{1}{j\omega}\mathbf{u}_n(\mathbf{r}) \times \int_S \nabla'_s \cdot \mathbf{J}_{ms}(\mathbf{r}')\nabla'G_0(\mathbf{r}, \mathbf{r}')dS(\mathbf{r}')$$

$$+\mathbf{u}_n(\mathbf{r}) \times \mu_0\mathbf{H}^{in}(\mathbf{r}),$$

$$-\frac{1}{2}\varepsilon \mathbf{J}_{ms}(\mathbf{r}) = \mathbf{u}_n(\mathbf{r}) \times \int_S j\omega\mu\varepsilon \mathbf{J}_s(\mathbf{r}')G(\mathbf{r},\mathbf{r}')dS(\mathbf{r}')$$

$$+ \mathbf{u}_n(\mathbf{r}) \times \int_S \mathbf{J}_{ms}(\mathbf{r}') \times \varepsilon\nabla'G(\mathbf{r},\mathbf{r}')dS(\mathbf{r}') \tag{5.44}$$

$$+ \frac{1}{j\omega}\mathbf{u}_n(\mathbf{r}) \times \int_S \nabla'_s \cdot \mathbf{J}_s(\mathbf{r}')\nabla'G(\mathbf{r},\mathbf{r}')dS(\mathbf{r}'),$$

$$\frac{1}{2}\mu \mathbf{J}_s(\mathbf{r}) = \mathbf{u}_n(\mathbf{r}) \times \int_S j\omega\mu\varepsilon \mathbf{J}_{ms}(\mathbf{r}')G(\mathbf{r},\mathbf{r}')dS(\mathbf{r}')$$

$$- \mathbf{u}_n(\mathbf{r}) \times \int_S \mathbf{J}_s(\mathbf{r}') \times \mu\nabla'G(\mathbf{r},\mathbf{r}')dS(\mathbf{r}') \tag{5.45}$$

$$+ \frac{1}{j\omega}\mathbf{u}_n(\mathbf{r}) \times \int_S \nabla'_s \cdot \mathbf{J}_{ms}(\mathbf{r}')\nabla'G(\mathbf{r},\mathbf{r}')dS(\mathbf{r}').$$

Adding (5.42) and (5.44) gives

$$-\frac{1}{2}(\varepsilon_0 + \varepsilon)\mathbf{J}_{ms}(\mathbf{r}) + j\mathbf{u}_n(\mathbf{r}) \times \int_S \left[k_0^2 G_0(\mathbf{r},\mathbf{r}') - k^2 G(\mathbf{r},\mathbf{r}')\right]\frac{1}{\omega}\mathbf{J}_s(\mathbf{r}')dS(\mathbf{r}')$$

$$+ \mathbf{u}_n(\mathbf{r}) \times \int_S \mathbf{J}_{ms}(\mathbf{r}') \times \left[\varepsilon_0\nabla'G_0(\mathbf{r},\mathbf{r}') - \varepsilon\nabla'G(\mathbf{r},\mathbf{r}')\right]dS(\mathbf{r}')$$

$$+ \frac{1}{j\omega}\mathbf{u}_n(\mathbf{r}) \times \int_S \nabla'_s \cdot \mathbf{J}_s(\mathbf{r}')\left[\nabla'G_0(\mathbf{r},\mathbf{r}') - \nabla'G(\mathbf{r},\mathbf{r}')\right]dS(\mathbf{r}') \tag{5.46}$$

$$= \mathbf{u}_n(\mathbf{r}) \times \varepsilon_0\mathbf{E}^{in}(\mathbf{r}).$$

Adding (5.43) and (5.45) gives

$$\frac{1}{2}(\mu_0 + \mu)\mathbf{J}_s(\mathbf{r}) + j\mathbf{u}_n(\mathbf{r}) \times \int_S \left[k_0^2 G_0(\mathbf{r},\mathbf{r}') - k^2 G(\mathbf{r},\mathbf{r}')\right]\frac{1}{\omega}\mathbf{J}_{ms}(\mathbf{r}')dS(\mathbf{r}')$$

$$+ \mathbf{u}_n(\mathbf{r}) \times \int_S \mathbf{J}_s(\mathbf{r}') \times \left[\mu\nabla'G(\mathbf{r},\mathbf{r}') - \mu_0\nabla'G_0(\mathbf{r},\mathbf{r}')\right]dS(\mathbf{r}')$$

$$+ \frac{1}{j\omega}\mathbf{u}_n(\mathbf{r}) \times \int_S \nabla'_s \cdot \mathbf{J}_{ms}(\mathbf{r}')\left[\nabla'G_0(\mathbf{r},\mathbf{r}') - \nabla'G(\mathbf{r},\mathbf{r}')\right]dS(\mathbf{r}') \tag{5.47}$$

$$= \mathbf{u}_n(\mathbf{r}) \times \mu_0\mathbf{H}^{in}(\mathbf{r}).$$

Making use of the relation $\int_S \nabla_s \cdot \mathbf{F}(\mathbf{r}) dS(\mathbf{r}) = 0$ for an arbitrary vector field $\mathbf{F}(\mathbf{r})$, the last integral in (5.46) and (5.47) may be written as

$$\int_S \nabla'_s \cdot \mathbf{J}_s(\mathbf{r}') \left[\nabla' G_0(\mathbf{r}, \mathbf{r}') - \nabla' G(\mathbf{r}, \mathbf{r}') \right] dS(\mathbf{r}')$$

$$= -\nabla \int_S \nabla'_s \cdot \mathbf{J}_s(\mathbf{r}') \left[G_0(\mathbf{r}, \mathbf{r}') - G(\mathbf{r}, \mathbf{r}') \right] dS(\mathbf{r}')$$

$$= -\int_S \left[\mathbf{J}_s(\mathbf{r}') \cdot \nabla' \right] \nabla' \left[G_0(\mathbf{r}, \mathbf{r}') - G(\mathbf{r}, \mathbf{r}') \right] dS(\mathbf{r}').$$

Therefore, (5.46) and (5.47) become

$$-\frac{1}{2}(\varepsilon_0 + \varepsilon) \mathbf{J}_{ms}(\mathbf{r}) + j \mathbf{u}_n(\mathbf{r}) \times \int_S \left[k_0^2 G_0(\mathbf{r}, \mathbf{r}') - k^2 G(\mathbf{r}, \mathbf{r}') \right] \frac{1}{\omega} \mathbf{J}_s(\mathbf{r}') dS(\mathbf{r}')$$

$$+ \mathbf{u}_n(\mathbf{r}) \times \int_S \mathbf{J}_{ms}(\mathbf{r}') \times \left[\varepsilon_0 \nabla' G_0(\mathbf{r}, \mathbf{r}') - \varepsilon \nabla' G(\mathbf{r}, \mathbf{r}') \right] dS(\mathbf{r}')$$

$$-\frac{1}{j\omega} \mathbf{u}_n(\mathbf{r}) \times \int_S \left[\mathbf{J}_s(\mathbf{r}') \cdot \nabla' \right] \left[\nabla' G_0(\mathbf{r}, \mathbf{r}') - \nabla' G(\mathbf{r}, \mathbf{r}') \right] dS(\mathbf{r}')$$

$$= \mathbf{u}_n(\mathbf{r}) \times \varepsilon_0 \mathbf{E}^{in}(\mathbf{r}),$$

(5.48)

$$\frac{1}{2}(\mu_0 + \mu) \mathbf{J}_s(\mathbf{r}) + j \mathbf{u}_n(\mathbf{r}) \times \int_S \left[k_0^2 G_0(\mathbf{r}, \mathbf{r}') - k^2 G(\mathbf{r}, \mathbf{r}') \right] \frac{1}{\omega} \mathbf{J}_{ms}(\mathbf{r}') dS(\mathbf{r}')$$

$$+ \mathbf{u}_n(\mathbf{r}) \times \int_S \mathbf{J}_s(\mathbf{r}') \times \left[\mu \nabla' G(\mathbf{r}, \mathbf{r}') - \mu_0 \nabla' G_0(\mathbf{r}, \mathbf{r}') \right] dS(\mathbf{r}')$$

$$-\frac{1}{j\omega} \mathbf{u}_n(\mathbf{r}) \times \int_S \left[\mathbf{J}_{ms}(\mathbf{r}') \cdot \nabla' \right] \left[\nabla' G_0(\mathbf{r}, \mathbf{r}') - \nabla' G(\mathbf{r}, \mathbf{r}') \right] dS(\mathbf{r}')$$

$$= \mathbf{u}_n(\mathbf{r}) \times \mu_0 \mathbf{H}^{in}(\mathbf{r}).$$

(5.49)

Equations (5.48) and (5.49) are the integral equations for an arbitrary dielectric scatterer, which may be rewritten as

$$-\frac{1}{2}\left(\frac{k_0}{\eta_0} + \frac{k}{\eta} \right) \mathbf{J}_{ms}(\mathbf{r}) + j \mathbf{u}_n(\mathbf{r}) \times \int_S \left[k_0^2 G_0(\mathbf{r}, \mathbf{r}') - k^2 G(\mathbf{r}, \mathbf{r}') \right] \mathbf{J}_s(\mathbf{r}') dS(\mathbf{r}')$$

$$+ \mathbf{u}_n(\mathbf{r}) \times \int_S \mathbf{J}_{ms}(\mathbf{r}') \times \left[\frac{k_0}{\eta_0} \nabla' G_0(\mathbf{r}, \mathbf{r}') - \frac{k}{\eta} \nabla' G(\mathbf{r}, \mathbf{r}') \right] dS(\mathbf{r}')$$

$$-\frac{1}{j} \mathbf{u}_n(\mathbf{r}) \times \int_S \left[\mathbf{J}_s(\mathbf{r}') \cdot \nabla' \right] \left[\nabla' G_0(\mathbf{r}, \mathbf{r}') - \nabla' G(\mathbf{r}, \mathbf{r}') \right] dS(\mathbf{r}')$$

(5.50)

$$= \mathbf{u}_n(\mathbf{r}) \times \frac{k_0}{\eta_0} \mathbf{E}^{in}(\mathbf{r}),$$

$$\frac{1}{2}(k_0\eta_0 + k\eta)\mathbf{J}_s(\mathbf{r}) + j\mathbf{u}_n(\mathbf{r}) \times \int_S \left[k_0^2 G_0(\mathbf{r}, \mathbf{r}') - k^2 G(\mathbf{r}, \mathbf{r}')\right]\mathbf{J}_{ms}(\mathbf{r}')dS(\mathbf{r}')$$

$$+ \mathbf{u}_n(\mathbf{r}) \times \int_S \mathbf{J}_s(\mathbf{r}') \times \left[k\eta\nabla' G(\mathbf{r}, \mathbf{r}') - k_0\eta_0\nabla' G_0(\mathbf{r}, \mathbf{r}')\right]dS(\mathbf{r}')$$

$$\quad (5.51)$$

$$- \frac{1}{j}\mathbf{u}_n(\mathbf{r}) \times \int_S \left[\mathbf{J}_{ms}(\mathbf{r}') \cdot \nabla'\right]\left[\nabla' G_0(\mathbf{r}, \mathbf{r}') - \nabla' G(\mathbf{r}, \mathbf{r}')\right]dS(\mathbf{r}')$$

$$= \mathbf{u}_n(\mathbf{r}) \times k_0\eta_0\mathbf{H}^{in}(\mathbf{r}),$$

where $\eta_0 = \sqrt{\mu_0/\varepsilon_0}$, $\eta = \sqrt{\mu/\varepsilon}$. The integral equations (5.50) and (5.51) for the dielectric scatterer can also be used to determine the interior resonant frequencies of an isolated dielectric resonator by assuming the right-hand side of both equations to be zero (Geyi and Hongshi, 1988(b)). At these frequencies the scattering solution of the dielectric object is not unique.

Remark 5.3: Several methods have been proposed to overcome the non-uniqueness of integral equations for the scattering problems (Mautz and Harrington, 1978; Poggio and Miller, 1973; Peterson, 1990; Correia, 1993). Since the matrix equations obtained from the discretization of the integral equations become ill-conditioned at the interior resonant frequencies, the matrix condition number can be used to detect the degree of ill-conditioning, thus providing an indicator for the interior resonant frequencies (Klein and Mittra, 1973). Another method of avoiding the non-uniqueness problem is to utilize the **extended boundary condition** (EBC). The EBC is defined as the requirement that a set of field quantities vanish over an observation domain in the zero-field region. The observation domain can be a closed surface, a portion of plane, or a portion of line in the zero-field region.   □

5.5.2   Two-Dimensional Scatterers

If the fields are independent of one coordinate, say, the z coordinate, Maxwell equations can be expressed by the sum of two different kinds of fields, called **transverse electric** (TE) field and **transverse magnetic** (TM) field. The TM field has no z-component of the magnetic field while the TE field has no z-component of the electric field. The fundamental equations for the TM field can be written as

$$\mathbf{u}_z \times \nabla E_z(\boldsymbol{\rho}) = j\omega\mu\mathbf{H}_t(\boldsymbol{\rho}),$$

$$\quad (5.52)$$

$$\mathbf{u}_z \cdot \nabla \times \mathbf{H}_t(\boldsymbol{\rho}) = j\omega\varepsilon E_z(\boldsymbol{\rho}) + J_z(\boldsymbol{\rho}),$$

where the subscript t denotes the transverse part of the vector field and $\boldsymbol{\rho} = (x, y)$ is the two-dimensional position vector. The fundamental equations for the TE field can be expressed as

$$\mathbf{u}_z \cdot \nabla \times \mathbf{E}_t(\boldsymbol{\rho}) = -j\omega\mu H_z(\boldsymbol{\rho}),$$

$$\quad (5.53)$$

$$-\mathbf{u}_z \times \nabla H_z(\boldsymbol{\rho}) = j\omega\varepsilon\mathbf{E}_t(\boldsymbol{\rho}) + \mathbf{J}_t(\boldsymbol{\rho}).$$

Eliminating $\mathbf{H}_t$ from (5.52) and $\mathbf{E}_t$ from (5.53) yields

$$\nabla_t^2 E_z(\boldsymbol{\rho}) + k^2 E_z(\boldsymbol{\rho}) = j\omega\mu J_z(\boldsymbol{\rho}), \text{ TM},$$

$$\quad (5.54)$$

$$\nabla_t^2 H_z(\boldsymbol{\rho}) + k^2 H_z(\boldsymbol{\rho}) = \nabla_t \cdot [\mathbf{u}_z \times \mathbf{J}_t(\boldsymbol{\rho})], \text{ TE}.$$

Let us consider a two-dimensional scatterer shown in Figure 5.5. On the boundary of the scatterer, the z-component of the TM fields must satisfy

$$(E_z)_+ = (E_{dz})_- , \quad \left(\frac{1}{jk_0\eta_0} \frac{\partial E_z}{\partial n} \right)_+ = \left(\frac{1}{jk\eta} \frac{\partial E_{dz}}{\partial n} \right)_- , \tag{5.55}$$

where the subscript d indicates the total field inside the scatterer. If the scatterer is a perfect conductor, the above boundary conditions reduce to $E_z = 0$. For a perfect conductor, the current distribution on the scatterer for the TM fields is then given by

$$\mathbf{J}_\Gamma = \mathbf{u}_z J_\Gamma = \mathbf{u}_z \frac{1}{jk_0\eta_0} \frac{\partial E_z}{\partial n}. \tag{5.56}$$

The boundary conditions for TE fields on the scatterer are

$$(H_{dz})_- - (H_z)_+ = \mathbf{J}_\Gamma \cdot \mathbf{u}_l, \quad \left(\frac{\eta}{jk} \frac{\partial H_{dz}}{\partial n} \right)_- = \left(\frac{\eta_0}{jk_0} \frac{\partial H_z}{\partial n} \right)_+ . \tag{5.57}$$

If the scatterer is a perfect conductor, the external field satisfies $(\partial H_z/\partial n)_+ = 0$ and the surface current for the TE fields is given by

$$\mathbf{J}_\Gamma = J_\Gamma \mathbf{u}_l = -H_z \mathbf{u}_l. \tag{5.58}$$

5.5.2.1 Conducting cylinder

For the TE incidence with the incident field generated by a nearby current source confined in Ω_0, the total z-component of the magnetic field in the region bounded by Γ_∞ and Γ may be obtained by using Green's identity for a scalar field

$$H_z(\boldsymbol{\rho}) = -\int_{\Omega_0} G_0(\boldsymbol{\rho}, \boldsymbol{\rho}') \nabla_t' \cdot \left[\mathbf{u}_z \times \mathbf{J}_t(\boldsymbol{\rho}') \right] d\Omega(\boldsymbol{\rho}')$$

$$- \int_\Gamma \left[G_0(\boldsymbol{\rho}, \boldsymbol{\rho}') \frac{\partial H_z(\boldsymbol{\rho}')}{\partial n(\boldsymbol{\rho}')} - H_z(\boldsymbol{\rho}') \frac{\partial G_0(\boldsymbol{\rho}, \boldsymbol{\rho}')}{\partial n(\boldsymbol{\rho}')} \right] d\Gamma(\boldsymbol{\rho}')$$

$$+ \int_{\Gamma_\infty} \left[G_0(\boldsymbol{\rho}, \boldsymbol{\rho}') \frac{\partial H_z(\boldsymbol{\rho}')}{\partial n(\boldsymbol{\rho}')} - H_z(\boldsymbol{\rho}') \frac{\partial G_0(\boldsymbol{\rho}, \boldsymbol{\rho}')}{\partial n(\boldsymbol{\rho}')} \right] d\Gamma(\boldsymbol{\rho}'),$$

where $G_0(\boldsymbol{\rho}, \boldsymbol{\rho}') = H_0^{(2)}(k_0 |\boldsymbol{\rho} - \boldsymbol{\rho}'|)/4j$. The integral on Γ_∞ vanishes as Γ_∞ approaches infinity because of the radiation condition. From the boundary condition on a perfect conductor, it follows that

$$H_z(\boldsymbol{\rho}) = -\int_{\Omega_0} G_0(\boldsymbol{\rho}, \boldsymbol{\rho}') \nabla_t' \cdot \left[\mathbf{u}_z \times \mathbf{J}_t(\boldsymbol{\rho}') \right] d\Omega(\boldsymbol{\rho}') + \int_\Gamma H_z(\boldsymbol{\rho}') \frac{\partial G_0(\boldsymbol{\rho}, \boldsymbol{\rho}')}{\partial n(\boldsymbol{\rho}')} d\Gamma(\boldsymbol{\rho}')$$

$$= H_z^{in}(\boldsymbol{\rho}) + \int_\Gamma H_z(\boldsymbol{\rho}') \frac{\partial G_0(\boldsymbol{\rho}, \boldsymbol{\rho}')}{\partial n(\mathbf{r}')} d\Gamma(\boldsymbol{\rho}'),$$

where $H_z^{in}(\boldsymbol{\rho}) = -\int_{V_0} G_0(\boldsymbol{\rho}, \boldsymbol{\rho}')\nabla_t' \cdot [\mathbf{u}_z \times \mathbf{J}_t(\boldsymbol{\rho}')]d\Omega(\boldsymbol{\rho}')$ is the incident field. As the observation point $\boldsymbol{\rho}$ approaches a point on the boundary Γ, the jump relation gives

$$\frac{1}{2}J_\Gamma(\boldsymbol{\rho}) - \int_\Gamma J_\Gamma(\boldsymbol{\rho}')\frac{\partial G_0(\boldsymbol{\rho}, \boldsymbol{\rho}')}{\partial n(\boldsymbol{\rho}')}d\Gamma(\boldsymbol{\rho}') = -H_z^{in}(\boldsymbol{\rho}), \tag{5.59}$$

where $J_\Gamma = -H_z$. This agrees with (5.34).

For the TM incidence, the z-component of the electric field in the region bounded by Γ_∞ and Γ may be expressed as

$$E_z(\boldsymbol{\rho}) = j\omega\mu_0 \int_{\Omega_0} G_0(\boldsymbol{\rho}, \boldsymbol{\rho}')J_z(\boldsymbol{\rho}')d\Omega(\boldsymbol{\rho}') - \int_\Gamma \left[G_0(\boldsymbol{\rho}, \boldsymbol{\rho}')\frac{\partial E_z(\boldsymbol{\rho}')}{\partial n(\boldsymbol{\rho}')} - E_z(\boldsymbol{\rho}')\frac{\partial G_0(\boldsymbol{\rho}, \boldsymbol{\rho}')}{\partial n(\boldsymbol{\rho}')} \right]d\Gamma(\boldsymbol{\rho}')$$
$$+ \int_{\Gamma_\infty} \left[G_0(\boldsymbol{\rho}, \boldsymbol{\rho}')\frac{\partial E_z(\boldsymbol{\rho}')}{\partial n(\boldsymbol{\rho}')} - E_z(\boldsymbol{\rho}')\frac{\partial G_0(\boldsymbol{\rho}, \boldsymbol{\rho}')}{\partial n(\boldsymbol{\rho}')} \right]d\Gamma(\boldsymbol{\rho}').$$

The integral on Γ_∞ must vanish as Γ_∞ approaches infinity. Introducing the boundary condition on the perfect conductor, we have

$$E_z(\boldsymbol{\rho}) = E_z^{in}(\boldsymbol{\rho}) - \int_\Gamma G_0(\boldsymbol{\rho}, \boldsymbol{\rho}')\frac{\partial E_z(\boldsymbol{\rho}')}{\partial n(\boldsymbol{\rho}')}d\Gamma(\boldsymbol{\rho}'), \tag{5.60}$$

where $E_z^{in}(\boldsymbol{\rho}) = j\omega\mu_0 \int_{V_0} G_0(\boldsymbol{\rho}, \boldsymbol{\rho}')J_z(\boldsymbol{\rho}')d\Omega(\boldsymbol{\rho}')$ is the incident field. If we let the observation point approach the boundary Γ, Equation (5.60) becomes

$$\int_\Gamma G_0(\boldsymbol{\rho}, \boldsymbol{\rho}')J_\Gamma(\boldsymbol{\rho}')d\Gamma(\boldsymbol{\rho}') = \frac{1}{jk_0\eta_0}E_z^{in}(\boldsymbol{\rho}), \tag{5.61}$$

where $J_\Gamma(\boldsymbol{\rho}) = (jk_0\eta_0)^{-1}\partial E_z(\boldsymbol{\rho})/\partial n(\boldsymbol{\rho})$ is the surface current. This agrees with (5.33).

Equation (5.61) is an integral equation of first kind. An integral equation of second kind for the TM incidence can be obtained by taking the normal derivative of (5.60) so that

$$\frac{\partial E_z(\boldsymbol{\rho})}{\partial n(\boldsymbol{\rho})} = \frac{\partial E_z^{in}(\boldsymbol{\rho})}{\partial n(\boldsymbol{\rho})} - \int_\Gamma \frac{\partial G_0(\boldsymbol{\rho}, \boldsymbol{\rho}')}{\partial n(\boldsymbol{\rho})}\frac{\partial E_z(\boldsymbol{\rho}')}{\partial n(\boldsymbol{\rho}')}d\Gamma(\boldsymbol{\rho}').$$

Letting the observation point $\boldsymbol{\rho}$ approach the boundary Γ and using the jump relation, we obtain

$$\frac{1}{2}J_\Gamma(\boldsymbol{\rho}) + \int_\Gamma J_\Gamma(\boldsymbol{\rho}')\frac{\partial G_0(\boldsymbol{\rho}, \boldsymbol{\rho}')}{\partial n(\boldsymbol{\rho})}d\Gamma(\boldsymbol{\rho}') = \frac{1}{jk_0\eta_0}\frac{\partial E_z^{in}(\boldsymbol{\rho})}{\partial n(\boldsymbol{\rho})}. \tag{5.62}$$

This is an integral equation of the second kind. Both (5.61) and (5.62) can be used to solve the scattering problem for the TM fields.

To determine the uniqueness of (5.59), we assume that there exists $J_{\Gamma p} \neq 0$, which satisfies the homogenous equation

$$\frac{1}{2} J_{\Gamma p}(\boldsymbol{\rho}) - \int_\Gamma J_{\Gamma p}(\boldsymbol{\rho}') \frac{\partial G_0(\boldsymbol{\rho}, \boldsymbol{\rho}')}{\partial n(\boldsymbol{\rho}')} d\Gamma(\boldsymbol{\rho}') = 0. \tag{5.63}$$

We may construct a single-layer potential function using $J_{\Gamma p}(\boldsymbol{\rho})$

$$u_p(\boldsymbol{\rho}) = \int_\Gamma J_{\Gamma p}(\boldsymbol{\rho}') G_0(\boldsymbol{\rho}, \boldsymbol{\rho}') d\Gamma(\boldsymbol{\rho}'),$$

and it satisfies the two-dimensional Helmholtz equation

$$\nabla_t^2 u_p + k_0^2 u_p = 0, \tag{5.64}$$

in the (x, y)-plane. It follows from the jump relation that

$$\left[\frac{\partial u_p(\boldsymbol{\rho})}{\partial n(\boldsymbol{\rho})} \right]_- = \frac{1}{2} J_{\Gamma p}(\boldsymbol{\rho}) + \int_\Gamma J_{\Gamma p}(\boldsymbol{\rho}') \frac{\partial G_0(\boldsymbol{\rho}, \boldsymbol{\rho}')}{\partial n(\boldsymbol{\rho})} d\Gamma(\boldsymbol{\rho}')$$

$$= \frac{1}{2} J_{\Gamma p}(\boldsymbol{\rho}) - \int_\Gamma J_{\Gamma p}(\boldsymbol{\rho}') \frac{\partial G_0(\boldsymbol{\rho}, \boldsymbol{\rho}')}{\partial n(\boldsymbol{\rho}')} d\Gamma(\boldsymbol{\rho}') = 0.$$

Hence the uniqueness of (5.63) fails at those values of k_0, which are the eigenvalues of the interior Neumann problem. The study of the uniqueness of (5.62) is similar to that of (5.59).

For the uniqueness of (5.61), we can assume that the integral equation

$$\int_\Gamma G_0(\boldsymbol{\rho}, \boldsymbol{\rho}') J_{\Gamma p}(\boldsymbol{\rho}') d\Gamma(\boldsymbol{\rho}') = 0, \; \boldsymbol{\rho} \in \Gamma \tag{5.65}$$

has a non-trivial solution $J_{\Gamma p}(\boldsymbol{\rho})$, and construct a single-layer function

$$u_p(\boldsymbol{\rho}) = \int_\Gamma J_{\Gamma p}(\boldsymbol{\rho}') G_0(\boldsymbol{\rho}, \boldsymbol{\rho}') d\Gamma(\boldsymbol{\rho}'),$$

which also satisfies the Helmholtz equation (5.64) in the (x, y)-plane. By the continuity of the single-layer function and (5.65), we have

$$[u_p(\boldsymbol{\rho})]_- = \int_\Gamma J_{\Gamma p}(\boldsymbol{\rho}') G_0(\boldsymbol{\rho}, \boldsymbol{\rho}') d\Gamma(\boldsymbol{\rho}') = 0.$$

So the uniqueness of (5.61) fails at those values of k_0, which are the eigenvalues of interior Dirichlet problem.

5.5.2.2 Dielectric cylinder

Similarly, the total z-component of the magnetic field for the TE incidence in the region bounded by Γ_∞ and Γ may be represented by

$$H_z(\boldsymbol{\rho}') = H_z^{in}(\boldsymbol{\rho}) - \int_\Gamma \left[G_0(\boldsymbol{\rho}, \boldsymbol{\rho}') \frac{\partial H_z(\boldsymbol{\rho}')}{\partial n(\boldsymbol{\rho}')} - H_z(\boldsymbol{\rho}') \frac{\partial G_0(\boldsymbol{\rho}, \boldsymbol{\rho}')}{\partial n(\boldsymbol{\rho}')} \right] d\Gamma(\boldsymbol{\rho}'), \qquad (5.66)$$

where the integral on Γ_∞ has been ignored because of the radiation condition and

$$H_z^{in}(\boldsymbol{\rho}) = - \int_{\Omega_0} G_0(\boldsymbol{\rho}', \boldsymbol{\rho}') \nabla_t' \cdot \left[\mathbf{u}_z \times \mathbf{J}_t(\boldsymbol{\rho}') \right] d\Omega(\boldsymbol{\rho}')$$

is the incident field. The integral expression for the total field inside Ω is

$$H_{dz}(\boldsymbol{\rho}) = \int_\Gamma \left[G(\boldsymbol{\rho}, \boldsymbol{\rho}') \frac{\partial H_{dz}(\boldsymbol{\rho}')}{\partial n(\boldsymbol{\rho}')} - H_{dz}(\boldsymbol{\rho}') \frac{\partial G(\boldsymbol{\rho}, \boldsymbol{\rho}')}{\partial n(\boldsymbol{\rho}')} \right] d\Gamma(\boldsymbol{\rho}'), \boldsymbol{\rho} \in \Omega$$

where $G(\boldsymbol{\rho}, \boldsymbol{\rho}') = H_0^{(2)}(k |\boldsymbol{\rho} - \boldsymbol{\rho}'|)/4j$. As the observation point $\boldsymbol{\rho}$ approaches the boundary Γ, the jump relations yield

$$\frac{1}{2} H_z(\boldsymbol{\rho}) = H_z^{in}(\boldsymbol{\rho}) - \int_\Gamma \left[G_0(\boldsymbol{\rho}, \boldsymbol{\rho}') \frac{\partial H_z(\boldsymbol{\rho}')}{\partial n(\boldsymbol{\rho}')} - H_z(\boldsymbol{\rho}') \frac{\partial G_0(\boldsymbol{\rho}, \boldsymbol{\rho}')}{\partial n(\boldsymbol{\rho}')} \right] d\Gamma(\boldsymbol{\rho}'),$$

$$\frac{1}{2} H_{dz}(\boldsymbol{\rho}) = \int_\Gamma \left[G(\boldsymbol{\rho}, \boldsymbol{\rho}') \frac{\partial H_{dz}(\boldsymbol{\rho}')}{\partial n(\boldsymbol{\rho}')} - H_{dz}(\boldsymbol{\rho}') \frac{\partial G(\boldsymbol{\rho}, \boldsymbol{\rho}')}{\partial n(\boldsymbol{\rho}')} \right] d\Gamma(\boldsymbol{\rho}').$$

Introducing the surface electric current $\mathbf{J}_\Gamma = \mathbf{u}_z J_\Gamma$ and magnetic current $\mathbf{J}_{m\Gamma} = \mathbf{u}_z J_{m\Gamma}$

$$J_\Gamma = -H_z = -H_{dz}, \quad J_{m\Gamma} = \frac{\eta_0}{jk_0} \frac{\partial H_z}{\partial n} = \frac{\eta}{jk} \frac{\partial H_{dz}}{\partial n},$$

and making use of the boundary condition (5.57), we have

$$-\frac{1}{2} J_\Gamma(\boldsymbol{\rho}) + \int_\Gamma \left[j \frac{k_0}{\eta_0} J_{m\Gamma}(\boldsymbol{\rho}') G_0(\boldsymbol{\rho}, \boldsymbol{\rho}') + J_\Gamma(\boldsymbol{\rho}') \frac{\partial G_0(\boldsymbol{\rho}, \boldsymbol{\rho}')}{\partial n(\boldsymbol{\rho}')} \right] d\Gamma(\boldsymbol{\rho}') = H_z^{in}(\boldsymbol{\rho}),$$

$$\tag{5.67}$$

$$\frac{1}{2} J_\Gamma(\boldsymbol{\rho}) + \int_\Gamma \left[j \frac{k}{\eta} J_{m\Gamma}(\boldsymbol{\rho}) G(\boldsymbol{\rho}, \boldsymbol{\rho}') + J_\Gamma(\boldsymbol{\rho}) \frac{\partial G(\boldsymbol{\rho}, \boldsymbol{\rho}')}{\partial n(\boldsymbol{\rho}')} \right] d\Gamma(\boldsymbol{\rho}') = 0.$$

These are the integral equations for the TE incidence. For the TM incidence, the z-component of the electric field in the region bounded by Γ_∞ and Γ may be expressed as

$$E_z(\boldsymbol{\rho}) = E_z^{in}(\boldsymbol{\rho}) - \int_\Gamma \left[G_0(\boldsymbol{\rho}, \boldsymbol{\rho}') \frac{\partial E_z(\boldsymbol{\rho}')}{\partial n(\boldsymbol{\rho}')} - E_z(\boldsymbol{\rho}') \frac{\partial G_0(\boldsymbol{\rho}, \boldsymbol{\rho}')}{\partial n(\boldsymbol{\rho}')} \right] d\Gamma(\boldsymbol{\rho}'),$$

where

$$E_z^{in}(\boldsymbol{\rho}) = j\omega\mu_0 \int_{\Omega_0} G_0(\boldsymbol{\rho}, \boldsymbol{\rho}') J_z(\boldsymbol{\rho}') d\Omega(\boldsymbol{\rho}')$$

is the incident field. The total field inside Ω can be expressed as

$$E_{dz}(\boldsymbol{\rho}) = \int_\Gamma \left[G(\boldsymbol{\rho}, \boldsymbol{\rho}') \frac{\partial E_{dz}(\boldsymbol{\rho}')}{\partial n(\boldsymbol{\rho}')} - E_{dz}(\boldsymbol{\rho}') \frac{\partial G(\boldsymbol{\rho}, \boldsymbol{\rho}')}{\partial n(\boldsymbol{\rho}')} \right] d\Gamma(\boldsymbol{\rho}'), \, \boldsymbol{\rho} \in \Omega.$$

Letting the observation point $\boldsymbol{\rho}$ approach the boundary Γ and making use of the jump relations yields

$$\frac{1}{2} E_z(\boldsymbol{\rho}) = E_z^{in}(\boldsymbol{\rho}) - \int_\Gamma \left[G_0(\boldsymbol{\rho}', \boldsymbol{\rho}') \frac{\partial E_z(\boldsymbol{\rho}')}{\partial n(\boldsymbol{\rho}')} - E_z(\boldsymbol{\rho}') \frac{\partial G_0(\boldsymbol{\rho}, \boldsymbol{\rho}')}{\partial n(\boldsymbol{\rho}')} \right] d\Gamma(\boldsymbol{\rho}'),$$

$$\frac{1}{2} E_{dz}(\boldsymbol{\rho}') = \int_\Gamma \left[G(\boldsymbol{\rho}, \boldsymbol{\rho}') \frac{\partial E_{dz}(\boldsymbol{\rho}')}{\partial n(\boldsymbol{\rho}')} - E_{dz}(\boldsymbol{\rho}') \frac{\partial G(\boldsymbol{\rho}, \boldsymbol{\rho}')}{\partial n(\boldsymbol{\rho}')} \right] d\Gamma(\boldsymbol{\rho}').$$

Introducing the surface electric current $\mathbf{J}_\Gamma = \mathbf{u}_z J_\Gamma$ and magnetic current $\mathbf{J}_{m\Gamma} = \mathbf{u}_l J_{m\Gamma}$

$$J_\Gamma = \frac{1}{jk_0\eta_0} \frac{\partial E_z}{\partial n} = \frac{1}{jk\eta} \frac{\partial E_{dz}}{\partial n}, \, J_{m\Gamma} = -E_z = -E_{dz},$$

and making use of the boundary conditions, we obtain the integral equations for TM incidence

$$-\frac{1}{2} J_{m\Gamma}(\boldsymbol{\rho}) + \int_\Gamma \left[jk_0\eta_0 G_0(\boldsymbol{\rho}, \boldsymbol{\rho}') J_\Gamma(\boldsymbol{\rho}') + J_{m\Gamma}(\boldsymbol{\rho}') \frac{\partial G_0(\boldsymbol{\rho}, \boldsymbol{\rho}')}{\partial n(\boldsymbol{\rho}')} \right] d\Gamma(\boldsymbol{\rho}') = E_z^{in}(\boldsymbol{\rho}),$$

$$\frac{1}{2} J_{m\Gamma}(\boldsymbol{\rho}) + \int_\Gamma \left[jk\eta G(\boldsymbol{\rho}, \boldsymbol{\rho}') J_\Gamma(\boldsymbol{\rho}') + J_{m\Gamma}(\boldsymbol{\rho}') \frac{\partial G(\boldsymbol{\rho}, \boldsymbol{\rho}')}{\partial n(\boldsymbol{\rho}')} \right] d\Gamma(\boldsymbol{\rho}') = 0.$$

$$(5.68)$$

The above integral equations for dielectric cylinder can also be obtained from the general integral equations (5.50) and (5.51) by using (5.32). The integral equations for dielectric cylinder also have the defect that the solution is not unique at the interior resonant frequencies.

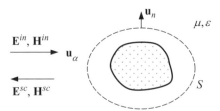

Figure 5.6   Arbitrary scatterer.

5.5.3   Scattering Cross-Section

The main purpose of studying the scattering problem is to extract the properties of the scattering object, such as its shape and size, from the information contained in the scattered waves. One of the important quantities characterizing the scattering object is the scattering cross-section. The **scattering cross-section** of a three-dimensional object is defined by

$$\sigma_{sc} = \frac{P^{sc}}{p^{in}},$$

where p^{in} denotes the time-average incident power density, and P^{sc} the time-averaged scattered power given by

$$P^{sc} = \frac{1}{2}\mathrm{Re}\int_S (\mathbf{E}^{sc} \times \bar{\mathbf{H}}^{sc}) \cdot \mathbf{u}_n dS,$$

where S is an arbitrary surface enclosing the scatterer, as shown in Figure 5.6. The time-averaged power absorbed by the scatterer is

$$P^{abs} = -\frac{1}{2}\mathrm{Re}\int_S (\mathbf{E} \times \bar{\mathbf{H}}) \cdot \mathbf{u}_n dS,$$

where $\mathbf{E}, \mathbf{H}$ are the total fields $\mathbf{E} = \mathbf{E}^{in} + \mathbf{E}^{sc}$, $\mathbf{H} = \mathbf{H}^{in} + \mathbf{H}^{sc}$. If we let P^{ext} denote the total power extracted from the incident power by the scatterer, we have

$$P^{ext} = P^{abs} + P^{sc} = -\frac{1}{2}\mathrm{Re}\int_S (\mathbf{E} \times \bar{\mathbf{H}} - \mathbf{E}^{sc} \times \bar{\mathbf{H}}^{sc}) \cdot \mathbf{u}_n dS$$

$$= -\frac{1}{2}\mathrm{Re}\int_S (\mathbf{E}^{in} \times \bar{\mathbf{H}}^{sc} + \mathbf{E}^{sc} \times \bar{\mathbf{H}}^{in}) \cdot \mathbf{u}_n dS, \tag{5.69}$$

where we have used the relation $\frac{1}{2}\mathrm{Re}\int_S (\mathbf{E}^{in} \times \bar{\mathbf{H}}^{in}) \cdot \mathbf{u}_n dS = 0$. The **absorption cross-section** is defined by

$$\sigma_{abs} = \frac{P^{abs}}{p^{in}}.$$

The sum of scattering cross-section and absorption cross-section is called **extinction cross-section**

$$\sigma_{ext} = \frac{P^{ext}}{P^{in}} = \sigma_{abs} + \sigma_{sc}.$$

5.5.4 Low Frequency Solutions of Integral Equations

The low frequency solution of electromagnetic scattering problem can be traced back to the work of Stevenson (Stevenson, 1953). Due to the lack of powerful calculating instruments at that time, Stevenson's idea did not attract much attention until late sixties. An alternative approach to the low frequency problems is based on Neumann's method in functional analysis (Bladel, 1977; Kleimann, 1978; Geyi, 1995). Consider the following operator equation

$$(\hat{I} - \lambda \hat{K}_k)u = f, \tag{5.70}$$

where $\hat{K}_k$ is an integral operator dependent of the wavenumber k; $\hat{I}$ is the unit operator, f is a known source function; λ is a parameter, and u is the unknown to be determined. We assume that both u and f belong to a Hilbert space H equipped with a norm denoted by $\|\cdot\|$. The **radius of spectrum** of $\hat{K}_k$ is defined by $r_\sigma(\hat{K}_k) = \sup_{\lambda \in \sigma(\hat{K}_k)} \{1/|\lambda|\}$, where $\sigma(\hat{K}_k)$ stands for the spectrum of $\hat{K}_k$, that is, all values of λ such that the inverse of $\hat{I} - \lambda \hat{K}_k$ does not exist or is unbounded, or the operator $\hat{I} - \lambda \hat{K}_k$ is not dense in H. If $r_\sigma(\hat{K}_k) < 1$, the solution of (5.70) can be expressed as a convergent **Neumann series**

$$u = \sum_{n=0}^{\infty} (\hat{K}_k)^n f. \tag{5.71}$$

Furthermore, if $r_\sigma(\hat{K}_k) \leq \rho < 1$, then $\|\hat{K}_k - \hat{K}_0\| \to 0$ as $|k| \to 0$. This implies that (5.71) is convergent for small k.

The Neumann's method can be used to solve the low frequency scattering problem. We will use the integral equations for scattering of conducting cylinder to illustrate the procedure. In order to apply the Neumann's method, the integral equations (5.59) and (5.61) must be regularized in the sense that all the kernels of integral equations are analytic with respect to the wavenumber k_0. For the TE incidence, Equation (5.59) may be written as

$$\frac{1}{2} J_\Gamma(\boldsymbol{\rho}) - \int_\Gamma J_\Gamma(\boldsymbol{\rho}') \left[\frac{\partial G_0(\boldsymbol{\rho}, \boldsymbol{\rho}')}{\partial n(\boldsymbol{\rho}')} - \frac{\partial \tilde{G}_0(\boldsymbol{\rho}, \boldsymbol{\rho}')}{\partial n(\boldsymbol{\rho}')} \right] d\Gamma(\boldsymbol{\rho}')$$
$$- \int_\Gamma J_\Gamma(\boldsymbol{\rho}') \frac{\partial \tilde{G}_0(\boldsymbol{\rho}, \boldsymbol{\rho}')}{\partial n(\boldsymbol{\rho}')} d\Gamma(\boldsymbol{\rho}') = -H_z^{in}(\boldsymbol{\rho}), \tag{5.72}$$

where $\tilde{G}_0(\boldsymbol{\rho}, \boldsymbol{\rho}') = -(2\pi)^{-1} \ln |\boldsymbol{\rho} - \boldsymbol{\rho}'|$ satisfying $\nabla^2 \tilde{G}_0(\boldsymbol{\rho}, \boldsymbol{\rho}') = -\delta(\boldsymbol{\rho} - \boldsymbol{\rho}')$ and

$$-\frac{1}{2} = \int_\Gamma \frac{\partial \tilde{G}_0(\boldsymbol{\rho}, \boldsymbol{\rho}')}{\partial n(\boldsymbol{\rho}')} d\Gamma(\boldsymbol{\rho}').$$

Multiplying both sides of the above equation by $J_\Gamma(\rho)$ and adding the result to (5.72) yield the regularized equation

$$(\hat{I} - \hat{K}_{k_0}^{TE})J_\Gamma(\rho) = -H_z^{in}(\rho),$$

(5.73)

where $\hat{K}_{k_0}^{TE}$ is an integral operator

$$\hat{K}_{k_0}^{TE}[J_\Gamma(\rho)] = \int_\Gamma \left[J_\Gamma(\rho') - J_\Gamma(\rho) \right] \frac{\partial \tilde{G}_0(\rho, \rho')}{\partial n(\rho')} d\Gamma(\rho')$$

$$+ \int_\Gamma J_\Gamma(\rho') \left[\frac{\partial G_0(\rho, \rho')}{\partial n(\rho')} - \frac{\partial \tilde{G}_0(\rho, \rho')}{\partial n(\rho')} \right] d\Gamma(\rho').$$

The Neumann solution of (5.73) is

$$J_\Gamma(\rho) = \sum_{n=0}^{\infty} (\hat{K}_{k_0}^{TE})^n [-H_z^{in}(\rho)].$$

(5.74)

This is a simple summation and can be easily carried out numerically.

To regularize (5.61) for the TM incidence, we assume that the origin of the coordinate system is located inside Ω. Then the value of the incident field at the origin is

$$E^{in}(0) = \int_\Gamma \left[G_0(\rho', 0) \frac{\partial E^{in}(\rho')}{\partial n(\rho')} - E^{in}(\rho') \frac{\partial G_0(\rho', 0)}{\partial n(\rho)} \right] d\Gamma(\rho')$$

$$= \int_\Gamma \left[G_0(\rho', 0) \frac{\partial E(\rho')}{\partial n(\rho')} - E(\rho') \frac{\partial G_0(\rho', 0)}{\partial n(\rho)} \right] d\Gamma(\rho')$$

(5.75)

$$= jk_0 \int_\Gamma J_s(\rho') G_0(\rho', 0) d\Gamma(\rho'),$$

where we have used the relation

$$\int_\Gamma \left[G_0(\rho', 0) \frac{\partial E^{sc}(\rho')}{\partial n(\rho')} - E^{sc}(\rho') \frac{\partial G_0(\rho', 0)}{\partial n(\rho)} \right] d\Gamma(\rho') = 0,$$

and the boundary conditions on the conductor surface. Equation (5.75) is equivalent to

$$\frac{1}{jk_0} E^{in}(0) \frac{\pi}{\ln k_0 \rho} \mathbf{u}_n(\rho) \cdot \nabla G_0(\rho, 0)$$

$$= \int_\Gamma J_\Gamma(\rho') \frac{\pi}{\ln k_0 \rho} \mathbf{u}_n(\rho) \cdot \nabla G_0(\rho, 0) G_0(\rho', 0) d\Gamma(\rho'),$$

(5.76)

where $\rho = |\boldsymbol{\rho}|$. Adding (5.61), (5.75) and (5.76), we obtain the following regularized integral equation

$$(\hat{I} - \hat{K}_{k_0}^{TM})J_{\Gamma}(\boldsymbol{\rho}) = -F^{in}(\boldsymbol{\rho}),$$

where

$$\hat{K}_{k_0}^{TM}[J_{\Gamma}(\boldsymbol{\rho})] = -2\int_{\Gamma} J(\boldsymbol{\rho}')\left[\mathbf{u}_n(\boldsymbol{\rho})\cdot\nabla G_0(\boldsymbol{\rho},\boldsymbol{\rho}') - \frac{\pi}{\ln k_0\rho}\mathbf{u}_n(\boldsymbol{\rho})\cdot\nabla G_0(\boldsymbol{\rho},0)G_0(\boldsymbol{\rho}',0)\right]d\Gamma(\boldsymbol{\rho}'),$$

$$F^{in}(\boldsymbol{\rho}) = \frac{2}{jk_0}\left[\frac{\partial E^{in}(\boldsymbol{\rho})}{\partial n(\boldsymbol{\rho})} + \frac{\pi}{\ln k_0\rho}E^{in}(0)\mathbf{u}_n(\boldsymbol{\rho})\cdot\nabla G_0(\boldsymbol{\rho},0)\right].$$

The Neumann's series solution is then given by

$$J_{\Gamma}(\boldsymbol{\rho}) = \sum_{n=0}^{\infty}(\hat{K}_{k_0}^{TM})^n[F^{in}(\boldsymbol{\rho})].$$

The above approach is similar to Colton and Kleimann (1980). The difference is that we use $\ln k_0\rho$ in (5.76) instead of $\ln k_0$. Utilizing $\ln k_0$ would lead to an inconsistency of dimension in the numerical calculations.

5.6   Multiple Metal Antenna System

When integral equations are applied to solve antenna problems, approximations of the source region are usually adopted by ignoring the antenna feeding waveguides. For example, a linear antenna is usually characterized by an integral equation in which the source region is usually modeled by a delta gap or a magnetic ring current. Such an approximation gives rise to a serious problem that the solution obtained cannot be checked experimentally because a feeding waveguide is always involved in every experimental set-up. In addition, the integral equation based on the delta gap is only valid for thin wires or low frequency problems. When the frequency is high or wire is thick, the integral equation cannot produce reasonable results, especially for the antenna input impedance. For this reason, a practical integral equation formulation for the antenna system must consider the influences of the feeding waveguides (Geyi, 2006b).[1]

Let us assume that the antenna system consists of N metal antennas. To get a universal integral equation for any operating conditions, the metal antenna system is assumed to include all possible sources, as shown in Figure 5.7. Each antenna may be in transmitting mode, receiving mode or in a mode that the antenna transmits and receives at the same time (for example, the antenna is in the transmitting mode but interfered with by an arbitrary incident field from the outside of the antenna). The source region $V_0^{(q)}(q = 1, 2, \cdots, N)$ of the i th antenna is chosen in such a way that its boundary $\partial V_0^{(q)}$ is coincident with the antenna surface,

---

[1] W, Geyi, "New magnetic field integral equation for antenna system", *Progress in Electromagnetics Research*, PIER 63, 153–76, 2006. Reproduced by permission of © 2006 The Electromagnetics Academy & EMW Publishing.

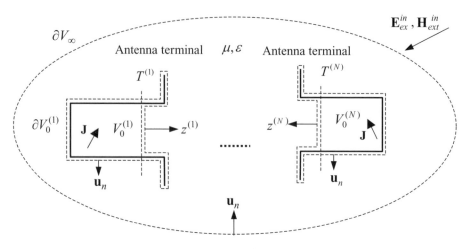

Figure 5.7   Multiple metal antenna system

which is assumed to be a perfect conductor (except for a cross-sectional portion $\Omega^{(q)}$ where $\partial V_0^{(q)}$ crosses the antenna input terminal). Let ∂V_∞ be a large surface that encloses the whole antenna system. From the representation theorem for electromagnetic fields, the total magnetic field in the region bounded by $\partial V_0 = \sum_{i=1}^{N} \partial V_0^{(q)}$ and ∂V_∞ can then be expressed as

$$
\mathbf{H}(\mathbf{r}) = -j\frac{k}{\eta} \int_{\partial V_0} G(\mathbf{r}, \mathbf{r}') \mathbf{J}_{ms}(\mathbf{r}') dS(\mathbf{r}') + \int_{\partial V_0} \mathbf{J}_s(\mathbf{r}') \times \nabla' G(\mathbf{r}, \mathbf{r}') dS(\mathbf{r}')
$$

$$
-\frac{1}{jk\eta} \int_{\partial V_0} \nabla'_s \cdot \mathbf{J}_{ms}(\mathbf{r}') \nabla' G(\mathbf{r}, \mathbf{r}') dS(\mathbf{r}') + \mathbf{H}^{in}_{ext}(\mathbf{r}),
$$

where $\eta = \sqrt{\mu/\varepsilon}$; $\mathbf{J}_s = \mathbf{u}_n \times \mathbf{H}$; $\mathbf{J}_{ms} = -\mathbf{u}_n \times \mathbf{E}$; $G(\mathbf{r}, \mathbf{r}') = e^{-jk|\mathbf{r}-\mathbf{r}'|}/4\pi \, |\mathbf{r} - \mathbf{r}'|$; ∇_s represents the surface divergence; and

$$
\mathbf{H}^{in}_{ext}(\mathbf{r}) = -j\frac{k}{\eta} \int_{\partial V_\infty} G(\mathbf{r}, \mathbf{r}') \mathbf{J}_{ms}(\mathbf{r}') dS(\mathbf{r}') + \int_{\partial V_\infty} \mathbf{J}_s(\mathbf{r}') \times \nabla' G(\mathbf{r}, \mathbf{r}') dS(\mathbf{r}')
$$

$$
-\frac{1}{jk\eta} \int_{\partial V_\infty} \nabla'_s \cdot \mathbf{J}_{ms}(\mathbf{r}') \nabla' G(\mathbf{r}, \mathbf{r}') dS(\mathbf{r}')
$$

stands for the external incident magnetic field. Letting the observation point $\mathbf{r}$ approach the boundary of the source region ∂V_0 from the interior of $\partial V_0 + \partial V_\infty$ and using the jump relations,

we obtain

$$
\mathbf{H}(\mathbf{r}) = -j\frac{k}{\eta} \int_{\partial V_0} G(\mathbf{r}, \mathbf{r}') \mathbf{J}_{ms}(\mathbf{r}') dS(\mathbf{r}') + \int_{\partial V_0} \mathbf{J}_s(\mathbf{r}') \times \nabla' G(\mathbf{r}, \mathbf{r}') dS(\mathbf{r}')
$$

$$
- \frac{1}{jk\eta} \int_{\partial V_0} \nabla'_s \cdot \mathbf{J}_{ms}(\mathbf{r}') \nabla' G(\mathbf{r}, \mathbf{r}') dS(\mathbf{r}') + \mathbf{H}^{in}_{ext}(\mathbf{r})
$$

$$
+ \frac{1}{2} \mathbf{J}_s(\mathbf{r}) \times \mathbf{u}_n(\mathbf{r}) - \frac{1}{j2k\eta} \mathbf{u}_n(\mathbf{r}) \nabla_s \cdot \mathbf{J}_{ms}(\mathbf{r}).
$$

Multiplying both sides of the above equations by $\mathbf{u}_n$ gives

$$
\frac{1}{2} \mathbf{J}_s(\mathbf{r}) = -j\frac{k}{\eta} \mathbf{u}_n(\mathbf{r}) \times \int_{\partial V_0} G(\mathbf{r}, \mathbf{r}') \mathbf{J}_{ms}(\mathbf{r}') dS(\mathbf{r}')
$$

$$
+ \mathbf{u}_n(\mathbf{r}) \times \int_{\partial V_0} \mathbf{J}_s(\mathbf{r}') \times \nabla' G(\mathbf{r}, \mathbf{r}') dS(\mathbf{r}')
$$

$$
- \frac{1}{jk\eta} \mathbf{u}_n(\mathbf{r}) \times \int_{\partial V_0} \nabla'_s \cdot \mathbf{J}_{ms}(\mathbf{r}') \nabla' G(\mathbf{r}, \mathbf{r}') dS(\mathbf{r}')
$$

$$
+ \mathbf{u}_n(\mathbf{r}) \times \mathbf{H}^{in}_{ext}(\mathbf{r}).
$$

Making use of the boundary conditions on the metal part of the antenna, the above equation can be written as

$$
-\frac{1}{2} \mathbf{J}_s(\mathbf{r}) + \mathbf{u}_n(\mathbf{r}) \times \int_{\partial V_0} \mathbf{J}_s(\mathbf{r}') \times \nabla' G(\mathbf{r}, \mathbf{r}') dS(\mathbf{r}')
$$

$$
= -\mathbf{u}_n(\mathbf{r}) \times [\mathbf{H}^{in}_{int}(\mathbf{r}) + \mathbf{H}^{in}_{ext}(\mathbf{r})],
$$

(5.77)

where

$$
\mathbf{H}^{in}_{int}(\mathbf{r}) = -j\frac{k}{\eta} \sum_{q=1}^{N} \int_{\Omega^{(q)}} G(\mathbf{r}, \mathbf{r}') \mathbf{J}_{ms}(\mathbf{r}') d\Omega(\mathbf{r}')
$$

$$
- \frac{1}{jk\eta} \sum_{q=1}^{N} \int_{\Omega^{(q)}} \nabla'_s \cdot \mathbf{J}_{ms}(\mathbf{r}') \nabla' G(\mathbf{r}, \mathbf{r}') d\Omega(\mathbf{r}')
$$

(5.78)

is determined by the equivalent surface magnetic current $\mathbf{J}_m = -\mathbf{u}_{z^{(q)}} \times \mathbf{E}$ on the antenna input terminals $\Omega^{(q)}$ ($q = 1, 2, \cdots, N_A$). In order to determine the equivalent magnetic current on the antenna input terminals, we can make use of the field expressions in the waveguide (see (3.71))

$$-\mathbf{u}_{z^{(q)}} \times \mathbf{E}(\mathbf{r}^{(q)}) = -\sum_{n=1}^{\infty} \mathbf{u}_{z^{(q)}} \times \mathbf{e}_n^{(q)}(\mathbf{r}^{(q)}) V_n^{(q)}(z^{(q)}),$$

(5.79)

$$\mathbf{u}_{z^{(q)}} \times \mathbf{H}(\mathbf{r}^{(q)}) = -\sum_{n=1}^{\infty} \mathbf{e}_n^{(q)}(\mathbf{r}^{(q)}) I_n^{(q)}(z^{(q)}),$$

where $\mathbf{r}^{(q)} = \mathbf{r} - \mathbf{r}_0$ is the local coordinate system for the q th feeding waveguide and $\mathbf{r}^{(q)} \in \Omega^{(q)}$, as shown in Figure 5.8; and

$$V_n^{(q)}(z^{(q)}) = A_n^{(q)} e^{-j\beta_n^{(q)} z^{(q)}} + B_n^{(q)} e^{j\beta_n^{(q)} z^{(q)}},$$

$$I_n^{(q)}(z^{(q)}) = \frac{1}{Z_{wn}^{(q)}} (A_n^{(q)} e^{-j\beta_n^{(q)} z^{(q)}} - B_n^{(q)} e^{j\beta_n^{(q)} z^{(q)}}),$$

$$\beta_n^{(q)} = \begin{cases} k, \text{ TEM mode} \\ \sqrt{k^2 - k_{cn}^{(q)2}}, \text{ TE or TM modes} \end{cases}, \quad Z_{wn}^{(q)} = \begin{cases} \eta, \text{ TEM mode} \\ \eta k/\beta_n^{(q)}, \text{ TE modes} \\ \eta \beta_n^{(q)}/k, \text{ TM modes} \end{cases}.$$

Assume that the feeding waveguides of antennas are in single-mode operation. The modal voltages and currents may be written as

$$V_1^{(q)}(z^{(q)}) = \delta^{(q)} e^{-j\beta_1^{(q)} z^{(q)}} + B_1^{(q)} e^{j\beta_1^{(q)} z^{(q)}},$$

$$V_n^{(q)}(z^{(q)}) = B_n^{(q)} e^{j\beta_n^{(q)} z^{(q)}}, n \geq 2,$$

$$I_1^{(q)}(z^{(q)}) = \frac{1}{Z_{w1}^{(q)}} (\delta^{(q)} e^{-j\beta_1^{(q)} z^{(q)}} - B_1^{(q)} e^{j\beta_1^{(q)} z^{(q)}}),$$

$$I_n^{(q)}(z^{(q)}) = -\frac{1}{Z_{wn}^{(q)}} B_n^{(q)} e^{j\beta_n^{(q)} z^{(q)}}, n \geq 2,$$

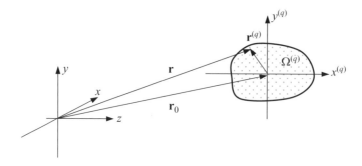

Figure 5.8   Coordinate systems

where $\delta^{(q)} = 1$ if the q th antenna is in transmitting mode and excited by the dominant mode of unit amplitude, and $\delta^{(q)} = 0$ if the q th antenna is in receiving mode. Thus, on the input terminal $\Omega^{(q)}$ ($z(q) = 0$), Equation (5.79) may be written as

$$\mathbf{J}_{ms}(\mathbf{r}^{(q)}) = -\mathbf{u}_{z^{(q)}} \times \mathbf{e}_1^{(q)}(\mathbf{r}^{(q)})(\delta^{(q)} + B_1^{(q)}) - \sum_{n=2}^{\infty} \mathbf{u}_{z^{(q)}} \times \mathbf{e}_n^{(q)}(\mathbf{r}^{(q)})B_n^{(q)},$$

$$\mathbf{J}_s(\mathbf{r}^{(q)}) = -\mathbf{e}_1^{(q)}(\mathbf{r}^{(q)})\frac{\delta^{(q)} - B_1^{(q)}}{Z_{w1}^{(q)}} + \sum_{n=2}^{\infty} \mathbf{e}_n^{(q)}(\mathbf{r}^{(q)})\frac{B_n^{(q)}}{Z_{wn}^{(q)}}.$$

The expansion coefficients can be determined by the second equation of the above equations

$$B_1^{(q)} = \delta^{(q)} + Z_{w1}^{(q)} \int_{\Omega^{(q)}} \mathbf{J}_s(\mathbf{r}^{(q)}) \cdot \mathbf{e}_1^{(q)}(\mathbf{r}^{(q)})d\Omega,$$

$$B_n^{(q)} = Z_{wn}^{(q)} \int_{\Omega^{(q)}} \mathbf{J}_s(\mathbf{r}^{(q)}) \cdot \mathbf{e}_n^{(q)}(\mathbf{r}^{(q)})d\Omega.$$

The equivalent magnetic current on the reference plane $T^{(q)}$ may thus be expressed by

$$\mathbf{J}_{ms}(\mathbf{r}^{(q)}) = -2\delta^{(q)}\mathbf{u}_{z^{(q)}} \times \mathbf{e}_1^{(q)}(\mathbf{r}^{(q)})$$

$$- \sum_{n=1}^{\infty} \mathbf{u}_{z^{(q)}} \times \mathbf{e}_n^{(q)}(\mathbf{r}^{(q)})Z_{wn}^{(q)} \int_{\Omega^{(q)}} \mathbf{J}_s(\mathbf{r}^{(q)}) \cdot \mathbf{e}_n^{(q)}(\mathbf{r}^{(q)})d\Omega(\mathbf{r}^{(q)}).$$

(5.80)

Inserting this into (5.78) yields

$$\mathbf{H}_{int}^{in}(\mathbf{r}) = \sum_{q=1}^{N} \left[2\delta^{(q)}\mathbf{G}_1^{(q)}(\mathbf{r}) + \sum_{n=1}^{\infty} Z_{wn}^{(q)}\mathbf{G}_n^{(q)}(\mathbf{r}) \int_{\Omega^{(q)}} \mathbf{J}_s(\mathbf{r}^{(q)}) \cdot \mathbf{e}_n^{(q)}(\mathbf{r}^{(q)})d\Omega(\mathbf{r}^{(q)}) \right], \quad (5.81)$$

where

$$\mathbf{G}_n^{(q)}(\mathbf{r}) = \frac{jk}{\eta} \int_{\Omega^{(q)}} G(\mathbf{r}, \mathbf{r}')\mathbf{u}_{z^{(q)}} \times \mathbf{e}_n^{(q)}(\mathbf{r}^{(q)})d\Omega(\mathbf{r}')$$

$$+ \frac{1}{jnk} \int_{\Omega^{(q)}} \nabla_s' \cdot \left[\mathbf{u}_{z^{(q)}} \times \mathbf{e}_n^{(q)}(\mathbf{r}^{(q)}) \right] \nabla' G(\mathbf{r}, \mathbf{r}')d\Omega(\mathbf{r}').$$

From (5.77) and (5.81), one may obtain the following modified MFIE

$$-\frac{1}{2}\mathbf{J}_s(\mathbf{r}) + \mathbf{u}_n(\mathbf{r}) \times \int_{\partial V_0} \mathbf{J}_s(\mathbf{r}') \times \nabla' G(\mathbf{r}, \mathbf{r}')dS(\mathbf{r}')$$

$$+ \sum_{q=1}^{N} \left[\sum_{n=1}^{\infty} Z_{wn}^{(q)}\mathbf{u}_n(\mathbf{r}) \times \mathbf{G}_n^{(q)}(\mathbf{r}) \int_{\Omega^{(q)}} \mathbf{J}_s(\mathbf{r}^{(q)}) \cdot \mathbf{e}_n^{(q)}(\mathbf{r}^{(q)})d\Omega\left(\mathbf{r}^{(q)}\right) \right]$$

(5.82)

$$= \sum_{q=1}^{N} \left[-2\delta^{(q)}\mathbf{u}_n(\mathbf{r}) \times \mathbf{G}_1^{(q)}(\mathbf{r}) \right] - \mathbf{u}_n(\mathbf{r}) \times \mathbf{H}_{ext}^{in}(\mathbf{r}), \mathbf{r} \in \partial V_0.$$

As pointed out before, the non-uniqueness problem occurs in the integral equation formulations when it is used to describe an isolated scatterer. When the antenna input terminals $\Omega^{(q)}$ exist, the electromagnetic energy is exchanged between the source region (enclosed by ∂V_0) and the exterior region (outside of ∂V_0) of the antennas. As a result, the physical conditions for interior resonance no longer exist, and the solution of (5.82) is unique.

5.7 Numerical Methods

Experiment, theory and computation form a tripod in modern scientific research. The traditional numerical methods for solving differential equations include finite element method and finite difference method. These types of numerical methods are called the **domain method** since the governing equation has to be solved over the entire defining region of the problem. On the other hand, the integral equations are defined on the boundary of the defining region and the numerical methods used to solve them are called the **boundary method**, such as the boundary element method. The domain method can easily be applied to nonlinear, inhomogeneous and time-varying problems. The numerical accuracy of domain methods is generally lower than the boundary method because the discretization error is limited only on the boundary for the latter. When the numerical methods are used to solve the integral equations, the integration region must be divided into sub-areas, called elements. Each element is usually approximated by a straight line for two-dimensional problems and a planar triangle or a planar rectangle for three-dimensional problems. The unknown function on each element is then approximated by a linear combination of some known basis functions. To get a general picture of this procedure, we may begin with a universal frame for the numerical methods, called the projection method.

5.7.1 Projection Method

Let U and W be two subspaces of a linear space E. The **sum** of U and W, denoted by $U + W$ is defined as the set of all vectors of the form $u + w$ with $u \in U$ and $w \in W$. The linear space E is said to be the **direct sum** of U and W, written as $E = U \oplus W$, if and only if, (1) $E = U + W$; (2) $U \cap W = 0$. (This means that the only common element is the zero vector.) W is called a **direct complement** of U in E.

If E has a direct sum decomposition $E = U \oplus W$, we may introduce a linear operator $\hat{P} : E \rightarrow U$, called the **projection** of E onto U along W, which satisfies (1) $\hat{P}u = u$, for all $u \in U$; (2) $\hat{P}w = 0$, for all $w \in W$. Apparently $\hat{I} - \hat{P}$ is the projection of E onto W along U. For all $h \in E$, we may write $h = u + w = \hat{P}h + w$ (Figure 5.9).

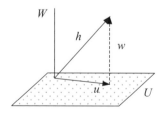

Figure 5.9   Projection of vector $h = u + w$

Let us consider an operator equation

$$\hat{T}x = y, x, y \in H, \tag{5.83}$$

where $\hat{T} : D(T) \subset H \rightarrow H$ is a linear operator (differential operator or integral operator) and H is a Hilbert space. Let $\{X^N\}$ and $\{Y^N\}$ be two given sequences of subspaces with $X^N \subset D(\hat{T})$ and $Y^N \subset H$, and let $\{\hat{P}^N\}$ be a sequence of projections of H onto Y^N. Whenever an exact solution of (5.83) is not available, we have to seek a numerical solution. Suppose x^N is an approximate solution in X^N. In general, $\hat{T}x^N - y$ is not zero. The **projection method** requires that the projection of $\hat{T}x^N - y$ onto Y^N vanishes

$$\hat{P}^N(\hat{T}x^N - y) = 0. \tag{5.84}$$

If $\{u^i \mid i = 1, 2, \cdots, N\}$ is an orthonormal basis of Y^N, the projection operator may be expressed by

$$\hat{P}^N(\cdot) = \sum_{i=1}^{N} (\cdot, u^i) u^i, \tag{5.85}$$

where $(\cdot, \cdot)$ stands for the inner product in H. If $\{v^i \mid i = 1, 2, \cdots, N\}$ is an orthonormal basis of X^N, we have the expansion

$$x^N = \sum_{i=1}^{N} a_i v^i. \tag{5.86}$$

Introducing (5.85) and (5.86) into (5.84) gives

$$\sum_{i=1}^{N} a_i (\hat{T}v^i, u^j) = (y, u^j), j = 1, 2, \cdots, N. \tag{5.87}$$

The projection method is also called the **method of weighted residuals** if the inner product is defined as an integral.

5.7.2 Moment Method

If we choose $u^i = \hat{S}v^i$, where $\hat{S}$ is an operator, the projection method reduces to the **moment method**. In this case, Equation (5.87) becomes

$$\sum_{i=1}^{N} a_i (\hat{T}v^i, \hat{S}v^j) = (y, \hat{S}v^j), j = 1, 2, \cdots, N. \tag{5.88}$$

Especially if we choose $\hat{S} = \hat{T}$, this becomes

$$\sum_{i=1}^{N} a_i (\hat{T}v^i, \hat{T}v^j) = (y, \hat{T}v^j), j = 1, 2, \cdots, N. \tag{5.89}$$

The same equation as (5.89) may be obtained if we minimize the error functional $\left\| \hat{T}x^N - y \right\|$ using (5.86). Thus, (5.89) is equivalent to the **method of least squares**. If we choose $\hat{S} = \hat{I}$, Equation (5.88) reduces to

$$\sum_{i=1}^{N} a_i(\hat{T}v^i, v^j) = (y, v^j), j = 1, 2, \cdots, N.$$

This is referred to as **Galerkin's method**, named after the Russian mathematician Boris Grigoryevich Galerkin (1871–1945).

5.7.3   Construction of Approximating Subspaces

The practical implementation of numerical methods depends on how to construct the approximating subspaces X^N and Y^N. For most applications, the solution of the operator equation (5.83) is defined in a region $\Omega \in R^m (m = 1, 2, 3)$. We may choose a set of points $\{\mathbf{r}^i | i = 1, 2, \cdots, N\}$ with $\mathbf{r}^i = (x_1^i, \cdots, x_m^i)$, called **global nodes**. The node numbering system is called the **global numbering system**. Consider a set of functions $\{l^i(\mathbf{r}) | i = 1, 2, \cdots, N\}$, which satisfies:

1. For each i, there exists a positive number ε_i such that

$$l^i(\mathbf{r}) = \begin{cases} \neq 0, & \left| \mathbf{r} - \mathbf{r}^i \right| \leq \varepsilon_i, \\ = 0, & \left| \mathbf{r} - \mathbf{r}^i \right| > \varepsilon_i. \end{cases}$$

2. $l^i(\mathbf{r})$ $(i = 1, 2, \cdots, N)$ are continuous and $l^i(\mathbf{r}^j) = \delta_{ij}$.

It is easy to show that the set $\{l^i(\mathbf{r}) | i = 1, 2, \cdots, N\}$ is linearly independent. Thus we let

$$X^N = Y^N = \text{span}\left\{ l^1(\mathbf{r}), l^2(\mathbf{r}), \cdots, l^N(\mathbf{r}) \right\}.$$

The approximate solution may be written as

$$x^N(\mathbf{r}) = \sum_{i=1}^{N} a_i l^i(\mathbf{r}).$$

The set $\{l^i(\mathbf{r}) | i = 1, 2, \cdots, N\}$ forms a global basis for X^N and Y^N. To construct these global basis functions, we divide the region Ω into n subregions (called **elements**) Ω_e ($e = 1, 2, \cdots, n$) such that the intersection of any two elements is either empty or consists of a common boundary curve or points (Figure 5.10). For each element, we choose N_e nodes $\mathbf{r}^\alpha$ ($\alpha = 1, 2, \cdots, N_e$) (the node numbering system α is called the **local numbering system**) and introduce the **Lagrange shape functions** l_e^α, which are smooth and satisfy

$$l_e^\alpha(\mathbf{r}) = 0, \mathbf{r} \notin \Omega_e,$$
$$l_e^\alpha(\mathbf{r}^\beta) = \delta_{\alpha\beta}, \alpha, \beta = 1, 2, \cdots, N_e.$$

Figure 5.10   Discretization of the solution region

The nodes that are not on the boundaries of elements are called **internal nodes**. Otherwise they are called **boundary nodes**. If m elements meet at $\mathbf{r}$, we say that $\mathbf{r}$ has m-multiplicity, denoted by $m(\mathbf{r})$. Let $\mathbf{r}^i$ be a node of m-multiplicity, that is, there exist m elements $\Omega_{(e_j)}$ $(j = 1, 2, \cdots, m)$ that meet at $\mathbf{r}^i$. Then the global basis functions can be constructed as follows

$$l^i(\mathbf{r}) = \sum_{j=1}^{m(\mathbf{r}^i)} \frac{1}{m(\mathbf{r})} l_{e_j}^{\alpha_j}(\mathbf{r}), \quad i = 1, 2, \cdots, N$$

where α_j is the local numbering for the node $\mathbf{r}^i$.

5.7.3.1 Lagrange shape function for line element

The construction of Lagrange shape function is an interpolation process. A line element is shown in Figure 5.11. The global coordinates of the two end points p_1 and p_2 are denoted by x_1 and x_2 respectively. Let p be an arbitrary point in the element with coordinate x. We may introduce the local coordinate system $\lambda = |p_1 p| / |p_1 p_2|$ (also called the **natural coordinate system**). Evidently

$$x = (1 - \lambda)x_1 + \lambda x_2. \tag{5.90}$$

Therefore an arbitrary line element in x-coordinate system is transformed into a standard line element in the natural coordinate system through (5.90). If we apply the linear interpolation to the standard element, the Lagrange shape functions are

$$l_e^1(\lambda) = 1 - \lambda, \, l_e^2(\lambda) = \lambda, 0 \le \lambda \le 1.$$

To better represent the unknowns, we use higher-order interpolation. For a quadratic interpolation, one more node p_3 must be introduced in the middle of the standard line element

(a)   (b)

Figure 5.11   Linear element

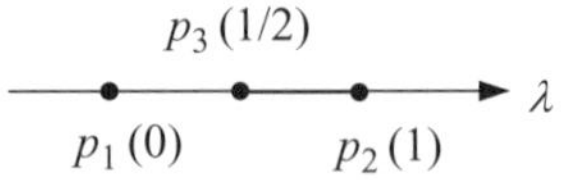

$$p_3 \, (1/2)$$

$$\xrightarrow{\quad\quad\quad\quad\quad} \lambda$$

$$p_1 \, (0) \qquad p_2 \, (1)$$

Figure 5.12   Quadratic element

(Figure 5.12). The Lagrange shape functions for the quadratic element are then given by

$$l_e^1(\lambda) = (\lambda - 1)(2\lambda - 1),\, l_e^2(\lambda) = \lambda(2\lambda - 1),\, l_e^3(\lambda) = 4\lambda(1 - \lambda),\, 0 \le \lambda \le 1.$$

5.7.3.2 Lagrange shape function for triangular element

Consider a planar triangular element $\Delta p_1 p_2 p_3$ shown in Figure 5.13 (a), whose area is denoted by Δ. The global coordinates for vertex p_i are denoted by (x_i, y_i) $(i = 1, 2, 3)$. The triangle is then divided into three small triangles using an arbitrary point p inside the triangle as a common vortex. We introduce the local coordinate system (natural or area coordinate system)

$$\lambda_1 = \Delta_1/\Delta,\, \lambda_2 = \Delta_2/\Delta,\, \lambda_3 = \Delta_3/\Delta, \tag{5.91}$$

where Δ_1, Δ_2 and Δ_3 are areas of the subtriangle $\Delta p_2 p_3 p$, $\Delta p_3 p_1 p$ and $\Delta p_1 p_2 p$ respectively. Note that $\sum_{i=1}^{3} \lambda_i = 1, 0 \le \lambda_i \le 1$. Therefore, only two natural coordinates are independent. It is easy to show that the global coordinate system is related to the local coordinate system by

$$\begin{bmatrix} x \\ y \\ 1 \end{bmatrix} = \begin{bmatrix} x_1 & x_2 & x_3 \\ y_1 & y_2 & y_3 \\ 1 & 1 & 1 \end{bmatrix} \begin{bmatrix} \lambda_1 \\ \lambda_2 \\ \lambda_3 \end{bmatrix}.$$

This transforms an arbitrary triangle in the global coordinate system into a standard right triangle shown in Figure 5.13(b). Therefore, it is only necessary to quote results for the standard right triangle element. If we apply the linear interpolation to the standard right triangle, the Lagrange shape functions are

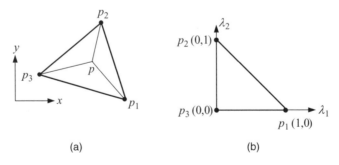

(a)   (b)

Figure 5.13   Linear triangular element

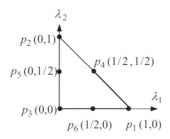

Figure 5.14   Quadratic triangular element

$$l_e^i(\lambda_1, \lambda_2) = \lambda_i, 0 \leq \lambda_i \leq 1, i = 1, 2, 3.$$

To achieve higher accuracy, a higher-order interpolation can be used. In this case more nodes other than the vertices must be inserted to the triangle. For example if we use quadratic interpolation, the mid-points 4, 5, and 6 of the sides of the standard right triangle may be introduced (Figure 5.14). The Lagrange shape functions are then given by

$$\begin{aligned}
l_e^1 &= \lambda_1(2\lambda_1 - 1), & l_e^2 &= \lambda_2(2\lambda_2 - 1), & l_e^3 &= \lambda_3(2\lambda_3 - 1),\\
l_e^4 &= 4\lambda_1\lambda_2, & l_e^5 &= 4\lambda_2\lambda_3, & l_e^6 &= 4\lambda_1\lambda_3.
\end{aligned}$$

When the above approximating subspaces are applied to integral equations, the procedure is called the **boundary element method**. The coefficient matrix of the algebraic system (5.87) resulting from an integral equation is not sparse in general. As the size of the matrix grows, iterative methods or the conjugate gradient method may be used to solve the algebraic system to save the computational time. Other acceleration techniques such as multi-pole expansion may also be used.

Remark 5.4: The integral equation method has been used in electromagnetic engineering for years and is usually applied to a source-free region. When the region contains sources, an integral equation, where both volume integral and boundary integral are involved, may be obtained. To numerically solve the integral equation, both the boundary and the source region have to be discretized, and the integration must be performed over both of them. In this case, the integral equation method loses its main advantages. In addition, the generation of mesh in three dimensions is not an easy task and very time-consuming even when an automatic mesh generator is available. To maintain the advantages of the integral equation method when it is applied to a source region, we must transform the volume integrals into boundary integrals. According to Huygens' principle, this transformation is physically possible.   □

It came as a complete surprise, when, in a short note published in 1900, the Swedish mathematician Ivar Fredholm showed that the general theory of all integral equations considered prior to him was, in fact, extremely simple.
              —Jean Alexandre Eugène Dieudonné (French mathematician, 1906–1992)

6

Network Formulations

The most practical solution is a good theory.

—Albert Einstein

Microwave field theory is an important branch of applied electromagnetics, and it studies the structures with dimensions being of the order of the wavelength. In a broad sense, the microwave field theory applies to the problems of guided waves, resonances, radiations and scattering. In many situations, a microwave field problem can be reduced to a network or circuit problem, which allows us to apply the circuit and network methods to solve the original field problem. The network formulation has eliminated unnecessary details in the field theory while reserving useful global information, such as the terminal voltages and currents.

A microwave network consists of waveguides, passive devices, and active devices. The waveguides at microwave frequencies are counterparts to connecting wires in low-frequency circuits, which interconnect different parts of the microwave network. The ports of the microwave network are a set of waveguide cross-sections at reference planes, through which the energy flows into or out of the circuit. The port is typically chosen to be far away from waveguide discontinuities so that only the dominant mode is propagating in the neighborhood of the reference planes. Instead of the circuit voltages and currents in low-frequency circuits, the modal voltages and currents are often used in a microwave network. The irregularities (or discontinuities) bounded by the waveguide ports will excite a number of higher-order modes, which are assumed to die out at the reference planes. As a result, the effects of these discontinuities may be considered as lumped. A microwave network may be characterized by a network matrix, such as the impedance matrix, the admittance matrix and the scattering matrix. To find the elements of the network matrix, we need to solve the Maxwell equations subject to various boundary conditions.

6.1 Transmission Line Theory

Transmission line theory is the cornerstone of electromagnetic engineering. The early history of the transmission lines has been summarized by Packard and Oliner (Packard, 1984; Oliner, 1984). The essential basis of modern transmission line theory was developed by Oliver

Foundations of Applied Electrodynamics   Geyi Wen
© 2010 John Wiley & Sons, Ltd

Heaviside in the late nineteenth century, who considered various possibilities for waves along wire lines and found that a single conductor line was not feasible, and a guided wave needs two wires. Heaviside also introduced the term '**impedance**', which is defined as the ratio of voltage to current in a circuit. The concept of the impedance was then extended to fields and waves by Schelkunoff in 1938 in a systematic way. The impedance is regarded as the characteristic of the field as well as the medium, and has a direction. In 1897, Rayleigh showed that waves could propagate within a hollow conducting cylinder and found that such waves existed only in a set of well-defined normal modes, and to support the modes in the hollow cylinder, the operating frequency must exceed the cut-off frequencies of the corresponding modes. The theory of dielectric waveguide was first studied by Sommerfeld in 1899 and then extended by the Greek physicist Demetrius Hondros (1882–1962) in 1909. The guided wave in a single dielectric rod is based on the fact that the discontinuity surface between two different media is likely to bind the wave to that surface, thus guiding the wave. The possible use of hollow waveguides was investigated during the 1930s by the American radio engineers George Clark Southworth (1890–1972) and Wilmer Lanier Barrow (1903–1975). Most of the important results on waveguide theory obtained in the first half of the last century have been included in the *Waveguide Handbook* (Marcuvitz, 1951). Nowadays it is well understood that any guided wave structure can be represented by a transmission line, and the term 'transmission line' is used in a broad sense, which may indicate the traditional two-wire transmission line, coaxial cable, a hollow waveguide, a dielectric waveguide or any other complicated structure as long as it supports a guided wave. The concept of transmission lines has also been generalized to study three-dimensional guided waves since only the transmission direction is important. The terms 'transmission line' and 'waveguide' are used interchangeably.

6.1.1   Transmission Line Equations

The transmission lines are used to transmit microwave signals. They are also used extensively in microwave circuit designs, such as directional couplers, filters, and power dividers. In the time domain, the voltage and current along a transmission line, as shown in Figure 6.1, satisfy the **transmission line equations**:

$$\frac{\partial v(z, t)}{\partial z} = -Ri(z, t) - L\frac{\partial i(z, t)}{\partial t}, \quad \frac{\partial i(z, t)}{\partial z} = -Gv(z, t) - C\frac{\partial v(z, t)}{\partial t},$$

where R, L, G and C are the resistance, inductance, conductance and capacitance per unit length of the transmission line respectively. For time-harmonic fields, these equations reduce to

$$\frac{dV}{dz} = -I Z_{unit}, \quad \frac{dI}{dz} = -V Y_{unit}, \tag{6.1}$$

where V and I are phasors, and $Z_{unit} = R + j\omega L$ and $Y_{unit} = G + j\omega C$ are the series impedance and shunt admittance per unit length of the transmission line. From the above

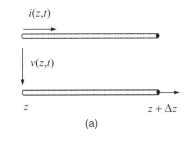

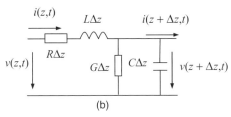

Figure 6.1   (a)Transmission line. (b) Equivalent circuit

equations, we obtain

$$\frac{d^2 V}{dz^2} - \gamma^2 V = 0, \frac{d^2 I}{dz^2} - \gamma^2 I = 0. \tag{6.2}$$

The quantity $\gamma = \sqrt{Z_{unit} Y_{unit}} = \alpha + j\beta$ is called the **propagation constant**. The solutions for the voltage and current can be obtained from (6.1) and (6.2) as

$$V = V^+ + V^- = Ae^{-\gamma z} + Be^{\gamma z}, I = I^+ - I^- = \frac{1}{Z_c}(Ae^{-\gamma z} - Be^{\gamma z}), \tag{6.3}$$

where $V^+ = Ae^{-\gamma z}$, $V^- = Be^{\gamma z}$, $I^+ = Ae^{-\gamma z}/Z_c$, and $I^- = Be^{\gamma z}/Z_c$ are the incident voltage wave, the reflected voltage wave, the incident current wave and the reflected current wave respectively; and

$$Z_c = \sqrt{\frac{Z_{unit}}{Y_{unit}}} = \frac{V^+}{I^+} = -\frac{V^-}{I^-}$$

is called the **characteristic impedance**. When the time factor is restored, we have

$$V^+ = Ae^{-\alpha z} e^{j(\omega t - \beta z)},$$

which stands for a wave moving along the positive z direction with an exponential damping factor determined by the attenuation constant α. The phase velocity is the speed of points of

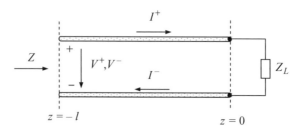

Figure 6.2   Transmission line terminated in a load

constant phase and is given by $v_p = \omega/\beta$. As a result, $\beta = 2\pi/\lambda$, where λ is the wavelength. The **reflection coefficient** Γ at position z is defined by

$$\Gamma = \frac{V^-}{V^+} = \frac{B}{A}e^{2\gamma z} = \Gamma_L e^{2\gamma z},$$

where $\Gamma_L = B/A$ is the reflection coefficient at $z = 0$ (Figure 6.2), called the **load reflection coefficient**. The input impedance at z can be obtained from (6.3)

$$Z = \frac{V}{I} = \frac{V^+ + V^-}{I^+ - I^-} = \frac{V^+}{I^+}\frac{1 + V^-/V^+}{1 - I^-/I^+} = Z_c\frac{1 + \Gamma}{1 - \Gamma} = Z_c\frac{1 + \Gamma_L e^{2\gamma z}}{1 - \Gamma_L e^{2\gamma z}}. \tag{6.4}$$

The reflection coefficient is

$$\Gamma = \frac{Z - Z_c}{Z + Z_c}. \tag{6.5}$$

It follows from (6.4) and (6.5) that

$$Z = Z_c\frac{Z_L + Z_c\tanh(-\gamma z)}{Z_c + Z_L\tanh(-\gamma z)}. \tag{6.6}$$

For a lossless transmission line, the input impedance at $z = -l$ becomes

$$Z = Z_c\frac{Z_L + jZ_c\tan(\beta l)}{Z_c + jZ_L\tan(\beta l)}. \tag{6.7}$$

Example 6.1 For the matched case: $Z_L = Z_c$, we have $Z = Z_c$. For the open circuit: $Z_L = \infty$, we have $Z = Z_c/j\tan\beta l$. For the short circuit: $Z_L = 0$, we have $Z = jZ_c\tan\beta l$. When $l = \lambda/4$, we have $Z = Z_c^2/Z_L$, which is called the **quarter wavelength transform**.   □

In Section 3.3.1, we showed that the modal voltage and the modal current in a waveguide satisfy

$$\frac{dV_n}{dz} = -j\beta_n Z_{wn}I_n(z), \quad \frac{dI_n}{dz} = -j\beta_n Y_{wn}V_n(z). \tag{6.8}$$

Here $Z_{wn} = 1/Y_{wn}$ is the wave impedance for the nth mode, and β_n and Z_{wn} are given by (3.67). From (6.8), we obtain

$$\frac{d^2 V_n}{dz^2} + \beta_n^2 V_n(z) = 0, \quad \frac{d^2 I_n}{dz^2} + \beta_n^2 I_n(z) = 0.$$

It is easy to see the following correspondences between (6.1) and (6.8):

$$Z_{unit} \leftrightarrow j\beta_n Z_{wn}, \; Y_{unit} \leftrightarrow j\beta_n Y_{wn}, \; Z_c \leftrightarrow Z_{wn}, \; \gamma \leftrightarrow j\beta_n.$$

6.1.2 Signal Propagations in Transmission Lines

Propagation of signals along transmission lines constitutes a vital part of a communication system. The types of signals handled by a modern communications system can be categorized either as an analogue signal or a digital signal. The analogue signal is the actual waveform that is related to a physical quantity while the digital signal is just pulses of the same shape.

According to Fourier analysis, any signal can be regarded as the sum of individual sine waves. After traveling a distance z in the transmission line, an individual sine wave $V_0 \sin \omega t$ will appear at the other end as $V_0 \exp(-\alpha z) \sin \omega (t - z/v_p)$, where $v_p = \omega/\beta$ is the phase speed. So the sine wave suffers a change of amplitude determined by the attenuation, and delay of z/v_p. If a signal, which is a sum of sine waves, passes the transmission line, it will be distorted in general. The distortion will be small if the attenuation does not vary very much over the frequency range and the phase speed v_p is independent of the frequency.

A signal $a(t)$ can be multiplied by a carrier $\cos \omega_c t$ to become a double-sided modulated signal $s(t) = a(t) \cos \omega_c t$. The Fourier transform of the signal $s(t)$ is of the form

$$\tilde{s}(\omega) = \frac{1}{2} \left[\tilde{a}(\omega - \omega_c) + \tilde{a}(\omega + \omega_c) \right],$$

where $\tilde{a}(\omega)$ is the Fourier transform of $a(t)$. The inverse transform is

$$s(t) = \frac{1}{2\pi} \int_{-\infty}^{\infty} \tilde{s}(\omega) e^{j\omega t} d\omega = \text{Re} \frac{1}{2\pi} \int_{0}^{\infty} \tilde{a}(\omega - \omega_c) e^{j\omega t} d\omega.$$

After traveling a distance z in a lossless transmission line, the modulated signal $s(t)$ becomes

$$s(z, t) = \text{Re} \frac{1}{2\pi} \int_{0}^{\infty} \tilde{a}(\omega - \omega_c) e^{j(\omega t - \beta z)} d\omega.$$

If we assume that $a(t)$ is a narrowband signal whose spectrum is zero outside the range $|\omega| > \omega_m$ with $\omega_m \ll \omega_c$, we can expand β as a Taylor series:

$$\beta = \beta_c + \omega_m \left(\frac{d\beta}{d\omega}\right)_c + \frac{\omega_m^2}{2}\left(\frac{d^2\beta}{d\omega^2}\right)_c + \cdots,$$

where the subscript c denotes that the function is evaluated $\omega = \omega_c$. If the range of ω_m is small enough, we can make use of the first-order approximation such that

$$s(z,t) = \text{Re}\frac{1}{2\pi}\int_0^\infty \tilde{a}(\omega - \omega_c)e^{j(\omega t - \beta z)}d\omega$$

$$= \text{Re}\frac{1}{2\pi}e^{j(\omega_c t - \beta_c z)}\int_0^\infty \tilde{a}(\omega - \omega_c)e^{j(\omega - \omega_c)[t - z(d\beta/d\omega)_c]}d\omega$$

$$= a\left[t - z\left(\frac{d\beta}{d\omega}\right)_c\right]\cos(\omega_c t - \beta_c z).$$

The term $\omega_c t - z\beta_c$ represents the phase-shifted carrier. After demodulation, we obtain the output $a[t - z(d\beta/d\omega)_c]$. Thus, each frequency component of the demodulated signal is delayed by the same amount. This delay is called **group delay** and is given by $\tau = z(d\beta/d\omega)_c$. The group speed is the velocity with which the envelope of the signal composed of a group of frequency components propagates (see Section 1.4). So the group speed is given by $v_g = (d\omega/d\beta)_c$. If the phase speed is constant so that $\omega = v_p\beta$, the group speed v_g is equal to the phase speed v_p.

6.2 Scattering Parameters for General Circuits

When the voltages and currents are defined at the reference planes of a microwave circuit, relations exist between the voltages and currents. For a linear microwave circuit, these relations are characterized by impedance or admittance matrices. In microwave engineering, the concept of power is more fundamental than the concepts of voltage and current since the latter are not easily measurable at microwave frequencies. For this reason, the scattering parameters are often introduced and are defined in such a way that the power relationship in the circuit can be expressed in a simple and straightforward manner. Scattering parameters originated in the theory of transmission lines and exist for all linear passive time-invariant systems.

6.2.1 One-Port Network

Let us consider a one-port network with input impedance Z as shown in Figure 6.3. The one-port network is connected to a voltage source V_s with source impedance Z_s. The **incident voltage** and the **incident current** are defined as the terminal voltage and current when the

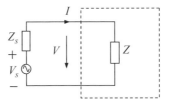

Figure 6.3   A one-port network

one-port network is conjugately matched to the source (that is, $Z = \bar{Z}_s$):

$$V^+ = \frac{V_s \bar{Z}_s}{Z_s + \bar{Z}_s} = \frac{V_s \bar{Z}_s}{2\mathrm{Re}Z_s}, \, I^+ = \frac{V_s}{Z_s + \bar{Z}_s} = \frac{V_s}{2\mathrm{Re}Z_s}.$$

So we have $V^+ = \bar{Z}_s I^+$. In this case, the load Z receives the maximum available power, denoted P^a, from the source

$$P^a = \frac{1}{2}\mathrm{Re}(V\bar{I}) = \frac{|V_s|^2}{8\mathrm{Re}(Z_s)} = \frac{|V^+|^2 \mathrm{Re}(Z_s)}{2|\bar{Z}_s|^2}.$$

The incident voltage and current are determined by the source only. The source impedance Z_s is called the **reference impedance** of the network. In general, the input impedance Z may not be conjugately matched to the source. The **reflected voltage** and the **reflected current** are then defined by

$$V^- = V - V^+, \, -I^- = I - I^+.$$

The minus sign in front of I^- implies that the reference direction of I^- is opposite the reference direction of I^+ (see Figure 6.2). The **normalized incident voltage wave** a and the **normalized reflected voltage wave** b are defined by

$$a = \frac{V^+ \sqrt{\mathrm{Re}Z_s}}{\bar{Z}_s}, b = \frac{V^- \sqrt{\mathrm{Re}Z_s}}{Z_s},$$

which can also be expressed as

$$a = I^+ \sqrt{\mathrm{Re}Z_s}, b = I^- \sqrt{\mathrm{Re}Z_s}.$$

The terminal voltage and current are thus given by

$$V = V^+ + V^- = \frac{1}{\sqrt{\mathrm{Re}Z_s}}(\bar{Z}_s a + Z_s b),$$

$$I = I^+ - I^- = \frac{1}{\sqrt{\mathrm{Re}Z_s}}(a - b),$$

from which we obtain

$$a = \frac{V + Z_s I}{2\sqrt{\mathrm{Re}\,Z_s}}, b = \frac{V - \bar{Z}_s I}{2\sqrt{\mathrm{Re}\,Z_s}}.$$

The voltage reflection coefficient and current reflection coefficient are

$$\Gamma^V = \frac{V^-}{V^+} = \frac{Z_s(Z - \bar{Z}_s)}{\bar{Z}_s(Z + \bar{Z}_s)}, \Gamma^I = \frac{I^-}{I^+} = \frac{Z - \bar{Z}_s}{Z + \bar{Z}_s}.$$

In general Γ^V is not equal to Γ^I. In microwave engineering, the reference impedance Z_s is usually assumed to be real, and so we have $\Gamma^V = \Gamma^I$. The ratio of the normalized reflection wave and the normalized incident wave is the reflection coefficient

$$\Gamma = \frac{b}{a} = \frac{Z - \bar{Z}_s}{Z + \bar{Z}_s} = \Gamma^I.$$

6.2.2 Multi-Port Network

For an n-port network with port number $i = 1, 2, \cdots, n$ shown in Figure 6.4, we may introduce the normalized incident wave and reflected wave at each port:

$$a_i = \frac{V_i + Z_{si} I_i}{2\sqrt{\mathrm{Re}\,Z_{si}}}, b_i = \frac{V_i - \bar{Z}_{si} I_i}{2\sqrt{\mathrm{Re}\,Z_{si}}}. \qquad (6.9)$$

For a linear network, the normalized reflected wave must be linearly related to the normalized incident wave:

$$[b] = [S][a], \qquad (6.10)$$

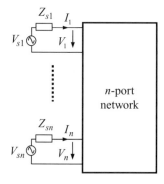

Figure 6.4 n-port network

where

$$[b] = \begin{bmatrix} b_1 \\ b_2 \\ \vdots \\ b_n \end{bmatrix}, [a] = \begin{bmatrix} a_1 \\ a_2 \\ \vdots \\ a_n \end{bmatrix}, [S] = \begin{bmatrix} S_{11} & S_{12} & \cdots & S_{1n} \\ S_{21} & S_{22} & \cdots & S_{2n} \\ \vdots & \vdots & \ddots & \vdots \\ S_{n1} & S_{n2} & \cdots & S_{nn} \end{bmatrix}$$

and $S_{ij}(i, j = 1, 2, \cdots, n)$ are called **scattering parameters**. From (6.10), we obtain

$$S_{ii} = \frac{b_i}{a_i}\bigg|_{a_l=0, l\neq i}, S_{ij} = \frac{b_i}{a_j}\bigg|_{a_l=0, l\neq j}.$$

Remark 6.1: For an n-port microwave network, the reference plane of each port is assumed to be in the single-mode region of the transmission line with real characteristic impedance $Z_{ci}(i = 1, 2, \cdots, n)$. In this case, the normalized incident wave and reflected wave at each port are defined by

$$a_i = \frac{V_i + Z_{ci}I_i}{2\sqrt{Z_{ci}}}, b_i = \frac{V_i - Z_{ci}I_i}{2\sqrt{Z_{ci}}}. \tag{6.11}$$

Thus,

$$V_i = \sqrt{Z_{ci}}(a_i + b_i), I_i = \frac{1}{\sqrt{Z_{ci}}}(a_i - b_i). \tag{6.12}$$

The normalized incident wave and reflected wave defined by (6.9) are a generalized version of (6.11) for an arbitrary circuit. □

Example 6.2 (Lossless condition): Consider an n-port network, the power delivered to the network is

$$P = \frac{1}{2}\text{Re}[V]^T[\bar{I}] = \frac{1}{2}([a]^T[\bar{a}] - [b]^T[\bar{b}]) = \frac{1}{2}[a]^T([1] - [S]^T[\bar{S}])[\bar{a}].$$

If the network is lossless, then $P = 0$. This gives the lossless condition

$$[1] - [S]^T[\bar{S}] = 0, \tag{6.13}$$

where [1] denotes the identity matrix. □

Example 6.3 (Power gain): Consider a two-port network shown in Figure 6.5. The input reflection coefficient is

$$\Gamma^{in} = \frac{b_1}{a_1} = S_{11} + \frac{S_{12}S_{21}\Gamma^L}{1 - S_{22}\Gamma^L}, \tag{6.14}$$

Figure 6.5 Two-port network

where $\Gamma^L = a_2/b_2$ is the reflection coefficient of the load. The output reflection coefficient is

$$\Gamma^{out} = \frac{b_2}{a_2} = S_{22} + \frac{S_{12}S_{21}\Gamma^S}{1 - S_{11}\Gamma^S},\tag{6.15}$$

where $\Gamma^S = a_1/b_1$ is the reflection coefficient of the source. The input power to the network is

$$P^{in} = \frac{1}{2}\mathrm{Re}\,V_1\bar{I}_1 = \frac{1}{2}(|a_1|^2 - |b_1|^2) = \frac{1}{2}|a_1|^2(1 - |\Gamma^{in}|^2).$$

The power absorbed by the load Z_L is

$$P^L = -\frac{1}{2}\mathrm{Re}\,V_2\bar{I}_2 = \frac{1}{2}(|b_2|^2 - |a_2|^2) = \frac{1}{2}|b_2|^2(1 - |\Gamma^L|^2).$$

The **power gain** is defined as the ratio of P^L over P^{in}:

$$G_P = \frac{P^L}{P^{in}} = \frac{\frac{1}{2}|b_2|^2(1 - |\Gamma^L|^2)}{\frac{1}{2}|a_1|^2(1 - |\Gamma^{in}|^2)} = \left|\frac{S_{21}}{1 - S_{22}\Gamma_L}\right|^2 \frac{1 - |\Gamma^L|^2}{1 - |\Gamma^{in}|^2}.\tag{6.16}$$

$\square$

6.3   Waveguide Junctions

Major advances of microwave network theory were made during World War II by a number of scientists, among whom the American physicists Schelkunoff, Julian Seymour Schwinger (1918–1994) and Nathan Marcuvitz played an important role. The main research topic of the microwave network theory was the representation of waveguide discontinuities by lumped circuit elements or network parameters. In most applications, the waveguide supports a single dominant propagating mode. When a discontinuity exists, such as discontinuity in cross-sectional shape or an obstacle in the waveguide, an infinite number of non-propagating modes will be excited in the vicinity of the discontinuity by the incident dominant propagating mode. A typical n-port waveguide discontinuity is shown in Figure 6.6 (a), which consists of n uniform waveguides and a discontinuity (a junction). The reference planes $T_1, T_2, \cdots$, and T_n

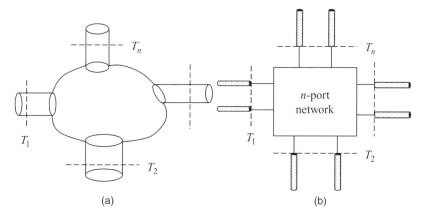

Figure 6.6 (a) Waveguide junction. (b) Equivalent circuit

are assumed to be far away from the discontinuity so that only the dominant modes exist at the reference planes.

The modal voltage V and the modal current I at a reference plane are proportional to the transverse electric field and transverse magnetic field in the waveguide respectively. The uniqueness theorem indicates that the modal voltages at the reference planes $V_1, V_2, \cdots, V_n$ can be determined by the modal currents $I_1, I_2, \cdots, I_n$ at the reference planes. If the medium is linear, the modal voltages and currents are linearly related. So we may write

$$[V] = [Z][I], \tag{6.17}$$

where

$$[V] = \begin{bmatrix} V_1 \\ V_2 \\ \vdots \\ V_n \end{bmatrix}, [I] = \begin{bmatrix} I_1 \\ I_2 \\ \vdots \\ I_n \end{bmatrix}, [Z] = \begin{bmatrix} Z_{11} & Z_{12} & \cdots & Z_{1n} \\ Z_{21} & Z_{22} & \cdots & Z_{2n} \\ \vdots & \vdots & \ddots & \vdots \\ Z_{n1} & Z_{n2} & \cdots & Z_{nn} \end{bmatrix},$$

and $Z_{ij}(i, j = 1, 2, \cdots, n)$ are called **impedance parameters**. It follows from (6.17) that

$$Z_{ii} = \frac{V_i}{I_i}\bigg|_{I_l=0, l \neq i}, Z_{ij} = \frac{V_i}{I_j}\bigg|_{I_l=0, l \neq j}.$$

Hence the impedance parameters are also called open circuit parameters. If the power delivered into the network, denoted P, is zero:

$$P = \frac{1}{4}[I]^T[Z^T + \bar{Z}][\bar{I}] = 0,$$

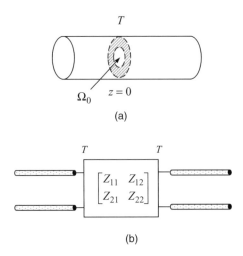

T

Ω_0 $z = 0$

(a)

T T

$$\begin{bmatrix} Z_{11} & Z_{12} \\ Z_{21} & Z_{22} \end{bmatrix}$$

(b)

Figure 6.7   (a) A waveguide discontinuity. (b) Equivalent circuit

the network is lossless and satisfies the lossless condition

$$[Z^T + \bar{Z}] = 0. \tag{6.18}$$

To determine the network parameters, the field distribution in the waveguide junction must be known. There are a number of analytical methods, which can be applied to solve the waveguide junction problems (for example, Collin, 1991; Schwinger and Saxon, 1968; Lewin, 1951). The variational method is the most commonly used one that can handle a large variety of discontinuity problems.

Example 6.4 Consider a waveguide junction with an arbitrarily shaped thin iris placed at $z = 0$ in a uniform waveguide, as shown in Figure 6.7 (a). Only the dominant mode is assumed to be propagating and all higher-order modes are evanescent. If the dominant mode of unit amplitude is incident from the left of the iris, a number of higher-order modes will be excited. The transverse electromagnetic fields in the region $z < 0$ may be expanded in terms of the complete orthonormal set $\{\mathbf{e}_n\}$ in the waveguide (see (3.71)):

$$\mathbf{E}_t^- = (e^{-j\beta_1 z} + \Gamma e^{j\beta_1 z})\mathbf{e}_1 + \sum_{n=2}^{\infty} V_n e^{j\beta_n z}\mathbf{e}_n,$$

$$\mathbf{H}_t^- = (e^{-j\beta_1 z} - \Gamma e^{j\beta_1 z})Z_{w1}^{-1}\mathbf{u}_z \times \mathbf{e}_1 - \sum_{n=2}^{\infty} V_n Z_{wn}^{-1} e^{j\beta_n z}\mathbf{u}_z \times \mathbf{e}_n,$$

where V_n are the modal voltages; Γ is the reflection coefficient for the dominant mode at $z = 0$; and β_n and Z_{wn} are given by (3.69). Similarly the fields in the region $z > 0$ have the

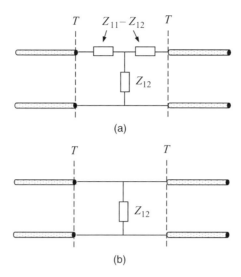

Figure 6.8   Equivalent circuits for the waveguide discontinuity

following expansions

$$\mathbf{E}_t^+ = V_1' e^{-j\beta_1' z} \mathbf{e}_1 + \sum_{n=2}^{\infty} \mathbf{e}_n V_n' e^{-j\beta_n' z},$$

$$\mathbf{H}_t^+ = V_1' e^{-j\beta_1' z} Z_{w1}^{-1} \mathbf{u}_z \times \mathbf{e}_1 + \sum_{n=2}^{\infty} \mathbf{u}_z \times \mathbf{e}_n V_n' Z_{wn}^{-1} e^{-j\beta_n' z}.$$

The tangential electric field must be continuous at $z = 0$. Thus

$$1 + \Gamma = V_1' = \int_{\Omega_0} \mathbf{E}_t(0) \cdot \mathbf{e}_1 d\Omega, \ V_n = V_n' = \int_{\Omega_0} \mathbf{E}_t(0) \cdot \mathbf{e}_n d\Omega, \ (n \geq 2), \tag{6.19}$$

where Ω_0 is the aperture at $z = 0$. Considering the symmetry property of the structure and that the tangential electric field must be continuous at $z = 0$, we have $Z_{11} = Z_{22}$ and the equivalent circuit shown in Figure 6.7 (b) can be simplified to a T-type circuit shown in Figure 6.8 (a). The first expression of (6.19) indicates that the two terminal voltages of the equivalent circuit are equal, which implies $Z_{11} = Z_{12}$, and the final equivalent circuit is shown in Figure 6.8 (b). Note that the tangential magnetic field must also be continuous at the aperture:

$$(1 - \Gamma) Z_{w1}^{-1} \mathbf{u}_z \times \mathbf{e}_1 - \sum_{n=2}^{\infty} \mathbf{u}_z \times \mathbf{e}_n V_n Z_{wn}^{-1} = V_1' Z_{w1}^{-1} \mathbf{u}_z \times \mathbf{e}_1 + \sum_{n=2}^{\infty} \mathbf{u}_z \times \mathbf{e}_n V_n' Z_{wn}^{-1}.$$

Substitution of (6.19) into the above equation gives

$$\mathbf{e}_1 = \mathbf{e}_1 \int_{\Omega_0} \mathbf{E}_t(0) \cdot \mathbf{e}_1 d\Omega + \sum_{n=2}^{\infty} \mathbf{e}_n Z_{w1} Z_{wn}^{-1} \int_{\Omega_0} \mathbf{E}_t(0) \cdot \mathbf{e}_n d\Omega, \ \text{in} \Omega_0. \tag{6.20}$$

This is an integral equation that can be used to determine the aperture field $\mathbf{E}_t(0)$. The input admittance is given by

$$Y = \frac{1}{Z_{w1}} + \frac{1}{Z_{12}} = \frac{1}{Z_{w1}} \frac{1-\Gamma}{1+\Gamma}.$$

Thus, it follows from the first equation of (6.19) that

$$\frac{1}{Z_{12}} = \frac{1}{Z_{w1}} \frac{-2\Gamma}{1+\Gamma} = \frac{1}{Z_{w1}} \frac{2\left(1 - \displaystyle\int_{\Omega_0} \mathbf{E}_t(0) \cdot \mathbf{e}_1 d\Omega\right)}{\displaystyle\int_{\Omega_0} \mathbf{E}_t(0) \cdot \mathbf{e}_1 d\Omega}. \tag{6.21}$$

Multiplying both sides of (6.20) by $\bar{\mathbf{E}}_t(0)$ and taking the integration over Ω_0, we have

$$1 - \int_{\Omega_0} \mathbf{E}_t(0) \cdot \mathbf{e}_1 d\Omega = \frac{\displaystyle\sum_{n=2}^{\infty} \mathbf{e}_n Z_{w1} Z_{wn}^{-1} \left|\int_{\Omega_0} \mathbf{E}_t(0) \cdot \mathbf{e}_n d\Omega\right|^2}{\displaystyle\int_{\Omega_0} \bar{\mathbf{E}}_t(0) \cdot \mathbf{e}_1 d\Omega}.$$

Substituting this into (6.21) gives

$$\frac{1}{Z_{12}} = \frac{1}{Z_{w1}} \frac{-2\Gamma}{1+\Gamma} = \frac{1}{Z_{w1}} \frac{2\displaystyle\sum_{n=2}^{\infty} \mathbf{e}_n Z_{w1} Z_{wn}^{-1} \left|\int_{\Omega_0} \mathbf{E}_t(0) \cdot \mathbf{e}_n d\Omega\right|^2}{\left|\displaystyle\int_{\Omega_0} \mathbf{E}_t(0) \cdot \mathbf{e}_1 d\Omega\right|^2}. \tag{6.22}$$

This is a variational expression, whose functional derivative with respect to the aperture electric field $\mathbf{E}_t(0)$ is zero (Kurokawa, 1969). □

6.4   Multiple Antenna System

A multiple antenna system is different from a waveguide junction. The former is an open system whose energy occupies the whole free space, while the latter is a closed system whose energy is confined in a finite region. From the viewpoint of the network theory, a multiple antenna system is also equivalent to a multi-port microwave network, and its parameters can be determined by solving the Maxwell equations subject to boundary conditions.

6.4.1   Impedance Matrix

Consider a system consisting of n antennas contained in a region V_∞ bounded by S_∞. Let the fields generated by antenna i $(i = 1, 2, \cdots, n)$ when antenna j $(j \neq i)$ are receiving be denoted by $\mathbf{E}_i$ and $\mathbf{H}_i$. We use $V_0^{(i)}$ to denote the source region for antenna i. The source region is chosen in such a way that its boundary, denoted by $\partial V_0^{(i)}$, is coincident with the metal surface of the antennas except for a portion $\Omega^{(i)}$ where the boundary crosses the antenna reference plane. This state of operation is illustrated in Figure 6.9 (a). Figure 6.9 (b) is the corresponding equivalent network representation. Taking the integration of Poynting theorem in frequency domain and using the divergence theorem over the region $V_\infty - \sum_{l=1}^{n} V_0^{(l)}$ with medium parameters σ, μ, and ε, as shown in Figure 6.9 (a), we have

$$\frac{1}{2} \int_{S_\infty} (\mathbf{E}_i \times \bar{\mathbf{H}}_i) \cdot \mathbf{u}_n dS + \frac{1}{2} \int_{\sum_{l=1}^{n} \partial V_0^{(l)}} (\mathbf{E}_i \times \bar{\mathbf{H}}_i) \cdot \mathbf{u}_n dS$$

$$\tag{6.23}$$

$$= -j2\omega \int_{V_\infty - \sum_{l=1}^{n} V_0^{(l)}} (w_{mi} - w_{ei}) dV - \frac{1}{2} \int_{V_\infty - \sum_{l=1}^{n} V_0^{(l)}} \sigma \mathbf{E}_i \cdot \bar{\mathbf{E}}_i dV,$$

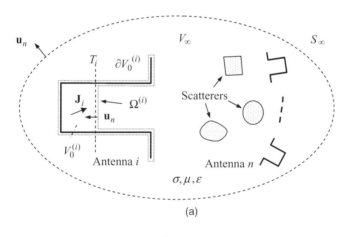

(a)

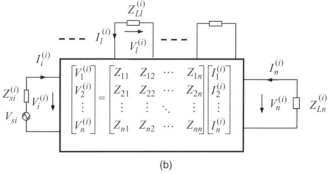

(b)

Figure 6.9 (a) Multiple antenna system. (b) Equivalent circuit

where $w_{mi} = \mu \mathbf{H}_i \cdot \bar{\mathbf{H}}_i / 4$ and $w_{ei} = \varepsilon \mathbf{E}_i \cdot \bar{\mathbf{E}}_i / 4$ are the magnetic and electric energy densities respectively; $\mathbf{u}_n$ is the unit outward normal on the surface enclosing the region $V_\infty - \sum_{l=1}^{n} V_0^{(l)}$, as shown in Figure 6.9. If the antenna surface is perfectly conducting, $\mathbf{E}_i \times \bar{\mathbf{H}}_i$ vanishes everywhere on $\partial V_0^{(i)}$ except over the antenna input terminal $\Omega^{(i)}$. For the single-mode feeding waveguides, we have

$$\int_{\partial V_0^{(l)}} (\mathbf{E}_i \times \bar{\mathbf{H}}_i) \cdot \mathbf{u}_n dS = -V_l^{(i)} \bar{I}_l^{(i)}, l = 1, 2, \cdots, n. \tag{6.24}$$

In deriving the above equations, the following relations have been used

$$\mathbf{E}_i(\mathbf{r}) = V_l^{(i)} \mathbf{e}_{1l}(x, y), \mathbf{H}_i(\mathbf{r}) = I_l^{(i)} \mathbf{u}_z \times \mathbf{e}_{1l}(x, y), \tag{6.25}$$

where $\mathbf{e}_{1l}$ denotes the vector modal function of the dominant mode in the feeding waveguide of antenna l, and $V_l^{(i)}$ and $I_l^{(i)}$ represent the modal voltage and modal current at the reference plane of antenna l when antenna i is transmitting and other antennas are receiving. Introducing (6.24) into (6.23), we obtain

$$\frac{1}{2} V_i^{(i)} \bar{I}_i^{(i)} + \frac{1}{2} \sum_{l=1, l \neq i}^{n} V_l^{(i)} \bar{I}_l^{(i)}$$

$$= \frac{1}{2} \int_{S_\infty} (\mathbf{E}_i \times \bar{\mathbf{H}}_i) \cdot \mathbf{u}_n dS + \frac{1}{2} \int_{V_\infty - \sum_{l=1}^{n} V_0^{(l)}} \sigma \mathbf{E}_i \cdot \bar{\mathbf{E}}_i dV + j2\omega \int_{V_\infty - \sum_{l=1}^{n} V_0^{(l)}} (w_{mi} - w_{ei}) dV. \tag{6.26}$$

If all other antennas $l(l \neq i)$ are in an open-circuit state while antenna i is transmitting, that is, $I_l^{(i)} = 0$ for $l \neq i$, the self-impedance can be expressed as

$$Z_{ii} = \frac{V_i^{(i)}}{I_i^{(i)}} \bigg|_{I_l^{(i)} = 0, l \neq i} = \frac{1}{\left| I_i^{(i)} \right|^2} \int_{S_\infty} (\mathbf{E}_i \times \bar{\mathbf{H}}_i) \cdot \mathbf{u}_n dS$$

$$+ \frac{1}{\left| I_i^{(i)} \right|^2} \int_{V_\infty - \sum_{l=1}^{n} V_0^{(l)}} \sigma \mathbf{E}_i \cdot \bar{\mathbf{E}}_i dV + \frac{j4\omega}{\left| I_i^{(i)} \right|^2} \int_{V_\infty - \sum_{l=1}^{n} V_0^{(l)}} (w_{mi} - w_{ei}) dV \tag{6.27}$$

from (6.26). Note that the fields $\mathbf{E}_i, \mathbf{H}_i$ in (6.27) are calculated with antenna i transmitting while the rest remain open. To calculate the mutual impedance Z_{ij}, we may use the frequency-domain reciprocity theorem

$$\int_S (\mathbf{E}_i \times \mathbf{H}_j - \mathbf{E}_j \times \mathbf{H}_i) \cdot \mathbf{u}_n dS = 0, \tag{6.28}$$

where S is an arbitrary closed surface that does not contain any impressed sources and $\mathbf{u}_n$ is the outward unit normal. Choosing $S = S_\infty + \sum_{l=1}^{n} \partial V_0^{(l)}$ in (6.28) yields

$$
\sum_{l=1}^{n} \int_{\partial V_0^{(l)}} (\mathbf{E}_i \times \mathbf{H}_j - \mathbf{E}_j \times \mathbf{H}_i) \cdot \mathbf{u}_n dS + \int_{S_\infty} (\mathbf{E}_i \times \mathbf{H}_j - \mathbf{E}_j \times \mathbf{H}_i) \cdot \mathbf{u}_n dS
$$
$$
= \sum_{l=1}^{n} \left[V_l^{(j)} I_l^{(i)} - V_l^{(i)} I_l^{(j)} \right] = 0,
\tag{6.29}
$$

where (6.25) has been used. This is the well-known reciprocity theorem in network theory. If we assume that all other antennas are in the state of open circuit when antenna i (or j) is transmitting, the above equation reduces to $V_i^{(j)} I_i^{(i)} = V_j^{(i)} I_j^{(j)}$, or

$$
Z_{ij} = \left. \frac{V_i^{(j)}}{I_j^{(j)}} \right|_{I_l^{(j)}=0, l \neq j} = \left. \frac{V_j^{(i)}}{I_i^{(i)}} \right|_{I_l^{(i)}=0, l \neq i} = Z_{ji}.
\tag{6.30}
$$

Therefore the impedance matrix is symmetric. To express Z_{ij} in terms of the field quantities, we may choose $S = S_i' + \partial V_0^{(i)}$ in (6.28), where S_i' is a closed surface containing antenna i only. Then

$$
\int_{\partial V_0^{(i)}} (\mathbf{E}_i \times \mathbf{H}_j - \mathbf{E}_j \times \mathbf{H}_i) \cdot \mathbf{u}_n dS + \int_{S_i'} (\mathbf{E}_i \times \mathbf{H}_j - \mathbf{E}_j \times \mathbf{H}_i) \cdot \mathbf{u}_n dS = 0.
$$

This implies

$$
V_i^{(i)} I_i^{(j)} - V_i^{(j)} I_i^{(i)} = \int_{S_i'} (\mathbf{E}_i \times \mathbf{H}_j - \mathbf{E}_j \times \mathbf{H}_i) \cdot \mathbf{u}_n dS.
\tag{6.31}
$$

Similarly,

$$
V_j^{(j)} I_j^{(i)} - V_j^{(i)} I_j^{(j)} = \int_{S_j'} (\mathbf{E}_j \times \mathbf{H}_i - \mathbf{E}_i \times \mathbf{H}_j) \cdot \mathbf{u}_n dS,
\tag{6.32}
$$

where S_j' is a closed surface containing antenna j only. The right-hand sides of (6.31) and (6.32) can be shown to be equal by choosing $S = S_i' + S_j' + \sum_{l=1, l \neq i, j}^{n} \partial V_0^{(l)} + S_\infty$ in (6.28).

When antenna i (or j) is transmitting with all other antennas being open, we have

$$V_i^{(j)} I_i^{(i)} = -\int_{S_i'} (\mathbf{E}_i \times \mathbf{H}_j - \mathbf{E}_j \times \mathbf{H}_i) \cdot \mathbf{u}_n dS$$

$$= -\int_{S_j'} (\mathbf{E}_j \times \mathbf{H}_i - \mathbf{E}_i \times \mathbf{H}_j) \cdot \mathbf{u}_n dS = V_j^{(i)} I_j^{(j)}. \tag{6.33}$$

By definition, the mutual impedance of the two-antenna system can be written as

$$Z_{ij} = \left. \frac{V_i^{(j)}}{I_j^{(j)}} \right|_{I_l^{(j)}=0, l \neq j} = -\frac{\displaystyle\int_{S_i'} (\mathbf{E}_i \times \mathbf{H}_j - \mathbf{E}_j \times \mathbf{H}_i) \cdot \mathbf{u}_n dS}{I_i^{(i)} I_j^{(j)}} = -\frac{\displaystyle\int_{V_0^{(i)}} \mathbf{J}_i \cdot \mathbf{E}_j dV}{I_i^{(i)} I_j^{(j)}}, \tag{6.34}$$

where use is made of the following reciprocity theorem

$$\int_{V_0^{(j)}} \mathbf{J}_j \cdot \mathbf{E}_i dV = \int_{S_j'} (\mathbf{E}_j \times \mathbf{H}_i - \mathbf{E}_i \times \mathbf{H}_j) \cdot \mathbf{u}_n dS$$

$$= \int_{S_i'} (\mathbf{E}_i \times \mathbf{H}_j - \mathbf{E}_j \times \mathbf{H}_i) \cdot \mathbf{u}_n dS = \int_{V_0^{(i)}} \mathbf{J}_i \cdot \mathbf{E}_j dV \tag{6.35}$$

Equation (6.34) may be regarded as an exact expression of Huygens' principle in a symmetrical form, and it is generally applicable to an inhomogeneous medium.

6.4.2 Scattering Matrix

Let Z_{sl} be the reference impedance for the input terminal of antenna l. Introducing

$$V_l^{(i)} = \frac{\bar{Z}_{sl}}{\sqrt{\mathrm{Re} Z_{sl}}} a_l^{(i)} + \frac{Z_{sl}}{\sqrt{\mathrm{Re} Z_{sl}}} b_l^{(i)}, \ I_l^{(i)} = \frac{1}{\sqrt{\mathrm{Re} Z_{sl}}} a_l^{(i)} - \frac{1}{\sqrt{\mathrm{Re} Z_{sl}}} b_l^{(i)} \tag{6.36}$$

into (6.29), we obtain

$$\sum_{l=1}^{n} \left[a_l^{(i)} b_l^{(j)} - a_l^{(j)} b_l^{(i)} \right] = 0. \tag{6.37}$$

If we assume that all other antennas are matched when antenna i (or j) is transmitting, Equation (6.37) reduces to $a_i^{(i)} b_i^{(j)} = a_j^{(j)} b_j^{(i)}$, which gives the symmetric property of scattering matrix

$$S_{ij} = \left. \frac{b_i^{(j)}}{a_j^{(j)}} \right|_{a_l^{(j)}=0, l \neq j} = \left. \frac{b_j^{(i)}}{a_i^{(i)}} \right|_{a_l^{(i)}=0, l \neq i} = S_{ji}.$$

In terms of incident and reflected power waves, Equations (6.31) and (6.32) can be written as

$$b_i^{(i)} a_i^{(j)} - b_i^{(j)} a_i^{(i)} = \frac{1}{2} \int_{S_i'} (\mathbf{E}_i \times \mathbf{H}_j - \mathbf{E}_j \times \mathbf{H}_i) \cdot \mathbf{u}_n dS, \tag{6.38}$$

$$b_j^{(j)} a_j^{(i)} - b_j^{(i)} a_j^{(j)} = \frac{1}{2} \int_{S_j'} (\mathbf{E}_j \times \mathbf{H}_i - \mathbf{E}_i \times \mathbf{H}_j) \cdot \mathbf{u}_n dS. \tag{6.39}$$

If all other antennas are matched when antenna i (or j) is transmitting, it follows from (6.35) that

$$
\begin{aligned}
S_{ij} = \left. \frac{b_i^{(j)}}{a_j^{(j)}} \right|_{a_l^{(j)}=0, l \neq j} &= -\frac{1}{2 a_i^{(i)} a_j^{(j)}} \int_{S_i'} (\mathbf{E}_i \times \mathbf{H}_j - \mathbf{E}_j \times \mathbf{H}_i) \cdot \mathbf{u}_n dS \\
&= -\frac{1}{2 a_i^{(i)} a_j^{(j)}} \int_{V_0^{(i)}} \mathbf{J}_i \cdot \mathbf{E}_j dV.
\end{aligned}
\tag{6.40}
$$

6.4.3 Antenna System with Large Separations

So far the separations between antennas are arbitrary. We now assume that the antennas are located in the far field region of each other. Determining the fields $\mathbf{E}_i$ and $\mathbf{H}_i$ produced by the antenna i with antennas $j(j \neq i)$ in place is not an easy task. Therefore, the following simplification is made: the calculation of fields $\mathbf{E}_i$ and $\mathbf{H}_i$ is carried out with the antennas j $(j \neq i)$ removed. Physically, this assumption is equivalent to neglecting the reflections between the antennas. To derive the expressions of the impedance parameters Z_{ij} when the antenna i and antenna j are far apart, two different coordinate systems for antenna i and antenna j may be used. The origins of the coordinate systems are chosen to be the geometrical center of the current distributions and the separation between antenna i and antenna j satisfies $kr_j \gg 1$, $r_j \gg 2a_j, r_j \gg 2a_i$ where $r_j = |\mathbf{r}_j|$ is the distance between antenna j and an arbitrary point of the circumscribing sphere (denoted by S_i') of antenna i, as shown in Figure 6.10. Let $\mathbf{r}_i'$ be the arbitrary point chosen on the circumscribing sphere of antenna i, and $\mathbf{r}_{i,j} = r_{i,j} \mathbf{u}_{r_{i,j}}$, where $r_{i,j}$ is the distance between the two origins and $\mathbf{u}_{r_{i,j}}$ is a unit vector directed from antenna i to antenna j. Thus the far field of antenna j at antenna i can be expressed as (see (4.7) and (4.8))

$$\mathbf{E}_j(\mathbf{r}_j) \approx -\frac{jk\eta I_j^{(j)} e^{-jkr_j}}{4\pi r_j} \mathbf{L}_j(\mathbf{u}_{r_j}), \quad \mathbf{H}_j(\mathbf{r}_j) \approx \frac{1}{\eta} \mathbf{u}_{r_j} \times \mathbf{E}_j(\mathbf{r}_j), \tag{6.41}$$

where $\mathbf{r}_j = \mathbf{r}_i' - \mathbf{r}_{i,j}$ is assumed to be a point on the sphere S_i' and

$$\mathbf{L}_j(\mathbf{u}_{r_j}) = \frac{1}{I_j^{(j)}} \int_{V_0^{(j)}} \left[\mathbf{J}_j - (\mathbf{J}_j \cdot \mathbf{u}_{r_j}) \mathbf{u}_{r_j} \right] e^{jk\mathbf{r}_j' \cdot \mathbf{u}_{r_j}} dV(\mathbf{r}_j').$$

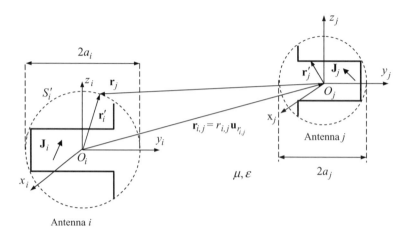

Figure 6.10   Coupling between two distant antennas

is the antenna's effective vector length. Since $\mathbf{r}'_i$ is very small compared to $r_{i,j}$ in magnitude, we can make the approximation $r_j = \left| \mathbf{r}'_i - \mathbf{r}_{i,j} \right| \approx r_{i,j} - \mathbf{u}_{r_{i,j}} \cdot \mathbf{r}'_i$. The field $\mathbf{E}_j$ in the coordinate system O_i can then be represented by

$$\mathbf{E}_j(\mathbf{r}_j) \approx -\frac{jk\eta I_j^{(j)} e^{-jkr_{i,j}} e^{jk\mathbf{u}_{r_{i,j}} \cdot \mathbf{r}'_i}}{4\pi r_{i,j}} \mathbf{L}_j(-\mathbf{u}_{r_{i,j}}),$$

$$(6.42)$$

$$\mathbf{H}_j(\mathbf{r}_j) \approx -\frac{1}{\eta} \mathbf{u}_{r_{i,j}} \times \mathbf{E}_j(\mathbf{r}_j).$$

Then

$$\int_{S'_i} (\mathbf{E}_i \times \mathbf{H}_j - \mathbf{E}_j \times \mathbf{H}_i) \cdot \mathbf{u}_n dS = \int_{S'_i} \left[-\eta^{-1} \mathbf{E}_i \times (\mathbf{u}_{r_{i,j}} \times \mathbf{E}_j) - \mathbf{E}_j \times \mathbf{H}_i \right] \cdot \mathbf{u}_n dS$$

$$(6.43)$$

$$= \int_{S'_i} \mathbf{E}_j \cdot \left[-\eta^{-1} \mathbf{u}_{r_{i,j}} \times (\mathbf{E}_i \times \mathbf{u}_n) - \mathbf{H}_i \times \mathbf{u}_n \right] dS = \int_{S'_i} \mathbf{E}_j \cdot (\mathbf{J}_{is} - \eta^{-1} \mathbf{u}_{r_{i,j}} \times \mathbf{J}_{ims}) dS,$$

where $\mathbf{J}_{is} = \mathbf{u}_n \times \mathbf{H}_i$, $\mathbf{J}_{ims} = -\mathbf{u}_n \times \mathbf{E}_i$ are the equivalent electric current and magnetic current on the surface S'_i respectively. Substituting (6.42) into (6.43), we obtain

$$\int_{S'_i} (\mathbf{E}_i \times \mathbf{H}_j - \mathbf{E}_j \times \mathbf{H}_i) \cdot \mathbf{u}_n dS \approx -\frac{4\pi r_{i,j} e^{jkr_{i,j}}}{jk\eta} \mathbf{E}_i(\mathbf{u}_{r_{i,j}}) \cdot \mathbf{E}_j(-\mathbf{u}_{r_{i,j}})$$

$$(6.44)$$

$$= \frac{-jk\eta I_i^{(i)} I_j^{(j)} e^{-jkr_{i,j}}}{4\pi r_{i,j}} \mathbf{L}_i(\mathbf{u}_{r_{i,j}}) \cdot \mathbf{L}_j(-\mathbf{u}_{r_{i,j}}).$$

Here we have used the far field expression of antenna i at antenna j

$$
\mathbf{E}_i(\mathbf{r}_{i,j}) = \frac{-jk\eta e^{-jkr_{i,j}}}{4\pi r_{i,j}} \int_{S_i'} e^{jk\mathbf{u}_{r_{i,j}} \cdot \mathbf{r}_i} \left[\mathbf{J}_{is}(\mathbf{r}'_i) - \mathbf{u}_{r_{i,j}} \times \eta^{-1} \mathbf{J}_{ims}(\mathbf{r}'_i) \right] dS
$$

$$
= \frac{-jk\eta I_i^{(i)} e^{-jkr_{i,j}}}{4\pi r_{i,j}} \mathbf{L}_i(\mathbf{u}_{r_{i,j}}).
$$

(6.45)

It follows from (6.44) that the mutual impedance Z_{ij} is given by

$$
Z_{ij} = \frac{V_i^{(j)}}{I_j^{(j)}} \bigg|_{I_i^{(j)}=0} = \frac{jk\eta e^{-jkr_{i,j}}}{4\pi r_{i,j}} \mathbf{L}_i(\mathbf{u}_{r_{i,j}}) \cdot \mathbf{L}_j(-\mathbf{u}_{r_{i,j}}).
$$

If the antennas are far apart, the self-impedance Z_{ii} is approximately equal to the input impedance of the antenna i when it is isolated (denoted by Z_i), and (6.27) reduces to

$$
Z_{ii} = \frac{V_i^{(i)}}{I_i^{(i)}} \bigg|_{I_j^{(i)}=0, j \neq i} \approx Z_i.
$$

(6.46)

The above simplification is valid for a multiple antenna system with large separation. If the transmitting performance of antenna i (assuming only antenna i is transmitting) is the major concern, the influence on antenna i due to antenna j ($j \neq i$) is negligible. However, if we are interested in the receiving performance of antenna j, the coupling from the transmitting antenna i to the receiving antenna j cannot be ignored. It is through this coupling that the receiving antenna collects electromagnetic energy from the transmitting antenna.

The scattering parameters for the multiple antenna system with large separation can be simplified in a similar manner and they are

$$
S_{ij} \approx \frac{jk\eta I_i^{(i)} I_j^{(j)} e^{-jkr_{i,j}}}{8\pi r_{i,j} a_i^{(i)} a_j^{(j)}} \mathbf{L}_i(\mathbf{u}_{r_{i,j}}) \cdot \mathbf{L}_j(-\mathbf{u}_{r_{i,j}})
$$

$$
= \frac{jk\eta e^{-jkr_{i,j}}}{8\pi r_{i,j}} \frac{(1 - \Gamma_i^{(i)})(1 - \Gamma_j^{(j)})}{\sqrt{\operatorname{Re} Z_{si}} \sqrt{\operatorname{Re} Z_{sj}}} \mathbf{L}_i(\mathbf{u}_{r_{i,j}}) \cdot \mathbf{L}_j(-\mathbf{u}_{r_{i,j}}),
$$

(6.47)

where $\Gamma_j^{(j)}$ is the reflection coefficient at the reference plane of antenna j.

Example 6.5 (Coupling between two small dipoles): A small two-dipole system is shown in Figure 6.11. The two dipoles are assumed to be identical and are located in the far field region of each other and separated by $r_{1,2}$. The radius of the dipoles is a_0 and their length is $2a$. The current distributions on the dipole surface may be assumed to be

$$
\mathbf{J}_i(\mathbf{r}) = \mathbf{u}_{z_i} \frac{I_i^{(i)}}{2\pi a_0} \left(1 - \frac{|z_i|}{a} \right), \quad -a < z_i < a, i = 1, 2.
$$

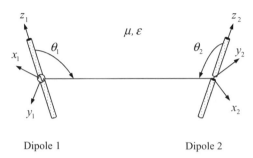

Dipole 1   Dipole 2

Figure 6.11   Two-dipole system

From the far field expressions, the vector effective lengths for dipole 1 and 2 can easily be found as

$$L_1(\mathbf{u}_{r_{1,2}}) = -\mathbf{u}_{\theta_1}(\mathbf{u}_{r_{1,2}})a\sin\theta_1(\mathbf{u}_{r_{1,2}}),$$
$$L_2(-\mathbf{u}_{r_{1,2}}) = -\mathbf{u}_{\theta_2}(-\mathbf{u}_{r_{1,2}})a\sin\theta_2(-\mathbf{u}_{r_{1,2}}),$$

and the mutual impedance is then given by

$$Z_{12} = \frac{jk\eta a^2 e^{-jkr_{1,2}}}{4\pi r_{1,2}}\sin\theta_1(\mathbf{u}_{r_{1,2}})\sin\theta_2(-\mathbf{u}_{r_{1,2}})\mathbf{u}_{\theta_1}(\mathbf{u}_{r_{1,2}})\cdot\mathbf{u}_{\theta_2}(-\mathbf{u}_{r_{1,2}}).$$

If dipole 1 and 2 are perpendicular or collinear, the mutual impedance is zero. The coupling is maximized when the two dipoles are placed in parallel.   □

Example 6.6 (Coupling between two small loop antennas): Consider two identical circular small loop antennas, separated by a distance $r_{1,2}$ and located in the far field region of each other as shown in Figure 6.12. The radius of the loop is assumed to be a and the radius of the wire is a_0. Since the loop is very small, the current density on the loop can be assumed to be constant:

$$\mathbf{J}_i(\mathbf{r}) = \mathbf{u}_{\varphi_i}\frac{I_i^{(i)}}{2\pi a_0}, i = 1, 2,$$

where φ_i is the polar angle in (x_i, y_i)-plane. The vector effective lengths for loop 1 and 2 are easily found to be

$$L_1(\mathbf{u}_{r_{1,2}}) = jk\pi a^2\mathbf{u}_{\varphi_1}(\mathbf{u}_{r_{1,2}})\sin\theta_1(\mathbf{u}_{r_{1,2}}),$$
$$L_2(-\mathbf{u}_{r_{1,2}}) = jk\pi a^2\mathbf{u}_{\varphi_2}(-\mathbf{u}_{r_{1,2}})\sin\theta_2(-\mathbf{u}_{r_{1,2}})$$

and the mutual impedance is

$$Z_{12} = -\frac{j\eta k^3\pi a^4 e^{-jkr_{1,2}}}{4r_{1,2}}\sin\theta_1(\mathbf{u}_{r_{1,2}})\sin\theta_2(-\mathbf{u}_{r_{1,2}})\mathbf{u}_{\phi_1}(\mathbf{u}_{r_{1,2}})\cdot\mathbf{u}_{\phi_2}(-\mathbf{u}_{r_{1,2}}).$$

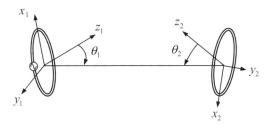

Figure 6.12   Two-loop system

If the two loops are perpendicular or placed face to face the coupling is zero. The coupling is maximized when the two loops lie in the same plane.   □

6.5   Power Transmission Between Antennas

The wireless power transmission may be considered as a three-step process. The first step is to convert direct-current power into radio-frequency power; the second step is to transmit the radio-frequency power to some distant point; the third step is to collect the radio-frequency power and convert it back to direct-current power at the receiving point. The last step is often accomplished by a rectenna, which is an antenna combined with a rectifier.

Wireless power transmission has been a research topic for years. Many applications can benefit from the research, such as microwave imaging, radar and directed energy weapons. The basic theory for the power transmission between two antennas was investigated in 1960s (Goubao and Schwinger, 1961; Kay, 1960; Sherman, 1962; Borgiotti, 1966), and it has found wide applications in many fields (Brown, 1984). Theoretically, power transmission efficiency of almost 100% is attainable by increasing the sizes of the antennas. For a given power transmission efficiency over a given distance between the transmitting and receiving antenna, there exists an optimum antenna aperture distribution which can minimize the transmitting and receiving aperture sizes. To achieve the maximum transmission efficiency, the transmitting antenna must be focused at the receiving antenna. In other words, the radiated electromagnetic energy must be focused in the vicinity of the axis of the transmitting and receiving antenna apertures as it propagates.

6.5.1   Universal Power Transmission Formula

Let us consider the power transmission between antenna i and antenna j ($j \neq i$) when antenna i is transmitting and antenna j is receiving. It follows from Figure 6.9 (b) that

$$V_i^{(i)} = Z_i^{(i)} I_i^{(i)}, \quad V_i^{(j)} = -I_i^{(j)} Z_{Li}^{(j)},$$
$$V_j^{(j)} = Z_j^{(j)} I_j^{(j)}, \quad V_j^{(i)} = -I_j^{(i)} Z_{Lj}^{(i)}.$$

$$(6.48)$$

Here $Z_i^{(i)}$ is the input impedance of antenna i when antenna i is transmitting and all other antennas are receiving. Substituting (6.48) into (6.33), we obtain

$$I_j^{(j)} I_j^{(i)} (Z_j^{(j)} + Z_{Lj}^{(i)}) = \int\limits_{S_i' \text{or} S_j'} (\mathbf{E}_i \times \mathbf{H}_j - \mathbf{E}_j \times \mathbf{H}_i) \cdot \mathbf{u}_n dS$$

$$= I_i^{(i)} I_i^{(j)} (Z_i^{(i)} + Z_{Li}^{(j)}). \tag{6.49}$$

Multiplying (6.49) by its conjugate, we obtain

$$\left| I_j^{(j)} \right|^2 \left| I_j^{(i)} \right|^2 \left| Z_j^{(j)} + Z_{Lj}^{(i)} \right|^2 = \left| \int\limits_{S_i' \text{or} S_{ji}'} (\mathbf{E}_i \times \mathbf{H}_j - \mathbf{E}_j \times \mathbf{H}_i) \cdot \mathbf{u}_n dS \right|^2$$

$$= \left| I_i^{(i)} \right|^2 \left| I_i^{(j)} \right|^2 \left| Z_i^{(i)} + Z_{Li}^{(j)} \right|^2. \tag{6.50}$$

If the antenna i and antenna j are conjugately matched, that is

$$\bar{Z}_j^{(j)} = Z_{Lj}^{(i)}, \ \bar{Z}_i^{(i)} = Z_{Li}^{(j)},$$

Equation (6.50) can be written as

$$\frac{\frac{1}{2} \left| I_j^{(i)} \right|^2 \operatorname{Re} Z_{Lj}^{(i)}}{\frac{1}{2} \left| I_i^{(i)} \right|^2 \operatorname{Re} Z_i^{(i)}} = \frac{\left| \int\limits_{S_i' \text{or} S_{ji}'} (\mathbf{E}_i \times \mathbf{H}_j - \mathbf{E}_j \times \mathbf{H}_i) \cdot \mathbf{u}_n dS \right|^2}{4 \left| I_i^{(i)} \right|^2 \operatorname{Re} Z_i^{(i)} \left| I_j^{(j)} \right|^2 \operatorname{Re} Z_j^{(i)}} = \frac{\frac{1}{2} \left| I_i^{(j)} \right|^2 \operatorname{Re} Z_{Li}^{(j)}}{\frac{1}{2} \left| I_j^{(j)} \right|^2 \operatorname{Re} Z_j^{(j)}}. \tag{6.51}$$

This implies

$$T_{ij} = \frac{P_j^{(i)}}{P_i^{(i)}} = \frac{\left| \int\limits_{S_i' \text{or} S_j'} (\mathbf{E}_i \times \mathbf{H}_j - \mathbf{E}_j \times \mathbf{H}_i) \cdot \mathbf{u}_n dS \right|^2}{4 \operatorname{Re} \int\limits_{S_i'} (\mathbf{E}_i \times \bar{\mathbf{H}}_i) \cdot \mathbf{u}_n dS \operatorname{Re} \int\limits_{S_j'} (\mathbf{E}_j \times \bar{\mathbf{H}}_j) \cdot \mathbf{u}_n dS} = \frac{P_i^{(j)}}{P_j^{(j)}} = T_{ji}, \tag{6.52}$$

where $P_i^{(i)}$ is the transmit power of antenna i when all other antennas are receiving and $P_j^{(i)}$ is the power received by antenna j when antenna i is transmitting. Equation (6.52) indicates that the ratio of the power received by antenna j to the transmitting power of antenna i (known as the **power transmission efficiency**, denoted T_{ij}) is equal to the ratio of the power received by antenna i to the transmitting power of antenna j (denoted T_{ji}). It also indicates that the radiation pattern of the antenna for reception is identical with that for transmission. Equation (6.52) is the theoretical foundation for the wireless power transmission in free space, and is the starting point for optimizing the aperture distribution to achieve the maximum possible power transmission efficiency. Evidently the power transmission efficiency is maximized if

$$\mathbf{E}_i = \bar{\mathbf{E}}_j, \mathbf{H}_i = -\bar{\mathbf{H}}_j \tag{6.53}$$

hold on some closed surface that encloses either antenna i or j. If the separation between antenna i and j is large enough, we may use (6.44) to obtain

$$\left| \int_{S_i' \text{or} S_j'} (\mathbf{E}_i \times \mathbf{H}_j - \mathbf{E}_j \times \mathbf{H}_i) \cdot \mathbf{u}_n dS \right|^2 \approx \frac{(4\pi r_{i,j})^2}{\eta^2 k^2} \left| \mathbf{E}_i(\mathbf{r}_{i,j}) \cdot \mathbf{E}_j(-\mathbf{r}_{i,j}) \right|^2$$

$$= \left(\frac{4\lambda}{r_{i,j}} \right)^2 U_i(\mathbf{u}_{r_{i,j}}) U_j(-\mathbf{u}_{r_{i,j}}) \cos \theta_{ij}, \tag{6.54}$$

where U_i and U_j are the radiation intensity of antenna i and j respectively, and θ_{ij} is the angle between $\mathbf{E}_i(\mathbf{r}_{i,j})$ and $\mathbf{E}_j(-\mathbf{r}_{i,j})$. Substituting (6.54) into (6.52), we obtain the well-known **Friis transmission formula**

$$\frac{P_j^{(i)}}{P_i^{(i)}} = \left(\frac{\lambda}{4\pi r_{i,j}} \right)^2 \frac{4\pi U_i(\mathbf{u}_{r_{i,j}}) 4\pi U_j(-\mathbf{u}_{r_{i,j}}) \cos \theta_{ij}}{\frac{1}{2}\text{Re} \int_{S_i'} (\mathbf{E}_i \times \bar{\mathbf{H}}_i) \cdot \mathbf{u}_n dS \frac{1}{2}\text{Re} \int_{S_j'} (\mathbf{E}_j \times \bar{\mathbf{H}}_j) \cdot \mathbf{u}_n dS}$$

$$= \left(\frac{\lambda}{4\pi r_{i,j}} \right)^2 G_i(\mathbf{u}_{r_{i,j}}) G_j(-\mathbf{u}_{r_{i,j}}) \cos \theta_{ij}, \tag{6.55}$$

where G_i and G_j are the gains of the antenna i and antenna j respectively. For a two-antenna system, Equation (6.55) may be written as

$$P_2^{(1)} = \frac{\text{EIRP}}{L_s} G_2(-\mathbf{u}_{r_{1,2}}) \cos \theta_{12}, \tag{6.56}$$

where $L_s = (4\pi r_{1,2}/\lambda)^2$ is known as **free-space path loss**, and EIRP stands for the **effective isotropic radiated power** defined by EIRP $= P_1^{(1)} G_1(\mathbf{u}_{r_{1,2}})$. The **received isotropic power** is defined as EIRP$/L_s$, which is the power received by an isotropic antenna ($G_2 = 1$).

Remark 6.2: In a wireless communication system, channel impairments are caused by several mechanisms, which are path loss, blockage, fast fading (multi-path effect), shadowing, random frequency modulation (relative motion of transmitter and receiver) and delay spread (multiple signals arrive with a slight additional delay which spreads the received signal and causes each symbol to overlap with proceeding and following symbols, producing inter-symbol interference). As a result, the propagation losses can be significantly higher than the free-space path loss. In practice, the path loss (called **propagation model**) for both indoor and outdoor applications should be modified as $L_s = (4\pi r_{1,2}/\lambda)^2 \times$ correction factors. The correction factors depend on the propagation environments and are usually based on measured data. Therefore, many propagation models are of semi-empirical type. The predictions from these models may have a large deviation from the actually measured data. □

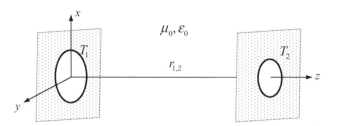

Figure 6.13   Two-planar aperture system

6.5.2   Power Transmission Between two Planar Apertures

If a two-antenna system is used to transmit electric power, the antenna geometries and their current distributions should be chosen properly in order that the electromagnetic power delivered from one antenna to the other is maximized. Let us consider the maximum power transmission between two-planar apertures. The configuration of a two-planar aperture system in free space is shown in Figure 6.13, where both apertures are in an infinite conducting screen so that the tangential electric field outside the aperture is zero. When the aperture i ($i = 1, 2$) is used as a transmitting antenna, the aperture field is assumed to be

$$\mathbf{E}_i = \mathbf{u}_x E_i, \mathbf{H}_i = \mathbf{u}_y \frac{1}{\eta_0} E_i,$$

where $\eta_0 = \sqrt{\mu_0/\varepsilon_0}$ is the wave impedance in free space. We will use the same notations for the aperture field distribution and the field produced by the aperture, and this will not cause any confusion. By means of equivalence theorem and image principle, the electric field produced by aperture 1 may be represented by

$$\mathbf{E}_1(\mathbf{r}) = \frac{1}{2\pi} \int_{T_1} \mathbf{u}_y \times \mathbf{u}_R \left(jk_0 + \frac{1}{|\mathbf{r} - \mathbf{r}'|} \right) e^{-jk_0|\mathbf{r} - \mathbf{r}'|} E_1(\mathbf{r}') dx' dy', \tag{6.57}$$

where $\mathbf{u}_R = (\mathbf{r} - \mathbf{r}')/|\mathbf{r} - \mathbf{r}'|$, $k_0 = \omega\sqrt{\mu_0\varepsilon_0}$. In deriving the above expression, we have neglected the multiple scattering between the apertures. If the apertures are located in the Fresnel region of each other and the observation point $\mathbf{r}$ is on the aperture 2, the following approximations can be made

$$\mathbf{u}_y \times \mathbf{u}_R \approx \mathbf{u}_x, \; |\mathbf{r} - \mathbf{r}'| \approx r_{1,2} + \frac{1}{2r_{1,2}}[(x - x')^2 + (y - y')^2].$$

From (6.57), we obtain

$$\mathbf{E}_1(\mathbf{r}) = \mathbf{u}_x E_1(\mathbf{r}) \approx \mathbf{u}_x \frac{je^{-jk_0 r_{1,2}}}{\lambda r_{1,2}} \int_{T_1} E_1 e^{-jk_0[(x-x')^2+(y-y')^2]/2r_{1,2}} dx' dy',$$

$$\tag{6.58}$$

$$\mathbf{H}_1(\mathbf{r}) = \mathbf{u}_y \frac{1}{\eta_0} E_1(\mathbf{r}).$$

Substituting these into (6.52) gives

$$T_{12} = \left(\frac{1}{\lambda r_{1,2}}\right)^2 \frac{\left|\int\limits_{T_2} \tilde{m}_1 m_2 dx dy\right|^2}{\int\limits_{T_1} |m_1|^2 dx dy \int\limits_{T_2} |m_2|^2 dx dy}, \qquad (6.59)$$

where

$$m_1(x, y) = E_1 e^{-jk_0(x^2+y^2)/2r_{1,2}},$$
$$m_2(x, y) = E_2 e^{-jk_0(x^2+y^2)/2r_{1,2}},$$
$$\tilde{m}_1(x, y) = \int\limits_{T_1} m_1(x', y') e^{jk_0(xx'+yy')/r_{1,2}} dx' dy',$$
$$\tilde{m}_2(x, y) = \int\limits_{T_2} m_2(x', y') e^{jk_0(xx'+yy')/r_{1,2}} dx' dy'.$$

Note that

$$\int\limits_{T_1} m_1 \tilde{m}_2 dx dy = \int\limits_{T_2} \tilde{m}_1 m_2 dx dy.$$

This is equivalent to $T_{12} = T_{21}$. We may introduce the power transmission efficiency between two ideal apertures

$$T_{12}^{ideal} = \frac{\text{Re} \int\limits_{T_2} (\mathbf{E}_1 \times \bar{\mathbf{H}}_1) \cdot \mathbf{u}_z dx dy}{\text{Re} \int\limits_{T_1} (\mathbf{E}_1 \times \bar{\mathbf{H}}_1) \cdot \mathbf{u}_z dx dy}.$$

Then

$$T_{12}^{ideal} = \left(\frac{1}{\lambda r_{1,2}}\right)^2 \frac{\int\limits_{T_2} |\tilde{m}_1|^2 dx dy}{\int\limits_{T_1} |m_1|^2 dx dy}. \qquad (6.60)$$

Thus (6.59) may be expressed as

$$T_{12} = T_{12}^{ideal} \cdot U,$$

where

$$
U = \frac{\left| \displaystyle\int_{T_2} \tilde{m}_1 m_2 dx dy \right|^2}{\displaystyle\int_{T_2} |\tilde{m}_1|^2 \, dx dy \displaystyle\int_{T_2} |m_2|^2 \, dx dy}.
$$

The power transmission efficiency T_{12} reaches maximum if both T_{12}^{ideal} and U are maximized. From Cauchy-Schwartz inequality, we have $\max U = 1$, which can be reached by letting $m_2(x, y) = c_1 \bar{\tilde{m}}_1(x, y)$, $(x, y) \in T_2$, that is,

$$
E_2(x, y) = c_2 \bar{E}_1(x, y), (x, y) \in T_2. \tag{6.61}
$$

Here both c_1 and c_2 are arbitrary complex numbers. The above equation implies that the aperture distribution of antenna 2 is equal to the complex conjugate of the field produced by antenna 1 at antenna 2. We now consider the condition for maximizing T_{12}^{ideal}. Equation (6.60) can be rewritten as

$$
T_{12}^{ideal} = \frac{(\hat{T}m_1, m_1)}{(m_1, m_1)}
$$

where $(\cdot, \cdot)$ denotes the inner product defined by $(u, v) = \int_{T_1} u \bar{v} dx dy$ for two arbitrary functions u and v, and $\hat{T}$ is a self-adjoint operator defined by

$$
\hat{T}m_1 \left(\xi', \varsigma' \right) = \int_{T_1} K_2(\xi, \varsigma; \xi', \varsigma') m_1(\xi, \varsigma) d\xi d\varsigma
$$

with

$$
K_2(\xi, \varsigma; \xi', \varsigma') = \left(\frac{1}{\lambda r_{1,2}} \right)^2 \int_{T_1} e^{jk_0[(\xi - \xi')x + (\varsigma - \varsigma')y]/r_{1,2}} dx dy.
$$

If the condition (6.61) is met, we have

$$
T_{12} = T_{12}^{ideal} = \frac{(\hat{T}m_1, m_1)}{(m_1, m_1)}. \tag{6.62}
$$

This is a variational expression (Rayleigh quotient), and attains an extremum when m_1 satisfies

$$
\hat{T}m_1(x, y) = T_{12} m_1(x, y). \tag{6.63}
$$

Therefore, the power transmission between planar apertures is maximized if the aperture field distributions satisfy (6.61) and (6.63) simultaneously. Equation (6.63) is an eigenvalue problem and its largest eigenvalue is the maximum possible value for the power transmission efficiency.

Table 6.1 Eigenvalues.

c_1	T_{12}
0.5	0.30969
1	0.57258
2	0.88056
4	0.99589
8	1.00000

Equation (6.63) may be used first to determine the aperture distribution of antenna 1, and the aperture distribution of antenna 2 can then be determined from (6.61).

Example 6.7: Let us consider the power transmission between two identical rectangular apertures with $T_1 = T_2 = [-a, a] \times (-\infty, +\infty)$, Equation (6.63) reduces to

$$\int_{-a}^{a} m_1(x') \frac{\sin[ka(x - x')/r_{1,2}]}{\pi(x - x')} dx' = T_{12} m_1(x)$$

where $m_1(x) = E_1(x)e^{-jkx^2/2r_{1,2}}$. The above eigenvalue problem also appears in signal theory and has been solved by Slepian and Pollak (Slepian and Pollak, 1961). The solutions are

$$T_{12} = \frac{2c_1}{\pi}[R_{00}^{(1)}(c_1, 1)]^2,$$

$$E_1(x) = S_{00}(c_1, x/a)e^{-jkx^2/2r_{1,2}},$$

where $c_1 = ka^2/r_{1,2}$, $R_{00}^{(1)}$ is the radial prolate spheroidal function, and S_{00} is the angular prolate spheroidal function. Some values of T_{12} are listed in Table 6.1. Observe that the power transmisson efficiency of 100% can be achieved by increasing the parameter c_1.   □

6.5.3   Power Transmission Between two Antenna Arrays

The proceeding discussions on the power transmission between two antennas can be generalized to two antenna arrays. Consider an $n_t + n_r$ antenna system shown in Figure 6.14, in which antennas $1 \sim n_t$ are transmitting while antennas $n_t + 1 \sim n_t + n_r$ are receiving. This system can be described as an $n_t + n_r$ network and can be characterized by

$$\begin{bmatrix} [b_t] \\ [b_r] \end{bmatrix} = \begin{bmatrix} [S_{tt}] & [S_{tr}] \\ [S_{rt}] & [S_{rr}] \end{bmatrix} \begin{bmatrix} [a_t] \\ [a_r] \end{bmatrix}, \tag{6.64}$$

where the normalized incident and reflected waves for transmitting antenna array and receiving antenna array are respectively given by

$$[a_t] = [a_1, a_2, \cdots, a_{n_t}]^T,$$
$$[b_t] = [b_1, b_2, \cdots, b_{n_t}]^T,$$
$$[a_r] = [a_{n_t+1}, a_{n_t+2}, \cdots, a_{n_t+n_r}]^T,$$
$$[b_r] = [b_{n_t+1}, b_{n_t+2}, \cdots, b_{n_t+n_r}]^T.$$

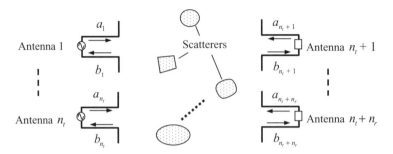

Figure 6.14 Power transmission between two antenna arrays

The power transmission efficiency between the two antenna arrays is defined as the ratio of the power delivered to the loads of the receiving array to the input power to the transmitting antenna array

$$T_{array} = \frac{\frac{1}{2}(|[b_r]|^2 - |[a_r]|^2)}{\frac{1}{2}(|[a_t]|^2 - |[b_t]|^2)}. \tag{6.65}$$

Assume that the receiving antenna array is matched so that $[a_r] = 0$. Making use of (6.64), Equation (6.65) can be written as

$$T_{array} = \frac{([A][a_t], [a_t])}{([B][a_t], [a_t])}, \tag{6.66}$$

where $(\cdot, \cdot)$ denotes the usual inner product of two column vectors, $[A]$ and $[B]$ are two matrices defined by

$$[A] = [S_{tr}]^T S_{rt}, [B] = [1] - [S_{tt}]^T [S_{tt}].$$

If the power transmission efficiency T_{array} reaches the maximum at $[a_t]$, then we have

$$[A][a_t] = T_{array}[B][a_t]. \tag{6.67}$$

Therefore, the maximum possible value of T_{array} is the largest eigenvalue of (6.67) and can be found numerically. From (6.67), we obtain

$$T_{array} = \frac{([B]^{-1}[A][a_t], [a_t])}{([a_t], [a_t])} \leq \left\| [B]^{-1}[A] \right\|,$$

where $\|\cdot\|$ denotes the matrix norm. The right-hand side is the maximum possible power transmission coefficient for a system with the receiving antenna array matched.

If both transmitting antenna array and receiving antenna array are matched, Equation (6.65) reduces to

$$T_{array} = \frac{([A][a_t], [a_t])}{([a_t], [a_t])} \le \|[A]\| .$$

(6.68)

The right-hand side is the maximum possible power transmission efficiency between two matched antenna arrays.

Remark 6.3: Let $[A]$ be a complex $m \times n$ matrix

$$[A] = \begin{bmatrix} a_{11} & a_{12} & \cdots & a_{1n} \\ a_{21} & a_{22} & \cdots & a_{2n} \\ \vdots & \vdots & \ddots & \vdots \\ a_{m1} & a_{m2} & \cdots & a_{mn} \end{bmatrix} .$$

One of the matrix norms can be defined as

$$\|[A]\| = \left(\sum_{i=1}^{m} \sum_{j=1}^{n} |a_{ij}|^2 \right)^{1/2} .$$

□

6.6 Network Parameters in a Scattering Environment

The network parameters vary with the environments. Predicting this variation is important in practice. For example, an embedded antenna in a handset is surrounded by a number of circuit components, any changes of these components, such as their locations, sizes and electrical properties, will affect the system performance. In this section, we present a perturbation method to predict how the system performance varies with the environments.

6.6.1 Compensation Theorem for Time-Harmonic Fields

In circuit theory, any element can be replaced by an idea current source of the same current intensity as in the element. This property is called the compensation theorem. The general form of the compensation theorem in electromagnetics says that the influence of substance on the fields can partly or completely be compensated by appropriate distribution of impressed currents (see also Section 1.3.2). Let us consider a closed surface S filled with a linear, isotropic medium and free of impressed source, as shown in Figure 6.15. The medium inside S may be inhomogeneous, with a permittivity $\varepsilon(\mathbf{r})$, permeability $\mu(\mathbf{r})$ and conductivity $\sigma(\mathbf{r})$. Thus, we may write

$$\nabla \times \mathbf{E}(\mathbf{r}) = -j\omega\mu(\mathbf{r})\mathbf{H}(\mathbf{r}), \ \nabla \times \mathbf{H}(\mathbf{r}) = [\sigma(\mathbf{r}) + j\omega\varepsilon(\mathbf{r})]\mathbf{E}(\mathbf{r}).$$

(6.69)

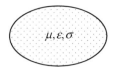

Figure 6.15 A region free of impressed source

If the medium parameters $\mu(\mathbf{r})$, $\varepsilon(\mathbf{r})$ and $\sigma(\mathbf{r})$ in S are changed to $\mu'(\mathbf{r})$, $\varepsilon'(\mathbf{r})$ and $\sigma'(\mathbf{r})$, the corresponding fields will be governed by

$$\nabla \times \mathbf{E}'(\mathbf{r}) = -j\omega\mu'(\mathbf{r})\mathbf{H}'(\mathbf{r}), \nabla \times \mathbf{H}'(\mathbf{r}) = [\sigma'(\mathbf{r}) + j\omega\varepsilon'(\mathbf{r})]\mathbf{E}'(\mathbf{r}). \tag{6.70}$$

This can be written as

$$\nabla \times \mathbf{E}'(\mathbf{r}) = -j\omega\mu(\mathbf{r})\mathbf{H}'(\mathbf{r}) - \mathbf{J}'_{m,imp}(\mathbf{r}),$$
$$\nabla \times \mathbf{H}'(\mathbf{r}) = [\sigma(\mathbf{r}) + j\omega\varepsilon(\mathbf{r})]\mathbf{E}'(\mathbf{r}) + \mathbf{J}'_{imp}(\mathbf{r}), \tag{6.71}$$

where

$$\mathbf{J}'_{m,imp}(\mathbf{r}) = j\omega\mathbf{H}'(\mathbf{r})[\mu'(\mathbf{r}) - \mu(\mathbf{r})],$$
$$\mathbf{J}'_{imp}(\mathbf{r}) = \{\sigma'(\mathbf{r}) - \sigma(\mathbf{r}) + j\omega[\varepsilon'(\mathbf{r}) - \varepsilon(\mathbf{r})]\}\mathbf{E}'(\mathbf{r}) \tag{6.72}$$

are the equivalent impressed current sources introduced in the region bounded by S. So the perturbed fields can be determined by introducing the equivalent current sources as if the medium parameters had not changed. This is one of the forms of the compensation theorem in electromagnetic field theory. The differential fields $\Delta\mathbf{E} = \mathbf{E}' - \mathbf{E}$, $\Delta\mathbf{H} = \mathbf{H}' - \mathbf{H}$ satisfy the following equations

$$\nabla \times \Delta\mathbf{E}(\mathbf{r}) = -j\omega\mu(\mathbf{r})\Delta\mathbf{H}(\mathbf{r}) - \mathbf{J}'_{m,imp}(\mathbf{r}),$$
$$\nabla \times \Delta\mathbf{H}(\mathbf{r}) = [\sigma(\mathbf{r}) + j\omega\varepsilon(\mathbf{r})]\Delta\mathbf{E}(\mathbf{r}) + \mathbf{J}'_{imp}(\mathbf{r}). \tag{6.73}$$

Hence the equivalent current sources in (6.72) generate the differential fields.

6.6.2 Scattering Parameters in a Scattering Environment

The influences of the changes of medium parameters can be studied by means of the compensation theorem. Let us consider the perturbation of the scattering parameters in a scattering environment. Figure 6.16 shows two antenna elements together with a region V_p enclosed by S_p, where the changes of medium parameters take place. It is assumed that the medium outside S_p is lossless. Two scenarios will be considered:

1) *Scenario 1*: The permeability, permittivity and conductivity in the region V_p are μ, ε and σ respectively, which may vary from point to point. The antenna i produces the fields $\mathbf{E}_i$ and $\mathbf{H}_i$ when antennas $j(j \neq i)$ are receiving. The normalized transmission coefficient between antenna i and antenna j is denoted by S_{ij}.

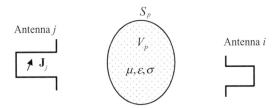

Figure 6.16   Multiple antenna system in a scattering environment

2) *Scenario 2*: The medium parameters μ, ε and σ in the region V_p are changed to μ', ε' and σ' respectively. The antenna i produces the fields $\mathbf{E}'_i$ and $\mathbf{H}'_i$ when antennas $j(j \neq i)$ are receiving. The normalized transmission coefficient between antenna i and antenna j is denoted by S'_{ij}.

From (6.40) and the reciprocity theorem in a region with impressed sources, the transmission coefficient for scenario 1 can be expressed as

$$S_{ij} = -\frac{1}{2a_i^{(i)}a_j^{(j)}} \int_{S'_i} (\mathbf{E}_i \times \mathbf{H}_j - \mathbf{E}_j \times \mathbf{H}_i) \cdot \mathbf{u}_n dS = -\frac{1}{2a_i^{(i)}a_j^{(j)}} \int_{V'_j} \mathbf{J}_j \cdot \mathbf{E}_i dV, \qquad (6.74)$$

where S'_i is the surface enclosing antenna i only (S_p is not included in S'_i), V'_j is the region enclosed by S'_j, which includes antenna j as well as V_p, and $\mathbf{J}_j$ is the current distribution of antenna j. Similarly the perturbed transmission coefficient for scenario 2 can be expressed as (assuming that the impressed current $\mathbf{J}_j$ remains unchanged)

$$S'_{ij} = -\frac{1}{2a_i^{(i)}a_j^{(j)}} \int_{S'_i} (\mathbf{E}'_i \times \mathbf{H}'_j - \mathbf{E}'_j \times \mathbf{H}'_i) \cdot \mathbf{u}_n dS = -\frac{1}{2a_i^{(i)}a_j^{(j)}} \int_{V'_j} \mathbf{J}_j \cdot \mathbf{E}'_i dV. \qquad (6.75)$$

Subtracting (6.74) from (6.75) gives

$$S'_{ij} - S_{ij} = -\frac{1}{2a_i^{(i)}a_j^{(j)}} \int_{V'_j} \mathbf{J}_j \cdot (\mathbf{E}'_i - \mathbf{E}_i) dV. \qquad (6.76)$$

Since V'_j contains the region V_p, from the reciprocity theorem it follows that

$$S'_{ij} - S_{ij} = -\frac{1}{2a_i^{(i)}a_j^{(j)}} \int_{V'_j} \mathbf{J}_j \cdot (\mathbf{E}'_i - \mathbf{E}_i) dV = -\frac{1}{2a_i^{(i)}a_j^{(j)}} \int_{V_p} \mathbf{J}'_{imp} \cdot \mathbf{E}_j - \mathbf{J}'_{m,imp} \cdot \mathbf{H}_j dV$$

$$\qquad (6.77)$$

$$= \frac{1}{2a_i^{(i)}a_j^{(j)}} \int_{V_p} \left\{ j\omega(\mu' - \mu)\mathbf{H}'_i \cdot \mathbf{H}_j - \left[\sigma' - \sigma + j\omega(\varepsilon' - \varepsilon)\right]\mathbf{E}'_i \cdot \mathbf{E}_j \right\} dV.$$

If we assume that the effect of the changes in permittivity and permeability has negligible effect on the fields, Equation (6.77) can be approximated by

$$S'_{ij} - S_{ij} = \frac{1}{2a_i^{(i)} a_j^{(j)}} \int_{V_p} \left\{ j\omega(\mu' - \mu)\mathbf{H}_i \cdot \mathbf{H}_j - \left[\sigma' - \sigma + j\omega(\varepsilon' - \varepsilon) \right] \mathbf{E}_i \cdot \mathbf{E}_j \right\} dV. \quad (6.78)$$

One of the applications of (6.78) is to study the reflections in a waveguide due to a slight inhomogeneity of the filled dielectric materials. Another application of (6.78) is to study the scattering caused by atmospheric inhomogeneity between two antennas (Monteath, 1973).

Equation (6.77) is useful to study the effect of changes in permittivity and permeability of the medium in a finite volume. But it is not convenient to study the changes in highly conducting bodies where the fields are confined to a shallow surface layer. In this case, a surface integral is more appropriate. Making use of reciprocity again, Equation (6.77) may be expressed as

$$S'_{ij} - S_{ij} = -\frac{1}{2a_i^{(i)} a_j^{(j)}} \int_{V_p} \mathbf{J}'_{imp} \cdot \mathbf{E}_j - \mathbf{J}'_{m,imp} \cdot \mathbf{H}_j dV$$

$$= -\frac{1}{2a_i^{(i)} a_j^{(j)}} \int_{S_p} \left[(\mathbf{E}'_i - \mathbf{E}_i) \times \mathbf{H}_j - \mathbf{E}_j \times (\mathbf{H}'_i - \mathbf{H}_i) \right] \cdot \mathbf{u}_n dS \quad (6.79)$$

$$= \frac{1}{2a_i^{(i)} a_j^{(j)}} \int_{S_p} (\mathbf{E}_j \times \mathbf{H}'_i - \mathbf{E}'_i \times \mathbf{H}_j) \cdot \mathbf{u}_n dS.$$

Suppose that the boundary S_p has a surface impedance Z_s in scenario 1 and Z'_s in scenario 2. Then

$$\mathbf{E}_{jt}(\mathbf{r}) = Z_s \mathbf{u}_n \times \mathbf{H}_{jt}(\mathbf{r}), \mathbf{E}'_{it}(\mathbf{r}) = Z'_s \mathbf{u}_n \times \mathbf{H}'_{jt}(\mathbf{r}), \mathbf{r} \in S_p,$$

where the subscript t denotes the tangential component. Introducing these into (6.79) yields

$$S'_{ij} - S_{ij} = \frac{1}{2a_i^{(i)} a_j^{(j)}} \int_{S_p} \left[\mathbf{H}'_i \cdot (\mathbf{u}_n \times \mathbf{E}_j) - \mathbf{H}_j \cdot (\mathbf{u}_n \times \mathbf{E}'_i) \right] dS$$

$$= \frac{1}{2a_i^{(i)} a_j^{(j)}} \int_{S_p} (Z'_s - Z_s)\mathbf{H}'_{it} \cdot \mathbf{H}_{jt} dS. \quad (6.80)$$

If there are m scatterers and each scatter occupies a region $V_p(p = 1, 2, \cdots, m)$, the integrals in Equations (6.77)–(6.80) become a summation of integrals over each scatterer. For instance, Equation (6.77) may be written as

$$S'_{ij} = S_{ij}$$

$$+ \frac{1}{2a_i^{(i)} a_j^{(j)}} \sum_{p=1}^{m} \int_{V_p} \left\{ j\omega(\mu' - \mu)\mathbf{H}'_i \cdot \mathbf{H}_j - \left[\sigma' - \sigma + j\omega(\varepsilon' - \varepsilon) \right] \mathbf{E}'_i \cdot \mathbf{E}_j \right\} dV. \quad (6.81)$$

The first term on the right-hand side corresponds to the contribution due to the direct path from antenna i to antenna j. The second term represents the m multi-path components contributed by the m scatterers. The presence of significant scatterers in the propagation medium guarantees that the waves from different paths will add differently at each receiving antenna element so that the receiving signals of different receiving antennas are independent.

Remark 6.4 (MIMO channel modeling): It is known that the performance of a wireless multiple-input and multiple-output (MIMO) system depends on the propagation channel. The propagation channel models can generally be divided into two different groups: the statistical models based on information theory and the site-specific models based on measurement or numerical simulation. It can be shown that the channel matrix can be identified as the scattering matrix, and (6.77) provides a deterministic approach to the MIMO channel prediction (Geyi, 2007(b)). □

6.6.3 Antenna Input Impedance in a Scattering Environment

The perturbation problems for the antenna input impedance can be studied in a similar manner. We still consider two scenarios for a single antenna system shown in Figure 6.17:

1) *Scenario 1*: The permeability, permittivity and conductivity in the region V_p bounded by S_p are μ, ε and σ respectively, which may vary from point to point. The antenna produces the fields $\mathbf{E}$ and $\mathbf{H}$. The antenna input impedance is Z.
2) *Scenario 2*: The medium parameters μ, ε and σ in the region V_p are changed to μ', ε' and σ' respectively. The antenna produces the fields $\mathbf{E}'$ and $\mathbf{H}'$. The antenna input impedance is Z'.

It will be shown later that the antenna input impedance can be expressed as [see (6.91)]

$$Z = -\frac{1}{I^2} \int_{V_0} \mathbf{E} \cdot \mathbf{J} dV + Z_{internal} \tag{6.82}$$

for scenario 1 and

$$Z' = -\frac{1}{I^2} \int_{V_0} \mathbf{E}' \cdot \mathbf{J} dV + Z_{internal} \tag{6.83}$$

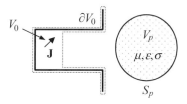

Figure 6.17 Perturbed antenna

for scenario 2. In (6.82) and (6.83), $Z_{internal}$ stands for the internal impedance of the antenna, and V_0 is the source region bounded by ∂V_0, which is coincident with the metal surface of the antenna except for the antenna input terminal. Both the excitation current $\mathbf{J}$ and the internal impedance $Z_{internal}$ are assumed to be constant while medium properties change. Subtracting (6.82) from (6.83), we obtain

$$Z' - Z = \frac{1}{I^2} \int_{V_0} (\mathbf{E}' - \mathbf{E}) \cdot \mathbf{J} dV. \tag{6.84}$$

Let V be the volume bounded by S, which encloses both the region V_p and the antenna. According to (6.72) and the frequency domain reciprocity theorem, Equation (6.84) can be written as

$$Z' - Z = -\frac{1}{I^2} \int_{V} (\mathbf{E}' - \mathbf{E}) \cdot \mathbf{J} dV = -\frac{1}{I^2} \int_{V_p} (\mathbf{J}'_{imp} \cdot \mathbf{E} - \mathbf{J}'_{m,imp} \cdot \mathbf{H}) dV$$

$$= \frac{1}{I^2} \int_{V_p} \left\{ j\omega(\mu' - \mu)\mathbf{H}' \cdot \mathbf{H} - \left[(\sigma' - \sigma + j\omega(\varepsilon' - \varepsilon)) \right] \mathbf{E}' \cdot \mathbf{E} \right\} dV. \tag{6.85}$$

Similarly if we assume that the effect of the changes in permittivity and permeability has negligible effect on the fields, Equation (6.85) can then be approximated by

$$Z' - Z = \frac{1}{I^2} \int_{V_p} \left\{ j\omega(\mu' - \mu)\mathbf{H} \cdot \mathbf{H} - \left[(\sigma' - \sigma + j\omega(\varepsilon' - \varepsilon)) \right] \mathbf{E} \cdot \mathbf{E} \right\} dV. \tag{6.86}$$

6.7   RLC Equivalent Circuits

A RLC circuit is also known as a resonant circuit, and has been widely used in radio engineering. Any one-port electrical network is equivalent to a RLC circuit.

6.7.1   RLC Equivalent Circuit for a One-Port Microwave Network

Consider a one-port network fed by a single-mode waveguide. The one-port microwave network is enclosed by a conducting surface S. We introduce a region V_0 such that its surface ∂V_0 coincides with S except a portion Ω where ∂V_0 crosses the waveguide at the reference plane T, as shown in Figure 6.18 (a). Applying the Poynting theorem to the region V_0 with medium parameters σ, μ, and ε, we have

$$\frac{1}{2} \int_{\partial V_0 - \Omega} (\mathbf{E} \times \bar{\mathbf{H}}) \cdot \mathbf{u}_n dS + \frac{1}{2} \int_{\Omega} (\mathbf{E} \times \bar{\mathbf{H}}) \cdot \mathbf{u}_n dS = -j2\omega(W_m - W_e) - P^{loss}, \tag{6.87}$$

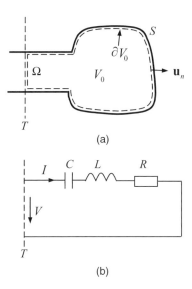

(a)

(b)

Figure 6.18   (a) A one-port network. (b) RLC equivalent circuit

where

$$P^{loss} = \frac{1}{2} \int_{V_0} \sigma \mathbf{E} \cdot \bar{\mathbf{E}} dV, \; W_e = \frac{1}{4} \int_{V_0} \varepsilon \mathbf{E} \cdot \bar{\mathbf{E}} dV, \; W_m = \frac{1}{4} \int_{V_0} \mu \mathbf{H} \cdot \bar{\mathbf{H}} dV \qquad (6.88)$$

are the power loss, total electric energy and total magnetic energy inside V_0 respectively. The first term on the left-hand side is zero due to the boundary condition. For a single-mode waveguide, the second term on the left-hand side of (6.87) is $-V\bar{I}/2$, where V and I are equivalent modal voltage and current at the reference plane. Thus, (6.87) becomes

$$\frac{1}{2}V\bar{I} = P^{loss} + j2\omega(W_m - W_e).$$

The input impedance Z is

$$Z = R^{loss} + j\left(\omega L - \frac{1}{\omega C}\right),$$

where

$$R^{loss} = \frac{2P^{loss}}{|I|^2}, \; L = \frac{4\tilde{W}_m}{|I|^2}, \; C = \frac{|I|^2}{4\omega^2 \tilde{W}_e},$$

are equivalent circuit elements. The equivalent RLC circuit of the one-port microwave network is shown in Figure 6.18 (b).

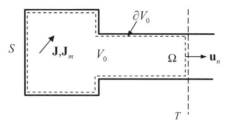

Figure 6.19   An arbitrary source

Remark 6.5: The above discussion is only valid for a bounded microwave system where the electromagnetic fields are confined in the finite region V_0. For an open system, such as an antenna, where the electromagnetic fields radiate into infinity, the total energies W_e and W_m defined by (6.88) become infinite. In this case, we must use stored energies instead of the total energies to derive the RLC equivalent circuit for the open system (see Section 4.4.2).   □

6.7.2   RLC Equivalent Circuits for Current Sources

From the equivalent circuits in network theory, developed by Helmholtz, Léon Charles Thévenin (French engineer, 1857–1926) and Edward Lawry Norton (American engineer, 1898–1983), an arbitrary network may be replaced by an equivalent voltage source in series with an equivalent impedance or alternatively by an equivalent current source in parallel with an equivalent admittance. In the following, we show that these equivalents also exist for general electromagnetic sources. Consider an arbitrary source connected to a feeding waveguide bounded by a conducting surface S, as shown in Figure 6.19. We introduce a source region V_0 such that its surface ∂V_0 coincides with S except a portion Ω where ∂V_0 crosses the waveguide at the reference plane T. All discontinuities and the exciting sources are assumed to be far from the reference plane T so that only the dominant mode exists at the reference plane. In the source region V_0, the generalized Maxwell equations can be expressed as

$$\nabla \times \mathbf{H} = (j\omega\varepsilon + \sigma)\mathbf{E} + \mathbf{J},$$
$$\nabla \times \mathbf{E} = -j\omega\mu\mathbf{H} - \mathbf{J}_m,$$

where μ, ε and σ are the medium parameters. From the above equations, we obtain

$$\nabla \cdot (\mathbf{E} \times \bar{\mathbf{H}}) = -j\omega\mu\,|\mathbf{H}|^2 + j\omega\varepsilon\,|\mathbf{E}|^2 - \sigma\,|\mathbf{E}|^2 - \bar{\mathbf{H}} \cdot \mathbf{J}_m - \mathbf{E} \cdot \bar{\mathbf{J}}. \qquad (6.89)$$

Taking the integration of (6.89) over the region V_0 and using the divergence theorem yield

$$\frac{1}{2}\int_{\partial V_0} (\mathbf{E} \times \bar{\mathbf{H}}) \cdot \mathbf{u}_n dS = -j2\omega(W_m - W_e) - P^{loss} - \frac{1}{2}\int_{V_0} \bar{\mathbf{H}} \cdot \mathbf{J}_m dV - \frac{1}{2}\int_{V_0} \mathbf{E} \cdot \bar{\mathbf{J}} dV,$$

$$(6.90)$$

where the notations are defined in (6.88). For a single-mode feeding waveguide, we have
$\frac{1}{2} \int_{\partial V_0} (\mathbf{E} \times \bar{\mathbf{H}}) \cdot \mathbf{u}_n dS = \frac{1}{2} V \bar{I}$, where V and I are the equivalent modal voltage and current at
the reference plane. From (6.90), we obtain

$$\frac{1}{2} V \bar{I} + \frac{1}{2} R_s |I|^2 + \frac{1}{2} j\omega L_s |I|^2 - \frac{1}{2} j \frac{|I|^2}{\omega C_s} = -\frac{1}{2} \int_{V_0} \bar{\mathbf{H}} \cdot \mathbf{J}_m dV - \frac{1}{2} \int_{V_0} \mathbf{E} \cdot \bar{\mathbf{J}} dV, \quad (6.91)$$

or

$$\frac{1}{2} V \bar{I} + \frac{1}{2} G_p |V|^2 - \frac{1}{2} j\omega C_p |V|^2 - \frac{1}{2} \frac{|V|^2}{j\omega L_p} = -\frac{1}{2} \int_{V_0} \bar{\mathbf{H}} \cdot \mathbf{J}_m dV - \frac{1}{2} \int_{V_0} \mathbf{E} \cdot \bar{\mathbf{J}} dV. \quad (6.92)$$

Here the circuit elements are defined by

$$R_s = \frac{2 P^{loss}}{|I|^2}, L_s = \frac{4 W_m}{|I|^2}, C_s = \frac{|I|^2}{4\omega^2 W_e},$$

$$G_p = \frac{2 P^{loss}}{|V|^2}, L_p = \frac{|V|^2}{4\omega^2 W_m}, C_p = \frac{4 W_e}{|V|^2}.$$

Introducing the voltage and current sources

$$V_s = -\frac{1}{\bar{I}} \int_{V_0} \bar{\mathbf{H}} \cdot \mathbf{J}_m dV, \bar{I}_s = -\frac{1}{V} \int_{V_0} \mathbf{E} \cdot \bar{\mathbf{J}} dV, \quad (6.93)$$

Equations (6.91) and (6.92) can be rewritten as

$$V + R_s I + j\omega L_s I + \frac{1}{j\omega C_s} I = V_s + V \frac{\bar{I}_s}{\bar{I}}, \quad (6.94)$$

$$I + G_p V + j\omega C_p V + \frac{V}{j\omega L_p} = \frac{\bar{V}_s}{\bar{V}} I + I_s. \quad (6.95)$$

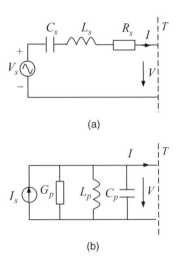

(a)

(b)

Figure 6.20 Equivalent circuits for an arbitrary source

respectively. If there is no electric current source inside the source region, the equivalent circuit can be constructed from (6.94), as shown in Figure 6.20 (a). If there is no magnetic current source inside the source region, an equivalent circuit can be constructed from (6.95), as shown in Figure 6.20 (b).

Example 6.8: For an electric source $\mathbf{J} = \mathbf{u}_z \tilde{I}_s \delta(x)\delta(y)(a < z < b)$, the equivalent current source is

$$\bar{I}_s = -\frac{1}{V}\int_{V_0} \mathbf{E} \cdot \bar{\mathbf{J}} dV = -\frac{\bar{\tilde{I}}_s}{V}\int_a^b \mathbf{E} \cdot \mathbf{u}_z dz = \bar{I}_s.$$

For a ring magnetic source $\mathbf{J}_m = -\mathbf{u}_\varphi \tilde{V}_s \delta(\rho - \rho_0)$, the equivalent voltage source is

$$V_s = -\frac{1}{I}\int_{V_0} \bar{\mathbf{H}} \cdot \mathbf{J}_m dV = \frac{\tilde{V}_s}{I}\int_{\rho=\rho_0} \bar{\mathbf{H}} \cdot \mathbf{u}_\varphi dc = \tilde{V}_s.$$

where $dc = \rho_0 d\varphi$ □

The equivalent circuit for the one-port microwave network and the equivalent circuit for the source can be combined to form the complete equivalent circuit for the whole microwave system, as shown in Figure 6.21.

It is to be noted that all the laws in circuit theory can be derived from Maxwell field equations (Ramo and Whinnery, 1953). There is a correspondence between circuit concepts and field concepts (Harrington, 1961). For example, Kirchhoff's voltage law in circuit theory is a result

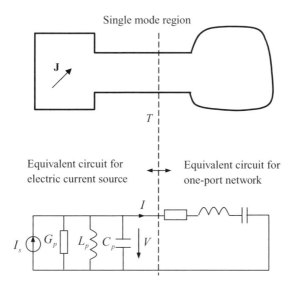

Figure 6.21   Equivalent circuit for a microwave system

of Faraday's law, while Kirchhoff's current law corresponds to the continuity equation. The circuit theory is the simplification and specialization of the field theory.

> The underlying physical laws necessary for the mathematical theory of a large part of physics and the whole of chemistry are thus completely known, and the difficulty is only that the exact application of these laws leads to equations much too complicated to be soluble.
> —Paul Adrien Maurice Dirac (British physicist, 1902–1984)

7

Fields in Inhomogeneous Media

The great success which these eminent men (Gauss, Weber, Riemann, J. & C. Neumann, Lorentz, etc.) attained in the application of mathematics to electrical phenomena, gives, as is natural, additional weight to their theoretical speculations.

—James Maxwell

An electromagnetic medium is called **inhomogeneous** if its electrical parameters are functions of space position, and the wave speed in the medium varies with position. Many transmission media in practice are inhomogeneous, such as the earth's atmosphere and optical fibers. In an inhomogeneous medium, the electromagnetic fields satisfy a partial differential equation with variable coefficients, which is more difficult to solve than a partial differential equation with constant coefficients.

An inhomogeneous waveguide refers to a guided structure, whose medium properties vary across its cross-section but remain constant in the direction of propagation. The reasons for introducing inhomogeneities in a waveguide are twofold. One is to provide mechanical support and the other is to obtain the performances that are not obtainable with homogeneous waveguides. An important feature of the inhomogeneous waveguide is that the free modes are of the hybrid type: they have both longitudinal components of the electric field and the magnetic field and the modes are generally neither transverse electric (TE) nor transverse magnetic (TM). The determination of the dispersion relationship is not as straightforward as the homogeneous waveguide, which requires the solution of a rather involved transcendental equation by numerical methods. When the waveguide is bounded by a metal surface, the number of modes is countable and the waveguide has a discrete spectrum. When the waveguide is open, additional radiating modes exist, which are non-countable. Typical inhomogeneous waveguides include microstrip lines, coplanar waveguides, and dielectric waveguides (optical fibers). In order to obtain a complete picture of the modes in the inhomogeneous waveguides, a sophisticated tool called spectral analysis in operator theory is a necessity.

Most of the time, the differential operator characterizing the electromagnetic boundary value problem can be extended to a self-adjoint operator so that the functional calculus for self-adjoint operators may apply. Equivalently, a variational formulation may be used by introducing a bilinear form associated with the self-adjoint operator. The spectral properties

Foundations of Applied Electrodynamics   Geyi Wen
© 2010 John Wiley & Sons, Ltd

of the self-adjoint operator can be obtained from the knowledge of the corresponding bilinear form.

7.1 Foundations of Spectral Analysis

The spectral theory has wide applications in physics. In quantum mechanics, for example, an observable can only assume values in the spectrum of its corresponding operator. The spectral theory may be applied to study the metal waveguides and metal cavity resonators (see Chapter 3). To deal with more complicated problems in electromagnetic theory, we need several more notions from the spectral theory.

7.1.1 The Spectrum

Let $\hat{A}_\lambda = \lambda \hat{I} - \hat{A}$, and $N(\hat{A}_\lambda)$ and $R(\hat{A}_\lambda)$ be the kernel and range of $\hat{A}_\lambda$ respectively. We recall that the spectrum of an operator $\hat{A}$ defined in a Hilbert space H consists of the point spectrum $\sigma_p(\hat{A})$, the residual spectrum $\sigma_r(\hat{A})$ and the continuous spectrum $\sigma_c(\hat{A})$:

$$\sigma_p(\hat{A}) = \left\{ \lambda \in C \mid N(\hat{A}_\lambda) \neq 0 \right\},$$

$$\sigma_r(\hat{A}) = \left\{ \lambda \in C \mid N(\hat{A}_\lambda) = 0, \overline{R(\hat{A}_\lambda)} \neq H \right\},$$

$$\sigma_c(\hat{A}) = \left\{ \lambda \in C \mid N(\hat{A}_\lambda) = 0, \overline{R(\hat{A}_\lambda)} = H, \hat{A}_\lambda^{-1} \text{ is unbounded} \right\},$$

where C is the set of all complex numbers. If $\hat{A}$ is self-adjoint, we have $\sigma_r(\hat{A}) = 0$. If $\hat{A}$ is a self-adjoint operator, we may introduce

$$\sigma_{ess}(\hat{A}) = \left\{ \lambda \in \sigma(\hat{A}) \mid \lambda \in \sigma_c(\hat{A}), \text{ or } \lambda \in \sigma_p(\hat{A}) \text{ with } \dim N(\hat{A}_\lambda) = +\infty \right\},$$

$$\sigma_{disc}(\hat{A}) = \left\{ \lambda \in \sigma_p(\hat{A}) \mid \dim N(\hat{A}_\lambda) < +\infty \right\},$$

which are called the **essential spectrum** and the **discrete spectrum** of operator $\hat{A}$ respectively. Some basic properties of the resolvent and the spectrum of a linear operator $\hat{A}$ defined in a Hilbert space H with a norm $\|\cdot\| = (\cdot, \cdot)^{1/2}$ are summarized below:

1. $\rho(\hat{A})$ is an open set in the complex plane.
2. $\sigma(\hat{A})$ is a closed set in the complex plane.
3. If $\hat{A}$ is self-adjoint, all nonreal numbers are in $\rho(\hat{A})$.
4. If $\hat{A}$ is self-adjoint, a real number λ is in $\sigma(\hat{A})$ if and only if there exists a sequence $\{u_n\} \subset D(\hat{A})$ (the domain of $\hat{A}$) such that $\|u_n\| = 1$ and $\|\hat{A}_\lambda u_n\| \to 0$ as $n \to \infty$.

Example 7.1 (Position operator): The position operator $\hat{A} = x$ is defined by $\hat{A}u(x) = xu(x)$ whose domain $D(\hat{A})$ is the set of those $u \in L^2(R)$ such that $xu \in L^2(R)$. It is easy to show that $D(\hat{A})$ is dense in $L^2(R)$ and $\hat{A}$ is self-adjoint. Consider the eigenvalue equation $(\hat{A} - \lambda\hat{I})u = (x - \lambda)u = 0$. This equation holds only if $u = 0$ for $x \neq \lambda$, which implies $u = 0$ almost everywhere, and there are no eigenvalues. Let λ be any real number, and we can try the sequence

$u_n = c_n \exp[-n^2(x - \lambda)^2/2]$ where c_n is chosen so that $\|u_n\| = 1$. Thus, $c_n = \sqrt{n}\pi^{-1/4}$. Since

$$\|(x - \lambda)u_n\|^2 = c_n^2 n^{-3} \int_{-\infty}^{\infty} x^2 e^{-x^2} dx \to 0,$$

we have $\lambda \in \sigma(\hat{A})$. Thus $\sigma(\hat{A})$ consists of the whole real axis. □

Let H be a Hilbert space. The sequence $\{x_n\}$ is said to converge weakly to x (**weak convergence**) if $\lim_{n\to\infty} (x_n, y) = (x, y)$ for all $y \in H$. This relation is denoted by $x_n \xrightarrow[n\to\infty]{w} x$. If a linear map $\hat{A} : H \to H$ is continuous, it is weakly continuous, that is, if $x_n \xrightarrow[n\to\infty]{w} x$, then $\hat{A}x_n \xrightarrow[n\to\infty]{w} \hat{A}x$. If $\{x_n\}$ is a bounded sequence in a separable Hilbert space, there exists a subsequence $\{x_{n_k}\}$ and a vector $x \in H$ such that $x_{n_k} \xrightarrow[k\to\infty]{w} x$.

The following theorems are important in studying the spectrum of a self-adjoint operator (for example, Schechter, 1981).

Theorem 7.1: *If $\hat{A}$ is self-adjoint, a number λ is in $\sigma_{ess}(\hat{A})$ if and only if there exists a sequence $\{u_n\} \subset D(\hat{A})$ (the domain of $\hat{A}$) such that*

1. $\|u_n\| = 1$,
2. $\{u_n\}$ has no convergent subsequence (or equivalently u_n converges to zero weakly), and
3. $\|\hat{A}_\lambda u_n\| \to 0$ as $n \to \infty$.

The sequence in the above theorem is called the **singular sequence**. An operator $\hat{B}$ is called **compact relative to operator** $\hat{A}$ or $\hat{A}$-compact if $D(\hat{A}) \subset D(\hat{B})$ and $\|u_n\| + \|\hat{A}u_n\| \le c_1$ (c_1 is a constant) implies that $\{\hat{B}u_n\}$ has a convergent subsequence.

Weyl theorem (named after the German mathematician Hermann Klaus Hugo Weyl, 1885–1955): *If $\hat{A}$ is a self-adjoint and $\hat{B}$ is symmetric and $\hat{A}$-compact, then*

1. $\hat{A} + \hat{B}$ is self-adjoint, and
2. $\sigma_{ess}(\hat{A} + \hat{B}) = \sigma_{ess}(\hat{A})$.

7.1.2 Spectral Theorem

An operator $\hat{P}$ is called an **orthogonal projection** if it is bounded and symmetric, and satisfies $\hat{P}^2 = \hat{P}$. A self-adjoint operator may be decomposed into the sum of the orthogonal projections.

Spectral theorem: *Let $\hat{A}$ be a self-adjoint operator on a Hilbert space H. Then there is a family $\{\hat{E}(\lambda)\}$ of orthogonal projections depending on a real parameter λ, called the **spectral family** of $\hat{A}$, such that*

1. If $\lambda_1 < \lambda_2$, then $\hat{E}(\lambda_1)\hat{E}(\lambda_2) = \hat{E}(\lambda_1)$.
2. For each $u \in H$ and real λ, $\hat{E}(\lambda + \varepsilon)u \to \hat{E}(\lambda)u$ as $\varepsilon \to 0^+$.

3. $u \in D(\hat{A})$ if and only if $\int\limits_{-\infty}^{\infty} \lambda^2 d \left\| \hat{E}(\lambda)u \right\|^2 < \infty.$

4. For $u \in D(\hat{A})$ and $v \in H$, we have $(\hat{A}u, v) = \int\limits_{-\infty}^{\infty} \lambda d(\hat{E}(\lambda)u, v).$

5. For any complex-valued function $f(\lambda)$, the operator

$$f(\hat{A}) = \int\limits_{-\infty}^{\infty} f(\lambda)d\hat{E}(\lambda) \qquad (7.1)$$

is defined on the set $D[f(\hat{A})]$ consisting of those $u \in H$ such that

$$\int\limits_{-\infty}^{\infty} |f(\lambda)|^2 d \left\| \hat{E}(\lambda)u \right\|^2 < \infty.$$

6. For $u \in D[f(\hat{A})], v \in D[g(\hat{A})]$

$$(f(\hat{A})u, g(\hat{A})v) = \int\limits_{-\infty}^{\infty} f(\lambda)\overline{g(\lambda)}d(\hat{E}(\lambda)u, v).$$

7.
$$f(\hat{A})^* = \int\limits_{-\infty}^{\infty} \overline{f(\lambda)}d\hat{E}(\lambda).$$

8. If I is the interval $(-\infty, \lambda]$, then $\hat{E}(\lambda) = \chi_I(\hat{A})$, χ_I is the characteristic function of the interval I

$$\chi_I(\lambda) = \begin{cases} 1, \lambda \in I \\ 0, \lambda \notin I \end{cases}.$$

7.1.3 Generalized Eigenfunctions of Self-Adjoint Operators

Let $\hat{A}$ be a self-adjoint operator on Hilbert space H, and $u \in H$. If its eigenfunctions u_n form a complete orthonormal set, we have the eigenfunction expansions

$$u = \sum_n (u, u_n)u_n, \quad \hat{A}u = \sum_n \lambda_n(u, u_n)u_n. \qquad (7.2)$$

It should be noted that such expansions do not always exist for a general self-adjoint operator. However, the following spectral decomposition always holds for a general self-adjoint operator

$$\hat{A} = \int\limits_{-\infty}^{\infty} \lambda d\hat{E}(\lambda). \qquad (7.3)$$

In the theory of differential equations, the eigenfunction expansions (7.2) are more convenient than the spectral decomposition (7.3), and it would be useful to extend (7.2) to a general self-adjoint operator. In the following, we only discuss the situation where $\hat{A}$ is a differential operator.

Let $\hat{A}$ be a self-adjoint operator in $L^2(R^N)$. Its domain $D(\hat{A})$ contains the fundamental space $D(R^N)$. We further assume that $\hat{A}$ is continuous from $D(R^N)$ to $D(R^N)$, and is real, that is, $\hat{A}u = \overline{\hat{A}u}$. Let $D'(R^N)$ denote the dual space of $D(R^N)$. We introduce a conjugate operator $\hat{A}'$ from $D'(R^N)$ to $D'(R^N)$ such that, for $T \in D'(R^N)$ and all $u \in D(R^N)$, we have

$$\langle \hat{A}'T, u \rangle = \langle T, \hat{A}u \rangle,$$

where $\langle T, u \rangle = T(u)$. Especially for $T \in D(\hat{A})$, we have

$$\langle \hat{A}'T, u \rangle = \langle T, \hat{A}u \rangle = \int_{R^N} T(\mathbf{x})(\hat{A}u)(\mathbf{x})d\mathbf{x}.$$

Since $\hat{A}$ is real and self-adjoint in $L^2(R^N)$, this can be rewritten as

$$\langle \hat{A}'T, u \rangle = \int_{R^N} T(\mathbf{x})(\hat{A}u)(\mathbf{x})d\mathbf{x} = \int_{R^N} (\hat{A}T)(\mathbf{x})u(\mathbf{x})d\mathbf{x} = \langle \hat{A}T, u \rangle.$$

Thus $\hat{A}'T = \hat{A}T$ for $T \in D(\hat{A})$, and $\hat{A}'$ is an extension of $\hat{A}$, $\hat{A} \subset \hat{A}'$. The generalized function $T \in D'(R^N)$ is called a **generalized eigenfunction** of operator $\hat{A}$ with the eigenvalue λ if $\hat{A}'T = \lambda T$. Apparently, the generalized eigenfunction becomes the eigenfunction of operator $\hat{A}$ when $T \in D(\hat{A})$.

Assume that $\hat{A}$ has a finite or countable set of eigenvalues $\{\lambda_n\}$ and the corresponding orthonormal eigenfunctions are denoted by $\{u_n\} \subset D(\hat{A})$. Furthermore, we assume that for each natural number n, there is a set $B_n \subset R$ such that the operator $\hat{A}$ has a generalized eigenfunction u_λ corresponding to each $\lambda \in B_n$. The system $\{u_n\} \cup \{u_\lambda | \lambda \in B_n, n = 1, 2, \cdots \}$ is said to form a complete orthonormal set of generalized eigenfunctions if for all $u, v \in D(R^N)$, we have

$$\int_{R^N} u(\mathbf{x})\bar{v}(\mathbf{x})d\mathbf{x} = \sum_n \langle u_n, u \rangle \overline{\langle u_n, v \rangle} + \sum_n \int_{B_n} \langle u_\lambda, u \rangle \overline{\langle u_\lambda, v \rangle}d\lambda. \tag{7.4}$$

Here some or all B_n may be empty, and they may have intersections. Formally (7.4) can be written as

$$u(\mathbf{x}) = \sum_n \langle u_n, u \rangle u_n + \sum_n \int_{B_n} \langle u_\lambda, u \rangle u_\lambda d\lambda, \tag{7.5}$$

or

$$\delta(\mathbf{x} - \boldsymbol{\xi}) = \sum_n u_n(\mathbf{x})u_n(\boldsymbol{\xi}) + \sum_n \int_{B_n} u_\lambda(\mathbf{x})u_\lambda(\boldsymbol{\xi})d\lambda. \tag{7.6}$$

It follows that

$$\hat{A}u(\mathbf{x}) = \sum_n \langle u_n, u \rangle \lambda_n u_n + \sum_n \int_{B_n} \langle u_\lambda, u \rangle u_\lambda \lambda d\lambda. \tag{7.7}$$

7.1.4   Bilinear Forms

Let H be a Hilbert space with inner product $(\cdot, \cdot)$ and the induced norm $\|\cdot\| = (\cdot, \cdot)^{1/2}$. A **bilinear form** $b(u, v)$ is a scalar function linear in u and conjugate linear in v defined for u, v in some subspace $D(b)$ of H

$$b(\alpha x + \beta y, v) = \alpha b(x, v) + \beta b(y, v),$$
$$b(x, \alpha u + \beta v) = \bar{\alpha}b(x, u) + \bar{\beta}b(x, v),$$

where α and β are complex numbers. The subspace $D(b)$ is called the **domain** of b. We will denote $b(u, u)$ by $b(u)$. A bilinear form is said to be **closed** if

$$u_n \to u, b(u_n - u_m) \to 0, \{u_n\} \subset D(b)$$

implies $u \in D(b)$ and $b(u_n - u) \to 0$. A bilinear form is called **Hermitian** (after French mathematician Charles Hermite, 1822–1901) if

$$b(u, v) = \overline{b(v, u)}, u, v \in D(b).$$

If $b(u, u) \geq 0$ for all $u \in D(b)$, b is called **positive**. If $b(u, u) \geq -c_1 \|u\|^2$ holds for all $u \in D(b)$ for some c_1, b is said to be **bounded from below**. Notice that b is automatically Hermitian if H is complex and b is bounded from below.

If the domain of a bilinear form is dense in H, we can define an operator $\hat{B}$ as follows. We say that $u \in D(\hat{B})$ and $\hat{B}u = f$ if and only if $u \in D(b)$ and

$$b(u, v) = (f, v), v \in D(b).$$

The operator $\hat{B}$ is linear and is called the **operator associated with** b.

Theorem 7.2: *Let b be a closed Hermitian bilinear form with dense domain in H. If there exists some constant c_1 such that $b(u) \geq c_1\|u\|^2$, $u \in D(b)$, the operator $\hat{B}$ associated with b is self-adjoint and $\sigma(\hat{B}) \subset [c_1, \infty)$.*

Theorem 7.3: *Let a be a closed Hermitian bilinear form with dense domain in H and b be a Hermitian bilinear form such that $D(a) \subset D(b)$ and $|b(u)| \leq c_1 a(u), u \in D(a)$. Assume that* *every sequence $\{u_n\} \subset D(a)$ satisfying $\|u_n\|^2 + a(u_n) \leq c_2$ has a subsequence $\{v_j\}$ such that $b(v_j - v_k) \to 0$, $j, k \to \infty$. Assume also that, for such a subsequence $\{v_j\}$, $v_j \to 0$ implies that $b(v_j) \to 0$. Let $c(u) = a(u) + b(u)$ and $\hat{B}$ and $\hat{C}$ be operators associated with b and c respectively. Then $\sigma_{ess}(\hat{C}) = \sigma_{ess}(\hat{A})$.*

Proofs of the above theorems can be found in (Schechter, 1981).

Example 7.2 (Dirichlet problem for second-order elliptical equation): Let Ω be a bounded region in R^N and we consider the second-order operator defined in Ω

$$\hat{L} = -\sum_{i=1}^{N}\sum_{j=1}^{N} \frac{\partial}{\partial x_j}\left(a_{ij}(\mathbf{x})\frac{\partial}{\partial x_i}\right) + \sum_{i=1}^{N} b_i(\mathbf{x})\frac{\partial}{\partial x_i} + c(\mathbf{x}), \tag{7.8}$$

where $a_{ij}(\mathbf{x})$ are symmetric, that is, $a_{ij}(\mathbf{x}) = a_{ji}(\mathbf{x})$, and $a_{ij}(\mathbf{x}) \in C^1(\bar{\Omega})$, $b_i(\mathbf{x})$, $c(\mathbf{x}) \in C(\bar{\Omega})$. If there exists a positive constant γ such that

$$\sum_{i=1}^{N}\sum_{j=1}^{N} a_{ij}(\mathbf{x})\xi_i\xi_j \leq \gamma \sum_{i=1}^{N} \xi_i^2, \mathbf{x} \in \bar{\Omega}, \, \xi = (\xi_1, \xi_2, \cdots, \xi_N) \in R^N,$$

the operator $\hat{L}$ is said to be uniformly elliptic. Let $\hat{L}$ be uniformly elliptic and consider the following Dirichlet problem

$$\hat{L}u(\mathbf{x}) = f(\mathbf{x}), \mathbf{x} \in \Omega, \, u(\mathbf{x})|_\Gamma = 0, \tag{7.9}$$

where Γ is the boundary of Ω, and $f \in C(\Omega)$. The set of classical solutions of the Dirichlet problem is

$$D(\hat{L}) = \left\{u \mid u \in C^2(\Omega) \cap C(\bar{\Omega}), \, u|_\Gamma = 0\right\}. \tag{7.10}$$

For a classical solution u, we may introduce the bilinear form on $D(\hat{L})$

$$b(u, v) = \int_\Omega \hat{L}u \cdot v d\mathbf{x}$$

for all $v \in D(\hat{L})$. It follows from integration by parts that

$$b(u, v) = \int_\Omega \left[\sum_{i=1}^{N}\sum_{j=1}^{N} a_{ij}(\mathbf{x})\frac{\partial u}{\partial x_i}\frac{\partial v}{\partial x_j} + \sum_{i=1}^{N} b_i(\mathbf{x})\frac{\partial u}{\partial x_i}v + c(\mathbf{x})uv\right] d\mathbf{x} = \int_\Omega f v d\mathbf{x} \tag{7.11}$$

for all $v \in D(\hat{L})$. Thus a classical solution satisfies the above relation. If u is not a classical solution or $f \notin C(\Omega)$, u may still satisfy the above equation. A natural extension of the set of classical solution is $H_0^1(\Omega)$. Let $f \in L^2(\Omega)$. If there exists $u \in H_0^1(\Omega)$ such that

$$\int_\Omega \left[\sum_{i=1}^N \sum_{j=1}^N a_{ij}(\mathbf{x}) \frac{\partial u}{\partial x_i} \frac{\partial v}{\partial x_j} + \sum_{i=1}^n b_i(\mathbf{x}) \frac{\partial u}{\partial x_i} v + c(\mathbf{x}) uv \right] dx = \int_\Omega f v dx \qquad (7.12)$$

for all $v \in H_0^1(\Omega)$, the function u is a weak solution of the Dirichlet problem. $\qquad \square$

Example 7.3: Let $H = [L^2(R^2)]^3$ and consider the bilinear form

$$b(\mathbf{h}_1, \mathbf{h}_2) = \int_{R^2} \left(\sum_{i=x,y,z} \nabla_t h_{1i} \cdot \nabla_t \bar{h}_{2i} + \beta^2 \mathbf{h}_1 \cdot \bar{\mathbf{h}}_2 \right) dxdy,$$

where $\nabla_t = \mathbf{u}_x \partial/\partial x + \mathbf{u}_y \partial/\partial y$ denotes the two-dimensional gradient and all the derivatives are understood in the generalized sense, $\mathbf{h}_i = (h_{ix}, h_{iy}, h_{iz})(i = 1, 2)$ are vector functions defined in (x, y)-plane, and β is a positive constant. If we choose $D(b)$ as the Sobolev space $[H^1(R^2)]^3$, then b is a closed Hermitian bilinear form with dense domain. $\qquad \square$

The functions in Sobolev spaces can be approximated by smooth functions. Let Ω be a bounded open set. Then

1. $C^\infty(\Omega)$ is dense in $H^m(\Omega)(m \geq 0)$.
2. $C_0^\infty(\Omega)$ is dense in $L^p(\Omega)$.
3. $C_0^\infty(R^N)$ is dense in $H^m(R^N)(m \geq 0)$.
4. $H_0^m(R^N) = H^m(R^N)$.

Note that $C_0^\infty(\Omega)$ is not dense in $H^m(\Omega)$ if Ω is a bounded open set.

7.1.5 Min-Max Principle

Let $\hat{B}$ be a self-adjoint operator that is bounded from below $(\hat{B}u, u) \geq c_1 \|u\|^2$. Define

$$\lambda_n(\hat{B}) = \sup_{u_1, u_2, \cdots, u_{n-1}} \quad \inf_{\substack{u \in D(\hat{B}), \|u\|=1 \\ u \in [u_1, u_2, \cdots, u_{n-1}]^\perp}} \quad (\hat{B}u, u), \qquad (7.13)$$

where $[u_1, u_2, \cdots, u_m]^\perp = \{u | (u, u_i) = 0, i = 1, 2, \cdots, m\}$. Note that the u_i are not necessarily independent. For each fixed n, the **min-max principle** states that

either

1. there are n eigenvalues (counting degenerate eigenvalues a number of times equal to their multiplicity) below the bottom of the essential spectrum, and $\lambda_n(\hat{B})$ is the nth eigenvalue;

or

2. λ_n is the bottom of the essential spectrum, that is, $\lambda_n = \inf\{\lambda \mid \lambda \in \sigma_{ess}(\hat{B})\}$ and in this case $\lambda_n = \lambda_{n+1} = \lambda_{n+2} = \cdots$, and there are at most $n - 1$ eigenvalues (counting multiplicity) below λ_n.

The min-max principle can be used to compare the eigenvalues of operators, to locate where σ_{ess} begins, and to prove the existence of eigenvalues.

7.1.6 A Bilinear Form for Maxwell Equations

Let $V \in R^3$ be a bounded domain with boundary S. Let $L^2(V)$ be the space of square integrable functions defined in V with the usual inner product $(u, v)_V = \int_V u\bar{v}\,dV$ and $L^2(S)$ the space of square integrable function defined on the boundary S with the inner product $(u, v)_S = \int_S u\bar{v}\,dS$. For vector functions $\mathbf{a}, \mathbf{b} \in [L^2(V)]^3$, the inner product is defined by $(\mathbf{a}, \mathbf{b})_V = \int_V \mathbf{a} \cdot \bar{\mathbf{b}}\,dV$. Similarly we can define $(\mathbf{a}, \mathbf{b})_S = \int_S \mathbf{a} \cdot \bar{\mathbf{b}}\,dS$, $\mathbf{a}, \mathbf{b} \in [L^2(S)]^3$. The Green's identities can thus be written as

$$(\nabla \times \mathbf{a}, \mathbf{b})_V - (\mathbf{a}, \nabla \times \mathbf{b})_V = (\mathbf{u}_n \times \mathbf{a}, \mathbf{b})_S,$$
$$(\nabla \cdot \mathbf{a}, \varphi)_V + (\mathbf{a}, \nabla\varphi)_V = (\mathbf{u}_n \cdot \mathbf{a}, \varphi)_S. \tag{7.14}$$

where $\mathbf{u}_n$ is the unit outward normal of S. If the region V is source free, the electric field in V satisfies

$$\nabla \times \overset{\leftrightarrow}{\mu}^{-1} \cdot \nabla \times \mathbf{E}(\mathbf{r}) - \omega^2 \overset{\leftrightarrow}{\varepsilon} \cdot \mathbf{E}(\mathbf{r}) = 0,$$

where $\overset{\leftrightarrow}{\mu}(\mathbf{r})$ and $\overset{\leftrightarrow}{\varepsilon}(\mathbf{r})$ are permeability and permittivity tensors (3×3 matrices) of the medium in V respectively, and they are functions of position. Since $\nabla \cdot (\overset{\leftrightarrow}{\varepsilon} \cdot \mathbf{E}) = 0$, the above equation can be regularized as

$$\hat{B}(\mathbf{E}) = \nabla \times \overset{\leftrightarrow}{\mu}^{-1} \cdot \nabla \times \mathbf{E}(\mathbf{r}) - \overset{\leftrightarrow}{\varepsilon}^* \nabla \tau \nabla \cdot (\overset{\leftrightarrow}{\varepsilon} \cdot \mathbf{E}) = \omega^2 \overset{\leftrightarrow}{\varepsilon} \cdot \mathbf{E}(\mathbf{r}),$$

where $\hat{B} = \nabla \times \overset{\leftrightarrow}{\mu}^{-1} \cdot \nabla \times \cdot - \overset{\leftrightarrow}{\varepsilon}^* \nabla \tau \nabla \cdot (\overset{\leftrightarrow}{\varepsilon}\cdot)$; $*$ denotes the adjoint; and τ is an arbitrary function. For $\mathbf{E}, \mathbf{F} \in [C^2(V)]^3$, we have

$$(\hat{B}(\mathbf{E}), \mathbf{F})_V = b(\mathbf{E}, \mathbf{F}) + (\mathbf{u}_n \times (\overset{\leftrightarrow}{\mu}^{-1} \cdot \nabla \times \mathbf{E}), \mathbf{F})_S - (\tau \nabla \cdot (\overset{\leftrightarrow}{\varepsilon} \cdot \mathbf{F}), \mathbf{u}_n \cdot (\overset{\leftrightarrow}{\varepsilon} \cdot \mathbf{F}))_S, \tag{7.15}$$

where

$$b(\mathbf{E}, \mathbf{F}) = \int_V \left\{ (\overset{\leftrightarrow}{\mu}^{-1} \cdot \nabla \times \mathbf{E}) \cdot (\nabla \times \bar{\mathbf{F}}) + \tau[\nabla \cdot (\overset{\leftrightarrow}{\varepsilon} \cdot \mathbf{E})][\nabla \cdot \overline{(\overset{\leftrightarrow}{\varepsilon} \cdot \mathbf{F})}] \right\} dV$$

is a bilinear form. It can be shown that this bilinear form is bounded from below (Costabel, 1991). Let $H = \{\mathbf{E} \in [H^1(V)]^3 \,|\, \mathbf{u}_n \times \mathbf{E} = 0 \text{ on } S\}$. For the operator equation

$$\hat{B}(\mathbf{E}) = \mathbf{f}, \mathbf{E} \in [C^2(V)]^3, \mathbf{f} \in [L^2(V)]^3,$$
$$\mathbf{u}_n \times \mathbf{E} = 0 \text{ on } S,$$

we may introduce the weak formulation (variational formulation): Find $\mathbf{E} \in H$ such that

$$b(\mathbf{E}, \mathbf{F}) = (\mathbf{f}, \mathbf{F}) \tag{7.16}$$

for $\mathbf{f} \in [L^2(V)]^3$ and for all $\mathbf{F} \in H$. From (7.15), we can see that $\mathbf{E}$ satisfies the natural boundary condition

$$\nabla \cdot (\ddot{\varepsilon} \cdot \mathbf{E}) = 0, \text{ on } S. \tag{7.17}$$

7.2   Plane Waves in Inhomogeneous Media

The calculation of fields in an inhomogeneous medium is a very difficult task. Exact solutions can be obtained only in some simple situations, one of which is when the medium parameters depend on one of the coordinates only.

7.2.1   Wave Equations in Inhomogeneous Media

Assume that the medium is inhomogeneous and isotropic so that $\mathbf{D} = \varepsilon \mathbf{E}$ and $\mathbf{B} = \mu \mathbf{H}$ in frequency domain. The wave equations for the time-harmonic fields in a source-free region are

$$\nabla \times \mu^{-1} \nabla \times \mathbf{E}(\mathbf{r}) - \omega^2 \varepsilon \mathbf{E}(\mathbf{r}) = 0,$$
$$\nabla \times \varepsilon^{-1} \nabla \times \mathbf{H}(\mathbf{r}) - \omega^2 \mu \mathbf{H}(\mathbf{r}) = 0. \tag{7.18}$$

Let us consider the case where the medium parameters depend on one coordinate only, say, z. We may write $\mu = \mu(z)$ and $\varepsilon = \varepsilon(z)$. If the electric field has only one component such that

$$\mathbf{E} = (E_x, E_y, E_z) = (0, E_y, 0),$$

we have $\nabla \cdot \mathbf{D} = \partial(\varepsilon E_y)/\partial y = 0$, which is equivalent to $\partial E_y/\partial y = 0$. From the first equation of (7.18) we obtain

$$\left[\frac{\partial^2}{\partial x^2} + \mu(z)\frac{\partial}{\partial z}\mu^{-1}(z)\frac{\partial}{\partial z} + \omega^2 \mu(z)\varepsilon(z) \right] E_y = 0. \tag{7.19}$$

Similarly if the magnetic field has only one component such that

$$\mathbf{H} = (H_x, H_y, H_z) = (0, H_y, 0),$$

we obtain

$$\left[\frac{\partial^2}{\partial x^2} + \varepsilon(z)\frac{\partial}{\partial z}\varepsilon^{-1}(z)\frac{\partial}{\partial z} + \omega^2\mu(z)\varepsilon(z)\right]H_y = 0. \tag{7.20}$$

7.2.2 Waves in Slowly Varying Layered Media and WKB Approximation

Both (7.19) and (7.20) are differential equations with variable coefficients. By the method of separation of variables, we may let

$$E_y(x, z) = e_y(z)e^{\pm jk_x x}, \ H_y(x, z) = h_y(z)e^{\pm jk_x x}.$$

Introducing these into (7.19) and (7.20) yields

$$\left[\mu(z)\frac{d}{dz}\mu^{-1}(z)\frac{d}{\partial z} + \omega^2\mu(z)\varepsilon(z) - k_x^2\right]e_y = 0,$$
$$\left[\varepsilon(z)\frac{\partial}{\partial z}\varepsilon^{-1}(z)\frac{\partial}{\partial z} + \omega^2\mu(z)\varepsilon(z) - k_x^2\right]h_y = 0. \tag{7.21}$$

If $\mu(z)$ and $\varepsilon(z)$ are slowly varying functions of z, Equations (7.21) may be approximated by

$$\frac{d^2\psi}{dz^2} + k^2(z)\psi = 0 \tag{7.22}$$

where ψ denotes e_y or h_y, and $k^2(z) = \omega^2\mu(z)\varepsilon(z) - k_x^2$ is a slowly varying function of z. We assume that the solution is of the form

$$\psi(z) = A(z)e^{-j\phi(z)} \tag{7.23}$$

and substitute it into (7.22), to obtain

$$A''(z) + \left\{k^2(z) - [\phi'(z)]^2\right\}A(z) - 2j\phi'(z)A'(z) - j\phi''(z)A(z) = 0,$$

where the prime denotes the derivative with respect to z. The real and imaginary parts of the left-hand side of the above equation must be zero, yielding

$$A''(z) + \{k^2(z) - [\phi'(z)]^2\}A(z) = 0,$$
$$\phi''(z)A(z) + 2\phi'(z)A'(z) = 0. \tag{7.24}$$

We further assume that A''/A is much smaller than the difference $k^2 - (\phi')^2$, and the upper equation of (7.24) may be approximated by

$$k^2(z) - [\phi'(z)]^2 = 0.$$

The solution is

$$\phi(z) = \pm \int_{z_0}^{z} k(z)dz, \tag{7.25}$$

where z_0 is some initial value. Using (7.25), the lower equation of (7.24) can be written as

$$\frac{A'}{A} = -\frac{\phi''}{2\phi'} = -\frac{k'}{2k}.$$

The solution of the equation is

$$A(z) = A(z_0)\sqrt{\frac{k(z_0)}{k(z)}}. \tag{7.26}$$

Substituting (7.25) and (7.26) into (7.23) gives the approximate solution

$$\psi(z) = \frac{1}{\sqrt{k(z)}}\left\{ C_0 \exp\left[-j\int_{z_0}^{z} k(z)dz\right] + D_0 \exp\left[j\int_{z_0}^{z} k(z)dz\right]\right\}, \tag{7.27}$$

where C_0, D_0 and z_0 are constants. The above procedure is known as **WKB approximation**, named after the German physicist Gregor Wentzel (1898–1978), the Dutch physicist Hendrik Anthony Kramers (1894–1952), and the French physicist Léon Brillouin, who developed the method in 1926.

7.2.3   High Frequency Approximations and Geometric Optics

The time-harmonic Maxwell equations in an isotropic inhomogeneous medium take the form

$$\nabla \times \mathbf{H}(\mathbf{r}) = j\omega\varepsilon(\mathbf{r})\mathbf{E}(\mathbf{r}),$$
$$\nabla \times \mathbf{E}(\mathbf{r}) = -j\omega\mu(\mathbf{r})\mathbf{H}(\mathbf{r}), \tag{7.28}$$
$$\nabla \cdot [\varepsilon(\mathbf{r})\mathbf{E}(\mathbf{r})] = 0, \nabla \cdot [\varepsilon(\mathbf{r})\mathbf{H}(\mathbf{r})] = 0.$$

The **refractive index** n of the medium is defined by $n = \sqrt{\mu\varepsilon/\mu_0\varepsilon_0}$, where μ_0 and ε_0 are the permeability and permittivity in free space. The wavenumber in free space will be denoted by $k_0 = \omega\sqrt{\mu_0\varepsilon_0}$. Similar to (7.23), we assume that

$$\mathbf{E} = \mathbf{E}_0(\mathbf{r})e^{-jk_0L(\mathbf{r})}, \mathbf{H} = \mathbf{H}_0(\mathbf{r})e^{-jk_0L(\mathbf{r})}. \tag{7.29}$$

The function $L(\mathbf{r})$ is known as **eikonal**. The **wavefronts** are defined as the surfaces of constant phase: $L(\mathbf{r}) = \text{const}$. Substituting (7.29) into (7.28), we obtain

$$
\begin{aligned}
&\mathbf{H}_0(\mathbf{r}) \times \nabla L(\mathbf{r}) - \frac{\omega \varepsilon(\mathbf{r})}{k_0} \mathbf{E}_0(\mathbf{r}) = j \frac{1}{k_0} \nabla \times \mathbf{H}_0(\mathbf{r}), \\
&\mathbf{E}_0(\mathbf{r}) \times \nabla L(\mathbf{r}) + \frac{\omega \mu(\mathbf{r})}{k_0} \mathbf{H}_0(\mathbf{r}) = j \frac{1}{k_0} \nabla \times \mathbf{E}_0(\mathbf{r}), \\
&\mathbf{E}_0(\mathbf{r}) \cdot \nabla L(\mathbf{r}) = \frac{1}{jk_0} \left[\mathbf{E}_0(\mathbf{r}) \cdot \nabla \ln \varepsilon(\mathbf{r}) + \nabla \cdot \mathbf{E}_0(\mathbf{r}) \right], \\
&\mathbf{H}_0(\mathbf{r}) \cdot \nabla L(\mathbf{r}) = \frac{1}{jk_0} \left[\mathbf{H}_0(\mathbf{r}) \cdot \nabla \ln \mu(\mathbf{r}) + \nabla \cdot \mathbf{H}_0(\mathbf{r}) \right].
\end{aligned}
\tag{7.30}
$$

If the frequency is very high, k_0 becomes very large and the right-hand side of (7.30) can be equated to zero. There results

$$
\begin{aligned}
&\mathbf{H}_0(\mathbf{r}) \times \nabla L(\mathbf{r}) - \frac{\omega \varepsilon(\mathbf{r})}{k_0} \mathbf{E}_0(\mathbf{r}) = 0, \\
&\mathbf{E}_0(\mathbf{r}) \times \nabla L(\mathbf{r}) + \frac{\omega \mu(\mathbf{r})}{k_0} \mathbf{H}_0(\mathbf{r}) = 0, \\
&\mathbf{E}_0(\mathbf{r}) \cdot \nabla L(\mathbf{r}) = 0, \\
&\mathbf{H}_0(\mathbf{r}) \cdot \nabla L(\mathbf{r}) = 0.
\end{aligned}
\tag{7.31}
$$

The last two equations show that $\mathbf{E}_0$ and $\mathbf{H}_0$ are transverse to ∇L, that is, transverse to the direction of propagation of the wavefront. From the first two equations of (7.31), it is easy to see that $\mathbf{E}_0 \cdot \mathbf{H}_0 = 0$. Therefore, the field is locally a plane wave. If $\mathbf{H}_0(\mathbf{r})$ is eliminated from the first two equations of (7.31), then

$$
n^2(\mathbf{r})\mathbf{E}_0(\mathbf{r}) + [\nabla L(\mathbf{r}) \cdot \mathbf{E}_0(\mathbf{r})]\nabla L(\mathbf{r}) - [\nabla L(\mathbf{r})]^2 \mathbf{E}_0(\mathbf{r}) = 0.
$$

The second term is zero due to the third equation of (7.31). Thus, if $\mathbf{E}_0$ is not identically zero, it is necessary that

$$
[\nabla L(\mathbf{r})]^2 = n^2(\mathbf{r}).
\tag{7.32}
$$

This is called an **eikonal equation**.

Making use of the second equation of (7.31), the Poynting vector may be written as

$$
\frac{1}{2}\text{Re}(\mathbf{E} \times \bar{\mathbf{H}}) = \frac{1}{2}\text{Re}\frac{k_0}{\omega\bar{\mu}} |\mathbf{E}_0(\mathbf{r})|^2 \nabla \bar{L}(\mathbf{r})e^{-jk_0(L-\bar{L})}.
$$

For real $L(\mathbf{r})$, we have

$$
\frac{1}{2}\text{Re}(\mathbf{E} \times \bar{\mathbf{H}}) = \frac{1}{2}\text{Re}\frac{k_0}{\omega\bar{\mu}} |\mathbf{E}_0(\mathbf{r})|^2 \nabla L(\mathbf{r}).
$$

So the direction of energy flow is normal to the wavefront. The curves whose tangent at each point is the direction of energy flow of the field are known as **rays**. In optics, the rays are used to model the propagation of light through an optical system, by representing the light field in terms of discrete rays. The ray optics can be used to study light reflections and refractions. Since the rays are normal to the wavefront, we may introduce a unit tangent vector to the rays:

$$\mathbf{s}(\mathbf{r}) = \frac{1}{n(\mathbf{r})} \nabla L(\mathbf{r}), \tag{7.33}$$

Let $\mathbf{r}$ be a point P on a ray and s be the arc length measured along the ray. Then $d\mathbf{r}/ds = \mathbf{s}$, and

$$n(\mathbf{r}) \frac{d\mathbf{r}}{ds} = \nabla L(\mathbf{r}).$$

Taking the derivative with respect to s and making use of the relation $\frac{d}{ds} = \frac{d\mathbf{r}}{ds} \cdot \nabla$, we obtain

$$\frac{d}{ds} n(\mathbf{r}) \frac{d\mathbf{r}}{ds} = \nabla n(\mathbf{r}). \tag{7.34}$$

This is the differential equation for the rays, called the **ray equation**, which can be solved numerically with initial data to determine the rays in a region.

Example 7.4 (Square-law distribution): Consider a square-law medium described by

$$n(x) = n_1[1 - \Delta(x/a)^2],$$

where n_1, Δ and a are constant with $\Delta(x/a)^2 \ll 1$. Since $\Delta(x/a)^2 \ll 1$, the above equation may be written as

$$n^2(x) = n_1^2[1 - 2\Delta(x/a)^2].$$

The solution of (7.34) in (x, z)-plane can be obtained by solving the following scalar differential equations

$$\frac{d}{ds} n(x) \frac{dx}{ds} = \frac{dn(x)}{dx}, \quad \frac{d}{ds} n(x) \frac{dz}{ds} = 0.$$

From the second equation we obtain

$$\frac{dz}{ds} = \frac{A}{n(x)},$$

where A is a constant that can be determined by the initial condition. Assume that the ray passes through the point (x_0, z_0). Then

$$\frac{dz}{ds}\bigg|_{(x_0,z_0)} = A \frac{1}{n(x_0)} = \cos\theta_0.$$

Here θ_0 is the angle between the ray and z-axis at (x_0, z_0), and $A = n(x_0)\cos\theta_0$. The z-component of the wavenumber $\mathbf{k}$ is

$$k_z(x_0, z_0) = k_0 n(x_0)\cos\theta_0 = k_z(x, z) = k_0 A.$$

Hence we may write

$$\frac{dz}{ds} = \frac{k_z}{k_0 n(x)}. \tag{7.35}$$

Since the ray is confined in the (x, z)-plane, we have

$$\frac{d\mathbf{r}}{ds} \cdot \frac{d\mathbf{r}}{ds} = \left(\frac{dx}{ds}\right)^2 + \left(\frac{dz}{ds}\right)^2 = 1.$$

It follows that

$$\frac{dx}{ds} = \left[1 - \left(\frac{dz}{ds}\right)^2\right]^{1/2}. \tag{7.36}$$

Combining (7.35) and (7.36) gives

$$\frac{dz}{dx} = \frac{ak_z}{[u^2 - v^2(x/a)^2]^{1/2}}, \tag{7.37}$$

where $u^2 = a^2(k_0^2 n_1^2 - k_z^2)$, $v^2 = a^2 k_0^2 2\Delta n_1^2$. Integrating (7.37) gives the equation for the ray

$$x = x_0 \cos\left(\frac{v}{a^2 k_z}z\right) + \frac{a^2 k_z}{v}\tan\theta_0 \sin\left(\frac{v}{a^2 k_z}z\right).$$

This indicates that the x-coordinate of the ray is a periodic function of its z-coordinate. □

Let κ be the curvature of the ray and $\mathbf{u}_c$ be the unit vector in the direction of the radius of curvature. Then $\kappa\mathbf{u}_c = d\mathbf{s}/ds = (\mathbf{s}\cdot\nabla)\mathbf{s} = -\mathbf{s}\times\nabla\times\mathbf{s}$. Thus

$$\kappa = -\mathbf{u}_c \cdot (\mathbf{s}\times\nabla\times\mathbf{s}) = \mathbf{u}_c \cdot \nabla\ln n.$$

The above equation indicates that the rays in a homogeneous medium are straight lines. In an inhomogeneous medium, the rays will bend towards the region of higher refractive index, that is, towards the region of lower speed of light.

The behavior of the magnitude $\mathbf{E}_0$ can be determined by the Maxwell equations. Introducing (7.29) into (7.18) yields

$$\frac{1}{jk_0}\left[(\nabla L \cdot \nabla \ln \mu - \nabla^2 L)\mathbf{E}_0 - 2(\nabla L \cdot \nabla)\mathbf{E}_0 - (\mathbf{E}_0 \cdot \nabla \ln \mu)\nabla L + (\nabla \cdot \mathbf{E}_0)\nabla L\right]$$

$$+\left[(\nabla L)^2 - n^2\right]\mathbf{E}_0 + \frac{1}{(jk_0)^2}\left[\nabla^2\mathbf{E}_0 + \nabla \ln \mu \times (\nabla \times \mathbf{E}_0) - \nabla(\nabla \cdot \mathbf{E}_0)\right] = 0. \tag{7.38}$$

The second term is zero due to the eikonal equation and the third term can be ignored for large k_0. From $\nabla \cdot \mathbf{D} = 0$ we have $\nabla \cdot \mathbf{E}_0 = -\mathbf{E}_0 \cdot \nabla \ln \varepsilon$. Thus, (7.38) may be written as

$$(\nabla L \cdot \nabla)\mathbf{E}_0 + \frac{1}{2}(\nabla^2 L)\mathbf{E}_0 + (\mathbf{E}_0 \cdot \nabla \ln n)\nabla L - \frac{1}{2}(\nabla L \cdot \nabla \ln \mu)\mathbf{E}_0 = 0. \tag{7.39}$$

This is the differential equation for the amplitude $\mathbf{E}_0$, called the **transport equation**. The amplitude $\mathbf{H}_0$ of the magnetic field satisfies the similar transport equation

$$(\nabla L \cdot \nabla)\mathbf{H}_0 + \frac{1}{2}(\nabla^2 L)\mathbf{H}_0 + (\mathbf{H}_0 \cdot \nabla \ln n)\nabla L - \frac{1}{2}(\nabla L \cdot \nabla \ln \varepsilon)\mathbf{H}_0 = 0. \tag{7.40}$$

Taking the scalar product of (7.39) with $\bar{\mathbf{E}}_0$ and adding the resultant equation to its conjugate, we obtain

$$n\frac{d}{ds}|\mathbf{E}_0|^2 + \mu |\mathbf{E}_0|^2 \nabla \cdot \left(\frac{1}{\mu}\nabla L\right) = 0.$$

The ratio of the field intensity at s_2 of a ray to s_1 is then given by

$$\frac{|\mathbf{E}_0|^2_{s_2}}{|\mathbf{E}_0|^2_{s_1}} = \exp\left[-\int_{s_1}^{s_2}\sqrt{\frac{\mu}{\varepsilon}}\nabla \cdot \left(\frac{1}{\mu}\nabla L\right)ds\right].$$

Similarly, we have

$$\frac{|\mathbf{H}_0|^2_{s_2}}{|\mathbf{H}_0|^2_{s_1}} = \exp\left[-\int_{s_1}^{s_2}\sqrt{\frac{\varepsilon}{\mu}}\nabla \cdot \left(\frac{1}{\varepsilon}\nabla L\right)ds\right].$$

A detailed study about the theory and applications of geometric optics can be found in (Kline and Kay, 1965; Jones, 1979).

7.2.4   Reflection and Transmission in Layered Media

If the fields are independent of one Cartesian coordinate, say, the y coordinate, Maxwell equations in an isotropic and homogeneous medium can be decoupled into two sets of equations

$$H_x = \frac{1}{j\omega\mu}\frac{\partial E_y}{\partial z}, \ H_z = -\frac{1}{j\omega\mu}\frac{\partial E_y}{\partial x},$$

$$\left(\frac{\partial^2}{\partial x^2} + \frac{\partial^2}{\partial z^2} + \omega^2\mu\varepsilon\right)E_y = 0,$$

(7.41)

and

$$E_x = -\frac{1}{j\omega\varepsilon}\frac{\partial H_y}{\partial z}, \ E_z = \frac{1}{j\omega\varepsilon}\frac{\partial H_y}{\partial x},$$

$$\left(\frac{\partial^2}{\partial x^2} + \frac{\partial^2}{\partial z^2} + \omega^2\mu\varepsilon\right)H_y = 0.$$

(7.42)

The fields determined by (7.41) are called transverse electric (TE) fields, which consist of one electric field component E_y and two magnetic field components H_x and H_z. The fields determined by (7.42) are called transverse magnetic (TM) fields, which contain one magnetic field component H_y and two electric field components E_x and E_z.

Let us consider a TE plane wave $\mathbf{E}^{in} = \mathbf{u}_y E_1 e^{-ik_z z - jk_x x}$, which is incident upon a layered medium from layer 1 as shown in Figure 7.1. Assume that the medium parameters μ and ε depend on the layer number $l(l = 1, 2, \cdots, m)$ but are constant inside each layer. The total fields in the layer l may be expressed as

$$E_y(x, z) = \left[A_l e^{-jk_{zl}(z-z_l)} + B_l e^{jk_{zl}(z-z_l)}\right]e^{-jk_x x},$$

$$H_x(x, z) = -\frac{k_{zl}}{\omega\mu_l}\left[A_l e^{-jk_{zl}(z-z_l)} - B_l e^{jk_{zl}(z-z_l)}\right]e^{-jk_x x},$$

$$H_z(x, z) = \frac{k_x}{\omega\mu_l}\left[A_l e^{-jk_{zl}(z-z_l)} + B_l e^{jk_{zl}(z-z_l)}\right]e^{-jk_x x},$$

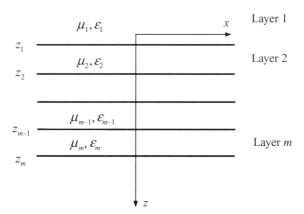

Figure 7.1   A layered medium

where the x-component of the wavenumber $\mathbf{k}$ has the same value as the incident field. At $z = z_l$, the tangential fields are given by

$$E_y(x, z_l) = (A_l + B_l)e^{-jk_xx}, \; H_x(x, z_l) = \frac{1}{\eta_l}(B_l - A_l)e^{-jk_xx},$$

where $\eta_l = \omega\mu_l/k_{zl}$. The tangential fields must be continuous across the interface. As a result, the fields at $z = z_{l-1}$ can be expressed as

$$E_y(x, z_{l-1}) = E_y(x, z_l)\cos\xi_l - j\eta_l H_x(x, z_l)\sin\xi_l,$$
$$H_y(x, z_{l-1}) = H_x(x, z_l)\cos\xi_l - j\frac{1}{\eta_l}E_y(x, z_l)\sin\xi_l,$$
(7.43)

where $\xi_l = k_{zl}(z_l - z_{l-1})$. Introducing the matrices

$$[F_l] = \begin{bmatrix} E_y(x, z_l) \\ H_x(x, z_l) \end{bmatrix}, \; [K_l] = \begin{bmatrix} \cos\xi_l & -j\eta_l\sin\xi_l \\ -j\dfrac{1}{\eta_l}\sin\xi_l & \cos\xi_l \end{bmatrix},$$

Equation (7.43) may be rewritten as

$$[F_{l-1}] = [K_l] \cdot [F_l].$$

The relationship between the fields at the lower boundary of layer 1 and the fields at the lower boundary of layer m is

$$[F_1] = [K] \cdot [F_m],$$
(7.44)

where

$$[K] = [K_2] \cdot [K_3] \cdots [K_m] = \begin{bmatrix} k_{11} & k_{12} \\ k_{21} & k_{22} \end{bmatrix}.$$

Since $\det[K_l] = 1$, we have $\det[K] = 1$ and

$$[K]^{-1} = \begin{bmatrix} k_{22} & -k_{21} \\ -k_{12} & k_{11} \end{bmatrix}.$$

We may introduce the reflection coefficient R_l and the surface impedance Z_l at $z = z_l$

$$R_l = \frac{B_l}{A_l}, \; Z_l = -\frac{E_y(x, z_l)}{H_x(x, z_l)}.$$

Then

$$R_l = \frac{Z_l - \eta_l}{Z_l + \eta_l}, \; Z_l = \eta_l\frac{1 + R_l}{1 - R_l}.$$

It follows from (7.44) that

$$Z_1 = \frac{k_{11}Z_m - k_{12}}{k_{22} - k_{11}Z_m}.$$

If the lower boundary of layer m is at $z = +\infty$, we have $B_m = 0$, $R_m = 0$ and $Z_m = \eta_m$. The study for TM incidence can be carried out similarly.

7.3 Inhomogeneous Metal Waveguides

Inhomogeneously filled waveguides, such as a rectangular waveguide partially filled with dielectric slabs, are used in a number of waveguide components. The determination of the propagation constants of the modes in the waveguides is the major focus of our interest.

7.3.1 General Field Relationships

Consider a metal waveguide, which is uniform along the z-axis. The cross-section of the waveguide is denoted by Ω and its boundary is assumed to be a perfect conductor and is denoted by $\Gamma = \Gamma_1 + \Gamma_2$, as shown in Figure 7.2. The waveguide is filled with an inhomogeneous medium in which μ and ε are functions of transverse positions but are constant along the z-axis. Assume that the fields in the waveguide have a z dependence of the form $e^{-j\beta z}$

$$\mathbf{E}(\mathbf{r}) = \mathbf{e}(\boldsymbol{\rho})e^{-j\beta z}, \ \mathbf{H}(\mathbf{r}) = \mathbf{h}(\boldsymbol{\rho})e^{-j\beta z}, \tag{7.45}$$

where $\boldsymbol{\rho} = (x, y) \in \Omega$ denotes the transverse position. Introducing these into Maxwell equations, we obtain

$$\nabla_\beta \times \mathbf{h} = j\omega\varepsilon\mathbf{e}, \ \nabla_\beta \times \mathbf{e} = -j\omega\mu\mathbf{h},$$
$$\nabla_\beta \cdot \varepsilon\mathbf{e} = 0, \ \nabla_\beta \cdot \mu\mathbf{h} = 0. \tag{7.46}$$

Here $\nabla_\beta = \nabla_t - j\beta\mathbf{u}_z$ denotes an operator obtained from ∇ by replacing the derivative with respect to z with multiplication by $-j\beta$, and ∇_t is transverse gradient operator. For an arbitrary

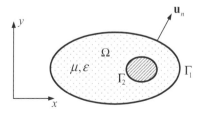

Figure 7.2 Inhomogeneous waveguide

vector function $\mathbf{f}(\boldsymbol{\rho})$ and a scalar function $u(\boldsymbol{\rho})$, we have

$$
\begin{aligned}
&\nabla_\beta \cdot (\nabla_\beta \times \mathbf{f}) = 0, \\
&\nabla_\beta \cdot (u\mathbf{f}) = u\nabla_\beta \cdot \mathbf{f} + \mathbf{f} \cdot \nabla_t u, \\
&\nabla_\beta \cdot (\nabla_\beta u) = \nabla_t^2 u - \beta^2 u, \\
&\nabla_\beta \times (\nabla_\beta u) = 0, \\
&\nabla_\beta \times \nabla_\beta \times \mathbf{f} = -\nabla_t^2 \mathbf{f} + \beta^2 \mathbf{f} + \nabla_\beta(\nabla_\beta \cdot \mathbf{f}).
\end{aligned}
\tag{7.47}
$$

It follows from (7.46) that

$$
\begin{aligned}
&\nabla_\beta \times \varepsilon_r^{-1}\nabla_\beta \times \mathbf{h} = k_0^2 \mu_r \mathbf{h}, \\
&\nabla_\beta \times \mu_r^{-1}\nabla_\beta \times \mathbf{e} = k_0^2 \varepsilon_r \mathbf{e}, \\
&\nabla_\beta \cdot \varepsilon_r \mathbf{e} = 0,\ \nabla_\beta \cdot \mu_r \mathbf{h} = 0,
\end{aligned}
\tag{7.48}
$$

where $\mu_r = \mu/\mu_0$, $\varepsilon_r = \varepsilon/\varepsilon_0$ and $k_0 = \omega\sqrt{\mu_0\varepsilon_0}$. A solution of (7.48) is called a **guided mode** of the waveguide if the field is non-trivial and has finite energy:

$$
(\beta, k_0) \in R^2,\ (\mathbf{e}, \mathbf{h}) \neq 0,\ \text{and } \mathbf{e}, \mathbf{h} \in [L^2(\Omega)]^3.
$$

7.3.2 Symmetric Formulation

It follows from (7.48) that the fields satisfy

$$
\begin{aligned}
&\nabla_\beta \times \varepsilon_r^{-1}\nabla_\beta \times \mathbf{h} = k_0^2 \mu_r \mathbf{h},\ \boldsymbol{\rho} \in \Omega, \\
&\mathbf{u}_n \cdot \mu_r \mathbf{h} = 0,\ \mathbf{u}_n \times \varepsilon_r^{-1}\nabla_\beta \times \mathbf{h} = 0,\ \boldsymbol{\rho} \in \Gamma,
\end{aligned}
\tag{7.49}
$$

and

$$
\begin{aligned}
&\nabla_\beta \times \mu_r^{-1}\nabla_\beta \times \mathbf{e} = k_0^2 \varepsilon_r \mathbf{e},\ \boldsymbol{\rho} \in \Omega, \\
&\mathbf{u}_n \times \mathbf{e} = 0,\ \mathbf{u}_n \cdot \mu_r^{-1}\nabla_\beta \times \mathbf{e} = 0,\ \boldsymbol{\rho} \in \Gamma,
\end{aligned}
\tag{7.50}
$$

where $\mathbf{u}_n$ is the unit outward normal on Γ. Let $L_q^2(\Omega)$ denote the space of square integrable functions defined in Ω with inner product defined by $(u, v)_q = \int_\Omega qu\bar{v}d\Omega$ and the corresponding norm is denoted by $\|\cdot\|_q = \sqrt{(\cdot, \cdot)_q}$. We also introduce the product space $H_q = [L_q^2(\Omega)]^3$ with an inner product $(\mathbf{p}_1, \mathbf{p}_2)_q = \int_\Omega q\mathbf{p}_1 \cdot \bar{\mathbf{p}}_2 d\Omega$, $\mathbf{p}_1, \mathbf{p}_2 \in H_q$. The corresponding norm is still denoted by $\|\cdot\|_q = \sqrt{(\cdot, \cdot)_q}$.

In (7.49) and (7.50), the propagation constant is considered as a parameter while the wavenumber k_0 is taken as the eigenvalue that is a function of β. For a given β, both (7.49) and (7.50) define a symmetric eigenvalue problem in H_{μ_r} and H_{ε_r} respectively. From (7.49)

we obtain

$$k_0^2 \int_\Omega \mu_r |\mathbf{h}|^2 \, d\Omega = \int_\Omega \frac{1}{n^2} (\nabla_\beta \times \mathbf{h}) \cdot (\overline{\nabla_\beta \times \mathbf{h}}) d\Omega$$

$$\geq \frac{1}{n_+^2} \int_\Omega (\nabla_\beta \times \mathbf{h}) \cdot (\overline{\nabla_\beta \times \mathbf{h}}) d\Omega$$

$$= \frac{1}{n_+^2} \int_\Omega (\nabla_\beta \times \nabla_\beta \times \mathbf{h}) \cdot \bar{\mathbf{h}} d\Omega,$$

where $n = \sqrt{\varepsilon_r}$ and $n_+ = \max_{\rho \in \Omega} n(\rho)$. Making use of the last equation of (7.47) and integration by parts, we have

$$k_0^2 \int_\Omega \mu_r |\mathbf{h}|^2 \, d\Omega \geq \frac{1}{n_+^2} \int_\Omega (-\nabla_t^2 \mathbf{h} + \beta^2 \mathbf{h} + \nabla_\beta \nabla_\beta \cdot \mathbf{h}) \cdot \bar{\mathbf{h}} d\Omega$$

$$= \frac{1}{n_+^2} \int_\Omega (|\nabla_t \times \mathbf{h}|^2 + |\nabla_t \cdot \mathbf{h}|^2 - |\nabla_\beta \cdot \mathbf{h}|^2) d\Omega + \frac{\beta^2}{n_+^2} \int_\Omega |\mathbf{h}|^2 \, d\Omega.$$

If μ_r is a constant, we have $\nabla_\beta \cdot \mathbf{h} = 0$ and the above is equivalent to

$$\int_\Omega (|\nabla_t \times \mathbf{h}|^2 + |\nabla_t \cdot \mathbf{h}|^2) d\Omega + (\beta^2 - k_0^2 \mu_r n_+^2) \int_\Omega |\mathbf{h}|^2 \, d\Omega \leq 0. \tag{7.51}$$

As a result, if $|\beta| \geq k_0 \sqrt{\mu_r} n_+$, then $\mathbf{h} = 0$ and (7.49) has a trivial solution. In other words, no guided modes exist in this case. Hence the solution (β, k_0) of (7.49) or (7.50) must satisfy

$$k_0 > \frac{|\beta|}{\sqrt{\mu_r} n_+}. \tag{7.52}$$

This is the **guidance condition** for an inhomogeneously filled waveguide.

7.3.3 Asymmetric Formulation

In engineering, the propagation constant β is usually considered as the eigenvalue while the frequency or the wavenumber k_0 is taken as a parameter. This arrangement often yields a non-symmetric eigenvalue problem, which is more difficult to study. The guided modes in the waveguide may be decomposed into a transverse and a longitudinal component

$$\mathbf{E}(\mathbf{r}) = [\mathbf{e}_t(\rho) + \mathbf{u}_z e_z(\rho)] e^{-j\beta z}, \quad \mathbf{H}(\mathbf{r}) = [\mathbf{h}_t(\rho) + \mathbf{u}_z h_z(\rho)] e^{-j\beta z}. \tag{7.53}$$

Introducing these into Maxwell equations, we obtain

$$\nabla \times \mathbf{h}_t = j\omega\varepsilon\mathbf{u}_z e_z, \nabla \times \mathbf{e}_t = -j\omega\mu\mathbf{u}_z h_z,$$

$$j\beta\mathbf{u}_z \times \mathbf{h}_t + \mathbf{u}_z \times \nabla h_z = -j\omega\varepsilon\mathbf{e}_t,$$

$$j\beta\mathbf{u}_z \times \mathbf{e}_t + \mathbf{u}_z \times \nabla e_z = j\omega\mu\mathbf{h}_t,$$ \qquad(7.54)

$$\nabla \cdot \varepsilon\mathbf{e}_t = j\beta\varepsilon e_z, \nabla \cdot \mu\mathbf{h}_t = j\beta\mu h_z.$$

By eliminating $\mathbf{h}_t$, e_z and h_z, we have the following eigenvalue problem

$$\mu\nabla \times \mu^{-1}\nabla \times \mathbf{e}_t - \nabla\varepsilon^{-1}\nabla \cdot \varepsilon\mathbf{e} - (\omega^2\mu\varepsilon - \beta^2)\mathbf{e}_t = 0, \rho \in \Omega,$$

$$\mathbf{u}_n \times \mathbf{e}_t = 0, \nabla \cdot \varepsilon\mathbf{e}_t = 0, \rho \in \Gamma.$$ \qquad(7.55)

In (7.55), β^2 is taken as the eigenvalue and ω^2 as the parameter. The differential operator in (7.55) is not symmetric. Let $\mathbf{e}_{tm}$ and $\mathbf{e}_{tn}$ be two different eigenfunctions corresponding to the eigenvalues β_m^2 and β_n^2 respectively. Then

$$\mu\nabla \times \mu^{-1}\nabla \times \mathbf{e}_{tm} - \nabla\varepsilon^{-1}\nabla \cdot \varepsilon\mathbf{e}_{tm} - (\omega^2\mu\varepsilon - \beta_m^2)\mathbf{e}_{tm} = 0.$$

Taking the scalar product of the above equation with $\nabla \times \mu^{-1}\nabla \times \mathbf{e}_{tn} - \omega^2\varepsilon\mathbf{e}_{tn}$ and integrating the resultant equation over Ω yields

$$\int_\Omega \mu(\nabla \times \mu^{-1}\nabla \times \mathbf{e}_{tm} - \omega^2\varepsilon\mathbf{e}_{tm}) \cdot (\nabla \times \mu^{-1}\nabla \times \mathbf{e}_{tn} - \omega^2\varepsilon\mathbf{e}_{tn})d\Omega$$

$$-\int_\Omega \omega^2\varepsilon^{-1}(\nabla \cdot \varepsilon\mathbf{e}_{tm})(\nabla \cdot \varepsilon\mathbf{e}_{tn})d\Omega + \beta_m^2 \int_\Omega (\mu^{-1}\nabla \times \mathbf{e}_{tm} \cdot \nabla \times \mathbf{e}_{tn} - \omega^2\varepsilon\mathbf{e}_{tm} \cdot \mathbf{e}_{tn})d\Omega = 0.$$

\qquad(7.56)

Interchanging m and n and subtracting the result from (7.56) gives

$$(\beta_m^2 - \beta_n^2)\int_\Omega (\mu^{-1}\nabla \times \mathbf{e}_{tm} \cdot \nabla \times \mathbf{e}_{tn} - \omega^2\varepsilon\mathbf{e}_{tm} \cdot \mathbf{e}_{tn})d\Omega = 0.$$ \qquad(7.57)

This implies the following orthogonality relation

$$\int_\Omega (\mu^{-1}\nabla \times \mathbf{e}_{tm} \cdot \nabla \times \mathbf{e}_{tn} - \omega^2\varepsilon\mathbf{e}_{tm} \cdot \mathbf{e}_{tn})d\Omega = 0,$$ \qquad(7.58)

if $\beta_m^2 \neq \beta_n^2$. From (7.54), the transverse magnetic field can be expressed in terms of the transverse electric field

$$\mathbf{u}_z \times \mathbf{h}_t = \frac{1}{\omega\beta}(\nabla \times \mu^{-1}\nabla \times \mathbf{e}_t - \omega^2\varepsilon\mathbf{e}_t),$$

and the orthogonality relation (7.58) can be written as

$$\int_{\Omega} (\mathbf{e}_{tm} \times \mathbf{h}_{tn}) \cdot \mathbf{u}_z d\Omega = 0, m \neq n. \tag{7.59}$$

This is the most general form of the orthogonality relation in a waveguide. The modes in a waveguide filled with homogeneous medium can be classified into TEM, TE and TM modes. In an inhomogeneous waveguide, such a classification is impossible since the modes contain both e_z and h_z.

7.4 Optical Fibers

An **optical fiber** consists of a core of dielectric material surrounded by a cladding of another dielectric material which has lower refractive index than that of the core. The electromagnetic fields are confined in the core region due to the total internal reflection and the fiber acts as a waveguide. Optical fibers have been widely used in fiber-optic communication, and they carry much more information and travel longer distances than conventional metal wires. Moreover, they are immune to electromagnetic interferences.

7.4.1 Circular Optical Fiber

In a cylindrical system, the total fields can be decomposed into a sum of the transverse component and longitudinal component

$$\mathbf{E} = \mathbf{E}_t + \mathbf{u}_z E_z, \mathbf{H} = \mathbf{H}_t + \mathbf{u}_z H_z.$$

If the fields have a z dependence of the form $e^{-j\beta z}$, we have the decomposition $\nabla = \nabla_t - j\beta\mathbf{u}_z$. From Maxwell equations, the transverse fields may be expressed in terms of the longitudinal fields as

$$\mathbf{E}_t = \frac{1}{k_c^2} [-j\omega\mu\nabla_t \times (\mathbf{u}_z H_z) - j\beta\nabla_t E_z],$$

$$\mathbf{H}_t = \frac{1}{k_c^2} [j\omega\varepsilon\nabla_t \times (\mathbf{u}_z E_z) - j\beta\nabla_t H_z], \tag{7.60}$$

where $k_c^2 = \omega^2\mu\varepsilon - \beta^2$. The longitudinal components satisfy the two-dimensional Helmholtz equation

$$(\nabla_t^2 + k_c^2)E_z = 0, (\nabla_t^2 + k_c^2)H_z = 0. \tag{7.61}$$

Consider the circular optical fiber shown in Figure 7.3. The core is the circular region of radius a with medium parameters $\mu_0, \varepsilon_{r1}\varepsilon_0$. The external region is the cladding with medium parameters $\mu_0, \varepsilon_{r2}\varepsilon_0$. The refractive indices $n_i = \sqrt{\varepsilon_{ri}}(i = 1, 2)$ are assumed to be constants.

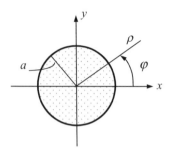

Figure 7.3   Circular optical fiber

The solutions of (7.61) in the core region must be finite, and may be written as

$$E_{z1} = A_1 \frac{J_m(k_{c1}\rho)}{J_m(k_{c1}a)} e^{jm\varphi} e^{-j\beta z} = A_1 \frac{J_m(u\rho')}{J_m(u)} e^{jm\varphi} e^{-j\beta z},$$

$$H_{z1} = B_1 \frac{J_m(k_{c1}\rho)}{J_m(k_{c1}a)} e^{jm\varphi} e^{-j\beta z} = B_1 \frac{J_m(u\rho')}{J_m(u)} e^{jm\varphi} e^{-j\beta z},$$

where $k_{c1}^2 = k_0^2 n_1^2 - \beta^2$; $k_0^2 = \omega^2 \mu_0 \varepsilon_0$; $u = k_{c1}a$; $\rho' = \rho/a$; A_1 and B_1 are constants to be determined by boundary conditions. The solutions of (7.61) in the cladding region must decrease as ρ increases to guarantee that the fields are square integrable (that is, the energy is finite), and they are given by

$$E_{z2} = A_2 \frac{K_m(k_{c2}\rho)}{K_m(k_{c2}a)} e^{jm\varphi} e^{-j\beta z} = A_2 \frac{K_m(v\rho')}{K_m(v)} e^{jm\varphi} e^{-j\beta z},$$

$$H_{z2} = B_2 \frac{K_m(k_{c2}\rho)}{K_m(k_{c2}a)} e^{jm\varphi} e^{-j\beta z} = B_2 \frac{K_m(v\rho')}{K_m(v)} e^{jm\varphi} e^{-j\beta z},$$

where K_m are modified Bessel functions of the second kind, and $k_{c2}^2 = \beta^2 - k_0^2 n_2^2$ and $v = k_{c2}a$. The transverse field components can then be determined from (7.60). In the core region, the fields are

$$E_{\rho 1} = -j \left(\frac{a}{u}\right)^2 \left[\frac{u\beta J_m'(u\rho')}{a J_m(u)} A_1 + \frac{j\omega\mu_0 m J_m(u\rho')}{\rho J_n(u)} B_1 \right] e^{jm\varphi} e^{-j\beta z},$$

$$E_{\varphi 1} = -j \left(\frac{a}{u}\right)^2 \left[\frac{j\beta m J_m(u\rho')}{\rho J_m(u)} A_1 - \frac{\omega\mu_0 u J_m'(u\rho')}{a J_m(u)} B_1 \right] e^{jm\varphi} e^{-j\beta z},$$

$$H_{\rho 1} = -j \left(\frac{a}{u}\right)^2 \left[\frac{u\beta J_m'(u\rho')}{a J_m(u)} B_1 - \frac{j\omega\varepsilon_0 n_1^2 m J_m(u\rho')}{\rho J_m(u)} A_1 \right] e^{jm\varphi} e^{-j\beta z},$$

$$H_{\varphi 1} = -j \left(\frac{a}{u}\right)^2 \left[\frac{j\beta m J_m(u\rho')}{\rho J_m(u)} B_1 + \frac{\omega\varepsilon_0 n_1^2 u J_m'(u\rho')}{a J_m(u)} A_1 \right] e^{jm\varphi} e^{-j\beta z}.$$

In the cladding region, the fields are

$$E_{\rho 2} = j\left(\frac{a}{v}\right)^2 \left[\frac{v\beta K'_m(v\rho')}{aK_m(v)}A_2 + \frac{j\omega\mu_0 m K_n(v\rho')}{\rho K_m(v)}B_2\right] e^{jm\varphi}e^{-j\beta z},$$

$$E_{\varphi 2} = j\left(\frac{a}{v}\right)^2 \left[\frac{j\beta m K_m(v\rho')}{\rho K_m(v)}A_2 - \frac{\omega\mu_0 v K'_m(v\rho')}{aK_m(v)}B_2\right] e^{jm\varphi}e^{-j\beta z},$$

$$H_{\rho 2} = j\left(\frac{a}{v}\right)^2 \left[\frac{v\beta K'_m(v\rho')}{aK_m(v)}B_2 - \frac{j\omega\varepsilon_0 n_2^2 m K_m(v\rho')}{\rho K_m(v)}A_2\right] e^{jm\varphi}e^{-j\beta z},$$

$$H_{\varphi 2} = j\left(\frac{a}{v}\right)^2 \left[\frac{j\beta m K_m(v\rho')}{\rho K_m(v)}B_2 + \frac{\omega\varepsilon_0 n_2^2 v K'_m(v\rho')}{aK_m(v)}A_2\right] e^{jm\varphi}e^{-j\beta z}.$$

The boundary conditions at $\rho = a$ require that the tangential fields must be continuous, which leads to

$$A_1 = A_2, \ B_1 = B_2,$$

$$-\left(\frac{a}{u}\right)^2 \left[\frac{j\beta m}{\rho}A_1 - \frac{\omega\mu_0 u J'_m(u)}{aJ_m(u)}B_1\right] = \left(\frac{a}{v}\right)^2 \left[\frac{j\beta m}{\rho}A_2 - \frac{\omega\mu_0 v K'_m(v)}{aK_m(v)}B_2\right],$$

$$-\left(\frac{a}{u}\right)^2 \left[\frac{j\beta m}{\rho}B_1 + \frac{\omega\varepsilon_0 n_1^2 u J'_n(u)}{aJ_m(u)}A_1\right] = \left(\frac{a}{v}\right)^2 \left[\frac{j\beta m}{\rho}B_2 + \frac{\omega\varepsilon_0 n_2^2 v K'_m(v)}{aK_m(v)}A_2\right].$$

This set of linear equations can be reduced to

$$A_1\left(\frac{1}{u^2} + \frac{1}{v^2}\right)\frac{j\beta m}{a} - B_1\frac{\omega\mu_0}{a}\left[\frac{1}{u}\frac{J'_m(u)}{J_m(u)} + \frac{1}{v}\frac{K'_m(v)}{K_m(v)}\right] = 0,$$

$$A_1\frac{\omega\varepsilon_0}{a}\left[\frac{n_1^2}{u}\frac{J'_m(u)}{J_m(u)} + \frac{n_2^2}{v}\frac{K'_m(v)}{K_m(v)}\right] + B_1\frac{j\beta m}{a}\left(\frac{1}{u^2} + \frac{1}{v^2}\right) = 0.$$

A non-trivial solution of the above set of equations requires that the determinant of the coefficient matrix vanishes, yielding

$$\left[\frac{1}{u}\frac{J'_m(u)}{J_m(u)} + \frac{1}{v}\frac{K'_m(v)}{K_m(v)}\right]\left[\frac{n_1^2}{u}\frac{J'_m(u)}{J_m(u)} + \frac{n_2^2}{v}\frac{K'_m(v)}{K_m(v)}\right] = \frac{m^2\beta^2}{k_0^2}\left(\frac{1}{u^2} + \frac{1}{v^2}\right),$$

which can be used to determine the propagation constant β. For the guided modes, both k_{c1} and k_{c2} must be positive. This requires

$$k_0 n_2 \leq \beta \leq k_0 n_1. \tag{7.62}$$

When $\beta = k_0 n_2$, we have $k_{c2} = 0$, which is called the **cut-off condition**. Note that the propagation constant β is not equal to zero when the optical fiber is at cut-off. This is different from a hollow metal waveguide. If $k_{c2} < 0$, the fields will radiate in ρ direction, and at same time, they still propagate along z direction. Such field distributions are called **radiation modes**.

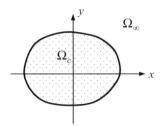

Figure 7.4 An optical fiber

7.4.2 *Guidance Condition*

Let Ω_c denote the cross-section of the core region of an arbitrary fiber, and the exterior region (the cladding) be denoted by Ω_∞, as shown in Figure 7.4. The medium parameters of the fiber are given by μ_0, $\varepsilon_r \varepsilon_0$. The refractive index of the fiber is denoted by $n(\rho) = \sqrt{\varepsilon_r}$, which is a positive function of the transverse coordinates $\rho = (x, y)$ only. We assume that the fiber cladding is homogeneous and extends infinitely in the transverse (x, y)-plane. This assumption is reasonable since the radius of the core is very small compared to the radius of the cladding in practice. The refractive index of the cladding is thus a constant, denoted by $n(\rho) = n_\infty$. If the index n is piecewise constant, the fiber is called a **step-index fiber**. If the index n is a continuous function, the fiber is called a **graded-index fiber**. For an optical fiber, Equation (7.48) reduces to

$$\nabla_\beta \times n^{-2}\nabla_\beta \times \mathbf{h} = k_0^2 \mathbf{h},$$
$$\nabla_\beta \times \nabla_\beta \times \mathbf{e} = k_0^2 n^2 \mathbf{e}, \tag{7.63}$$
$$\nabla_\beta \cdot n^2 \mathbf{e} = 0, \ \nabla_\beta \cdot \mathbf{h} = 0.$$

Multiplying both sides of the first equation by $\bar{\mathbf{h}}$ and integrating over (x, y)-plane by use of integration by parts, we obtain

$$k_0^2 \int_{R^2} |\mathbf{h}|^2 \, d\Omega = \int_{R^2} \frac{1}{n^2}(\nabla_\beta \times \mathbf{h}) \cdot (\overline{\nabla_\beta \times \mathbf{h}}) d\Omega$$
$$\geq \frac{1}{n_+^2} \int_{R^2} (\nabla_\beta \times \mathbf{h}) \cdot (\overline{\nabla_\beta \times \mathbf{h}}) d\Omega$$
$$= \frac{1}{n_+^2} \int_{R^2} (\nabla_\beta \times \nabla_\beta \times \mathbf{h}) \cdot \bar{\mathbf{h}} d\Omega,$$

where $n_+ = \max_{\rho \in R^2} n(\rho)$. Making use of the last equation of (7.47) and integration by parts again, we have

$$k_0^2 \int_{R^2} |\mathbf{h}|^2 \, d\Omega \geq \frac{1}{n_+^2} \int_{R^2} (-\nabla_t^2 \mathbf{h} + \beta^2 \mathbf{h}) \cdot \bar{\mathbf{h}} d\Omega$$
$$= \frac{1}{n_+^2} \int_{R^2} (|\nabla_t \times \mathbf{h}|^2 + |\nabla_t \cdot \mathbf{h}|^2) d\Omega + \frac{\beta^2}{n_+^2} \int_{R^2} |\mathbf{h}|^2 \, d\Omega.$$

This is equivalent to

$$\int\limits_{R^2} (|\nabla_t \times \mathbf{h}|^2 + |\nabla_t \cdot \mathbf{h}|^2)d\Omega + (\beta^2 - k_0^2 n_+^2) \int\limits_{R^2} |\mathbf{h}|^2\, d\Omega \leq 0. \qquad (7.64)$$

If $|\beta| \geq k_0 n_+$ the above inequality implies $\mathbf{h} = \mathbf{e} = 0$, and no guided modes exist in the optical fiber. Therefore, the guided solution (β, k) of (7.63) must satisfy

$$k_0 > |\beta|/n_+. \qquad (7.65)$$

In the region Ω_∞, the first equation of (7.63) becomes

$$\nabla_t^2 \mathbf{h} + (k_0^2 n_\infty^2 - \beta^2)\mathbf{h} = 0,\ \rho \in \Omega_\infty.$$

From the uniqueness theorem for Helmholtz equation, the above relation implies that $\mathbf{h}$ must be zero in $R^2 - D$ (D is a disk containing the core region Ω_c) in order that $\mathbf{h} \in [L^2(R^2)]^3$ if $k_0 > |\beta|/n_\infty$, and hence $\mathbf{h}$ must be identically zero in R^2. Therefore, the guided solution (β, k_0) of (7.63) must also satisfy

$$k_0 \leq |\beta|/n_\infty. \qquad (7.66)$$

Combining (7.65) and (7.66) gives

$$|\beta|/n_+ < k_0 \leq |\beta|/n_\infty. \qquad (7.67)$$

The expression is similar to (7.62) and implies

$$n_+ > n_\infty. \qquad (7.68)$$

This is called the **guidance condition** for the optical fiber.

7.4.3   Eigenvalues and Essential Spectrum

Consider the optical fiber shown in Figure 7.4 and let $H = [L^2(R^2)]^3$. The inner product of H is defined by $(\mathbf{h}_1, \mathbf{h}_2) = \int\limits_{R^2} \mathbf{h}_1 \cdot \bar{\mathbf{h}}_2 d\Omega$. Considering the last equation of (7.63), the first equation of (7.63) can be modified as follows

$$\nabla_\beta \times \frac{1}{n^2}\nabla_\beta \times \mathbf{h} - \frac{1}{n_\infty^2}\nabla_\beta(\nabla_\beta \cdot \mathbf{h}) = \hat{B}(\mathbf{h}) = k^2 \mathbf{h}, \qquad (7.69)$$

where $\hat{B} = \nabla_\beta \times n^{-2}\nabla_\beta \times (\cdot) - n_\infty^{-2}\nabla_\beta\nabla_\beta(\cdot)$. Equation (7.69) is a two-dimensional eigen-value problem. For convenience, we let $D(\hat{B}) = [C_0^\infty(R^2)]^3$. For all $\mathbf{h}_1, \mathbf{h}_2 \in D(\hat{B})$, we have

$$
(\hat{B}(\mathbf{h}_1), \mathbf{h}_2) = \int_{R^2} \left[\nabla_\beta \times \frac{1}{n^2}\nabla_\beta \times \mathbf{h}_1 - \frac{1}{n_\infty^2}\nabla_\beta(\nabla_\beta \cdot \mathbf{h}_1) \right] \cdot \bar{\mathbf{h}}_2 d\Omega,
$$

$$
= \int_{R^2} \left(\frac{1}{n^2}\nabla_\beta \times \mathbf{h}_1 \cdot \overline{\nabla_\beta \times \mathbf{h}_2} + \frac{1}{n_\infty^2}\nabla_\beta \cdot \mathbf{h}_1 \overline{\nabla_\beta \cdot \mathbf{h}_2} \right) d\Omega. \tag{7.70}
$$

Hence $\hat{B}$ is a symmetric operator. Let n_- be defined by $n_- = \min\limits_{\rho \in R^2} n(\rho)$. It follows from (7.70) that

$$
(\hat{B}(\mathbf{h}_1), \mathbf{h}_2) \geq \frac{1}{n_+^2} \int_{R^2} (\nabla_\beta \times \mathbf{h}_1 \cdot \overline{\nabla_\beta \times \mathbf{h}_2} + \nabla_\beta \cdot \mathbf{h}_1 \overline{\nabla_\beta \cdot \mathbf{h}_2}) d\Omega,
$$

$$
(\hat{B}(\mathbf{h}_1), \mathbf{h}_2) \leq \frac{1}{n_-^2} \int_{R^2} (\nabla_\beta \times \mathbf{h}_1 \cdot \overline{\nabla_\beta \times \mathbf{h}_2} + \nabla_\beta \cdot \mathbf{h}_1 \overline{\nabla_\beta \cdot \mathbf{h}_2}) d\Omega. \tag{7.71}
$$

For all $\mathbf{h}_1, \mathbf{h}_2 \in D(\hat{B})$, we have

$$
\int_{R^2} (\nabla_\beta \times \mathbf{h}_1 \cdot \overline{\nabla_\beta \times \mathbf{h}_2} + \nabla_\beta \cdot \mathbf{h}_1 \overline{\nabla_\beta \cdot \mathbf{h}_2}) d\Omega
$$

$$
= \int_{R^2} (-\bar{\mathbf{h}}_2 \cdot \nabla_t^2 \mathbf{h}_1 + \beta^2 \mathbf{h}_1 \cdot \bar{\mathbf{h}}_2 + \bar{\mathbf{h}}_2 \cdot \nabla_\beta(\nabla_\beta \cdot \mathbf{h}_1) + \nabla_\beta \cdot \mathbf{h}_1 \overline{\nabla_\beta \cdot \mathbf{h}_2}) d\Omega \tag{7.72}
$$

$$
= \int_{R^2} (-\bar{\mathbf{h}}_2 \cdot \nabla_t^2 \mathbf{h}_1 + \beta^2 \mathbf{h}_1 \cdot \bar{\mathbf{h}}_2) d\Omega,
$$

and (7.71) implies

$$
(\hat{B}(\mathbf{h}), \mathbf{h}) \geq \frac{1}{n_+^2} \int_{R^2} (|\nabla_t \times \mathbf{h}|^2 + |\nabla_t \cdot \mathbf{h}|^2 + \beta^2 |\mathbf{h}|^2) d\Omega,
$$

$$
(\hat{B}(\mathbf{h}), \mathbf{h}) \leq \frac{1}{n_-^2} \int_{R^2} (|\nabla_t \times \mathbf{h}|^2 + |\nabla_t \cdot \mathbf{h}|^2 + \beta^2 |\mathbf{h}|^2) d\Omega. \tag{7.73}
$$

Hence $\hat{B}$ is positive-bounded-below if $\beta \neq 0$ and we may introduce the energy product $(\mathbf{h}_1, \mathbf{h}_2)_{\hat{B}} = (\hat{B}(\mathbf{h}_1), \mathbf{h}_2)$. The completion of $D(\hat{B})$ with respect to the energy norm $\|\cdot\|_{\hat{B}}$ is the energy space $H_{\hat{B}}$. We now show that $H_{\hat{B}}$ is isomorphic to the space

$$
H(\nabla_t \times, \nabla_t \cdot) = \{\mathbf{h} \in [L^2(R^2)]^3 | \nabla_t \times \mathbf{h} \in [L^2(R^2)]^3, \nabla_t \cdot \mathbf{h} \in L^2(R^2)\}
$$

equipped with the norm

$$\|\mathbf{h}\|_{H(\nabla_t \times, \nabla_t \cdot)} = (\|\mathbf{h}\|^2 + \|\nabla_t \times \mathbf{h}\|^2 + \|\nabla_t \cdot \mathbf{h}\|^2)^{1/2},$$

where all derivatives are understood in the generalized sense. Let $\mathbf{h} \in H_{\hat{B}}$. Then there exists admissible sequence $\{\mathbf{h}_n \in D(\hat{B})\}$ for $\mathbf{h}$ such that

$$\mathbf{h}_n \xrightarrow[n \to \infty]{} \mathbf{h} \tag{7.74}$$

in H and $\{\mathbf{h}_n\}$ is a Cauchy sequence in $H_{\hat{B}}$. From (7.73), we obtain

$$\frac{1}{n_-^2} \int_{R^2} (|\nabla_t \times \mathbf{h}_n - \nabla_t \times \mathbf{h}_m|^2 + |\nabla_t \cdot \mathbf{h}_n - \nabla_t \cdot \mathbf{h}_m|^2 + \beta^2 |\mathbf{h}_n - \mathbf{h}_m|^2) d\Omega \leq \|\mathbf{h}_n - \mathbf{h}_m\|_{\hat{B}}^2$$

$$\leq \frac{1}{n_+^2} \int_{R^2} (|\nabla_t \times \mathbf{h}_n - \nabla_t \times \mathbf{h}_m|^2 + |\nabla_t \cdot \mathbf{h}_n - \nabla_t \cdot \mathbf{h}_m|^2 + \beta^2 |\mathbf{h}_n - \mathbf{h}_m|^2) d\Omega. \tag{7.75}$$

Consequently, $\{\nabla_t \times \mathbf{h}_n\}$ and $\{\nabla_t \cdot \mathbf{h}_n\}$ are Cauchy sequences in $[L^2(R^2)]^3$ and $L^2(R^2)$ respectively. As a result, there exist $\mathbf{e} \in [L^2(R^2)]^3$, and $\rho \in L^2(R^2)$ such that

$$\nabla_t \times \mathbf{h}_n \xrightarrow[n \to \infty]{} \mathbf{e}, \nabla_t \cdot \mathbf{h}_n \xrightarrow[n \to \infty]{} \rho. \tag{7.76}$$

From

$$\int_{R^2} \nabla_t \times \mathbf{h}_n \cdot \boldsymbol{\varphi} d\Omega = \int_{R^2} \mathbf{h}_n \cdot \nabla_t \times \boldsymbol{\varphi} d\Omega, \boldsymbol{\varphi} \in [C_0^\infty(R^2)]^2,$$

$$\int_{R^2} (\nabla_t \cdot \mathbf{h}_n) \varphi d\Omega = - \int_{R^2} \mathbf{h}_n \cdot \nabla_t \varphi d\Omega, \varphi \in C_0^\infty(R^2),$$

we obtain

$$\int_{R^2} \mathbf{e} \cdot \boldsymbol{\varphi} d\Omega = \int_{R^2} \mathbf{h} \cdot \nabla_t \times \boldsymbol{\varphi} d\Omega, \boldsymbol{\varphi} \in [C_0^\infty(R^2)]^2,$$

$$\int_{R^2} \rho \varphi d\Omega = - \int_{R^2} \mathbf{h} \cdot \nabla_t \varphi d\Omega, \varphi \in C_0^\infty(R^2).$$

Therefore, $\nabla_t \times \mathbf{h} = \mathbf{e}$ and $\nabla_t \cdot \mathbf{h} = \rho$ hold in the generalized sense. By (7.74) and (7.76), we have $\mathbf{h} \in H(\nabla_t \times, \nabla_t \cdot)$. On the other hand, if we assume that $\mathbf{h} \in H(\nabla_t \times, \nabla_t \cdot)$, then there exists a sequence $\{\mathbf{h}_n\} \subset D(\hat{B})$ such that (7.74) and (7.76) hold (since $D(\hat{B})$ is dense in $H(\nabla_t \times, \nabla_t \cdot)$). Thus, $\{\mathbf{h}_n\}$ is an admissible sequence for $\mathbf{h}$ from (7.75). Therefore, $\mathbf{h} \in H_{\hat{B}}$. The Friedrichs extension of $\hat{B}$ is denoted by $\hat{B}_F$. Instead of solving (7.69), we consider the following generalized eigenvalue problem

$$\hat{B}_F(\mathbf{h}) = \nabla_\beta \times \frac{1}{n^2} \nabla_\beta \times \mathbf{h} - \frac{1}{n_\infty^2} \nabla_\beta(\nabla_\beta \cdot \mathbf{h}) = k_0^2 \mathbf{h}, \tag{7.77}$$

where all derivatives are understood in the generalized sense. The bilinear form associated with $\hat{B}_F$ is

$$
\begin{aligned}
b(\mathbf{h}_1, \mathbf{h}_2) &= (\hat{B}_F(\mathbf{h}_1), \mathbf{h}_2) \\
&= \int_{R^2} \left(\frac{1}{n^2} \nabla_\beta \times \mathbf{h}_1 \cdot \overline{\nabla_\beta \times \mathbf{h}_2} + \frac{1}{n_\infty^2} \nabla_\beta \cdot \mathbf{h}_1 \overline{\nabla_\beta \cdot \mathbf{h}_2} \right) d\Omega,
\end{aligned}
\tag{7.78}
$$

which is a closed Hermitian bilinear form with $D(b) = H(\nabla_t \times, \nabla_t \cdot)$. From Theorem 7.2 and (7.73), we obtain

$$\sigma(\hat{B}_F) \subset [\beta^2/n_+^2, +\infty). \tag{7.79}$$

It follows from (7.67) that the eigenvalues k_0^2 of $\hat{B}_F$ satisfy

$$\sigma_p(\hat{B}_F) \subset (|\beta|^2/n_+^2, |\beta|^2/n_\infty^2]. $$

The continuous part of the spectrum is the essential spectrum denoted by $\sigma_{ess}(\hat{B}_F)$, which corresponds to the radiation modes. It can be shown that

$$\sigma_{ess}(\hat{B}_F) = [\beta^2/n_\infty^2, +\infty). \tag{7.80}$$

To prove this, one may rewrite (7.78) as (Bamberger and Bonnet, 1990)

$$b(\mathbf{h}_1, \mathbf{h}_2) = d(\mathbf{h}_1, \mathbf{h}_2) + \beta d_1(\mathbf{h}_1, \mathbf{h}_2) + \beta^2 d_2(\mathbf{h}_1, \mathbf{h}_2) \tag{7.81}$$

with

$$
\begin{aligned}
d(\mathbf{h}_1, \mathbf{h}_2) = &\int_{R^2} \left(\frac{1}{n^2} \nabla_t \times \mathbf{h}_{t1} \cdot \nabla_t \times \bar{\mathbf{h}}_{t2} + \frac{1}{n_\infty^2} \nabla_t \cdot \mathbf{h}_{t1} \nabla_t \cdot \bar{\mathbf{h}}_{t2} + \frac{1}{n^2} \nabla_t h_{z1} \cdot \nabla_t \bar{h}_{z2} \right) d\Omega \\
&+ \frac{\beta^2}{n_\infty^2} \int_{R^2} \mathbf{h}_1 \cdot \bar{\mathbf{h}}_2 d\Omega,
\end{aligned}
$$

$$d_1(\mathbf{h}_1, \mathbf{h}_2) = j \int_{R^2} (n^{-2} - n_\infty^{-2})(\nabla_t h_{z1} \cdot \bar{\mathbf{h}}_{t2} - \mathbf{h}_{t1} \cdot \nabla_t \bar{h}_{z2}) d\Omega,$$

$$d_2(\mathbf{h}_1, \mathbf{h}_2) = \int_{R^2} (n^{-2} - n_\infty^{-2}) \mathbf{h}_{t1} \cdot \bar{\mathbf{h}}_{t2} d\Omega.$$

It is easy to show that d_0, d_1 and d_2 are Hermitian and continuous on $H(\nabla_t \times, \nabla_t \cdot)$. We have

$$d\,(\mathbf{h}, \mathbf{h}) \geq \frac{1}{n_+^2} \int_{R^2} (|\nabla_t \times \mathbf{h}_t|^2 + |\nabla_t \cdot \mathbf{h}_t|^2 + |\nabla_t h_z|^2) d\Omega + \frac{\beta^2}{n_\infty^2} \int_{R^2} |\mathbf{h}|^2 \, d\Omega$$

$$\geq \min\left\{ \frac{1}{n_+^2}, \frac{\beta^2}{n_\infty^2} \right\} \int_{R^2} (|\nabla_t \times \mathbf{h}|^2 + |\nabla_t \cdot \mathbf{h}|^2 + |\mathbf{h}|^2 + |\nabla_t h_z|^2) d\Omega$$

$$\geq \min\left\{ \frac{1}{n_+^2}, \frac{\beta^2}{n_\infty^2} \right\} \left[\int_{R^2} (|\nabla_t \times \mathbf{h}|^2 + |\nabla_t \cdot \mathbf{h}|^2 + |\mathbf{h}|^2) d\Omega \right]$$

$$= \min\left\{ \frac{1}{n_+^2}, \frac{\beta^2}{n_\infty^2} \right\} \|\mathbf{h}\|_{H(\nabla_t \times, \nabla_t \cdot)}^2 .$$

(7.82)

Let $c_i (i = 1, 2, \cdots)$ denote constants. We now show that

$$|d_1(\mathbf{h}, \mathbf{h})| \leq c_1 d(\mathbf{h}, \mathbf{h}), \ |d_2(\mathbf{h}, \mathbf{h})| \leq c_2 d(\mathbf{h}, \mathbf{h}). \tag{7.83}$$

Since $n^{-2} - n_\infty^{-2}$ is zero outside Ω_c, we may write

$$|d_1(\mathbf{h}, \mathbf{h})| \leq c_3 \left| \int_{\Omega_c} (\nabla_t h_z \cdot \bar{\mathbf{h}}_t - \mathbf{h}_t \cdot \nabla_t \bar{h}_z) d\Omega \right|$$

$$\leq c_3 \int_{\Omega_c} (|\nabla_t h_z \cdot \bar{\mathbf{h}}_t| + |\mathbf{h}_t \cdot \nabla_t \bar{h}_z|) d\Omega \leq 2c_3 \int_{\Omega_c} |\nabla_t h_z| \cdot |\bar{\mathbf{h}}_t| \, d\Omega.$$

Therefore,

$$|d_1(\mathbf{h}, \mathbf{h})| \leq c_3 \int_{\Omega_c} (|\nabla_t h_z|^2 + |\mathbf{h}_t|^2) d\Omega, \tag{7.84}$$

$$|d_1(\mathbf{h}, \mathbf{h})| \leq c_4 \int_{\Omega_c} |\nabla_t h_z|^2 \, d\Omega \int_{\Omega_c} |\mathbf{h}_t|^2 \, d\Omega. \tag{7.85}$$

It is readily found that

$$d_2(\mathbf{h}, \mathbf{h}) \leq c_5 \int_{\Omega_c} |\mathbf{h}|^2 \, d\Omega. \tag{7.86}$$

Thus (7.83) holds due to (7.84) and (7.86). Consider a sequence $\{\mathbf{h}_n\} \subset H(\nabla_t \times, \nabla_t \cdot)$ such that

$$\|\mathbf{h}_n\|^2 + d(\mathbf{h}_n, \mathbf{h}_n) \leq c_6.$$

This inequality implies

$$\|\mathbf{h}_n\|_{H(\nabla_t \times, \nabla_t \cdot)} \le c_7.$$

It is easy to see that $H(\nabla_t \times, \nabla_t \cdot)$ is compactly embedded in $[L^2(R^2)]^3$. As a result, there is a subsequence, still denoted by $\{\mathbf{h}_n\}$ such that

$$\|\mathbf{h}_m - \mathbf{h}_n\| \to 0 \tag{7.87}$$

as $m, n \to \infty$. It follows from (7.85), (7.86) and (7.87) that

$$|d_1(\mathbf{h}_m - \mathbf{h}_n, \mathbf{h}_m - \mathbf{h}_n)| \to 0,$$
$$|d_2(\mathbf{h}_m - \mathbf{h}_n, \mathbf{h}_m - \mathbf{h}_n)| \to 0.$$

By Theorem 7.3, we have

$$\sigma_{ess}(\hat{B}_F) = \sigma_{ess}(\hat{D}),$$

where $\hat{D}$ is the operator associated with the form d. We now show that $\sigma_{ess}(\hat{D}) = [\beta^2/n_\infty^2, +\infty)$. To this purpose, we only need to show that $\sigma_{ess}(\hat{D}_0) = [0, +\infty)$, where $\hat{D}_0$ is the operator associated with d_0 defined by

$$d_0(\mathbf{h}_1, \mathbf{h}_2) = \int\limits_{R^2} \left(\frac{1}{n^2} \nabla_t \times \mathbf{h}_{t1} \cdot \nabla_t \times \bar{\mathbf{h}}_{t2} + \frac{1}{n_\infty^2} \nabla_t \cdot \mathbf{h}_{t1} \nabla_t \cdot \bar{\mathbf{h}}_{t2} + \frac{1}{n^2} \nabla_t h_{z1} \cdot \nabla_t \bar{h}_{z2} \right) d\Omega.$$

Making use of (7.79) with $\beta = 0$, we have $\sigma_{ess}(\hat{D}_0) \subset [0, +\infty)$. We only need to prove that $(0, +\infty) \subset \sigma_{ess}(\hat{D}_0)$ since $\sigma_{ess}(\hat{D}_0)$ is closed. Let us consider the sequence $\{\mathbf{h}_n\}$ with (Bamberger and Bonnet, 1990)

$$\mathbf{h}_n(\rho) = \frac{1}{\sqrt{n}} \psi(\rho/n) J_0(\sqrt{\gamma}\rho)\mathbf{h}_0, \rho \in R^2,$$

where ψ is a function of $C_0^\infty(R^2)$ which vanishes in Ω_c, $\mathbf{h}_0$ is a vector in C^3 (C is the complex plane), $\gamma \in (0, +\infty)$ is an arbitrary positive real number. By properly choosing $\mathbf{h}_0$, it is easy to show that

1. $\|\mathbf{h}_n\| = 1$.
2. $\|\hat{D}_0\mathbf{h}_n - \gamma\mathbf{h}_n\| \to 0, n \to \infty$.
3. $\mathbf{h}_n \xrightarrow[n \to \infty]{w} 0$ in H.

By Theorem 7.1, we have $(0, +\infty) \subset \sigma_{ess}(\hat{D}_0)$. Thus, we have proved (7.80).

In order to find the dispersion relation for the guided modes, we must find all pairs (β, k_0) with $\beta > 0$, $k_0 > 0$ such that there exists $\mathbf{h} \neq 0$ satisfying (7.77) or

$$b(\mathbf{h}, \mathbf{h}) = k_0^2(\mathbf{h}, \mathbf{h}). \tag{7.88}$$

The eigenvalues k_0^2 as a function of β can be obtained by applying the min-max principle. Let

$$\lambda_m(\beta) = \sup_{\mathbf{h}_1, \mathbf{h}_2, \cdots, \mathbf{h}_{m-1} \in [L^2(R^2)]^3} \inf_{\substack{\mathbf{h} \in [L(R^2)]^3, \|\mathbf{h}\|=1 \\ \mathbf{h} \in [\mathbf{h}_1, \mathbf{h}_2, \cdots, \mathbf{h}_{m-1}]^\perp}} b(\mathbf{h}, \mathbf{h}). \tag{7.89}$$

The solutions of (β, k_0) of (7.88) are the roots of the dispersion relation

$$k_0^2 = \lambda_m(\beta), m = 1, 2, \cdots. \tag{7.90}$$

Note that we may use (7.81) to write (7.88) as

$$\beta^2 \left[d_2(\mathbf{h}, \mathbf{h}) + \frac{1}{n_\infty^2}(\mathbf{h}, \mathbf{h}) \right] + \beta d_1(\mathbf{h}, \mathbf{h}) + d_0(\mathbf{h}, \mathbf{h}) - k_0^2(\mathbf{h}, \mathbf{h}) = 0. \tag{7.91}$$

Given the wavenumber k_0, the quadratic equation has two solutions $\beta_{1,2}(k_0)$.

Remark 7.1 (**Weakly guiding** approximation): Denote $\tau = 1 - n_\infty^2/n_+^2$. An optical fiber is called weakly guiding if $\tau \ll 1$. Let $\xi(\boldsymbol{\rho}) = (1 - n^2/n_+^2)/\tau$. Then, $0 \leq \xi(\boldsymbol{\rho}) \leq 1$. It follows from (7.54) that

$$\nabla_t^2 \mathbf{e}_t + \nabla_t(\mathbf{e}_t \cdot \nabla_t \ln n^2) + k_0^2 n^2 \mathbf{e}_t = \beta^2 \mathbf{e}_t, \boldsymbol{\rho} \in R^2. \tag{7.92}$$

The second term is negligible if the optical fiber is weakly guiding since

$$-\nabla_t \ln n^2 = -\nabla_t \ln(1 - \tau\xi) = \tau\nabla_t\xi + \tau^2\nabla_t(\xi^2) + \cdots.$$

In weak guidance, Equation (7.92) reduces to

$$\nabla_t^2 \mathbf{e}_t + k_0^2 n^2 \mathbf{e}_t = \beta^2 \mathbf{e}_t.$$

The equation can be solved using k_0^2 as the parameter and β^2 as the eigenvalue.   □

7.5 Inhomogeneous Cavity Resonator

Inhomogeneous cavity resonators are often found in microwave engineering. For example, a dielectric resonator enclosed by a metal shield to prevent radiation forms an inhomogeneous cavity resonator. Another example is to use a metal cavity to measure the dielectric constant of a sample placed inside the cavity.

7.5.1 Mode Theory

By eliminating $\mathbf{H}$ from the time-harmonic Maxwell equations in an inhomogeneous metal cavity, which occupies a region V bounded by a conducting surface S, we may find that the electric field $\mathbf{E}$ satisfies the following eigenvalue equation

$$\nabla \times \mu_r^{-1} \nabla \times \mathbf{E}(\mathbf{r}) = k_e^2 \varepsilon_r \mathbf{E}(\mathbf{r}), \mathbf{r} \in V,$$
$$\mathbf{u}_n \times \mathbf{E}(\mathbf{r}) = 0, \mathbf{r} \in S,$$
(7.93)

where $\mu_r = \mu / \mu_0$ and $\varepsilon_r = \varepsilon / \varepsilon_0$ are relative permeability and permittivity of the medium filled in the cavity respectively, and k_e^2 is the eigenvalue to be determined. The above equation implies that $\nabla \cdot \varepsilon_r \mathbf{E}(\mathbf{r}) = 0$. Instead of solving (7.93), we may consider the following regularized eigenvalue problem

$$\varepsilon_r^{-1} \nabla \times \mu_r^{-1} \nabla \times \mathbf{E} - \nabla \nabla \cdot \varepsilon_r \mathbf{E} - k_e^2 \mathbf{E} = 0, \mathbf{r} \in V,$$
$$\mathbf{u}_n \times \mathbf{E} = 0, \nabla \cdot \varepsilon_r \mathbf{E} = 0, \mathbf{r} \in S.$$
(7.94)

Since the first equation does not imply the divergence-free condition $\nabla \cdot \varepsilon_r \mathbf{E}(\mathbf{r}) = 0$, the boundary condition $\nabla \cdot \varepsilon_r \mathbf{E} = 0$ has been introduced in (7.94). The advantage of (7.94) is that it is an elliptical differential equation of second order. Similarly, we can construct the following eigenvalue problem for the magnetic field

$$\mu_r^{-1} \nabla \times \varepsilon_r^{-1} \nabla \times \mathbf{H} - \nabla \nabla \cdot \mu_r \mathbf{H} - k_h^2 \mathbf{H} = 0, \mathbf{r} \in V,$$
$$\mathbf{u}_n \cdot \mu_r \mathbf{H} = 0, \mathbf{u}_n \times \varepsilon_r^{-1} \nabla \times \mathbf{H} = 0, \mathbf{r} \in S.$$
(7.95)

Introducing the operator $\hat{B}_e = \varepsilon_r^{-1} \nabla \times \mu_r^{-1} \nabla \times - \nabla \nabla \cdot \varepsilon_r$, Equation (7.94) can be written as

$$\hat{B}_e(\mathbf{E}) = k_e^2 \mathbf{E}, \mathbf{r} \in V,$$
$$\mathbf{u}_n \times \mathbf{E} = 0, \nabla \cdot \varepsilon_r \mathbf{E} = 0, \mathbf{r} \in S.$$
(7.96)

The domain of definition of the operator $\hat{B}_e$ is

$$D(\hat{B}_e) = \left\{ \mathbf{E} | \mathbf{E} \in [C^\infty(V)]^3, \mathbf{u}_n \times \mathbf{E} = 0, \nabla \cdot \varepsilon_r \mathbf{E} = 0, \mathbf{r} \in S \right\}.$$
(7.97)

Let $L_{\varepsilon_r}^2(V)$ denote the space of square integrable functions defined in V with inner product defined by $(u, v)_{\varepsilon_r} = \int_V \varepsilon_r u v dV$ and the corresponding norm is denoted by $\|\cdot\|_{\varepsilon_r} = (\cdot, \cdot)_{\varepsilon_r}^{1/2}$.
We also introduce the inner product space $H_{\varepsilon_r} = [L_{\varepsilon_r}^2(V)]^3$ whose inner product is defined by

$$(\mathbf{E}_1, \mathbf{E}_2)_{\varepsilon_r} = \int_V \varepsilon_r \mathbf{E}_1 \cdot \mathbf{E}_2 dV, \mathbf{E}_1, \mathbf{E}_2 \in H_{\varepsilon_r}.$$

The corresponding norm is still denoted by $\|\cdot\|_{\varepsilon_r} = \sqrt{(\cdot, \cdot)_{\varepsilon_r}}$. For all $\mathbf{E}_1, \mathbf{E}_2 \in D(\hat{B}_e)$, we have

$$(\hat{B}_e(\mathbf{E}_1), \mathbf{E}_2)_{\varepsilon_r} = \int_V (\nabla \times \mu_r^{-1} \nabla \times \mathbf{E}_1 - \varepsilon_r \nabla \nabla \cdot \varepsilon_r \mathbf{E}_1) \cdot \mathbf{E}_2 dV$$

$$= \int_V \left[\mu_r^{-1} \nabla \times \mathbf{E}_1 \cdot \nabla \times \mathbf{E}_2 + (\nabla \cdot \varepsilon_r \mathbf{E}_1)(\nabla \cdot \varepsilon_r \mathbf{E}_2) \right] dV$$

from integration by parts. Therefore, $\hat{B}_e$ is symmetric and positive definite. Introducing a positive parameter ξ, Equation (7.96) can be modified as

$$\begin{aligned} \hat{A}_e(\mathbf{E}) &= (k_e^2 + \xi)\mathbf{E}, \mathbf{r} \in V, \\ \mathbf{u}_n \times \mathbf{E} &= 0, \nabla \cdot \varepsilon_r \mathbf{E} = 0, \mathbf{r} \in S, \end{aligned} \tag{7.98}$$

where $\hat{A}_e = \hat{B}_e + \xi \hat{I}$ is symmetric and positive-bounded-below with $D(\hat{A}_e) = D(\hat{B}_e)$. For all $\mathbf{E}_1, \mathbf{E}_2 \in D(\hat{A}_e)$, we may introduce the energy inner product

$$\begin{aligned} (\mathbf{E}_1, \mathbf{E}_2)_{\hat{A}_e} &= (\hat{A}_e(\mathbf{E}_1), \mathbf{E}_2)_{\varepsilon_r} \\ &= \int_V \left[\mu_r^{-1} \nabla \times \mathbf{E}_1 \cdot \nabla \times \mathbf{E}_2 + (\nabla \cdot \varepsilon_r \mathbf{E}_1)(\nabla \cdot \varepsilon_r \mathbf{E}_2) + \xi \varepsilon_r \mathbf{E}_1 \cdot \mathbf{E}_2 \right] dV. \end{aligned}$$

The completion of $D(\hat{A}_e)$ with respect to the norm $\|\cdot\|_{\hat{A}_e} = (\cdot, \cdot)_{\hat{A}_e}$ is denoted by $H_{\hat{A}_e}$. Let $\mathbf{E} \in H_{\hat{A}_e}$. Then there exists an admissible sequence $\{\mathbf{E}_n \in D(\hat{A}_e)\}$ such that

$$\|\mathbf{E}_n - \mathbf{E}\|_{\varepsilon_r} \xrightarrow[n \to \infty]{} 0.$$

Since $\{\mathbf{E}_n\}$ is a Cauchy sequence in $H_{\hat{A}_e}$, we have

$$\|\mathbf{E}_n - \mathbf{E}_m\|_{\hat{A}_e} \xrightarrow[n,m \to \infty]{} 0,$$

which implies

$$\|\nabla \times \mathbf{E}_n - \nabla \times \mathbf{E}_m\|_{\varepsilon_r} \xrightarrow[n,m \to \infty]{} 0,$$

$$\|\nabla \cdot \varepsilon_r \mathbf{E}_n - \nabla \cdot \varepsilon_r \mathbf{E}_m\|_{\varepsilon_r} \xrightarrow[n,m \to \infty]{} 0.$$

Hence there exist $\mathbf{H}' \in H_{\varepsilon_r}$ and $\rho \in L_{\varepsilon_r}^2(V)$ such that

$$\|\nabla \times \mathbf{E}_n - \mathbf{H}'\|_{\varepsilon_r} \xrightarrow[n,m \to \infty]{} 0,$$

$$\|\nabla \cdot \varepsilon_r \mathbf{E}_n - \rho\|_{\varepsilon_r} \xrightarrow[n,m \to \infty]{} 0.$$

From

$$\int_V \nabla \times \mathbf{E}_n \cdot \boldsymbol{\varphi} d\Omega = \int_V \mathbf{E}_n \cdot \nabla \times \boldsymbol{\varphi} d\Omega, \; \boldsymbol{\varphi} \in [C_0^\infty(V)]^3,$$

$$\int_V (\nabla \cdot \varepsilon_r \mathbf{E}_n) \varphi dV = -\int_V \varepsilon_r \mathbf{E}_n \cdot \nabla \varphi dV, \; \varphi \in C_0^\infty(V),$$

we may let $n \to \infty$ to obtain

$$\int_V \mathbf{H}' \cdot \boldsymbol{\varphi} dV = \int_V \mathbf{E} \cdot \nabla \times \boldsymbol{\varphi} dV,$$

$$\int_V \rho \varphi dV = -\int_V \varepsilon_r \mathbf{E} \cdot \nabla \varphi dV.$$

Therefore, $\nabla \times \mathbf{E} = \mathbf{H}'$ and $\nabla \cdot \varepsilon_r \mathbf{E} = \rho$ hold in the generalized sense. For arbitrary $\mathbf{E}_1, \mathbf{E}_2 \in H_{\hat{A}_e}$ there are admissible sequences $\{\mathbf{E}_{1n}\}$ and $\{\mathbf{E}_{2n}\}$ such that

$$\|\mathbf{E}_{1n} - \mathbf{E}_1\|_{\varepsilon_r} \xrightarrow[n \to \infty]{} 0, \; \|\mathbf{E}_{2n} - \mathbf{E}_2\|_{\varepsilon_r} \xrightarrow[n \to \infty]{} 0.$$

We define

$$(\mathbf{E}_1, \mathbf{E}_2)_{\hat{A}_e} = \lim_{n \to \infty} (\mathbf{E}_{1n}, \mathbf{E}_{2n})_{\hat{A}_e}$$

$$= \int_V \left[\mu_r^{-1} \nabla \times \mathbf{E}_1 \cdot \nabla \times \mathbf{E}_2 + (\nabla \cdot \varepsilon_r \mathbf{E}_1)(\nabla \cdot \varepsilon_r \mathbf{E}_2) + \xi \varepsilon_r \mathbf{E}_1 \cdot \mathbf{E}_2 \right] dV,$$

where the derivatives are understood in the generalized sense. We now prove that the embedding $H_{\hat{A}_e} \subset H_{\varepsilon_r}$ is compact. Let $\hat{J}(\mathbf{E}) = \mathbf{E}, \mathbf{E} \in H_{\hat{A}_e}$. Then the linear operator $\hat{J} : H_{\hat{A}_e} \to H_{\varepsilon_r}$ is continuous since

$$\left\| \hat{J}(\mathbf{E}) \right\|_{\varepsilon_r}^2 = \|\mathbf{E}\|_{\varepsilon_r}^2 = \int_V \varepsilon_r \mathbf{E} \cdot \mathbf{E} dV \leq \xi^{-1} \|\mathbf{E}\|_{\hat{A}_e}^2 .$$

A bounded sequence $\{\mathbf{E}_n\} \subset H_{\hat{A}_e}$ implies

$$\|\mathbf{E}_n\|_{\hat{A}_e}^2 = \int_V \left[\mu_r^{-1} \nabla \times \mathbf{E}_n \cdot \nabla \times \mathbf{E}_n + (\nabla \cdot \varepsilon_r \mathbf{E}_n)(\nabla \cdot \varepsilon_r \mathbf{E}_n) + \xi \varepsilon_r \mathbf{E}_n \cdot \mathbf{E}_n \right] dV \leq c_1,$$

where c_1 is a constant. The compactness of the operator $\hat{J}$ follows from Rellich's theorem. From the general eigenvalue theory discussed in section 3.2.1, Equation (7.96) has an infinite set of eigenvalues $0 \leq k_{e1}^2 \leq k_{e2}^2 \leq \cdots \leq k_{en}^2 \leq \cdots$, and $k_{en}^2 \to \infty$ as $n \to \infty$. The corresponding set of eigenfunctions $\{\mathbf{E}_n\}$ constitutes a complete set in H_{ε_r}. The different eigenfunctions corresponding to different eigenvalues are orthogonal and can be normalized as follows

$$\int_V \varepsilon_r \mathbf{E}_m \cdot \mathbf{E}_n dV = \begin{cases} 0 & (m \neq n) \\ 1 & (m = n) \end{cases}.$$

Each eigenfunction $\mathbf{E}_n$ belongs to one of the following four categories:

1. $\nabla \times \mathbf{E}_n = 0, \nabla \cdot \varepsilon_r \mathbf{E}_n = 0$.
2. $\nabla \times \mathbf{E}_n \neq 0, \nabla \cdot \varepsilon_r \mathbf{E}_n = 0$.
3. $\nabla \times \mathbf{E}_n = 0, \nabla \cdot \varepsilon_r \mathbf{E}_n \neq 0$.
4. $\nabla \times \mathbf{E}_n \neq 0, \nabla \cdot \varepsilon_r \mathbf{E}_n \neq 0$.

We now show that a complete set of eigenfunctions can be derived from the first three categories. Assuming that $\mathbf{E}_n$ belongs to the category 4, we can introduce two new functions

$$\mathbf{E}' = A\varepsilon_r^{-1} \nabla \times \mu_r^{-1} \nabla \times \mathbf{E}_n, \ \mathbf{E}'' = B\nabla \nabla \cdot \varepsilon_r \mathbf{E}_n,$$

where A and B are constants. By use of (7.96), $\mathbf{E}_n$ can be expressed as a linear combination of $\mathbf{E}'$ and $\mathbf{E}''$

$$\mathbf{E}_n = k_{en}^{-2}(A^{-1}\mathbf{E}' - B^{-1}\mathbf{E}''). \tag{7.99}$$

Since $\mathbf{E}_n$ belongs to category 4, we have $k_{en}^2 \neq 0$, $\mathbf{E}' \neq 0$ and $\mathbf{E}'' \neq 0$. Applying $\varepsilon_r^{-1}\nabla \times \mu_r^{-1}\nabla\times$ to (7.96), we obtain

$$\varepsilon_r^{-1} \nabla \times \mu_r^{-1} \nabla \times \mathbf{E}' = k_{en}^2 \mathbf{E}'.$$

Since $\nabla \cdot \varepsilon_r \mathbf{E}' = 0$, we may add $\nabla \nabla \cdot \varepsilon_r \mathbf{E}'$ to the left-hand side of the above equation

$$\varepsilon_r^{-1} \nabla \times \mu_r^{-1} \nabla \times \mathbf{E}' - \nabla \nabla \cdot \varepsilon_r \mathbf{E}' = k_{en}^2 \mathbf{E}'.$$

Therefore $\mathbf{E}'$ satisfies the same differential equation as the eigenfunction $\mathbf{E}_n$. The tangential component of $\mathbf{E}'$ on the boundary of the cavity is

$$\begin{aligned} \mathbf{u}_n \times \mathbf{E}' &= A\mathbf{u}_n \times \varepsilon_r^{-1} \nabla \times \mu_r^{-1} \nabla \times \mathbf{E}_n \\ &= A\mathbf{u}_n \times \nabla \nabla \cdot \varepsilon_r \mathbf{E}_n + A\mathbf{u}_n \times k_{en}^2 \mathbf{E}_n \\ &= A\mathbf{u}_n \times \left(\mathbf{u}_n \frac{\partial}{\partial n} + \mathbf{u}_t \frac{\partial}{\partial t} \right) \nabla \cdot \varepsilon_r \mathbf{E}_n = 0, \end{aligned}$$

where $\mathbf{u}_t$ is the unit tangent vector of S and use is made of (7.96). By definition, $\nabla \cdot \varepsilon_r \mathbf{E}'$ vanishes on S. Therefore, $\mathbf{E}'$ satisfies (7.96) and $\mathbf{E}'$ is an eigenfunction belonging to category

2. It follows from (7.99) that $\mathbf{E}''$ is also an eigenfunction belonging to category 3. Furthermore $\mathbf{E}'$ and $\mathbf{E}''$ are orthogonal. In fact,

$$\int_V \varepsilon_r \mathbf{E}' \cdot \mathbf{E}'' d\Omega = \int_V AB(\nabla \times \mu_r^{-1}\nabla \times \mathbf{E}_n) \cdot (\nabla\nabla \cdot \varepsilon_r \mathbf{E}_n) dV$$

$$= -\int_V AB\nabla \cdot \varepsilon_r \mathbf{E}_n \nabla \cdot (\nabla \times \mu_r^{-1}\nabla \times \mathbf{E}_n) dV$$

$$+ \int_V AB\nabla \cdot \left[(\nabla \times \mu_r^{-1}\nabla \times \mathbf{E}_n)(\nabla \cdot \varepsilon_r \mathbf{E}_n)\right] dV = 0.$$

The eigenvalue problem (7.95) may be studied in a similar manner. Equation (7.95) has an infinite set of eigenvalues $0 \leq k_{h1}^2 \leq k_{h2}^2 \leq \cdots \leq k_{hn}^2 \leq \cdots$, and $k_{hn}^2 \to \infty$ as $n \to \infty$. The corresponding set of eigenfunctions $\{\mathbf{H}_n\}$ constitutes a complete set in H_{μ_r}. The different eigenfunctions corresponding to different eigenvalues are orthogonal and can be normalized as follows

$$\int_V \mu_r \mathbf{H}_m \cdot \mathbf{H}_n dV = \begin{cases} 0 & (m \neq n) \\ 1 & (m = n) \end{cases}.$$

Also each eigenfunction $\mathbf{H}_n$ can be chosen from one of the following three categories:

  1. $\nabla \times \mathbf{H}_n = 0, \nabla \cdot \mathbf{H}_n = 0.$
  2. $\nabla \times \mathbf{H}_n \neq 0, \nabla \cdot \mathbf{H}_n = 0.$
  3. $\nabla \times \mathbf{H}_n = 0, \nabla \cdot \mathbf{H}_n \neq 0.$

The eigenfunctions belonging to category 2 in the two sets of eigenfunctions $\{\mathbf{E}_n\}$ and $\{\mathbf{H}_n\}$ are related. To show this, let $\mathbf{E}_n$ belong to category 2. Then $k_{en} \neq 0$, and we can define a function $\mathbf{H}_n$ through

$$\nabla \times \mathbf{E}_n = k_{en}\mu_r \mathbf{H}_n. \qquad (7.100)$$

Obviously $\mathbf{H}_n$ belongs to category 2. In addition

$$\mu_r^{-1}\nabla \times \varepsilon_r^{-1}\nabla \times \mathbf{H}_n - k_{en}^2 \mathbf{H}_n = k_{en}^{-1}\mu_r^{-1}\nabla \times \left(\varepsilon_r^{-1}\nabla \times \mu_r^{-1}\nabla \times \mathbf{E}_n - k_{en}^2 \mathbf{E}_n\right) = 0$$

and on the boundary S, we have

$$\mathbf{u}_n \times \nabla \times \mathbf{H}_n = k_{en}^{-1}\mathbf{u}_n \times \nabla \times \mu_r^{-1}\nabla \times \mathbf{E}_n = k_{en}^{-1}\mathbf{u}_n \times k_{en}^2 \varepsilon_r \mathbf{E}_n = 0.$$

Consider the integration of $\mathbf{u}_n \cdot \mathbf{H}_n$ over an arbitrary part of S, denoted ΔS

$$\int_{\Delta S} \mathbf{u}_n \cdot \mu_r \mathbf{H}_n dS = k_{en}^{-1} \int_{\Delta S} \mathbf{u}_n \cdot \nabla \times \mathbf{E}_n dS = k_{en}^{-1} \int_{\Delta\Gamma} \mathbf{u}_\Gamma \cdot \mathbf{E}_n d\Gamma$$

where $\Delta\Gamma$ is the closed contour around ΔS and $\mathbf{u}_\Gamma$ is the unit tangent vector along the contour. The right-hand side vanishes for an arbitrary ΔS, which implies $\mathbf{u}_n \cdot \mathbf{H}_n = 0$, $\mathbf{r} \in S$. Therefore $\mathbf{H}_n$ satisfies (7.95) and the corresponding eigenvalue is k_{en}^2. If $\mathbf{H}_m$ is another eigenfunction corresponding to $\mathbf{E}_m$ belonging to category 2, then

$$\int_V \mu_r \mathbf{H}_m \cdot \mathbf{H}_n d\Omega = (k_{em}k_{en})^{-1} \int_V \mu_r^{-1} \nabla \times \mathbf{E}_m \cdot \nabla \times \mathbf{E}_n dV$$

$$= (k_{em}k_{en})^{-1} \int_V \nabla \times \mu_r^{-1} \nabla \times \mathbf{E}_m \cdot \mathbf{E}_n dV + (k_{em}k_{en})^{-1}$$

$$\times \int_V \nabla \cdot \left(\mathbf{E}_n \times \mu_r^{-1} \nabla \times \mathbf{E}_m\right) dV$$

$$= (k_{em}k_{en})^{-1} k_{em}^2 \int_V \varepsilon_r \mathbf{E}_m \cdot \mathbf{E}_n dV + (k_{em}k_{en})^{-1}$$

$$\times \int_S \mathbf{u}_n \cdot \left(\mathbf{E}_n \times \mu_r^{-1} \nabla \times \mathbf{E}_m\right) dS$$

$$= \frac{k_{em}}{k_{en}} \int_V \varepsilon_r \mathbf{E}_m \cdot \mathbf{E}_n dV.$$

So the eigenfunctions $\mathbf{H}_n$ defined by (7.100) are orthogonal to each other, and the eigenfunctions $\mathbf{H}_n$ in category 2 can be derived from the eigenfunction $\mathbf{E}_n$ in category 2. Conversely if $\mathbf{H}_n$ is in category 2, we can define $\mathbf{E}_n$ by

$$\nabla \times \mathbf{H}_n = k_{hn}\varepsilon_r \mathbf{E}_n, \tag{7.101}$$

and a similar discussion indicates that $\mathbf{E}_n$ is an eigenfunction of (7.94) with k_{hn} being the eigenvalue. So the completeness of the two sets are still guaranteed if the eigenfunctions belonging to category 2 in $\{\mathbf{E}_n\}$ and $\{\mathbf{H}_n\}$ are related through either (7.100) or (7.101). From now on, Equations (7.100) and (7.101) will be assumed to hold, and $k_{en} = k_{hn}$ will be denoted by k_n. The complete set $\{\mathbf{E}_n\}$ is most appropriate for the expansion of electric field, and $\{\mathbf{H}_n\}$ is most appropriate for the expansion of the magnetic field.

7.5.2   Field Expansions

If the cavity contains an impressed electric current source $\mathbf{J}$ and a magnetic current source $\mathbf{J}_m$, the fields excited by these sources satisfy the Maxwell equations

$$\varepsilon_r^{-1} \nabla \times \mathbf{H}(\mathbf{r}) = (\sigma + j\omega\varepsilon_0)\mathbf{E}(\mathbf{r}) + \mathbf{J}(\mathbf{r}),$$
$$\mu_r^{-1} \nabla \times \mathbf{E}(\mathbf{r}) = -j\omega\mu_0 \mathbf{H}(\mathbf{r}) - \mathbf{J}_m(\mathbf{r}), \tag{7.102}$$

and can be expanded in terms of the vector eigenfunctions

$$\mathbf{E} = \sum_n v_n \mathbf{E}_n + \sum_v v_v \mathbf{E}_v, \mathbf{H} = \sum_n i_n \mathbf{H}_n + \sum_\tau i_\tau \mathbf{H}_\tau, \qquad (7.103)$$

$$\mu_r^{-1} \nabla \times \mathbf{E} = \sum_n \mathbf{H}_n \int_V \nabla \times \mathbf{E} \cdot \mathbf{H}_n dV + \sum_\tau \mathbf{H}_\tau \int_V \nabla \times \mathbf{E} \cdot \mathbf{H}_\tau dV,$$

$$\varepsilon_r^{-1} \nabla \times \mathbf{H} = \sum_n \mathbf{E}_n \int_V \nabla \times \mathbf{H} \cdot \mathbf{E}_n dV + \sum_v \mathbf{E}_v \int_V \nabla \times \mathbf{H} \cdot \mathbf{E}_v dV, \qquad (7.104)$$

where the subscript n denotes the eigenfunctions belonging to category 2, and the Greek subscript v and τ for the eigenfunctions belonging to category 1 or 3, and

$$v_{n(v)} = \int_V \varepsilon_r \mathbf{E} \cdot \mathbf{E}_{n(v)} dV, i_{n(\tau)} = \int_V \mu_r \mathbf{H} \cdot \mathbf{H}_{n(\tau)} dV. \qquad (7.105)$$

Considering the following calculations

$$\int_V \nabla \times \mathbf{E} \cdot \mathbf{H}_n dV = \int_V \mathbf{E} \cdot \nabla \times \mathbf{H}_n dV + \int_V \nabla \cdot (\mathbf{E} \times \mathbf{H}_n) dV = k_n v_n,$$

$$\int_V \nabla \times \mathbf{E} \cdot \mathbf{H}_\tau dV = \int_V \mathbf{E} \cdot \nabla \times \mathbf{H}_\tau dv + \int_V \nabla \cdot (\mathbf{E} \times \mathbf{H}_\tau) dV = 0,$$

$$\int_V \nabla \times \mathbf{H} \cdot \mathbf{E}_n dV = \int_V \mathbf{H} \cdot \nabla \times \mathbf{E}_n dV + \int_V \nabla \cdot (\mathbf{H} \times \mathbf{E}_n) dV = k_n i_n,$$

$$\int_V \nabla \times \mathbf{H} \cdot \mathbf{E}_v dV = \int_V \mathbf{H} \cdot \nabla \times \mathbf{E}_v dV + \int_V \nabla \cdot (\mathbf{H} \times \mathbf{E}_v) dV = 0,$$

Equation (7.104) can be written as

$$\mu_r^{-1} \nabla \times \mathbf{E} = \sum_n k_n v_n \mathbf{H}_n, \varepsilon_r^{-1} \nabla \times \mathbf{H} = \sum_n k_n i_n \mathbf{E}_n. \qquad (7.106)$$

Substituting the above expansions into (7.102) leads to

$$\sum_n k_n v_n \mathbf{H}_n = -j\omega\mu_0 \left(\sum_n i_n \mathbf{H}_n + \sum_\tau i_\tau \mathbf{H}_\tau \right) - \mathbf{J}_m,$$

$$\sum_n k_n i_n \mathbf{E}_n = (\sigma + j\omega\varepsilon_0) \left(\sum_n v_n \mathbf{E}_n + \sum_v v_v \mathbf{E}_v \right) + \mathbf{J}. \qquad (7.107)$$

Thus the expansion coefficients can easily be determined by

$$k_n v_n = -j\omega\mu_0 i_n - \int_V \mu_r \mathbf{J}_m \cdot \mathbf{H}_n dV,$$

$$j\omega\mu_0 i_\tau + \int_V \mu_r \mathbf{J}_m \cdot \mathbf{H}_\tau dV = 0,$$

$$k_n i_n = (\sigma + j\omega\varepsilon_0) v_n + \int_V \varepsilon_r \mathbf{J}(\mathbf{r}) \cdot \mathbf{E}_n dV, \quad (7.108)$$

$$(\sigma + j\omega\varepsilon_0) v_\nu + \int_\Omega \varepsilon_r \mathbf{J} \cdot \mathbf{E}_\nu d\Omega = 0.$$

The wave propagation in inhomogeneous media is a very complicated process. In most situations, we have to adopt the numerical methods, such as finite difference and finite element, to fully understand the physical process.

> The mathematical facts worthy of being studied are those which, by their analogy with other facts, are capable of leading us to the knowledge of a physical law.
> —Henri Poincaré

8

Time-domain Theory

> Since Maxwell's time, physical reality has been thought of as represented by continuous fields, and not capable of any mechanical interpretation. This change in the conception of reality is the most profound and the most fruitful that physics has experienced since the time of Newton.
>
> —Albert Einstein

The ever-increasing interest in ultra-wideband techniques and high speed devices has made time-domain analysis an important research field. The short pulses may be used to obtain high resolution and high accuracy in radar and to increase information transmission rate in communication systems. An important feature of the short pulses is that the rate of decay of electromagnetic energy can be slowed down when the pulse shape is properly chosen. A very short intense electromagnetic pulse can result in irreversible damages to electronic equipments, such as computers and radio receivers.

Compared to the voluminous literature on time-harmonic theory of electromagnetics, time-domain electromagnetics is still a virgin land to be cultivated. In the time-domain theory, the fields are assumed to start at a finite instant of time and Maxwell equations are solved subject to initial conditions, boundary conditions, excitation conditions and causality. According to the linear system theory and Fourier analysis, the response of the system to an arbitrary pulse can be obtained by superimposing its responses to all the real frequencies. In other words, the solution to the time-domain problem can be expressed in terms of the time-harmonic solution through the use of the Fourier transform. This process can be assisted by the fast Fourier transform and has been used extensively in studying the transient responses of electromagnetic systems. The procedure, however, is not always most effective and is not a trivial exercise since the time-harmonic problem must be solved for a large range of frequencies, and only an approximate time-harmonic solution valid over a finite frequency band can be obtained.

Moreover, the time-harmonic solution may not be able to give the correct physical picture in some situations. The time-harmonic field theory is founded on the assumption that a monotonic electromagnetic source turns on at $t = -\infty$ and the initial conditions of the fields produced by the source are ignored. This assumption does not cause any problem if the system has dissipation or radiation loss. When the system is lossless, the assumption may lead to physically unacceptable solutions. For example, the time-harmonic theory predicts that the

Foundations of Applied Electrodynamics   Geyi Wen
© 2010 John Wiley & Sons, Ltd

field response of a lossless metal cavity is sinusoidal if the excitation source is sinusoidal. The time-domain theory, however, shows that a sinusoidal response can be built up only if the cavity is excited by a sinusoidal source whose frequency coincides with one of the resonant frequencies. In addition, the field responses in a lossless cavity predicted by the time-harmonic theory are singular everywhere inside the cavity if the frequency of the sinusoidal excitation source coincides with one of the resonant frequencies of the cavity, while the time-domain theory always gives finite field responses. Therefore, we are forced to seek a solution in the time domain in some situations.

8.1   Time-domain Theory of Metal Waveguides[1]

In high-speed circuits, the signal frequency spectrum of a pulse may extend to terahertz regime and signal integrity problems may occur, which requires a deep understanding of the propagation characteristics of the transients in a waveguide. One of the research topics is to determine the response of the waveguide to an arbitrary input signal. If the input signal of the waveguide is $x(t)$, after traveling a distance z, the output from the waveguide is given by the Fourier integral

$$y(t) = \frac{1}{2\pi} \int\limits_{-\infty}^{\infty} X(\omega) e^{j(\omega t - \beta z)} d\omega. \tag{8.1}$$

Here $X(\omega)$ is the Fourier transform of $x(t)$; $\beta = v^{-1}\sqrt{\omega^2 - \omega_c^2}$ is the propagation constant; ω_c is the cut-off frequency of the propagating mode; and $v = 1/\sqrt{\mu\varepsilon}$ is the signal speed, with μ and ε being the permeability and permittivity of the medium filling the waveguide.

Several methods have been proposed to evaluate the Fourier integral in (8.1), such as the saddle point integration method (Namiki and Horuchi, 1952) and the stationary phase method (Ito, 1965). A serious drawback to the stationary phase method is that it contradicts the physical realizability as well as causality (that is, the response appears before the input signal is launched). A more rigorous approach is based on impulse response function for a lossless waveguide, which is defined as the inverse Fourier transform of the transfer function $e^{-j\beta z}$. The impulse response function can be expressed as an exact closed form and has been applied to study transient responses of a waveguide to various input signals (for example, Schulz-Dubois, 1970).

The response given by (8.1) is, however, hardly realistic when describing the propagation of a very short pulse or an ultra-wideband signal since it is based on an assumption that the waveguide is in a single-mode operation. This assumption is reasonable only for a narrow band signal but fails for a short pulse that covers a very wide range of frequency spectrums and thus will excite a number of higher-order modes in the waveguide. Many authors have approached the transient responses of waveguides for various input signals, such as a step function, a rectangular pulse or even a δ impulse, tacitly assuming the waveguide is in a

[1] W. Geyi, 'A time-domain theory of waveguide', *Progress in Electromagnetics Research*, PIER 59, 267–97, 2006. Reproduced by permission of ©2006 The Electromagnetics Academy & EMW Publishing.

single-mode operation. This approach has oversimplified the problem and the results obtained cannot accurately describe the transient process in the waveguide.

In order to find the real picture of the transient process in the waveguide, the time-domain Maxwell equations must be solved subject to initial conditions, boundary conditions and excitation conditions, and the higher-order mode effects must be taken into account.

8.1.1   Field Expansions

Assume that the medium in the waveguide is homogeneous and isotropic with medium parameters μ, ε and σ. The cross-section of the waveguide is denoted by Ω and its boundary by Γ, which is assumed to be a perfect conductor. The transient electrical field in a source-free region of the waveguide satisfies the equation

$$\nabla^2 \mathbf{E}(\mathbf{r}, t) - \frac{1}{v^2} \frac{\partial^2 \mathbf{E}(\mathbf{r}, t)}{\partial t^2} - \sigma \frac{\eta}{v} \frac{\partial \mathbf{E}(\mathbf{r}, t)}{\partial t} = 0, \mathbf{r} \in \Omega,$$

$$\nabla \cdot \mathbf{E}(\mathbf{r}, t) = 0, \mathbf{r} \in \Omega, \tag{8.2}$$

$$\mathbf{u}_n \times \mathbf{E}(\mathbf{r}, t) = 0, \mathbf{r} \in \Gamma,$$

where $v = 1/\sqrt{\mu \varepsilon}$ and $\mathbf{u}_n$ is the unit outward normal to the boundary Γ. The solution of (8.2) can be expressed as the sum of a transverse component and a longitudinal component, both of which are separable functions of transverse coordinates ρ and the longitudinal coordinate z with time

$$\mathbf{E}(\mathbf{r}, t) = [\mathbf{e}(\rho) + \mathbf{u}_z e_z(\rho)] u(z, t). \tag{8.3}$$

Inserting (8.3) into (8.2) and taking the boundary condition into account, we obtain

$$\nabla \times \nabla \times \mathbf{e}(\rho) - \nabla \nabla \cdot \mathbf{e}(\rho) - k_c^2 \mathbf{e}(\rho) = 0, \rho \in \Omega,$$

$$\mathbf{u}_n \times \mathbf{e}(\rho) = \nabla \cdot \mathbf{e}(\rho) = 0, \rho \in \Gamma, \tag{8.4}$$

where k_c^2 is the separation constant. The function $u(z, t)$ satisfies the modified Klein-Gordon equation

$$\frac{\partial^2 u(z, t)}{\partial z^2} - \frac{1}{v^2} \frac{\partial^2 u(z, t)}{\partial t^2} - \sigma \frac{\eta}{v} \frac{\partial u(z, t)}{\partial t} - k_c^2 u(z, t) = 0. \tag{8.5}$$

When $\sigma = 0$, Equation (8.5) reduces to the **Klein–Gordon equation**, named after the Swedish physicist Oskar Benjamin Klein (1894–1977) and German physicist Walter Gordon (1893–1939), who proposed the equation in 1927. There exists a complete orthonormal system of vector modal functions $\{\mathbf{e}_n | n = 1, 2, \cdots\}$ satisfying (8.4) (see Section 3.3.1). These vector modal functions can be used to expand the fields in both frequency and time domain. The transient electromagnetic fields in the waveguide can then be expressed as

$$\mathbf{E}(\mathbf{r}, t) = \sum_{n=1}^{\infty} v_n(z, t) \mathbf{e}_n(\rho) + \mathbf{u}_z \sum_{n=1}^{\infty} e'_{zn}(z, t) \frac{\nabla \cdot \mathbf{e}_n(\rho)}{k_{cn}},$$

$$\mathbf{H}(\mathbf{r}, t) = \sum_{n=1}^{\infty} i_n(z, t)\mathbf{u}_z \times \mathbf{e}_n(\boldsymbol{\rho}) + \mathbf{u}_z \frac{1}{\sqrt{\Omega}} \int_{\Omega} \frac{\mathbf{u}_z \cdot \mathbf{H}(\mathbf{r}, t)}{\sqrt{\Omega}} d\Omega \tag{8.6}$$

$$+ \sum_{n=1}^{\infty} h'_{zn}(z, t) \frac{\nabla \times \mathbf{e}_n(\boldsymbol{\rho})}{k_{cn}},$$

$$\nabla \times \mathbf{E} = \sum_{n=1}^{\infty} \left(\frac{\partial v_n}{\partial z} + k_{cn} e'_{zn} \right) \mathbf{u}_z \times \mathbf{e}_n + \sum_{n=1}^{\infty} k_{cn} v_n \frac{\nabla \times \mathbf{e}_n}{k_{cn}},$$

$$\nabla \times \mathbf{H} = \sum_{n=1}^{\infty} \left(-\frac{\partial i_n}{\partial z} + k_{cn} h'_{zn} \right) \mathbf{e}_n + \mathbf{u}_z \sum_{n=1}^{\infty} k_{cn} i_n \frac{\nabla \cdot \mathbf{e}_n}{k_{cn}},$$

$$\tag{8.7}$$

where v_n and i_n are the **time-domain modal voltage** and the **time-domain modal current** defined by

$$v_n(z, t) = \int_{\Omega} \mathbf{E}(\mathbf{r}, t) \cdot \mathbf{e}_n(\boldsymbol{\rho}) d\Omega,$$

$$i_n(z, t) = \int_{\Omega} \mathbf{H}(\mathbf{r}, t) \cdot \mathbf{u}_z \times \mathbf{e}_n(\boldsymbol{\rho}) d\Omega,$$

$$\tag{8.8}$$

and e'_{zn} and h'_{zn} are given by

$$e'_{zn}(z, t) = \int_{\Omega} \mathbf{u}_z \cdot \mathbf{E}(\mathbf{r}, t) \frac{\nabla \cdot \mathbf{e}_n(\boldsymbol{\rho})}{k_{cn}} d\Omega,$$

$$h'_{zn}(z, t) = \int_{\Omega} \mathbf{H}(\mathbf{r}, t) \cdot \frac{\nabla \times \mathbf{e}_n(\boldsymbol{\rho})}{k_{cn}} d\Omega.$$

Substituting (8.6) and (8.7) into the generalized Maxwell equations

$$\nabla \times \mathbf{E}(\mathbf{r}, t) = -\mu \frac{\partial \mathbf{H}(\mathbf{r}, t)}{\partial t} - \mathbf{J}_m(\mathbf{r}, t),$$

$$\nabla \times \mathbf{H}(\mathbf{r}, t) = \varepsilon \frac{\partial \mathbf{E}(\mathbf{r}, t)}{\partial t} + \mathbf{J}(\mathbf{r}, t) + \sigma \mathbf{E}(\mathbf{r}, t),$$

and comparing the transverse and longitudinal components, we obtain

$$-\frac{\partial i_n}{\partial z} + k_{cn} h'_{zn} = \varepsilon \frac{\partial v_n}{\partial t} + \sigma v_n + \int_{\Omega} \mathbf{J} \cdot \mathbf{e}_n d\Omega, \tag{8.9}$$

$$k_{cn} i_n = \varepsilon \frac{\partial e'_{zn}}{\partial t} + \sigma e'_{zn} + \int_{\Omega} \mathbf{u}_z \cdot \mathbf{J} \frac{\nabla \cdot \mathbf{e}_n}{k_{cn}} d\Omega, \text{ for TM modes only}, \tag{8.10}$$

$$\frac{\partial v_n}{\partial z} + e'_{zn} k_{cn} = -\mu \frac{\partial i_n}{\partial t} - \int_{\Omega} \mathbf{J}_m \cdot \mathbf{u}_z \times \mathbf{e}_n d\Omega, \tag{8.11}$$

$$k_{cn}v_n = -\mu \frac{\partial h'_{zn}}{\partial t} - \int_\Omega \mathbf{u}_z \cdot \mathbf{J}_m \frac{\mathbf{u}_z \cdot \nabla \times \mathbf{e}_n}{k_{cn}} d\Omega, \text{ for TE modes only,} \qquad (8.12)$$

$$-\mu \frac{\partial}{\partial t} \int_\Omega \frac{\mathbf{H} \cdot \mathbf{u}_z}{\sqrt{\Omega}} d\Omega = \int_\Omega \frac{\mathbf{u}_z \cdot \mathbf{J}_m}{\sqrt{\Omega}} d\Omega, \text{ for TE modes only.} \qquad (8.13)$$

For the TEM modes, the modal voltages and modal currents satisfy

$$\frac{\partial v_n^{TEM}}{\partial z} = -\mu \frac{\partial i_n^{TEM}}{\partial t} - \int_\Omega \mathbf{J}_m \cdot \mathbf{u}_z \times \mathbf{e}_n d\Omega,$$

$$\frac{\partial i_n^{TEM}}{\partial z} = -\varepsilon \frac{\partial v_n^{TEM}}{\partial t} - \sigma v_n^{TEM} - \int_\Omega \mathbf{J} \cdot \mathbf{e}_n d\Omega,$$

$$\qquad (8.14)$$

from (8.9) and (8.11). The modal voltages for the TEM modes satisfy the wave equation

$$\frac{\partial^2 v_n^{TEM}}{\partial z^2} - \frac{1}{v^2} \frac{\partial^2 v_n^{TEM}}{\partial t^2} - \sigma \frac{\eta}{v} \frac{\partial v_n^{TEM}}{\partial t} = \mu \frac{\partial}{\partial t} \int_\Omega \mathbf{J} \cdot \mathbf{e}_n d\Omega - \frac{\partial}{\partial z} \int_\Omega \mathbf{J}_m \cdot \mathbf{u}_z \times \mathbf{e}_n d\Omega. \qquad (8.15)$$

The modal currents i_n^{TEM} can be determined by the time integration of v_n^{TEM}.
  For the TE modes, we have

$$\frac{\partial v_n^{TE}}{\partial z} = -\mu \frac{\partial i_n^{TE}}{\partial t} - \int_\Omega \mathbf{J}_m \cdot \mathbf{u}_z \times \mathbf{e}_n d\Omega,$$

$$\frac{\partial i_n^{TE}}{\partial z} - k_{cn} h'_{zn} = -\varepsilon \frac{\partial v_n^{TE}}{\partial t} - \sigma v_n^{TE} - \int_\Omega \mathbf{J} \cdot \mathbf{e}_n d\Omega, \qquad (8.16)$$

$$\mu \frac{\partial h'_{zn}}{\partial t} = -k_{cn} v_n^{TE} - \int_\Omega \mathbf{u}_z \cdot \mathbf{J}_m \frac{\mathbf{u}_z \cdot \nabla \times \mathbf{e}_n}{k_{cn}} d\Omega,$$

from (8.9), (8.11) and (8.12). The modal voltages v_n^{TE} satisfy the modified Klein–Gordon equation

$$\frac{\partial^2 v_n^{TE}}{\partial z^2} - \frac{1}{v^2} \frac{\partial^2 v_n^{TE}}{\partial t^2} - \sigma \frac{\eta}{v} \frac{\partial v_n^{TE}}{\partial t} - k_{cn}^2 v_n^{TE}$$

$$= \mu \frac{\partial}{\partial t} \int_\Omega \mathbf{J} \cdot \mathbf{e}_n d\Omega - \frac{\partial}{\partial z} \int_\Omega \mathbf{J}_m \cdot \mathbf{u}_z \times \mathbf{e}_n d\Omega + k_{cn} \int_\Omega \mathbf{u}_z \cdot \mathbf{J}_m \frac{\mathbf{u}_z \cdot \nabla \times \mathbf{e}_n}{k_{cn}} d\Omega.$$

$$\qquad (8.17)$$

The modal currents i_n^{TE} can be determined by the time integration of $\partial v_n^{TE}/\partial z$.

For the TM modes, the modal voltages and modal currents satisfy

$$\frac{\partial v_n^{TM}}{\partial z} + k_{cn}e'_{zn} = -\mu\frac{\partial i_n^{TM}}{\partial t} - \int_{\Omega} \mathbf{J}_m \cdot \mathbf{u}_z \times \mathbf{e}_n d\Omega,$$

$$\frac{\partial i_n^{TM}}{\partial z} = -\varepsilon\frac{\partial v_n^{TM}}{\partial t} - \sigma v_n^{TM} - \int_{\Omega} \mathbf{J} \cdot \mathbf{e}_n d\Omega, \tag{8.18}$$

$$\varepsilon\frac{\partial \bar{e}_{zn}}{\partial t} = k_{cn}i_n^{TM} - \sigma e'_{zn} - \int_{\Omega} \mathbf{u}_z \cdot \mathbf{J}\frac{\nabla \cdot \mathbf{e}_n}{k_{cn}} d\Omega,$$

from (8.9), (8.10) and (8.11). The modal currents i_n^{TM} also satisfy the modified Klein–Gordon equation

$$\frac{\partial^2 i_n^{TM}}{\partial z^2} - \frac{1}{v^2}\frac{\partial^2 i_n^{TM}}{\partial t^2} - \sigma\frac{\eta}{v}\frac{\partial i_n^{TM}}{\partial t} - k_{cn}^2 i_n^{TM}$$

$$= -\frac{\partial}{\partial z}\int_{\Omega} \mathbf{J} \cdot \mathbf{e}_n d\Omega - k_{cn}\int_{\Omega} \mathbf{u}_z \cdot \mathbf{J}\frac{\nabla \cdot \mathbf{e}_n}{k_{cn}} d\Omega \tag{8.19}$$

$$+ \varepsilon\frac{\partial}{\partial t}\int_{\Omega} \mathbf{J}_m \cdot \mathbf{u}_z \times \mathbf{e}_n d\Omega + \sigma\int_{\Omega} \mathbf{J}_m \cdot \mathbf{u}_z \times \mathbf{e}_n d\Omega.$$

The modal voltages v_n^{TM} can then be determined by a time integration of $\partial i_n^{TM}/\partial z$. The excitation problem in the waveguide is now reduced to the solution of a series of inhomogeneous modified Klein–Gordon equations.

8.1.2   Solution of the Modified Klein–Gordon Equation

To find the complete solution of the transient fields in the waveguide, we need to solve the modified Klein–Gordon equation. This can be done by using the retarded Green's function. The retarded Green's function of the modified Klein–Gordon equation is defined by

$$\left(\frac{\partial^2}{\partial z^2} - \frac{1}{v^2}\frac{\partial^2}{\partial t^2} - \sigma\frac{\eta}{v}\frac{\partial}{\partial t} - k_{cn}^2\right) G_n(z, t; z', t') = -\delta(z - z')\delta(t - t'),$$

$$G_n(z, t; z', t')\big|_{t<t'} = 0. \tag{8.20}$$

The second equation represents the causality condition. From the Fourier transform pair

$$\tilde{G}_n(p, \omega; z', t') = \int_{-\infty}^{\infty}\int_{-\infty}^{\infty} G_n(z, t; z', t')e^{-jpz-j\omega t} dz dt,$$

$$\tag{8.21}$$

$$G_n(z, t; z', t') = \frac{1}{(2\pi)^2}\int_{-\infty}^{\infty}\int_{-\infty}^{\infty} \tilde{G}_n(p, \omega; z', t')e^{jpz+j\omega t} dp d\omega,$$

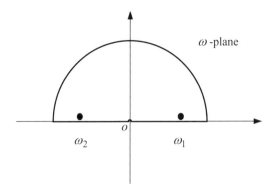

Figure 8.1   Integration contour

we find that

$$\tilde{G}_n(p, \omega; z', t') = \frac{-v^2 e^{-jpz' - j\omega t'}}{\omega^2 - p^2 v^2 - k_{cn}^2 v^2 - j\omega\sigma/\varepsilon}.$$

Substituting this into the second equation of (8.21) yields

$$G_n(z, t; z', t') = -\frac{v^2}{(2\pi)^2} \int\limits_{-\infty}^{\infty} e^{jp(z-z')} dp \int\limits_{-\infty}^{\infty} \frac{e^{j\omega(t-t')}}{\omega^2 - p^2 v^2 - k_{cn}^2 v^2 - j\omega\sigma/\varepsilon} d\omega.$$

To calculate the integral with respect to ω, we may extend ω to the complex plane and use the residue theorem in complex variable analysis. There are two simple poles in the integrand: $\omega_{1,2} = j\gamma \pm \sqrt{p^2 v^2 + k_{cn}^2 v^2 - \gamma^2}$, where $\gamma = \sigma/2\varepsilon$. To satisfy the causality condition, we only need to consider the integral along a closed contour consisting of the real axis from $-\infty$ to ∞ and an infinite semicircle in the upper half plane, as shown in Figure 8.1. The contour integral along the large semicircle is zero for $t > t'$. Using the residue theorem and the relation (Gradsheyn and Ryzhik, 1994)

$$\int\limits_{0}^{\infty} \frac{\sin q\sqrt{x^2 + a^2}}{\sqrt{x^2 + a^2}} \cos bx\, dx = \frac{\pi}{2} J_0(a\sqrt{q^2 - b^2}) H(q - b) \qquad (8.22)$$

$$a > 0, q > 0, b > 0,$$

we obtain the retarded Green's function:

$$G_n(z, t; z', t') = \frac{v}{2} e^{-\gamma(t-t')} H[(t - t') - |z - z'|/v]$$

$$\cdot J_0\left[(k_{cn}^2 v^2 - \gamma^2)^{1/2} \sqrt{(t - t')^2 - |z - z'|^2 / v^2} \right], \qquad (8.23)$$

where $J_0(x)$ is the Bessel function of first kind and $H(x)$ is the unit step function. The retarded Green's function can now be used to solve the modified Klein–Gordon equation with the known source function $f(z, t)$:

$$\left(\frac{\partial^2}{\partial z^2} - \frac{1}{v^2} \frac{\partial^2}{\partial t^2} - \sigma \frac{\eta}{v} \frac{\partial}{\partial t} - k_{cn}^2 \right) u_n(z, t) = f(z, t).$$

The above equation and (8.20) can be transformed into the frequency domain by using the Fourier transform as follows

$$\left(\frac{\partial^2}{\partial z^2} + \beta_{cn}^2 \right) \tilde{u}_n(z, \omega) = \tilde{f}(z, \omega), \tag{8.24}$$

$$\left(\frac{\partial^2}{\partial z^2} + \beta_{cn}^2 \right) \tilde{G}_n(z, \omega; z', t') = -\delta(z - z')e^{-j\omega t'}, \tag{8.25}$$

where $\beta_{cn}^2 = k^2 - k_{cn}^2 - j\sigma k\eta$, $k = \omega/v$. Multiplying (8.24) and (8.25) by $\tilde{G}_n$ and $\tilde{u}$ respectively and then subtracting the resultant equations yields

$$\tilde{u}_n(z, \omega) \frac{\partial^2 \tilde{G}_n(z, \omega; z', t')}{\partial z^2} - \tilde{G}_n(z, \omega; z', t') \frac{\partial^2 \tilde{u}_n(z, \omega)}{\partial z^2}$$

$$= -\delta(z - z')\tilde{u}_n(z, \omega)e^{-j\omega t'} - \tilde{f}(z, \omega)\tilde{G}_n(z, \omega; z', t'). \tag{8.26}$$

We assume that the source function $f(z, t)$ is limited in a finite interval (a, b), as shown in Figure 8.2. Taking the integration of the above equation over the interval $[a, b]$ and then taking the inverse Fourier transform, we obtain the solution

$$u_n(z, t) = \int_{-\infty}^{\infty} G_n(z, t; z', t') \frac{\partial u_n(z', t')}{\partial z'} dt' \Big|_{z=a}^{b}$$

$$- \int_{-\infty}^{\infty} u_n(z', t') \frac{\partial G_n(z, t; z', t')}{\partial z'} dt' \Big|_{z=a}^{b} \tag{8.27}$$

$$- \int_{a}^{b} \int_{-\infty}^{\infty} f(z', t')G_n(z, t; z', t')dt'dz, \ z \in (a, b),$$

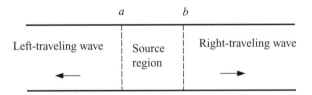

Figure 8.2   Left-traveling wave and right-traveling wave in a waveguide

where the symmetry of Green's function about z and z' has been used. If we let $a \to -\infty$ and $b \to \infty$, the above expression becomes

$$u_n(z, t) = -\int\limits_{-\infty}^{\infty} \int\limits_{-\infty}^{\infty} f(z', t')G_n(z, t; z', t')dt'dz', \; z \in (-\infty, \infty). \qquad (8.28)$$

The solution in the region $(z_+, +\infty)$ $(z_+ > b)$ and $(-\infty, z_-)$ $(z_- < a)$ may be expressed in terms of its boundary values at z_+ and z_-. Without loss of generality, we assume that $a < 0$ and $b > 0$, and the medium in the waveguide is lossless. Taking the integration of (8.26) over $[z_+, +\infty)$ with $z_+ > b$ and using integration by parts, we obtain

$$\tilde{G}_n(z_+, \omega; z', t')\frac{\partial \tilde{u}_n(z_+, \omega)}{\partial z} - \tilde{u}_n(z_+, \omega)\frac{\partial \tilde{G}_n(z_+, \omega; z', t')}{\partial z} = -\tilde{u}_n(z', \omega)e^{-j\omega t'},$$

where the radiation condition at $z = +\infty$ has been used. Taking the inverse Fourier transform and letting $z' = z_+$ leads to

$$u_n(z_+, t - t') + v\int\limits_{-\infty}^{t-t'} J_0\left[k_{cn}v(t - \tau - t')\right]\frac{\partial u_n(z_+, \tau)}{\partial z}d\tau = 0.$$

Since z_+ and t' are arbitrary, the above equation can be written as

$$u_n(z, t) + v J_0(k_{cn}vt)H(t) * \frac{\partial u_n(z, t)}{\partial z} = 0, \; z \geq b > 0. \qquad (8.29)$$

This equation is called the **right-traveling condition** of the wave (Kristensson, 1995). Similarly, taking the integration of (8.26) over $(-\infty, z_-]$ with $z_- < a$, we obtain the **left-traveling condition**

$$u_n(z, t) - v J_0(k_{cn}vt)H(t) * \frac{\partial u_n(z, t)}{\partial z} = 0, \; z \leq a < 0. \qquad (8.30)$$

Both (8.29) and (8.30) are integral-differential equations. If the source is turned on at $t = 0$, all the fields must be zero when $t < 0$, and (8.29) and (8.30) can be solved by the single-sided Laplace transform defined by $\tilde{u}_n(z, s) = \int\limits_{0}^{\infty} u_n(z, t)e^{-st}dt$. Thus

$$\tilde{u}_n(z, s) + [(s/v)^2 + k_{cn}^2]^{-1/2}\frac{\partial \tilde{u}_n(z, s)}{\partial z} = 0, \; z \geq b > 0,$$

$$u_n(z, t) - [(s/v)^2 + k_{cn}^2]^{-1/2}\frac{\partial \tilde{u}_n(z, s)}{\partial z} = 0, \; z \leq a < 0.$$

The solutions of the above equations are

$$\tilde{u}_n^+(z,s) = \tilde{u}_n^+(b,s)e^{-\sqrt{(s/v)^2+k_{cn}^2}(z-b)}, z \geq b > 0,$$

$$\tilde{u}_n^-(z,s) = \tilde{u}_n^-(a,s)e^{\sqrt{(s/v)^2+k_{cn}^2}(z-a)}, z \leq a < 0.$$

By means of the inverse Laplace transform, the solutions of (8.29) and (8.30) can be expressed as

$$u_n(z,t) = u_n\left(b, t - \frac{z-b}{v}\right) - ck_{cn}(z-b)$$

$$\times \int_0^{t-\frac{z-b}{v}} \frac{J_1\left[k_{cn}v\sqrt{(t-\tau)^2-(z-b)^2/v^2}\right]}{\sqrt{(t-\tau)^2-(z-b)^2/v^2}} u_n(b,\tau)d\tau, z \geq b > 0,$$

$$u_n(z,t) = u_n\left(a, t + \frac{z-a}{v}\right) + ck_{cn}(z-a)$$

$$\times \int_0^{t+\frac{z-a}{v}} \frac{J_1\left[k_{cn}v\sqrt{(t-\tau)^2-(z-a)^2/v^2}\right]}{\sqrt{(t-\tau)^2-(z-a)^2/v^2}} u_n(a,\tau)d\tau, z \leq a < 0.$$

Once the input signal is known, the output signal after traveling a certain distance in the waveguide can be determined by the convolution integral.

8.1.3 Excitation of Waveguides

A wideband pulse in the waveguide will excite a number of higher-order modes and the field distributions are determined by (8.6). The bandwidth of the excitation pulse must be adjusted properly in order to control the number of modes excited in the waveguide. For example, if the spectrum of the baseband signal is limited to the range $[0, \omega_f]$ and ω_0 is the carrier frequency with $\omega_f \ll \omega_0$, the frequency spectrum range of the modulated signal will be $[\omega_0 - \omega_f, \omega_0 + \omega_f]$. If ω_f is properly chosen so that $\omega_{c1} < \omega_0 - \omega_f$ and $\omega_0 + \omega_f < \omega_{c2}$, where ω_{c1} and ω_{c2} are respectively the cut-off frequencies of the dominant mode and the first higher-order mode, only the dominant mode will propagate in the waveguide and the radio signal will be transmitted without distortion.

Example 8.1 (Rectangular waveguide): Let us consider a rectangular waveguide of width a and height b as depicted in Figure 8.3. The waveguide is excited by a line current extending across the waveguide located at $x = x_0$, and the current density is given by

$$\mathbf{J}(\mathbf{r},t) = \mathbf{u}_y\delta(x-x_0)\delta(z-z_0)f(t). \tag{8.31}$$

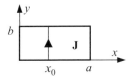

Figure 8.3 Rectangular waveguide

Since the line current is uniform in y direction, the fields excited by the current are independent of y. As a consequence, only TE_{n0} modes will be excited and we have

$$k_{cn} = \frac{n\pi}{a}, \mathbf{e}_n(x, y) = \mathbf{e}_{n0}^{TE}(x, y) = -\mathbf{u}_y \left(\frac{2}{ab}\right)^{1/2} \sin \frac{n\pi x}{a}, n = 1, 2, 3 \cdots. \qquad (8.32)$$

From (8.28), the modal voltages may be written as

$$v_n^{TE}(z, t) = \frac{b\eta}{2} \left(\frac{2}{ab}\right)^{1/2} \sin \frac{n\pi}{a} x_0 \int_{-\infty}^{t-|z-z_0|/v} \frac{df(t')}{dt'} J_0 \left[k_{cn}v\sqrt{(t - t')^2 - |z - z_0|^2 /v^2}\right] dt',$$

$$(8.33)$$

where $\eta = \sqrt{\mu/\varepsilon}$. Thus the time-domain voltages v_n^{TE} for $n = 2, 4, 6, \cdots$ vanish. The total electric field in the waveguide is given by the first equation of (8.6)

$$\mathbf{E} = \mathbf{u}_y E_y = -\mathbf{u}_y \left(\frac{2}{ab}\right)^{1/2} \sum_{n=1}^{\infty} v_n^{TE} \sin \frac{n\pi x}{a}. \qquad (8.34)$$

For the time-domain response to a continuous sinusoidal wave turned on at $t = 0$, we may expect that the time-domain response approaches the well-known steady-state response as time goes to infinity. Let $x_0 = a/2, z_0 = 0$, and $f(t) = H(t) \sin \omega t$ in (8.34). Equation (8.33) may be expressed as the sum of two parts:

$$v_n^{TE}(z, t) = v_n^{TE}(z, t)\big|_{\text{steady}} + v_n^{TE}(z, t)\big|_{\text{transient}}, t > |z|/v,$$

where $v_n^{TE}(z, t)\big|_{\text{steady}}$ and $v_n^{TE}(z, t)\big|_{\text{transient}}$ represent the steady-state part and the transient part of the response respectively:

$$v_n^{TE}(z, t)\big|_{\text{steady}} = \frac{b\eta}{2} \left(\frac{2}{ab}\right)^{1/2} ka \sin \frac{n\pi}{2}$$

$$\times \int_{|z|/a}^{\infty} \cos ka(vt/a - u) J_0 \left[k_{cn}a\sqrt{u^2 - |z|^2 /a^2}\right] du,$$

$$v_n^{TE}(z,t)\big|_{\text{transient}} = -\frac{b\eta}{2}\left(\frac{2}{ab}\right)^{1/2} ka \sin\frac{n\pi}{2}$$

$$\times \int_{vt/a}^{\infty} \cos ka(vt/a - u) J_0\left[k_{cn}a\sqrt{u^2 - |z|^2/a^2}\right] du.$$

The transient part of the response approaches zero as $t \to \infty$. By means of the following relations (Gradsheyn and Ryzhik, 1994)

$$\int_a^{\infty} J_0(b\sqrt{x^2 - a^2}) \sin dxdx = \begin{cases} 0, 0 < d < b \\ \cos(a\sqrt{d^2 - b^2})/\sqrt{d^2 - b^2}, 0 < b < d \end{cases}$$

$$\int_a^{\infty} J_0(b\sqrt{x^2 - a^2}) \cos dxdx = \begin{cases} \exp(-a\sqrt{b^2 - d^2})/\sqrt{b^2 - d^2}, 0 < d < b \\ -\sin(a\sqrt{d^2 - b^2})/\sqrt{d^2 - b^2}, 0 < b < d \end{cases}$$

(8.35)

the steady-state response is found to be

$$v_n^{TE}(z,t)\big|_{\text{steady}} = \frac{b\eta}{2}\left(\frac{2}{ab}\right)^{1/2} \frac{ka \sin\frac{n\pi}{2}}{\sqrt{|(ka)^2 - (k_{cn}a)^2|}}$$

$$\times \begin{cases} \sin\left(ka\frac{vt}{a} - \frac{|z|}{a}\sqrt{(ka)^2 - (k_{cn}a)^2}\right), k > k_{cn} \\ \cos\left(ka\frac{vt}{a}\right)\exp\left[-\left|\frac{z}{a}\right|\sqrt{(k_{cn}a)^2 - (ka)^2}\right], k < k_{cn} \end{cases}.$$

Thus as $|z|$ increases, the modal voltages decrease rapidly when $k < k_{cn}$. In other words, only those modes satisfying $k > k_{cn}$ propagate in the steady state. When $v_n^{TE}\big|_{\text{steady}}$ are inserted into (8.34), it can be found that the steady-state response of the electric field agrees with the traditional time-harmonic theory of waveguides.

If the excitation is a unit step pulse $f(t) = H(t)$, Equation (8.33) becomes

$$v_n^{TE}(z,t) = \frac{\eta b}{2}\left(\frac{2}{ab}\right)^{1/2} \sin\frac{n\pi}{2} J_0\left[k_{cn}a\sqrt{\left(\frac{vt}{a}\right)^2 - \frac{|z|^2}{a^2}}\right] H\left(\frac{vt}{a} - \frac{|z|}{a}\right).$$

The plots of the time-domain voltages v_n^{TE} indicate that the voltages for the higher-order modes cannot be ignored for a unit step pulse. The time responses of the fields are different from the original excitation pulse since a hollow waveguide is essentially a high pass filter and blocks all the low frequency components below the first cut-off frequency (Geyi, 2006(a)). Therefore, a hollow metal waveguide is not an ideal medium to transmit a wideband signal. Instead, one must use multi-conductor transmission lines, which support a TEM mode whose cut-off frequency is zero.   □

Example 8.2 (Coaxial waveguide): To see how a pulse propagates in a TEM transmission line as well as the effects of the higher-order modes, let us consider a coaxial line consisting

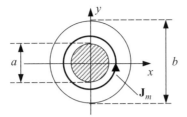

Figure 8.4 Coaxial waveguide

of an inner conductor of radius a and an outer conductor of radius b, as shown in Figure 8.4. We assume that the coaxial line is excited by a magnetic current located at $z = z_0$

$$\mathbf{J}_m(\mathbf{r}, t) = \mathbf{u}_\varphi f(t)\delta(z - z_0)\delta(\rho - \rho_0), a < \rho_0 < b \tag{8.36}$$

where (ρ, φ, z) are the polar coordinates and $\mathbf{u}_\varphi$ is the unit vector in φ direction. According to the symmetry, only the TEM mode and those TM_{0n} modes independent of φ will be excited. The orthonormal vector modal functions for these modes are given by (Marcuvitz, 1951)

$$\mathbf{e}_{t1}(\rho, \varphi) = \mathbf{u}_\rho e_1(\rho), k_{c1} = 0, e_1(\rho) = \frac{1}{\rho\sqrt{2\pi \ln c_1}},$$

$$\mathbf{e}_{tn}(\rho, \varphi) = \mathbf{u}_\rho e_n(\rho), k_{cn} = \frac{\chi_n}{a},$$

$$e_n(\rho) = \frac{\sqrt{\pi}}{2} \frac{\chi_n}{a} \frac{J_1(\chi_n \rho/a)N_0(\chi_n) - N_1(\chi_n \rho/a)J_0(\chi_n)}{\sqrt{J_0^2(\chi_n)/J_0^2(c_1\chi_n) - 1}}, n \geq 2, \tag{8.37}$$

where $c_1 = b/a$, $\mathbf{u}_\rho$ is the unit vector in ρ direction, and χ_n is the nth nonvanishing root of the equation $J_0(\chi_n c_1)N_0(\chi_n) - N_0(\chi_n c_1)J_0(\chi_n) = 0$. From (8.28), the time-domain modal currents may be expressed as

$$\frac{\eta i_1^{TEM}}{a} = -\frac{\pi}{a\sqrt{2\pi \ln c_1}} f(t - |z - z_0|/v),$$

$$\frac{\eta i_n^{TM}}{a} = -\frac{\pi \rho_0}{a} e_n(\rho_0) \tag{8.38}$$

$$\cdot \int_{-\infty}^{t-|z-z_0|/v} \frac{df(t')}{dt'} J_0\left[k_{cn}v\sqrt{(t - t')^2 - |z - z_0|^2/v^2}\right] dt', n \geq 2.$$

Equation (8.38) indicates that the signal can be transmitted without distortion in a coaxial line

if the highest frequency component of the excitation pulse is below the cut-off frequency of
the first higher order mode. Otherwise the higher-order modes will be excited. The magnetic
field in the coaxial cable may be obtained from the second equation of (8.6) as follows

$$\eta \mathbf{H} = \mathbf{u}_\varphi \eta H_\varphi = \mathbf{u}_\varphi \frac{\eta i_n^{TEM}}{a} \frac{1}{(\rho/a)\sqrt{2\pi \ln c_1}}$$

$$+ \mathbf{u}_\varphi \sum_{n=2}^{\infty} \frac{\eta i_n^{TM}}{a} \frac{\sqrt{\pi}\chi_n}{2} \frac{J_1(\chi_n\rho/a)N_0(\chi_n) - N_1(\chi_n\rho/a)J_0(\chi_n)}{\sqrt{J_0^2(\chi_n)/J_0^2(c_1\chi_n) - 1}}.$$

□

8.2 Time-domain Theory of Metal Cavity Resonators[2]

A metal cavity resonator constitutes a typical eigenvalue problem in electromagnetic theory
and the study of the transient process in a metal cavity may be carried out by the field
expansions in terms of the vector modal functions studied in Section 3.3.2. When these
expansions are introduced into the time-domain Maxwell equations, we may find that the
expansion coefficients satisfy the ordinary differential equations of second order, which can
be easily solved once the initial conditions and the excitations are known.

8.2.1 Field in Arbitrary Cavities

Consider a metal cavity with a perfectly conducting wall, and assume that the medium in
the cavity is homogeneous and isotropic with medium parameters σ, μ and ε. The volume
occupied by the cavity is denoted by V and its boundary by S. If the cavity contains an
impressed electric current source $\mathbf{J}$ and a magnetic current source $\mathbf{J}_m$, the fields excited by
these sources satisfy the Maxwell equations in the cavity:

$$\nabla \times \mathbf{H}(\mathbf{r}, t) = \varepsilon \frac{\partial \mathbf{E}(\mathbf{r}, t)}{\partial t} + \sigma \mathbf{E}(\mathbf{r}, t) + \mathbf{J}(\mathbf{r}, t),$$

$$\nabla \times \mathbf{E}(\mathbf{r}, t) = -\mu \frac{\partial \mathbf{H}(\mathbf{r}, t)}{\partial t} - \mathbf{J}_m(\mathbf{r}, t),$$

(8.39)

with the boundary conditions $\mathbf{u}_n \times \mathbf{E} = 0$ and $\mathbf{u}_n \cdot \mathbf{H} = 0$ on the boundary S. Here $\mathbf{u}_n$ is the
unit outward normal to the boundary. The fields inside the cavity can be expanded in terms of
its vector modal functions as follows

$$\mathbf{E}(\mathbf{r}, t) = \sum_n V_n(t)\mathbf{e}_n(\mathbf{r}) + \sum_\nu V_\nu(t)\mathbf{e}_\nu(\mathbf{r}),$$

$$\mathbf{H}(\mathbf{r}, t) = \sum_n I_n(t)\mathbf{h}_n(\mathbf{r}) + \sum_\tau I_\tau(t)\mathbf{h}_\tau(\mathbf{r}),$$

(8.40)

[2] W. Geyi, 'Time-domain theory of metal cavity resonator', *Progress in Electromagnetics Research*, PIER 78, 219–53, 2008. Reproduced by permission of ©2008 The Electromagnetics Academy & EMW Publishing.

$$\nabla \times \mathbf{E}(\mathbf{r}, t) = \sum_n \mathbf{h}_n(\mathbf{r}) \int_V \nabla \times \mathbf{E}(\mathbf{r}, t) \cdot \mathbf{h}_n(\mathbf{r}) dV + \sum_\tau \mathbf{h}_\tau(\mathbf{r}) \int_V \nabla \times \mathbf{E}(\mathbf{r}, t) \cdot \mathbf{h}_\tau(\mathbf{r}) dV,$$

$$\nabla \times \mathbf{H}(\mathbf{r}, t) = \sum_n \mathbf{e}_n(\mathbf{r}) \int_V \nabla \times \mathbf{H}(\mathbf{r}, t) \cdot \mathbf{e}_n(\mathbf{r}) dV + \sum_\nu \mathbf{e}_\nu(\mathbf{r}) \int_V \nabla \times \mathbf{H}(\mathbf{r}, t) \cdot \mathbf{e}_\nu(\mathbf{r}) dV,$$

$$(8.41)$$

where the subscript n denotes the vector modal functions belonging to category 2, and the Greek subscript ν and τ for the vector modal functions belonging to category 1 or 3, and

$$V_{n(\nu)}(t) = \int_V \mathbf{E}(\mathbf{r}, t) \cdot \mathbf{e}_{n(\nu)}(\mathbf{r}) dV, \quad I_{n(\tau)}(t) = \int_V \mathbf{H}(\mathbf{r}, t) \cdot \mathbf{h}_{n(\tau)}(\mathbf{r}) dV. \qquad (8.42)$$

Considering the following calculations

$$\int_V \nabla \times \mathbf{E} \cdot \mathbf{h}_n dV = \int_V \mathbf{E} \cdot \nabla \times \mathbf{h}_n dV + \int_S (\mathbf{E} \times \mathbf{h}_n) \cdot \mathbf{u}_n dS = k_n V_n,$$

$$\int_V \nabla \times \mathbf{E} \cdot \mathbf{h}_\tau dV = \int_V \mathbf{E} \cdot \nabla \times \mathbf{h}_\tau dV + \int_S (\mathbf{E} \times \mathbf{h}_\tau) \cdot \mathbf{u}_n dS = 0,$$

$$\int_V \nabla \times \mathbf{H} \cdot \mathbf{e}_n dS = \int_V \mathbf{H} \cdot \nabla \times \mathbf{e}_n dV + \int_S (\mathbf{H} \times \mathbf{e}_n) \cdot \mathbf{u}_n dS = k_n I_n,$$

$$\int_V \nabla \times \mathbf{H} \cdot \mathbf{e}_\nu dS = \int_V \mathbf{H} \cdot \nabla \times \mathbf{e}_\nu dV + \int_S (\mathbf{H} \times \mathbf{e}_\nu) \cdot \mathbf{u}_n dS = 0,$$

Equation (8.41) can be written as

$$\nabla \times \mathbf{E} = \sum_n k_n V_n \mathbf{h}_n, \quad \nabla \times \mathbf{H} = \sum_n k_n I_n \mathbf{e}_n.$$

Substituting the above expansions into (8.39) and equating the expansion coefficients of the vector modal functions, we obtain

$$\frac{\partial V_n}{\partial t} + \frac{\sigma}{\varepsilon} V_n - \frac{k_n}{\varepsilon} I_n = -\frac{1}{\varepsilon} \int_V \mathbf{J} \cdot \mathbf{e}_n dV,$$

$$\frac{\partial V_\nu}{\partial t} + \frac{\sigma}{\varepsilon} V_\nu = -\frac{1}{\varepsilon} \int_V \mathbf{J} \cdot \mathbf{e}_\nu dV,$$

$$\frac{\partial I_n}{\partial t} + \frac{k_n}{\mu} V_n = -\frac{1}{\mu} \int_V \mathbf{J}_m \cdot \mathbf{h}_n dV,$$

$$(8.43)$$

$$\frac{\partial I_\tau}{\partial t} = -\frac{1}{\mu} \int_V \mathbf{J}_m \cdot \mathbf{h}_\tau dV.$$

From these equations, we may find that

$$\frac{\partial^2 I_n}{\partial t^2} + 2\gamma \frac{\partial I_n}{\partial t} + \omega_n^2 I_n = \omega_n S_n^I,$$

$$\frac{\partial^2 V_n}{\partial t^2} + 2\gamma \frac{\partial V_n}{\partial t} + \omega_n^2 V_n = \omega_n S_n^V,$$

(8.44)

where $\omega_n = k_n v$, $\gamma = \sigma/2\varepsilon$ and

$$S_n^I = v \int_V \mathbf{J} \cdot \mathbf{e}_n dV - \frac{1}{k_n \eta} \frac{\partial}{\partial t} \int_V \mathbf{J}_m \cdot \mathbf{h}_n dV - \frac{\sigma v}{k_n} \int_V \mathbf{J}_m \cdot \mathbf{h}_n dV,$$

$$S_n^V = -\frac{\eta}{k_n} \frac{\partial}{\partial t} \int_V \mathbf{J} \cdot \mathbf{e}_n dV - v \int_V \mathbf{J}_m \cdot \mathbf{h}_n dV.$$

The expansion coefficients I_n and V_n may be determined by use of the retarded Green's function defined by

$$\frac{\partial^2 G_n(t, t')}{\partial t^2} + 2\gamma \frac{\partial G_n(t, t')}{\partial t} + \omega_n^2 G_n(t, t') = -\delta(t - t'),$$

$$G_n(t, t')\big|_{t<t'} = 0.$$

(8.45)

The solution of (8.45) is readily found to be

$$G_n(t, t') = -\frac{e^{-\gamma(t-t')}}{\sqrt{\omega_n^2 - \gamma^2}} \sin \sqrt{\omega_n^2 - \gamma^2}(t - t') H(t - t').$$

(8.46)

Therefore, the general solution of I_n may be written as

$$I_n(t) = -\int_{-\infty}^{\infty} G_n(t, t') \omega_n S_n^I(t') dt' + e^{-\gamma t} \left(c_1 \cos \sqrt{\omega_n^2 - \gamma^2} t + c_2 \sin \sqrt{\omega_n^2 - \gamma^2} t \right),$$

(8.47)

where c_1 and c_2 are two arbitrary constants. If the source is turned on at $t = 0$, both $V_n(0^-)$ and $I_n(0^-)$ may be assumed to be zero due to causality. Considering the third equation of (8.43), the second term of (8.47) vanishes. Thus

$$I_n(t) = \frac{\omega_n}{\sqrt{\omega_n^2 - \gamma^2}} \int_{0^-}^{t} e^{-\gamma(t-t')} \sin \sqrt{\omega_n^2 - \gamma^2}(t - t')$$

$$\times \left[v \int_V \mathbf{J} \cdot \mathbf{e}_n dV - \frac{1}{k_n \eta} \frac{\partial}{\partial t'} \int_V \mathbf{J}_m \cdot \mathbf{h}_n dV - \frac{\sigma v}{k_n} \int_V \mathbf{J}_m \cdot \mathbf{h}_n dV \right] dt'.$$

(8.48)

Similarly, we have

$$
V_n(t) = \frac{\omega_n}{\sqrt{\omega_n^2 - \gamma^2}} \int_{0^-}^{t} e^{-\gamma(t-t')} \sin\sqrt{\omega_n^2 - \gamma^2}(t - t')
$$

$$
\times \left[-\frac{\eta}{k_n} \frac{\partial}{\partial t'} \int_V \mathbf{J} \cdot \mathbf{e}_n dV - \upsilon \int_V \mathbf{J}_m \cdot \mathbf{h}_n dV \right] dt'.
$$

(8.49)

and

$$
V_\upsilon(t) = -\frac{1}{\varepsilon} e^{-2\gamma t} \int_{0^-}^{t} e^{2\gamma t'} dt' \int_V \mathbf{J} \cdot \mathbf{e}_\upsilon dV,
$$

$$
I_\tau(t) = -\frac{1}{\mu} \int_{0^-}^{t} dt' \int_V \mathbf{J}_m \cdot \mathbf{h}_\tau dV.
$$

(8.50)

Substituting (8.48), (8.49) and (8.50) into (8.40), we may find out the field distributions inside the metal cavity.

Example 8.3: Assume that the current source is sinusoidal and is turned on at $t = 0$

$$
\mathbf{J}(\mathbf{r}, t) = \mathbf{J}'(\mathbf{r}) H(t) \sin \omega t
$$

(8.51)

and $\mathbf{J}_m(\mathbf{r}, t) = 0$. It follows from (8.48), (8.49), and (8.50) that

$$
I_n(t) = \omega_n \upsilon \int_V \mathbf{J}' \cdot \mathbf{e}_n dV \left[\frac{-(\omega_n^2 - \omega^2)\sin\omega t + 2\omega\gamma\cos\omega t}{(\omega_n^2 - \omega^2)^2 + 4\omega^2\gamma^2} \right.
$$

$$
\left. + \frac{1}{\beta_n} \frac{-(\omega_n^2 - \omega^2)\omega\sin\beta_n t + 2\omega\gamma(\beta_n\cos\omega t + \gamma\sin\beta_n t)}{(\omega_n^2 - \omega^2)^2 + 4\omega^2\gamma^2} e^{-\gamma t} \right],
$$

(8.52)

$$
V_n(t) = -\frac{\eta\omega\omega_n}{k_n} \int_V \mathbf{J}' \cdot \mathbf{e}_n dV \left[\frac{(\omega_n^2 - \omega^2)\cos\omega t + 2\omega\gamma\sin\omega t}{(\omega_n^2 - \omega^2)^2 + 4\omega^2\gamma^2} \right.
$$

$$
\left. + \frac{1}{\beta_n} \frac{-(\omega_n^2 - \omega^2)(\gamma\sin\beta_n t + \beta_n\cos\beta_n t) - 2\omega^2\gamma\sin\beta_n t}{(\omega_n^2 - \omega^2)^2 + 4\omega^2\gamma^2} e^{-\gamma t} \right],
$$

(8.53)

$$
V_\upsilon(t) = -\frac{1}{\varepsilon} \int_V \mathbf{J}' \cdot \mathbf{e}_\upsilon dV \left[\frac{2\gamma\sin\omega t - \omega\cos\omega t}{\omega^2 + 4\gamma^2} + \frac{\omega e^{-2\gamma t}}{\omega^2 + 4\gamma^2} \right],
$$

$$
I_\tau(t) = 0.
$$

The time-domain electromagnetic fields are given by

$$
\begin{aligned}
\mathbf{E}(\mathbf{r}, t) = &-\sum_n \frac{\eta \omega \omega_n}{k_n} \mathbf{e}_n(\mathbf{r}) \int_V \mathbf{J}' \cdot \mathbf{e}_n dV \left[\frac{(\omega_n^2 - \omega^2) \cos \omega t + 2\omega\gamma \sin \omega t}{(\omega_n^2 - \omega^2)^2 + 4\omega^2\gamma^2} \right. \\
&\left. - \frac{1}{\beta_n} \frac{(\omega_n^2 - \omega^2)(\gamma \sin \beta_n t + \beta_n \cos \beta_n t) + 2\omega^2\gamma \sin \beta_n t}{(\omega_n^2 - \omega^2)^2 + 4\omega^2\gamma^2} e^{-\gamma t} \right] \\
&+ \sum_v \frac{1}{\varepsilon} \mathbf{e}_v(\mathbf{r}) \int_V \mathbf{J}' \cdot \mathbf{e}_v dV \left[\frac{\omega \cos \omega t - 2\gamma \sin \omega t}{\omega^2 + 4\gamma^2} + \frac{\omega e^{-2\gamma t}}{\omega^2 + 4\gamma^2} \right],
\end{aligned}
\tag{8.54}
$$

$$
\begin{aligned}
\mathbf{H}(\mathbf{r}, t) = &\sum_n \omega_n v \mathbf{h}_n(\mathbf{r}) \int_V \mathbf{J}' \cdot \mathbf{e}_n dV \left[\frac{(\omega_n^2 - \omega^2) \sin \omega t - 2\omega\gamma \cos \omega t}{(\omega_n^2 - \omega^2)^2 + 4\omega^2\gamma^2} \right. \\
&\left. + \frac{1}{\beta_n} \frac{-(\omega_n^2 - \omega^2)\omega \sin \beta_n t + 2\omega\gamma(\beta_n \cos \omega t + \gamma \sin \beta_n t)}{(\omega_n^2 - \omega^2)^2 + 4\omega^2\gamma^2} e^{-\gamma t} \right].
\end{aligned}
\tag{8.55}
$$

Compared to the time-harmonic solutions (3.91) and (3.92) for a sinusoidal excitation $\mathbf{J}(\mathbf{r}, t) = \mathbf{J}'(\mathbf{r}) \sin \omega t$, Equations (8.54) and (8.55) have additional terms with exponential factor, which tend to zero with increasing time and may be viewed as the transient response for a lossy system. Hence the response in a metal cavity resonator can be separated into the sum of a steady-state response and a transient response if the medium is lossy. The time-domain solutions (8.54) and (8.55) approach the time-harmonic solutions (3.91) and (3.92) for a lossy system as time goes to infinity.

For a lossless cavity, Equations (8.54) and (8.55) become

$$
\begin{aligned}
\mathbf{E}(\mathbf{r}, t) = &-\sum_n \frac{\eta \omega \omega_n}{k_n} \frac{\cos \omega t - \cos \omega_n t}{\omega_n^2 - \omega^2} \mathbf{e}_n(\mathbf{r}) \int_V \mathbf{J}' \cdot \mathbf{e}_n dV \\
&+ \sum_v \frac{1 + \cos \omega t}{\omega \varepsilon} \mathbf{e}_v(\mathbf{r}) \int_V \mathbf{J}' \cdot \mathbf{e}_v dV,
\end{aligned}
\tag{8.56}
$$

$$
\mathbf{H}(\mathbf{r}, t) = \sum_n v \frac{\omega_n \sin \omega t - \omega \sin \omega_n t}{\omega_n^2 - \omega^2} \mathbf{h}_n(\mathbf{r}) \int_V \mathbf{J}' \cdot \mathbf{e}_n dV.
\tag{8.57}
$$

The time-domain solutions (8.56) and (8.57) do not agree with the time-harmonic solutions (3.93) and (3.94). When the cavity is lossless, the time-domain solutions are not sinusoidal even if the time approaches infinity, in contrast to the fact that the time-harmonic solutions are always sinusoidal. When ω approaches ω_n, Equations (8.56) and (8.57) may be rewritten as

$$
\begin{aligned}
\mathbf{E}(\mathbf{r}, t) = &-\sum_n \frac{\eta v}{2} t \sin \omega_n t \mathbf{e}_n(\mathbf{r}) \int_V \mathbf{J}' \cdot \mathbf{e}_n dV \\
&+ \sum_v \frac{1 + \cos \omega t}{\omega \varepsilon} \mathbf{e}_v(\mathbf{r}) \int_V \mathbf{J}' \cdot \mathbf{e}_v dV,
\end{aligned}
\tag{8.58}
$$

$$
\mathbf{H}(\mathbf{r}, t) = \sum_n v \left(-\frac{1}{2} t \cos \omega_n t + \frac{1}{2\omega_n} \sin \omega_n t \right) \mathbf{h}_n(\mathbf{r}) \int_V \mathbf{J}' \cdot \mathbf{e}_n dV.
\tag{8.59}
$$

Therefore the time-domain solutions (8.58) and (8.59) are finite for a finite time t, and there is no infinity problem that occurs in the time-harmonic solutions (3.90) and (3.91).

The above phenomenon can be explained by the uniqueness theorem of electromagnetic fields. In a bounded region, such as a cavity, the time-harmonic Maxwell equations have a unique solution if and only if the system is lossy, while the time-domain Maxwell equations always have a unique solution even if the system is lossless.   □

Remark 8.1: A cavity resonator is similar to an RLC circuit in low-frequency circuit, which is the most fundamental circuit system discussed in introductory physics courses. The RLC circuit is often used as an equivalent circuit to represent a single-port network. Figure 8.5 shows an RLC circuit whose element values are assumed to be independent of frequency. We now investigate this circuit in both frequency domain and time domain with a sinusoidal excitation. In frequency-domain analysis, the voltage source is assumed to be

$$v_s(t) = V \sin \omega t = \text{Re} V_s e^{j\omega t}, \tag{8.60}$$

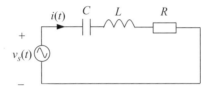

Figure 8.5 RLC circuit

where $V_s = V e^{-j\pi/2}$ is the phasor of the voltage source. According to the phasor arithmetic, the phasor of the current is

$$I(\omega) = \frac{V_s \omega}{L} \frac{2\omega\gamma - j(\omega^2 - \omega_r^2)}{4\omega^2\gamma^2 + (\omega^2 - \omega_r^2)^2},$$

where $\omega_r = 1/\sqrt{LC}$ is the resonant frequency of the circuit, and $\gamma = R/2L$ is the attenuation constant. Thus, the current in the time domain is given by

$$i(t) = \text{Re} I(\omega) e^{j\omega t} = \frac{V\omega}{L} \frac{(\omega_r^2 - \omega^2)\cos \omega t + 2\omega\gamma \sin \omega t}{(\omega_r^2 - \omega^2)^2 + 4\omega^2\gamma^2}. \tag{8.61}$$

If the loss R is sent to zero, the limit of current for a resonant LC circuit becomes

$$i(t) = \frac{V}{L} \frac{\omega \cos \omega t}{\omega_r^2 - \omega^2}. \tag{8.62}$$

We now re-examine the same problem in the time domain by means of Laplace transform. Assume that the RLC circuit is excited by a sinusoidal source turned on at $t = 0$

$$v_s(t) = V H(t) \sin \omega t,$$

where $H(t)$ is the unit step function. The current $i(t)$ satisfies the following differential equation in time domain

$$Ri(t) + L\frac{di(t)}{dt} + \frac{1}{C}\int_{0^-}^{t} i(\tau)d\tau = v_s(t).$$

Taking the Laplace transform yields

$$I(s) = V\frac{s}{Ls^2 + Rs + 1/C}\frac{\omega}{s^2 + \omega^2}.$$

This expression has four poles: $s_{1,2} = \pm j\omega$, $s_{3,4} = -\gamma \pm j\sqrt{\omega_r^2 - \gamma^2}$ and one zero at origin. The inverse Laplace transform gives the time-domain current, which can be split into the sum of two parts

$$i(t) = \frac{V\omega}{L}\frac{(\omega_r^2 - \omega^2)\cos\omega t + 2\omega\gamma\sin\omega t}{(\omega_r^2 - \omega^2)^2 + 4\omega^2\gamma^2}$$
$$-\frac{V\omega}{L\beta_r}\frac{\gamma(\omega^2 + \omega_r^2)\sin\beta_r t + \beta_r(\omega_r^2 - \omega^2)\cos\beta_r t}{(\omega_r^2 - \omega^2)^2 + 4\omega^2\gamma^2}e^{-\gamma t}. \qquad (8.63)$$

The first term on the right-hand side of (8.63) stands for the steady-state response of the system, which coincides with the frequency-domain solution (8.61). The second term may be viewed as the transient response of the lossy system, which approaches zero as time goes to infinity. For a lossless system ($\gamma = 0$), Equation (8.63) reduces to

$$i(t) = \frac{V}{L}\frac{\omega\cos\omega t}{(\omega_r^2 - \omega^2)} - \frac{V}{L}\frac{\omega\cos\omega_r t}{(\omega_r^2 - \omega^2)}. \qquad (8.64)$$

By comparing (8.62) against (8.64), we may find that the time-domain analysis disagrees with the frequency-domain analysis for a lossless system. The second term in the time-domain solution (8.64) does not appear in the time-harmonic solution (8.62). The second term in (8.64) is originated from the second term in (8.63), which is the transient part when loss exists and is no longer transient when loss disappears. From (8.62) and (8.64), two key differences between the frequency-domain and time-domain analysis for a lossless system can be found:

1. The frequency-domain analysis and time-domain analysis give different results for a lossless system. The time-harmonic solution (8.62) is sinusoidal while the time-domain solution (8.64) is not.
2. When ω approaches the resonant frequency ω_r, the time-harmonic solution (8.62) tends to infinity while the time-domain solution (8.64) reduces to

$$i(t) = \frac{V}{2L}t\sin\omega_r t. \qquad (8.65)$$

$\square$

From the above analysis, we conclude that the frequency-domain analysis fails when the system is lossless. The time-harmonic fields in a lossless medium cannot be considered the limit of the corresponding fields in a lossy medium as the loss goes to zero. For a lossless system we have to rely on the time-domain analysis to find a reasonable solution.

8.2.2   Fields in Waveguide Cavities

Evaluating the vector modal functions in an arbitrary metal cavity is not an easy task. When the metal cavity consists of a section of a uniform metal waveguide, the analysis of the transient process in the metal cavity can be carried out by means of the time-domain theory of waveguide.

8.2.2.1 Field expansions

Consider a waveguide cavity with a perfect electric wall of length L, as shown in Figure 8.6. The transient electromagnetic fields inside the waveguide cavity with current source $\mathbf{J}$ and $\mathbf{J}_m$ can be expanded in terms of the transverse vector modal functions $\mathbf{e}_n$ in the waveguide

$$
\begin{aligned}
\mathbf{E}(\mathbf{r}, t) &= \sum_{n=1}^{\infty} v_n(z, t)\mathbf{e}_n(\boldsymbol{\rho}) + \mathbf{u}_z \sum_{n=1}^{\infty} \frac{\nabla \cdot \mathbf{e}_n(\boldsymbol{\rho})}{k_{cn}} e'_{zn}, \\
\mathbf{H}(\mathbf{r}, t) &= \sum_{n=1}^{\infty} i_n(z, t)\mathbf{u}_z \times \mathbf{e}_n(\boldsymbol{\rho}) + \mathbf{u}_z \frac{1}{\sqrt{\Omega}} \int_{\Omega} \frac{\mathbf{u}_z \cdot \mathbf{H}}{\sqrt{\Omega}} d\Omega + \sum_{n=1}^{\infty} \frac{\nabla \times \mathbf{e}_n(\boldsymbol{\rho})}{k_{cn}} h'_{zn},
\end{aligned}
\tag{8.66}
$$

where $\boldsymbol{\rho} = (x, y)$ is the position vector in the waveguide cross-section Ω, and

$$
v_n(z, t) = \int_{\Omega} \mathbf{E} \cdot \mathbf{e}_n d\Omega, \; i_n(z, t) = \int_{\Omega} \mathbf{H} \cdot \mathbf{u}_z \times \mathbf{e}_n d\Omega,
$$

$$
h'_{zn}(z, t) = \int_{\Omega} \mathbf{H} \cdot \left(\frac{\nabla \times \mathbf{e}_{tn}}{k_{cn}} \right) d\Omega, \; e'_{zn}(z, t) = \int_{\Omega} \mathbf{u}_z \cdot \mathbf{E} \left(\frac{\nabla \cdot \mathbf{e}_{tn}}{k_{cn}} \right) d\Omega.
$$

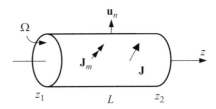

Figure 8.6   A metal cavity formed by a waveguide

Similar to the time-domain theory of waveguide, the modal voltages v_n^{TEM} and currents i_n^{TEM} for the TEM modes satisfy the wave equation

$$\frac{\partial^2 v_n^{TEM}}{\partial z^2} - \frac{1}{v^2}\frac{\partial^2 v_n^{TEM}}{\partial t^2} - \sigma\frac{\eta}{v}\frac{\partial v_n^{TEM}}{\partial t}$$

$$= \frac{\eta}{v}\frac{\partial}{\partial t}\int_\Omega \mathbf{J}\cdot\mathbf{e}_n d\Omega - \frac{\partial}{\partial z}\int_\Omega \mathbf{J}_m\cdot\mathbf{u}_z\times\mathbf{e}_n d\Omega,$$

$$\frac{\partial^2 i_n^{TEM}}{\partial z^2} - \frac{1}{v^2}\frac{\partial^2 i_n^{TEM}}{\partial t^2} - \sigma\frac{\eta}{v}\frac{\partial i_n^{TEM}}{\partial t}$$

$$= \sigma\int_\Omega \mathbf{J}_m\cdot\mathbf{u}_z\times\mathbf{e}_n d\Omega - \frac{\partial}{\partial z}\int_\Omega \mathbf{J}\cdot\mathbf{e}_n d\Omega + \frac{1}{\eta v}\frac{\partial}{\partial t}\int_\Omega \mathbf{J}_m\cdot\mathbf{u}_z\times\mathbf{e}_n d\Omega.$$

(8.67)

Once v_n^{TEM} (or i_n^{TEM}) are determined, i_n^{TEM} (or v_n^{TEM}) can be determined by the time integration of v_n^{TEM} (or i_n^{TEM}). The modal voltages v_n^{TE} for TE modes satisfy the modified Klein–Gordon equation

$$\frac{\partial^2 v_n^{TE}}{\partial z^2} - \frac{1}{v^2}\frac{\partial^2 v_n^{TE}}{\partial t^2} - \sigma\frac{\eta}{v}\frac{\partial v_n^{TE}}{\partial t} - k_{cn}^2 v_n^{TE}$$

$$= \frac{\eta}{v}\frac{\partial}{\partial t}\int_\Omega \mathbf{J}\cdot\mathbf{e}_n d\Omega - \frac{\partial}{\partial z}\int_\Omega \mathbf{J}_m\cdot\mathbf{u}_z\times\mathbf{e}_n d\Omega$$

$$+ k_{cn}\int_\Omega (\mathbf{u}_z\cdot\mathbf{J}_m)\left(\frac{\mathbf{u}_z\cdot\nabla\times\mathbf{e}_n}{k_{cn}}\right)d\Omega.$$

(8.68)

The modal currents i_n^{TE} for TE modes can be determined by the time integration of $\partial v_n^{TE}/\partial z$:

$$i_n^{TE}(z,t) = -\frac{\eta}{v}\int_{-\infty}^t \frac{\partial v_n^{TE}(z,t')}{\partial z}dt'$$

$$-\frac{\eta}{v}\int_{-\infty}^t \left\{\int_\Omega \mathbf{J}_m(\mathbf{r},t')\cdot[\mathbf{u}_z\times\mathbf{e}_n(\boldsymbol{\rho})]d\Omega(\boldsymbol{\rho})\right\}dt'.$$

(8.69)

The modal currents i_n^{TM} for TM modes also satisfy the modified Klein–Gordon equation

$$\frac{\partial^2 i_n^{TM}}{\partial z^2} - \frac{1}{v^2}\frac{\partial^2 i_n^{TM}}{\partial t^2} - \sigma\frac{\eta}{v}\frac{\partial i_n^{TM}}{\partial t} - k_{cn}^2 i_n^{TM}$$

$$= \sigma\int_\Omega \mathbf{J}_m\cdot\mathbf{u}_z\times\mathbf{e}_n d\Omega - \frac{\partial}{\partial z}\int_\Omega \mathbf{J}\cdot\mathbf{e}_n d\Omega$$

$$+ \frac{1}{\eta v}\frac{\partial}{\partial t}\int_\Omega \mathbf{J}_m\cdot\mathbf{u}_z\times\mathbf{e}_n d\Omega - k_{cn}\int_\Omega \mathbf{u}_z\cdot\mathbf{J}\left(\frac{\nabla\cdot\mathbf{e}_n}{k_{cn}}\right)d\Omega.$$

(8.70)

The modal voltages v_n^{TM} can then be determined by the time integration of $\partial i_n^{TM}/\partial z$:

$$v_n^{TM}(z,t) = -\eta v \int_{-\infty}^{t} \frac{\partial i_n^{TM}(z,t')}{\partial z} dt' - \eta v \int_{-\infty}^{t} \left[\int_{\Omega} \mathbf{J}(\mathbf{r},t') \cdot \mathbf{e}_n(\boldsymbol{\rho}) d\Omega(\boldsymbol{\rho}) \right] dt'. \tag{8.71}$$

8.2.2.2 Solutions of the modified Klein–Gordon equations

Since the tangential electric field on the electric conductor must be zero, the time-domain voltages satisfy the homogeneous Dirichlet boundary conditions

$$v_n(z,t)|_{z=z_1} = v_n(z,t)|_{z=z_2} = 0. \tag{8.72}$$

Considering (8.9) and the boundary condition that the normal component of the magnetic field on an electric conductor must be zero, the time-domain currents must satisfy the homogeneous Neumann boundary conditions

$$\left.\frac{\partial i_n(z,t)}{\partial z}\right|_{z=z_1} = \left.\frac{\partial i_n(z,t)}{\partial z}\right|_{z=z_2} = 0. \tag{8.73}$$

In order to solve (8.67), (8.68) and (8.70) subject to the boundary conditions (8.72) and (8.73), we may introduce the retarded Green's functions

$$\left(\frac{\partial^2}{\partial z^2} - \frac{1}{v^2}\frac{\partial^2}{\partial t^2} - \sigma\frac{\eta}{v}\frac{\partial}{\partial t} - k_{cn}^2 \right) G_n^v(z,t;z',t') = -\delta(z-z')\delta(t-t'),$$

$$G_n^v(z,t;z',t')|_{t<t'} = 0, \tag{8.74}$$

$$G_n^v(z,t;z',t')|_{z=z_1} = G_n^v(z,t;z',t')|_{z=z_2} = 0,$$

for the modal voltages and

$$\left(\frac{\partial^2}{\partial z^2} - \frac{1}{v^2}\frac{\partial^2}{\partial t^2} - \sigma\frac{\eta}{v}\frac{\partial}{\partial t} - k_{cn}^2 \right) G_n^i(z,t;z',t') = -\delta(z-z')\delta(t-t'),$$

$$G_n^i(z,t;z',t')|_{t<t'} = 0, \tag{8.75}$$

$$\left.\frac{\partial G_n^i(z,t;z',t')}{\partial z}\right|_{z=z_1} = \left.\frac{\partial G_n^i(z,t;z',t')}{\partial z}\right|_{z=z_2} = 0,$$

for the modal currents. Taking the Fourier transform of the Green's functions with respect to time t

$$\tilde{G}_n^{v,i}(z,\omega;z',t') = \int_{-\infty}^{\infty} G_n^{v,i}(z,t;z',t')e^{-j\omega t} dt$$

gives

$$\left(\frac{\partial^2}{\partial z^2} + \beta_n^2\right) \tilde{G}_n^{v,i}(z, \omega; z', t') = -e^{-j\omega t'} \delta(z - z'), \tag{8.76}$$

where $\beta_n^2 = k^2 - k_{cn}^2 - j\sigma k\eta$, $k = \omega/v$. The above equations can be solved by the method of eigenfunctions and the solutions are

$$\tilde{G}_n^v(z, \omega; z', t') = \sum_{m=1}^{\infty} \frac{-1}{\beta_n^2 - (m\pi/L)^2} \frac{2}{L} \sin\frac{m\pi}{L}(z - z_1) \sin\frac{m\pi}{L}(z' - z_1)e^{-j\omega t'},$$

$$\tilde{G}_n^i(z, \omega; z', t') = \sum_{m=0}^{\infty} \frac{-1}{\beta_n^2 - (m\pi/L)^2} \frac{\varepsilon_m}{L} \cos\frac{m\pi}{L}(z - z_1) \cos\frac{m\pi}{L}(z' - z_1)e^{-j\omega t'}.$$

Taking the inverse Fourier transform and making use of the residue theorem, we obtain

$$G_n^v(z, t; z', t') = \sum_{m=1}^{\infty} \frac{2v}{L} \sin\frac{m\pi}{L}(z - z_1) \sin\frac{m\pi}{L}(z' - z_1)$$

$$\cdot e^{-\gamma(t-t')} \frac{\sin\left[v(t - t')\sqrt{k_{cn}^2 + (m\pi/L)^2 - (\gamma/v)^2}\right]}{\sqrt{k_{cn}^2 + (m\pi/L)^2 - (\gamma/v)^2}} H(t - t'), \tag{8.77}$$

$$G_n^i(z, t; z', t') = \sum_{m=0}^{\infty} \frac{\varepsilon_m v}{L} \cos\frac{m\pi}{L}(z - z_1) \cos\frac{m\pi}{L}(z' - z_1)$$

$$\cdot e^{-\gamma(t-t')} \frac{\sin\left[v(t - t')\sqrt{k_{cn}^2 + (m\pi/L)^2 - (\gamma/v)^2}\right]}{\sqrt{k_{cn}^2 + (m\pi/L)^2 - (\gamma/v)^2}} H(t - t'), \tag{8.78}$$

where $\gamma = \sigma/2\varepsilon$. If one of the ends of the waveguide cavity extends to infinity, say, $z_2 \to \infty$, the discrete values $m\pi/L$ become a continuum. In this case, Equations (8.77) and (8.78) become

$$G_n^v(z, t; z', t')\big|_{z_2 \to \infty} = -\frac{v}{\pi} e^{-\gamma(t-t')}$$

$$\cdot \int_0^{\infty} [\cos k(z + z' - 2z_1) - \cos k(z - z')] \frac{\sin\left[v(t - t')\sqrt{k_{cn}^2 + k^2 - (\gamma/v)^2}\right]}{\sqrt{k_{cn}^2 + k^2 - (\gamma/v)^2}} dk,$$

$$G_n^i(z, t; z', t')\big|_{z_2 \to \infty} = \frac{v}{\pi} e^{-\gamma(t-t')}$$

$$\cdot \int_0^{\infty} [\cos k(z + z' - 2z_1) + \cos k(z - z')] \frac{\sin\left[v(t - t')\sqrt{k_{cn}^2 + k^2 - (\gamma/v)^2}\right]}{\sqrt{k_{cn}^2 + k^2 - (\gamma/v)^2}} dk.$$

These integrations may be carried out by using (8.22), and the retarded Green's functions are given by

$$e^{\gamma(t-t')}G_n^v(z,t;z',t')\Big|_{z_2 \to \infty} =$$

$$-\frac{v}{2}J_0\left[(k_{cn}^2v^2-\gamma^2)^{1/2}\sqrt{(t-t')^2-|z+z'-2z_1|^2/v^2}\right]H[v(t-t')-|z+z'-2z_1|]$$

$$+\frac{v}{2}J_0\left[(k_{cn}^2v^2-\gamma^2)^{1/2}\sqrt{(t-t')^2-|z-z'|^2/v^2}\right]H[v(t-t')-|z-z'|], \qquad (8.79)$$

$$e^{\gamma(t-t')}G_n^i(z,t;z',t')\Big|_{z_2 \to \infty}$$

$$=\frac{v}{2}J_0\left[(k_{cn}^2v^2-\gamma^2)^{1/2}\sqrt{(t-t')^2-|z+z'-2z_1|^2/v^2}\right]H[v(t-t')-|z+z'-2z_1|]$$

$$+\frac{v}{2}J_0\left[(k_{cn}^2v^2-\gamma^2)^{1/2}\sqrt{(t-t')^2-|z-z'|^2/v^2}\right]H[v(t-t')-|z-z'|]. \qquad (8.80)$$

The retarded Green's functions can be used to solve the following modified Klein–Gordon equations

$$\left(\frac{\partial^2}{\partial z^2}-\frac{1}{v^2}\frac{\partial^2}{\partial t^2}-\sigma\frac{\eta}{v}\frac{\partial}{\partial t}-k_{cn}^2\right)v_n(z,t)=f(z,t), z_1 < z < z_2,$$

$$\left(\frac{\partial^2}{\partial z^2}-\frac{1}{v^2}\frac{\partial^2}{\partial t^2}-\sigma\frac{\eta}{v}\frac{\partial}{\partial t}-k_{cn}^2\right)i_n(z,t)=g(z,t), z_1 < z < z_2,$$

subject to the boundary conditions (8.72) and (8.73). The solutions of the above equations are

$$v_n(z,t)=-\int_{z_1}^{z_2}dz'\int_{-\infty}^{\infty}f(z',t')G_n^v(z,t;z',t')dt', z_1 < z < z_2,$$

$$i_n(z,t)=-\int_{z_1}^{z_2}dz'\int_{-\infty}^{\infty}g(z',t')G_n^i(z,t;z',t')dt', z_1 < z < z_2. \qquad (8.81)$$

Thus the solutions of (8.68) and (8.70) can be expressed as

$$v_n^{TE}(z,t)=-\frac{\eta}{v}\int_{z_1}^{z_2}dz'\int_{-\infty}^{\infty}G_n^v(z,t;z',t')dt'\int_{\Omega}\frac{\partial}{\partial t'}\mathbf{J}(\boldsymbol{\rho}',z',t')\cdot\mathbf{e}_n(\boldsymbol{\rho}')d\Omega(\boldsymbol{\rho}')$$

$$-\int_{z_1}^{z_2}dz'\int_{-\infty}^{\infty}\frac{\partial G_n^v(z,t;z',t')}{\partial z'}dt'\int_{\Omega}\mathbf{J}_m(\boldsymbol{\rho}',z',t')\cdot\mathbf{u}_z\times\mathbf{e}_n(\boldsymbol{\rho}')d\Omega(\boldsymbol{\rho}') \qquad (8.82)$$

$$-k_{cn}\int_{z_1}^{z_2}dz'\int_{-\infty}^{\infty}G_n^v(z,t;z',t')dt'\int_{\Omega}\mathbf{u}_z\cdot\mathbf{J}_m(\boldsymbol{\rho}',z',t')\frac{\mathbf{u}_z\cdot\nabla\times\mathbf{e}_n(\boldsymbol{\rho}')}{k_{cn}}d\Omega(\boldsymbol{\rho}'),$$

$$i_n^{TM}(z,t) = -\int_{z_1}^{z_2} dz' \int_{-\infty}^{\infty} \frac{\partial G_n^i(z,t;z',t')}{\partial z'} dt' \int_{\Omega} \mathbf{J}(\boldsymbol{\rho}',z',t') \cdot \mathbf{e}_n(\boldsymbol{\rho}') d\Omega(\boldsymbol{\rho}')$$

$$-\frac{1}{\eta v}\int_{z_1}^{z_2} dz' \int_{-\infty}^{\infty} G_n^i(z,t;z',t') dt' \int_{\Omega} \frac{\partial}{\partial t'} \mathbf{J}_m(\boldsymbol{\rho}',z',t') \cdot \mathbf{u}_z \times \mathbf{e}_n(\boldsymbol{\rho}') d\Omega(\boldsymbol{\rho}') \qquad (8.83)$$

$$+k_{cn}\int_{z_1}^{z_2} dz' \int_{-\infty}^{\infty} G_n^i(z,t;z',t') dt' \int_{\Omega} \mathbf{u}_z \cdot \mathbf{J}(\boldsymbol{\rho}',z',t') \frac{\nabla \cdot \mathbf{e}_n(\boldsymbol{\rho}')}{k_{cn}} d\Omega(\boldsymbol{\rho}').$$

In deriving these expressions, it has been assumed that all sources are confined inside the cavity. It should be notified that the time-domain voltage and current do not satisfy the homogeneous boundary conditions (8.72) and (8.73) at $z = z_1$ or $z = z_2$ if the magnetic current $\mathbf{J}_m$ is tightly pressed on the electric wall $z = z_1$ or $z = z_2$.

8.2.2.3 Excitation of waveguide cavities

The time-domain response inside a metal cavity resonator is uniquely determined by the boundary conditions, initial conditions, and source conditions, regardless of whether the cavity involves loss or not. A wideband signal source in a waveguide cavity will excite an infinite number of waveguide modes, and the total fields in the cavity are the linear combination of these modes and are determined by (8.66), in which each expansion coefficient represents the contribution from the corresponding mode and can be determined by (8.82) or (8.83). We now give some examples to illustrate the transient processes inside the waveguide cavities.

Example 8.4 (A shorted rectangular waveguide): Consider a shorted rectangular waveguide shown in Figure 8.7. The shorted waveguide is excited by a line current extending across the waveguide centered at $x = x_0 = a/2$, $z = z_0$, which is given by (8.31). By the symmetry of the structure and excitation, only TE_{n0} modes will be excited and the vector modal functions are shown in (8.32). Assuming $f(t) = H(t)\sin \omega t$ and ignoring the heat loss, the time-domain voltages may be found from (8.79) and (8.82) as follows

$$v_n^{TE}(z,t) = \frac{b\eta}{2}\left(\frac{2}{ab}\right)^{1/2} ka\sin\frac{n\pi}{a}x_0$$

$$\cdot \left\{ \int_{|z-z_0|/a}^{vt/a} \cos ka(vt/a - u)J_0\left[k_{cn}a\sqrt{u^2 - |z - z_0|^2/a^2}\right]du \right.$$

$$\left. - \int_{|z+z_0|/a}^{vt/a} \cos ka(vt/a - u)J_0\left[k_{cn}a\sqrt{u^2 - |z + z_0|^2/a^2}\right]du \right\}.$$

Figure 8.7   A shorted rectangular waveguide excited by a centered current source

Due to the existence of radiation loss in the shorted waveguide, the time-domain responses may be divided into the sum of a steady-state part and a transient part

$$v_n^{TE}(z,t) = v_n^{TE}(z,t)\big|_{steady} + v_n^{TE}(z,t)\big|_{transient}$$

where

$$v_n^{TE}(z,t)\big|_{steady} = \frac{b\eta}{2}\left(\frac{2}{ab}\right)^{1/2} ka \sin\frac{n\pi}{a} x_0$$

$$\cdot\left\{\int_{|z-z_0|/a}^{\infty} \cos ka(vt/a - u) J_0\left[k_{cn}a\sqrt{u^2 - |z-z_0|^2/a^2}\right] du\right.$$

$$\left. - \int_{|z+z_0|/a}^{\infty} \cos ka(vt/a - u) J_0\left[k_{cn}a\sqrt{u^2 - |z+z_0|^2/a^2}\right] du\right\},$$

$$v_n^{TE}(z,t)\big|_{transient} = -\frac{b\eta}{2}\left(\frac{2}{ab}\right)^{1/2} ka \sin\frac{n\pi}{a} x_0$$

$$\cdot\left\{\int_{vt/a}^{\infty} \cos ka(vt/a - u) J_0\left[k_{cn}a\sqrt{u^2 - |z-z_0|^2/a^2}\right] du\right.$$

$$\left. - \int_{vt/a}^{\infty} \cos ka(vt/a - u) J_0\left[k_{cn}a\sqrt{u^2 - |z+z_0|^2/a^2}\right] du\right\}.$$

The transient part approaches zero as $t \to \infty$. The integrals in the steady-state part can be carried out by use of (8.35). Thus

$$v_n^{TE}(z,t)\big|_{steady} = \frac{b\eta}{2}\left(\frac{2}{ab}\right)^{1/2} \frac{ka}{\sqrt{|(ka)^2 - (k_{cn}a)^2|}} \sin\frac{n\pi}{2}$$

$$\times \begin{cases} \sin\left(ka\dfrac{vt}{a} - \dfrac{|z-z_0|}{a}\sqrt{(ka)^2 - (k_{cn}a)^2}\right) \\[2mm] -\sin\left(ka\dfrac{vt}{a} - \dfrac{|z+z_0|}{a}\sqrt{(ka)^2 - (k_{cn}a)^2}\right), \ k > k_{cn} \\[2mm] \cos\left(ka\dfrac{vt}{a}\right)\exp\left[-\dfrac{|z-z_0|}{a}\sqrt{(k_{cn}a)^2 - (ka)^2}\right] \\[2mm] -\cos\left(ka\dfrac{vt}{a}\right)\exp\left[-\dfrac{|z+z_0|}{a}\sqrt{(k_{cn}a)^2 - (ka)^2}\right], \ k < k_{cn} \end{cases}.$$

In the region $0 < z < z_0$, the steady-state response may be rewritten as

$$v_n^{TE}(z,t)\big|_{\text{steady}} = \frac{1}{\sqrt{2}}\left(\frac{b}{a}\right)^{1/2}\frac{\eta k}{\beta_n}\sin\frac{n\pi}{2}\cdot\begin{cases} 2\sin(\beta_n z)\cos(\omega t - \beta_n z_0),\ k > k_{cn} \\[2mm] \cos(\omega t)\exp[\beta_n(z-z_0)]- \\[1mm] \cos(\omega t)\exp[-\beta_n(z+z_0)],\ k < k_{cn} \end{cases},$$

where $\beta_n = (|k^2 - k_{cn}^2|)^{1/2}$. The time-domain voltages for the TE_{n0} modes in the shorted waveguide are a standing wave if the operating frequency is higher than the cut-off frequency of the TE_{n0} mode. The time-domain currents can be determined by (8.69) as

$$i_n^{TE}(z,t) = \left(\frac{2b}{a}\right)^{1/2}\sin\frac{n\pi}{2}\left[-\frac{1}{2}\sin\omega(t - |z+z_0|/v) - \frac{1}{2}\sin\omega(t - |z-z_0|/v)\right]$$

$$+\left(\frac{2b}{a}\right)^{1/2}\sin\frac{n\pi}{2}\left\{\frac{k_{cn}(z+z_0)}{2}\int_0^{t-|z+z_0|/v}\frac{J_1\left[k_{cn}v\sqrt{(t-t')^2 - |z+z_0|^2/v^2}\right]}{\sqrt{(t-t')^2 - |z+z_0|^2/v^2}}\sin\omega t'\,dt'\right.$$

$$\left.-\frac{k_{cn}(z-z_0)}{2}\int_0^{t-|z-z_0|/v}\frac{J_1\left[k_{cn}v\sqrt{(t-t')^2 - |z-z_0|^2/v^2}\right]}{\sqrt{(t-t')^2 - |z-z_0|^2/v^2}}\sin\omega t'\,dt'\right\}.$$

The steady-state part of $i_n^{TE}(z,t)$ is

$$i_n^{TE}(z,t)\big|_{\text{steady}}$$

$$= \left(\frac{2b}{a}\right)^{1/2}\sin\frac{n\pi}{2}\left[-\frac{1}{2}\sin\omega(t - |z+z_0|/v) - \frac{1}{2}\sin\omega(t - |z-z_0|/v)\right]$$

$$+\left(\frac{2b}{a}\right)^{1/2}\sin\frac{n\pi}{2}\left\{\frac{k_{cn}(z+z_0)}{2}\int_{|z+z_0|/v}^{\infty}\frac{J_1\left[k_{cn}v\sqrt{u^2 - |z+z_0|^2/v^2}\right]}{\sqrt{u^2 - |z+z_0|^2/v^2}}\sin\omega(t-u)\,du\right.$$

$$\left.-\frac{k_{cn}(z-z_0)}{2}\int_{|z-z_0|/v}^{\infty}\frac{J_1\left[k_{cn}v\sqrt{u^2 - |z-z_0|^2/v^2}\right]}{\sqrt{u^2 - |z-z_0|^2/v^2}}\sin\omega(t-u)\,du\right\}.$$

Assuming that $k > k_{cn}$ and making use of the following calculations

$$\int_a^\infty \frac{\sin dx}{\sqrt{x^2 - a^2}} J_\nu(b\sqrt{x^2 - a^2})dx$$

$$= \frac{\pi}{2} J_{\nu/2}\left[\frac{a}{2}(d - \sqrt{d^2 - b^2})\right] J_{-\nu/2}\left[\frac{a}{2}(d + \sqrt{d^2 - b^2})\right],$$

$$\int_a^\infty \frac{\cos dx}{\sqrt{x^2 - a^2}} J_\nu(b\sqrt{x^2 - a^2})dx$$

$$= -\frac{\pi}{2} J_{\nu/2}\left[\frac{a}{2}(d - \sqrt{d^2 - b^2})\right] N_{-\nu/2}\left[\frac{a}{2}(d + \sqrt{c^2 - b^2})\right],$$

$$(a > 0, 0 < b < d),$$

we obtain

$$i_n^{TE}(z, t)\big|_{steady} = -\frac{2}{\sqrt{2}}\left(\frac{b}{a}\right)^{1/2} \sin\frac{n\pi}{2} \cos(\beta_n z) \sin(\omega t - \beta_n z_0)$$

for $0 < z < z_0$. Let $V_n^{TE}(z)$ and $I_n^{TE}(z)$ be the phasors of $v_n^{TE}(z, t)\big|_{steady}$ and $i_n^{TE}(z, t)\big|_{steady}$ respectively, then

$$V_n^{TE}(z) = \frac{2}{\sqrt{2}}\left(\frac{b}{a}\right)^{1/2} \frac{\eta k}{\beta_n} \sin\frac{n\pi}{2} \sin(\beta_n z)e^{-j\beta_n z_0}, k > k_{cn},$$

$$I_n^{TE}(z) = j\frac{2}{\sqrt{2}}\left(\frac{b}{a}\right)^{1/2} \sin\frac{n\pi}{2} \cos(\beta_n z)e^{-j\beta_n z_0}, k > k_{cn}.$$

Since the currents are assumed to be in positive z-direction, the impedances for the TE modes at $z \in (0, z_0)$ are given by

$$Z_n(z) = \frac{V_n^{TE}(z)}{-I_n^{TE}(z)} = j\frac{\eta k}{\beta_n} \tan(\beta_n z), k > k_{cn}.$$

This is a well-known result in time-harmonic theory.   □

Example 8.5 (A rectangular waveguide cavity): A rectangular waveguide cavity is obtained by letting the shorted waveguide be closed by a perfect conducting wall at $z = L$ with $L > z_0$, as shown in Figure 8.8. For the same excitation source (8.31), only TE_{n0} mode will be excited in the cavity. It follows from (8.77) and (8.82) that

$$v_n^{TE}(z, t) = \frac{2\eta}{L}\left(\frac{2b}{a}\right)^{1/2} \sin\frac{n\pi x_0}{a} \sum_{m=1}^\infty \frac{\sin\frac{m\pi}{L}z \sin\frac{m\pi}{L}z_0}{\sqrt{(n\pi/a)^2 + (m\pi/L)^2}}$$

$$\cdot \int_{-\infty}^t \frac{df(t')}{dt'} \sin\left[c(t - t')\sqrt{\left(\frac{n\pi}{a}\right)^2 + \left(\frac{m\pi}{L}\right)^2}\right] dt'. \tag{8.84}$$

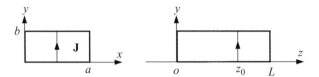

Figure 8.8   A rectangular waveguide cavity excited by a current source

If the cavity is excited by a sinusoidal signal turned on at $t = 0$, that is $f(t) = H(t) \sin \omega t$, the time-domain modal voltage may be written as

$$
v_n^{TE}(z, t) = -\frac{2\eta k}{L}\left(\frac{2b}{a}\right)^{1/2} \sin\frac{n\pi x_0}{a} \sum_{m=1}^{\infty} \sin\frac{m\pi}{L} z \sin\frac{m\pi}{L} z_0
$$
$$
\cdot \frac{\cos kct - \cos\left[vt\sqrt{(n\pi/a)^2 + (m\pi/L)^2}\right]}{k^2 - (n\pi/a)^2 - (m\pi/L)^2}.
$$
(8.85)

It can be seen that the time-domain response cannot be divided into a transient part and a steady-state part due to the lossless assumption. Numerical plots indicate that the response is not sinusoidal as $t \to \infty$ if k is not equal to any resonant wavenumber $\sqrt{(n\pi/a)^2 + (m\pi/L)^2}$ (Geyi, 2008). Notice that the response (8.85) is finite as k approaches any resonant wavenumber, and in this case, a sinusoidal wave gradually builds up as $t \to \infty$. Therefore, the response of a lossless metal cavity is sinusoidal if and only if the frequency of the exciting sinusoidal wave coincides with one of the resonant frequencies of the metal cavity.

If the excitation waveform is a unit step function $f(t) = H(t)$, Equation (8.84) becomes

$$
v_n^{TE}(z, t) = \frac{2\eta}{L}\left(\frac{2b}{a}\right)^{1/2} \sin\frac{n\pi x_0}{a}
$$
$$
\cdot \sum_{m=1}^{\infty} \sin\frac{m\pi}{L} z \sin\frac{m\pi}{L} z_0 \frac{\sin\left[ct\sqrt{(n\pi/a)^2 + (m\pi/L)^2}\right]}{\sqrt{(n\pi/a)^2 + (m\pi/L)^2}}.
$$
(8.86)

and the response of the metal cavity is no longer a unit step function.   □

Example 8.6 (A coaxial waveguide): A lossless coaxial waveguide cavity of length L consisting of an inner conductor of radius a and an outer conductor of radius b is shown in Figure 8.9. The coaxial waveguide is excited by a magnetic ring current located at $z = z_0$, which is given by (8.36). According to the symmetry, only the TEM mode and those TM_{0q} modes that are independent of φ will be excited and the vector modal functions for these modes are given

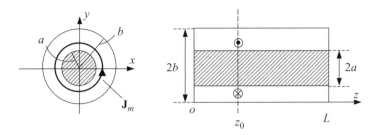

Figure 8.9 Cross-section of a coaxial waveguide

by (8.37). It follows from (8.78) and (8.83) that

$$i_n^{TM}(z, t) = -\frac{2\pi}{\eta L} \rho_0 e_n(\rho_0) \sum_{m=0}^{\infty} \varepsilon_m \frac{\cos(m\pi z/L)\cos(m\pi z_0/L)}{\sqrt{(\chi_n/a)^2 + (m\pi/L)^2}}$$

$$\cdot \int_{-\infty}^{t} \frac{df(t')}{dt'} \sin\left[v(t - t')\sqrt{(\chi_n/a)^2 + (m\pi/L)^2}\right] dt'. \tag{8.87}$$

For $f(t) = H(t)\sin\omega t$, the above expression becomes

$$i_n^{TM}(z, t) = \frac{2k\pi}{\eta L} \rho_0 e_n(\rho_0) \sum_{m=0}^{\infty} \varepsilon_m \cos\frac{m\pi}{L} z \cos\frac{m\pi}{L} z_0$$

$$\cdot \frac{\cos kct - \cos\left[vt\sqrt{(\chi_n/a)^2 + (m\pi/L)^2}\right]}{k^2 - (\chi_n/a)^2 - (m\pi/L)^2}. \tag{8.88}$$

In the case where k does not coincide with any resonant wavenumber $\sqrt{(\chi_n/a)^2 + (m\pi/L)^2}$, the field response determined from (8.66) is not sinusoidal. When the frequency of the excitation waveform approaches one of the resonant frequencies, a sinusoidal wave will gradually build up inside the coaxial waveguide cavity.

If the coaxial waveguide cavity is excited by the unit step waveform $f(t) = H(t)$, Equation (8.87) becomes

$$i_n^{TM}(z, t) = -\frac{2\pi}{\eta L} \rho_0 e_n(\rho_0)$$

$$\cdot \sum_{m=0}^{\infty} \varepsilon_m \frac{\cos(m\pi z/L)\cos(m\pi z_0/L)}{\sqrt{(\chi_n/a)^2 + (m\pi/L)^2}} \sin\left[vt\sqrt{(\chi_n/a)^2 + (m\pi/L)^2}\right].$$

The field response is no longer a unit step waveform. This is completely different from the time-domain response in a coaxial waveguide, which retains the original excitation waveform although it is distorted by the higher-order modes.   □

8.3 Spherical Wave Expansions in Time-domain

In the spherical coordinate system (r, θ, φ), the electromagnetic fields can be decomposed into the transverse components and the radial components

$$\mathbf{E} = \mathbf{E}_t + \mathbf{u}_r E_r, \quad \mathbf{H} = \mathbf{H}_t + \mathbf{u}_r H_r.$$

If these decompositions are introduced into Maxwell equations, the radial components can be eliminated to get the equations for the transverse components.

8.3.1 Transverse Field Equations

Taking the vector and scalar product of Maxwell equations

$$\nabla \times \mathbf{H}(\mathbf{r}, t) = \varepsilon \frac{\partial \mathbf{E}(\mathbf{r}, t)}{\partial t} + \mathbf{J}(\mathbf{r}, t),$$

$$\nabla \times \mathbf{E}(\mathbf{r}, t) = -\mu \frac{\partial \mathbf{H}(\mathbf{r}, t)}{\partial t} - \mathbf{J}_m(\mathbf{r}, t),$$

with the vector $\mathbf{r}$, we obtain

$$\varepsilon \frac{\partial}{\partial t}(\mathbf{r} \times \mathbf{E}) + \mathbf{r} \times \mathbf{J} = \nabla(\mathbf{r} \cdot \mathbf{H}) - (\mathbf{r} \cdot \nabla)\mathbf{H} - \mathbf{H},$$

$$-\mu \frac{\partial}{\partial t}(\mathbf{r} \times \mathbf{H}) - \mathbf{r} \times \mathbf{J}_m = \nabla(\mathbf{r} \cdot \mathbf{E}) - (\mathbf{r} \cdot \nabla)\mathbf{E} - \mathbf{E}, \tag{8.89}$$

$$-\nabla \cdot (\mathbf{r} \times \mathbf{H}_t) = \varepsilon \frac{\partial(\mathbf{r} \cdot \mathbf{E})}{\partial t} + \mathbf{r} \cdot \mathbf{J},$$

$$\nabla \cdot (\mathbf{r} \times \mathbf{E}_t) = \mu \frac{\partial(\mathbf{r} \cdot \mathbf{H})}{\partial t} + \mathbf{r} \cdot \mathbf{J}_m. \tag{8.90}$$

For an arbitrary vector $\mathbf{F}$, its directional derivative along $\mathbf{r}$ is

$$(\mathbf{r} \cdot \nabla)\mathbf{F} = r \frac{\partial \mathbf{F}}{\partial r} = r \frac{\partial}{\partial r}(F_r \mathbf{u}_r + F_\theta \mathbf{u}_\theta + F_\varphi \mathbf{u}_\varphi)$$

$$= r \mathbf{u}_r \frac{\partial}{\partial r} F_r + r \frac{\partial}{\partial r}(F_\theta \mathbf{u}_\theta + F_\varphi \mathbf{u}_\varphi) = r \mathbf{u}_r \frac{\partial}{\partial r} F_r + r \frac{\partial}{\partial r} \mathbf{F}_t.$$

Considering this equation and comparing the transverse components of (8.89), we obtain

$$\frac{1}{r} \nabla_{\theta\varphi}(\mathbf{r} \cdot \mathbf{H}) - r \frac{\partial \mathbf{H}_t}{\partial r} - \mathbf{H}_t = \varepsilon \frac{\partial}{\partial t}(\mathbf{r} \times \mathbf{E}_t) + \mathbf{r} \times \mathbf{J}_t,$$

$$\frac{1}{r} \nabla_{\theta\varphi}(\mathbf{r} \cdot \mathbf{E}) - r \frac{\partial \mathbf{E}_t}{\partial r} - \mathbf{E}_t = -\mu \frac{\partial}{\partial t}(\mathbf{r} \times \mathbf{H}_t) - \mathbf{r} \times \mathbf{J}_{mt}, \tag{8.91}$$

where $\nabla_{\theta\varphi} = \mathbf{u}_\theta \frac{\partial}{\partial\theta} + \mathbf{u}_\varphi \frac{1}{\sin\theta}\frac{\partial}{\partial\varphi}$. The radial components in (8.91) may be eliminated by using (8.90), to obtain the equations for the transverse fields

$$-\mu\frac{\partial}{\partial r}\left[\frac{\partial(r\mathbf{H}_t)}{\partial t}\right] + \frac{1}{r^2}\nabla_{\theta\varphi}\nabla_{\theta\varphi}\cdot(\mathbf{r}\times\mathbf{E}_t) - \mu\varepsilon\frac{\partial^2}{\partial t^2}(\mathbf{r}\times\mathbf{E}_t)$$

$$= \mu\frac{\partial}{\partial t}(\mathbf{r}\times\mathbf{J}_t) + \frac{1}{r}\nabla_{\theta\varphi}(\mathbf{r}\cdot\mathbf{J}_m),$$

$$-\varepsilon\frac{\partial}{\partial r}\left[\frac{\partial(r\mathbf{E}_t)}{\partial t}\right] - \frac{1}{r^2}\nabla_{\theta\varphi}\nabla_{\theta\varphi}\cdot(\mathbf{r}\times\mathbf{H}_t) + \mu\varepsilon\frac{\partial^2}{\partial t^2}(\mathbf{r}\times\mathbf{H}_t)$$

$$= -\varepsilon\frac{\partial}{\partial t}(\mathbf{r}\times\mathbf{J}_{mt}) + \frac{1}{r}\nabla_{\theta\varphi}(\mathbf{r}\cdot\mathbf{J}).$$

(8.92)

8.3.2   Spherical Transmission Line Equations

Similar to (3.102), the transverse electromagnetic fields in time domain may be represented by

$$r\mathbf{E}_t(\mathbf{r}, t) = \sum_{n,m,l}\left[V_{nml}^{TM}(r, t)\mathbf{e}_{nml}(\theta, \varphi) + V_{nml}^{TE}(r, t)\mathbf{h}_{nml}(\theta, \varphi)\right],$$

$$r\mathbf{H}_t(\mathbf{r}, t) = \sum_{n,m,l}\left[I_{nml}^{TM}(r, t)\mathbf{h}_{nml}(\theta, \varphi) - I_{nml}^{TE}(r, t)\mathbf{e}_{nml}(\theta, \varphi)\right].$$

Thus

$$\mathbf{r}\times\mathbf{E}_t(\mathbf{r}, t) = \sum_{n,m,l}\left[V_{nml}^{TM}(r, t)\mathbf{h}_{nml}(\theta, \varphi) - V_{nml}^{TE}(r, t)\mathbf{e}_{nml}(\theta, \varphi)\right],$$

$$\mathbf{r}\times\mathbf{H}_t(\mathbf{r}, t) = \sum_{n,m,l}\left[-I_{nml}^{TM}(r, t)\mathbf{e}_{nml}(\theta, \varphi) - I_{nml}^{TE}(r, t)\mathbf{h}_{nml}(\theta, \varphi)\right].$$

Introducing the following calculations

$$\nabla_{\theta\varphi}\nabla_{\theta\varphi}\cdot(\mathbf{r}\times\mathbf{E}_t) = \sum_{n,m,l}V_{nml}^{TE}n(n+1)\mathbf{e}_{nml},$$

$$\nabla_{\theta\varphi}\nabla_{\theta\varphi}\cdot(\mathbf{r}\times\mathbf{H}_t) = \sum_{n,m,l}I_{nml}^{TM}n(n+1)\mathbf{e}_{nml},$$

$$\frac{\partial}{\partial r}\left[\frac{\partial(r\mathbf{E}_t)}{\partial t}\right] = \sum_{n,m,l}\left(\frac{\partial^2 V_{nml}^{TM}}{\partial r\partial t}\mathbf{e}_{nml} + \frac{\partial^2 V_{nml}^{TE}}{\partial r\partial t}\mathbf{h}_{nml}\right),$$

$$\frac{\partial}{\partial r}\left[\frac{\partial(r\mathbf{H}_t)}{\partial t}\right] = \sum_{n,m,l}\left(\frac{\partial^2 I_{nml}^{TM}}{\partial r\partial t}\mathbf{h}_{nml} - \frac{\partial^2 I_{nml}^{TE}}{\partial r\partial t}\mathbf{e}_{nml}\right),$$

$$\frac{\partial^2}{\partial t^2}(\mathbf{r}\times\mathbf{E}_t) = \sum_{n,m,l}\left(\frac{\partial^2 V_{nml}^{TM}}{\partial t^2}\mathbf{h}_{nml} - \frac{\partial^2 V_{nml}^{TE}}{\partial t^2}\mathbf{e}_{nml}\right),$$

$$\frac{\partial^2}{\partial t^2}(\mathbf{r}\times\mathbf{H}_t) = \sum_{n,m,l}\left(-\frac{\partial^2 I_{nml}^{TM}}{\partial t^2}\mathbf{e}_{nml} - \frac{\partial^2 I_{nml}^{TE}}{\partial t^2}\mathbf{h}_{nml}\right),$$

into (8.92) yields

$$
-\mu \sum_{n,m,l} \left(\frac{\partial^2 I_{nml}^{TM}}{\partial r \partial t} \mathbf{h}_{nml} - \frac{\partial^2 I_{nml}^{TE}}{\partial r \partial t} \mathbf{e}_{nml} \right) + \frac{1}{r^2} \sum_{n,m,l} V_{nml}^{TE} n(n+1) \mathbf{e}_{nml}
$$

$$
-\mu\varepsilon \sum_{n,m,l} \left(\frac{\partial^2 V_{nml}^{TM}}{\partial t^2} \mathbf{h}_{nml} - \frac{\partial^2 V_{nml}^{TE}}{\partial t^2} \mathbf{e}_{nml} \right)
$$

$$
= \mu \frac{\partial}{\partial t} (\mathbf{r} \times \mathbf{J}_t) + \frac{1}{r} \nabla_{\theta\varphi} (\mathbf{r} \cdot \mathbf{J}_m),
$$

$$
-\varepsilon \sum_{n,m,l} \left(\frac{\partial^2 V_{nml}^{TM}}{\partial r \partial t} \mathbf{e}_{nml} + \frac{\partial^2 V_{nml}^{TE}}{\partial r \partial t} \mathbf{h}_{nml} \right) - \frac{1}{r^2} \sum_{n,m,l} I_{nml}^{TM} n(n+1) \mathbf{e}_{nml}
$$

$$
+\mu\varepsilon \sum_{n,m,l} \left(-\frac{\partial^2 I_{nml}^{TM}}{\partial t^2} \mathbf{e}_{nml} - \frac{\partial^2 I_{nml}^{TE}}{\partial t^2} \mathbf{h}_{nml} \right)
$$

$$
= -\varepsilon \frac{\partial}{\partial t} (\mathbf{r} \times \mathbf{J}_{mt}) + \frac{1}{r} \nabla_{\theta\varphi} (\mathbf{r} \cdot \mathbf{J}).
$$

Equating the coefficients before the vector basis functions gives

$$
\varepsilon \frac{\partial^2 V_{nml}^{TM}}{\partial t^2} + \frac{\partial^2 I_{nml}^{TM}}{\partial r \partial t} = - \int_{S'} \left[\frac{\partial}{\partial t} (\mathbf{r} \times \mathbf{J}_t) + \frac{1}{\mu r} \nabla_{\theta\varphi} (\mathbf{r} \cdot \mathbf{J}_m) \right] \cdot \mathbf{h}_{nml} d\Omega,
$$

$$
-\mu \frac{\partial^2 I_{nml}^{TM}}{\partial t^2} - \frac{\partial^2 V_{nml}^{TM}}{\partial r \partial t} - \frac{1}{\varepsilon r^2} I_{nml}^{TM} n(n+1) = \int_{S'} \left[-\frac{\partial}{\partial t} (\mathbf{r} \times \mathbf{J}_{mt}) + \frac{1}{\varepsilon r} \nabla_{\theta\varphi} (\mathbf{r} \cdot \mathbf{J}) \right] \cdot \mathbf{e}_{nml} d\Omega,
$$

$$
-\mu \frac{\partial^2 I_{nml}^{TE}}{\partial t^2} - \frac{\partial^2 V_{nml}^{TE}}{\partial r \partial t} = \int_{S'} \left[-\frac{\partial}{\partial t} (\mathbf{r} \times \mathbf{J}_{mt}) + \frac{1}{\varepsilon r} \nabla_{\theta\varphi} (\mathbf{r} \cdot \mathbf{J}) \right] \cdot \mathbf{h}_{nml} d\Omega,
$$

$$
\varepsilon \frac{\partial^2 V_{nml}^{TE}}{\partial t^2} + \frac{\partial^2 I_{nml}^{TE}}{\partial r \partial t} + \frac{1}{\mu r^2} V_{nml}^{TE} n(n+1) = \int_{S'} \left[\frac{\partial}{\partial t} (\mathbf{r} \times \mathbf{J}_t) + \frac{1}{\mu r} \nabla_{\theta\varphi} (\mathbf{r} \cdot \mathbf{J}_m) \right] \cdot \mathbf{e}_{nml} d\Omega,
$$

where S' is a sphere enclosing the source and $d\Omega$ is the differential element of the solid angle. After some manipulations, we obtain

$$
\left[\frac{\partial^2}{\partial r^2} - \frac{1}{v^2} \frac{\partial^2}{\partial t^2} - \frac{n(n+1)}{r^2} \right] I_{nml}^{TM}
$$

$$
= \varepsilon \int_{S'} \left[-\frac{\partial}{\partial t} (\mathbf{r} \times \mathbf{J}_{mt}) + \frac{1}{\varepsilon r} \nabla_{\theta\varphi} (\mathbf{r} \cdot \mathbf{J}) \right] \cdot \mathbf{e}_{nml} d\Omega
$$

$$
- \frac{\partial}{\partial r} \int_{-\infty}^{t} \left\{ \int_{S'} \left[\frac{\partial}{\partial t'} (\mathbf{r} \times \mathbf{J}_t) + \frac{1}{\mu r} \nabla_{\theta\varphi} (\mathbf{r} \cdot \mathbf{J}_m) \right] \cdot \mathbf{h}_{nml} d\Omega \right\} dt',
$$

$$
\left[\frac{\partial^2}{\partial r^2} - \frac{1}{v^2} \frac{\partial^2}{\partial t^2} - \frac{n(n+1)}{r^2} \right] V_{nml}^{TE}
$$

$$
= -\mu \int_{S'} \left[\frac{\partial}{\partial t} (\mathbf{r} \times \mathbf{J}_t) + \frac{1}{\mu r} \nabla_{\theta\varphi} (\mathbf{r} \cdot \mathbf{J}_m) \right] \cdot \mathbf{e}_{nml} d\Omega
$$

$$
- \frac{\partial}{\partial r} \int_{-\infty}^{t} \left\{ \int_{S'} \left[-\frac{\partial}{\partial t'} (\mathbf{r} \times \mathbf{J}_{mt}) + \frac{1}{\varepsilon r} \nabla_{\theta\varphi} (\mathbf{r} \cdot \mathbf{J}) \right] \cdot \mathbf{h}_{nml} d\Omega \right\} dt'.
$$

These are the **time-domain spherical transmission line equations**.

8.4 Radiation and Scattering in Time-domain

The application of short pulses in radar and telecommunication has made the analysis and design of ultra-wideband antenna in the time-domain an active research field. The ultra-wideband systems have some advantages over the traditional narrow band system because of the use of very short pulses, such as efficient transfer of localized electromagnetic energy and high-resolution interrogation of targets.

8.4.1 Radiation From an Arbitrary Source

In the time domain, the vector and scalar potential in free space satisfy the wave equations

$$
\left(\nabla^2 - \mu_0 \varepsilon_0 \frac{\partial^2}{\partial t^2} \right) \mathbf{A}(\mathbf{r}, t) = -\mu \mathbf{J}(\mathbf{r}, t),
$$

$$
\left(\nabla^2 - \mu_0 \varepsilon_0 \frac{\partial^2}{\partial t^2} \right) \phi(\mathbf{r}, t) = -\frac{\rho(\mathbf{r}, t)}{\varepsilon}.
$$

For the radiation problem in free space, the solutions are

$$
\mathbf{A}(\mathbf{r}, t) = \int_V \frac{\mu_0 \mathbf{J}(\mathbf{r}', T)}{4\pi R} dV(\mathbf{r}'), \ \phi(\mathbf{r}, t) = \int_V \frac{\rho(\mathbf{r}', T)}{4\pi \varepsilon_0 R} dV(\mathbf{r}'),
$$

where $R = |\mathbf{r} - \mathbf{r}'|$, $T = t - R/c$, and $c = 1/\sqrt{\mu_0 \varepsilon_0}$. In time-domain electromagnetics, the difference between the time of emission and the time of observation is the basis for the existence of electromagnetic radiation. When the observation point is very close to the sources, the changes of the sources ρ and $\mathbf{J}$ over a time scale of R/c are not significant and retardation effects can be ignored. In this case, the time of emission T is approximately equal to the time of observation t. In other words the potentials are approximately those occurring in statics. On the contrary, retardation effects become significant when the observation point is far from the source. Suppose that the origin of the coordinates lies inside the source distribution, and the source distribution has a characteristic dimension d. We can make the following approximations:

$$R = |\mathbf{r} - \mathbf{r}'| = \sqrt{r^2 - 2\mathbf{r} \cdot \mathbf{r}' + r'^2} \approx r - \mathbf{u}_r \cdot \mathbf{r}',$$

$$T \approx t_r = t - r/c + \mathbf{u}_r \cdot \mathbf{r}'/c$$

(8.93)

for $r \gg d$. The last term in t_r denotes the amount of time it takes for the radiation to propagate across the source. Making use of (8.93), the potentials can then be rewritten as

$$\mathbf{A}(\mathbf{r}, t) = \frac{\mu_0}{4\pi r} \int_V \mathbf{J}\left(\mathbf{r}', t - \frac{r}{c} + \frac{1}{c}\mathbf{u}_r \cdot \mathbf{r}'\right) dV(\mathbf{r}'),$$

$$\phi(\mathbf{r}, t) = \frac{1}{4\pi \varepsilon_0 r} \int_V \rho\left(\mathbf{r}', t - \frac{r}{c} + \frac{1}{c}\mathbf{u}_r \cdot \mathbf{r}'\right) dV(\mathbf{r}').$$

By use of the following approximations

$$\nabla \frac{f(t - R/c)}{R} \approx -\frac{\mathbf{u}_r}{c} \frac{\partial}{\partial t} \frac{f(t_r)}{r} + o\left(\frac{1}{r^2}\right),$$

$$\nabla \times \frac{\mathbf{J}(t - R/c)}{R} \approx -\frac{\mathbf{u}_r \times \mathbf{J}'(t_r)}{cr} + o\left(\frac{1}{r}\right),$$

the far fields produced by the source may be expressed as

$$\mathbf{E}(\mathbf{r}, t) = -\nabla\phi - \frac{\partial \mathbf{A}}{\partial t} \approx -\frac{\mu_0}{4\pi r} \int_V \frac{\partial \mathbf{J}(\mathbf{r}', t_r)}{\partial t} dV(\mathbf{r}') + \frac{\mathbf{u}_r \eta_0}{4\pi r} \int_V \frac{\partial \rho(\mathbf{r}', t_r)}{\partial t} dV(\mathbf{r}'),$$

$$\mathbf{H}(\mathbf{r}, t) = \nabla \times \mathbf{A} \approx -\frac{\mathbf{u}_r}{4\pi cr} \times \int_V \frac{\partial \mathbf{J}(\mathbf{r}', t_r)}{\partial t} dV(\mathbf{r}'),$$

where $\eta_0 = \sqrt{\mu_0/\varepsilon_0}$. To get rid of the charge distribution, we may use the continuity equation

$$\nabla'_{\mathbf{r}'} \cdot \mathbf{J}(\mathbf{r}', t_r) = -\frac{\partial \rho(\mathbf{r}', t_r)}{\partial t},$$

where $\nabla'_{\mathbf{r}'}$ denotes the operation applied to the spatial argument only. Since

$$\nabla' \cdot \mathbf{J}(\mathbf{r}', t_r) = \nabla'_{\mathbf{r}'} \cdot \mathbf{J}(\mathbf{r}', t_r) - \frac{1}{c}\mathbf{u}_R \cdot \frac{\partial \mathbf{J}(\mathbf{r}', t_r)}{\partial t}$$

$$\approx \nabla'_{\mathbf{r}'} \cdot \mathbf{J}(\mathbf{r}', t_r) - \frac{1}{c}\mathbf{u}_r \cdot \frac{\partial \mathbf{J}(\mathbf{r}', t_r)}{\partial t},$$

the continuity equation can be written as

$$\frac{\partial \rho(\mathbf{r}', t_r)}{\partial t} = \nabla' \cdot \mathbf{J}(\mathbf{r}', t_r) + \frac{1}{c}\mathbf{u}_r \cdot \frac{\partial \mathbf{J}(\mathbf{r}', t_r)}{\partial t}.$$

(8.94)

From $\int_V \nabla' \cdot \mathbf{J}(\mathbf{r}', t_r) dV(\mathbf{r}') = 0$ and (8.94), the radiated electric field can be represented by

$$\mathbf{E}(\mathbf{r}, t) \approx -\frac{\mu_0}{4\pi r} \int_V \frac{\partial \mathbf{J}(\mathbf{r}', t_r)}{\partial t} dV(\mathbf{r}') + \frac{\mu_0}{4\pi r} \mathbf{u}_r \int_V \mathbf{u}_r \cdot \frac{\partial \mathbf{J}(\mathbf{r}', t_r)}{\partial t} dV(\mathbf{r}')$$

$$= \mathbf{u}_r \times \frac{\mu_0}{4\pi r} \int_V \mathbf{u}_r \times \frac{\partial \mathbf{J}(\mathbf{r}', t_r)}{\partial t} dV(\mathbf{r}') = -\eta_0 \mathbf{u}_r \times \mathbf{H}(\mathbf{r}, t).$$

(8.95)

Thus, the far field generated by any transient sources is a transverse electromagnetic wave relative to the radial direction. The rate of energy radiated per unit area is

$$\frac{d P^{rad}}{dS} = \mathbf{u}_r \cdot (\mathbf{E} \times \mathbf{H}) = \eta_0 |\mathbf{H}|^2$$

$$= \frac{\eta_0}{16\pi^2 c^2 r^2} \left| \mathbf{u}_r \times \int_V \frac{\partial \mathbf{J}(\mathbf{r}', t_r)}{\partial t} dV(\mathbf{r}') \right|^2 ,$$

where dS stands for the differential area of the sphere of radius r. The rate of energy radiated per unit solid angle (angular distribution of radiated power) is

$$\frac{d P^{rad}}{d\Omega} = \frac{d P^{rad}}{r^2 dS} = \frac{\eta}{16\pi^2 c^2} \left| \mathbf{u}_r \times \int_V \frac{\partial \mathbf{J}(\mathbf{r}', t_r)}{\partial t} dV(\mathbf{r}') \right|^2 .$$

8.4.2 Radiation From Elementary Sources

The time-domain analysis of radiation mechanism by elementary electric and magnetic sources is helpful for understanding the physical process of radiation, which involves the exchange of energies between the source and the radiated fields.

8.4.2.1 Radiation by accelerated charged particle

Any accelerated charges will radiate electromagnetic energy. Let $\mathbf{c}(t')$ denote the position of a charged particle with charge q moving with velocity $\mathbf{v}$, small compared with the speed of light, $|\mathbf{v}|/c \ll 1$. The current density is

$$\mathbf{J}(\mathbf{r}', t') = q\mathbf{v}(t')\delta[\mathbf{r}' - \mathbf{c}(t')].$$

The time of emission of radiation may be approximated by

$$t_r = t - \frac{r}{c} + \frac{1}{c}\mathbf{u}_r \cdot \mathbf{c}(t') \approx t - \frac{r}{c} = t_e$$

for $|\mathbf{v}|/c \ll 1$. Then

$$\int_V \frac{\partial \mathbf{J}(\mathbf{r}', t_r)}{\partial t} dV(\mathbf{r}') \approx \frac{d}{dt} \int_V \mathbf{J}(\mathbf{r}', t_e) dV(\mathbf{r}') = q \frac{d\mathbf{v}(t_e)}{dt_e},$$

which indicates that the radiation is produced whenever the charged particle is accelerated. The angular distribution of the radiated power is

$$\frac{d P^{rad}}{d\Omega} = \frac{\eta_0 q^2}{16\pi^2 c^2} \left| \mathbf{u}_r \times \frac{d\mathbf{v}(t_e)}{dt_e} \right|^2 = \frac{\eta_0 q^2}{16\pi^2 c^2} \left| \frac{d\mathbf{v}(t_e)}{dt_e} \right|^2 \sin^2 \theta(t_e),$$

where $\theta(t_e)$ is the angle between the direction of observation $\mathbf{u}_r$ and the direction of the acceleration at the emission time t_e. Note that there is no radiation along the direction of acceleration. Since $\int \sin^2 \theta d\Omega = 8\pi/3$, the total radiated power is

$$P^{rad} = \frac{\eta q^2}{6\pi c^2} \left| \frac{d\mathbf{v}(t_e)}{dt_e} \right|^2.$$

This is the well-known **Larmor formula**, named after Irish physicist Joseph Larmor (1857–1942).

8.4.2.2 Radiation from electric dipole and magnetic dipole

An electric dipole consists of two point charges of equal magnitude $q(t)$ but opposite sign separated by a fixed distance l, as shown in Figure 8.10. Let $I(t)$ be the current flowing in the direction of $\mathbf{u}_d$ and $\mathbf{u}_d$ is a unit vector from the negative charge to the positive charge. The electric dipole moment $\mathbf{p}$ is defined by

$$\mathbf{p}(t) = p(t)\mathbf{u}_d = q(t)l\mathbf{u}_d.$$

An infinitesimal dipole is obtained when the length of the dipole goes to zero and charges become infinity in such a way that the dipole moment remains finite. The radiated electromagnetic fields of the dipole can be obtained from (8.95)

$$\mathbf{E}(\mathbf{r}, t) \approx \frac{\mu_0}{4\pi r} \{\mathbf{u}_r \times [\mathbf{u}_r \times \ddot{\mathbf{p}}(t_r)]\}_{t_r=t-r/c} = \frac{\mu_0}{4\pi r} [\ddot{p}(t_r)]_{t_r=t-r/c} \sin\theta \mathbf{u}_\theta,$$

$$\mathbf{H}(\mathbf{r}, t) \approx -\frac{1}{4\pi rc} [\mathbf{u}_r \times \ddot{\mathbf{p}}(t_r)]_{t_r=t-r/c} = \frac{1}{4\pi rc} [\ddot{p}(t_r)]_{t_r=t-r/c} \sin\theta \mathbf{u}_\varphi.$$

$$(8.96)$$

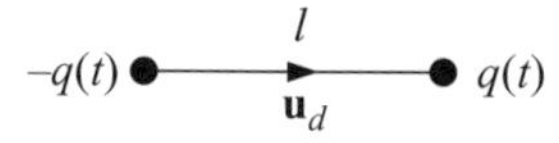

Figure 8.10 A dipole

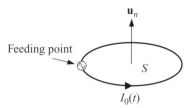

Figure 8.11 A loop

At large radial distance from the dipole, the rate of energy radiated per unit area is

$$\frac{dP^{rad}}{dS} = \frac{1}{(4\pi)^2 \varepsilon_0 c^3} \frac{\sin^2 \theta}{r^2} \ddot{p}\left(t - \frac{r}{c}\right).$$

The definition of a magnetic dipole is exactly the same as the electric dipole and we only need to replace $q(t)$ by $q_m(t)$, and $I(t)$ by $I_m(t)$. A magnetic dipole is a fictitious source that is equivalent to a small electric current loop in the sense that they both give the same fields outside the source region. It can be shown that the electromagnetic fields produced by the magnetic dipole are the same as that produced by a small electric current loop if we define the magnetic dipole element by $\mathbf{m}(t) = \mu_0 I_0(t) S \mathbf{u}_n$, where S is the area of the loop, $I_0(t)$ is the current in the loop, and $\mathbf{u}_n$ is the unit normal whose direction is determined by the right-hand rule as shown in Figure 8.11. When the loop is electrically small, the current is approximately uniform with the value $I_0(t)$. The fields produced by the loop can be obtained by duality through (8.96) as

$$\mathbf{E}(\mathbf{r}, t) \approx -\frac{1}{4\pi c r} [\dot{m}(t_r)]_{t_r = t - r/c} \sin\theta \, \mathbf{u}_\varphi,$$

$$\mathbf{H}(\mathbf{r}, t) \approx \frac{\varepsilon_0}{4\pi r} [\dot{m}(t_r)]_{t_r = t - r/c} \sin\theta \, \mathbf{u}_\theta.$$

The radiated fields of the small loop is proportional to $\dot{m}(t)$.

8.4.3   Enhancement of Radiation

The vector wave equations for the electromagnetic fields in free space can be rewritten as

$$\nabla^2 \mathbf{E} - \mu_0 \varepsilon_0 \frac{\partial^2 \mathbf{E}}{\partial t^2} = \mu_0 \frac{\partial \mathbf{J}}{\partial t} + \nabla \times \mathbf{J}_m + \frac{\nabla \rho}{\varepsilon_0},$$

$$\nabla^2 \mathbf{H} - \mu_0 \varepsilon_0 \frac{\partial^2 \mathbf{H}}{\partial t^2} = -\nabla \times \mathbf{J} + \varepsilon_0 \frac{\partial \mathbf{J}_m}{\partial t} + \frac{\nabla \rho_m}{\mu_0}.$$

If all sources are within a finite volume V, the solutions of these equations are

$$\mathbf{E}(\mathbf{r}, t) = -\int_V \frac{\mathbf{S}_E(\mathbf{r}', t - |\mathbf{r} - \mathbf{r}'|/v)}{4\pi R} dV(\mathbf{r}'),$$

$$\mathbf{H}(\mathbf{r}, t) = -\int_V \frac{\mathbf{S}_H(\mathbf{r}', t - |\mathbf{r} - \mathbf{r}'|/v)}{4\pi R} dV(\mathbf{r}'),$$

(8.97)

where

$$S_E(\mathbf{r}, t) = \mu_0 \frac{\partial \mathbf{J}(\mathbf{r}, t)}{\partial t} + \nabla \times \mathbf{J}_m(\mathbf{r}, t) + \frac{\nabla \rho(\mathbf{r}, t)}{\varepsilon_0},$$

$$S_H(\mathbf{r}, t) = -\nabla \times \mathbf{J}(\mathbf{r}, t) + \varepsilon_0 \frac{\partial \mathbf{J}_m(\mathbf{r}, t)}{\partial t} + \frac{\nabla \rho_m(\mathbf{r}, t)}{\mu_0}.$$

It can be seen that the contributions of the sources to the fields are not directly through the sources themselves but through their time variation and space variations. As a result, the wire antennas with concentrated loadings along their length radiate more efficiently than unloaded wires and significant contributions to the radiation fields come from the ends of the wire antennas. Physically the loading and the discontinuity increase the gradient of charges along the wire antennas.

The radiation can also be enhanced by decreasing the rise-time of the pulse (Geyi, 1996). To demonstrate this point, we may consider the current distribution

$$\mathbf{J}(\mathbf{r}, t) = \mathbf{J}(\mathbf{r}) f(t) \delta(z), \mathbf{r} \in \Omega,$$

and its radiated electric field on the z-axis:

$$\mathbf{E}(0, 0, z, t) = -\frac{\mu}{4\pi z} \frac{df(t - z/c)}{dt} \int_\Omega \mathbf{J}(\mathbf{r}') d\Omega(\mathbf{r}').$$

The time-integrated Poynting vector of the fields is

$$S(0, 0, z) = \frac{1}{\eta_0} \int_{-\infty}^{\infty} |\mathbf{E}(0, 0, z, t)|^2 \, dt$$

$$= \eta_0 \left(\frac{1}{4\pi z c}\right)^2 \left|\int_\Omega \mathbf{J}(\mathbf{r}') d\Omega(\mathbf{r}')\right|^2 \int_{-\infty}^{\infty} \left|\frac{df(t)}{dt}\right|^2 dt. \tag{8.98}$$

This indicates that the shorter the rise time of the exciting pulse, the slower is the decay of the radiating energy. The property is very important to carrier-free radar, and it implies that the radar range can be increased by a shorter exciting pulse. Let us consider an interesting case where the exciting pulse is a modulated signal with a finite duration D and a carrier whose cycle is $D_0 = D/n(n > 1)$:

$$f(t) = Ag(t) \sin\left(\frac{\pi t}{D_0}\right), 0 < t < D. \tag{8.99}$$

If the energy of the pulse is normalized, that is,

$$\int_0^D g^2(t) dt = 1, \int_0^D f^2(t) dt = 1,$$

we have $A \geq 1$. Substituting (8.99) into (8.98) gives

$$S(0, 0, z) = \eta_0 \left(\frac{1}{4\pi zc}\right)^2 \left|\int_\Omega \mathbf{J}d\Omega\right|^2$$

$$\cdot \left\{\left(\frac{n\pi}{D}\right)^2 + \frac{A^2}{2}\int_0^D \left[\frac{dg(t)}{dt}\right]^2 dt - \frac{A^2}{2}\int_0^D \left[\frac{dg(t)}{dt}\right]^2 \cos\left(\frac{2n\pi t}{D}\right) dt\right\}.$$

$$(8.100)$$

The last term in the curved brackets decreases rapidly as n increases. As a result, the time integrated energy on the z-axis increases as n increases. In other words, the decay of the energy density of the radiated electromagnetic pulse can be slowed down by increasing the frequency of the carrier.

Example 8.7: Consider three types of pulses with unit energy: a triangular pulse, a single-cycle sinusoidal pulse and a single-cycle sinusoidal pulse with a carrier, respectively defined by

$$f_1(t) = \sqrt{2/D}(1 - 2|t|/D), \ |t| < D/2,$$

$$f_2(t) = \sqrt{2/D}\sin(\pi t/D), 0 < t < D,$$

$$f_3(t) = 2\sqrt{1/D}\sin(\pi t/D)\sin(\pi t/D_0), \ D = nD_0, n > 1, 0 < t < D.$$

The corresponding time-integrated Poynting vectors are

$$S_1(0, 0, z) = \eta_0 \frac{8}{D^2}\left(\frac{1}{4\pi zc}\right)^2 \left|\int_\Omega \mathbf{J}(\mathbf{r}')d\Omega(\mathbf{r}')\right|^2,$$

$$S_2(0, 0, z) = \eta_0 \frac{\pi^2}{D^2}\left(\frac{1}{4\pi zc}\right)^2 \left|\int_\Omega \mathbf{J}(\mathbf{r}')d\Omega(\mathbf{r}')\right|^2,$$

$$S_3(0, 0, z) = \eta_0 \frac{\pi^2}{D^2}(1 + n^2)\left(\frac{1}{4\pi zc}\right)^2 \left|\int_\Omega \mathbf{J}(\mathbf{r}')d\Omega(\mathbf{r}')\right|^2,$$

and we have $S_1 < S_2 < S_3$.   □

8.4.4   Time-domain Integral Equations

The time-domain integral equations can be used to analyze the transient radiation characteristics of complex antennas. The derivation of integral equations in the time domain is very similar to that in the frequency domain.

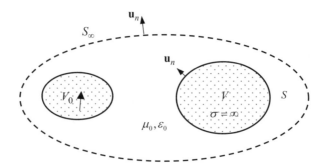

Figure 8.12 An arbitrary metal antenna

8.4.4.1 Integral equations for metal antenna

An arbitrary metal antenna is shown in Figure 8.12. It is assumed that the scatterer is a perfect conductor and occupies a finite region V bounded by S. A nearby current source $\mathbf{J}$ is located in the region V_0. Let S_∞ be a closed surface, which is large enough to enclose both V_0 and the scatterer V. From the representation theorems, the total fields inside the region bounded by S and S_∞ can be expressed as

$$\mathbf{E}(\mathbf{r},t) = \nabla \times \int_{S_\infty} \mathbf{J}_{ms}(\mathbf{r}',T)G_0(\mathbf{r},\mathbf{r}')dS(\mathbf{r}') + \frac{1}{\varepsilon_0}\nabla \int_{S_\infty} \rho_s(\mathbf{r}',T)G_0(\mathbf{r},\mathbf{r}')dS(\mathbf{r}')$$

$$+ \mu_0 \frac{\partial}{\partial t} \int_{S_\infty} \mathbf{J}_s(\mathbf{r}',T)G_0(\mathbf{r},\mathbf{r}')dS(\mathbf{r}') - \nabla \times \int_{S} \mathbf{J}_{ms}(\mathbf{r}',T)G_0(\mathbf{r},\mathbf{r}')dS(\mathbf{r}')$$

$$- \frac{1}{\varepsilon_0}\nabla \int_{S} \rho_s(\mathbf{r}',T)G_0(\mathbf{r},\mathbf{r}')dS(\mathbf{r}') - \mu_0 \frac{\partial}{\partial t} \int_{S} \mathbf{J}_s(\mathbf{r}',T)G_0(\mathbf{r},\mathbf{r}')dS(\mathbf{r}')$$

$$+ \mathbf{E}^{in}(\mathbf{r},t),$$

$$\mathbf{H}(\mathbf{r},t) = -\nabla \times \int_{S_\infty} \mathbf{J}_s(\mathbf{r}',T)G_0(\mathbf{r},\mathbf{r}')dS(\mathbf{r}') + \frac{1}{\mu_0}\nabla \int_{S_\infty} \rho_{ms}(\mathbf{r}',T)G_0(\mathbf{r},\mathbf{r}')dS(\mathbf{r}')$$

$$+ \varepsilon_0 \frac{\partial}{\partial t} \int_{S_\infty} \mathbf{J}_{ms}(\mathbf{r}',T)G_0(\mathbf{r},\mathbf{r}')dS(\mathbf{r}') + \nabla \times \int_{S} \mathbf{J}_s(\mathbf{r}',T)G_0(\mathbf{r},\mathbf{r}')dS(\mathbf{r}')$$

$$- \frac{1}{\mu_0}\nabla \int_{S} \rho_{ms}(\mathbf{r}',T)G_0(\mathbf{r},\mathbf{r}')dS(\mathbf{r}') - \varepsilon_0 \frac{\partial}{\partial t} \int_{S} \mathbf{J}_{ms}(\mathbf{r}',T)G_0(\mathbf{r},\mathbf{r}')dS(\mathbf{r}')$$

$$+ \mathbf{H}^{in}(\mathbf{r},t),$$

where $\mathbf{J}_s = \mathbf{u}_n \times \mathbf{H}, \quad \mathbf{J}_{ms} = -\mathbf{u}_n \times \mathbf{E}, \quad \rho_{ms} = \mu \mathbf{u}_n \cdot \mathbf{H}, \quad \rho_s = \varepsilon \mathbf{u}_n \cdot \mathbf{E}, \quad G_0(\mathbf{r},\mathbf{r}') =$
$1/4\pi \left|\mathbf{r}-\mathbf{r}'\right|, T = t - \left|\mathbf{r}-\mathbf{r}'\right|/c, c = 1/\sqrt{\mu_0\varepsilon_0}$, and

$$\mathbf{E}^{in}(\mathbf{r},t) = -\frac{1}{\varepsilon_0}\nabla \int_{V_0} \rho_s(\mathbf{r}',T)G_0(\mathbf{r},\mathbf{r}')dS(\mathbf{r}') - \mu_0 \frac{\partial}{\partial t} \int_{V_0} \mathbf{J}_s(\mathbf{r}',T)G_0(\mathbf{r},\mathbf{r}')dS(\mathbf{r}'),$$

$$\mathbf{H}^{in}(\mathbf{r},t) = \nabla \times \int_{V_0} \mathbf{J}_s(\mathbf{r}',T)G_0(\mathbf{r},\mathbf{r}')dS(\mathbf{r}'),$$

are the incident fields. The integrals over S_∞ must be zero as S_∞ approaches infinity. Let $\mathbf{r}$ approach S from the outside. From the jump relations, it follows that

$$\mathbf{E}(\mathbf{r}, t) = \frac{1}{2}\mathbf{u}_n(\mathbf{r}) \times \mathbf{J}_{ms}(\mathbf{r}, t) - \nabla \times \int_S \mathbf{J}_{ms}(\mathbf{r}', T)G_0(\mathbf{r}, \mathbf{r}')dS(\mathbf{r}')$$

$$+ \frac{1}{2\varepsilon_0}\mathbf{u}_n(\mathbf{r})\rho_s(\mathbf{r}, t) - \frac{1}{\varepsilon_0}\nabla \int_S \rho_s(\mathbf{r}', T)G_0(\mathbf{r}, \mathbf{r}')dS(\mathbf{r}')$$

$$- \mu_0\frac{\partial}{\partial t}\int_S \mathbf{J}_s(\mathbf{r}', T)G_0(\mathbf{r}, \mathbf{r}')dS(\mathbf{r}') + \mathbf{E}^{in}(\mathbf{r}, t),$$

$$\mathbf{H}(\mathbf{r}, t) = -\frac{1}{2}\mathbf{u}_n(\mathbf{r}) \times \mathbf{J}_s(\mathbf{r}, t) + \nabla \times \int_S \mathbf{J}_s(\mathbf{r}', T)G_0(\mathbf{r}, \mathbf{r}')dS(\mathbf{r}')$$

$$+ \frac{1}{2\mu_0}\mathbf{u}_n(\mathbf{r})\rho_{ms}(\mathbf{r}, t) - \frac{1}{\mu_0}\nabla \int_S \rho_{ms}(\mathbf{r}', T)G_0(\mathbf{r}, \mathbf{r}')dS(\mathbf{r}')$$

$$- \varepsilon_0\frac{\partial}{\partial t}\int_S \mathbf{J}_{ms}(\mathbf{r}', T)G_0(\mathbf{r}, \mathbf{r}')dS(\mathbf{r}') + \mathbf{H}^{in}(\mathbf{r}, t).$$

Multiplying the above equations vectorially by $\mathbf{u}_n(\mathbf{r})$, we obtain

$$\mathbf{u}_n(\mathbf{r}) \times \mu_0\frac{\partial}{\partial t}\int_S \mathbf{J}_s(\mathbf{r}', T)G_0(\mathbf{r}, \mathbf{r}')dS(\mathbf{r}')$$

$$+ \mathbf{u}_n(\mathbf{r}) \times \frac{1}{\varepsilon_0}\nabla \int_S \rho_s(\mathbf{r}', T)G_0(\mathbf{r}, \mathbf{r}')dS(\mathbf{r}') = \mathbf{u}_n(\mathbf{r}) \times \mathbf{E}^{in}(\mathbf{r}, t), \qquad (8.101)$$

and

$$\frac{1}{2}\mathbf{J}_s(\mathbf{r}, t) - \mathbf{u}_n(\mathbf{r}) \times \nabla \times \int_S \mathbf{J}_s(\mathbf{r}', T)G_0(\mathbf{r}, \mathbf{r}')dS(\mathbf{r}') = \mathbf{u}_n(\mathbf{r}) \times \mathbf{H}^{in}(\mathbf{r}, t). \qquad (8.102)$$

Considering

$$\nabla \left[\rho_s(\mathbf{r}', T)G_0(\mathbf{r}, \mathbf{r}')\right]$$

$$= \rho_s(\mathbf{r}', T)\nabla G_0(\mathbf{r}, \mathbf{r}') - \mathbf{u}_R G_0(\mathbf{r}, \mathbf{r}')\frac{\partial\rho_s(\mathbf{r}', T)}{c\partial t} \qquad (8.103)$$

$$= \left[\rho_s(\mathbf{r}', T) + |\mathbf{r} - \mathbf{r}'|\frac{\partial\rho_s(\mathbf{r}', T)}{c\partial t}\right]\nabla G_0(\mathbf{r}, \mathbf{r}'),$$

$$\nabla \times \left[\mathbf{J}_s(\mathbf{r}', T)G_0(\mathbf{r}, \mathbf{r}')\right]$$

$$= \nabla G_0(\mathbf{r}, \mathbf{r}') \times \mathbf{J}_s(\mathbf{r}', T) - G_0(\mathbf{r}, \mathbf{r}')\mathbf{u}_R \times \frac{\partial\mathbf{J}_s(\mathbf{r}', T)}{c\partial t} \qquad (8.104)$$

$$= \nabla G_0(\mathbf{r}, \mathbf{r}') \times \left[\mathbf{J}_s(\mathbf{r}', T) + |\mathbf{r} - \mathbf{r}'|\frac{\partial\mathbf{J}_s(\mathbf{r}', T)}{c\partial t}\right],$$

Equations (8.101) and (8.102) can be written as

$$
\mathbf{u}_n(\mathbf{r}) \times \mu_0 \frac{\partial}{\partial t} \int_S \mathbf{J}_s(\mathbf{r}', T) G_0(\mathbf{r}, \mathbf{r}') dS(\mathbf{r}')
$$

$$
-\mathbf{u}_n(\mathbf{r}) \times \frac{1}{\varepsilon_0} \int_S \left[\rho_s(\mathbf{r}', T) + |\mathbf{r} - \mathbf{r}'| \frac{\partial \rho_s(\mathbf{r}', T)}{c \partial t} \right] \nabla' G_0(\mathbf{r}, \mathbf{r}') dS(\mathbf{r}') \tag{8.105}
$$

$$
= \mathbf{u}_n(\mathbf{r}) \times \mathbf{E}^{in}(\mathbf{r}, t),
$$

$$
\frac{1}{2}\mathbf{J}_s(\mathbf{r}, t) - \mathbf{u}_n(\mathbf{r}) \times \int_S \left[\mathbf{J}_s(\mathbf{r}', T) + |\mathbf{r} - \mathbf{r}'| \frac{\partial \mathbf{J}_s(\mathbf{r}', T)}{c \partial t} \right] \times \nabla' G_0(\mathbf{r}, \mathbf{r}') dS(\mathbf{r}')
$$
$$
\tag{8.106}
$$

$$
= \mathbf{u}_n(\mathbf{r}) \times \mathbf{H}^{in}(\mathbf{r}, t).
$$

These are the **time-domain integral equations** for the metal antenna. The surface currents and charge densities satisfy the continuity equations

$$
\nabla_s \cdot \mathbf{J}_s(\mathbf{r}, t) + \frac{\partial \rho(\mathbf{r}, t)}{\partial t} = 0, \ \nabla_s \cdot \mathbf{J}_{ms}(\mathbf{r}, t) + \frac{\partial \rho_{ms}(\mathbf{r}, t)}{\partial t}, \tag{8.107}
$$

which can be used to eliminate the surface charge density in (8.105).

Remark 8.2 (Method of Laplace transform): Laplace transform has been the traditional method used for the analysis of transient phenomena. By taking the Laplace transform of (8.105) and (8.106), we obtain the integral equations in the complex frequency domain. After discretization and matrix inversion, the time-domain response can be obtained by taking the inverse Laplace transform. The time response of the scatterer is usually represented as a series of exponentials. This indicates that the field response may be characterized by the singularities of the Laplace transform. The **singularity expansion method** (SEM) is based on this idea (Baum, 1976; Marin, 1973). The singularities are often simple poles and are dependent on the geometry of the scatterer only. Therefore, they can be used to characterize the scatterer.   □

8.4.4.2 Integral equations for dielectric antenna

An arbitrary dielectric antenna is shown in Figure 8.13. It is assumed that the dielectric scatterer is finite and homogeneous, and occupies a region V bounded by S. When the fields generated by a source $\mathbf{J}$ are incident upon the dielectric scatterer, they will generate scattered fields outside S and a transmitted field inside S. The tangential components of the total fields must be continuous across S

$$
(\mathbf{u}_n \times \mathbf{E})_+ = (\mathbf{u}_n \times \mathbf{E}_d)_-, \ (\mathbf{u}_n \times \mathbf{H})_+ = (\mathbf{u}_n \times \mathbf{H}_d)_-, \tag{8.108}
$$

where the subscript d stands for the transmitted fields inside the dielectric scatterer. Ignoring the integral over S_∞ (S_∞ approaches infinity), the total fields inside the region bounded by S

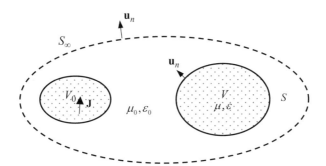

Figure 8.13   An arbitrary dielectric antenna

and S_∞ can be represented by

$$\mathbf{E}(\mathbf{r}, t) = -\nabla \times \int_S \mathbf{J}_{ms}(\mathbf{r}', T)G_0(\mathbf{r}, \mathbf{r}')dS(\mathbf{r}') - \frac{1}{\varepsilon_0}\nabla \int_S \rho_s(\mathbf{r}', T)G_0(\mathbf{r}, \mathbf{r}')dS(\mathbf{r}')$$

$$- \mu_0\frac{\partial}{\partial t} \int_S \mathbf{J}_s(\mathbf{r}', T)G_0(\mathbf{r}, \mathbf{r}')dS(\mathbf{r}') + \mathbf{E}^{in}(\mathbf{r}, t),$$

$$\mathbf{H}(\mathbf{r}, t) = \nabla \times \int_S \mathbf{J}_s(\mathbf{r}', T)G_0(\mathbf{r}, \mathbf{r}')dS(\mathbf{r}') - \frac{1}{\mu_0}\nabla \int_S \rho_{ms}(\mathbf{r}', T)G_0(\mathbf{r}, \mathbf{r}')dS(\mathbf{r}')$$

$$- \varepsilon_0\frac{\partial}{\partial t} \int_S \mathbf{J}_{ms}(\mathbf{r}', T)G_0(\mathbf{r}, \mathbf{r}')dS(\mathbf{r}') + \mathbf{H}^{in}(\mathbf{r}, t),$$

where

$$\mathbf{E}^{in}(\mathbf{r}, t) = -\frac{1}{\varepsilon_0}\nabla \int_{V_0} \rho_s(\mathbf{r}', T)G_0(\mathbf{r}, \mathbf{r}')dV(\mathbf{r}') - \mu_0\frac{\partial}{\partial t} \int_{V_0} \mathbf{J}_s(\mathbf{r}', T)G_0(\mathbf{r}, \mathbf{r}')dV(\mathbf{r}'),$$

$$\mathbf{H}^{in}(\mathbf{r}, t) = \nabla \times \int_{V_0} \mathbf{J}_s(\mathbf{r}', T)G_0(\mathbf{r}, \mathbf{r}')dV(\mathbf{r}'),$$

are the incident fields. The total fields inside S can be expressed as

$$\mathbf{E}_d(\mathbf{r}, t) = \nabla \times \int_S \mathbf{J}_{ms}(\mathbf{r}', T_d)G_0(\mathbf{r}, \mathbf{r}')dS(\mathbf{r}') + \frac{1}{\varepsilon}\nabla \int_S \rho_s(\mathbf{r}', T_d)G_0(\mathbf{r}, \mathbf{r}')dS(\mathbf{r}')$$

$$+ \mu\frac{\partial}{\partial t} \int_S \mathbf{J}_s(\mathbf{r}', T_d)G_0(\mathbf{r}, \mathbf{r}')dS(\mathbf{r}'),$$

$$\mathbf{H}_d(\mathbf{r}, t) = -\nabla \times \int_S \mathbf{J}_s(\mathbf{r}', T_d)G_0(\mathbf{r}, \mathbf{r}')dS(\mathbf{r}') + \frac{1}{\mu}\nabla \int_S \rho_{ms}(\mathbf{r}', T_d)G_0(\mathbf{r}, \mathbf{r}')dS(\mathbf{r}')$$

$$+ \varepsilon\frac{\partial}{\partial t} \int_S \mathbf{J}_{ms}(\mathbf{r}', T_d)G_0(\mathbf{r}, \mathbf{r}')dS(\mathbf{r}'),$$

where $T_d = t - |\mathbf{r} - \mathbf{r}'|/v$, $v = 1/\sqrt{\mu\varepsilon}$ and we have used (8.108). Let the observation point approach the boundary S from the outside. Making use of the jump relations, we may find that

$$\mathbf{E}(\mathbf{r}, t) = \frac{1}{2}\mathbf{u}_n(\mathbf{r}) \times \mathbf{J}_{ms}(\mathbf{r}, t) - \nabla \times \int_S \mathbf{J}_{ms}(\mathbf{r}', T)G_0(\mathbf{r}, \mathbf{r}')dS(\mathbf{r}')$$

$$+ \frac{1}{2\varepsilon_0}\mathbf{u}_n(\mathbf{r})\rho_s(\mathbf{r}, t) - \frac{1}{\varepsilon_0}\nabla \int_S \rho_s(\mathbf{r}', T)G_0(\mathbf{r}, \mathbf{r}')dS(\mathbf{r}')$$

$$- \mu_0\frac{\partial}{\partial t}\int_S \mathbf{J}_s(\mathbf{r}', T)G_0(\mathbf{r}, \mathbf{r}')dS(\mathbf{r}') + \mathbf{E}^{in}(\mathbf{r}, t),$$

$$\mathbf{H}(\mathbf{r}, t) = -\frac{1}{2}\mathbf{u}_n(\mathbf{r}) \times \mathbf{J}_s(\mathbf{r}, t) + \nabla \times \int_S \mathbf{J}_s(\mathbf{r}', T)G_0(\mathbf{r}, \mathbf{r}')dS(\mathbf{r}')$$

$$+ \frac{1}{2\mu_0}\mathbf{u}_n(\mathbf{r})\rho_{ms}(\mathbf{r}, t) - \frac{1}{\mu_0}\nabla \int_S \rho_{ms}(\mathbf{r}', T)G_0(\mathbf{r}, \mathbf{r}')dS(\mathbf{r}')$$

$$- \varepsilon_0\frac{\partial}{\partial t}\int_S \mathbf{J}_{ms}(\mathbf{r}', T)G_0(\mathbf{r}, \mathbf{r}')dS(\mathbf{r}') + \mathbf{H}^{in}(\mathbf{r}, t),$$

$$\mathbf{E}_d(\mathbf{r}, t) = \frac{1}{2}\mathbf{u}_n(\mathbf{r}) \times \mathbf{J}_{ms}(\mathbf{r}, t) + \nabla \times \int_S \mathbf{J}_{ms}(\mathbf{r}', T_d)G_0(\mathbf{r}, \mathbf{r}')dS(\mathbf{r}')$$

$$+ \frac{1}{2\varepsilon}\mathbf{u}_n(\mathbf{r})\rho_s(\mathbf{r}, t) + \frac{1}{\varepsilon}\nabla \int_S \rho_s(\mathbf{r}', T_d)G_0(\mathbf{r}, \mathbf{r}')dS(\mathbf{r}')$$

$$+ \mu\frac{\partial}{\partial t}\int_S \mathbf{J}_s(\mathbf{r}', T_d)G_0(\mathbf{r}, \mathbf{r}')dS(\mathbf{r}'),$$

$$\mathbf{H}_d(\mathbf{r}, t) = -\frac{1}{2}\mathbf{u}_n(\mathbf{r}) \times \mathbf{J}_s(\mathbf{r}, t) - \nabla \times \int_S \mathbf{J}_s(\mathbf{r}', T_d)G_0(\mathbf{r}, \mathbf{r}')dS(\mathbf{r}')$$

$$+ \frac{1}{2\mu}\mathbf{u}_n(\mathbf{r})\rho_{ms}(\mathbf{r}, t) + \frac{1}{\mu}\nabla \int_S \rho_{ms}(\mathbf{r}', T_d)G_0(\mathbf{r}, \mathbf{r}')dS(\mathbf{r}')$$

$$+ \varepsilon\frac{\partial}{\partial t}\int_S \mathbf{J}_{ms}(\mathbf{r}', T_d)G_0(\mathbf{r}, \mathbf{r}')dS(\mathbf{r}').$$

Multiplying these equations vectorially by $\mathbf{u}_n$ yields

$$-\frac{1}{2}\varepsilon_0\mathbf{J}_{ms}(\mathbf{r}, t) = -\mu_0\varepsilon_0\mathbf{u}_n(\mathbf{r}) \times \frac{\partial}{\partial t}\int_S \mathbf{J}_s(\mathbf{r}', T)G_0(\mathbf{r}, \mathbf{r}')dS(\mathbf{r}')$$

$$-\mathbf{u}_n(\mathbf{r}) \times \nabla \times \int_S \mathbf{J}_{ms}(\mathbf{r}', T)\varepsilon_0 G_0(\mathbf{r}, \mathbf{r}')dS(\mathbf{r}') \qquad (8.109)$$

$$-\mathbf{u}_n(\mathbf{r}) \times \nabla \int_S \rho_s(\mathbf{r}', T)G_0(\mathbf{r}, \mathbf{r}')dS(\mathbf{r}') + \mathbf{u}_n(\mathbf{r}) \times \varepsilon_0\mathbf{E}^{in}(\mathbf{r}, t),$$

$$\frac{1}{2}\mu_0 \mathbf{J}_s(\mathbf{r}, t) = -\mu_0 \varepsilon_0 \mathbf{u}_n(\mathbf{r}) \times \frac{\partial}{\partial t} \int_S \mathbf{J}_{ms}(\mathbf{r}', T) G_0(\mathbf{r}, \mathbf{r}') dS(\mathbf{r}')$$

$$+ \mathbf{u}_n(\mathbf{r}) \times \nabla \times \int_S \mathbf{J}_s(\mathbf{r}', T) \mu_0 G_0(\mathbf{r}, \mathbf{r}') dS(\mathbf{r}')$$

$$- \mathbf{u}_n(\mathbf{r}) \times \nabla \int_S \rho_{ms}(\mathbf{r}', T) G_0(\mathbf{r}, \mathbf{r}') dS(\mathbf{r}') + \mathbf{u}_n(\mathbf{r}) \times \mu_0 \mathbf{H}^{in}(\mathbf{r}, t),$$

(8.110)

$$-\frac{1}{2}\varepsilon \mathbf{J}_{ms}(\mathbf{r}, t) = \mu\varepsilon \mathbf{u}_n(\mathbf{r}) \times \frac{\partial}{\partial t} \int_S \mathbf{J}_s(\mathbf{r}', T_d) G_0(\mathbf{r}, \mathbf{r}') dS(\mathbf{r}')$$

$$+ \mathbf{u}_n(\mathbf{r}) \times \nabla \times \int_S \mathbf{J}_{ms}(\mathbf{r}', T_d) \varepsilon G_0(\mathbf{r}, \mathbf{r}') dS(\mathbf{r}')$$

$$+ \mathbf{u}_n(\mathbf{r}) \times \nabla \int_S \rho_s(\mathbf{r}', T_d) G_0(\mathbf{r}, \mathbf{r}') dS(\mathbf{r}'),$$

(8.111)

$$\frac{1}{2}\mu \mathbf{J}_s(\mathbf{r}, t) = \mu\varepsilon \mathbf{u}_n(\mathbf{r}) \times \frac{\partial}{\partial t} \int_S \mathbf{J}_{ms}(\mathbf{r}', T_d) G_0(\mathbf{r}, \mathbf{r}') dS(\mathbf{r}')$$

$$- \mathbf{u}_n(\mathbf{r}) \times \nabla \times \int_S \mathbf{J}_s(\mathbf{r}', T_d) \mu G_0(\mathbf{r}, \mathbf{r}') dS(\mathbf{r}')$$

(8.112)

$$+ \mathbf{u}_n(\mathbf{r}) \times \nabla \int_S \rho_{ms}(\mathbf{r}', T_d) G_0(\mathbf{r}, \mathbf{r}') dS(\mathbf{r}').$$

Adding (8.109) and (8.111) gives

$$-\frac{1}{2}(\varepsilon_0 + \varepsilon) \mathbf{J}_{ms}(\mathbf{r}, t)$$

$$+ \mathbf{u}_n(\mathbf{r}) \times \int_S \left[\mu_0 \varepsilon_0 \frac{\partial}{\partial t} \mathbf{J}_s(\mathbf{r}', T) - \mu\varepsilon \frac{\partial}{\partial t} \mathbf{J}_s(\mathbf{r}', T_d) \right] G_0(\mathbf{r}, \mathbf{r}') dS(\mathbf{r}')$$

$$+ \mathbf{u}_n(\mathbf{r}) \times \nabla \times \int_S \left[\varepsilon_0 \mathbf{J}_{ms}(\mathbf{r}', T) - \varepsilon \mathbf{J}_{ms}(\mathbf{r}', T_d) \right] G_0(\mathbf{r}, \mathbf{r}') dS(\mathbf{r}')$$

(8.113)

$$+ \mathbf{u}_n(\mathbf{r}) \times \nabla \int_S \left[\rho_s(\mathbf{r}', T) - \rho_s(\mathbf{r}', T_d) \right] G_0(\mathbf{r}, \mathbf{r}') dS(\mathbf{r}') = \mathbf{u}_n(\mathbf{r}) \times \varepsilon_0 \mathbf{E}^{in}(\mathbf{r}, t).$$

Adding (8.110) and (8.112) gives

$$\frac{1}{2}(\mu_0 + \mu) \mathbf{J}_s(\mathbf{r}, t)$$

$$+ \mathbf{u}_n(\mathbf{r}) \times \int_S \left[\mu_0 \varepsilon_0 \frac{\partial}{\partial t} \mathbf{J}_{ms}(\mathbf{r}', T) - \mu\varepsilon \frac{\partial}{\partial t} \mathbf{J}_{ms}(\mathbf{r}', T_d) \right] G_0(\mathbf{r}, \mathbf{r}') dS(\mathbf{r}')$$

$$+ \mathbf{u}_n(\mathbf{r}) \times \nabla \times \int_S \left[\mu \mathbf{J}_s(\mathbf{r}', T_d) - \mu_0 \mathbf{J}_s(\mathbf{r}', T) \right] G_0(\mathbf{r}, \mathbf{r}') dS(\mathbf{r}')$$

$$+ \mathbf{u}_n(\mathbf{r}) \times \nabla \int_S \left[\rho_{ms}(\mathbf{r}', T) - \rho_{ms}(\mathbf{r}', T_d) \right] G_0(\mathbf{r}, \mathbf{r}') dS(\mathbf{r}') = \mathbf{u}_n(\mathbf{r}) \times \mu_0 \mathbf{H}^{in}(\mathbf{r}, t).$$

(8.114)

By use of (8.103) and (8.104), Equations (8.113) and (8.114) can be rewritten as

$$\frac{1}{2}(\mu_0 + \mu)\mathbf{J}_s(\mathbf{r}, t)$$

$$+ \mathbf{u}_n(\mathbf{r}) \times \int_S \left[\mu_0 \varepsilon_0 \frac{\partial}{\partial t} \mathbf{J}_{ms}(\mathbf{r}', T) - \mu \varepsilon \frac{\partial}{\partial t} \mathbf{J}_{ms}(\mathbf{r}', T_d) \right] G_0(\mathbf{r}, \mathbf{r}') dS(\mathbf{r}')$$

$$+ \mathbf{u}_n(\mathbf{r}) \times \int_S \left[\mathbf{J}_s(\mathbf{r}', T_d) + |\mathbf{r} - \mathbf{r}'| \frac{\partial \mathbf{J}_s(\mathbf{r}', T_d)}{v \partial t} \right] \times \mu \nabla' G_0(\mathbf{r}, \mathbf{r}') dS(\mathbf{r}')$$

$$- \mathbf{u}_n(\mathbf{r}) \times \int_S \left[\mathbf{J}_s(\mathbf{r}', T) + |\mathbf{r} - \mathbf{r}'| \frac{\partial \mathbf{J}_s(\mathbf{r}', T)}{c \partial t} \right] \times \mu_0 \nabla' G_0(\mathbf{r}, \mathbf{r}') dS(\mathbf{r}')$$

$$- \mathbf{u}_n(\mathbf{r}) \times \int_S \left[\rho_{ms}(\mathbf{r}', T) + |\mathbf{r} - \mathbf{r}'| \frac{\partial \rho_{ms}(\mathbf{r}', T)}{c \partial t} \right] \nabla' G_0(\mathbf{r}, \mathbf{r}') ds(\mathbf{r}')$$ (8.115)

$$+ \mathbf{u}_n(\mathbf{r}) \times \int_S \left[\rho_{ms}(\mathbf{r}', T_d) + |\mathbf{r} - \mathbf{r}'| \frac{\partial \rho_{ms}(\mathbf{r}', T_d)}{v \partial t} \right] \nabla' G_0(\mathbf{r}, \mathbf{r}') dS(\mathbf{r}')$$

$$= \mathbf{u}_n(\mathbf{r}) \times \mu_0 \mathbf{H}^{in}(\mathbf{r}, t),$$

$$-\frac{1}{2}(\varepsilon_0 + \varepsilon)\mathbf{J}_{ms}(\mathbf{r}, t)$$

$$+ \mathbf{u}_n(\mathbf{r}) \times \int_S \left[\mu_0 \varepsilon_0 \frac{\partial}{\partial t} \mathbf{J}_s(\mathbf{r}', T) - \mu \varepsilon \frac{\partial}{\partial t} \mathbf{J}_s(\mathbf{r}', T_d) \right] G_0(\mathbf{r}, \mathbf{r}') dS(\mathbf{r}')$$

$$+ \mathbf{u}_n(\mathbf{r}) \times \int_S \left[\mathbf{J}_{ms}(\mathbf{r}', T) + |\mathbf{r} - \mathbf{r}'| \frac{\partial \mathbf{J}_{ms}(\mathbf{r}', T)}{c \partial t} \right] \times \varepsilon_0 \nabla' G_0(\mathbf{r}, \mathbf{r}') dS(\mathbf{r}')$$

$$- \mathbf{u}_n(\mathbf{r}) \times \int_S \left[\mathbf{J}_{ms}(\mathbf{r}', T_d) + |\mathbf{r} - \mathbf{r}'| \frac{\partial \mathbf{J}_{ms}(\mathbf{r}', T_d)}{v \partial t} \right] \times \varepsilon \nabla' G_0(\mathbf{r}, \mathbf{r}') dS(\mathbf{r}')$$ (8.116)

$$- \mathbf{u}_n(\mathbf{r}) \times \int_S \left[\rho_s(\mathbf{r}', T) + |\mathbf{r} - \mathbf{r}'| \frac{\partial \rho_s(\mathbf{r}', T)}{c \partial t} \right] \nabla' G_0(\mathbf{r}, \mathbf{r}') dS(\mathbf{r}')$$

$$+ \mathbf{u}_n(\mathbf{r}) \times \int_S \left[\rho_s(\mathbf{r}', T_d) + |\mathbf{r} - \mathbf{r}'| \frac{\partial \rho_s(\mathbf{r}', T_d)}{v \partial t} \right] \nabla' G_0(\mathbf{r}, \mathbf{r}') dS(\mathbf{r}')$$

$$= \mathbf{u}_n(\mathbf{r}) \times \varepsilon_0 \mathbf{E}^{in}(\mathbf{r}, t).$$

Equations (8.115) and (8.116) are the **time-domain integral equations** for the dielectric antenna. Equations (8.107) can be used to eliminate the surface charge densities in the above equations. In recent years, there has been increasing interest in the use of the time-domain integral equations to numerically solve the scattering problems, which is very efficient for a homogeneous scattering environment. When the environment gets complicated, we must consider the domain method to tackle the time-domain problems. One popular method is the **finite-difference time-domain** (FDTD), which is a numerical scheme proposed by Yee in 1966 to solve Maxwell equations (Yee, 1966). Compared with the time-domain integral equations, the FDTD method is very powerful in dealing with inhomogeneous or nonlinear problems. When FDTD is applied to solve a problem with infinite computational domain, a truncated boundary must be introduced to make the number of unknowns in the domain finite. To simulate the original infinite domain problem, absorbing boundary conditions must be carefully designed to minimize the reflections from the truncated boundary (Umashankar and Taflove, 1993).

And the continuity of our science has not been affected by all these turbulent happenings, as the older theories have always been included as limiting cases in the new ones.

—Max Born (German physicist, 1882–1970)

9

Relativity

The special theory of relativity owes its origins to Maxwell equations of the electromagnetic field.

—Albert Einstein

Newton's first law states that a body at rest or in uniform motion will remain at rest or in uniform motion unless some external force is applied to it. It is evident that Newton's first law does not hold for all coordinate systems. Those coordinate systems where Newton's first law is applicable are called **inertial systems**.

A physical law can be described as a relation between some physical quantities, which are generally the functions of space and time in a given coordinate system. A common belief is that the physical laws should retain the same functional form in different inertial coordinate systems. This is called the **principle of relativity**. Newton believed the existence of an absolute reference frame and all other inertial frames are at rest or in uniform motion with respect to it. Based on the assumption of the existence of a universal time, the transformation laws between different inertial coordinate systems could be derived. This is called the **Galilean transformation**, named after the Italian scientist Galileo Galilei (1564–1642). The principle of relativity based on the Galilean transformation is called the **principle of Galilean relativity**, and Newton's law of motion is invariant under Galilean transformation, that is, it preserves the functional form under Galilean transformation. An important feature that Galilean transformation predicts is that the velocity of a particle with respect to one inertial system is simply the algebraic addition of the velocity of the particle with respect to another inertial system and the relative velocity of the two inertial systems.

Maxwell first formulated the mathematical equations for describing the electromagnetic phenomena in 1861 based on a model of ether to give a mechanical explanation of the electromagnetic waves. The ether is an imaginative medium whose mechanical vibration forms electromagnetic waves. The speed of light predicted by Maxwell equations would be the speed with respect to the ether, which fills all the space and thus provides an absolute reference frame. If we assume that Maxwell equations hold for all inertial frames, which are related by Galilean transformations, the light speed would be the same in all inertial frames, which contradicts the simple addition of velocities mentioned above. It can also be shown

Foundations of Applied Electrodynamics   Geyi Wen
© 2010 John Wiley & Sons, Ltd

that Maxwell equations are not invariant under Galilean transformation. Therefore, either Maxwell's theory or Galilean transformation was subject to a major revision.

In 1881, American physicists Albert Abraham Michelson (1852–1931) and Edward Williams Morley (1838–1923) performed a famous experiment in attempt to find any motion of the earth relative to the ether. However, their experiment gave a null result and thus overturned the assumption about the existence of ether. This experiment motivated Einstein to abandon the concept of ether and Galilean transformation and to adopt the Lorentz transformation, a generalization of Galilean transformation, and led him to form a complete theory of relativity.

The relativity theory can be divided into special theory, which was systematically studied by Einstein in 1905, and general theory presented by Einstein in 1916. The special theory studies the physical phenomena perceived by different observers traveling at a constant speed relative to each other while the general theory studies the phenomena perceived by different observers traveling at an arbitrary relative speed. The tensor algebra and tensor analysis provide a natural mathematical language for describing the relativity.

9.1 Tensor Algebra on Linear Spaces

It is well known that the classical vector analysis enables one to express a physical law in a concise manner that does not depend on a coordinate system. However, the concept of vector, which is uniquely determined by its three components relative to some coordinate system, is too limited in practice. Some quantities in mathematics and physics must be described by a tensor, which requires more than three components for a complete specification. For example, one must introduce the stress tensor to fully specify the stress in elasticity. The tensor analysis reserves the major advantages of vector analysis and allows for an effortless transition from a given coordinate system to another. A physical theory is not considered complete if it cannot be expressed in tensorial form.

9.1.1 Tensor Algebra

Let E be an n dimensional linear space and E^* its dual space. If $\{e_1, e_2, \cdots, e_n\}$ is an ordered basis of E, there is a unique ordered basis of E^*, the **dual basis** $\{e^1, e^2, \cdots, e^n\}$, such that $e^j(e_i) = \delta^j{}_i$, where $\delta^j{}_i = 1$ if $j = i$ and 0 otherwise. Let $L^k(E_1, E_2, \cdots, E_k; F)$ denote the vector space of continuous k-multilinear maps of $E_1 \times E_2 \times \cdots \times E_k$ to F, where $E_i (i = 1, 2, \cdots, k)$ and F are vector spaces. For all $x \in E$ and $\alpha \in E^*$, we have the following expansions

$$x = \sum_{i=1}^{n} e^i(x)e_i, \alpha = \sum_{i=1}^{n} \alpha(e_i)e^i.$$

We put $T^r{}_s(E) = L^{r+s}(E^*, \cdots, E^*, E, \cdots, E; R)$ (r copies of E^* and s copies of E). Elements of $T^r{}_s(E)$ are called **tensors** on E, contravariant of order r and covariant of order s; or simply type (r, s). Given $T_1 \in T^{r_1}{}_{s_1}(E)$ and $T_2 \in T^{r_2}{}_{s_2}(E)$, the **tensor product** of T_1 and T_2 is a tensor $T_1 \otimes T_2 \in T^{r_1+r_2}{}_{s_1+s_2}(E)$ defined by

$$T_1 \otimes T_2\left(\beta^1, \cdots, \beta^{r_1}, \gamma^1, \cdots \gamma^{r_2}, f_1, \cdots, f_{s_1}, g_1, \cdots, g_{s_2}\right)$$
$$= T_1\left(\beta^1, \cdots, \beta^{r_1}, f_1, \cdots, f_{s_1}\right) T_2\left(\gamma^1, \cdots \gamma^{r_2}, g_1, \cdots, g_{s_2}\right),$$

where $\beta^j, \gamma^j \in E^*$ and $f_j, g_j \in E$. By convention, we let $T^0{}_0(E) = R$. We always identify E^{**} (the dual space of E^*) with E. Thus, $e_i(e^j) = e^j(e_i)$ and $T^1{}_0(E) = L(E^*, R) = E^{**} = E$. If $\{e_1, \cdots, e_n\}$ is a basis of E and $\{e^1, \cdots, e^n\}$ is the dual basis, then

$$\left\{ e_{i_1} \otimes \cdots \otimes e_{i_r} \otimes e^{j_1} \otimes \cdots \otimes e^{j_s} \,\middle|\, i_1, \cdots, i_r, j_1, \cdots, j_s = 1, \cdots, n \right\}$$

is a basis of $T^r{}_s(E)$. Thus, $\dim[T^r{}_s(E)] = n^{r+s}$. Note that the definition of tensor does not involve any coordinate system.

We use **summation convention** in this chapter: summation is implied when an index is repeated on upper and lower levels. Thus any tensor $T \in T^r{}_s(E)$ may be represented by

$$T = T^{i_1 \cdots i_r}{}_{j_1 \cdots j_s} e_{i_1} \otimes \cdots \otimes e_{i_r} \otimes e^{j_1} \otimes \cdots \otimes e^{j_s}, \tag{9.1}$$

where $T^{i_1 \cdots i_r}{}_{j_1 \cdots j_s} = T(e^{i_1}, \cdots, e^{i_r}, e_{j_1}, \cdots e_{j_s})$ are called the **components** of T relative to the basis $\{e_1, \cdots, e_n\}$. If the components of tensors T_1 and T_2 are $T_1^{i_1 \cdots i_{r_1}}{}_{j_1 \cdots j_{s_1}}$ and $T_2^{h_1 \cdots h_{r_2}}{}_{k_1 \cdots k_{s_2}}$ respectively, the components of the tensor product $T_1 \otimes T_2$ are

$$(T_1 \otimes T_2)^{i_1 \cdots i_{r_1} h_1 \cdots h_{r_2}}{}_{j_1 \cdots j_{s_1} k_1 \cdots k_{s_2}} = T_1^{i_1 \cdots i_{r_1}}{}_{j_1 \cdots j_{s_1}} T_2^{h_1 \cdots h_{r_2}}{}_{k_1 \cdots k_{s_2}}. \tag{9.2}$$

An $(r, 0)$-tensor is called **symmetric** if

$$T(\alpha^1, \cdots, \alpha^r) = T(\alpha^{\sigma(1)}, \cdots, \alpha^{\sigma(r)})$$

holds for all permutations σ of $\{1, \cdots, r\}$ and all elements $\alpha^1, \cdots \alpha^r \in E^*$. An $(r, 0)$-tensor is called antisymmetric if

$$T(\alpha^1, \cdots, \alpha^r) = \operatorname{sgn} \sigma \, T(\alpha^{\sigma(1)}, \cdots, \alpha^{\sigma(r)})$$

holds for all permutations σ of $\{1, \cdots, r\}$ and all elements $\alpha^1, \cdots \alpha^r \in E^*$. Here $\operatorname{sgn} \sigma = 1$ for even permutation and $\operatorname{sgn} \sigma = -1$ for odd permutation. Similar definitions hold for $(0, s)$-tensor.

The (k, l)-**contraction** map $C^k{}_l : T^r{}_s(E) \to T^{r-1}{}_{s-1}(E)$ is defined by

$$C^k{}_l\left(T^{i_1 \cdots i_r}{}_{j_1 \cdots j_s} e_{i_1} \otimes \cdots \otimes e_{i_r} \otimes e^{j_1} \otimes \cdots \otimes e^{j_s}\right)$$

$$= T^{i_1 \cdots, i_{k-1} q i_{k+1} \cdots i_r}{}_{j_1 \cdots j_{l-1} q j_{l+1} \cdots j_s} e_{i_1} \otimes \cdots \otimes \breve{e}_{i_k} \otimes \cdots \otimes e_{i_r} \otimes e^{j_1} \otimes \cdots \otimes \breve{e}^{j_l} \otimes \cdots \otimes e^{j_s},$$

where $\breve{e}$ means that the term e is omitted.

Example 9.1: An inner product $(\cdot, \cdot)$ on E is a symmetric $(0,2)$-tensor. Its matrix has components $g_{ij} = (e_i, e_j)$. Thus the matrix $[g_{ij}]$ is symmetric and positive definite. The components of the inverse matrix are denoted by g^{ij}. We define the **index lowering operator** $^b : E \to E^*$ by $x \mapsto (x, \cdot)$. Its inverse is denoted by $^\# : E^* \to E$ and is called **index raising operator**. For $x = x^i e_i$ and $\alpha = \alpha_i e^i$, we have

$$(x^b)_i = g_{ji} x^j, \quad (\alpha^\#)^i = g^{ij} \alpha_j.$$

The inner product can be expressed as

$$(x, y) = x^b(y) = g_{ij} x^i y^j.$$

The index raising and lowering operator can be applied to tensors to produce new ones. □

The components of a tensor depend on the choice of the basis. Consider another basis $\{\tilde{e}_i\}$. Each basis vector e_i can be expressed as a linear combination of the new basis vectors $\tilde{e}_i$

$$e_i = A^j{}_i \tilde{e}_j. \tag{9.3}$$

Similarly, for the dual basis we can write

$$e^i = B^i{}_j \tilde{e}^j. \tag{9.4}$$

It is easy to see that $[B^i{}_j]$ is the inverse matrix of $[A^j{}_i]$. Thus we have

$$\tilde{e}_i = B^j{}_i e_j, \tag{9.5}$$

and

$$\tilde{e}^i = A^i{}_j e^j. \tag{9.6}$$

In the new basis vectors, we can express an arbitrary tensor as

$$T = \tilde{T}^{i_1 \cdots i_r}{}_{j_1 \cdots j_s} \tilde{e}_{i_1} \otimes \cdots \otimes \tilde{e}_{i_r} \otimes \tilde{e}^{j_1} \otimes \cdots \otimes \tilde{e}^{j_s}. \tag{9.7}$$

Inserting (9.3) and (9.4) into (9.1) gives

$$T = A^{i_1}{}_{h_1} \cdots A^{i_r}{}_{h_r} T^{h_1 \cdots h_r}{}_{k_1 \cdots k_s} B^{k_1}{}_{j_1} \cdots B^{k_s}{}_{j_s} \tilde{e}_{i_1} \otimes \cdots \otimes \tilde{e}_{i_r} \otimes \tilde{e}^{j_1} \otimes \cdots \otimes \tilde{e}^{j_s}.$$

Comparing the above expression with (9.7) gives

$$\tilde{T}^{i_1 \cdots i_r}{}_{j_1 \cdots j_s} = A^{i_1}{}_{h_1} \cdots A^{i_r}{}_{h_r} B^{k_1}{}_{j_1} \cdots B^{k_s}{}_{j_s} T^{h_1 \cdots h_r}{}_{k_1 \cdots k_s}. \tag{9.8}$$

This is known as the **tensoriality criterion** and is used to define a tensor in classical analysis. As special cases, a covariant tensor of order s obeys the transformation rule

$$\tilde{T}_{j_1 \cdots j_s} = B^{i_1}{}_{j_1} \cdots B^{i_s}{}_{j_s} T_{i_1 \cdots i_s},$$

and a contravariant tensor of order r obeys the following transformation rule

$$\tilde{T}^{i_1 \cdots i_r} = A^{i_1}{}_{j_1} \cdots A^{i_r}{}_{j_r} T^{j_1 \cdots j_r}.$$

A vector x in E has the following expansions

$$x = x^i e_i = \tilde{x}^i \tilde{e}_i.$$

It follows that

$$\tilde{x}^i = A^i{}_j x^j, \, x^i = B^i{}_j \tilde{x}^j. \tag{9.9}$$

Therefore a vector is a $(1,0)$-tensor. Based on (9.9), the tensor transformation rule (9.8) can be written in a somewhat different form. From (9.9) it follows that

$$\frac{\partial \tilde{x}^i}{\partial x^j} = A^i{}_j, \, \frac{\partial x^i}{\partial \tilde{x}^j} = B^i{}_j. \tag{9.10}$$

Thus (9.8) becomes

$$\widetilde{T}^{i_1 \cdots i_r}{}_{j_1 \cdots j_s} = \frac{\partial \tilde{x}^{i_1}}{\partial x^{h_1}} \cdots \frac{\partial \tilde{x}^{i_r}}{\partial x^{h_r}} \frac{\partial x^{k_1}}{\partial \tilde{x}^{j_1}} \cdots \frac{\partial x^{k_s}}{\partial \tilde{x}^{j_s}} T^{h_1 \cdots h_r}{}_{k_1 \cdots k_s}. \tag{9.11}$$

A **coordinate system** on E is a bijection ϕ from the n dimensional linear space E to R^n, denoted by $\phi : p \in E \rightarrow (x^1, x^2, \cdots, x^n) \in R^n$ or (E, ϕ). Let ψ be another coordinate system on E, $\psi : p \in E \rightarrow (\tilde{x}^1, \tilde{x}^2, \cdots, \tilde{x}^n) \in R^n$. Thus a vector $x \in E$ may be denoted by $(x^1, x^2, \cdots, x^n)$ and $(\tilde{x}^1, \tilde{x}^2, \cdots, \tilde{x}^n)$ respectively. These two n-tuples of real numbers are related to each other by the transformation equations

$$\tilde{x}^1 = f^1(x^1, x^2, \cdots, x^n),$$
$$\tilde{x}^2 = f^2(x^1, x^2, \cdots, x^n),$$
$$\vdots$$
$$\tilde{x}^n = f^n(x^1, x^2, \cdots, x^n),$$

where $f^1, \cdots,$ and f^n are n distinct functions of n variables. In the sequel, we will denote the n-tuple $(x^1, x^2, \cdots, x^n)$ by a single letter x^h and the above transformation equations will be written in the compressed form

$$\tilde{x}^j = \tilde{x}^j(x^h), \tag{9.12}$$

and the inverse of the above will be expressed in the form

$$x^h = x^h(\tilde{x}^j). \tag{9.13}$$

It follows from (9.12), (9.13) and the chain rule that

$$\frac{\partial \tilde{x}^j}{\partial x^h}\frac{\partial x^h}{\partial \tilde{x}^l} = \delta^j{}_l, \quad \frac{\partial x^h}{\partial \tilde{x}^j}\frac{\partial \tilde{x}^j}{\partial x^k} = \delta^h{}_k. \tag{9.14}$$

For the linear space E, it is always possible to find a single coordinate system to cover the whole space.

9.1.2 Tangent Space, Cotangent Space and Tensor Space

In classical calculus, vectors are considered as arrows characterized by a direction and a length and they are independent of their location in space. Such vectors are called free vectors. In both mathematics and physics, we need to introduce the notion of vectors, which depend on locations. For example, a vector representing the electric field depends on the space point at which the electric field is defined. Another example is the normal vector defined on a surface. As position changes, the normal vector changes its directions.

Let E be an n-dimensional linear space. A **tangent vector** X_p at $p \in E$ is an ordered pair (p, X), where $X \in E$. The point p is called the **base point** of the tangent vector X_p. The set of all tangent vectors at a point $p \in E$ is called the **tangent space** at p and is denoted by $T_p(E)$. Given two tangent vectors X_p and Y_p and a constant a, we can define new tangent vectors at p by $(X + Y)_p = X_p + Y_p$ and $(aX)_p = aX_p$. The tangent space $T_p(E)$ with above addition and scalar multiplication constitutes a linear vector space. Note that two tangent vectors at different points cannot be added. Basically it is not defined. Let X_p be a tangent vector at $p \in U \subset E$ where U is open in E, and let $f : U \subset E \to R$ be smooth function defined on U. The **directional derivative** of f at the point p in the direction of X_p is defined by

$$(X_p f)(p) = Df(p) \cdot X_p, \tag{9.15}$$

where $Df(p)$ is the derivative of f at p(see Section 2.7.1).

Equation (9.15) implies that a tangent vector may be viewed as an operator on the smooth functions defined in a neighborhood of the point. The operator assigns to a function the directional derivative of the function in the direction of the vector. This understanding is very different from the usual concept of a vector. Let C_p^∞ denote the set of smooth functions defined in the neighborhood of p. For $f, g \in C_p^\infty$ and $a, b \in R$, we have

$$X_p(af + bg) = aX_p(f) + bX_p(g),$$
$$X_p(fg) = f(p)X_p(g) + g(p)X_p(f). \tag{9.16}$$

Let ϕ be a coordinate system on E and $p \in E$. For any $f \in C_p^\infty$, the function $g = f \circ \phi^{-1} :$ $R^n \to R$ is a function of the coordinates. The **partial derivatives** of f with respect to x^j are defined by

$$\frac{\partial f}{\partial x^j} = \frac{\partial (f \circ \phi^{-1})}{\partial x^j} \circ \phi, \ j = 1, 2, \cdots, n. \tag{9.17}$$

For $f, g \in C_p^\infty$, and $a, b \in R$, we have

$$\frac{\partial}{\partial x^j}(af + bg) = a\frac{\partial f}{\partial x^j} + b\frac{\partial g}{\partial x^j},$$

$$\frac{\partial}{\partial x^j}(fg) = f\frac{\partial g}{\partial x^j} + g\frac{\partial f}{\partial x^j}.$$

It follows from (9.15) and (9.17) that

$$(X_p f)(p) = D(f \circ \phi^{-1} \circ \phi)(p) \cdot X_p = D(f \circ \phi^{-1}) \circ D\phi(p) \cdot X_p.$$

Since $D\phi(p) \cdot X_p \in R^n$, the above equation may be written as

$$(X_p f)(p) = a_j \frac{\partial f}{\partial x^j}.$$

where $a_j(j = 1, 2, \cdots, n)$ are constants. It is natural to use the following identification

$$X_p = a_j \frac{\partial}{\partial x^j},$$

which implies $a^j = X_p(x^j)$, called the **components** of the tangent vector X_p. The n operators $\partial/\partial x^j$, $j = 1, 2, \cdots, n$ are linearly independent. Actually if $a_j \partial/\partial x^j = 0$ then $(a_j \partial/\partial x^j)x^i = a^i = 0$. Consequently the n operators $\partial/\partial x^j$, $j = 1, 2, \cdots, n$ constitute a basis of the tangent space $T_p(E)$. The basis $\{\partial/\partial x^j\}$ is called the **coordinate basis** induced by the coordinate system (E, ϕ). Let $(\tilde{x}^1, \tilde{x}^2, \cdots, \tilde{x}^n)$ be another coordinate system. Then

$$X_p = X^h \frac{\partial}{\partial x^h} = \tilde{X}^j \frac{\partial}{\partial \tilde{x}^j}.$$

It is readily found that

$$\tilde{X}^j = \frac{\partial \tilde{x}^j}{\partial x^h} X^h, \ \frac{\partial}{\partial x^h} = \frac{\partial \tilde{x}^j}{\partial x^h}\frac{\partial}{\partial \tilde{x}^j}.$$

The dual space of $T_p(E)$ is denoted by $T_p^*(E)$, called the **cotangent space**. The element in $T_p^*(E)$ is called the **covector**. For any $f \in C_p^\infty$ we can define a unique element $df \in T_p^*(E)$ by

$$df(X_p) = X_p(f), \ X_p \in T_p(E),$$

from which we obtain

$$dx^h \left(\frac{\partial}{\partial x^k} \right) = \delta^h{}_k.$$

Thus, the set $\{dx^k\}$ is a basis of the cotangent space $T_p^*(E)$ that is dual to the basis $\{\partial/\partial x^h\}$ of $T_p(E)$. Any $\omega_p \in T_p^*(E)$ can be expressed as $\omega_p = \omega_k dx^k$. Especially

$$df = \frac{\partial f}{\partial x^k} dx^k,$$

which is consistent with the usual expression of differential of a function. If $(\tilde{x}^1, \tilde{x}^2, \cdots, \tilde{x}^n)$ is another coordinate system, we have

$$\omega_h = \frac{\partial \tilde{x}^k}{\partial x^h} \tilde{\omega}_k, \, d\tilde{x}^j = \frac{\partial \tilde{x}^j}{\partial x^h} dx^h.$$

We introduce the **tensor space** at $p \in E$, denoted by

$$T_{p\ s}^r(E) = L^{r+s}[T_p^*(E), \cdots, T_p^*(E), T_p(E), \cdots, T_p(E); R]$$

with r copies of $T_p^*(E)$ and s copies of $T_p(E)$. An element $t \in T_{p\ s}^r(E)$ is called a **tensor of contravariant order** r **and covariant order** s **at** p. An element $t \in T_{p\ s}^r(E)$ can be expressed as

$$T = T^{i_1 \cdots i_r}{}_{j_1 \cdots j_s} \frac{\partial}{\partial x^{i_1}} \otimes \cdots \otimes \frac{\partial}{\partial x^{i_r}} \otimes dx^{j_1} \otimes \cdots \otimes dx^{j_s}. \tag{9.18}$$

Under the coordinate transformation (9.12), the components of the above tensor change according to (9.11).

Let $U \subset E$ be an open subset. The set $T(U) = \bigcup_{p \in U} T_p(E)$ is called the **tangent bundle** restricted to U. The tangent bundle is not a vector space but it has more structure than just a set. Similarly, we can define the **cotangent bundle** $T^*(U) = \bigcup_{p \in U} T_p^*(E)$ and rhe **tensor bundle** $T^r{}_s(E) = \bigcup_{p \in U} T_{p\ s}^r(E)$. If for all points $p \in U$, a unique element is prescribed in each $T_{p\ s}^r(E)$ we obtain a **tensor field** on U. A **vector field** on $U \subset E$ is a (1,0)-tensor field. A **scalar field** on $U \subset E$ is a (0,0)-tensor field. Relative to a coordinate system x^j, the components of tensor fields are defined in the same way as (9.18). In this case, the components are functions of the coordinates x^j of p. The tensor field is said to be smooth if these coefficients are smooth. All operations about the tensors can be extended to tensor fields pointwise. For example, the directional derivatives can be extended to vector fields by defining $X(f)(p) = X_p(f)$. Once the coordinate system is chosen, a tensor field T is simply denoted by its components $T^{i_1 \cdots i_r}{}_{j_1 \cdots j_s}$.

9.1.3 Metric Tensor

Let X and Y be two vector fields. A **metric tensor field** on a linear space E is a symmetric (0,2)-tensor field, denoted by $\langle \cdot, \cdot \rangle$, such that if $\langle X, Y \rangle = 0$ for all Y then $X = 0$. A linear space E equipped with a metric tensor field is called a **Riemannian space**, named after the German mathematician Georg Friedrich Bernhard Riemann (1826–1866). $\langle X, Y \rangle$ is called the inner product between X and Y. In terms of the coordinate system, the inner product is given by

$$\langle X, Y \rangle = X^i Y^j \left\langle \frac{\partial}{\partial x^i}, \frac{\partial}{\partial x^j} \right\rangle = g_{ij} X^i Y^j,$$

where $g_{ij} = \langle \partial/\partial x^i, \partial/\partial x^j \rangle$ are the coefficients of the metric tensor field. We define the square norm of X by

$$\|X\|^2 = \langle X, X \rangle = g_{ij} X^i X^j.$$

X is respectively called **timelike, spacelike**, and **null** if $\|X\|^2 < 0$, $\|X\|^2 > 0$ and $\|X\|^2 = 0$.

Example 9.2: Let $E = R^4$ and $g_{ij} = \eta_{ij}$ with

$$[\eta_{ij}] = \begin{bmatrix} 1 & 0 & 0 & 0 \\ 0 & 1 & 0 & 0 \\ 0 & 0 & 1 & 0 \\ 0 & 0 & 0 & -1 \end{bmatrix}. \tag{9.19}$$

This metric is called the **Minkowski metric**, named after the German mathematician Hermann Minkowski (1864–1909). The space R^4 equipped with the Minkowski metric is called the **Minkowski space**. □

In classical analysis, the metric tensor is introduced by coordinate transformations starting from a rectangular coordinate system. Let $\bar{x}^i, i = 1, 2, \cdots, n$ be the rectangular coordinates and $x^j, j = 1, 2, \cdots, n$ be the new (curvilinear) coordinates, both being related by $\bar{x}^i = x^i(x^j)$, $i, j = 1, 2, \cdots, n$. Then we have $d\bar{x}^i = \frac{\partial \bar{x}^i}{\partial x^j} dx^j$. The differential length may be expressed as

$$ds^2 = \sum_{l=1}^{n} (d\bar{x}^l)^2 = \sum_{l=1}^{n} \frac{\partial \bar{x}^l}{\partial x^i} \frac{\partial \bar{x}^l}{\partial x^j} dx^i dx^j = g_{ij} dx^i dx^j$$

where $g_{ij} = \sum_{l=1}^{n} \frac{\partial \bar{x}^l}{\partial x^i} \frac{\partial \bar{x}^l}{\partial x^j}$ is the metric tensor in the new coordinate system x^j. It is easy to see that the metric tensor is symmetric $g_{ij} = g_{ji}$.

9.2   Einstein's Postulates for Special Relativity

Einstein's theory of relativity has changed our view of space and time. One of Einstein's starting points was the principle of relativity due to Galileo. This principle claims that all physical laws should retain their functional forms under Galilean transformation, and was widely accepted by physicists from the seventeenth until the middle of the nineteenth century when James Clerk Maxwell discovered Maxwell equations. Maxwell equations seemed not to obey Galileo's principle of relativity. That is, they do not retain their functional form under Galilean transformation. To solve this dilemma, Einstein abandoned Galileo's principle of relativity, which is based on Galilean transformation. Instead, he adopted the principle of relativity based on the Lorentz transformation.

9.2.1   Galilean Relativity Principle

In physics we want to record what is happening using clocks and measuring rulers. To do this we use the **reference frame**, which defines the time and spatial position. Associated with each reference frame is a four-dimensional Cartesian coordinate system. Different frames put different labels to the same event and different relationships between them. We will make no distinction between a reference frame and the coordinate system associated with it. A coordinate system in which a free particle will remain at rest or in uniform motion is called an **inertial reference frame**. In classical mechanics, we assume that the inertial reference frames do exist. Different inertial reference frames are connected by the **Galilean transformation**

$$\tilde{\mathbf{r}} = \mathbf{r} - \mathbf{R} = \mathbf{r} - \mathbf{v}t,$$
$$\tilde{t} = t. \tag{9.20}$$

Here $(\mathbf{r}, t)$ and $(\tilde{\mathbf{r}}, \tilde{t})$ are respectively the space–time coordinates in coordinate system S and $\tilde{S}$, of which the relative velocity is $\mathbf{v}$, and $\mathbf{R} = \mathbf{v}t$, as shown in Figure 9.1.

Newton's equation of motion is invariant under the Galilean transformation. Thus all inertial systems are equivalent in Newton's theory. However it is found that the wave equation for the

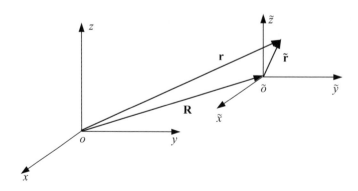

Figure 9.1   Galilean transformation

electric field is not invariant under the Galilean transformation. As a result, Maxwell equations only hold in a single coordinate system, and this violates the relativity principle.

If the relativity principle is considered a universal law, we only have two possibilities in order to solve the dilemma. One is to modify the Maxwell equations so that they are invariant under the Galilean transformation and the other is to modify the Galilean transformation so that the Maxwell equations are invariant under the new transformation. Einstein chose the latter.

9.2.2   Fundamental Postulates

Einstein's special theory of relativity is based on two postulates:

1. **Relativity principle:** The law of nature preserves its form in all inertial coordinate systems. This is also called invariance or covariance of physical laws.
2. **Invariance of light speed:** The speed of light is finite and independent of the motion of its source. In different inertial systems the light speed is the same.

The relativity principle generalizes the Galilean relativity principle for classical mechanics to all physical laws, including electromagnetism. It applies to everything! The invariance of light speed implies that light must travel at the same speed for all observers, regardless of their state of motion. In other words, Maxwell equations must be valid for all observers. The invariance of light speed is very much similar to the properties of sound waves. The propagation speed of sound wave is independent of the motion of the source.

9.3   The Lorentz Transformation

In special relativity, we work in four-dimensional flat space–time, three of space, one of time (in distinction to the curved space–time in general relativity). Therefore the space will be the four-dimensional linear space R^4. Furthermore, we will use orthonormal (Cartesian-like) coordinates with the metric tensor given by (9.19). The Lorentz transformation to be studied below had been derived by many physicists before Einstein established the special theory of relativity in 1905. Actually the Irish physicists George Francis FitzGerald (1851–1901) and Larmor, and Lorentz himself had all arrived at the Lorentz transformation by 1892. Poincaré also asserted before 1905 that all physical laws should retain their functional forms under Lorentz transformation.

From now on, all tensors will use Greek indices. Some Greek symbols will be used to stand for both physical quantities and indices. This will not cause any confusion, as one can easily identify what a Greek symbol implies from the context.

9.3.1   Intervals

We take the first three coordinates x^1, x^2, x^3 to be spatial, that is, $x^1 = x$, $x^2 = y$, $x^3 = z$ and the fourth one x^4 to be ct, where c is the light speed. Therefore the coordinate system $x^\alpha (\alpha = 1, 2, 3, 4)$ is our reference frame. A point (x^1, x^2, x^3, x^4) in R^4 is called an **event**.

All events constitute our **space–time**. A **physical process** is an ordered sequence of events in space–time. For example, the movement of a particle in space–time is a typical physical process. The path of the particle, $(x^1(\tau), x^2(\tau), x^3(\tau), x^4(\tau))$ for some parameter τ, is called the **world line** of the particle, which is a continuous curve according to the classical continuum assumption. The space–time **interval** of two events is defined by

$$s = \left[\left(x_2^1 - x_1^1\right)^2 + \left(x_2^2 - x_1^2\right)^2 + \left(x_2^3 - x_1^3\right)^2 - \left(x_2^4 - x_1^4\right)^2 \right]^{1/2}. \tag{9.21}$$

The intervals can be classified into three categories:

1. $s^2 = 0$, **light-like interval**.
2. $s^2 < 0$, **time-like interval**.
3. $s^2 > 0$, **space-like interval**.

Let us observe two events in coordinate system $S = (x^1, x^2, x^3, x^4)$. The first event is to send a light signal at t_1 from the point $\mathbf{r}_1 = (x_1^1, x_1^2, x_1^3)$, and the second event is that the signal arrives to the point $\mathbf{r}_2 = (x_2^1, x_2^2, x_2^3)$ at t_2. The interval between the two evens is zero since the light signal travels a distance $c(t_2 - t_1)$. Thus

$$\left(x_2^1 - x_1^1\right)^2 + \left(x_2^2 - x_1^2\right)^2 + \left(x_2^3 - x_1^3\right)^2 - \left(x_2^4 - x_1^4\right)^2 = 0.$$

If we observe the same events in a different coordinate system $\tilde{S} = (\tilde{x}^1, \tilde{x}^2, \tilde{x}^3, \tilde{x}^4)$, which is moving with a velocity $\mathbf{v}$ relative to S, we should also have

$$\left(\tilde{x}_2^1 - \tilde{x}_1^1\right)^2 + \left(\tilde{x}_2^2 - \tilde{x}_1^2\right)^2 + \left(\tilde{x}_2^3 - \tilde{x}_1^3\right)^2 - \left(\tilde{x}_2^4 - \tilde{x}_1^4\right)^2 = 0,$$

since the light signal travels at the same speed c in the inertial coordinate system $\tilde{S}$ by Einstein's postulates. The above observations show that if the interval is zero in one inertial coordinate system, it is zero in all other inertial coordinate systems. For two infinitesimally separated events in the inertial coordinate system S, the intervals can be written as

$$ds^2 = (dx^1)^2 + (dx^2)^2 + (dx^3)^2 - (dx^4)^2 = \eta_{\mu\nu} dx^\mu dx^\nu,$$

where $\eta_{\mu\nu}$ is the metric tensor given by (9.19). Similarly, in the inertial coordinate $\tilde{S}$, we have

$$d\tilde{s}^2 = (d\tilde{x}^1)^2 + (d\tilde{x}^2)^2 + (d\tilde{x}^3)^2 - (d\tilde{x}^4)^2 = \tilde{\eta}_{\mu\nu} d\tilde{x}^\mu d\tilde{x}^\nu.$$

In a flat space–time, we have $\eta_{\mu\nu} = \tilde{\eta}_{\mu\nu}$. Since ds and $d\tilde{s}$ have the same dimension (length) they must be related linearly and we can write $ds = ad\tilde{s} + b$, where a and b are constants. Since $ds = 0$ implies $d\tilde{s} = 0$ we have $b = 0$. Thus $ds = a(\mathbf{v})d\tilde{s}$. The constant a depends only on the magnitude of velocity for space and time are assumed to be homogeneous. Hence $a(\mathbf{v}) = a(v)$. By relativity principle, we also have $d\tilde{s} = a(v)ds$. It follows that $a^2(v) = 1$ or $a(v) = \pm 1$. As a special case, if $\tilde{S}$ is at rest relative to S we have $ds = d\tilde{s}$, which implies

$a(v) = 1$. Therefore ds^2 is an invariant under the transformation of two inertial systems

$$ds = d\tilde{s}, \tag{9.22}$$

which is the mathematical expression of invariance of light speed. The above equation can be written as

$$\eta_{\mu\nu}dx^{\mu}dx^{\nu} = \tilde{\eta}_{\mu\nu}d\tilde{x}^{\mu}d\tilde{x}^{\nu}. \tag{9.23}$$

9.3.2 Derivation of the Lorentz Transformation

In the special theory of relativity, we assume that there exists a collection of inertial reference frames in which the world line of a particle subject to no external force is a straight line. The transformation from one inertial reference frame S to another inertial reference frame $\tilde{S}$ is the **Lorentz transformation**. We write

$$\tilde{x}^{\mu} = L^{\mu}{}_{\nu}x^{\nu}. \tag{9.24}$$

Making use of (9.23) we obtain

$$\tilde{\eta}_{\mu\nu}d\tilde{x}^{\mu}d\tilde{x}^{\nu} = \tilde{\eta}_{\mu\nu}L^{\mu}{}_{\alpha}L^{\nu}{}_{\beta}dx^{\alpha}dx^{\beta} = \eta_{\alpha\beta}dx^{\alpha}dx^{\beta},$$

which implies $\tilde{\eta}_{\mu\nu}L^{\mu}{}_{\alpha}L^{\nu}{}_{\beta} = \eta_{\alpha\beta}$ in flat space–time. From the above equation it follows that $\det[L^{\mu}{}_{\nu}] = \pm 1$. As a result, we have

$$\sum_{\mu=1}^{3}\left(L^{\mu}{}_{\nu}\right)^2 - \left(L^{4}{}_{\nu}\right)^2 = 1, \nu = 1, 2, 3,$$

$$\sum_{\mu=1}^{3}\left(L^{\mu}{}_{4}\right)^2 - \left(L^{4}{}_{4}\right)^2 = -1, \tag{9.25}$$

$$\tilde{\eta}_{\mu\nu}L^{\mu}{}_{\alpha}L^{\nu}{}_{\beta} = 0, \alpha \neq \beta.$$

Let us consider a special case where the inertial reference frame $\tilde{S}$ is moving relative to the other frame S, and the coordinate axes of the two frames are parallel and oriented so that the frame $\tilde{S}$ is moving in the positive x direction with speed v as viewed from S. The two origins o and $\tilde{o}$ coincide at $t = \tilde{t} = 0$. Since coordinates x^2, x^3 do not change, Equation (9.24) may be written as

$$\begin{bmatrix} \tilde{x}^1 \\ \tilde{x}^2 \\ \tilde{x}^3 \\ \tilde{x}^4 \end{bmatrix} = \begin{bmatrix} L^1{}_1 & 0 & 0 & L^1{}_4 \\ 0 & 1 & 0 & 0 \\ 0 & 0 & 1 & 0 \\ L^4{}_1 & 0 & 0 & L^4{}_4 \end{bmatrix} \begin{bmatrix} x^1 \\ x^2 \\ x^3 \\ x^4 \end{bmatrix}$$

and Equation (9.25) reduces to

$$\left(L^1{}_1\right)^2 - \left(L^4{}_1\right)^2 = 1,$$
$$\left(L^1{}_4\right)^2 - \left(L^4{}_4\right)^2 = -1, \tag{9.26}$$
$$L^1{}_1 L^1{}_4 - L^4{}_1 L^4{}_4 = 0.$$

Since the origin of the frame $\tilde{S}$ viewed in frame S is $(vt, 0, 0, x^4)$, we have

$$L^1{}_1 v/c + L^1{}_4 = 0. \tag{9.27}$$

The four unknowns $L^1{}_1$, $L^4{}_1$, $L^1{}_4$, and $L^4{}_4$ can be obtained from (9.26) and (9.27). Finally, we obtain

$$\begin{aligned}
\tilde{x}^1 &= \gamma(x^1 - x^4 v/c), \\
\tilde{x}^2 &= x^2, \\
\tilde{x}^3 &= x^3, \\
\tilde{x}^4 &= \gamma(x^4 - x^1 v/c),
\end{aligned} \tag{9.28}$$

where $\gamma = 1/\sqrt{1 - (v/c)^2}$. The inverse transform is

$$\begin{aligned}
x^1 &= \gamma(\tilde{x}^1 + \tilde{x}^4 v/c), \\
x^2 &= \tilde{x}^2, \\
x^3 &= \tilde{x}^3, \\
x^4 &= \gamma(\tilde{x}^4 + \tilde{x}^1 v/c).
\end{aligned} \tag{9.29}$$

We denote

$$\left[L^i{}_k\right]_{x^1} = \begin{bmatrix} \gamma & 0 & 0 & -\gamma v/c \\ 0 & 1 & 0 & 0 \\ 0 & 0 & 1 & 0 \\ -\gamma v/c & 0 & 0 & \gamma \end{bmatrix}.$$

It is easy to see that $\det[L^\mu{}_v]_{x^1} = 1$.

In deriving the Lorentz transformation, we have assumed that the origin o and $\tilde{o}$ coincide at the time $t = \tilde{t} = 0$. Thus the Lorentz transformation is assumed to be homogeneous. If this assumption is abandoned, the resulting Lorentz transformation is inhomogeneous. If $\tilde{S}$ is moving relative to S with an arbitrary velocity $\mathbf{v}$ as viewed from S and two origins o and $\tilde{o}$ coincide at $t = \tilde{t} = 0$, the homogeneous Lorentz transformation may be expressed in a compact form:

$$\tilde{\mathbf{r}} = \overset{\leftrightarrow}{\alpha} \cdot \mathbf{r} - \gamma \boldsymbol{\beta} x^4, \quad \tilde{x}^4 = \gamma(x^4 - \boldsymbol{\beta} \cdot \mathbf{r}),$$

where

$$\tilde{\mathbf{r}} = (\tilde{x}^1, \tilde{x}^2, \tilde{x}^3), \mathbf{r} = (x^1, x^2, x^3),$$

$$\overset{\leftrightarrow}{\alpha} = \overset{\leftrightarrow}{\mathbf{I}} + (\gamma - 1)\frac{\boldsymbol{\beta}\boldsymbol{\beta}}{\beta^2}, \boldsymbol{\beta} = \mathbf{v}/c.$$

Under the Lorentz transformation, the differential volume element is an invariant:

$$d\tilde{x}^1 d\tilde{x}^2 d\tilde{x}^3 d\tilde{x}^4 = \frac{\partial(\tilde{x}^1, \tilde{x}^2, \tilde{x}^3, \tilde{x}^4)}{\partial(x^1, x^2, x^3, x^4)}dx^1 dx^2 dx^3 dx^4 = dx^1 dx^2 dx^3 dx^4.$$

9.3.3 Properties of Space–Time

A large number of physical theories can be simplified by combining space and time into a single continuum: space–time. Some properties of space–time may be derived from the Lorentz transformation.

9.3.3.1 Simultaneity in special relativity

Let us consider the special Lorentz transformation (9.28). Assume that $(x_1^1, x_1^2, x_1^3, x_1^4)$ and $(x_2^1, x_2^2, x_2^3, x_2^4)$ are two arbitrary points in frame S. The corresponding two points in frame $\tilde{S}$ are denoted by $(\tilde{x}_1^1, \tilde{x}_1^2, \tilde{x}_1^3, \tilde{x}_1^4)$ and $(\tilde{x}_2^1, \tilde{x}_2^2, \tilde{x}_2^3, \tilde{x}_2^4)$ respectively. From (9.28), we obtain

$$\tilde{x}_2^4 - \tilde{x}_1^4 = \frac{x_2^4 - x_1^4 - (x_2^1 - x_1^1)v/c}{\sqrt{1 - (v/c)^2}}. \tag{9.30}$$

The above equation shows that two observers in relative motion would disagree about the time interval between two events.

9.3.3.2 Fitzgerald-Lorentz contraction

It follows from (9.28) that

$$\tilde{x}_2^1 - \tilde{x}_1^1 = \gamma(x_2^1 - x_1^1) - \gamma\frac{v}{c}(x_2^4 - x_1^4). \tag{9.31}$$

Suppose we have a ruler, which is at rest in the frame $\tilde{S}$. It is natural to define the length of the ruler relative to any inertial system as the difference between simultaneous coordinate values of the end points. The length of the ruler measured by the observer in the frame $\tilde{S}$ is $l_0 = \tilde{x}_2^1 - \tilde{x}_1^1$, which is the length of the ruler as seen in the rest frame of the ruler and is called the **proper length**. Now we want to measure the length of the ruler in frame S. By definition, we may find the length l of the ruler relative to S as the difference between simultaneous coordinate values of the end points of the ruler. This can be done by placing the ruler to the x^1 axis and reading the coordinates of the two ends of the ruler at the same time (synchronously).

Thus we have $x_2^4 = x_1^4$, and (9.31) becomes

$$l = l_0\sqrt{1 - (v/c)^2},\tag{9.32}$$

where $l = x_2^1 - x_1^1$ is the length of the ruler measured by the observer in the frame S. The above equation indicates that a moving ruler is contracted in the direction of motion. This is called **Fitzgerald-Lorentz contraction**. In other words, the distance between two points in one coordinate system appears to be contracted to an observer in relative motion parallel to the line connecting these two points. This contraction is an effect caused by the operational definition of the measurement of length. Note that the proper length l_0 is an invariant.

It follows from (9.28) that a ruler which is placed perpendicular to the x-axis will have the same length in S and $\tilde{S}$. In general, a body, which moves relative to an inertial system, is contracted in the direction of its motion while its transverse dimension does not change. Consider a volume V_0 which is at rest in the frame $\tilde{S}$. The corresponding value of the volume measured synchronously in S is then given by

$$V = V_0\sqrt{1 - (v/c)^2}.\tag{9.33}$$

The transformation of surface element may be obtained similarly (see Bladel, 1984).

9.3.3.3 Time dilation

It follows from (9.29) that

$$x_2^4 - x_1^4 = \gamma\left(\tilde{x}_2^4 - \tilde{x}_1^4\right) + \gamma\frac{v}{c}\left(\tilde{x}_2^1 - \tilde{x}_1^1\right).$$

Suppose we have a clock, which is at rest in the frame $\tilde{S}$ and is fixed at some point ($\tilde{x}_2^1 = \tilde{x}_1^1$), and the time interval measured is $\tilde{t}_2 - \tilde{t}_1$. The above equation reduces to

$$t_2 - t_1 = \frac{\tilde{t}_2 - \tilde{t}_1}{\sqrt{1 - (v/c)^2}}.\tag{9.34}$$

This indicates that a moving clock appears to be running slow compared to the stationary clock. This effect is known as **time dilation**. The time $\tilde{t}$ as seen in the rest frame of the clock is called the **proper time**, denoted by τ. We rewrite (9.34) as

$$d\tau = dt\sqrt{1 - (v/c)^2}.\tag{9.35}$$

The interval in the frame $\tilde{S}$ is

$$d\tilde{s}^2 = -c^2 d\tau^2.$$

Since $d\tilde{s}$ is an invariant, the proper time $d\tau$ is also an invariant.

9.4 Relativistic Mechanics in Inertial Reference Frame

Newton's law of motion is invariant under Galilean transformation but not under Lorentz transformation. Thus Newton's law must be modified so that they are consistent with the special relativity.

9.4.1 Four-Velocity Vector

In the coordinate system x^μ, the **four-velocity vector** u^μ of a particle is defined by $u^\mu = dx^\mu/d\tau$, where $d\tau$ is the proper time. By definition, we have

$$ds^2 = (dx^1)^2 + (dx^2)^2 + (dx^3)^2 - (dx^4)^2 = -c^2 d\tau^2,$$

which implies $d\tau = dt\sqrt{1 - u^2/c^2}$ and

$$u^\mu = \left(\frac{\mathbf{u}}{\sqrt{1 - u^2/c^2}}, \frac{c}{\sqrt{1 - u^2/c^2}} \right),$$

where $\mathbf{u} = (dx^1/dt, dx^2/dt, dx^3/dt) = (u_x, u_y, u_z)$ is the velocity of the particle, and $u = |\mathbf{u}|$. Under the special Lorentz transformation (9.28), the four velocity vector u^μ changes according to $\tilde{u}^\mu = \frac{\partial \tilde{x}^\mu}{\partial x^\nu} u^\nu$. Hence

$$\tilde{u}_x = \frac{u_x - v}{(1 - vu_x/c^2)}, \tilde{u}_y = \frac{u_y}{(1 - vu_x/c^2)\gamma}, \tilde{u}_z = \frac{u_z}{(1 - vu_x/c^2)\gamma}.$$

9.4.2 Four-Momentum Vector

In the coordinate system x^μ, the **four-momentum vector** of a particle is defined by

$$p^\mu = m_0 u^\mu = \left(\frac{m_0 \mathbf{u}}{\sqrt{1 - u^2/c^2}}, \frac{m_0 c}{\sqrt{1 - u^2/c^2}} \right),$$

where m_0 is the mass of the particle at rest, called rest mass. The first three components form a vector

$$\mathbf{p} = \frac{m_0 \mathbf{u}}{\sqrt{1 - u^2/c^2}} = m\mathbf{u}, \tag{9.36}$$

where

$$m = \frac{m_0}{\sqrt{1 - (u/c)^2}} \tag{9.37}$$

is defined as the mass of a moving particle. Equation (9.36) reduces to $m_0\mathbf{u}$, the classical definition of momentum, when $u \ll c$. Therefore, $\mathbf{p}$ can be considered as the definition of momentum in relativity. If $u \ll c$, the fourth component of p^μ has the following expansion

$$p^4 = \frac{m_0 c}{\sqrt{1 - u^2/c^2}} = \frac{1}{c}\left(m_0 c^2 + \frac{1}{2}m_0 u^2 + \cdots\right).$$

It can be seen that the second term in the bracket is the kinetic energy in classical mechanics. Consequently, all quantities in the bracket represent energy. When the particle is at rest only the term $m_0 c^2$ is left. For this reason, the term $m_0 c^2$ stands for the rest energy of a particle. The energy of a free particle with mass m and velocity u is thus defined by

$$E = \frac{m_0 c^2}{\sqrt{1 - u^2/c^2}} = mc^2.$$

This is the famous **Einstein mass-energy relation**. The kinetic energy of the particle is then given by $T = E - m_0 c^2$. Thus the four momentum vector may be written as

$$p^\mu = \left(\frac{m_0 \mathbf{u}}{\sqrt{1 - u^2/c^2}}, \frac{1}{c}E\right).$$

Since $p_\mu p^\mu$ is invariant under coordinate transformation from S to $\tilde{S}$

$$p_\mu p^\mu = \tilde{p}_\mu \tilde{p}^\mu. \tag{9.38}$$

the rest mass is an invariant. If the particle is at rest in the frame $\tilde{S}$, then $\tilde{p}^\mu = (0, 0, 0, m_0 c^2/c)$, and from (9.38) it follows that

$$p^2 - E^2/c^2 = -m_0^2 c^2,$$

where $p = |\mathbf{p}|$. This gives

$$E = \sqrt{p^2 c^2 + m_0^2 c^4}.$$

When the speed of the particle increases, its mass, energy and momentum increase correspondingly. In classical mechanics, the energy can be determined up to a constant (the rest energy) and the constant can be ignored. In special relativity the constant cannot be ignored.

9.4.3   Relativistic Equation of Motion

In relativistic mechanics, Newton's equation must be modified to

$$\frac{d\mathbf{p}}{dt} = \mathbf{F}, \tag{9.39}$$

where $\mathbf{F}$ is the external force acting on the particle and $\mathbf{p}$ is the momentum defined by (9.36). If the velocity of the particle is $\mathbf{u}$, the work done by $\mathbf{F}$ per unit time is equal to the increase of the energy of the particle per unit time $dE/dt = \mathbf{F} \cdot \mathbf{u}$ or

$$\frac{d}{dt}\left(\frac{E}{c}\right) = \frac{\mathbf{F} \cdot \mathbf{u}}{c}. \tag{9.40}$$

Combining (9.39) and (9.40) yields

$$\frac{d}{dt}\left(\mathbf{p}, \frac{E}{c}\right) = \left(\mathbf{F}, \frac{\mathbf{F} \cdot \mathbf{u}}{c}\right).$$

In terms of proper time, this becomes

$$\frac{dp^\mu}{d\tau} = F^\mu, \tag{9.41}$$

where

$$F^\mu = \left(\frac{\mathbf{F}}{\sqrt{1 - u^2/c^2}}, \frac{\mathbf{F} \cdot \mathbf{u}}{c\sqrt{1 - u^2/c^2}}\right)$$

is called **four-force vector**. Under the special Lorentz transformation (9.28), the four-force F^μ changes according to $\tilde{F}^\mu = \frac{\partial \tilde{x}^\mu}{\partial x^\nu} F^\nu$. Thus

$$\tilde{F}_x = \frac{F_x - \mathbf{F} \cdot \mathbf{u}v/c^2}{1 - vu_x/c^2}, \quad \tilde{F}_y = \frac{F_y}{(1 - vu_x/c^2)\gamma}, \quad \tilde{F}_z = \frac{F_z}{(1 - vu_x/c^2)\gamma}. \tag{9.42}$$

Remark 9.1: Sometimes we need to consider the system in which the external forces produce a change in the rest mass of the particle. For example, the Joule heat energy produced in a conducting body due to the electromagnetic forces will contribute to the rest mass. In this case, in order to maintain (9.41), the four-force vector should be changed to

$$F^\mu = \left(\frac{\mathbf{F}}{\sqrt{1 - u^2/c^2}}, \frac{\mathbf{F} \cdot \mathbf{u} + Q}{c\sqrt{1 - u^2/c^2}}\right) \tag{9.43}$$

where Q is the amount of heat or non-mechanical energy developed per unit time in the body. Thus (9.40) becomes

$$\frac{dE}{dt} = \mathbf{F} \cdot \mathbf{u} + Q. \qquad \square$$

When the force $\mathbf{F}$ is conservative, we can write $\mathbf{F} = -\nabla\psi$, where ψ is the potential function. Equation (9.40) implies

$$E + \psi = W = \text{constant} \tag{9.44}$$

where W is the total energy.

9.4.4   Angular Momentum Tensor and Energy-Momentum Tensor

From the two four-vectors x^μ and p^ν we can form an antisymmetric (2,0)-tensor

$$M^{\mu\nu} = x^\mu p^\nu - x^\nu p^\mu.$$

This is called the **angular momentum tensor**. We also can introduce another antisymmetric (2,0)-tensor

$$D^{\mu\nu} = x^\mu F^\nu - x^\nu F^\mu.$$

Then it is easy to verify that

$$\frac{dM^{\mu\nu}}{d\tau} = D^{\mu\nu}.$$

Let us consider a continuous matter distribution, which moves with a local average velocity $\mathbf{u} = \mathbf{u}_x u$ with respect to the frame S. Let ΔV_0 be a differential volume at a point, which is at rest momentarily with respect to $\tilde{S}$. The corresponding differential volume ΔV in the frame S measured synchronously is given by $\Delta V = \Delta V_0 \sqrt{1-(u/c)^2}$. Let ρ_m be the volume density of mass and ρ_{m_0} be the density of rest mass, both being the functions of space coordinates and time. Then these two mass densities are related by

$$\rho_m = \frac{\rho_{m_0}}{\sqrt{1-(u/c)^2}} \tag{9.45}$$

due to (9.37). Considering a point inside the matter distribution at a given time, we assume that the matter at this point is momentarily at rest relative to $\tilde{S}$. In the frame $\tilde{S}$, Equation (9.45) reduces to $\tilde{\rho}_m = \tilde{\rho}_{m_0}$. Since the rest mass is invariant, we have $\rho_{m_0} \Delta V = \tilde{\rho}_{m_0} \Delta V_0$. Thus we get

$$\rho_m = \frac{\tilde{\rho}_{m_0}}{1-(u/c)^2}, \; \rho_{m_0} = \frac{\tilde{\rho}_{m_0}}{\sqrt{1-(u/c)^2}}.$$

We further assume that the interactions between the mass particles are negligible (that is we consider an incoherent matter, which is a mass distribution without pressure or viscosity). From (9.41) it follows that

$$\frac{d}{d\tau}(m_0 u^\mu) = \frac{d}{d\tau}(\rho_{m_0} \Delta V u^\mu) = \frac{d}{d\tau}(\tilde{\rho}_{m_0} \Delta V_0 u^\mu) = F^\mu. \tag{9.46}$$

Note that the four-force vector may be written as

$$F^\mu = \left(\frac{\mathbf{F}}{\sqrt{1-(u/c)^2}}, \frac{\mathbf{F} \cdot \mathbf{u}}{c\sqrt{1-(u/c)^2}} \right) = \left(\frac{\mathbf{f}\Delta V}{\sqrt{1-(u/c)^2}}, \frac{\mathbf{f}\Delta V \cdot \mathbf{u}}{c\sqrt{1-(u/c)^2}} \right)$$

$$= \left(\mathbf{f}, \frac{\mathbf{f} \cdot \mathbf{u}}{c} \right) \Delta V_0 = f^\mu \Delta V_0$$

where $f^\mu = (\mathbf{f}, \mathbf{f} \cdot \mathbf{u}/c)$ is the four-force density which is a four-vector since ΔV_0 is an invariant. Equation (9.46) may be expressed as

$$\tilde{\rho}_{m_0} u^\mu \frac{d(\Delta V_0)}{d\tau} + \Delta V_0 \frac{d(\tilde{\rho}_{m_0} u^\mu)}{d\tau} = f^\mu \Delta V_0. \tag{9.47}$$

When the rest mass is conserved, the first term on the left-hand side is zero. When the rest mass is not conserved, the four-force density should be changed to (see (9.43)) $f^\mu = \left(\mathbf{f}, \frac{\mathbf{f} \cdot \mathbf{u} + q}{c} \right)$, where q is the amount of non-mechanical energy developed per unit of volume and time.

In order to calculate the first term on the left-hand side of (9.47) when the rest mass is not conserved, let us consider matter distribution which occupies a volume V at the time t. At the time $t + \Delta t$ the volume of this matter will be increased by an amount

$$dV = dt \int_S \mathbf{u} \cdot \mathbf{u}_n dS = dt \int_V \nabla \cdot \mathbf{u} dV,$$

where S is the bounding surface of V and $\mathbf{u}_n$ is unit outward normal of S. The above equation must hold for every part of the material body. In particular if V is an infinitesimal volume element ΔV we have

$$\frac{1}{\Delta V} \frac{d\Delta V}{dt} = \nabla \cdot \mathbf{u}. \tag{9.48}$$

Since ΔV_0 and $d\tau$ are measured in the rest frame $\tilde{S}$, from (9.48) we have

$$\frac{d\Delta V_0}{d\tau} = \Delta V_0 \tilde{\nabla} \cdot \tilde{\mathbf{u}}, \tag{9.49}$$

where $\tilde{\nabla}$ means the differentiation with respect to the coordinates in $\tilde{S}$. In the rest system $\tilde{S}$ we have $\tilde{\mathbf{u}} = 0$. But its derivatives with respect to the coordinates need not be zero. Let $\partial_\alpha = \partial/\partial x^\alpha (\alpha = 1, 2, 3, 4)$. Since $\partial_\alpha u^\alpha$ is an invariant, we have

$$\partial_\alpha u^\alpha = \tilde{\partial}_\alpha \tilde{u}^\alpha = \tilde{\nabla} \cdot \tilde{\mathbf{u}}. \tag{9.50}$$

By use of (9.49) and (9.50), Equation (9.47) may be written as

$$\frac{d(\tilde{\rho}_{m_0} u^\mu)}{d\tau} + \tilde{\rho}_{m_0} u^\mu \partial_\alpha u^\alpha = f^\mu. \tag{9.51}$$

The first term on the left-hand side is

$$\frac{d(\tilde{\rho}_{m_0} u^\mu)}{d\tau} = \frac{\partial(\tilde{\rho}_{m_0} u^\mu)}{\partial x^\alpha} \cdot \frac{\partial x^\alpha}{\partial \tau} = u^\alpha \partial_\alpha (\tilde{\rho}_{m_0} u^\mu).$$

Thus (9.51) becomes

$$\partial_\nu T_m^{\mu\nu} = f^\mu. \tag{9.52}$$

The quantity $T_m^{\mu\nu} = \tilde{\rho}_{m_0} u^\mu u^\nu$ is a symmetric (2,0)-tensor, called the **energy-momentum tensor**. Introduce the momentum density vector $\mathbf{g} = \rho_m \mathbf{u}$. Then (9.52) implies

$$\frac{\partial e}{\partial t} + \nabla \cdot (e\mathbf{u}) = cf^4, \quad \frac{\partial g^\alpha}{\partial t} + \nabla \cdot (g^\alpha \mathbf{u}) = f^\alpha, \alpha = 1, 2, 3,$$

where $e = \rho_m c^2$ stands for the total energy. The first equation means the conservation of energy, and the second equations mean the conservation of momentum.

9.5 Electrodynamics in Inertial Reference Frame

It is required by relativity that all the physical laws must retain their forms under the Lorentz transformation. This is called invariance or covariance of physical laws. A physical law is expressed as an equation consisting of physical quantities. If both sides of the equation can be written in the form of tensors of the same type, it will satisfy the principle of relativity. For example, if a physical law in the reference frame S can be expressed as

$$A^\alpha = B^{\alpha\beta} C_\beta, \tag{9.53}$$

where A^α, $B^{\alpha\beta}$ and C_β are (1,0)-tensor, (2,0)-tensor and (0,1)-tensor respectively. Under the coordinate transformation $\tilde{x}^\alpha = \tilde{x}^\alpha(x^\beta)$, the above tensors change according to

$$\tilde{A}^\alpha = \frac{\partial \tilde{x}^\alpha}{\partial x^\beta} A^\beta, \ \tilde{C}_\beta = \frac{\partial x^\alpha}{\partial \tilde{x}^\beta} C_\alpha, \ \tilde{B}^{\mu\nu} = \frac{\partial \tilde{x}^\mu}{\partial x^\alpha} \frac{\partial \tilde{x}^\nu}{\partial x^\beta} B^{\alpha\beta}.$$

Then

$$\tilde{B}^{\mu\nu}\tilde{C}_\nu = \frac{\partial \tilde{x}^\mu}{\partial x^\alpha} \frac{\partial \tilde{x}^\nu}{\partial x^\lambda} \frac{\partial x^\beta}{\partial \tilde{x}^\nu} B^{\alpha\lambda} C_\beta = \frac{\partial \tilde{x}^\mu}{\partial x^\alpha} B^{\alpha\lambda} C_\lambda = \frac{\partial \tilde{x}^\mu}{\partial x^\alpha} A^\alpha = \tilde{A}^\mu.$$

Hence (9.53) is invariant under the coordinate transformation.

9.5.1 Covariance of Continuity Equation

Electric charge will be assumed to be invariant. This fact has been verified by a number of experiments. Introducing $J^\alpha = (J_x, J_y, J_z, c\rho)$, the continuity equation may be written in the form

$$\partial_\alpha J^\alpha = 0, \tag{9.54}$$

where $\partial_\alpha = \partial/\partial x^\alpha$ is formally a (0,1)-tensor. If we can prove that J^α is a (1,0)-tensor then $\partial_\alpha J^\alpha$ is a scalar, a (0,0)-tensor, and (9.54) is the covariant form required. Actually the charge in the differential volume element is $dq = \rho dx^1 dx^2 dx^3$. Thus

$$dq dx^4 = \rho dx^1 dx^2 dx^3 dx^4.$$

Since dq and $dx^1 dx^2 dx^3 dx^4$ are invariant under the Lorentz transformation, the current density ρ changes according to dx^4. The charge passing through $dx^2 dx^3$ during the time interval dt is

$$dq = J^1 dx^2 dx^3 dt = J^1 dx^2 dx^3 dx^4 / c$$

or

$$dx^1 c\, dq = J^1 dx^1 dx^2 dx^3 dx^4.$$

Hence J^1 changes according to dx^1. In general, J^α changes according to dx^α. Therefore we have proved that J^α is a contravariant vector, and the continuity equation (9.54) satisfies the principle of relativity.

9.5.2   Covariance of Maxwell Equations

The vector potential $\mathbf{A}$ and scalar potential ϕ in free space satisfy the following equations

$$\nabla^2 \mathbf{A} - \frac{1}{c^2}\frac{\partial^2 \mathbf{A}}{\partial t^2} = -\mu_0 \mathbf{J}, \ \nabla^2 \phi - \frac{1}{c^2}\frac{\partial^2 \phi}{\partial t^2} = -\frac{\rho}{\varepsilon_0}, \tag{9.55}$$

and the Lorentz gauge condition

$$\nabla \cdot \mathbf{A} + \frac{1}{c^2}\frac{\partial \phi}{\partial t} = 0. \tag{9.56}$$

If we introduce $A^\mu = (A_x, A_y, A_z, \phi/c)$, Equations (9.55) may be written as

$$\Box A^\mu = \mu_0 J^\mu,$$

where $\Box = \frac{\partial^2}{\partial t^2} - \frac{1}{c^2}\left(\frac{\partial^2}{\partial x^2} + \frac{\partial^2}{\partial y^2} + \frac{\partial^2}{\partial z^2}\right)$ is the wave operator. It is easy to show that the wave operator $\Box$ is an invariant under Lorentz transformation. Since J^μ is a contravariant vector A^μ is also a contravariant vector. We rewrite (9.56) as

$$\partial_\mu A^\mu = 0.$$

This is a covariant form. We now introduce an antisymmetric (0, 2)-tensor, called the **electromagnetic field-strength tensor**

$$F_{\mu\nu} = \partial_\mu A_\nu - \partial_\nu A_\mu = -F_{\nu\mu},$$

where $A_\mu = \eta_{\mu\nu} A^\nu$. Another field strength tensor may be constructed by the metric tensor

$$F^{\mu\nu} = \eta^{\mu\alpha}\eta^{\nu\beta} F_{\alpha\beta}.$$

The two inhomogeneous Maxwell equations

$$\nabla \times \mathbf{B} - \frac{1}{c^2} \frac{\partial \mathbf{E}}{\partial t} = \mu_0 \mathbf{J}, \ \nabla \cdot \mathbf{E} = \frac{\rho}{\varepsilon_0}$$

can be written in the form

$$\partial_\mu F^{\mu\nu} = -\mu_0 J^\nu, \tag{9.57}$$

and the two homogeneous Maxwell equations

$$\nabla \times \mathbf{E} + \frac{\partial \mathbf{B}}{\partial t} = 0, \ \nabla \cdot \mathbf{B} = 0$$

can be written in the form

$$\partial_\mu F_{\nu\alpha} + \partial_\nu F_{\alpha\mu} + \partial_\alpha F_{\mu\nu} = 0. \tag{9.58}$$

Here $(\mu, \nu, \alpha) = (1,2,3)$ gives $\nabla \cdot \mathbf{B} = 0$, and $(\mu, \nu, \alpha) = (4,2,3),(4,3,1),(4,1,2)$ give $\nabla \times \mathbf{E} + \partial \mathbf{B}/\partial t = 0$. Therefore, there are only four independent equations.

Equations (9.57) and (9.58) are the covariant form of Maxwell equations. Equation (9.58) can be written into a different form by using the **dual field-strength tensor** $G^{\alpha\beta}$ defined by

$$G^{\alpha\beta} = \frac{1}{2} \bar{\varepsilon}^{\alpha\beta\gamma\delta} F_{\gamma\delta},$$

where $\bar{\varepsilon}^{\alpha\beta\gamma\delta}$ is an antisymmetric (4, 0)-tensor, defined by

1. $\bar{\varepsilon}^{\alpha\beta\gamma\delta} = 0$ whenever any two indices are the same;
2. $\bar{\varepsilon}^{1234} = 1$;
3. $\bar{\varepsilon}^{\alpha\beta\gamma\delta} = 1$ if four indices are different and can be transformed to 1234 by an even permutation;
4. $\bar{\varepsilon}^{\alpha\beta\gamma\delta} = -1$ if four indices are different and can be transformed to 1234 by an odd permutation.

Then (9.58) is equivalent to $\partial_\alpha G^{\alpha\beta} = 0$. As a result, Maxwell equations can also be written as

$$\partial_\mu F^{\mu\nu} = -\mu_0 J^\nu,$$
$$\partial_\alpha G^{\alpha\beta} = 0.$$

9.5.3 Transformation of Electromagnetic Fields and Sources

Since $F^{\mu\nu}$ is a (2,0)-tensor, we have the following transformation relation

$$\tilde{F}^{\mu\nu} = \frac{\partial \tilde{x}^\mu}{\partial x^\alpha} \frac{\partial \tilde{x}^\nu}{\partial x^\beta} F^{\alpha\beta}. \tag{9.59}$$

Under the Lorentz transformation (9.28), Equation (9.59) gives the field transformations

$$\tilde{E}_x = E_x, \ \tilde{E}_y = \gamma(E_y - vB_z), \ \tilde{E}_z = \gamma(E_z + vB_y),$$
$$\tilde{B}_x = B_x, \ \tilde{B}_y = \gamma\left(B_y + E_z\frac{v}{c^2}\right), \ \tilde{B}_z = \gamma\left(B_z - E_y\frac{v}{c^2}\right). \tag{9.60}$$

The four current changes according to the rule $\tilde{J}^\mu = \frac{\partial \tilde{x}^\mu}{\partial x^\alpha} J^\alpha$, which gives the source transformations

$$\tilde{J}_x = \gamma(J_x - v\rho), \ \tilde{J}_y = J_y, \ \tilde{J}_z = J_z, \ c\tilde{\rho} = \gamma\left(c\rho - J_x\frac{v}{c}\right).$$

9.5.4   Covariant Forms of Electromagnetic Conservation Laws

It is readily found that the Lorentz force density equation $\mathbf{f} = \rho\mathbf{E} + \mathbf{J} \times \mathbf{B}$ can be written in the covariant form

$$f^\mu = F^{\mu\nu} J_\nu, \tag{9.61}$$

where $J_\mu = g_{\mu\nu} J^\nu = (J_x, J_y, J_z, -c\rho)$. The first three components of f^μ represent the force density $\mathbf{f}$. Making use of (9.57), Equation (9.61) may be expressed as

$$f^\mu = -\frac{1}{\mu_0} F^{\mu\nu} \partial_\alpha F^\alpha{}_\nu.$$

We recall that the conservation laws of electromagnetic energy and momentum are

$$\nabla \cdot \mathbf{S} + \frac{\partial w}{\partial t} = -\mathbf{J}_{imp} \cdot \mathbf{E} - \mathbf{J}_{ind} \cdot \mathbf{E} = -\mathbf{J} \cdot \mathbf{E},$$
$$\nabla \cdot \overset{\leftrightarrow}{\mathbf{T}} - \frac{\partial \mathbf{g}}{\partial t} = \mathbf{f},$$

where $w = \mathbf{E} \cdot \mathbf{D}/2 + \mathbf{H} \cdot \mathbf{B}/2$, $\mathbf{J} = \mathbf{J}_{imp} + \mathbf{J}_{ind}$ and $\overset{\leftrightarrow}{\mathbf{T}}$ is the Maxwell stress tensor. Introducing the (2,0)-tensor of second order, called the **electromagnetic energy–momentum tensor**

$$T^{\mu\nu}_{EM} = -\frac{1}{\mu_0}\left(\eta^{\mu\alpha} F_{\alpha\beta} F^{\beta\nu} + \frac{1}{4}\eta^{\mu\nu} F_{\alpha\beta} F^{\alpha\beta}\right),$$

the conservation laws of electromagnetic energy and momentum may be written as

$$\partial_\nu T^{\mu\nu}_{EM} = -f^\mu. \tag{9.62}$$

The covariant form of the conservation laws of electromagnetic angular momentum in the source-free region is

$$\partial_\alpha M^{\alpha\beta\gamma} = 0,$$

where $M^{\alpha\beta\gamma} = T^{\alpha\gamma}_{EM}x^\beta - T^{\alpha\beta}_{EM}x^\gamma$.

Remark 9.2 (Electromagnetic field invariants): Two invariants can be constructed from the electromagnetic field-strength tensor

$$F^{\mu\nu} F_{\mu\nu} = \text{invariant}, \quad \bar{\varepsilon}^{\alpha\beta\gamma\delta} F_{\alpha\beta} F_{\gamma\delta} = \text{invariant},$$

which imply

$$|\mathbf{B}|^2 - \frac{1}{c^2}|\mathbf{E}|^2 = \text{invariant}, \quad \frac{1}{c}\mathbf{E} \cdot \mathbf{B} = \text{invariant}. \tag{9.63}$$

$\square$

9.5.5   Total Energy-Momentum Tensor

If the incoherent matter contains charged particles subjected to electromagnetic forces, the four-force density is given by (9.62). For a closed system consisting of matter and fields, we rewrite (9.52) as

$$\partial_\beta T^{\alpha\beta} = 0, \tag{9.64}$$

where

$$T^{\alpha\beta} = T_m^{\alpha\beta} + T_{EM}^{\alpha\beta} \tag{9.65}$$

is the total energy-momentum tensor.

9.6   General Theory of Relativity

In the theory of special relativity, there is a preferred class of 'non-accelerating' global frames, called inertial frames, in which Newton's first law holds. All inertial frames are equivalent, and physical laws preserve their form under the coordinate transformation (characterized by the Lorentz transformation) from one inertial frame to another. Special relativity reveals that the space and time are not separate. A question may then be raised whether we can find an inertial frame after all. By definition we only need to check if a free particle travels in a straight line at constant speed relative to the frame. However, a completely free particle does not exist in reality since the gravity affects all matter equally. In other words, there are no particles that are free of all forces.

In 1916, Einstein published his important paper 'The foundations of the general theory of relativity' in *Annalen der Physik*. By accepting the experimental fact that the inertial mass is equal to gravitational mass, Einstein assumed that the gravity and acceleration are equivalent. As a result, a gravitational field can be simulated by an acceleration field, and a uniform gravitational field can be eliminated from a reference frame by suitable acceleration. If the gravitational field is not uniform, the gravity can be eliminated in a sufficiently small region in which the gravitational field can be treated as uniform. As an accelerated coordinate system is characterized by a metric tensor $g_{\alpha\beta}$, the gravitational field can also be characterized by $g_{\alpha\beta}$. Therefore the space–time is curved by the gravity. Given the sources of the gravitational field, the determination of this metric tensor $g_{\alpha\beta}$ is one of the major tasks in general relativity.

9.6.1   *Principle of Equivalence*

In Newton's theory there are two kinds of mass: inertial mass in his law of motion and gravitational mass in his law of gravity. In Newton's theory there is no reason why these two masses should be related to each other. Newton's second law states that the force exerted on a particle is proportional to its acceleration. Thus one may write

$$\mathbf{F} = m_i \mathbf{a}, \tag{9.66}$$

where m_i is a constant, called the **inertial mass**. Newton's law of universal gravitation states that the gravitational force of a particle of mass M exerted on a particle of mass m_g is proportional to the gradient of a scalar field Φ, known as the **gravitational potential**

$$\mathbf{F}_g = -\frac{GMm_g\mathbf{r}}{r^3} = -m_g\nabla\Phi, \tag{9.67}$$

where $r = |\mathbf{r}|$ is the distance between two particles, $G = 6.67 \times 10^{-11}$ is the constant of universal gravitation, and $\Phi = GM/r$. The mass m_g is called the **gravitational mass**. Many experiments have demonstrated that

$$m_i = m_g. \tag{9.68}$$

It follows from (9.66), (9.67) and (9.68) that

$$\mathbf{a} = -\nabla\Phi. \tag{9.69}$$

Therefore all particles accelerate in the same way, independent of their mass. Einstein generalized the fact (9.68) and assumed that there is no observable distinction between the effects of gravity and acceleration. This is referred to as the **principle of equivalence**. In a closed box, there is no physical experiment within the box that can tell if the box is at rest on the earth's surface or the box is accelerating at $1g$ (g is the acceleration due to the gravity) in empty space.

In a gravitational field, it is impossible to identify the global inertial frames in special relativity. But we can pick out local inertial frames in which gravity disappears (that is, frames in free-fall). Therefore the principle of equivalence can also be stated as: gravitation can be made to vanish locally through an appropriate choice of frames. This statement implies, in small enough regions of space–time, the laws of physics reduce to those of special relativity and it is impossible to detect the existence of a gravitational field. Mathematically, this statement implies that, given any point p in the four dimensional space of events, one can find a coordinate system x^μ (that is, local inertial frame) with its origin being at p, such that the metric becomes locally Minkowskian

$$g_{\mu\nu}(x^\alpha) = \eta_{\mu\nu} + o(x^\alpha)^2,$$

where $\eta_{\mu\nu}$ denotes the Minkowskian metric tensor in special relativity. In the local inertial frame, we have

$$g_{\mu\nu}(p) = \eta_{\mu\nu}, \ \frac{\partial g_{\mu\nu}}{\partial x^{\alpha}}(p) = 0.$$

9.6.2 Manifolds

The tensor transformation (9.11) is based on the linear coordinate transformation (9.9), and furthermore the space E on which the tensor is defined is assumed to be a linear space. In practice, it is often necessary to use curvilinear coordinate systems, the transition to which involves non-linear coordinate transformations. Recurrently, we have to consider the tensor analysis on a curved surface, which cannot be covered by a single coordinate system. For example, a sphere in R^3 cannot be covered by a single two-dimensional coordinate system. Manifolds are an abstraction of the idea of a smooth surface in Euclidean space and generalize the parameter representation of the surface. The abstraction strips away the containing space and makes the constructions intrinsic to the manifold itself (Abraham *et al.*, 1988).

An n-dimensional manifold is a point set M, which is covered completely by a countable set of neighborhoods (open sets) $U_1, U_2, \cdots$, such that each point $p \in M$ belongs to at least one of these neighborhoods. A **local coordinate system** on each U_i is a bijection $\phi_i : p \in U_i \to (x^1, x^2, \cdots, x^n) \in \phi_i(U_i) \subset R^n$, where $\phi_i(U_i)$ is an open set in R^n. Let U_i and U_j be any two coordinate neighborhoods such that $U_i \cap U_j \neq \emptyset$. For $p \in U_i \cap U_j$, we have two sets of coordinates $x^1, x^2, \cdots, x^n \in \phi_i(U_i)$ and $\tilde{x}^1, \tilde{x}^2, \cdots, \tilde{x}^n \in \phi_j(U_j)$ for p. If the overlap map

$$\phi_j \circ \phi_i^{-1} : (x^1, x^2, \cdots, x^n) \to (\tilde{x}^1, \tilde{x}^2, \cdots, \tilde{x}^n) \tag{9.70}$$

as shown in Figure 9.2, is smooth, the manifold is called the n-**dimensional differentiable manifold**. The pair (U_i, ϕ_i) is called a **chart** of dimension n with coordinate neighborhood

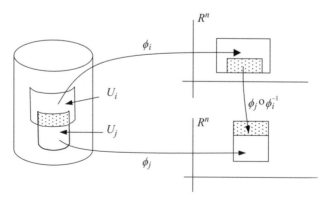

Figure 9.2   Local coordinate systems

U_i. The collection of all charts is called an **atlas** on M. The map (9.70) will be denoted by

$$\tilde{x}^j = \tilde{x}^j(x^h),$$

and the inverse of the above will be expressed in the compressed form

$$x^h = x^h(\tilde{x}^j).$$

An n manifold M immersed in R^m $(n \leq m)$ is characterized by

$$\begin{aligned}
y_1 &= y_1(x^1, x^2, \cdots, x^n), \\
y_2 &= y_2(x^1, x^2, \cdots, x^n), \\
&\quad \cdots\cdots \\
y_m &= y_m(x^1, x^2, \cdots, x^n).
\end{aligned} \tag{9.71}$$

We call $(x^1, x^2, \cdots, x^n)$ the local coordinates and $(y_1, y_2, \cdots, y_m)$ the global or ambient coordinates.

9.6.3 Tangent Bundles, Cotangent Bundles and Tensor Bundles

Let (U, ϕ) be a chart and $p \in U$. The set of all C^∞ functions $U \to R$ is denoted by C_p^∞. For any $f \in C_p^\infty$, the function $f \circ \phi^{-1} : R^n \to R$ is a function of the coordinates. The **partial derivatives** of f with respect to x^j are defined by

$$\frac{\partial f}{\partial x^j} = \frac{\partial(f \circ \phi^{-1})}{\partial x^j} \circ \phi.$$

Obviously we have the usual laws of partial differentiation: For $f, g \in C_p^\infty$, and $a, b \in R$

$$\frac{\partial}{\partial x^j}(af + bg) = a\frac{\partial f}{\partial x^j} + b\frac{\partial g}{\partial x^j},$$

$$\frac{\partial}{\partial x^j}(fg) = f\frac{\partial g}{\partial x^j} + g\frac{\partial f}{\partial x^j}.$$

The **tangent space** at a point p of a differentiable manifold M is denoted by $T_p(M)$, which is the set of all maps $X_p : C_p^\infty \to R$ that satisfy

$$\begin{aligned}
X_p(af + bg) &= aX_p(f) + bX_p(g), \\
X_p(fg) &= f(p)X_p(g) + g(p)X_p(f),
\end{aligned}$$

for all $f, g \in C_p^\infty$, and all $a, b \in R$, with vector space operations in $T_p(M)$ defined by

$$\begin{aligned}
(X_p + Y_p)f &= X_p(f) + Y_p(f), \quad X_p, Y_p \in T_p(M), \\
(aX_p)f &= a(X_p f), \quad X_p \in T_p(M), a \in R.
\end{aligned}$$

Any element $X_p \in T_p(M)$ is called a **tangent vector** of M at p.

Remark 9.3: The tangent vector may be introduced in a different way. Let M be an n dimensional manifold. A **curve** at $p \in M$ is defined as a C^1 map $c : I \to M$ from an interval $I \subset R$ into M with $0 \in I$ and $c(0) = p$. In a coordinate system, the curve is represented by $(x^1(t), x^2(t), \cdots, x^n(t))$. For any C^1 function f defined on M the map

$$X_p : f \to \frac{d}{dt} f[c(t)] \Big|_{t=0} = \frac{d}{dt} f \circ \varphi^{-1} \circ \varphi[c(t)] = \frac{dx^i}{dt} \Big|_{t=0} \frac{\partial f}{\partial x^i}$$

is called the **tangent vector** of curve c at p. Therefore,

$$X_p = \frac{dx^i}{dt} \Big|_{t=0} \frac{\partial}{\partial x^i}. \tag{9.72}$$

Here dx^i/dt are called the **components** of the tangent vector. Two curves c_1 and c_2 at $p \in E$ are said to be **tangent** at $c_1(0) = c_2(0) = p$ if they have the same components.   □

The dual space of $T_p(M)$ is denoted by $T_p^*(M)$, called the **cotangent space**. The element in $T_p^*(M)$ is called the **covector** or **1-form**. For any $f \in C_p^\infty$, we define a unique element $df \in T_p^*(M)$ by

$$df(X_p) = X_p(f), \ X_p \in T_p(M).$$

The **tensor space** at $p \in M$ is defined by

$$T_{p\ s}^r(M) = L^{r+s}[T_p^*(M), \cdots, T_p^*(M), T_p(M), \cdots, T_p(M); R]$$

with r copies of $T_p^*(M)$ and s copies of $T_p(M)$. An element $t \in T_{p\ s}^r(M)$ is called a **tensor of contravariant order r and covariant order** s.

All the properties of tensors defined on linear space discussed in Section 9.1.2 can be carried forward to the tensors defined on a chart of manifold. Similarly we can define the **tangent bundle** TM, the **cotangent bundle** $T^*(M)$ and the **tensor bundle** $T^r{}_s(M)$ restricted to a subset $U \subset M$:

$$T(M)|_U = \bigcup_{p \in U} T_p(M),$$

$$T^*(M)\big|_U = \bigcup_{p \in U} T_p^*(M),$$

$$T^r{}_s(M)\big|_U = \bigcup_{p \in U} T_{p\ s}^r(M).$$

If for any point $p \in U$, a unique element is prescribed in each $T_{p\ s}^r(M)$ we obtain a tensor field on U. More precisely a tensor field on U is a map from U to $T^r{}_s(M)|_U$. A vector field on U is defined as a map from U to $T(M)|_U$, and it is a $(1,0)$-tensor field. A scalar field is a $(0, 0)$-tensor field. Relative to a coordinate system x^j the components of tensor fields are defined in the same way as (9.18). In this case, the components are functions of the coordinates x^j of p. The tensor field is said to be smooth if these coefficients are smooth.

9.6.4   Riemannian Manifold

A Riemannian manifold is a differential manifold in which each tangent space is equipped with an inner product which varies smoothly from point to point. A smooth inner product or metric on a manifold M is a bilinear form $\langle \cdot, \cdot \rangle$ that associates a pair of smooth vector fields X and Y to a scalar field $\langle X, Y \rangle$, satisfying the following properties:

1. Symmetry: $\langle X, Y \rangle = \langle Y, X \rangle$.
2. Bilinearity:

$$\langle aX, bY \rangle = ab \langle X, Y \rangle, a, b \in R,$$
$$\langle X, Y + Z \rangle = \langle X, Y \rangle + \langle X, Z \rangle,$$
$$\langle X + Y, Z \rangle = \langle X, Z \rangle + \langle Y, Z \rangle.$$

3. Non-degeneracy: If $\langle X, Y \rangle = 0$ for all Y, then $X = 0$.

A manifold endowed with a smooth metric tensor field $\langle \cdot, \cdot \rangle$ is called a **Riemannian manifold**. In terms of a local coordinate system, the metric tensor field is given by

$$\langle X, Y \rangle = X^i Y^j \left\langle \frac{\partial}{\partial x^i}, \frac{\partial}{\partial x^j} \right\rangle = g_{ij} X^i Y^j,$$

where $g_{ij} = \langle \partial/\partial x^i, \partial/\partial x^j \rangle$ are the coefficients of the metric tensor field. If X is a vector field on M, we define the square norm of X by

$$\|X\|^2 = \langle X, X \rangle = g_{ij} X^i X^j.$$

Therefore the square norm of a vector field is an invariant. Notice that we have the following three situations:

1. X is timelike if $\|X\|^2 < 0$.
2. X is spacelike if $\|X\|^2 > 0$.
3. X is lightlike if $\|X\|^2 = 0$.

In general, the inequality $\|X + Y\| \leq \|X\| + \|Y\|$ does not hold. A Riemannian 4-manifold M is called **locally Minkowskian** if its metric in a local coordinate system is given by (9.19).

Let $c : I = (a, b) \to M$ be a curve, given by $x^i(t), t \in I, i = 1, 2, \cdots, n$ in a coordinate system. Let $X = \frac{dx^i}{dt} \frac{\partial}{\partial x^i}$ be the tangent vector. The curve is said to be **non-null** if $\|X\|^2 \neq 0$. The curve is said to be spacelike if $\|X\|^2 > 0$ and timelike if $\|X\|^2 < 0$. If the curve c is a non-null in M, its **length** is defined by

$$L(a, b) = \int_a^b \sqrt{\pm g_{ij} \frac{dx^i}{dt} \frac{dx^j}{dt}} dt, \tag{9.73}$$

where the sign ± 1 is chosen as $+1$ if the curve is spacelike and -1 if the curve is timelike. Equivalently the differential of arc-length is defined by

$$ds = \sqrt{\pm g_{ij} dx^i dx^j}. \tag{9.74}$$

Remark 9.4: A reference frame consists of a collection of material objects (clocks and rulers) with respect to which the observer relates his measurements (Bladel, 1984). The connection between coordinates and measurements is based on the form of the differential length given by (9.74) associated with the coordinates. In general, it is impossible to introduce on the surface a global Cartesian coordinates for which the differential arc-length is $ds = \sqrt{dx^2 + dy^2 + dz^2}$. The geometry on the surface is non-Euclidean but a Riemannian geometry. To determine the geometry on the surface, we use an intrinsic way without referring to the embedding of the surface to the three-dimensional Euclidean space. For an arbitrary coordinate system x^i on the surface, the distance between two neighboring points x^i and $x^i + dx^i$ can be measured by means of a measuring ruler. Then ds and dx^i are known numbers and the metric tensor g_{ij} can be determined by (9.74). In this way, the geometry on the surface becomes an empirical science subjected to the limitations arising from the limited measuring accuracy (Møller, 1952). □

9.6.5 Accelerated Reference Frames

An accelerated reference frame is usually relative to the inertial reference frame. Let us start with an inertial reference frame $\bar{x}^\alpha (\alpha = 1, 2, 3, 4)$ whose metric tensor is given by (9.19). The differential interval is thus given by

$$ds^2 = \eta_{\alpha\beta} d\bar{x}^\alpha d\bar{x}^\beta = (d\bar{x}^1)^2 + (d\bar{x}^2)^2 + (d\bar{x}^3)^2 - (d\bar{x}^4)^2.$$

We may introduce the general curvilinear coordinates $x^\alpha (\alpha = 1, 2, 3, 4)$ (the first three coordinates are used to represent spatial position and the last one is for time defined by $t = x^4/c$ at each spatial point) by means of the transformation

$$\bar{x}^\alpha = \bar{x}^\alpha (x^\beta), \quad \alpha, \beta = 1, 2, 3, 4, \tag{9.75}$$

then

$$d\bar{x}^\alpha = \frac{\partial \bar{x}^\alpha}{\partial x^\beta} dx^\beta = a^\alpha{}_\beta dx^\beta, \tag{9.76}$$

where $a^\alpha{}_\beta = \partial \bar{x}^\alpha / \partial x^\beta$. The differential length may be expressed in terms of the curvilinear coordinates

$$ds^2 = g_{\alpha\beta} dx^\alpha dx^\beta,$$

where

$$g_{\alpha\beta} = \sum_{\gamma=1}^{3} \frac{\partial \bar{x}^\gamma}{\partial x^\alpha} \frac{\partial \bar{x}^\gamma}{\partial x^\beta} - \frac{\partial \bar{x}^4}{\partial x^\alpha} \frac{\partial \bar{x}^4}{\partial x^\beta} = \sum_{\gamma=1}^{3} a^\gamma{}_\alpha a^\gamma{}_\beta - a^4{}_\alpha a^4{}_\beta \tag{9.77}$$

is the metric tensor in the new coordinate system x^α, which is symmetric $g_{\alpha\beta} = g_{\beta\alpha}$. Therefore, ds^2 is an invariant. In the general situation, different points in the coordinate system x^α may have largely varying velocities relative to the Minkowski coordinate system $\bar{x}^\alpha$. Therefore, the motion of different points in the coordinate system x^α relative to the Minkowski coordinate system $\bar{x}^\alpha$ is very similar to the motion of fluid. We will confine ourselves to such a picture, that is, each point in the coordinate system x^α flows like a real fluid with respect to the Minkowski coordinate system $\bar{x}^\alpha$. This means that the velocity of each point in x^α relative to $\bar{x}^\alpha$ must be smaller than the light speed c.

For a fixed point p in x^α, we have $dx^\alpha = 0 (\alpha = 1, 2, 3)$. It follows from (9.76) that the velocity components v^α of the point p relative to the inertial reference system $\bar{x}^\alpha$ are

$$\frac{v^\alpha}{c} = \frac{d\bar{x}^\alpha}{d\bar{x}^4} = \frac{a^\alpha{}_4}{a^4{}_4}. \tag{9.78}$$

The requirement that the velocity of the point must be smaller than light speed, that is, $\sum_{\alpha=1}^{3} (v^\alpha)^2 < c^2$ puts a restriction on the transformation (9.75)

$$g_{44} = \sum_{\alpha=1}^{3} a^\alpha{}_4 - \left(a^4{}_4\right)^2 < 0. \tag{9.79}$$

In order that the reference frame x^α defined by (9.75) is physically realizable, the admissive space–time transformation (9.75) must satisfy this condition.

If the physical process is required to be causal, further restrictions must be imposed on the transformation (9.75). Let us consider the following two events in the inertial reference frame $\bar{x}^\alpha$. The first event is to transmit a signal from a point $(\bar{x}^1, \bar{x}^2, \bar{x}^3)$ at the time $\bar{t}$, and the second event is that the signal arrives at the point $(\bar{x}^1 + d\bar{x}^1, \bar{x}^2 + d\bar{x}^2, \bar{x}^3 + d\bar{x}^3)$ at time $\bar{t} + d\bar{t}$. Since the signal speed must be smaller than the light speed, the differential interval of the two adjacent events $ds^2 = \sum_{\alpha=1}^{3} (d\bar{x}^\alpha)^2 - (cd\bar{t})^2$ must be smaller than or equal to zero. In the reference frame x^α, this implies $ds^2 = g_{\alpha\beta} dx^\alpha dx^\beta \leq 0$. For any two adjacent events, which are simultaneous in the reference frame x^α (that is, $dx^4 = 0$) and thus cannot be connected by a signal, we must have

$$\sum_{\alpha,\beta=1}^{3} g_{\alpha\beta} dx^{\alpha} dx^{\beta} > 0.$$

This must hold for arbitrary $dx^{\alpha} (\alpha = 1, 2, 3)$. Therefore, we have

$$g_{\alpha\alpha} > 0, \begin{vmatrix} g_{\alpha\alpha} & g_{\alpha\beta} \\ g_{\beta\alpha} & g_{\beta\beta} \end{vmatrix} > 0, \begin{vmatrix} g_{11} & g_{12} & g_{13} \\ g_{21} & g_{22} & g_{23} \\ g_{31} & g_{32} & g_{33} \end{vmatrix} > 0, (\alpha, \beta = 1, 2, 3). \qquad (9.80)$$

It follows from (9.79) and (9.80) that

$$g = |g_{\alpha\beta}| = \begin{vmatrix} g_{11} & g_{12} & g_{13} & g_{14} \\ g_{21} & g_{22} & g_{23} & g_{24} \\ g_{31} & g_{32} & g_{33} & g_{34} \\ g_{41} & g_{42} & g_{43} & g_{44} \end{vmatrix} < 0. \qquad (9.81)$$

Remark 9.5: The metric tensor (9.77) is based on a starting inertial reference frame $\bar{x}^{\alpha}$. Actually we can start with any other inertial reference frames, which will give the same metric tensor. To prove this, let us consider a different inertial frame $\bar{x}'^{\alpha}$. The metric tensor based on this new inertial frame is

$$g'_{\alpha\beta} = \sum_{\gamma=1}^{3} \frac{\partial \bar{x}'^{\gamma}}{\partial x^{\alpha}} \frac{\partial \bar{x}'^{\gamma}}{\partial x^{\beta}} - \frac{\partial \bar{x}'^{4}}{\partial x^{\alpha}} \frac{\partial \bar{x}'^{4}}{\partial x^{\beta}} = \sum_{\lambda=1}^{4} \left[\left(\sum_{\gamma=1}^{3} \frac{\partial \bar{x}'^{\gamma}}{\partial \bar{x}^{\lambda}} \frac{\partial \bar{x}'^{\gamma}}{\partial \bar{x}^{\lambda}} - \frac{\partial \bar{x}'^{4}}{\partial \bar{x}^{\lambda}} \frac{\partial \bar{x}'^{4}}{\partial \bar{x}^{\lambda}} \right) \frac{\partial \bar{x}^{\lambda}}{\partial x^{\alpha}} \frac{\partial \bar{x}^{\lambda}}{\partial x^{\beta}} \right]$$

$$= \sum_{\lambda=1}^{3} \frac{\partial \bar{x}^{\lambda}}{\partial x^{\alpha}} \frac{\partial \bar{x}^{\lambda}}{\partial x^{\beta}} - \frac{\partial \bar{x}^{4}}{\partial x^{\alpha}} \frac{\partial \bar{x}^{4}}{\partial x^{\beta}} = g_{\alpha\beta}.$$

Here we have used (9.25). This relation indicates that the metric tensor in a general coordinate system is independent of the inertial reference system that is used to define the metric tensor. □

Remark 9.6: It is known that we have to introduce fictitious forces, such as the centrifugal force and the Coriolis force, to describe a mechanical system in an accelerated reference frame. Such forces have no connection with the physical properties of the mechanical system itself, and in fact they depend on the accelerated system introduced relative to the inertial system. Therefore, Newton introduced the concept of absolute space (an ideal reference frame) relative to which the physical laws have the simplest and natural form. However, the theory of special relativity abandoned the concept of absolute space because it is impossible to determine by experiments which inertial system should be regarded as the absolute system. Einstein gave a new interpretation of these fictitious forces in an accelerated reference frame. According to this interpretation, these fictitious forces are treated as a kind of gravitational force. □

Example 9.3 (Rotating system of reference): We can introduce a rotating reference frame $x^\alpha = (r, \varphi, z, ct)$ by

$$
\begin{aligned}
\bar{x}^1 &= r\cos(\Omega t + \varphi), \\
\bar{x}^2 &= r\sin(\Omega t + \varphi), \\
\bar{x}^3 &= z, \\
\bar{x}^4 &= c\bar{t} = ct.
\end{aligned}
$$

Then

$$
g_{\alpha\beta} =
\begin{bmatrix}
1 & 0 & 0 & 0 \\
0 & r^2 & 0 & \dfrac{\Omega r^2}{c} \\
0 & 0 & 1 & 0 \\
0 & \dfrac{\Omega r^2}{c} & 0 & -\left(1 - \dfrac{\Omega^2 r^2}{c^2}\right)
\end{bmatrix}
\tag{9.82}
$$

□

9.6.6 Time and Length in Accelerated Reference Frame

The time and length measurements must be examined carefully in an accelerated system. Following Møller (Møller, 1952), we consider two points p and q whose space coordinates are $x^\alpha (\alpha = 1, 2, 3)$ and $x^\alpha + dx^\alpha$ $(\alpha = 1, 2, 3)$ respectively. The spatial distance dl between p and q at the time $t = x^4/c$ can be measured by a standard ruler, which is at rest relative to the point p. If dx^α is very small the point q will also be at rest relative to the ruler. In order to express dl in terms of the metric tensor $g_{\alpha\beta}$, we may introduce an inertial reference frame $\bar{x}^\alpha$ relative to which the point p (approximately point q) is at rest momentarily, which is called an **instantaneous co-moving inertial reference frame**. The transformation from x^α to $\bar{x}^\alpha$ is given by (9.75). Since the point p is at rest momentarily relative to the inertial reference frame $\bar{x}^\alpha$, from (9.78) we obtain

$$
a^\alpha{}_4 = 0, \alpha = 1, 2, 3
\tag{9.83}
$$

at the point p. The differences $d\bar{x}^\alpha$ $(\alpha = 1, 2, 3)$ corresponding to the two points p and q in the reference x^α are

$$
d\bar{x}^\alpha = \sum_{\beta=1}^{3} a^\alpha{}_\beta dx^\beta, \alpha = 1, 2, 3
$$

due to (9.83). Therefore

$$
\begin{aligned}
ds^2 &= \eta_{\alpha\beta} d\bar{x}^\alpha d\bar{x}^\beta = (d\bar{x}^1)^2 + (d\bar{x}^2)^2 + (d\bar{x}^3)^2 - (d\bar{x}^4)^2 \\
&= dl^2 - c^2 d\bar{t}^2 = g_{\alpha\beta} dx^\alpha dx^\beta.
\end{aligned}
\tag{9.84}
$$

Note that $d\bar{l}^2 = (d\bar{x}^1)^2 + (d\bar{x}^2)^2 + (d\bar{x}^3)^2$ only depends on dx^1, dx^2 and dx^3. Thus we may write $g_{\alpha\beta}dx^\alpha dx^\beta$ as (Bladel, 1984)

$$ds^2 = g_{\alpha\beta}dx^\alpha dx^\beta = \sum_{\alpha,\beta=1}^{3} g_{\alpha\beta}dx^\alpha dx^\beta + g_{44}(dx^4)^2 + 2dx^4 \sum_{\beta=1}^{3} g_{4\beta}dx^\beta$$

$$= \sum_{\alpha,\beta=1}^{3} (g_{\alpha\beta} - g_{4\alpha}g_{4\beta}/g_{44})dx^\alpha dx^\beta \tag{9.85}$$

$$- \left(\sqrt{-g_{44}}dx^4 - \sum_{\beta=1}^{3} g_{4\beta}dx^\beta \bigg/ \sqrt{-g_{44}} \right)^2 .$$

Comparing (9.84) and (9.85) we obtain

$$d\bar{l}^2 = \sum_{\alpha,\beta=1}^{3} \gamma_{\alpha\beta}dx^\alpha dx^\beta ,$$

$$c d\bar{t} = \sqrt{-g_{44}}dx^4 - \sum_{\beta=1}^{3} \gamma_\beta dx^\beta = \sqrt{-g_{44}}dx^4 \left(1 - \sum_{\beta=1}^{3} \gamma_\beta v^\beta \bigg/ c\sqrt{-g_{44}} \right), \tag{9.86}$$

where $v^\beta = dx^\beta/dt$ is the velocity and

$$\gamma_{\alpha\beta} = g_{\alpha\beta} + \gamma_\alpha \gamma_\beta, \gamma_\alpha = g_{4\alpha}\bigg/\sqrt{-g_{44}}. \tag{9.87}$$

It follows from (9.85) and (9.86) that

$$ds^2 = -c^2 d\tau^2 = \sum_{\alpha,\beta=1}^{3} \gamma_{\alpha\beta}dx^\alpha dx^\beta + g_{44}(dx^4)^2 \left(1 - \sum_{\beta=1}^{3} \gamma_\beta v^\beta \bigg/ c\sqrt{-g_{44}} \right)^2 . \tag{9.88}$$

Remark 9.7: We may assume that $d\bar{l} = dl$, where dl is the distance between the two neighboring points p and q measured in the coordinate system x^α (Møller, 1952). □

Example 9.4: Consider the rotating system of reference again, whose metric tensor is given by (9.82). From (9.87), we obtain

$$\gamma_{11} = 1, \gamma_{22} = \frac{r^2}{1 - \Omega^2 r^2/c^2}, \gamma_{33} = 1, \gamma_{\alpha\beta} = 0, \alpha \neq \beta.$$

Thus the distance dl between two neighboring points (r, φ, z) and $(r + dr, \varphi + d\varphi, z)$ measured in the rotating system of reference is

$$dl^2 = \sum_{\alpha, \beta = 1}^{3} \gamma_{\alpha\beta} dx^\alpha dx^\beta = dr^2 + \frac{r^2 d\varphi^2}{1 - \Omega^2 r^2 / c^2}. \qquad (9.89)$$

This relation shows that the geometrical theorems obtained from an observer in the reference frame, which is at rest relative to a rotating system, will be different from the theorems of Euclidean geometry. Let us consider an circle given by $r = $ constant. According to (9.89), the periphery of the this circle measured by the observer in the rotating system will be

$$\int_{0}^{2\pi} \frac{r d\varphi}{\sqrt{1 - \Omega^2 r^2 / c^2}} = \frac{2\pi r}{\sqrt{1 - \Omega^2 r^2 / c^2}}.$$

Therefore the periphery of the circle is larger than $2\pi r$. □

Now consider a clock fixed at a given reference point p of the reference frame x^α. From the second expression of (9.86), the time interval (that is, proper time interval obtained by an observer in the instantaneous co-moving inertial reference frame at p, which is at rest relative to the clock) of the trajectory of the clock is

$$c d\tau = c d\bar{t} = \sqrt{-g_{44}} dx^4 \qquad (9.90)$$

since $dx^\alpha = 0$ $(\alpha = 1, 2, 3)$ for a clock at rest. Note that $t = x^4/c$ denotes the time shown by the clock at p.

Example 9.5: For a rotating reference frame given by (9.82), Equation (9.90) becomes

$$d\tau = \frac{1}{c}\sqrt{-g_{44}} dx^4 = \sqrt{1 - \Omega^2 r^2/c^2} dt, \qquad (9.91)$$

where dt is the coordinate time interval, which is equal to the proper time interval only for points on the rotational axis. □

9.6.7 Covariant Derivative and Connection

In order to be able to relate tensors at distinct points, it is necessary to use a process of differentiation. However, the ordinary differential operation on a tensor does not yield a tensor. Therefore we need to use a different concept of differentiation, a **covariant derivative** to guarantee that the new tensor after the differentiation is still a tensor. The basic idea is to introduce a 'compensating field', called a **connection**, in the ordinary differential operation. This connection will be denoted by Γ. The covariant derivative of the scalar field is defined to be its ordinary derivative

$$D_\mu \phi = \partial_\mu \phi,$$

which is covariant. The covariant derivative of a contravariant vector field X^μ is defined by

$$D_\mu X^\nu = \partial_\mu X^\nu + \Gamma^\nu{}_{\lambda\mu} X^\lambda.$$

The covariant derivative of a covariant vector field X_μ is defined by

$$D_\mu X_\nu = \partial_\mu X_\nu - \Gamma^\lambda{}_{\nu\mu} X_\lambda.$$

In general, the covariant derivative of arbitrary tensor is defined by

$$D_\mu T^{\nu_1\cdots\nu_r}{}_{\sigma_1\cdots\sigma_s} = \partial_\mu T^{\nu_1\cdots\nu_r}{}_{\sigma_1\cdots\sigma_s} + \sum_{m=1}^{r} \Gamma^{\nu_m}{}_{\nu\mu} T^{\nu_1\cdots\nu_{m-1}\nu\nu_{m+1}\nu_r}{}_{\sigma_1\cdots\sigma_s}$$

$$- \sum_{n=1}^{s} \Gamma^\sigma{}_{\sigma_n\mu} T^{\nu_1\cdots\nu_r}{}_{\sigma_1\cdots\sigma_{n-1}\sigma\sigma_{n+1}\cdots\sigma_s}.$$

Under a coordinate transform $x^\mu \to \tilde{x}^\mu$, the connection Γ must transform according to

$$\tilde{\Gamma}^\mu{}_{\nu\lambda} = \frac{\partial\tilde{x}^\mu}{\partial x^\alpha}\frac{\partial x^\beta}{\partial\tilde{x}^\nu}\frac{\partial x^\chi}{\partial\tilde{x}^\lambda}\Gamma^\alpha{}_{\beta\chi} - \frac{\partial^2\tilde{x}^\mu}{\partial x^\beta\partial x^\chi}\frac{\partial x^\beta}{\partial\tilde{x}^\nu}\frac{\partial x^\chi}{\partial\tilde{x}^\lambda} \tag{9.92}$$

to make the covariant derivative of a (r, s)-tensor $T^{\nu_1\cdots\nu_r}{}_{\sigma_1\cdots\sigma_s}$ be a $(r, s+1)$-tensor. The transformation law (9.92) may be expressed in a different form as

$$\tilde{\Gamma}^\mu{}_{\nu\lambda} = \frac{\partial\tilde{x}^\mu}{\partial x^\alpha}\frac{\partial x^\beta}{\partial\tilde{x}^\nu}\frac{\partial x^\chi}{\partial\tilde{x}^\lambda}\Gamma^\alpha{}_{\beta\chi} + \frac{\partial\tilde{x}^\mu}{\partial x^\alpha}\frac{\partial^2 x^\alpha}{\partial\tilde{x}^\nu\partial\tilde{x}^\lambda}. \tag{9.93}$$

This can be proved by the chain rule.

If the covariant derivative of a tensor field is zero in a subset of a manifold, the tensor field is said to be **parallel-transported** in the subset. If a metric tensor $g_{\rho\sigma}$ is parallel-transported, that is, $D_\mu g_{\rho\sigma} = 0$, then

$$\partial_\mu g_{\rho\sigma} = \Gamma^\lambda{}_{\rho\mu} g_{\lambda\sigma} + \Gamma^\lambda{}_{\sigma\mu} g_{\rho\lambda} = \Gamma_{\sigma\rho\mu} + \Gamma_{\rho\sigma\mu} = 2\Gamma_{(\rho\sigma)\mu}. \tag{9.94}$$

This is called the **metricity condition**. Here $\Gamma_{\sigma\rho\mu} = \Gamma^\lambda{}_{\rho\mu} g_{\lambda\sigma}$ and $\Gamma_{(\rho\sigma)\mu}$ denote the symmetrized part $\Gamma_{(\rho\sigma)\mu} = (\Gamma_{\rho\sigma\mu} + \Gamma_{\sigma\rho\mu})/2$. The **antisymmetrized part** is defined by $\Gamma_{[\rho\sigma]\mu} = (\Gamma_{\rho\sigma\mu} - \Gamma_{\sigma\rho\mu})/2$.

The connection on a manifold is not unique. Given a metric $g_{\rho\sigma}$ in a coordinate system x^μ, there exist infinite connections that satisfy the metricity condition. One of these connections, called the **Levi-Civita connection**, named after the Italian mathematician Tullio Levi-Civita (1873–1941), is given by

$$\underline{\Gamma}^\lambda{}_{\rho\sigma} = \frac{1}{2}g^{\lambda\mu}(\partial_\rho g_{\sigma\mu} + \partial_\sigma g_{\rho\mu} - \partial_\mu g_{\rho\sigma}), \tag{9.95}$$

which satisfies the metricity condition

$$D_\mu g_{\rho\sigma} = \partial_\mu g_{\rho\sigma} - \underline{\Gamma}^\lambda{}_{\rho\mu} g_{\lambda\sigma} - \underline{\Gamma}^\lambda{}_{\sigma\mu} g_{\rho\lambda} = 0. \tag{9.96}$$

The components of the Levi-Civita connection are called **Christoffel symbols**, named after the German mathematician Elwin Bruno Christoffel (1829–1900). The Levi-Civita connection is symmetric in the last two indices. From now on, all quantities derived from $\underline{\Gamma}^\lambda{}_{\rho\sigma}$ will be underlined. Multiplying (9.96) by $g^{\rho\sigma}$ we obtain

$$\underline{\Gamma}^\lambda{}_{\lambda\mu} = \frac{1}{2} g^{\rho\sigma} \partial_\mu g_{\rho\sigma}. \tag{9.97}$$

Let the determinant of the matrix $[g_{\rho\sigma}]$ be denoted by $g = |g_{\rho\sigma}|$. Then

$$\frac{\partial g}{\partial x^\lambda} = g g^{\rho\sigma} \frac{\partial g_{\rho\sigma}}{\partial x^\lambda}.$$

Making use of this relation, we may write (9.97) as

$$\underline{\Gamma}^\lambda{}_{\lambda\mu} = \frac{1}{2g} \frac{\partial g}{\partial x^\mu} = \frac{\partial}{\partial x^\mu} \ln \sqrt{-g}. \tag{9.98}$$

The **Riemann curvature tensor** of a Levi-Civita connection is defined by

$$\underline{R}^\nu{}_{\rho\mu\sigma} = \partial_\mu \underline{\Gamma}^\nu{}_{\rho\sigma} - \partial_\sigma \underline{\Gamma}^\nu{}_{\rho\mu} + \underline{\Gamma}^\nu{}_{\lambda\mu} \underline{\Gamma}^\lambda{}_{\rho\sigma} - \underline{\Gamma}^\nu{}_{\lambda\sigma} \underline{\Gamma}^\lambda{}_{\rho\mu},$$

which has the symmetric properties:

$$\underline{R}_{\nu\rho\mu\sigma} = -\underline{R}_{\nu\rho\sigma\mu} = \underline{R}_{\rho\nu\mu\sigma} = \underline{R}_{\rho\nu\sigma\mu},$$
$$\underline{R}_{\nu\rho\mu\sigma} = \underline{R}_{\mu\sigma\nu\rho}.$$

Here $\underline{R}_{\nu\rho\mu\sigma} = g_{\mu\nu} \underline{R}^\mu{}_{\rho\mu\sigma}$. The **Ricci tensor**, named after the Italian mathematician Gregorio Ricci-Curbastro (1853–1925), is defined by $\underline{R}_{\rho\sigma} = \underline{R}^\mu{}_{\rho\mu\sigma}$ and is also symmetric

$$\underline{R}_{\rho\sigma} = \underline{R}_{\sigma\rho}.$$

The **scalar curvature** is defined by $\underline{R} = g^{\mu\nu} \underline{R}_{\mu\nu}$.

Let $\tilde{x}^\mu$ be another coordinate system. The transformation law of the metric tensor is

$$g_{\rho\sigma} = \tilde{g}_{\mu\nu} \frac{\partial \tilde{x}^\mu}{\partial x^\rho} \cdot \frac{\partial \tilde{x}^\nu}{\partial x^\sigma}.$$

By a lengthy calculation, it can be shown that the Levi-Civita connection changes according to

$$\underline{\Gamma}^\lambda{}_{\rho\sigma} = \frac{\partial x^\lambda}{\partial \tilde{x}^\mu} \left(\frac{\partial \tilde{x}^\alpha}{\partial x^\rho} \frac{\partial \tilde{x}^\beta}{\partial x^\sigma} \underline{\tilde{\Gamma}}^\mu{}_{\alpha\beta} + \frac{\partial^2 \tilde{x}^\mu}{\partial x^\rho \partial x^\sigma} \right). \tag{9.99}$$

9.6.8   Geodesics and Equation of Motion in Gravitational Field

Let us consider the length of a curve defined by (9.73). A **geodesic** is a curve which extremizes the length functional (9.73). Taking the variation of the functional, a geodesic $x^\mu(t)$ must satisfy the famous **geodesic equation**

$$\frac{d^2 x^\mu}{dt^2} + \Gamma^\mu_{\ \rho\sigma} \frac{dx^\rho}{dt} \frac{dx^\sigma}{dt} = 0, \tag{9.100}$$

if the parameter t is an affine parameter, that is, if it is related to the proper time by $t = a\tau + b$. If the proper time is used as the parameter of the curve the geodesic equation may be written as

$$\frac{du^\mu}{d\tau} + \Gamma^\mu_{\ \rho\sigma} u^\rho u^\sigma = 0, \tag{9.101}$$

where u^μ is the four velocity. The geodesics are locally straight in an inertial coordinate system.

Equation (9.101) is assumed to be the equation of motion of a test particle, not acted on by any forces except an arbitrary gravitational field, in an arbitrary coordinate system. It can be derived by the equivalence principle. By Einstein equivalence principle, there exists a free falling coordinate system $\bar{x}^\alpha$ where the equation of motion of a particle under the influence of purely gravitational force is

$$\frac{d^2 \bar{x}^\alpha}{d\tau^2} = 0, \tag{9.102}$$

where $d\tau$ is the proper time

$$-c^2 d\tau^2 = \bar{\eta}_{\alpha\beta} d\bar{x}^\alpha d\bar{x}^\beta. \tag{9.103}$$

Here $\bar{\eta}_{\alpha\beta}$ is used to represent the Minkowski metric. It can be shown that the transformation of (9.102) and (9.103) gives (9.100) with

$$g_{\mu\nu} = \frac{\partial \bar{x}^\alpha}{\partial x^\mu} \frac{\partial \bar{x}^\beta}{\partial x^\nu} \bar{\eta}_{\alpha\beta}.$$

In fact,

$$\begin{aligned}
\frac{d^2 \bar{x}^\alpha}{d\tau^2} &= \frac{d}{d\tau}\left(\frac{\partial \bar{x}^\alpha}{\partial x^\rho} \cdot \frac{dx^\rho}{d\tau} \right) = \frac{d}{d\tau} \frac{\partial \bar{x}^\alpha}{\partial x^\rho} \cdot \frac{dx^\rho}{d\tau} + \frac{\partial \bar{x}^\alpha}{\partial x^\mu} \cdot \frac{d^2 x^\mu}{d\tau^2} \\
&= \left(\frac{\partial^2 \bar{x}^\alpha}{\partial x^\rho \partial x^\sigma} \cdot \frac{\partial x^\mu}{\partial \bar{x}^\alpha} \cdot \frac{dx^\sigma}{d\tau} \cdot \frac{dx^\rho}{d\tau} + \frac{d^2 x^\mu}{d\tau^2} \right) \cdot \frac{\partial \bar{x}^\alpha}{\partial x^\mu} = 0.
\end{aligned}$$

Making use of (9.99) we obtain (9.100).

For a particle which is momentarily at rest in the gravitational field (that is, $u^\alpha = 0, \alpha = 1, 2, 3$), Equation (9.101) can be written as

$$\sum_{\mu=1}^{3} g_{\nu\mu} \frac{du^\mu}{d\tau} + \frac{d}{d\tau}(g_{\nu4}u^4) = \frac{1}{2}\frac{\partial g_{44}}{\partial x^\nu}u^4 u^4, \quad \nu = 1, 2, 3. \tag{9.104}$$

From (9.88), we obtain

$$-c^2\left(\frac{d\tau}{dx^4}\right)^2 = \sum_{\alpha,\beta=1}^{3} \gamma_{\alpha\beta}\frac{dx^\alpha}{dx^4}\frac{dx^\beta}{dx^4} + g_{44}\left(1 - \sum_{\beta=1}^{3}\frac{\gamma_\beta}{\sqrt{-g_{44}}}\frac{v^\beta}{c}\right)^2,$$

or

$$u^4 = c\left[-g_{44}\left(1 - \sum_{\beta=1}^{3}\frac{\gamma_\beta}{\sqrt{-g_{44}}}\frac{v^\beta}{c}\right)^2 - \frac{v^2}{c^2}\right]^{-1/2}, \tag{9.105}$$

where $v = \sqrt{\sum_{\alpha,\beta=1}^{3}\gamma_{\alpha\beta}\frac{dx^\alpha}{dt}\frac{dx^\beta}{dt}}$ is the magnitude of the particle velocity. Thus we have

$$g_{\nu4}u^4 = c\gamma_\nu\left[\left(1 - \sum_{\beta=1}^{3}\frac{\gamma_\beta}{\sqrt{-g_{44}}}\frac{v^\beta}{c}\right)^2 + \frac{v^2}{c^2 g_{44}}\right]^{-1/2},$$

and after some calculation we get

$$\frac{d}{d\tau}(g_{\nu4}u^4) = \frac{1}{\sqrt{-g_{44}}}\left(c^2\frac{d\gamma_\nu}{dx^4} + \sum_{\beta=1}^{3}\frac{\gamma_\nu\gamma_\beta}{\sqrt{-g_{44}}}\frac{dv^\beta}{dt}\right) \tag{9.106}$$

for a particle which is momentarily at rest. Substituting (9.106) and (9.105) into (9.104) we obtain

$$\sum_{\mu=1}^{3}\gamma_{\nu\mu}\frac{d^2x^\mu}{dt^2} = -\frac{\partial}{\partial x^\nu}\left(-\frac{c^2 g_{44}}{2}\right) - \sqrt{-g_{44}}c\frac{d\gamma_\nu}{dt}, \quad \nu = 1, 2, 3.$$

Therefore the particle which is momentarily at rest gets accelerated. The covariant components of the acceleration is

$$a_\nu = \sum_{\mu=1}^{3}\gamma_{\nu\mu}\frac{d^2x^\mu}{dt^2}.$$

If we let $g_{44} = -(1 + 2\Phi/c^2)$, the acceleration can be expressed as

$$a_\nu = -\frac{\partial \Phi}{\partial x^\nu} - c\sqrt{1 + \frac{2\Phi}{c^2}}\frac{d\gamma_\nu}{dt}, \nu = 1, 2, 3. \tag{9.107}$$

The metric tensor $g_{\alpha\beta}$ defines the motion of a test particle under the gravitational field. The influence of gravity on the particle has been included in the metric tensor $g_{\alpha\beta}$. Therefore we sometimes call the metric tensor $g_{\alpha\beta}$ the gravitational potential.

Remark 9.8 (Newton mechanics as a limit case): If we make the following assumptions:

1. The particle moves slowly, which implies we may neglect $dx^\mu/d\tau$ compared to $dt/d\tau$.
2. The gravitational field is static.
3. The gravitational field is weak.

we can get the Newton equation from (9.100). Based on the above assumptions, Equation (9.100) may be approximated by

$$\frac{d^2 x^\mu}{d\tau^2} + \underline{\Gamma}^\mu_{44}\frac{dx^4}{d\tau}\frac{dx^4}{d\tau} = 0. \tag{9.108}$$

Since the field is static, all the time derivatives of $g_{\alpha\beta}$ vanish. Thus we have

$$\underline{\Gamma}^\mu_{44} = -\frac{1}{2}g^{\mu\nu}\frac{\partial g_{44}}{\partial x^\nu}.$$

A weak gravitational field means that we can introduce a local coordinate system x^α in which the metric tensor can be written as

$$g_{\alpha\beta} = \eta_{\alpha\beta} + h_{\alpha\beta}, |h_{\alpha\beta}| \ll 1,$$

where $\eta_{\alpha\beta}$ is given by (9.19); $h_{\alpha\beta}$ and their derivatives are small quantities whose squares may be ignored. Therefore

$$\underline{\Gamma}^\mu_{44} = -\frac{1}{2}\eta^{\mu\nu}\frac{\partial h_{44}}{\partial x^\nu}, \tag{9.109}$$

and (9.108) reduces to

$$\frac{d^2 x^\mu}{d\tau^2} - \frac{c^2}{2}\frac{\partial h_{44}}{\partial x^\mu}\left(\frac{dt}{d\tau}\right)^2 = 0, \mu = 1, 2, 3,$$

$$\frac{d^2 t}{d\tau^2} = 0.$$

The last equation implies $dt/d\tau = $ constant. Thus the first equation can be written as

$$\frac{d^2 x^\mu}{dt^2} = \frac{1}{2}\frac{\partial\left(c^2 h_{44}\right)}{\partial x^\mu}, \mu = 1, 2, 3.$$

If we compare this equation to (9.69) we obtain

$$h_{44} = -2\Phi/c^2 + \text{constant}.$$

Since the coordinate system must become Minkowskian at great distance, h_{44} must vanish if we define the gravitational potential Φ to vanish at infinity. Then we have

$$h_{44} = -2\Phi/c^2, \tag{9.110}$$

and

$$g_{44} = -(1 + 2\Phi/c^2). \qquad \square$$

9.6.9 Bianchi Identities

Let us consider a vector field X^α. It can be shown that

$$[\underline{D}_\gamma, \underline{D}_\beta]X^\alpha = \underline{D}_\gamma \underline{D}_\beta X^\alpha - \underline{D}_\beta \underline{D}_\gamma X^\alpha = -\underline{R}^\alpha{}_{\nu\beta\gamma} X^\nu$$

where $[\underline{D}_\gamma, \underline{D}_\beta] = \underline{D}_\gamma \underline{D}_\beta - \underline{D}_\beta \underline{D}_\gamma$ is called the commutator of operators $\underline{D}_\gamma$ and $\underline{D}_\beta$. We also have the **first Bianchi identity**, named after the Italian mathematician Luigi Bianchi (1856–1928),

$$\underline{R}^\nu{}_{\rho\mu\sigma} + \underline{R}^\nu{}_{\mu\sigma\rho} + \underline{R}^\nu{}_{\sigma\rho\mu} = 0, \tag{9.111}$$

and the **second Bianchi identity**

$$\underline{D}_\mu \underline{R}_{\nu\lambda\rho\sigma} + \underline{D}_\sigma \underline{R}_{\nu\lambda\mu\rho} + \underline{D}_\rho \underline{R}_{\nu\lambda\sigma\mu} = 0. \tag{9.112}$$

Since the metric tensor $g^{\nu\rho}$ has zero covariant derivative, it can be inserted to the middle of each term in (9.112). By contraction, we obtain

$$\underline{D}_\mu \underline{R}_{\lambda\sigma} - \underline{D}_\sigma \underline{R}_{\lambda\mu} + \underline{D}_\rho \underline{R}^\rho{}_{\lambda\sigma\mu} = 0.$$

Similarly by inserting $g^{\lambda\sigma}$ into the above identity and contracting, we obtain

$$\underline{D}_\mu \underline{R} - \underline{D}_\sigma \underline{R}^\sigma{}_\mu - \underline{D}_\rho \underline{R}^\rho{}_\mu = \underline{D}_\mu \underline{R} - 2\underline{D}_\sigma \underline{R}^\sigma{}_\mu = 0.$$

This can also be written as

$$\underline{D}_\mu \left(\underline{R}^\mu{}_\nu - \frac{1}{2}\delta^\mu{}_\nu \underline{R} \right) = 0. \tag{9.113}$$

The **Einstein tensor** is defined by

$$G_{\mu\nu} = \underline{R}_{\mu\nu} - \frac{1}{2}g_{\mu\nu}\underline{R}.$$

Raising the first index and then contracting the above equation we obtain

$$G^{\mu}{}_{\mu} = -\underline{R}. \tag{9.114}$$

9.6.10   Principle of General Covariance and Minimal Coupling

The principle of equivalence can be used to study the effects of gravitation on physical systems. We can first write down the physical equations in the locally inertial coordinate system, that is, the equations of special relativity and then perform a coordinate transformation to find the corresponding equations in an arbitrary (accelerated) coordinate system. A different approach is based on the **principle of general covariance**. It states that a physical equation holds in a general gravitational field if the following conditions are met (Weinberg, 1972)

1. The equation holds in the absence of gravitation. In other words, it agrees with the laws of special relativity when the metric tensor $g_{\alpha\beta}$ equals the Minkowski tensor $\eta_{\alpha\beta}$ and when the connection coefficients $\Gamma^{\gamma}{}_{\alpha\beta}$ vanish.
2. The equation is generally covariant; that is, it retains its form under a general change of coordinates.

The principle of general covariance follows from the principle of equivalence. The principle of general covariance requires that the equations preserve their form under a general change of coordinates. Given a physical equation expressed in Minkowski space (that is, special relativity), we may get its version in the presence of gravitation field by introducing gravitation through the substitution of $g_{\alpha\beta}$ for $\eta_{\alpha\beta}$ and of $\underline{D}_{\mu}$ for ∂_{μ} in the physical equation written in special relativity in Minkowski coordinates. This procedure is called **minimal coupling**.

9.6.11   Einstein Field Equations

The gravitational potential Φ satisfies

$$\nabla^2\Phi = 4\pi G\tilde{\rho}_{m_0}, \tag{9.115}$$

where $\tilde{\rho}_{m_0}$ is the volume density of mass measured in rest frame, In relativity, we need an covariant analogue of (9.115), a form which is independent of the choices of coordinate system. By the mass-energy relation the mass density can be generalized to energy density. Since the energy density is just one component of the stress-energy tensor, we may need to use whole of $T^{\mu\nu}$. Furthermore, the gravitational potential Φ should be related to $g^{\mu\nu}$. Therefore, we may expect that the covariant analogue of (9.115) may be written in the form

$$\diamondsuit(g^{\mu\nu}) = \chi T^{\mu\nu}, \tag{9.116}$$

where $\diamond$ is a second order differential operator to be determined that generalizes ∇^2, and χ is some constant. $\diamond(g^{\mu\nu})$ should be some linear combination of $g^{\mu\nu}$, their first derivative and second derivative of $g^{\mu\nu}$, and must be symmetric since $T^{\mu\nu}$ is symmetric. Examples of such a tensors are Ricci tensors $\underline{R}^{\mu\nu}$, $g^{\mu\nu}\underline{R}$ and $g^{\mu\nu}$. Thus the left-hand side of (9.116) may be expressed as a linear combination of these quantities

$$\underline{R}^{\mu\nu} + ag^{\mu\nu}\underline{R} + \Lambda g^{\mu\nu} = \chi T^{\mu\nu}. \tag{9.117}$$

Applying the conservation laws $\underline{D}_\nu T^{\mu\nu} = 0$ we obtain

$$\underline{D}_\nu(\underline{R}^{\mu\nu} + ag^{\mu\nu}\underline{R}) = 0,$$

where we have used $\underline{D}_\nu g^{\mu\nu} = 0$. If this is compared to (9.113) we obtain $a = -1/2$. Hence (9.117) becomes

$$\underline{R}^{\mu\nu} - \frac{1}{2}g^{\mu\nu}\underline{R} + \Lambda g^{\mu\nu} = \chi T^{\mu\nu}. \tag{9.118}$$

These are the **Einstein field equations**, which relate the curvature of space–time with the mass, energy, and momentum within it.

The constant Λ is called **cosmological constant**, which is usually set to zero. Thus we have

$$\underline{R}^{\mu\nu} - \frac{1}{2}g^{\mu\nu}\underline{R} = \chi T^{\mu\nu}. \tag{9.119}$$

By lowering the second index and contracting, we obtain

$$G^\mu{}_\mu = \chi T^\mu{}_\mu = -\underline{R}$$

from (9.114). Thus Einstein field equation can also be written as

$$\underline{R}^{\mu\nu} = \chi \left(T^{\mu\nu} - \frac{1}{2}g^{\mu\nu}T^\mu{}_\mu \right). \tag{9.120}$$

Remark 9.9: The energy-momentum tensor describes not only matter distribution but also fields of all kinds, such as the electromagnetic field but it does not include the contributions from the gravitational field. □

Equation (9.118) is a set of non-linear partial differential equations in the metric coefficients $g^{\mu\nu}$ and it is not easy to solve mainly because the technique of superposition of solutions cannot be used.

Now we determine the constant χ. When the gravitational field is very weak, Equation (9.118) can be approximated by a set of linear differential equations. A weak field means that we can introduce a local coordinate system x^α in which the metric tensor can be written as

$$g_{\alpha\beta} = \eta_{\alpha\beta} + h_{\alpha\beta},$$

where $\eta_{\alpha\beta}$ is given by (9.19); $h_{\alpha\beta}$ and their derivatives are small quantities whose squares may be ignored. The Christoffel symbols are small of the first order. Therefore, we can neglect all terms depending on the square of Christoffel symbols in Ricci tensor to get

$$\underline{R}^{\mu\nu} = g^{\mu\rho} g^{\nu\sigma} \underline{R}_{\rho\sigma} \approx g^{\mu\rho} g^{\nu\sigma} \left(\partial_\mu \underline{\Gamma}^\mu_{\ \rho\sigma} - \partial_\sigma \underline{\Gamma}^\mu_{\ \rho\mu} \right)$$
$$\approx \eta^{\mu\rho} \eta^{\nu\sigma} \left(\partial_\mu \underline{\Gamma}^\mu_{\ \rho\sigma} - \partial_\sigma \underline{\Gamma}^\mu_{\ \rho\mu} \right).$$

Making use of the following calculations

$$\partial_\mu \underline{\Gamma}^\mu_{\ \rho\sigma} \approx \frac{1}{2} \eta^{\mu\lambda} \left(\frac{\partial^2 h_{\sigma\lambda}}{\partial x^\mu \partial x^\rho} + \frac{\partial^2 h_{\rho\lambda}}{\partial x^\mu \partial x^\sigma} - \frac{\partial^2 h_{\rho\sigma}}{\partial x^\mu \partial x^\lambda} \right),$$

$$\partial_\sigma \underline{\Gamma}^\mu_{\ \rho\mu} \approx \frac{1}{2} \eta^{\mu\lambda} \left(\frac{\partial^2 h_{\mu\lambda}}{\partial x^\sigma \partial x^\rho} + \frac{\partial^2 h_{\rho\lambda}}{\partial x^\sigma \partial x^\mu} - \frac{\partial^2 h_{\rho\mu}}{\partial x^\sigma \partial x^\lambda} \right),$$

we obtain

$$\underline{R}_{\rho\sigma} = \partial_\mu \underline{\Gamma}^\mu_{\ \rho\sigma} - \partial_\sigma \underline{\Gamma}^\mu_{\ \rho\mu}$$
$$= -\frac{1}{2} \eta^{\mu\lambda} \frac{\partial^2 h_{\rho\sigma}}{\partial x^\mu \partial x^\lambda} + \frac{1}{2} \eta^{\mu\lambda} \left(\frac{\partial^2 h_{\sigma\lambda}}{\partial x^\mu \partial x^\rho} + \frac{\partial^2 h_{\rho\mu}}{\partial x^\sigma \partial x^\lambda} - \frac{\partial^2 h_{\mu\lambda}}{\partial x^\sigma \partial x^\rho} \right).$$

Assuming the gravitational field is static, then all the derivatives of $h_{\mu\nu}$ with respect to time are zero. For the component $\underline{R}^{44}$, it follows from (9.109) and (9.110) that

$$\underline{R}^{44} \approx \eta^{4\rho} \eta^{4\sigma} \left(\partial_\mu \underline{\Gamma}^\mu_{\ 44} - \partial_4 \underline{\Gamma}^\mu_{\ 4\mu} \right) = \partial_\mu \underline{\Gamma}^\mu_{\ 44}$$
$$= -\frac{1}{2} \nabla^2 h_{44} = \nabla^2 \left(\frac{\Phi}{c^2} \right). \tag{9.121}$$

Now we consider an incoherent matter which is at rest. Its energy-momentum tensor is $T_m^{\alpha\beta} = \tilde{\rho}_{m_0} u^\alpha u^\beta$. Then we have

$$T_m^{44} = \rho_m c^2 = \tilde{\rho}_{m_0} c^2,$$
$$T_m^\alpha_{\ \alpha} = \tilde{\rho}_{m_0} u^\alpha u_\alpha = -\tilde{\rho}_{m_0} c^2. \tag{9.122}$$

Substituting (9.121) and (9.122) into (9.120) we obtain

$$\nabla^2 \Phi = \chi \left(\frac{1}{2} \tilde{\rho}_{m_0} c^4 \right).$$

Comparing this equation with (9.115) gives rise to $\chi = 8\pi G/c^4$. Finally, the Einstein field equations (9.119) may written as

$$\underline{R}^{\mu\nu} - \frac{1}{2} g^{\mu\nu} \underline{R} = \frac{8\pi G}{c^4} T^{\mu\nu}.$$

or

$$R^{\mu\nu} = \frac{8\pi G}{c^4}\left(T^{\mu\nu} - \frac{1}{2}g^{\mu\nu}T^{\mu}{}_{\mu}\right).$$
(9.123)

from (9.120).

The general relativity is the simplest theory on gravity and is found to be consistent with experiments. However, the general relativity has not been completely reconciled with quantum mechanics yet.

9.6.12   The Schwarzschild Solution

The Schwarzschild solution, named after the German physicist Karl Schwarzschild (1873–1916), is one of the most useful solutions of the Einstein field equations. Consider an inertial reference frame $\bar{x}^{\alpha}(\alpha = 1, 2, 3, 4)$ whose metric tensor is given by (9.19). The differential interval is thus given by

$$ds^2 = \eta_{\alpha\beta}d\bar{x}^{\alpha}d\bar{x}^{\beta} = (d\bar{x}^1)^2 + (d\bar{x}^2)^2 + (d\bar{x}^3)^2 - (d\bar{x}^4)^2.$$

In a spherical coordinate system (r, θ, φ, ct), this can be written as

$$ds^2 = dr^2 + r^2 d\theta^2 + r^2 \sin^2\theta d\varphi - c^2 t^2.$$
(9.124)

In a gravitational field, the differential interval will take a more complicated form

$$ds^2 = g_{\alpha\beta}dx^{\alpha}dx^{\beta}.$$

Consider an empty space surrounding a static star or planet. We use the spherical coordinate system (r, θ, φ, ct) with the origin at the center of the planet. If the planet does not rotate very fast, we may ignore the effects of rotation and assume that the gravitational field is spherically symmetric. In this case, the metric coefficients $g_{\alpha\beta}$ as a function of (r, θ, φ, ct) must satisfy

$$g_{\alpha\beta}(r, \theta, \varphi, ct) = g_{\alpha\beta}(r, \theta, \varphi, -ct),$$
$$g_{\alpha\beta}(r, \theta, \varphi, ct) = g_{\alpha\beta}(r, \theta, -\varphi, ct),$$
$$g_{\alpha\beta}(r, \theta, \varphi, ct) = g_{\alpha\beta}(r, -\theta, \varphi, ct),$$

which imply $g_{\alpha\beta} = 0(\alpha \neq \beta)$. Thus the differential interval may be written as

$$ds^2 = g_{11}dr^2 + g_{22}d\theta^2 + g_{33}d\varphi^2 - g_{44}c^2 dt^2,$$

where $g_{ii}(i = 1, 2, 3, 4)$ are independent of time. According to the spherical symmetry, both g_{11} and g_{44} only depend on r. Now if we let $dr = 0$ and $dt = 0$, the differential interval should reduce to differential length on the sphere of radius r

$$ds^2 = g_{22}d\theta^2 + g_{33}d\varphi^2 = r^2 d\theta^2 + r^2 \sin^2\theta d\varphi.$$

Therefore $g_{22} = r^2$, $g_{33} = r^2 \sin^2 \theta$, and the differential interval becomes

$$ds^2 = g_{11}(r)dr^2 + r^2 d\theta^2 + r^2 \sin^2 \theta d\varphi^2 - g_{44}(r)c^2 dt^2.$$

In order to determine g_{11} and g_{44}, we need to use the Einstein field equations. In the empty space we have $T^{\mu\nu} = 0$ and Einstein field equations reduce to

$$\underline{R}^{\mu\nu} = 0.$$

One can find g_{11} and g_{44} from these equations. Ignoring the details, the differential interval may be expressed as

$$ds^2 = \left(1 - \frac{2GM}{c^2 r^2}\right)^{-1} dr^2 + r^2 d\theta^2 + r^2 \sin^2 \theta d\varphi^2 - \left(1 - \frac{2GM}{c^2 r^2}\right) c^2 dt^2, \quad (9.125)$$

where M is the mass of the planet. This is the **Schwarzschild solution** outside the planet, which shows that the mass M introduces the curvature of the space–time. As $r \to \infty$, Equation (9.125) reduces to (9.124).

9.6.13　Electromagnetic Fields in An Accelerated System

We have already shown that the electromagnetic fields in an inertial system can be written as

$$\partial_\mu F^{\mu\nu} = -\mu_0 J^\nu,$$
$$\partial_\mu F_{\nu\alpha} + \partial_\nu F_{\alpha\mu} + \partial_\alpha F_{\mu\nu} = 0.$$

By minimal coupling, these equations can be generalized to an arbitrary coordinate system

$$\underline{D}_\mu F^{\mu\nu} = -\mu_0 J^\nu,$$
$$\underline{D}_\mu F_{\nu\alpha} + \underline{D}_\nu F_{\alpha\mu} + \underline{D}_\alpha F_{\mu\nu} = 0.$$
$$(9.126)$$

By means of the following calculations

$$\underline{D}_\mu F^{\mu\nu} = \frac{\partial F^{\mu\nu}}{\partial x^\mu} + \Gamma^\mu{}_{\alpha\mu} F^{\alpha\nu} + \Gamma^\nu{}_{\beta\mu} F^{\mu\beta},$$

$$\underline{D}_\alpha F_{\mu\nu} = \frac{\partial F_{\mu\nu}}{\partial x^\alpha} - \Gamma^\beta{}_{\mu\alpha} F_{\beta\nu} - \Gamma^\gamma{}_{\nu\alpha} F_{\mu\gamma},$$

$$\underline{D}_\nu F_{\alpha\mu} = \frac{\partial F_{\alpha\mu}}{\partial x^\nu} - \Gamma^\beta{}_{\alpha\nu} F_{\beta\mu} - \Gamma^\gamma{}_{\mu\nu} F_{\alpha\gamma},$$

$$\underline{D}_\mu F_{\nu\alpha} = \frac{\partial F_{\nu\alpha}}{\partial x^\mu} - \Gamma^\beta{}_{\nu\mu} F_{\beta\alpha} - \Gamma^\gamma{}_{\alpha\mu} F_{\nu\gamma},$$

and (9.98), Equation (9.126) can be written as

$$\frac{\partial F_{\mu\nu}}{\partial x^\alpha} + \frac{\partial F_{\nu\alpha}}{\partial x^\mu} + \frac{\partial F_{\alpha\mu}}{\partial x^\nu} = 0,$$

$$\frac{1}{\sqrt{-g}} \frac{\partial}{\partial x^\mu}\left(\sqrt{-g}\, F^{\mu\nu}\right) = -\mu_0 J^\nu.$$

(9.127)

These are most general form of Maxwell equations in an arbitrary coordinate system.

> Henceforth space by itself and time by itself are doomed to fade away into mere shadows, and only a kind of the union of the two will preserve an independent reality.
>
> —Hermann Minkowski

10

Quantization of Electromagnetic Fields

The true logic in this world lies in probability theory.

—James Maxwell

Reality is merely an illusion, albeit a very persistent one.

—Albert Einstein

In the late 1800s and early 1900s, there was an increase in the number of experimental observations that could not be explained by classical physics. Physical scientists began to believe that the physics needs a complete reformulation. This led to the birth of modern physics, which rests on three pillars: the quantum theory, special relativity and general relativity.

The quantum theory was invented to describe the physics of small systems, such as atoms and nuclei, and was pioneered by Max Planck (German physicist, 1858–1947), Albert Einstein, Niels Henrik David Bohr (Danish physicist, 1885–1962) and Louis de Broglie (French physicist, 1892–1987). It is the work by the German physicist Werner Heisenberg (1901–1976) and the Austrian physicist Erwin Rudolf Josef Alexander Schrödinger (1887–1961) that made the early quantum theory become what is known as quantum mechanics today. Quantum mechanics has replaced Newtonian mechanics as the correct description of small particle systems.

On the atomic and molecular scale, both the charged particles and electromagnetic fields must obey the laws of quantum mechanics, which leads to quantum electrodynamics. In classical electrodynamics, the field strengths may take arbitrary values. In quantum electrodynamics, however, the field strengths cannot be identically zero. Quantum electrodynamics is remarkable for its incredible accuracy in predicting small particle systems, and it has been widely used for calculating the interaction of electromagnetic radiation with atomic and molecular matter. It should be mentioned that a mathematically rigorous quantum field theory for describing the behavior of elementary particles is still not available.

Foundations of Applied Electrodynamics   Geyi Wen
© 2010 John Wiley & Sons, Ltd

10.1   Fundamentals of Quantum Mechanics

In classical mechanics, the motion of a particle can be determined in an exact manner. In quantum mechanics, we do not have this determinism. The small particle system is so delicate that any means of observation will interfere with the system and modify it. For this reason, the quantum mechanics is basically a theory about measurement, and it is based on a few postulates or axioms that we accept without proofs.

10.1.1   Basic Postulates of Quantum Mechanics

The mathematical foundation of quantum mechanics is the theory of Hilbert space. Any quantum system can be described by a complex Hilbert space H with inner product $(\cdot, \cdot)$ and norm $\|\cdot\| = (\cdot, \cdot)^{1/2}$. The basic postulates for the quantum mechanics are summarized as follows.

1. A physical system is characterized by a unit vector $\psi \in H$, called a **state** or a **wavefunction** that contains all the information of the system and satisfies the **Schrödinger equation**

$$\hat{H}\psi = j\hbar \frac{\partial \psi}{\partial t}, \qquad (10.1)$$

   where $\hbar = h/2\pi$, and $h = 6.62377 \times 10^{-34}$ are known as **Planck's constant**.

2. Each physical quantity A (such as position, momentum, angular momentum and energy) is associated with a self-adjoint operator (or Hermitian operator), which is denoted by $\hat{A}$: $D(\hat{A}) \subset H \rightarrow H$, and is called **observable**. In particular, the self-adjoint operator $\hat{H}$: $D(\hat{H}) \subset H \rightarrow H$ corresponding to the energy of the quantum system is called the **Hamiltonian** of the quantum system.

3. A measurement of the observable A in the state ψ results in a number that is statistical and belongs to one of the eigenvalues of $\hat{A}$. The number $\langle \hat{A} \rangle \equiv (\hat{A}\psi, \psi)$ is called the **mean value** of A. (Note that we use $\langle \cdot \rangle$ to denote the mean value and $(\cdot, \cdot)$ a bilinear form.) Let $\Delta \hat{A} = \hat{A} - \langle \hat{A} \rangle$, and the number $\langle \Delta \hat{A}^2 \rangle = \left\| \hat{A}\psi - \langle \hat{A} \rangle \psi \right\|^2$ is called the **dispersion** of the observable in the state ψ. The mean value is always real, and the dispersion is always positive.

Let $\{u_n\}$ be an orthonormal system of eigenfunctions of an observable $\hat{A}$

$$\hat{A}u_n = \lambda_n u_n.$$

For each state $\psi \in H$, we may write

$$\psi = \sum_n (\psi, u_n) u_n.$$

Thus

$$(\psi, \psi) = \sum_n |(\psi, u_n)|^2 = 1,$$

$$\langle \hat{A} \rangle = (\hat{A}\psi, \psi) = \sum_n \lambda_n |(\psi, u_n)|^2,$$

$$\langle \Delta \hat{A}^2 \rangle = \left\| \hat{A}\psi - \langle \hat{A} \rangle \psi \right\|^2 = \sum_n (\lambda_n - \langle \hat{A} \rangle)^2 |(\psi, u_n)|^2.$$

Especially if $\psi = u_m$, we have $(\psi, u_n) = 0$ $(n \neq m)$, and $\langle \hat{A} \rangle = \lambda_m$, $\langle \Delta \hat{A}^2 \rangle = 0$. The number

$$P_\psi(\lambda_n) = |(\psi, u_n)|^2 \tag{10.2}$$

is defined as the **probability** that the measurement of A will yield the value λ_n or the probability for realizing the state u_n. More generally, if the quantum system is in the state ψ and ϕ is another state, then $|(\psi, \phi)|^2$ is defined as the **probability** for realizing ϕ.

The operator $[\hat{A}, \hat{B}] = \hat{A}\hat{B} - \hat{B}\hat{A}$ is the **commutator** of the operators $\hat{A}$ and $\hat{B}$. The operators $\hat{A}$ and $\hat{B}$ are said to commute if $[\hat{A}, \hat{B}] = 0$. We may interpret $\hat{A}\hat{B}$ as the process of measuring B first followed by measuring A. The reverse order is denoted by $\hat{B}\hat{A}$. The result of measuring A and B depends on the order of measurement if $[\hat{A}, \hat{B}] \neq 0$. If two adjoint operators commute, they have common eigenfunctions, and vice versa.

10.1.2   Quantum Mechanical Operators

According to the basic postulates of quantum mechanics, a physical quantity is associated with a Hermitian operator. But we have not described any rules as to how the Hermitian operator should be constructed. The **correspondence principle** proposed by Bohr states that the quantum theory and classical theory must agree in the cases where quantum effects are not important. Therefore the operator equation in quantum mechanics must be reduced to its classical counterpart when the system is becoming very large. To fulfill this requirement, the quantum mechanical operator can be defined as the same as their classical counterparts.

The **position operators** $\hat{x}$, $\hat{y}$ and $\hat{z}$ are defined as

$$\hat{x}\psi(\mathbf{r}, t) = x\psi(\mathbf{r}, t), \ \hat{y}\psi(\mathbf{r}, t) = y\psi(\mathbf{r}, t), \ \hat{z}\psi(\mathbf{r}, t) = z\psi(\mathbf{r}, t).$$

In three dimensions, we write $\hat{\mathbf{r}} = \hat{x}\mathbf{u}_x + \hat{y}\mathbf{u}_y + \hat{z}\mathbf{u}_z$, where $\mathbf{u}_x$, $\mathbf{u}_y$ and $\mathbf{u}_z$ are unit vectors along the x, y and z axis.

Remark 10.1 (Particle-wave duality): **Particle-wave duality** proposes that all energy exhibits both wave-like and particle-like properties. A particle of energy E and momentum $\mathbf{p}$ in free space may be associated with a wavefunction $\psi(\mathbf{r}, t) = c_1 e^{-j(Et - \mathbf{p} \cdot \mathbf{r})/\hbar}$, where c_1 is a constant.   □

When the momentum operator $\hat{\mathbf{p}}$ is applied to the wavefunction $\psi(\mathbf{r}, t) = c_1 e^{-j(Et - \mathbf{p} \cdot \mathbf{r})/\hbar}$, it must yield the momentum $\mathbf{p}$. Thus the **momentum operator** $\hat{p}_x$, $\hat{p}_y$ and $\hat{p}_z$ may be defined as

$$\hat{p}_x = -j\hbar \frac{\partial}{\partial x}, \; \hat{p}_y = -j\hbar \frac{\partial}{\partial y}, \; \hat{p}_z = -j\hbar \frac{\partial}{\partial z}.$$

In three dimensions, we may write

$$\hat{\mathbf{p}} = -j\hbar \nabla = -j\hbar \left(\mathbf{u}_x \frac{\partial}{\partial x} + \mathbf{u}_y \frac{\partial}{\partial y} + \mathbf{u}_z \frac{\partial}{\partial z} \right).$$

The **angular momentum operator** is defined by $\hat{\mathbf{l}} = \hat{\mathbf{r}} \times \hat{\mathbf{p}}$ or componentwise

$$\hat{l}_x = \hat{y}\hat{p}_z - \hat{z}\hat{p}_y, \hat{l}_y = \hat{z}\hat{p}_x - \hat{x}\hat{p}_z, \hat{l}_z = \hat{x}\hat{p}_y - \hat{y}\hat{p}_x.$$

The mean value of a Hermitian operator $\hat{A}$ changes with time according to

$$\frac{d}{dt}\langle \hat{A} \rangle = \left(\hat{A}\psi, \frac{\partial \psi}{\partial t} \right) + \left(\hat{A}\frac{\partial \psi}{\partial t}, \psi \right) + \left(\frac{\partial \hat{A}}{\partial t}\psi, \psi \right) = \frac{1}{j\hbar}\langle [\hat{A}, \hat{H}] \rangle + \left\langle \frac{\partial \hat{A}}{\partial t} \right\rangle. \quad (10.3)$$

If $\hat{A}$ does not depend on time explicitly, we have

$$\frac{d}{dt}\langle \hat{A} \rangle = \frac{1}{j\hbar}\langle [\hat{A}, \hat{H}] \rangle. \quad (10.4)$$

This is called the **Heisenberg equation of motion**. If $\hat{A}$ and $\hat{H}$ commute, Equation (10.4) reduces to $\frac{d}{dt}\langle \hat{A} \rangle = 0$. In this case, the mean value of the Hermitian operator does not change with time.

10.1.3 The Uncertainty Principle

In quantum mechanics, some physical quantities cannot be measured simultaneously with arbitrary accuracy. Let $\hat{A}$ and $\hat{B}$ be two observables, and ψ be a state in the Hilbert space H. Let ξ be an arbitrary real number. Then

$$(\xi\hat{A}\psi + j\hat{B}\psi, \xi\hat{A}\psi + j\hat{B}\psi) = \xi^2(\hat{A}^2\psi, \psi) - j\xi([\hat{A}, \hat{B}]\psi, \psi) + (\hat{B}^2\psi, \psi) \geq 0.$$

Let $j\hat{C} = [\hat{A}, \hat{B}]$. Then $\hat{C}$ is a Hermitian operator. We may write the above as

$$\xi^2\langle \hat{A}^2 \rangle - \xi\langle \hat{C} \rangle + \langle \hat{B}^2 \rangle = \langle \hat{A}^2 \rangle \left(\xi - \frac{\langle \hat{C} \rangle}{2\langle \hat{A}^2 \rangle} \right)^2 + \left(\langle \hat{B}^2 \rangle - \frac{\langle \hat{C} \rangle^2}{4\langle \hat{A}^2 \rangle} \right) \geq 0.$$

Choosing $\xi = \langle \hat{C} \rangle / 2 \langle \hat{A}^2 \rangle$ yields

$$\langle \hat{A}^2 \rangle \langle \hat{B}^2 \rangle \geq \frac{1}{4} \langle \hat{C} \rangle^2 = \frac{1}{4} \langle [\hat{A}, \hat{B}] \rangle^2. \tag{10.5}$$

This is valid for any two Hermitian operators $\hat{A}$ and $\hat{B}$, and is thus also valid for $\Delta \hat{A}$ and $\Delta \hat{B}$. Therefore

$$\sqrt{\langle \Delta \hat{A}^2 \rangle} \sqrt{\langle \Delta \hat{B}^2 \rangle} \geq \frac{1}{2} \left| \langle [\hat{A}, \hat{B}] \rangle \right|, \tag{10.6}$$

where we have used the relation $\langle [\Delta \hat{A}, \Delta \hat{B}] \rangle = \langle [\hat{A}, \hat{B}] \rangle$. Equation (10.6) implies that if $\hat{A}$ and $\hat{B}$ do not commute, it is impossible to measure the two corresponding physical quantities simultaneously with arbitrary accuracy. This is called the **uncertainty principle**.

Example 10.1 (Heisenberg's uncertainty principle): Heisenberg's principle states that it is impossible to measure the position and momentum simultaneously with arbitrary accuracy. Actually,

$$[\hat{x}, \hat{p}_x] \psi = -j\hbar \left[x \frac{\partial \psi}{\partial x} - \frac{\partial}{\partial x} (x \psi) \right] = j\hbar \psi.$$

Thus $[\hat{x}, \hat{p}_x] = j\hbar$. In general, we have

$$[\hat{\alpha}, \hat{p}_\beta] = \begin{cases} j\hbar, \alpha = \beta \\ 0, \alpha \neq \beta \end{cases}, \alpha = x, y, z.$$

It follows from (10.6) that

$$\sqrt{\langle \Delta \hat{x}^2 \rangle} \sqrt{\langle \Delta \hat{p}_x^2 \rangle} \geq \frac{\hbar}{2}.$$

If we localize a particle more precisely (that is, $\sqrt{\langle \Delta \hat{x}^2 \rangle}$ is small), the velocity of the particle becomes more uncertain (that is, $\sqrt{\langle \Delta \hat{p}_x^2 \rangle}$ is large). Conversely if we measure the velocity more precisely, the position of the particle becomes more uncertain. □

Example 10.2: Let the Hamiltonian of the system be $\hat{H}$, and $\hat{A}$ be an arbitrary Hermitian operator which does not explicitly depend on time. By means of (10.6), we obtain

$$\sqrt{\langle \Delta \hat{H}^2 \rangle} \sqrt{\langle \Delta \hat{A}^2 \rangle} \geq \frac{1}{2} \left| \langle [\hat{A}, \hat{H}] \rangle \right|.$$

It follows from (10.4) that

$$\sqrt{\langle \Delta \hat{H}^2 \rangle} \sqrt{\langle \Delta \hat{A}^2 \rangle} \geq \frac{\hbar}{2} \left| \frac{d}{dt} \langle \hat{A} \rangle \right|.$$

Let $\tau_{\hat{A}} = \sqrt{\langle \Delta \hat{A}^2 \rangle}(d\langle \hat{A} \rangle/dt)^{-1}$ denote the time interval needed for $\langle \hat{A} \rangle$ to increase by $\sqrt{\langle \Delta \hat{A}^2 \rangle}$. In a given state, each physical quantity has a $\tau_{\hat{A}}$, among which we can choose a smallest one, denoted by Δt. Thus

$$\Delta E \cdot \Delta t \geq \hbar/2, \qquad (10.7)$$

where $\Delta E = \sqrt{\langle \Delta \hat{H}^2 \rangle}$. This equation indicates that the higher the precision of the measurement of the energy, the longer the time needed for the measurement. Essentially the uncertainty principle states that any measurement made on a physical system perturbs the system.   □

It must be mentioned that the uncertainty principle is not a limit set by the accuracy of measuring equipment. It is a fundamental requirement by the nature.

10.1.4   Quantization of Classical Mechanics

Quantization is a procedure for building quantum mechanics from classical mechanics, which consists of two main steps. The first step is to construct the Hamiltonian function in terms of the generalized coordinates and generalized momenta. The second step is to convert the classical Hamiltonian function to the quantum mechanical Hamiltonian operator by replacing the generalized coordinates and the generalized momenta with quantum mechanical operators subject to the commutation rules. In classical non-relativistic physics, the motion of a particle of mass m in R^3 is governed by the classical Newtonian equation

$$m\frac{d^2\mathbf{r}(t)}{dt^2} = \mathbf{F}[\mathbf{r}(t)],$$

where $\mathbf{F}$ is the force field. If the force field possesses a potential $V(\mathbf{r})$ such that $\mathbf{F} = -\nabla V$, the total energy of the particle is given by

$$E = \frac{\mathbf{p}^2}{2m} + V, \qquad (10.8)$$

where $\mathbf{p} = md\mathbf{r}/dt$ is the momentum vector at time t. The first term of the above equation denotes the kinetic energy and the second term the potential energy. The energy operator, the Hamiltonian, of a single particle can be obtained from (10.8) by replacing $\mathbf{p}$ with $\hat{\mathbf{p}}$ as

$$\hat{H} = \frac{\hat{\mathbf{p}}^2}{2m} + V(\mathbf{r}, t) = -\frac{\hbar^2}{2m}\nabla^2 + V(\mathbf{r}, t).$$

In quantum mechanics, the motion of a particle of mass m is described by the Schrödinger equation

$$-\frac{\hbar^2}{2m}\nabla^2\psi(\mathbf{r}, t) + V(\mathbf{r}, t)\psi(\mathbf{r}, t) = j\hbar\frac{\partial\psi(\mathbf{r}, t)}{\partial t}. \qquad (10.9)$$

The physical state $\psi(\mathbf{r}, t)$ of the particle satisfies $\int_{R^3} |\psi(\mathbf{r}, t)|^2 \, d\mathbf{r} = 1$. Since the expectation value of the position vector is given by $\langle \hat{\mathbf{r}} \rangle = \int_{R^3} \mathbf{r} |\psi(\mathbf{r}, t)|^2 \, d\mathbf{r}$, we may interpret $|\psi(\mathbf{r}, t)|^2 \, d\mathbf{r}$ as the probability of finding the particle inside the differential volume $d\mathbf{r}$ at time t. If $\hat{H}$ does not explicitly depend on time, the general solution of (10.9) may be written as

$$\psi(\mathbf{r}, t) = \sum_n a_n u_n(\mathbf{r}) e^{-j E_n t / \hbar},$$

where u_n is the eigenfunction of $\hat{H}$ with energy E_n

$$-\frac{\hbar^2}{2m} \nabla^2 u_n(\mathbf{r}) + V(\mathbf{r}) u_n(\mathbf{r}) = E_n u_n(\mathbf{r}).$$

This is called the **stationary Schrödinger equation**. The probability of finding the system in the state u_n with energy E_n is

$$P_\psi(E_n) = |(\psi, u_n)|^2 = |a_n|^2.$$

10.1.5   Harmonic Oscillator

The study of the harmonic oscillator is fundamental in quantum mechanics. In classical mechanics, an idealized harmonic oscillator is a point mass connected to the end of a frictionless idealized spring. The motion of the mass is governed by the ordinary differential equation

$$m \frac{d^2 x(t)}{dt^2} = -Kx$$

where $-Kx$ is the restoring force. Let $p_x = mdx/dt$. The total energy of the system is $E = p_x^2/2m + Kx^2/2$. In quantum mechanics, the harmonic oscillator is described by the Schrödinger equation

$$\hat{H}(\psi) = \left(\frac{\hat{p}_x^2}{2m} + \frac{1}{2} K \hat{x}^2 \right) \psi = -\frac{\hbar^2}{2m} \frac{\partial^2 \psi(x, t)}{\partial x^2} + \frac{1}{2} Kx^2 \psi(x, t)$$
$$= j\hbar \frac{\partial \psi(x, t)}{\partial t}.$$

The corresponding stationary Schrödinger equation is

$$-\frac{\hbar^2}{2m} \frac{d^2 u(x)}{dx^2} + \frac{1}{2} Kx^2 u(x) = Eu(x).$$

This eigenvalue equation can be solved by the method of power series expansion. The normalized harmonic oscillator wavefunction is [for example, Yariv, 1982]

$$u_n(x) = \left(\frac{\alpha}{\sqrt{\pi}n!2^n}\right)^{1/2} H_n(\alpha x)e^{-\alpha^2 x^2/2}, \quad n = 0, 1, 2, \cdots,$$

where H_n are Hermite polynomials:

$$H_n(\xi) = e^{\xi^2}(-1)^n \frac{d^n e^{-\xi^2}}{d\xi^n}$$

and $\alpha = (m\omega/\hbar)^{1/2}$, $\omega = (K/m)^{1/2}$. The corresponding eigenvalues are given by

$$E_n = \hbar\omega\left(n + \frac{1}{2}\right), \quad n = 0, 1, 2, \cdots.$$

As a result, the energy of the harmonic oscillator is quantized and is not zero even in its lowest energy state $n = 0$. We may introduce two new operators

$$\hat{a} = \frac{\alpha}{\sqrt{2}}\hat{x} + j\frac{1}{\sqrt{2}\hbar\alpha}\hat{p}_x, \quad \hat{a}^+ = \frac{\alpha}{\sqrt{2}}\hat{x} - j\frac{1}{\sqrt{2}\hbar\alpha}\hat{p}_x, \tag{10.10}$$

which are called the **annihilation operator** and the **creation operator** respectively. Then it is easy to show that

$$\hat{a}u_n = n^{1/2}u_{n-1}, \quad \hat{a}^+u_n = (n+1)^{1/2}u_{n+1}. \tag{10.11}$$

The operators $\hat{a}$ and $\hat{a}^+$ are not Hermitian, and we have

$$[\hat{a}, \hat{a}^+] = 1. \tag{10.12}$$

It follows from (10.10) that

$$\hat{x} = \frac{1}{\sqrt{2}\alpha}(\hat{a}^+ + \hat{a}), \quad \hat{p}_x = \frac{j\hbar\alpha}{\sqrt{2}}(\hat{a}^+ - \hat{a}).$$

Making use of these relations, we obtain

$$\hat{H} = \frac{\hbar\omega}{2}(\hat{a}\hat{a}^+ + \hat{a}^+\hat{a}) = \hbar\omega\left(\hat{a}^+\hat{a} + \frac{1}{2}\right).$$

Since

$$\hat{a}^+\hat{a}u_n = nu_n,$$

the operator $\hat{a}^+\hat{a}$ is referred to as the **particle number operator**.

10.1.6 Systems of Identical Particles

The principle of indistinguishability in quantum mechanics states that it is impossible to distinguish between n identical particles. Consider a system of two non-interacting identical particles, numbered as 1 and 2. Then

$$\hat{H}(1)\psi_\alpha(1) = E_\alpha\psi_\alpha(1), \ \hat{H}(2)\psi_\beta(2) = E_\beta\psi_\beta(2), \tag{10.13}$$

where ψ_α and ψ_β are any two eigenstates of the single-particle Hamiltonian. The subscripts α and β are used to denote the eigenstates while the numbers in parentheses denote the particle or its spatial and spin coordinates. Since the particles are assumed to be non-interacting, the total Hamiltonian of the system is

$$\hat{H}(1, 2) = \hat{H}(1) + \hat{H}(2). \tag{10.14}$$

From (10.13) and (10.14), we obtain

$$\hat{H}(1, 2)\psi(1, 2) = (E_\alpha + E_\beta)\psi(1, 2), \tag{10.15}$$

where $\psi(1, 2)$ could be one of the following eigenfunctions

$$\psi_\alpha(1)\psi_\beta(2), \ \psi_\alpha(2)\psi_\beta(1), \ \frac{1}{2}\left[\psi_\alpha(1)\psi_\beta(2) \pm \psi_\alpha(2)\psi_\beta(1)\right]$$

from the mathematical point of view. To find the correct eigenstate, the physical conditions must be taken into account. Let $\hat{P}_{12}$ denote the interchange of particles 1 and 2 with $\hat{P}_{12}\psi(1, 2) = \psi(2, 1)$. Since an interchange of particles 1 and 2 does not change the state of the system, we have

$$[\hat{P}_{12}, \hat{H}(1, 2)] = \hat{P}_{12}\hat{H}(1, 2) - \hat{H}(1, 2)\hat{P}_{12} = 0$$

from (10.15). This implies that $\hat{P}_{12}$ and $\hat{H}$ commute and they possess common eigenfunctions. Thus we may write

$$\hat{P}_{12}\psi(1, 2) = \lambda\psi(1, 2).$$

Since two permutations give the original state, we have

$$\hat{P}_{12}^2\psi(1, 2) = \lambda^2\psi(1, 2) = \psi(1, 2),$$

which yields $\lambda = \pm 1$. Hence the eigenfunction $\psi(1, 2)$ is either symmetric ($\lambda = 1$) or anti-symmetric ($\lambda = -1$) when two particles are interchanged. Let ψ_s and ψ_a denote the symmetric and antisymmetric wavefunctions respectively. For two identical non-interacting particles in the state ψ_α and ψ_β, the normalized symmetric and antisymmetric wavefunctions are given by

$$\psi_s = \frac{1}{\sqrt{2}}\left[\psi_\alpha(1)\psi_\beta(2) + \psi_\alpha(2)\psi_\beta(1)\right],$$

$$\psi_a = \frac{1}{\sqrt{2}}\left[\psi_\alpha(1)\psi_\beta(2) - \psi_\alpha(2)\psi_\beta(1)\right].$$

There exist two completely different kinds of elementary particles in nature. The particles with half-odd-integral spin (such as electrons, protons, and neutrons) are called **fermions**, named after the Italian physicist Enrico Fermi (1901–1954), and they are material particles. The particles with integral spin (such as photons) are called **bosons**, named after the Indian physicist Satyendra Nath Bose (1894–1974), and they are force carriers transferring interactions between fermions. Experiments show that the fermions are described by antisymmetric wavefunctions, and the bosons are described by symmetric wavefunctions.

The normalized antisymmetric wavefunctions of a system of more than two particles can be constructed in terms of Slater determinant

$$\psi_a(1, 2, \cdots, N) = \frac{1}{\sqrt{N!}}\begin{vmatrix} \psi_1(1) & \psi_1(2) & \cdots & \psi_1(N) \\ \psi_2(1) & \psi_2(2) & \cdots & \psi_2(N) \\ \vdots & \vdots & \vdots & \vdots \\ \psi_N(1) & \psi_N(2) & \cdots & \psi_N(N) \end{vmatrix}. \tag{10.16}$$

The symmetric wavefunctions ψ_s can be obtained by replacing all $(-)$ signs in the expansion of (10.16) by $(+)$. It can be seen that if any two fermions are placed in the same eigenstate, ψ_a will vanish. This fact is known as the **Pauli exclusion principle**, named after the Austrian physicist Wolfgang Ernst Pauli (1900–1958), which results from the postulate of fermions antisymmetry.

10.2   Quantization of Free Electromagnetic Fields

In Newton's time light was considered a beam of particles. The wavelike nature of light was disclosed during the first half of nineteenth century, and light was demonstrated to be an electromagnetic wave. The polarization of light was due to the vectorial character of electric field. However, the classical electromagnetic theory could not explain blackbody radiation, which led Max Planck in 1900 to postulate that the energy of an electromagnetic wave of frequency f is quantized and is an integral multiple of the smallest amount of energy hf. Therefore, the electromagnetic field of frequency f consists of a beam of particles, called photons, each of which carries energy hf. The particle parameters (the energy E and the momentum $\mathbf{p}$ of a photon) and the wave parameters (the angular frequency $\omega = 2\pi f$ and wave vector $\mathbf{k} = \mathbf{u}_k k$ with $k = 2\pi/\lambda$) of a photon are related by Planck–Einstein relations

$$E = \hbar\omega, \mathbf{p} = \hbar\mathbf{k}.$$

Quantum electrodynamics (QED) deals with the interaction of electromagnetic fields with atoms and molecules. On the atomic and molecular scale, both the particles and fields are subject to quantum conditions. To find the quantized field, the classical Hamiltonian must be converted to quantum mechanical Hamiltonian operator. The classical field variables may be treated as dynamical variables, called canonical coordinates, and their time derivatives are the canonical momenta. The canonical coordinates and canonical momenta are subject to the canonical commutation rules. The field is then converted to an operator through combinations of creation and annihilation operators. This procedure is known as quantization. The QED has been the most precise theory that agrees closely with experiments. The non-relativistic version of QED is also known as quantum optics or cavity QED, which is useful for studying the properties of light and its interaction with materials.

The quantization of the field is a very important concept in modern physics. The result from the quantization is the field quanta, which can be created and annihilated. The quanta of the electromagnetic fields are photons, and they are force-carrier particles in electromagnetic interactions.

10.2.1   Quantization in Terms of Plane Wave Functions

To overcome the difficulties of normalization, we first assume that the electromagnetic fields are confined in arbitrary volume V. For convenience, we take the volume V as a cube of sides L. Using the Coulomb gauge condition, the fields in source-free region may be expressed as

$$\mathbf{E} = -\frac{\partial \mathbf{A}}{\partial t}, \mathbf{B} = \nabla \times \mathbf{A},$$

where the vector potential $\mathbf{A}$ satisfies

$$\nabla^2 \mathbf{A}(\mathbf{r}, t) - \frac{1}{v^2} \frac{\partial^2 \mathbf{A}(\mathbf{r}, t)}{\partial t^2} = 0,$$

and $v = 1/\sqrt{\mu\varepsilon}$. By separation of variables, we may try a solution of the form

$$\mathbf{A}(\mathbf{r}, t) = a(t)\mathbf{a}(\mathbf{r}), \tag{10.17}$$

where $a(t)$ and $\mathbf{a}(\mathbf{r})$ satisfy

$$\ddot{a}(t) + \omega^2 a(t) = 0, \tag{10.18}$$

$$\nabla^2 \mathbf{a}(\mathbf{r}) + k^2 \mathbf{a}(\mathbf{r}) = 0. \tag{10.19}$$

Here $k = \omega/v$ is the separation constant and the dot denotes the derivative with respect to time. The solution of (10.19) may be taken as the plane wave solution

$$\mathbf{a}_k(\mathbf{r}) = \sqrt{1/V}\mathbf{u}(\mathbf{k})\exp(-j\mathbf{k} \cdot \mathbf{r}),$$

where $\mathbf{u}(\mathbf{k})$ is the unit polarization vector and $k = |\mathbf{k}|$. The Coulomb gauge condition requires that $\mathbf{u}(\mathbf{k}) \cdot \mathbf{k} = 0$. This is called the **transversality condition**, which indicates that the polarization vector is perpendicular to $\mathbf{k}$. The vector $\mathbf{u}(\mathbf{k})$ can be chosen as one of the mutually orthonormal directions transverse to $\mathbf{k}$, denoted by $\mathbf{e}^{(\lambda)}(\mathbf{k})$ ($\lambda = 1, 2$). We have

$$\mathbf{e}^{(1)}(\mathbf{k}) \cdot \mathbf{e}^{(2)}(\mathbf{k}) = 0, \; \mathbf{e}^{(1)}(\mathbf{k}) \cdot \mathbf{k} = \mathbf{e}^{(2)}(\mathbf{k}) \cdot \mathbf{k} = 0.$$

The set $\{\mathbf{e}^{(1)}(\mathbf{k}), \mathbf{e}^{(2)}(\mathbf{k}), \mathbf{k}\}$ forms a right-handed system. We further assume that the vector potential $\mathbf{A}$ satisfies periodic boundary conditions, that is, it takes the same value on opposite faces of the cube of sides L. Then $\mathbf{k}$ may take the following values

$$\mathbf{k} = (k_x, k_y, k_z) = \frac{2\pi}{L}(n_x, n_y, n_z), n_x, n_y, n_z = 0, \pm 1, \pm 2, \cdots \qquad (10.20)$$

except in the situation $n_x = n_y = n_z = 0$. Corresponding to each $\mathbf{k}$, the solution of (10.18) may be represented by $a_{\mathbf{k}}^{(\lambda)}(t) = b_{\mathbf{k}}^{(\lambda)} e^{j\omega_{\mathbf{k}} t}$, where $b_{\mathbf{k}}^{(\lambda)}$ is independent of time. Therefore the vector potential may be written in the form

$$\mathbf{A}(\mathbf{r}, t) = \sum_{\mathbf{k},\lambda} \left[b_{\mathbf{k}}^{(\lambda)} \mathbf{a}_{\mathbf{k}}^{(\lambda)}(\mathbf{r}) e^{j\omega_{\mathbf{k}} t} + \bar{b}_{\mathbf{k}}^{(\lambda)} \bar{\mathbf{a}}_{\mathbf{k}}^{(\lambda)}(\mathbf{r}) e^{-j\omega_{\mathbf{k}} t} \right]$$

where $\mathbf{a}_{\mathbf{k}}^{(\lambda)}(\mathbf{r}) = \sqrt{1/V} \mathbf{e}^{(\lambda)}(\mathbf{k}) \exp(-j\mathbf{k} \cdot \mathbf{r})$. The electromagnetic fields may thus be expressed as

$$\mathbf{E}(\mathbf{r}, t) = -\frac{\partial \mathbf{A}(\mathbf{r}, t)}{\partial t} = -j \sum_{\mathbf{k},\lambda} \omega_{\mathbf{k}} \left[b_{\mathbf{k}}^{(\lambda)} \mathbf{a}_{\mathbf{k}}^{(\lambda)}(\mathbf{r}) e^{j\omega_{\mathbf{k}} t} - \bar{b}_{\mathbf{k}}^{(\lambda)} \bar{\mathbf{a}}_{\mathbf{k}}^{(\lambda)}(\mathbf{r}) e^{-j\omega_{\mathbf{k}} t} \right],$$

$$\mathbf{B}(\mathbf{r}, t) = \nabla \times \mathbf{A}(\mathbf{r}, t) = \sum_{\mathbf{k},\lambda} \left[b_{\mathbf{k}}^{(\lambda)} \nabla \times \mathbf{a}_{\mathbf{k}}^{(\lambda)}(\mathbf{r}) e^{j\omega_{\mathbf{k}} t} + \bar{b}_{\mathbf{k}}^{(\lambda)} \nabla \times \bar{\mathbf{a}}_{\mathbf{k}}^{(\lambda)}(\mathbf{r}) e^{-j\omega_{\mathbf{k}} t} \right].$$

$$(10.21)$$

The total electromagnetic energy or Hamiltonian of the system is given by

$$H = \int_V \left(\frac{1}{2}\varepsilon |\mathbf{E}|^2 + \frac{1}{2\mu} |\mathbf{B}|^2 \right) d\mathbf{r} = \varepsilon \sum_{\mathbf{k},\lambda} \omega_{\mathbf{k}}^2 (b_{\mathbf{k}}^{(\lambda)} \bar{b}_{\mathbf{k}}^{(\lambda)} + \bar{b}_{\mathbf{k}}^{(\lambda)} b_{\mathbf{k}}^{(\lambda)}),$$

where we have used the following calculations

$$\int_V \mathbf{a}_{\mathbf{k}}^{(\lambda)}(\mathbf{r}) \cdot \bar{\mathbf{a}}_{\mathbf{k}'}^{(\lambda')}(\mathbf{r}) d\mathbf{r} = \delta_{\mathbf{k}\mathbf{k}'} \delta_{\lambda\lambda'},$$

$$\int_V \frac{1}{2}\varepsilon |\mathbf{E}|^2 d\mathbf{r} = \frac{1}{2}\varepsilon \sum_{\mathbf{k},\lambda} \omega_{\mathbf{k}}^2 (b_{\mathbf{k}}^{(\lambda)} \bar{b}_{\mathbf{k}}^{(\lambda)} + \bar{b}_{\mathbf{k}}^{(\lambda)} b_{\mathbf{k}}^{(\lambda)}),$$

$$\int_V \frac{1}{2\mu} |\mathbf{B}|^2 d\mathbf{r} = \frac{1}{2}\varepsilon \sum_{\mathbf{k},\lambda} \omega_{\mathbf{k}}^2 (b_{\mathbf{k}}^{(\lambda)} \bar{b}_{\mathbf{k}}^{(\lambda)} + \bar{b}_{\mathbf{k}}^{(\lambda)} b_{\mathbf{k}}^{(\lambda)}).$$

Introducing the transformation

$$q_{\mathbf{k}}^{(\lambda)} = \alpha(b_{\mathbf{k}}^{(\lambda)} + \bar{b}_{\mathbf{k}}^{(\lambda)}),$$

$$p_{\mathbf{k}}^{(\lambda)} = \alpha(\dot{b}_{\mathbf{k}}^{(\lambda)} + \dot{\bar{b}}_{\mathbf{k}}^{(\lambda)}) = j\omega_{\mathbf{k}}\alpha(b_{\mathbf{k}}^{(\lambda)} - \bar{b}_{\mathbf{k}}^{(\lambda)}),$$

where α is an arbitrary real constant, the Hamiltonian of the system admits the representation

$$H = \sum_{\mathbf{k},\lambda} \frac{\varepsilon}{2\alpha^2}(p_{\mathbf{k}}^{(\lambda)2} + \omega_{\mathbf{k}}^2 q_{\mathbf{k}}^{(\lambda)2}) = \sum_{\mathbf{k},\lambda} \frac{1}{2}(p_{\mathbf{k}}^{(\lambda)2} + \omega_{\mathbf{k}}^2 q_{\mathbf{k}}^{(\lambda)2}). \tag{10.22}$$

In the above, we have chosen $\alpha = \sqrt{\varepsilon}$. Equation (10.22) indicates that the total Hamiltonian of the electromagnetic field can be expressed as the sum of Hamiltonians of harmonic oscillators. Note that $q_{\mathbf{k}}^{(\lambda)}$ and $p_{\mathbf{k}}^{(\lambda)}$ are canonically conjugate. Actually

$$\frac{\partial H}{\partial p_{\mathbf{k}}^{(\lambda)}} = \dot{q}_{\mathbf{k}}^{(\lambda)}, \quad \frac{\partial H}{\partial q_{\mathbf{k}}^{(\lambda)}} = -\dot{p}_{\mathbf{k}}^{(\lambda)}.$$

The energy operator $\hat{H}$ can be obtained by replacing $p_{\mathbf{k}}^{(\lambda)}$ and $q_{\mathbf{k}}^{(\lambda)}$ with $\hat{p}_{\mathbf{k}}^{(\lambda)}$ and $\hat{q}_{\mathbf{k}}^{(\lambda)}$ respectively, together with the following canonical commutation relations

$$\left[\hat{q}_{\mathbf{k}}^{(\lambda)}, \hat{q}_{\mathbf{k}'}^{(\lambda')}\right] = 0, \quad \left[\hat{p}_{\mathbf{k}}^{(\lambda)}, \hat{p}_{\mathbf{k}'}^{(\lambda')}\right] = 0, \quad \left[\hat{q}_{\mathbf{k}}^{(\lambda)}, \hat{p}_{\mathbf{k}'}^{(\lambda')}\right] = j\hbar\delta_{\mathbf{k}\mathbf{k}'}\delta_{\lambda\lambda'}.$$

Thus

$$\hat{H} = \sum_{\mathbf{k},\lambda} \frac{1}{2}(\hat{p}_{\mathbf{k}}^{(\lambda)2} + \omega_{\mathbf{k}}^2 \hat{q}_{\mathbf{k}}^{(\lambda)2}). \tag{10.23}$$

Similar to (10.10), we introduce the new operators

$$\hat{a}_{\mathbf{k}}^{(\lambda)} = \sqrt{\frac{1}{2\hbar\omega_{\mathbf{k}}}}(j\hat{p}_{\mathbf{k}}^{(\lambda)} + \omega_{\mathbf{k}}\hat{q}_{\mathbf{k}}^{(\lambda)}),$$

$$\hat{a}_{\mathbf{k}}^{(\lambda)+} = \sqrt{\frac{1}{2\hbar\omega_{\mathbf{k}}}}(-j\hat{p}_{\mathbf{k}}^{(\lambda)} + \omega_{\mathbf{k}}\hat{q}_{\mathbf{k}}^{(\lambda)}). \tag{10.24}$$

Thus

$$\left[\hat{a}_{\mathbf{k}}^{(\lambda)}, \hat{a}_{\mathbf{k}'}^{(\lambda')}\right] = \left[\hat{a}_{\mathbf{k}}^{(\lambda)+}, \hat{a}_{\mathbf{k}'}^{(\lambda')+}\right] = 0,$$

$$\left[\hat{a}_{\mathbf{k}}^{(\lambda)}, \hat{a}_{\mathbf{k}'}^{(\lambda')+}\right] = \delta_{\mathbf{k}\mathbf{k}'}\delta_{\lambda\lambda'}. \tag{10.25}$$

It follows from (10.23) and (10.24) that

$$\hat{H} = \sum_{\mathbf{k},\lambda} \hbar\omega_{\mathbf{k}} \left(\hat{a}_{\mathbf{k}}^{(\lambda)+} \hat{a}_{\mathbf{k}}^{(\lambda)} + \frac{1}{2} \right). \tag{10.26}$$

In order to find the eigenvalues of $\hat{H}$, we only need to find the eigenvalues of $\hat{a}_{\mathbf{k}}^{(\lambda)+} \hat{a}_{\mathbf{k}}^{(\lambda)}$. Assuming that

$$\hat{a}_{\mathbf{k}}^{(\lambda)+} \hat{a}_{\mathbf{k}}^{(\lambda)} u_n(\mathbf{k}, \lambda) = \alpha_n(\mathbf{k}, \lambda) u_n(\mathbf{k}, \lambda), \tag{10.27}$$

the eigenvalues $\alpha_n(\mathbf{k}, \lambda)$ are then real and non-negative. Applying $\hat{a}_{\mathbf{k}}^{(\lambda)}$ to the above equation and making use of the last equation of (10.25) yields

$$\hat{a}_{\mathbf{k}}^{(\lambda)+} \hat{a}_{\mathbf{k}}^{(\lambda)} \left[\hat{a}_{\mathbf{k}}^{(\lambda)} u_n(\mathbf{k}, \lambda) \right] = [\alpha_n(\mathbf{k}, \lambda) - 1] \hat{a}_{\mathbf{k}}^{(\lambda)} u_n(\mathbf{k}, \lambda). \tag{10.28}$$

If this is repeated m times we may find

$$\hat{a}_{\mathbf{k}}^{(\lambda)+} \hat{a}_{\mathbf{k}}^{(\lambda)} \left[\hat{a}_{\mathbf{k}}^{(\lambda)m} u_n(\mathbf{k}, \lambda) \right] = [\alpha_n(\mathbf{k}, \lambda) - m] \hat{a}_{\mathbf{k}}^{(\lambda)m} u_n(\mathbf{k}, \lambda).$$

For a given $\alpha_n(\mathbf{k}, \lambda)$, the eigenvalue $\alpha_n(\mathbf{k}, \lambda) - m$ may become negative if m is very large. This situation is not allowed because the eigenvalues of $\hat{a}_{\mathbf{k}}^{(\lambda)+} \hat{a}_{\mathbf{k}}^{(\lambda)}$ must be non-negative. As a result, there must exist a non-negative integer $n(\mathbf{k}, \lambda)$ for which

$$\hat{a}_{\mathbf{k}}^{(\lambda)n(\mathbf{k},\lambda)} u_n(\mathbf{k}, \lambda) \neq 0, \quad \hat{a}_{\mathbf{k}}^{(\lambda)[n(\mathbf{k},\lambda)+1]} u_n(\mathbf{k}, \lambda) = 0.$$

Thus

$$\hat{a}_{\mathbf{k}}^{(\lambda)+} \hat{a}_{\mathbf{k}}^{(\lambda)} \left[\hat{a}_{\mathbf{k}}^{(\lambda)n(\mathbf{k},\lambda)} u_n(\mathbf{k}, \lambda) \right] = [\alpha_n(\mathbf{k}, \lambda) - n(\mathbf{k}, \lambda)] \hat{a}_{\mathbf{k}}^{(\lambda)n(\mathbf{k},\lambda)} u_n(\mathbf{k}, \lambda),$$

$$\hat{a}_{\mathbf{k}}^{(\lambda)} \left[\hat{a}_{\mathbf{k}}^{(\lambda)n(\mathbf{k},\lambda)} u_n(\mathbf{k}, \lambda) \right] = 0.$$

These equations imply that $\alpha_n(\mathbf{k}, \lambda) = n(\mathbf{k}, \lambda)$ and we have

$$\hat{a}_{\mathbf{k}}^{(\lambda)+} \hat{a}_{\mathbf{k}}^{(\lambda)} u_n(\mathbf{k}, \lambda) = n(\mathbf{k}, \lambda) u_n(\mathbf{k}, \lambda). \tag{10.29}$$

Similarly we may apply $\hat{a}_{\mathbf{k}}^{(\lambda)+}$ to (10.27) and use the last equation of (10.25) to get

$$\hat{a}_{\mathbf{k}}^{(\lambda)+} \hat{a}_{\mathbf{k}}^{(\lambda)} \left[\hat{a}_{\mathbf{k}}^{(\lambda)+} u_n(\mathbf{k}, \lambda) \right] = [n(\mathbf{k}, \lambda) + 1] \hat{a}_{\mathbf{k}}^{(\lambda)+} u_n(\mathbf{k}, \lambda). \tag{10.30}$$

Considering (10.28), (10.29) and (10.30), we may write

$$\hat{a}_{\mathbf{k}}^{(\lambda)} u_n(\mathbf{k}, \lambda) = c_{\mathbf{k}}^{(\lambda)} u_{n-1}(\mathbf{k}, \lambda),$$

$$\hat{a}_{\mathbf{k}}^{(\lambda)+} u_n(\mathbf{k}, \lambda) = d_{\mathbf{k}}^{(\lambda)} u_{n+1}(\mathbf{k}, \lambda).$$

Since all the eigenfunctions are assumed to be normalized, we have

$$n(\mathbf{k}, \lambda) = (\hat{a}_{\mathbf{k}}^{(\lambda)+}\hat{a}_{\mathbf{k}}^{(\lambda)}u_n(\mathbf{k}, \lambda), u_n(\mathbf{k}, \lambda))$$

$$= (\hat{a}_{\mathbf{k}}^{(\lambda)}u_n(\mathbf{k}, \lambda), \hat{a}_{\mathbf{k}}^{(\lambda)}u_n(\mathbf{k}, \lambda)) = \left|c_{\mathbf{k}}^{(\lambda)}\right|^2,$$

$$n(\mathbf{k}, \lambda) = ((\hat{a}_{\mathbf{k}}^{(\lambda)}\hat{\mathbf{a}}_{\mathbf{k}}^{(\lambda)+} - 1)u_n(\mathbf{k}, \lambda), u_n(\mathbf{k}, \lambda))$$

$$= (\hat{a}_{\mathbf{k}}^{(\lambda)+}u_n(\mathbf{k}, \lambda), \hat{a}_{\mathbf{k}}^{(\lambda)+}u_n(\mathbf{k}, \lambda)) - 1 = \left|d_{\mathbf{k}}^{(\lambda)}\right|^2.$$

Ignoring the phase factor, we obtain

$$c_{\mathbf{k}}^{(\lambda)} = \sqrt{n(\mathbf{k}, \lambda)}, d_{\mathbf{k}}^{(\lambda)} = \sqrt{n(\mathbf{k}, \lambda) + 1}.$$

Thus

$$\hat{a}_{\mathbf{k}}^{(\lambda)}u_n(\mathbf{k}, \lambda) = \sqrt{n(\mathbf{k}, \lambda)}u_{n-1}(\mathbf{k}, \lambda),$$

$$\hat{a}_{\mathbf{k}}^{(\lambda)+}u_n(\mathbf{k}, \lambda) = \sqrt{n(\mathbf{k}, \lambda) + 1}u_{n+1}(\mathbf{k}, \lambda).$$

The operator $\hat{a}_{\mathbf{k}}^{(\lambda)+}\hat{a}_{\mathbf{k}}^{(\lambda)}$ is the number operator. The eigenvalue $n(\mathbf{k}, \lambda)$ of the number operator is called the occupation number and represents the number of photons in the $(\mathbf{k}, \lambda)$ state. The operators $\hat{a}_{\mathbf{k}}^{(\lambda)+}$ and $\hat{a}_{\mathbf{k}}^{(\lambda)}$ are the creation and annihilation operators. The eigenvalues of the energy operator $\hat{H}$ are given by

$$\sum_{\mathbf{k},\lambda}\left[n(\mathbf{k}, \lambda) + \frac{1}{2}\right]\hbar\omega_{\mathbf{k}}. \tag{10.31}$$

By definition, the electromagnetic field momentum is

$$\mathbf{P} = \int_V \varepsilon\mathbf{E} \times \mathbf{B}d\mathbf{r} = \sum_{\mathbf{k},\lambda}\varepsilon\omega_{\mathbf{k}}(b_{\mathbf{k}}^{(\lambda)}\bar{b}_{\mathbf{k}}^{(\lambda)} + \bar{b}_{\mathbf{k}}^{(\lambda)}b_{\mathbf{k}}^{(\lambda)})$$

$$= \sum_{\mathbf{k},\lambda}\frac{\mathbf{k}}{2\omega_{\mathbf{k}}}(p_{\mathbf{k}}^{(\lambda)2} + \omega_{\mathbf{k}}^2 q_{\mathbf{k}}^{(\lambda)2}). \tag{10.32}$$

The operator $\hat{\mathbf{p}}$ can be obtained by replacing $p_{\mathbf{k}}^{(\lambda)}$ and $q_{\mathbf{k}}^{(\lambda)}$ with $\hat{p}_{\mathbf{k}}^{(\lambda)}$ and $\hat{q}_{\mathbf{k}}^{(\lambda)}$ respectively

$$\hat{\mathbf{p}} = \sum_{\mathbf{k},\lambda}\frac{1}{2}\frac{\mathbf{k}}{\omega_{\mathbf{k}}}(\hat{p}_{\mathbf{k}}^{(\lambda)2} + \omega_{\mathbf{k}}^2\hat{q}_{\mathbf{k}}^{(\lambda)2}) = \sum_{\mathbf{k},\lambda}\hbar\mathbf{k}\left(\hat{a}_{\mathbf{k}}^{(\lambda)+}\hat{a}_{\mathbf{k}}^{(\lambda)} + \frac{1}{2}\right). \tag{10.33}$$

The eigenvalues of $\hat{\mathbf{p}}$ are given by

$$\sum_{\mathbf{k},\lambda}\left[n(\mathbf{k}, \lambda) + \frac{1}{2}\right]\hbar\mathbf{k}.$$

The quantized electromagnetic field becomes a system of photons. The number of photons in the $(\mathbf{k}, \lambda)$ state is $n(\mathbf{k}, \lambda)$. Each photon in the $(\mathbf{k}, \lambda)$ state has energy $E_{\mathbf{k}} = \hbar \omega_{\mathbf{k}}$ and momentum $\mathbf{p_k} = \hbar \mathbf{k}$. Note that $E_{\mathbf{k}}^2 = |\mathbf{p_k}|^2 c^2$, which reveals that the rest mass of photon is zero. The ground state of the system corresponds to $n(\mathbf{k}, \lambda) = 0$ for all $(\mathbf{k}, \lambda)$. It follows from (10.31) that the energy of the ground state is $\sum_{\mathbf{k}, \lambda} \hbar \omega_{\mathbf{k}}/2$. Since the index $(\mathbf{k}, \lambda)$ has infinite number of choices, the energy of the ground state (or zero-point energy) is infinite. This is the well-known conceptual difficulty of quantized field theory. The zero-point energy represents the energy of vacuum fluctuations.

The operators $\hat{\mathbf{A}}$ for the vector potential and $\hat{\mathbf{E}}$ for the electric field can be obtained as follows

$$\hat{\mathbf{A}}(\mathbf{r}, t) = \sum_{\mathbf{k}, \lambda} \sqrt{\frac{\hbar}{2 \varepsilon \omega_{\mathbf{k}}}} \left[\hat{a}_{\mathbf{k}}^{(\lambda)+} \mathbf{a}_{\mathbf{k}}^{(\lambda)}(\mathbf{r}) e^{j \omega_{\mathbf{k}} t} + \hat{a}_{\mathbf{k}}^{(\lambda)} \bar{\mathbf{a}}_{\mathbf{k}}^{(\lambda)}(\mathbf{r}) e^{-j \omega_{\mathbf{k}} t} \right],$$

$$\hat{\mathbf{E}}(\mathbf{r}, t) = -j \sum_{\mathbf{k}, \lambda} \sqrt{\frac{\hbar \omega_{\mathbf{k}}}{2 \varepsilon}} \left[\hat{a}_{\mathbf{k}}^{(\lambda)+} \mathbf{a}_{\mathbf{k}}^{(\lambda)}(\mathbf{r}) e^{j \omega_{\mathbf{k}} t} - \hat{a}_{\mathbf{k}}^{(\lambda)} \bar{\mathbf{a}}_{\mathbf{k}}^{(\lambda)}(\mathbf{r}) e^{-j \omega_{\mathbf{k}} t} \right], \tag{10.34}$$

$$\hat{\mathbf{B}}(\mathbf{r}, t) = -j \sum_{\mathbf{k}, \lambda} \sqrt{\frac{\hbar}{2 \varepsilon \omega_{\mathbf{k}}}} \left[\hat{a}_{\mathbf{k}}^{(\lambda)+} \mathbf{k} \times \mathbf{a}_{\mathbf{k}}^{(\lambda)}(\mathbf{r}) e^{j \omega_{\mathbf{k}} t} - \hat{a}_{\mathbf{k}}^{(\lambda)} \mathbf{k} \times \bar{\mathbf{a}}_{\mathbf{k}}^{(\lambda)}(\mathbf{r}) e^{-j \omega_{\mathbf{k}} t} \right].$$

So far we have assumed that the electromagnetic fields are confined in a cube V of side L, and thus $\mathbf{k} = (k_x, k_y, k_z)$ is discrete and is determined by (10.20) with $k_\alpha = 2 \pi n_\alpha / L (\alpha = x, y, z)$. If L approaches infinity, the values of $k_\alpha = 2 \pi n_\alpha / L$ become very dense in $\mathbf{k}$-space. Let $\Delta k_x = \Delta k_y = \Delta k_z = 2\pi/L$. Thus $\Delta k_x \Delta k_y \Delta k_z = (2\pi)^3/V$.

Remark 10.2 (Mode density): The field inside the volume V is the superposition of plane wave modes whose wave vectors are given by (10.20). Each triplet of integers n_x, n_y and n_z defines two modes with different polarizations. Each mode may be associated with a volume $\Delta k_x \Delta k_y \Delta k_z = (2\pi)^3/V$. Let $k = |\mathbf{k}| = 2\pi f / v$. The number of modes whose magnitudes of $\mathbf{k}$ vectors lies between 0 and k is

$$N_f = 2 \cdot \frac{4\pi}{3} k^3 \bigg/ \left(\frac{2\pi}{L} \right)^3 = \frac{8\pi}{3} \frac{f^3 V}{v^3}. \tag{10.35}$$

The **mode density** is defined as the number of modes per unit volume per unit frequency interval, and may be expressed as

$$p(f) = \frac{1}{V} \frac{dN_f}{df} = \frac{8\pi f^2}{v^3}. \tag{10.36}$$

In the limit of $V \to \infty$, the sum over $\mathbf{k}$ may be replaced by an integral $\square$

$$\frac{1}{V} \sum_{\mathbf{k}} (\cdot) = \frac{1}{(2\pi)^3} \sum_{\mathbf{k}} (\cdot) \Delta k_x \Delta k_y \Delta k_z \xrightarrow[v \to \infty]{} \frac{1}{(2\pi)^3} \int d\mathbf{k} (\cdot).$$

The sum over the polarization vector can be obtained by the cosine sum rule

$$\sum_{\lambda=1,2} e_i^{(\lambda)}(\mathbf{k}) e_j^{(\lambda)}(\mathbf{k}) = \delta_{ij} - \frac{k_i k_j}{k^2}.$$

If the polarization vectors are complex, the above equation can be generalized to

$$\sum_{\lambda=1,2} e_i^{(\lambda)}(\mathbf{k}) \bar{e}_j^{(\lambda)}(\mathbf{k}) = \delta_{ij} - \frac{k_i k_j}{k^2}.$$

Introducing a polarization vector in the direction of the magnetic field

$$\mathbf{b}^{(\lambda)}(\mathbf{k}) = \frac{\mathbf{k}}{k} \times \mathbf{e}^{(\lambda)}(\mathbf{k}),$$

then we have

$$\sum_{\lambda=1,2} e_i^{(\lambda)}(\mathbf{k}) \bar{b}_j^{(\lambda)}(\mathbf{k}) = \varepsilon_{ijl} \frac{k_l}{k^2},$$

$$\sum_{\lambda=1,2} b_i^{(\lambda)}(\mathbf{k}) \bar{b}_j^{(\lambda)}(\mathbf{k}) = \delta_{ij} - \frac{k_i k_j}{k^2}.$$

10.2.2   *Quantization in Terms of Spherical Wavefunctions*

The electromagnetic fields can also be expanded in terms of spherical wavefunctions. By separation of variables, we still assume that (10.17), (10.18) and (10.19) hold. We further assume that the fields are confined in a sphere of radius r_0, denoted by V. Physically $\mathbf{a(r)}$ must be bounded in the sphere. Moreover the tangent component of $\mathbf{a(r)}$ on the spherical surface $r = r_0$ must be zero to guarantee that the electric field $\mathbf{E}$ only has a normal component on the spherical surface $r = r_0$. In this case $\mathbf{E} \times \mathbf{H} = 0$ on the spherical surface, that is, there is no electromagnetic energy flowing out of the sphere.

To find the spherical wave solution of (10.19) under the transversality condition, we start from the scalar Helmholtz equation

$$(\nabla^2 + k^2) u(\mathbf{r}) = 0. \tag{10.37}$$

The solution of the above equation in the sphere can be written as

$$u_{mn} = j_n(kr) Y_{mn}(\theta, \varphi),$$

where j_n is the spherical Bessel function, Y_{mn} is the spherical harmonics defined by

$$Y_{mn} = N_{mn} P_n^m(\cos\theta) e^{jm\varphi}, \; n = 0, 1, 2, \cdots, \; |m| = 0, 1, 2, \cdots, n,$$

where N_{mn} is normalization constant

$$N_{mn} = \left[\left(\frac{2m+1}{4\pi} \right) \frac{(n-|m|)!}{(n+|m|)!} \right]^{1/2}.$$

Let $\hat{\mathbf{l}}$ be the angular momentum operator. Since $[\hat{\mathbf{l}}, \nabla^2] = 0$, $\hat{\mathbf{l}} u_{mn}$ is a solution of (10.19) and $\nabla \cdot \hat{\mathbf{l}} u_{mn} = 0$. Introducing the notation $\mathbf{M}_{mn} = j c_n \hat{\mathbf{l}} u_{mn}$, where c_n is the normalization constant to be determined, then

$$(\nabla^2 + k^2)\mathbf{M}_{nm} = 0.$$

The tangential component of $\mathbf{M}_{mn}$ must vanish at $r = r_0$. Thus we have $j_n(kr_0) = 0$. Making use of the asymptotic expression

$$j_n(x) \xrightarrow[v \to \infty]{} \frac{\sin(x - n\pi/2)}{x},$$

we have $kr_0 - n\pi/2 = l\pi (l = 0, 1, 2, \cdots)$ for sufficiently large r_0, that is,

$$k = k_l = \left(l + \frac{n}{2} \right) \frac{\pi}{r_0}.$$

We may introduce another solution of (10.19)

$$\mathbf{N}_{mn} = \frac{c_n}{k} \nabla \times (\hat{\mathbf{l}} u_{mn}) = \frac{1}{jk} \nabla \times \mathbf{M}_{mn},$$

which satisfies transversality condition and the wave equation

$$(\nabla^2 + k^2)\mathbf{N}_{mn} = 0.$$

It is easy to show that

$$\nabla \times (\hat{\mathbf{l}} u_{mn}) = j\hbar \left\{ \frac{\partial}{\partial r} [r j_n(kr)] \right\} \nabla Y_{mn} - j\hbar j_n(kr) \mathbf{r} \nabla^2 Y_{mn},$$

where the first term is the tangent component. So we must let

$$\left. \frac{\partial}{\partial r} [r j_n(kr)] \right|_{r=r_0} = 0.$$

For sufficiently large r_0, the above equation implies $\cos(kr_0 - n\pi/2) = 0$, which gives

$$k = k_l = \left(l + \frac{n+1}{2} \right) \frac{\pi}{r_0}, l = 0, 1, 2 \cdots.$$

The normalization constant c_n can be chosen as

$$c_n = \sqrt{\frac{2}{n(n+1)r_0}\frac{\omega_l}{\hbar}}, \quad \omega_l = k_l c,$$

so that

$$\int_V \mathbf{M}_{mn} \cdot \bar{\mathbf{M}}_{m'n'} d\mathbf{r} = \delta_{mm'}\delta_{nn'}\delta_{k_l k_{l'}},$$

$$\int_V \mathbf{N}_{mn} \cdot \bar{\mathbf{N}}_{m'n'} d\mathbf{r} = \delta_{mm'}\delta_{nn'}\delta_{k_l k_{l'}},$$

$$\int_V \mathbf{M}_{mn} \cdot \bar{\mathbf{N}}_{m'n'} d\mathbf{r} = 0.$$

We now use $\lambda = 0$ to stand for $\mathbf{M}$ and $\lambda = 1$ for $\mathbf{N}$. We also use q to denote the multi-index $[l, m, n]$. Then $\mathbf{M}_{mn}$ and $\mathbf{N}_{mn}$ can be represented by a single notation $\mathbf{a}_q^{(\lambda)}$, and we have

$$\int_V \mathbf{a}_q^{(\lambda)} \cdot \bar{\mathbf{a}}_{q'}^{(\lambda')} d\mathbf{r} = \delta_{qq'}\delta_{\lambda\lambda'}.$$

Given (q, λ), the solution of (10.18) may be written as

$$a_q^{(\lambda)}(t) = b_q^{(\lambda)} e^{j\omega_q t}.$$

The vector potential $\mathbf{A}(\mathbf{r}, t)$ can thus be represented by

$$\mathbf{A}(\mathbf{r}, t) = \sum_{q,\lambda}\left[b_q^{(\lambda)}\mathbf{a}_q^{(\lambda)}(\mathbf{r})e^{j\omega_q t} + \bar{b}_q^{(\lambda)}\bar{\mathbf{a}}_q^{(\lambda)}(\mathbf{r})e^{-j\omega_q t}\right].$$

Following a similar approach leading to (10.26), we may find that the energy operator $\hat{H}$ of the system is

$$\hat{H} = \sum_{q,\lambda}\left(\hat{a}_q^{(\lambda)+}\hat{a}_q^{(\lambda)} + \frac{1}{2}\right)\hbar\omega_q$$

with

$$\left[\hat{a}_q^{(\lambda)}, \hat{a}_{q'}^{(\lambda')+}\right] = \delta_{qq'}\delta_{\lambda\lambda'},$$

$$\left[\hat{a}_q^{(\lambda)+}, \hat{a}_{q'}^{(\lambda')+}\right] = \left[\hat{a}_q^{(\lambda)}, \hat{a}_{q'}^{(\lambda')}\right] = 0.$$

10.3 Quantum Statistics

When a system has a large number of constituent particles, it is impossible to find the evolution of the system by solving the Schrödinger equation. In this case, we may adopt a similar treatment in classical statistical mechanics. Consider a physical system consisting of N identical particles confined in a finite volume V. If the interactions between particles are very weak, the total energy of the system and the number of the particles can be expressed as

$$E = \sum_{j=1}^{\infty} n_j e_j, \quad N = \sum_{j=1}^{\infty} n_j,$$

where n_j is the number of particles with energy e_j. The three parameters N, V and E define a **macrostate** of the system. In classical mechanics, the states of particles in a system are described by their positions and speeds, and all particles are considered distinguishable and can be labeled and tracked. The change of position of any particle results in a different configuration of the system. Each particle can be assumed to be in any state accessible to the system.

In quantum mechanics, the states of particles in a system are described by the wavefunctions. Some particles are indistinguishable from one another, and interchanging any two particles does not result in a new configuration of the system. This means that the wavefunction of the system is invariant with respect to the interchange of the constituent particles. Given the macrostate (N, V, E), there will be a large number of different ways to arrange the N particles in V so that the total energy of the system is E. Each of these different ways defines a **microstate** of the system. In other words, for a given macrostate, there exist a number of corresponding microstates. It is natural to assume that at any time the system in equilibrium is equally likely to be in any one of these microstates. This assumption is known as the **equal a priori probability postulate**, and is the backbone of statistical mechanics. The particles in quantum mechanics can be divided into three categories:

1. Identical but distinguishable particles. An example is a collection of harmonic oscillators which are distinguishable.
2. Identical indistinguishable particles of half-odd-integral spin (fermions). Fermions such as electrons and protons are material particles which obey the Pauli exclusion principle. Any two particles cannot occupy the same quantum state.
3. Identical indistinguishable particles of integral spin (Bosons). Bosons such as photons do not obey Pauli exclusion principle, and more than one particle can occupy the same quantum state.

10.3.1 Statistical States

Let Q be a quantum system composed of identical particles and H be a Hilbert space. A tuple $\Psi = (\psi_1, p_1; \psi_2, p_2; \cdots)$ is referred to as a **statistical state** of Q, where ψ_n satisfy the Schrödinger equation and form a complete orthonormal system in H, and p_n are real numbers with $0 \leq p_n \leq 1$, $n \geq 1$. The real number p_n is called the probability of finding the system Q in the physical state ψ_n. The statistical state Ψ is called a **pure state** if $p_{n_o} = 1$ for some fixed n_0 and $p_n = 0$ for all $n \neq n_0$. Otherwise Ψ is called a **mixed state**. Suppose that we measure the observable $\hat{A} : D(\hat{A}) \subset H \to H$ in the statistical state Ψ. The outcome of the measurement is statistical. The number

$$\langle \hat{A} \rangle_e = \sum_n p_n (\hat{A} \psi_n, \psi_n)$$

is called the **ensemble average** of $\hat{A}$. The number

$$\langle \Delta \hat{A}^2 \rangle_e = \sum_n p_n \left\| \hat{A} \psi_n - \langle \hat{A} \rangle_e \psi_n \right\|^2$$

is called the dispersion.

10.3.2   *Most Probable Distributions*

A single particle's quantum state is characterized by

$$\hat{H} \psi_j = e_j \psi_j,$$

where $\hat{H}$ is the Hamiltonian of the particle when it is in the volume V all by itself, and e_j are the energy eigenvalues and ψ_j are the corresponding eigenfunctions. Consider an N-particle system Q. A macrostate of the system is specified by

$$
\begin{array}{ccccc}
e_1 & e_2 & \cdots & e_j & \cdots, \\
n_1 & n_2 & \cdots & n_j & \cdots,
\end{array}
\tag{10.38}
$$

where n_j is the number of particles with energy $e_j (j = 1, 2, \cdots)$. Thus the total energy E and the particle number N are

$$E = \sum_{j=1}^{\infty} n_j e_j, \quad N = \sum_{j=1}^{\infty} n_j. \tag{10.39}$$

A macrostate of the system is specified once the numbers n_j are given. The probability of finding the system in the macrostate is proportional to the number of microstates, denoted by P, that realize the macrostate. The number of microstates that correspond to the macrostate $(n_1, n_2, \cdots n_j, \cdots)$ are [for example, Yariv, 1982]

1. $P = N! \prod\limits_{j=1}^{\infty} \frac{g_j^{n_j}}{n_j!}$ (Identical distinguishable particles).

2. $P = \prod\limits_{j=1}^{\infty} \frac{g_j!}{(g_j - n_j)! n_j}$ (Identical indistinguishable particles of half-odd-integral spin).

3. $P = \prod\limits_{j=1}^{\infty} \frac{(n_j + g_j - 1)!}{n_j!(g_j - 1)!}$ (Identical indistinguishable particles of integral spin).

where g_j is the degeneracy of state j. The most probable macrostate $(n_1, n_2, \cdots n_j, \cdots)$can be obtained by maximizing P subject to the constraints (10.39). Introducing the function

$$F = \ln P - \alpha \left(\sum_{j=1}^{\infty} n_j - N \right) - \beta \left(\sum_{j=1}^{\infty} n_j e_j - E \right)$$

and using the method of Lagrange multipliers, we obtain the expected number of particles in energy state j

1. $n_j = \frac{g_j}{e^{\alpha+\beta e_j}}$ (Identical distinguishable particles).
2. $n_j = \frac{g_j}{e^{\alpha+\beta e_j}+1}$ (Identical indistinguishable particles of half-odd-integral spin).
3. $n_j = \frac{g_j}{e^{\alpha+\beta e_j}-1}$ (Identical indistinguishable particles of integral spin).

They are called Maxwell-Boltzmann statistics, Fermi-Dirac statistics and Bose-Einstein statistics respectively. Here $\beta = (kT)^{-1}$, $k = 1.3807 \times 10^{-23}$ is the Boltzmann constant, and T is the temperature. The constants α can be determined by the nature of the system of the particles.

10.3.3   Blackbody Radiation

Consider the thermal radiation field inside a large cubic box of sides L. The system is assumed to be at thermal equilibrium at temperature T. The electromagnetic field is quantized using plane wave modes as shown in (10.26). A mode in state j has energy $e_j = (j + 1/2)\hbar\omega$. Since the modes are distinguishable, we use Maxwell-Boltzmann statistics. The probability of finding the mode in state j is

$$p_j = \frac{n_j}{N} = \frac{e^{-e_j/kT}}{\sum\limits_{l=1}^{\infty} e^{-e_l/kT}}.$$

The average thermal energy per mode is

$$E_{av} = \sum_{j=1}^{\infty} p_j e_j = \sum_{j=0}^{\infty} e_j \frac{e^{-e_j/kT}}{\sum\limits_{l=1}^{\infty} e^{-e_l/kT}} = \frac{\hbar\omega}{2} + \frac{\hbar\omega}{e^{\hbar\omega/kT} - 1}.$$

The first term on the right-hand side represents the zero point energy, which cannot be extracted, and thus can be ignored. Therefore

$$E_{av} = \frac{\hbar\omega}{e^{\hbar\omega/kT} - 1}.$$

The product of average thermal energy per mode with mode density (10.36) gives the blackbody energy density per unit frequency

$$\rho(f) = \frac{8\pi h f^3}{v^3(e^{hf/kT} - 1)}.$$

The blackbody spectral intensity is the blackbody energy density per unit frequency multiplied by the velocity of light and averaged over all directions (since the radiation is the same in all directions)

$$I(f) = \frac{\rho(f)}{4\pi} v = \frac{2hf^3}{v^2(e^{hf/kT} - 1)},$$

which is the well-known **Planck's law**.

10.4   Interaction of Electromagnetic Fields With the Small Particle System

So far, the dynamics of particles and fields have been discussed separately. To study the dynamical system consisting of both particles and fields, the fields and the particles are subject to quantum conditions. Quantum electrodynamics has been the most successful theory in studying the interaction between radiation fields and small particle system (atoms and molecules). In quantum electrodynamics, the Hamiltonian is of the form

$$\hat{H} = \hat{H}_{part} + \hat{H}_{rad} + \hat{H}_{int}, \tag{10.40}$$

where $\hat{H}_{part}$ and $\hat{H}_{rad}$ represent the contributions from the particles and electromagnetic fields respectively while $\hat{H}_{int}$ stands for the interaction between fields and particles. Equation (10.40) indicates that the electromagnetic fields can deliver energy to the particles or receive energy from them. When the electromagnetic field is very strong, the influence of the particles on the field is negligibly small. In this case, the field can be considered as an external field and treated classically. This process is called the **semi-classical method**.

10.4.1   The Hamiltonian Function of the Coupled System

The Hamiltonian function can be constructed from the Lagrangian function as indicated in Chapter 2. The Lagrangian function also consists of three parts, one each for the particles, fields and the interaction between them. The least action principle leads to the Lagrangian equations. It is known that the Lagrangian function for a system is not unique. A correct choice of Lagrangian function must guarantee that the Lagrangian equations are the equations of motion. The total Lagrangian function for the system of particles and fields in free space may be written as

$$L = L_{part} + L_{rad} + L_{int}, \tag{10.41}$$

where

$$L_{part} = \frac{1}{2} \sum_\alpha m_\alpha \dot{\mathbf{q}}_\alpha^2 - V(\mathbf{q}_1, \mathbf{q}_2, \cdots),$$

$$L_{rad} = \int \mathcal{L}_{rad} d\mathbf{r}, \quad L_{int} = \int \mathcal{L}_{int} d\mathbf{r},$$

$$\mathcal{L}_{rad} = \frac{\varepsilon_0}{2} \left[\left(\frac{\partial \mathbf{A}(\mathbf{r}, t)}{\partial t} \right)^2 - c^2 [\nabla \times \mathbf{A}(\mathbf{r}, t)]^2 \right],$$

$$\mathcal{L}_{int} = \mathbf{J}^\perp \cdot \mathbf{A}.$$

Here $\mathbf{J}^\perp = \sum_\alpha e_\alpha \dot{\mathbf{q}}_\alpha \cdot \overset{\leftrightarrow}{\delta}{}^\perp (\mathbf{r} - \mathbf{q}_\alpha)$ is the transverse part of the total current $\mathbf{J} = \sum_\alpha e_\alpha \dot{\mathbf{q}}_\alpha \delta(\mathbf{r} - \mathbf{q}_\alpha)$, and V is the potential function due to the electrostatic field, and the Hamiltonian gauge has been assumed. We have

$$\frac{\partial L}{\partial \dot{q}_{\alpha i}} = m_\alpha \dot{q}_{\alpha i} + e_\alpha A_i(\mathbf{q}_\alpha, t),$$

$$\frac{d}{dt} \frac{\partial L}{\partial \dot{q}_{\alpha i}} = m_\alpha \ddot{q}_{\alpha i} + e_\alpha \frac{\partial A_i(\mathbf{q}_\alpha, t)}{\partial t} + e_\alpha \sum_{j=1}^{3} \frac{\partial A_i(\mathbf{q}_\alpha, t)}{\partial q_{\alpha j}} \dot{q}_{\alpha j},$$

$$\frac{\partial L}{\partial q_{\alpha i}} = -\frac{\partial V}{\partial q_{\alpha i}} + e_\alpha \sum_{j=1}^{3} \frac{\partial A_j(\mathbf{q}_\alpha, t)}{\partial q_{\alpha j}} \dot{q}_{\alpha j},$$

$$\frac{d}{dt} \left(\frac{\partial \mathcal{L}}{\partial (\partial A_i / \partial t)} \right) = \varepsilon_0 \frac{\partial^2 A_i}{\partial t^2},$$

$$\frac{\partial \mathcal{L}}{\partial A_i} = \mathbf{J}^\perp,$$

$$\sum_{j=1}^{3} \frac{d}{dx_j} \left(\frac{\partial \mathcal{L}}{\partial (\partial A_i / \partial x_j)} \right) = -\varepsilon_0 c^2 \nabla^2 A_i,$$

where $\mathcal{L} = \mathcal{L}_{rad} + \mathcal{L}_{int}$, and $i = 1, 2, 3$. In deriving the first equation, we have used the following relation

$$\int \overset{\leftrightarrow}{\delta}{}^\| (\mathbf{r} - \mathbf{q}_\alpha) \cdot \mathbf{A}(\mathbf{r}, t) d\mathbf{r} = 0.$$

The Lagrangian equations for the particles are

$$m_\alpha \ddot{q}_{\alpha i} = -\frac{\partial V}{\partial q_{\alpha i}} - e_\alpha \frac{\partial A_i(\mathbf{q}_\alpha, t)}{\partial t} + e_\alpha \sum_{j=1}^{3} \left[\frac{\partial A_j(\mathbf{q}_\alpha, t)}{\partial q_{\alpha i}} - \frac{\partial A_i(\mathbf{q}_\alpha, t)}{\partial q_{\alpha j}} \right] \dot{q}_{\alpha j}$$

$$= -\frac{\partial V}{\partial q_{\alpha i}} + e_\alpha E_i(\mathbf{q}_\alpha, t) + e_\alpha [\mathbf{q}_\alpha \times \mathbf{B}(\mathbf{q}_\alpha, t)]_i.$$

and the Lagrangian equation for the field is

$$\left(\nabla^2 + \frac{1}{c^2}\frac{\partial^2}{\partial t^2}\right) \mathbf{A}(\mathbf{r}, t) = -\frac{1}{\varepsilon_0 c^2}\mathbf{J}^{\perp}(\mathbf{r}, t).$$

Therefore the Lagrangian function defined by (10.41) gives the correct equations of motion.
The generalized momenta conjugate to $\mathbf{q}$ and $\mathbf{A}$ are

$$\mathbf{p}_\alpha = \frac{\partial L}{\partial \dot{\mathbf{q}}_\alpha} = m_\alpha \dot{\mathbf{q}}_\alpha + e_\alpha \mathbf{A}(\mathbf{q}_\alpha, t),$$

$$\mathbf{\Pi} = \frac{\partial \mathcal{L}}{\partial \dot{\mathbf{A}}} = \varepsilon_0 \frac{\partial \mathbf{A}(\mathbf{r}, t)}{\partial t} = -\varepsilon_0 \mathbf{E}(\mathbf{r}, t).$$

Thus the Hamiltonian function of the system can be written as

$$H = \sum_\alpha \mathbf{p}_\alpha \cdot \dot{\mathbf{q}}_\alpha + \int \mathbf{\Pi} \cdot \frac{\partial \mathbf{A}}{\partial t} d\mathbf{r} - L$$

$$= \sum_\alpha \frac{1}{2m_\alpha}[\mathbf{p}_\alpha - e_\alpha \mathbf{A}(\mathbf{q}_\alpha, t)]^2 + V(\mathbf{q}_1, \mathbf{q}_2, \cdots) \qquad (10.42)$$

$$+ \frac{1}{2}\int \left[\frac{\mathbf{\Pi}^2}{\varepsilon_0} + \varepsilon_0 c^2 [\nabla \times \mathbf{A}(\mathbf{r}, t)]^2\right] d\mathbf{r}.$$

10.4.2 Quantization of the Coupled System

The quantization of the coupled system can be carried out by replacing the classical dynamical variables in the Hamiltonian with Hermitian operators in Hilbert space. The operators are subject to the commutation relations

$$\begin{cases} [\hat{q}_{\alpha i}, \hat{p}_{\beta j}] = j\hbar\delta_{ij}\delta_{\alpha\beta}, \\ [\hat{A}_i(\mathbf{r}, t), \hat{\Pi}_j(\mathbf{r}', t)] = j\hbar\delta_{ij}^{\perp}(\mathbf{r} - \mathbf{r}'), \end{cases} \quad i, j = 1, 2, 3; \alpha, \beta = 1, 2, \cdots,$$

where $\delta_{ij}^{\perp}$ are the components of the transverse δ-dyadic.

Consider an atomic system whose Hamiltonian in an electromagnetic field is given by (10.42). Making use of (10.22), Equation (10.42) may be written as

$$H = \sum_\alpha \frac{1}{2m_\alpha}[\mathbf{p}_\alpha - e_\alpha \mathbf{A}(\mathbf{q}_\alpha, t)]^2 + \frac{1}{4\pi\varepsilon_0}\sum_{\alpha<\beta}\frac{e_\alpha e_\beta}{|\mathbf{q}_\alpha - \mathbf{q}_\beta|} + \sum_{\mathbf{k},\lambda}\frac{1}{2}\left(p_{\mathbf{k}}^{(\lambda)2} + \omega_{\mathbf{k}}^2 q_{\mathbf{k}}^{(\lambda)2}\right).$$

The operator $\hat{H}$ can then be obtained by replacing canonical variables with corresponding operators

$$\hat{H} = \sum_\alpha \frac{1}{2m_\alpha}[\hat{\mathbf{p}}_\alpha - e_\alpha \hat{\mathbf{A}}(\mathbf{q}_\alpha, t)]^2 + \frac{1}{4\pi\varepsilon_0}\sum_{\alpha<\beta}\frac{e_\alpha e_\beta}{|\mathbf{q}_\alpha - \mathbf{q}_\beta|} + \sum_{\mathbf{k},\lambda}\frac{1}{2}\left(\hat{p}_{\mathbf{k}}^{(\lambda)2} + \omega_{\mathbf{k}}^2 \hat{q}_{\mathbf{k}}^{(\lambda)2}\right).$$

We rewrite this as

$$\hat{H} = \hat{H}_0 + \hat{H}',$$

where $\hat{H}_0$ is the Hamiltonian for the atomic system and the fields, and $\hat{H}'$ is the Hamiltonian for describing the interaction between the atomic system and fields

$$\hat{H}_0 = \sum_{\mathbf{k},\lambda} \frac{1}{2}\left(\hat{p}_{\mathbf{k}}^{(\lambda)2} + \omega_{\mathbf{k}}^2 \hat{q}_{\mathbf{k}}^{(\lambda)2}\right) + \frac{1}{4\pi\varepsilon_0}\sum_{\alpha<\beta}\frac{e_\alpha e_\beta}{|\mathbf{q}_\alpha - \mathbf{q}_\beta|} + \sum_\alpha \frac{\hat{\mathbf{p}}_\alpha^2}{2m_\alpha},$$

$$\hat{H}' = \sum_\alpha \frac{e_\alpha^2 \hat{\mathbf{A}}^2(\mathbf{q}_\alpha, t)}{2m_\alpha} - \sum_\alpha \frac{e_\alpha}{2m_\alpha}[\hat{\mathbf{p}}_\alpha \cdot \hat{\mathbf{A}}(\mathbf{q}_\alpha, t) + \hat{\mathbf{A}}(\mathbf{q}_\alpha, t) \cdot \hat{\mathbf{p}}_\alpha].$$

Making use of the commutation relation

$$\hat{\mathbf{p}}_\alpha \cdot \hat{\mathbf{A}}(\mathbf{q}_\alpha, t) - \hat{\mathbf{A}}(\mathbf{q}_\alpha, t) \cdot \hat{\mathbf{p}}_\alpha = 0$$

where Hamiltonian gauge is assumed, we have

$$\hat{H}' = \sum_\alpha \frac{e_\alpha^2 \hat{\mathbf{A}}^2(\mathbf{q}_\alpha, t)}{2m_\alpha} - \sum_\alpha \frac{e_\alpha}{m_\alpha}\hat{\mathbf{p}}_\alpha \cdot \hat{\mathbf{A}}(\mathbf{q}_\alpha, t). \tag{10.43}$$

When the wavelength of the radiation is long compared with the dimensions of the atoms, the variation of the electromagnetic field over the atoms in (10.43) can be neglected and the vector potential $\hat{\mathbf{A}}(\mathbf{q}_\alpha, t)$ can be approximated by $\hat{\mathbf{A}}(0, t)$, where 0 denotes the center of mass of the atomic system. We also assume that the radiation field strength is much smaller than the Coulomb fields within the atoms. So we may ignore the first term in (10.43) to get

$$\hat{H}' = -\left(\sum_\alpha \frac{e_\alpha}{m_\alpha}\hat{\mathbf{p}}_\alpha\right) \cdot \hat{\mathbf{A}}(0, t). \tag{10.44}$$

Furthermore we make the approximation $\hat{\mathbf{p}}_\alpha = m_\alpha \dot{\mathbf{q}}_\alpha + e_\alpha \hat{\mathbf{A}}(0, t) \approx m_\alpha \dot{\mathbf{q}}_\alpha$ to rewrite (10.44) as

$$\hat{H}' = -\left(\sum_\alpha e_\alpha \dot{\mathbf{q}}_\alpha\right) \cdot \hat{\mathbf{A}}(0, t) = -\dot{\hat{\boldsymbol{\mu}}} \cdot \hat{\mathbf{A}}(0, t), \tag{10.45}$$

where $\hat{\boldsymbol{\mu}} = \sum_\alpha e_\alpha \mathbf{q}_\alpha$ is the dipole moment of the atomic system. Since adding a term $d[\hat{\boldsymbol{\mu}} \cdot \hat{\mathbf{A}}(0, t)]/dt$ to (10.45) does not affect the final results, Equation (10.45) may be written as

$$\hat{H}' = \hat{\boldsymbol{\mu}} \cdot \dot{\hat{\mathbf{A}}}(0, t) = -\hat{\boldsymbol{\mu}} \cdot \mathbf{E}(0, t). \tag{10.46}$$

10.4.3 Perturbation Theory

Suppose that the Hamiltonian of a system can be written as

$$\hat{H} = \hat{H}_0 + \hat{H}', \tag{10.47}$$

where $\hat{H}_0$ is a Hamiltonian with known eigenvalues and eigenfunctions

$$\hat{H}_0 u_n = E_n u_n, \tag{10.48}$$

and $\hat{H}'$ is small compared to $\hat{H}_0$. We will use perturbation techniques to find the approximate eigenvalues and eigenfunctions of $\hat{H}$. By the postulate of quantum mechanics, the system state changes with time according to the Schrödinger equation

$$\hat{H}\psi(t) = j\hbar \frac{\partial \psi(t)}{\partial t}.$$

If $\hat{H}$ does not explicitly depend on time, we have

$$\psi(t) = \sum_n a_n u_n(\mathbf{r}) e^{-jE_n t/\hbar},$$

where a_n are independent of time and are the expansion coefficients of $\psi(0)$

$$\psi(0) = \sum_n a_n u_n(\mathbf{r})$$

and $\{u_n\}$ is the complete set of eigenfunctions of (10.48). If the system is in the state u_m with energy E_m at $t = 0$, that is, $\psi(0) = u_m$, we have $a_n = \delta_{nm}$, which holds for all subsequent times. Thus we have $\psi(t) = u_m(\mathbf{r}) e^{-jE_m t/\hbar}$, and the system is still in the state u_m.

We now consider a more interesting situation. Suppose that the system is in the state u_m at $t = 0$ with Hamiltonian $\hat{H}_0$, which does not explicitly depend on time. Then the system is perturbed by an external influence $\hat{H}'$, which may depend on time so that the total Hamiltonian is given by

$$\hat{H} = \hat{H}_0 + \lambda \hat{H}',$$

where $\lambda > 0$ is a small parameter. In this case, the system will not remain in the initial state u_m. The state $\psi(t)$ satisfies the Schrödinger equation

$$(\hat{H}_0 + \lambda \hat{H}')\psi(t) = j\hbar \frac{\partial \psi(t)}{\partial t}. \tag{10.49}$$

Since $\{u_n\}$ is a complete set we may write

$$\psi(t) = \sum_n a_n(t) u_n(\mathbf{r}) e^{-jE_n t/\hbar}. \tag{10.50}$$

The expansion coefficients a_n are now dependent of time. Substituting (10.50) into (10.49) and using the orthonormal condition for u_n we obtain

$$\dot{a}_n = -\frac{j}{\hbar} \sum_l \lambda a_l (\hat{H}' u_l, u_n) e^{j\omega_{nl}t} \tag{10.51}$$

where $\omega_{nl} = (E_n - E_l)/\hbar$. Since we have assumed that the system is in the state u_m at $t = 0$, Equation (10.51) is subject to the following initial condition

$$a_n(0) = \delta_{nm}. \tag{10.52}$$

It is very difficult to solve (10.49) for a general $\hat{H}'$. However, if $\hat{H}'$ is very small, that is, $\hat{H}' \ll \hat{H}_0$, we may adopt the perturbation method to find the approximate solutions. Assuming the following power series expansion for a_n

$$a_n = a_n^{(0)} + \lambda a_n^{(1)} + \lambda^2 a_n^{(2)} + \cdots \tag{10.53}$$

and substituting this into (10.51) yields

$$\dot{a}_n^{(0)} + \lambda \dot{a}_n^{(1)} + \lambda^2 \dot{a}_n^{(2)} + \cdots = -\frac{j}{\hbar} \sum_l \lambda \left(a_l^{(0)} + \lambda a_l^{(1)} + \lambda^2 a_l^{(2)} + \cdots \right) (\hat{H}' u_l, u_n) e^{j\omega_{nl}t}.$$

Equating the coefficients for the same powers of λ, we obtain

$$\dot{a}_n^{(0)} = 0,$$
$$\dot{a}_n^{(1)} = -\frac{j}{\hbar} \sum_l a_l^{(0)} (\hat{H}' u_l, u_n) e^{j\omega_{nl}t},$$
$$\vdots \tag{10.54}$$
$$\dot{a}_n^{(s)} = -\frac{j}{\hbar} \sum_l a_l^{(s-1)} (\hat{H}' u_l, u_n) e^{j\omega_{nl}t}.$$

The first equation gives $a_n^{(0)} = $ constant. We may let

$$a_n(0) = a_n^{(0)} = \delta_{nm}.$$

This corresponds to the situation of $\lambda = 0$ where no perturbation exists. Thus the second equation of (10.54) becomes

$$\dot{a}_n^{(1)} = -\frac{j}{\hbar} (\hat{H}' u_m, u_n) e^{j\omega_{nm}t}.$$

The first-order solution may be obtained by letting $\lambda = 1$ in (10.53)

$$a_n(t) = a_n^{(0)} + a_n^{(1)} = \delta_{nm} + a_n^{(1)}.$$

Usually the final state is different from the initial state. So we have

$$a_n(t) = a_n^{(1)} = -\frac{j}{\hbar} \int_0^t (\hat{H}' u_m, u_n) e^{j\omega_{nm}t} dt \, (n \neq m),$$
(10.55)

where the initial condition $a_n^{(1)}(0) = 0 (n \neq m)$ has been used. Therefore $\left|a_n^{(1)}\right|^2$ is the probability of finding the system at time t in the state u_n when the system is in the state u_m at $t = 0$, and is called the **transition probability** from state u_m to u_n. The **transition rate** per unit time is defined by

$$W_{m \to n}(t) = \frac{d}{dt} \left|a_n^{(1)}\right|^2.$$
(10.56)

The perturbation method can be used to obtain approximate solutions in many practical situations where the radiation field strengths are much smaller than Coulomb fields within the atoms.

Example 10.3 (Time-harmonic perturbation): The time-harmonic perturbation refers to the perturbation that varies sinusoidally with time

$$\hat{H}' = \hat{H}_s e^{j\omega t} + \hat{H}_s^* e^{-j\omega t}, \, t > 0,$$

where $\hat{H}_s = \hat{H}_s^*$ is independent of time. Then it is readily found that

$$a_n^{(1)} = -\frac{j}{\hbar} \int_0^t (\hat{H}' u_m, u_n) e^{j\omega_{nm}t} dt$$

$$= -\frac{1}{\hbar} \left[(\hat{H}_s u_m, u_n) \frac{e^{j(\omega_{nm}+\omega)t} - 1}{\omega_{nm} + \omega} + (\hat{H}_s^* u_m, u_n) \frac{e^{j(\omega_{nm}-\omega)t} - 1}{\omega_{nm} - \omega} \right].$$

We assume that the frequency of perturbation is very close to $|\omega_{nm}|$, that is, $\hbar\omega \approx |E_n - E_m|$. In this case, the transition probability may be written as

$$\left|a_n^{(1)}\right|^2 \approx \frac{\left|(\hat{H}_s u_m, u_n)\right|^2}{\hbar^2} \frac{\sin^2[(\omega_{nm} \pm \omega)t/2]}{[(\omega_{nm} \pm \omega)t/2]^2}$$
(10.57)

where $+$ is for the situation where $\omega_{mn} \approx \omega$, and $-$ is for $\omega_{nm} \approx \omega$. When t is very large, the transition rate is given by

$$W_{m \to n} = \frac{d}{dt} \left|a_n^{(1)}\right|^2 \approx \frac{2\pi \left|(\hat{H}_s u_m, u_n)\right|^2}{\hbar^2} \delta(\omega_{nm} \pm \omega)$$

$$= \frac{2\pi \left|(\hat{H}_s u_m, u_n)\right|^2}{\hbar} \delta(E_n - E_m \pm \hbar\omega),$$
(10.58)

where the following relation has been used

$$\frac{\sin^2(xt/2)}{(x/2)^2} \rightarrow 2\pi t \delta(x), t \rightarrow \infty.$$

We now calculate the transition probability from m to a group of final states clustered about the state n. Let $\rho(\omega_{nm})$ be the density of these final states per unit of ω_{nm}. Equation (10.57) can be expressed as

$$\left|a_n^{(1)}\right|^2 \approx \frac{1}{\hbar^2} \int_{-\infty}^{\infty} \left|(\hat{H}_s u_m, u_n)\right|^2 \frac{\sin^2\left[(\omega_{nm} \pm \omega)t/2\right]}{\left[(\omega_{nm} \pm \omega)/2\right]^2} \rho(\omega_{nm}) d\omega_{nm}.$$

When t is very large $\rho(\omega_{nm})$ is a slowly varying function compared to the rest of the integrand. Thus

$$\left|a_n^{(1)}\right|^2 \approx \frac{2\pi}{\hbar^2} \left|(\hat{H}_s u_m, u_n)\right|^2 \rho(\omega_{nm} \approx \mp \omega)t.$$

The transition rate is

$$W_{m \rightarrow n} = \frac{d}{dt} \left|a_n^{(1)}(t)\right|^2 = \frac{2\pi}{\hbar^2} \left|(\hat{H}_s u_m, u_n)\right|^2 \rho(\omega_{nm} \approx \mp \omega)$$

$$= \frac{2\pi}{\hbar^2} \left|(\hat{H}_s u_m, u_n)\right|^2 \rho(E = E_m \mp \hbar\omega).$$

This is called **Fermi's Golden Rule**, and it represents a transition rate from a single state m to a continuum of states n.   □

10.4.4   Induced Transition and Spontaneous Transition

An atomic system subject to electromagnetic fields may undergo transition to lower or higher states. This process is called **induced transition**. When the fields are not present, an atom may undergo transition to lower states. This process is called **spontaneous transition**.

10.4.4.1 Induced transition

Let us consider an atomic system that has two possible states u_1 and u_2 with $E_2 > E_1$ and $\Delta E = E_2 - E_1 = \hbar\omega_{21}$. The atomic system is perturbed by electromagnetic radiation with

$$\mathbf{E}(\mathbf{r}, t) = \mathbf{u} E_0(\mathbf{r}) \cos \omega t,$$

where $\mathbf{u}$ is a unit vector. In atomic system, the speed of electrons is much smaller than the light speed. This implies $|\mathbf{v} \times \mathbf{B}| / |\mathbf{E}| \sim v/c \ll 1$. For this reason, the influence of magnetic field on the electrons is much smaller than that of the electric field and can be ignored. The potential energy of an electron in the field $\mathbf{E}$ is

$$V = -e\mathbf{E} \cdot \mathbf{r} = -e\mathbf{u} \cdot \mathbf{r}E_0(\mathbf{r})\cos\omega t.$$

We may take the perturbation Hamiltonian due to the field as

$$\hat{H}' = -\frac{e\mathbf{u} \cdot \hat{\mathbf{r}}E_0(\mathbf{r})}{2}(e^{j\omega t} + e^{-j\omega t}) = \hat{H}_s(e^{j\omega t} + e^{-j\omega t}),$$

where $\hat{H}_s = -e\mathbf{u} \cdot \hat{\mathbf{r}}E_0(\mathbf{r})/2$. On the atomic scale, the electric field can be considered as uniform. By means of (10.58), we have

$$W_{1\to2} = \frac{\pi e^2 E_0^2}{2\hbar}|r_{12}|^2 \,\delta\,(\Delta E - \hbar\omega) = W_{2\to1} = W_i, \tag{10.59}$$

where $r_{12} = \int \mathbf{u} \cdot \mathbf{r}u_1 u_2^* d\mathbf{r}$. In real atomic systems, it is impossible to specify the exact value of $\Delta E = E_2 - E_1$. Thus we have to introduce a probability density function $g(E)$ with $g(E)dE$ being the probability of finding ΔE between E and $E + dE$ and $\int_{-\infty}^{\infty} g(E)dE = 1$. Thus, the induced transition rate W_i can be obtained by superimposing all possible values of ΔE with $g(E)$ as a weighting function

$$\begin{aligned}
W_i &= \frac{\pi e^2 E_0^2}{2\hbar}|r_{12}|^2 \int_{-\infty}^{\infty} \delta\,(\Delta E - \hbar\omega)\,g(\Delta E)d\Delta E \\
&= \frac{\pi e^2 E_0^2}{2\hbar}|r_{12}|^2\,g(\hbar\omega) = \frac{e^2 E_0^2}{4\hbar^2}|r_{12}|^2\,g(f),
\end{aligned} \tag{10.60}$$

where f stands for the frequency, and the relation $g(f)df = g(\hbar\omega)d\hbar\omega$ has been used. The power-flow density, denoted by I_f, is the product of energy density and energy velocity

$$I_f = \frac{1}{2}\varepsilon E_0^2 \times \frac{1}{\sqrt{\mu\varepsilon}} = \frac{1}{2}cn\varepsilon_0 E_0^2,$$

where $n = \sqrt{\varepsilon_r}$ is the index of refraction and $c = 1/\sqrt{\mu_0\varepsilon_0}$. Equation (10.60) may be rewritten as

$$W_i = \frac{e^2 |r_{12}|^2}{2\hbar^2 cn\varepsilon_0}g(f)I_f. \tag{10.61}$$

10.4.4.2 Spontaneous transition

We still consider an atom which has two possible states u_1 and u_2 with $E_2 > E_1$ and is subject to electromagnetic radiation $\mathbf{E}(\mathbf{r}, t) = \mathbf{u}E_0(\mathbf{r}, t)$, where $\mathbf{u}$ is a unit vector. We ignore the influence of magnetic field on the electrons. The potential energy of an electron in the field $\mathbf{E}$ is

$$V = -e\mathbf{E} \cdot \mathbf{r} = -e\mathbf{u} \cdot \mathbf{r}E_0(\mathbf{r}, t). \tag{10.62}$$

According to (10.34), the electric field may be quantized and expressed as a linear combination of radiation modes as follows

$$\hat{\mathbf{E}}(\mathbf{r}, t) = -j\mathbf{u} \sum_{\mathbf{k}} \sqrt{\frac{\hbar \omega_{\mathbf{k}}}{2V\varepsilon}} \left[\hat{a}_{\mathbf{k}}^{+} e^{j\omega_{\mathbf{k}}t} e^{-j\mathbf{k}\cdot\mathbf{r}} - \hat{a}_{\mathbf{k}} e^{-j\omega_{\mathbf{k}}t} e^{j\mathbf{k}\cdot\mathbf{r}} \right].$$

In order to get the transition rate from state u_2 to u_1, we may first derive the transition rate due to a single radiation mode and then sum up the transition rates over all radiation modes. Assume that the atom is initially in the excited state u_2 and the radiation mode is the state $u_n(\mathbf{k})$. The initial state of the combined system is characterized as $u_i = u_2 u_n(\mathbf{k})$. After the transition, the atom is in the lower state u_1 and the radiation mode is elevated to state $u_{n+1}(\mathbf{k})$ for gaining a quantum of radiation. The final state of the combined system is $u_f = u_1 u_{n+1}(\mathbf{k})$. The initial and final energies of the combined system are

$$E_i = E_2 + \hbar \omega_{\mathbf{k}} \left[n(\mathbf{k}) + \frac{1}{2} \right], \quad E_f = E_1 + \hbar \omega_{\mathbf{k}} \left[n(\mathbf{k}) + \frac{1}{2} + 1 \right].$$

The radiation mode is considered as a perturbation. Considering (10.62), the perturbation Hamiltonian can be taken as

$$\hat{H}' = \hat{H}_s e^{j\omega_{\mathbf{k}}t} + \hat{H}_s^{*} e^{-j\omega_{\mathbf{k}}t} = j e\mathbf{u} \cdot \mathbf{r} \sqrt{\frac{\hbar \omega_{\mathbf{k}}}{2V\varepsilon}} \left[\hat{a}_{\mathbf{k}}^{+} e^{j\omega_{\mathbf{k}}t} e^{-j\mathbf{k}\cdot\mathbf{r}} - \hat{a}_{\mathbf{k}} e^{-j\omega_{\mathbf{k}}t} e^{j\mathbf{k}\cdot\mathbf{r}} \right], \tag{10.63}$$

where $\hat{H}_s = j e \sqrt{\frac{\hbar \omega_{\mathbf{k}}}{2V\varepsilon}} \mathbf{u} \cdot \mathbf{r} \hat{a}_{\mathbf{k}}^{+} e^{-j\mathbf{k}\cdot\mathbf{r}}$. The transition rate is thus given by (10.58)

$$W(\mathbf{k}) = \frac{2\pi \left| (\hat{H}_s u_i, u_f) \right|^2}{\hbar} \delta (E_1 - E_2 + \hbar \omega_{\mathbf{k}}).$$

Note that

$$\hat{H}_s u_i = j e\mathbf{u} \cdot \mathbf{r}u_2 e^{-j\mathbf{k}\cdot\mathbf{r}} \sqrt{\frac{\hbar \omega_{\mathbf{k}}}{2V\varepsilon}} \sqrt{n(\mathbf{k}) + 1}u_{n+1}(\mathbf{k}),$$

and

$$(\hat{H}_s u_i, u_f) = j e e^{-j\mathbf{k}\cdot\mathbf{r}} r_{12} \sqrt{\frac{\hbar \omega_{\mathbf{k}}}{2V\varepsilon}} \sqrt{n(\mathbf{k}) + 1}.$$

where $r_{12} = (\mathbf{u} \cdot \mathbf{r}u_2, u_1)$. So the transition rate is

$$W(\mathbf{k}) = W_s(\mathbf{k}) + W_i(\mathbf{k}),$$

where $W_s(\mathbf{k})$ and $W_i(\mathbf{k})$ stand for the spontaneous transition rate and induced transition rate respectively

$$W_s(\mathbf{k}) = \frac{\pi e^2 \omega_{\mathbf{k}}}{V \varepsilon} |r_{12}|^2 \delta(E_1 - E_2 + \hbar \omega_{\mathbf{k}}),$$

$$W_i(\mathbf{k}) = \frac{\pi e^2 \omega_{\mathbf{k}}}{V \varepsilon} n(\mathbf{k}) |r_{12}|^2 \delta(E_1 - E_2 + \hbar \omega_{\mathbf{k}}).$$

The induced transition rate is thus proportional to the mode energy density $n(\mathbf{k})\hbar\omega(\mathbf{k})/V$ while the spontaneous transition rate is independent of the mode energy. The total spontaneous and induced transition rate can be obtained by summing over all radiation modes. Since the mode density is (see (10.36))

$$p(f_{\mathbf{k}}) = \frac{8\pi n^3 f_{\mathbf{k}}^2}{c^3},$$

where $c = 1/\sqrt{\mu_0 \varepsilon_0}$, the total spontaneous transition rate is given by

$$W_s = \frac{1}{t_s} = \int_0^\infty W_s(\mathbf{k}) p(f_{\mathbf{k}}) V df_{\mathbf{k}} = \frac{16\pi^3 n^3 e^2 f_0^3}{\varepsilon h c^3} |r_{12}|^2, \qquad (10.64)$$

where we have let $E_2 - E_1 = h f_0$ and t_s is the spontaneous lifetime.

10.4.5 Absorption and Amplification

When electromagnetic waves propagate in a medium, the medium absorbs the energy of the field (the energy of a photon is delivered to electrons of an atom) and causes the attenuation of the field along with dispersion in which the phase velocity of a wave depends on its frequency. It follows from (10.61) and (10.64) that the induced transition rate in a two-level atomic system may be written as

$$W_i = \frac{\lambda^3}{8\pi h n^2 c t_s} g(f) I_f \qquad (10.65)$$

where λ is the wavelength of the wave in free space. Let N_1 and N_2 be the atomic population densities (atoms/m^3) in levels 2 and 1 ($E_2 > E_1$) respectively. Consider a differential volume in the medium which is illuminated by a plane wave propagating in z direction. The differential volume has an area A in the (x, y)-plane and a thickness dz in z direction, as shown in Figure 10.1. The power-flow densities of the plane wave at z and $z + dz$ are denoted by $I_f(z)$ and $I_f(z + dz)$ respectively. Since the increase of the power flow should be equal to the power gain induced by the net transition from level 2 to level 1, we have

$$\frac{dI_f}{dz} = (N_2 - N_1) W_i h f = -\alpha I_f,$$

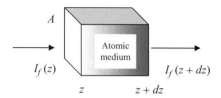

Figure 10.1 A differential volume of an atomic medium

where $\alpha = (N_1 - N_2)\frac{\lambda^2 g(f)}{8\pi n^2 t_s}$ is called the absorption coefficient. The solution of the equation is

$$I_f(z) = I_f(0)e^{-\alpha z}.$$

If $N_1 > N_2$, the medium absorbs energy from the incident field and the incident field is attenuated. If $N_2 > N_1$ (called the **population inversion**), we let $\gamma = -\alpha$, and it is called the **gain constant**. In the latter case, the incident wave gets energy from the medium and it is amplified while passing through the medium. This is the basis of laser.

10.4.6 Quantum Mechanical Derivation of Dielectric Constant

Consider a two-level atom with energies E_1 and E_2

$$\hat{H}_0\psi_n(\mathbf{r}) = E_n\psi_n(\mathbf{r})(n = 1, 2),$$

where $\hat{H}_0$ is the Hamiltonian of the unperturbed atom. When an electromagnetic field is applied to the atom, the total Hamiltonian of the two-level system may be expressed as

$$\hat{H} = \hat{H}_0 + \hat{H}'.$$

Here the interaction Hamiltonian is assumed to be of the dipole type

$$\hat{H}' = -\hat{\boldsymbol{\mu}} \cdot \mathbf{E}(t) = -\hat{\mu}E(t)$$

where $\hat{\mu}$ is the projection of the dipole operator along the electric field $\mathbf{E}$. The wave function ψ of the two-level system satisfies

$$\hat{H}\psi(\mathbf{r}, t) = j\hbar\frac{\partial\psi(\mathbf{r}, t)}{\partial t}, \tag{10.66}$$

and may be expanded in terms of ψ_n

$$\psi(\mathbf{r}, t) = \sum_{n=1,2} C_n(t)\psi_n(\mathbf{r}),$$

where $C_n(t) = (\psi, \psi_n)$. Introducing this into (10.66) yields

$$j\hbar \sum_{n=1,2} \frac{dC_n(t)}{dt} \psi_n(\mathbf{r}) = \sum_{n=1,2} C_n(t) \left[E_n \psi_n(\mathbf{r}) + \hat{H}' \psi_n(\mathbf{r}) \right].$$

Multiplying both sides by $\bar{\psi}_m$ and taking the integration, we obtain

$$\frac{dC_1(t)}{dt} = \frac{1}{j\hbar} C_1(t) \left[E_1 - E(t)\mu_{11} \right] - \frac{1}{j\hbar} C_2(t) E(t)\mu_{21},$$

$$\frac{dC_2(t)}{dt} = \frac{1}{j\hbar} C_2(t) \left[E_2 - E(t)\mu_{22} \right] - \frac{1}{j\hbar} C_1(t) E(t)\mu_{12},$$

(10.67)

where $\mu_{mn} = (\hat{\mu}\psi_m, \psi_n)$. It follows from (10.67) that

$$\frac{dC_1 \bar{C}_1}{dt} = -\frac{1}{j\hbar} C_1 \bar{C}_1 \left[E_1 - E(t)\bar{\mu}_{11} \right] + \frac{1}{j\hbar} C_1 \bar{C}_1 \left[E_1 - E(t)\mu_{11} \right]$$

$$+ \frac{1}{j\hbar} C_1 \bar{C}_2 E(t)\bar{\mu}_{21} - \frac{1}{j\hbar} \bar{C}_1 C_2 E(t)\mu_{21},$$

$$\frac{dC_2 \bar{C}_2}{dt} = -\frac{1}{j\hbar} C_2 \bar{C}_2 \left[E_2 - E(t)\bar{\mu}_{22} \right] + \frac{1}{j\hbar} C_2 \bar{C}_2 \left[E_2 - E(t)\mu_{22} \right]$$

$$+ \frac{1}{j\hbar} \bar{C}_1 C_2 E(t)\bar{\mu}_{12} - \frac{1}{j\hbar} C_1 \bar{C}_2 E(t)\mu_{12},$$

$$\frac{dC_1 \bar{C}_2}{dt} = -\frac{1}{j\hbar} C_1 \bar{C}_2 \left[E_2 - E(t)\bar{\mu}_{22} \right] + \frac{1}{j\hbar} C_1 \bar{C}_2 \left[E_1 - E(t)\mu_{11} \right]$$

$$+ \frac{1}{j\hbar} C_1 \bar{C}_1 E(t)\bar{\mu}_{12} - \frac{1}{j\hbar} C_2 \bar{C}_2 E(t)\mu_{21},$$

$$\frac{dC_2 \bar{C}_1}{dt} = \frac{1}{j\hbar} C_2 \bar{C}_1 \left[E_2 - E(t)\mu_{22} \right] - \frac{1}{j\hbar} C_2 \bar{C}_1 \left[E_1 - E(t)\bar{\mu}_{11} \right]$$

$$- \frac{1}{j\hbar} C_1 \bar{C}_1 E(t)\mu_{12} + \frac{1}{j\hbar} C_2 \bar{C}_2 E(t)\bar{\mu}_{21}.$$

The expectation of $\hat{\mu}$ is given by

$$\langle \hat{\mu} \rangle = (\hat{\mu}\psi, \psi) = \left(\sum_{m=1,2} C_m(t) \hat{\mu} \psi_m(\mathbf{r}), \sum_{n=1,2} C_n(t) \psi_n(\mathbf{r}) \right) = \sum_{m=1,2} \sum_{n=1,2} C_m \bar{C}_n \mu_{mn}.$$

If the system contains a large number of identical atoms, we may take the ensemble average of the dipole moment

$$\langle \hat{\mu} \rangle_e = \sum_{m=1,2} \sum_{n=1,2} \langle C_m \bar{C}_n \rangle_e \mu_{mn} = \sum_{m=1,2} \sum_{n=1,2} \rho_{mn} \mu_{mn},$$

where $\rho_{mn} = \langle C_m \bar{C}_n \rangle_e$. Physically ρ_{mn} are the density matrix elements. The diagonal elements ρ_{nn} denote the population of the state ψ_n, and the off-diagonal elements ρ_{mn} denote the coherence. Apparently ρ_{mn} satisfy the same equations as $C_m \bar{C}_n$. By symmetry we assume that

$$\mu_{11} = \mu_{22} = 0, \mu_{12} = \mu_{21} = \mu,$$

where μ is real. Then

$$\langle \hat{\mu} \rangle_e = \mu(\rho_{12} + \rho_{21}). \tag{10.68}$$

It is easy to show that

$$\frac{d\rho_{21}}{dt} = -j\rho_{21}\frac{(E_2 - E_1)}{\hbar} + j\frac{\mu E(t)}{\hbar}(\rho_{11} - \rho_{22}),$$
$$\frac{d}{dt}(\rho_{11} - \rho_{22}) = j\frac{2}{\hbar}\mu E(t)(\rho_{21} - \bar{\rho}_{21}).$$

Intuitively, once the excitation stops we may expect the population ρ_{nn} will tend toward their thermodynamic equilibrium levels $\rho_{nn}^{(0)}$ with a certain time constant resulting from stochastic interactions. This time constant, denoted by T_1 is called diagonal relaxation time or population lifetime. In the same way, we may expect the off-diagonal elements to lose coherence with a time constant T_2. Introducing these relaxation times, the equations for density matrix elements may be modified as follows

$$\frac{d\rho_{21}}{dt} = -j\rho_{21}\frac{(E_2 - E_1)}{\hbar} + j\frac{\mu E(t)}{\hbar}(\rho_{11} - \rho_{22}) - \frac{\rho_{21}}{T_2},$$
$$\frac{d}{dt}(\rho_{11} - \rho_{22}) = j\frac{2}{\hbar}\mu E(t)(\rho_{21} - \bar{\rho}_{21}) - \frac{(\rho_{11} - \rho_{22}) - (\rho_{11}^{(0)} - \rho_{22}^{(0)})}{T_1}.$$

Assuming $E(t) = E_0 \cos \omega t$, the solution of the off-diagonal element may be found as follows

$$\rho_{21} = \frac{(\omega_0 - \omega)\Omega T_2^2(\rho_{11}^{(0)} - \rho_{22}^{(0)})}{1 + (\omega - \omega_0)^2 T_2^2 + 4\Omega^2 T_1 T_2} + j\frac{\Omega T_2(\rho_{11}^{(0)} - \rho_{22}^{(0)})}{1 + (\omega - \omega_0)^2 T_2^2 + 4\Omega^2 T_1 T_2},$$

where $\omega_0 = (E_2 - E_1)/\hbar$ and $\Omega = \mu E_0/2\hbar$. Let N denote the number of atoms per unit volume. The magnitude of the polarization vector is

$$P(t) = N\langle \hat{\mu} \rangle_e = \frac{\mu^2(N_1^{(0)} - N_2^{(0)})T_2}{\hbar} \cdot \frac{(\omega_0 - \omega)T_2 E_0 \cos \omega t + E_0 \sin \omega t}{1 + (\omega - \omega_0)^2 T_2^2 + 4\Omega^2 T_1 T_2},$$

where $N_1^{(0)} = N\rho_{11}^{(0)}$, $N_2^{(0)} = N\rho_{22}^{(0)}$. Comparing the above equation with the following

$$P(t) = \text{Re}(\varepsilon_0 \chi_e E_0 e^{j\omega t}) = \varepsilon_0 \chi_e' E_0 \cos \omega t + \varepsilon_0 \chi_e'' E_0 \sin \omega t,$$

where $\chi_e = \chi'_e - j\chi''_e$, we obtain

$$\chi'_e = \frac{\mu^2 (N_1^{(0)} - N_2^{(0)})T_2}{\varepsilon_0 \hbar} \cdot \frac{1}{1 + (\omega - \omega_0)^2 T_2^2 + 4\Omega^2 T_1 T_2},$$

$$\chi''_e = \frac{\mu^2 (N_1^{(0)} - N_2^{(0)})T_2}{\varepsilon_0 \hbar} \cdot \frac{(\omega_0 - \omega)T_2}{1 + (\omega - \omega_0)^2 T_2^2 + 4\Omega^2 T_1 T_2}.$$

(10.69)

The dielectric constant is then given by

$$\varepsilon = \varepsilon' - j\varepsilon', \varepsilon' = \varepsilon_0(1 + \chi'_e), \varepsilon'' = \varepsilon_0 \chi''_e.$$

10.5 Relativistic Quantum Mechanics

The Schrödinger equation (10.9) is not covariant and thus must be modified so that it is compatible with special relativity.

10.5.1 The Klein–Gordon Equation

We may start with the following identity for a free particle in special relativity (see Chapter 9)

$$\mathbf{p}^2 c^2 + m_0^2 c^4 = E^2.$$

(10.70)

After quantization we obtain

$$[(-j\hbar\nabla)^2 c^2 + m_0^2 c^4]\psi = \left(j\hbar\frac{\partial}{\partial t}\right)^2 \psi.$$

Rearranging terms gives

$$\left(\nabla^2 - \frac{1}{c^2}\frac{\partial^2}{\partial t^2}\right)\psi = \frac{m_0^2 c^2}{\hbar^2}\psi.$$

(10.71)

This is the Klein–Gordon equation. The equation involves a second-order time derivative, so we must specify the initial values of both ψ and $\partial\psi/\partial t$ to solve the equation. This is unacceptable in quantum mechanics as the wavefunction ψ is supposed to contain all the information for the prediction of the system's behavior. Therefore the Klein–Gordon equation is not suitable for describing a single particle. In 1934, the Austrian physicists Pauli and Victor Frederick Weisskopf (1908–2002) reinterpreted the Klein–Gordon equation as a field equation (like Maxwell equations). Since the Klein–Gordon equation is a scalar field equation, it is used to describe the spinless particles.

10.5.2 The Dirac Equation

In order to overcome the difficulties encountered in the Klein–Gordon equation, Dirac proposed the relativistic wave equation for particles in 1928. After quantization, Equation (10.70) becomes

$$j\hbar\frac{\partial\psi}{\partial t} = \sqrt{\hat{\mathbf{p}}^2c^2 + m_0^2c^4}\,\psi. \tag{10.72}$$

This is not easy to work with because of the square root. Following Dirac, we may assume the equation sought is of the form, called the **Dirac equation**

$$j\hbar\frac{\partial\psi}{\partial t} = (c\alpha_x\hat{p}_x + c\alpha_y\hat{p}_y + c\alpha_z\hat{p}_z + \beta m_0c^2)\psi, \tag{10.73}$$

where α_i ($i = x, y, z$) and β are constants to be determined. It follows from (10.72) and (10.73) that

$$\hat{\mathbf{p}}^2c^2 + m_0^2c^4 = (c\alpha_x\hat{p}_x + c\alpha_y\hat{p}_y + c\alpha_z\hat{p}_z + \beta m_0c^2)^2.$$

Equating the similar terms gives

$$\alpha_x^2 + \alpha_y^2 + \alpha_z^2 = 1,$$
$$\alpha_x\alpha_y + \alpha_y\alpha_x = \alpha_z\alpha_y + \alpha_y\alpha_z = \alpha_x\alpha_z + \alpha_z\alpha_x = 0,$$
$$\alpha_x\beta + \beta\alpha_x = \alpha_y\beta + \beta\alpha_y = \alpha_z\beta + \beta\alpha_z = 0.$$

These conditions can be met only if α_i and β are matrices. One possible choice is

$$\alpha_i = \begin{bmatrix} 0 & \sigma_i \\ \sigma_i & 0 \end{bmatrix}, \beta = \begin{bmatrix} I & 0 \\ 0 & -I \end{bmatrix}, i = x, y, z \tag{10.74}$$

where

$$\sigma_x = \begin{bmatrix} 0 & 1 \\ 1 & 0 \end{bmatrix}, \sigma_y = \begin{bmatrix} 0 & -j \\ j & 0 \end{bmatrix}, \sigma_z = \begin{bmatrix} 1 & 0 \\ 0 & -1 \end{bmatrix} \tag{10.75}$$

are **Pauli spin matrices** and I is 2×2 unit matrix. Since α_i and β are matrices, ψ must be vector

$$\psi = \begin{bmatrix} \psi_1 & \psi_2 & \psi_3 & \psi_4 \end{bmatrix}^T = \begin{bmatrix} \psi_+ & \psi_- \end{bmatrix}^T,$$

and the Dirac equation (10.73) can be written as

$$j\hbar\frac{\partial\psi_+}{\partial t} = \sigma\cdot\hat{\mathbf{p}}c\psi_- + m_0c^2\psi_+,$$
$$j\hbar\frac{\partial\psi_-}{\partial t} = \sigma\cdot\hat{\mathbf{p}}c\psi_+ - m_0c^2\psi_-, \tag{10.76}$$

where $\boldsymbol{\sigma} = \sigma_x \mathbf{u}_x + \sigma_y \mathbf{u}_y + \sigma_z \mathbf{u}_z$, $\psi_+ = [\psi_1, \psi_2]^T$ and $\psi_- = [\psi_3, \psi_4]^T$. Note that any solution of the Dirac equation is also a solution of the Klein–Gordon equation, but the converse is not true.

If the particle described by the Dirac equation has a charge q, subject to an applied electromagnetic field characterized by the vector potential $\mathbf{A}$ and scalar ϕ, the Dirac equation (10.73) must be modified according to the correspondence principle by replacing $\hat{\mathbf{p}}$ with $\hat{\mathbf{p}} - q\mathbf{A}$

$$j\hbar\frac{\partial \psi}{\partial t} = \left[c\alpha_x(\hat{p}_x - qA_x) + c\alpha_y(\hat{p}_y - qA_y) + c\alpha_z(\hat{p}_z - qA_z) + \beta m_0 c^2 \right] \psi. \qquad (10.77)$$

Accordingly, Equation (10.76) must be modified as

$$j\hbar\frac{\partial \psi_+}{\partial t} = \boldsymbol{\sigma} \cdot (\hat{\mathbf{p}} - q\mathbf{A})c\psi_- + m_0 c^2 \psi_+,$$

$$(10.78)$$

$$j\hbar\frac{\partial \psi_-}{\partial t} = \boldsymbol{\sigma} \cdot (\hat{\mathbf{p}} - q\mathbf{A})c\psi_+ - m_0 c^2 \psi_-.$$

Finally, we note that quantum mechanics is a set of principles for describing small particles. We accept these principles simply because they are self-consistent and pragmatic, and can interpret the experiments with high degree of accuracy. Quantum mechanics provides a mathematical frame for many branches of physics and chemistry and has many successful applications such as in laser, semiconductors, magnetic resonance imaging, and electron microscope. Current active research topics are to explore the possibility of manipulating quantum states, which include quantum information and quantum computation, and are still in their infancy [see Nielsen and Chuang, 2000].

There is no quantum world. There is only an abstract physical description. It is wrong to think that the task of physics is to find out how nature is. Physics concerns what we can say about nature.

—Niels Bohr

Appendix A

Set Theory

The set theory is the foundation of modern mathematics, and is used in the definitions of all mathematical objects.

A.1 Basic Concepts

A set A is a collection of certain different objects. These objects are called the elements of the set. We use $a \in A$ (or $a \notin A$) to indicate that a is (or is not) an element of A. The set that has no element is called empty set, denoted by $\emptyset$. Sets can be represented by enumerating their elements in braces or by defining properties possessed by the elements. For example, the set of natural numbers can be expressed by

$$\{0, 1, 2, \cdots\} = \{x \,|\, x \text{ is a natural number}\}.$$

Two sets A and B are said to be identical or equal, denoted by $A = B$, if they have exactly the same elements. The set A is called a subset of B if all elements of A belong to B, denoted by $A \subset B$. The empty set is a subset of any set A.

A.2 Set Operations

The union of two sets A and B, denoted by $A \cup B$, is defined by

$$A \cup B = \{x \,|\, x \in A \text{ or } x \in B\}.$$

The intersection of two sets A and B, denoted by $A \cap B$, is defined by

$$A \cap B = \{x \,|\, x \in A \text{ and } x \in B\}.$$

If the intersection of two sets is empty, they are called disjoint.

Foundations of Applied Electrodynamics   Geyi Wen
© 2010 John Wiley & Sons, Ltd

The difference of two sets A and B, denoted by $A - B$, is defined as the set of elements that belong to A but not to B. Consider the subset A of a given set M. The complement of A with respect to M is defined as $M - A$.

The Cartesian product of two sets A and B is defined by

$$A \times B = \{(a, b)| a \in A \text{ and } b \in B\}.$$

The elements (a, b) are called ordered pairs. Similarly one can define n times Cartesian product.

A.3   Set Algebra

1. Associative laws

$$(A \cap B) \cap C = A \cap (B \cap C),$$
$$(A \cup B) \cup C = A \cup (B \cup C).$$

2. Commutative laws

$$A \cap B = B \cap A,$$
$$A \cup B = B \cup A.$$

3. Distributive laws

$$(A \cup B) \cap C = (A \cap C) \cup (B \cap C),$$
$$(A \cap B) \cup C = (A \cup C) \cap (B \cup C).$$

4. De Morgan laws

$$\overline{A \cap B} = \bar{A} \cup \bar{B},$$
$$\overline{A \cup B} = \bar{A} \cap \bar{B}.$$

Appendix B

Vector Analysis

Vector analysis studies the differentiation and integration of vector fields. It plays an important role in electromagnetic field theory.

B.1 Formulas from Vector Analysis

By definition, the gradient of a scalar function $\phi(\mathbf{r})$ at $\mathbf{r}$ is

$$\nabla\phi(\mathbf{r}) = \lim_{V \to 0} \frac{1}{V} \int_S \mathbf{u}_n \phi dS, \tag{B.1}$$

where V is a volume containing the point $\mathbf{r}$, S is its boundary and $\mathbf{u}_n$ is the unit outward normal of S. The gradient measures the rate and direction of change in a scalar field. The divergence of a vector function $\mathbf{A}$ is defined by

$$\nabla \cdot \mathbf{A} = \lim_{V \to 0} \frac{1}{V} \int_S \mathbf{u}_n \cdot \mathbf{A} dS. \tag{B.2}$$

The divergence measures the magnitude of the source of the vector field at a point. The rotation of a vector function $\mathbf{A}$ is defined by

$$\nabla \times \mathbf{A} = \lim_{V \to 0} \frac{1}{V} \int_S \mathbf{u}_n \times \mathbf{A} dS. \tag{B.3}$$

The rotation measures the tendency of the vector field to rotate about a point. Let $\mathbf{a}$, $\mathbf{b}$, $\mathbf{c}$ and $\mathbf{d}$ be vector functions; and ϕ and ψ be scalar functions. Then

1. $\mathbf{a} \cdot \mathbf{b} \times \mathbf{c} = \mathbf{b} \cdot \mathbf{c} \times \mathbf{a} = \mathbf{c} \cdot \mathbf{a} \times \mathbf{b}$.
2. $\mathbf{a} \times (\mathbf{b} \times \mathbf{c}) = (\mathbf{a} \cdot \mathbf{c})\mathbf{b} - (\mathbf{a} \cdot \mathbf{b})\mathbf{c}$.
3. $(\mathbf{a} \times \mathbf{b}) \cdot (\mathbf{c} \times \mathbf{d}) = (\mathbf{a} \cdot \mathbf{c})(\mathbf{b} \cdot \mathbf{d}) - (\mathbf{a} \cdot \mathbf{d})(\mathbf{b} \cdot \mathbf{c})$.

Foundations of Applied Electrodynamics Geyi Wen
© 2010 John Wiley & Sons, Ltd

4. $\nabla(\phi\psi) = \phi\nabla\psi + \psi\nabla\phi$.
5. $\nabla(\mathbf{a} \cdot \mathbf{b}) = (\mathbf{a} \cdot \nabla)\mathbf{b} + (\mathbf{b} \cdot \nabla)\mathbf{a} + \mathbf{a} \times (\nabla \times \mathbf{b}) + \mathbf{b} \times (\nabla \times \mathbf{a})$.
6. $\nabla \cdot (\phi\mathbf{a}) = \mathbf{a} \cdot \nabla\phi + \phi\nabla \cdot \mathbf{a}$.
7. $\nabla \cdot (\mathbf{a} \times \mathbf{b}) = \mathbf{b} \cdot \nabla \times \mathbf{a} - \mathbf{a} \cdot \nabla \times \mathbf{b}$.
8. $\nabla \times (\phi\mathbf{a}) = \nabla\phi \times \mathbf{a} + \phi\nabla \times \mathbf{a}$.
9. $\nabla \times (\mathbf{a} \times \mathbf{b}) = \mathbf{a}\nabla \cdot \mathbf{b} - \mathbf{b}\nabla \cdot \mathbf{a} + (\mathbf{b} \cdot \nabla)\mathbf{a} - (\mathbf{a} \cdot \nabla)\mathbf{b}$.
10. $\nabla \times \nabla \times \mathbf{a} = \nabla\nabla \cdot \mathbf{a} - \nabla^2\mathbf{a}$.

Let V be a volume bounded by the closed surface S and $\mathbf{u}_n$ be the unit outward normal of S. Then the following Gauss theorems hold

1. $\int_V \nabla\phi dV = \int_S \phi\mathbf{u}_n dS$.
2. $\int_V \nabla \cdot \mathbf{a} dV = \int_S \mathbf{u}_n \cdot \mathbf{a} dS$.
3. $\int_V \nabla \times \mathbf{a} dV = \int_S \mathbf{u}_n \times \mathbf{a} dS$.

Let S be an unclosed surface bounded by the contour Γ. Then we have the following Stokes theorems:

1. $\int_S \mathbf{u}_n \times \nabla\phi dS = \int_\Gamma \phi\mathbf{u}_l d\Gamma$.
2. $\int_S \mathbf{u}_n \cdot \nabla \times \mathbf{a} dS = \int_\Gamma \mathbf{a} \cdot \mathbf{u}_l d\Gamma$.

where $\mathbf{u}_l$ is the unit tangent vector along Γ in the positive sense with respect to $\mathbf{u}_n$.
   If $\mathbf{a}$ vanishes as rapidly as r^{-2} at infinity, we have Helmholtz identity

$$\mathbf{a}(\mathbf{r}) = -\frac{1}{4\pi}\nabla \int_{R^3} \frac{\nabla' \cdot \mathbf{a}(\mathbf{r}')}{|\mathbf{r} - \mathbf{r}'|} dV(\mathbf{r}') + \frac{1}{4\pi}\nabla \times \int_{R^3} \frac{\nabla' \times \mathbf{a}(\mathbf{r}')}{|\mathbf{r} - \mathbf{r}'|} dV(\mathbf{r}')$$

and

$$\int_{R^3} |\mathbf{a}(\mathbf{r})|^2 dV(\mathbf{r}) = \int_{R^3}\int_{R^3} \frac{\nabla \cdot \mathbf{a}(\mathbf{r})\nabla' \cdot \bar{\mathbf{a}}(\mathbf{r}') + \nabla \times \mathbf{a}(\mathbf{r}) \cdot \nabla' \times \bar{\mathbf{a}}(\mathbf{r}')}{4\pi |\mathbf{r} - \mathbf{r}'|} dV(\mathbf{r})dV(\mathbf{r}').$$

The Helmholtz identity indicates that a vector field is determined by its divergence and rotation.

B.2   Vector Analysis in Curvilinear Coordinate Systems

A curvilinear coordinate system is usually obtained from the standard Cartesian coordinate system by a non-linear transformation. In a curvilinear coordinate system, the coordinate lines are curved.

B.2.1 Curvilinear Coordinate Systems

Let $\Omega \in R^3$ be a connected region, and (x, y, z) be the rectangular coordinate system in R^3.
Let

$$v^1 = v^1(x, y, z), v^2 = v^2(x, y, z), v^3 = v^3(x, y, z), (x, y, z) \in \Omega \qquad (B.4)$$

be three independent, continuous and single-valued functions of the rectangular coordinates (x, y, z) defined in Ω. The range of the new variables (v^1, v^2, v^3) is denoted by D. If the above transform is invertible, (B.4) can be solved with respect to (x, y, z) to give

$$x = x(v^1, v^2, v^3), y = y(v^1, v^2, v^3), z = z(v^1, v^2, v^3), (v^1, v^2, v^3) \in D, \qquad (B.5)$$

which are also independent, continuous and single-valued functions in the region D. For each point (x, y, z) in Ω, there is associated a point (v^1, v^2, v^3) and vice versa. (v^1, v^2, v^3) is called the curvilinear coordinate system in Ω. Through each point (x, y, z), there pass three surfaces

$$v^i = \text{constant } (i = 1, 2, 3),$$

which are called the coordinate surfaces. On each coordinate surface, one coordinate is constant and the other two are variable. Two coordinate surfaces intersect in a curve, called a coordinate curve, along which two coordinates are constant and one is variable. A coordinate surface is designated by the constant coordinate and a coordinate curve is designated by variable coordinate, as shown in Figure B.1.

Let $\mathbf{r}$ denote the vector from the origin of the rectangular system to a variable point $P = (x, y, z)$

$$\mathbf{r} = x\mathbf{u}_x + y\mathbf{u}_y + z\mathbf{u}_z, \qquad (B.6)$$

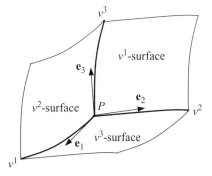

Figure B.1 Curvilinear coordinate system

where $\mathbf{u}_i$ is the unit vector along $i(i = x, y, z)$ direction. Considering (B.5), (B.6) may be written as

$$\mathbf{r} = x(v^1, v^2, v^3)\mathbf{u}_x + y(v^1, v^2, v^3)\mathbf{u}_y + z(v^1, v^2, v^3)\mathbf{u}_z.$$

The vectors

$$\mathbf{e}_1 = \frac{\partial \mathbf{r}}{\partial v^1}, \mathbf{e}_2 = \frac{\partial \mathbf{r}}{\partial v^2}, \mathbf{e}_3 = \frac{\partial \mathbf{r}}{\partial v^3}$$

are linearly independent and form a basis at the point P, which is called a local frame. Note that the base vector $\mathbf{e}_i$ is the tangent vector along the coordinate curve $v^i (i = 1, 2, 3)$ as shown in Figure B.1. The metric tensor is defined by

$$g_{ij} = \mathbf{e}_i \cdot \mathbf{e}_j = \frac{\partial x}{\partial v^i}\frac{\partial x}{\partial v^j} + \frac{\partial y}{\partial v^i}\frac{\partial y}{\partial v^j} + \frac{\partial z}{\partial v^i}\frac{\partial z}{\partial v^j}.$$

The symmetric matrix $[g_{ij}]$ possesses an inverse denoted by $[g^{ij}]$

$$\sum_{k=1}^{3} g_{ik}g^{kj} = \delta^j{}_i.$$

The dual basis is defined by

$$\mathbf{e}^i = \sum_{j=1}^{3} g^{ij}\mathbf{e}_j (i = 1, 2, 3).$$

The dual basis has the following properties

$$\mathbf{e}_i \cdot \mathbf{e}^j = \delta^j{}_i, \mathbf{e}^i \cdot \mathbf{e}^j = g^{ij},$$

$$\mathbf{e}^i = \frac{1}{\sqrt{g}}\mathbf{e}_j \times \mathbf{e}_k,$$

$$\mathbf{e}_i = \sqrt{g}\mathbf{e}^j \times \mathbf{e}^k,$$

where (i, j, k) is a permutation of $(1, 2, 3)$ and

$$g = \det[g_{ij}] = [\mathbf{e}_1 \cdot (\mathbf{e}_2 \times \mathbf{e}_3)]^2.$$

A vector function $\mathbf{A}$ at the point P may be expanded in terms of the basis $\{\mathbf{e}_1, \mathbf{e}_2, \mathbf{e}_3\}$ or the dual basis $\{\mathbf{e}^1, \mathbf{e}^2, \mathbf{e}^3\}$ at the point P

$$\mathbf{A} = \sum_{i=1}^{3} a^i\mathbf{e}_i = \sum_{i=1}^{3} a_i\mathbf{e}^i.$$

The differential $d\mathbf{r}$ is an infinitesimal displacement from the point (v^1, v^2, v^3) to a neighboring point $(v^1 + dv^1, v^2 + dv^2, v^3 + dv^3)$

$$d\mathbf{r} = \sum_{i=1}^{3} \frac{\partial \mathbf{r}}{\partial v^i} dv^i = \sum_{i=1}^{3} \mathbf{e}_i dv^i .$$

The magnitude of this displacement is denoted by ds

$$ds^2 = d\mathbf{r} \cdot d\mathbf{r} = \sum_{i,j=1}^{3} \mathbf{e}_i \cdot \mathbf{e}_j dv^i dv^j = \sum_{i,j=1}^{3} g_{ij} dv^i dv^j .$$

Especially an infinitesimal displacement at (v^1, v^2, v^3) along the v^i-curve is

$$d\mathbf{r}_i = \mathbf{e}_i dv^i$$

and the magnitude of the infinitesimal displacement along the v^i-curve is

$$ds_i = \sqrt{d\mathbf{r}_i \cdot d\mathbf{r}_i} = \sqrt{g_{ii}} dv^i .$$

Consider an infinitesimal parallelogram in the v^1-surface bounded by intersecting v^2- and v^3-curves as shown in Figure B.2. The area of the infinitesimal parallelogram in the v^1-surface is equal to

$$d\Sigma_1 = |d\mathbf{r}_2 \times d\mathbf{r}_3| = |\mathbf{e}_2 \times \mathbf{e}_3| \, dv^2 dv^3 = \sqrt{g_{22}g_{33} - g_{23}^2} dv^2 dv^3 .$$

Similarly for the areas of the infinitesimal parallelograms in the v^2- and v^3-surfaces

$$d\Sigma_2 = |d\mathbf{r}_1 \times d\mathbf{r}_3| = |\mathbf{e}_1 \times \mathbf{e}_3| \, dv^1 dv^3 = \sqrt{g_{11}g_{33} - g_{13}^2} dv^1 dv^3 ,$$

$$d\Sigma_3 = |d\mathbf{r}_1 \times d\mathbf{r}_2| = |\mathbf{e}_1 \times \mathbf{e}_2| \, dv^1 dv^2 = \sqrt{g_{11}g_{22} - g_{12}^2} dv^1 dv^2 .$$

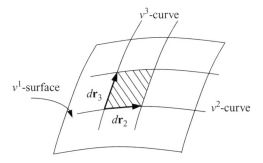

Figure B.2 $\;$ Element of area in v^1-surface

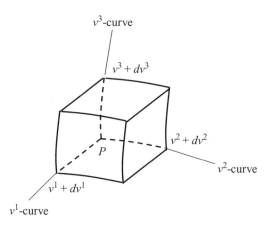

Figure B.3   A volume element in curvilinear coordinate system

The volume element bounded by coordinate surfaces is given by

$$dv = d\mathbf{r}_1 \cdot d\mathbf{r}_2 \times d\mathbf{r}_3 = \mathbf{e}_1 \cdot \mathbf{e}_2 \times \mathbf{e}_3 dv^1 dv^2 dv^3 = \sqrt{g}\,dv^1 dv^2 dv^3.$$

B.2.2 Differential Operators

Consider a volume element bounded by coordinate surfaces at point P as indicated in Figure B.3. For two end faces that lie in v^2-surfaces, the area of the first face at v^2 is $\mathbf{e}_1 \times \mathbf{e}_3 dv^1 dv^3$ and the area of the second face at $v^2 + dv^2$ is $\mathbf{e}_3 \times \mathbf{e}_1 dv^1 dv^3$.

B.2.2.1 Gradient

The net contribution of these two end faces to the integral in (B.1) is

$$(\phi \mathbf{e}_1 \times \mathbf{e}_3 dv^1 dv^3)_{v^2} + (\phi \mathbf{e}_3 \times \mathbf{e}_1 dv^1 dv^3)_{v^2 + dv^2}.$$

For sufficiently small dv^2, the above expression may be approximated by

$$\frac{\partial \phi}{\partial v^2}(\mathbf{e}_3 \times \mathbf{e}_1 dv^1 dv^2 dv^3) = \frac{\partial \phi}{\partial v^2}\left(\frac{1}{\sqrt{g}}\mathbf{e}_3 \times \mathbf{e}_1 \sqrt{g}\,dv^1 dv^2 dv^3\right)$$

$$= \frac{\partial \phi}{\partial v^2}(\mathbf{e}^2 \sqrt{g}\,dv^1 dv^2 dv^3).$$

Similar results may be obtained from the two remaining pairs of faces. Note that V in (B.1) can be approximated by $dv = \sqrt{g}\,dv^1 dv^2 dv^3$ as $V \to 0$. Therefore the gradient of ϕ is

$$\nabla \phi = \sum_{i=1}^{3} \mathbf{e}^i \frac{\partial \phi}{\partial v^i}.$$

B.2.2.2 Divergence

The net contribution of the two end faces in the v^2-surfaces to the integral in (B.2) is

$$(\mathbf{A} \cdot \mathbf{e}_1 \times \mathbf{e}_3 dv^1 dv^3)_{v^2} + (\mathbf{A} \cdot \mathbf{e}_3 \times \mathbf{e}_1 dv^1 dv^3)_{v^2+dv^2}.$$

For sufficiently small dv^2, this may be approximated by

$$\frac{\partial}{\partial v^2}(\mathbf{A} \cdot \mathbf{e}_3 \times \mathbf{e}_1 dv^1 dv^2 dv^3) = \frac{\partial}{\partial v^2}(\mathbf{A} \cdot \mathbf{e}^2 \sqrt{g} dv^1 dv^2 dv^3)$$

$$= \frac{\partial}{\partial v^2}(a^2 \sqrt{g}) dv^1 dv^2 dv^3.$$

Following a similar discussion, the divergence of a vector $\mathbf{A}$ referred to a curvilinear coordinate system is

$$\nabla \cdot \mathbf{A} = \frac{1}{\sqrt{g}} \sum_{i=1}^{3} \frac{\partial}{\partial v^i}(a^i \sqrt{g}).$$

B.2.2.3 Rotation

The net contribution of the two end faces in the v^2-surfaces to the integral in (B.3) is

$$(\mathbf{e}_1 \times \mathbf{e}_3 dv^1 dv^3 \times \mathbf{A})_{v^2} + (\mathbf{e}_3 \times \mathbf{e}_1 dv^1 dv^3 \times \mathbf{A})_{v^2+dv^2}.$$

For sufficiently small dv^2, this may be approximated by

$$\frac{\partial}{\partial v^2}(\mathbf{e}_3 \times \mathbf{e}_1 dv^1 dv^2 dv^3 \times \mathbf{A}) = \frac{\partial}{\partial v^2}(\mathbf{e}^2 \times \mathbf{A}\sqrt{g} dv^1 dv^2 dv^3).$$

The net contribution of other two remaining pairs of faces to the integral in (B.3) is

$$\frac{\partial}{\partial v^1}(\mathbf{e}^1 \times \mathbf{A}\sqrt{g} dv^1 dv^2 dv^3) + \frac{\partial}{\partial v^3}(\mathbf{e}^3 \times \mathbf{A}\sqrt{g} dv^1 dv^2 dv^3).$$

Note that

$$\mathbf{e}^1 \times \mathbf{A} = \mathbf{e}^1 \times \mathbf{e}^2 a_2 + \mathbf{e}^1 \times \mathbf{e}^3 a_3 = \frac{1}{\sqrt{g}} a_2 \mathbf{e}_3 - \frac{1}{\sqrt{g}} a_3 \mathbf{e}_2,$$

$$\mathbf{e}^2 \times \mathbf{A} = \mathbf{e}^2 \times \mathbf{e}^1 a_1 + \mathbf{e}^2 \times \mathbf{e}^3 a_3 = -\frac{1}{\sqrt{g}} a_1 \mathbf{e}_3 + \frac{1}{\sqrt{g}} a_3 \mathbf{e}_1,$$

$$\mathbf{e}^3 \times \mathbf{A} = \mathbf{e}^3 \times \mathbf{e}^1 a_1 + \mathbf{e}^3 \times \mathbf{e}^2 a_2 = \frac{1}{\sqrt{g}} a_1 \mathbf{e}_2 - \frac{1}{\sqrt{g}} a_2 \mathbf{e}_1.$$

Thus the rotation of a vector $\mathbf{A}$ with respect to a curvilinear coordinate system is

$$\nabla \times \mathbf{A} = \frac{1}{\sqrt{g}} \left[\left(\frac{\partial a_3}{\partial v^2} - \frac{\partial a_2}{\partial v^3} \right) \mathbf{e}_1 + \left(\frac{\partial a_1}{\partial v^3} - \frac{\partial a_3}{\partial v^1} \right) \mathbf{e}_2 + \left(\frac{\partial a_2}{\partial v^1} - \frac{\partial a_1}{\partial v^2} \right) \mathbf{e}_3 \right].$$

B.2.2.4 Laplace operator

Finally, we note that the Laplace operator ∇^2 can be expressed as

$$\nabla^2 \phi = \nabla \cdot \nabla \phi = \frac{1}{\sqrt{g}} \sum_{i=1}^{3} \sum_{j=1}^{3} \frac{\partial}{\partial v^i} \left(g^{ij} \sqrt{g} \frac{\partial \phi}{\partial v^j} \right).$$

B.2.3 Orthogonal Systems

The most useful curvilinear coordinate systems are orthogonal, and the base vectors are mutually perpendicular. Hence $g_{ij} = 0, i \neq j$. In this case, it is customary to introduce the metrical coefficients

$$h_i = \sqrt{g_{ii}} = \sqrt{1/g^{ii}}, i = 1, 2, 3$$

and an orthonormal basis $\{\mathbf{u}_1, \mathbf{u}_2, \mathbf{u}_3\}$

$$\mathbf{u}_i = \mathbf{e}_i / h_i = h_i \mathbf{e}^i, \mathbf{u}_i \cdot \mathbf{u}_j = \delta_{ij}.$$

Thus an arbitrary vector $\mathbf{A}$ has following expansion

$$\mathbf{A} = \sum_{i=1}^{3} A_i \mathbf{u}_i,$$

and we have

$$\nabla \cdot \mathbf{A} = \frac{1}{h_1 h_2 h_3} \left[\frac{\partial}{\partial v^1} (h_2 h_3 A_1) + \frac{\partial}{\partial v^2} (h_1 h_3 A_2) + \frac{\partial}{\partial v^3} (h_1 h_2 A_3) \right],$$

$$\nabla \times \mathbf{A} = \frac{1}{h_2 h_3} \left[\frac{\partial}{\partial v^2} (h_3 A_3) - \frac{\partial}{\partial v^3} (h_2 A_2) \right] \mathbf{u}_1 + \frac{1}{h_3 h_1} \left[\frac{\partial}{\partial v^3} (h_1 A_1) - \frac{\partial}{\partial v^1} (h_3 A_3) \right] \mathbf{u}_2$$

$$+ \frac{1}{h_1 h_2} \left[\frac{\partial}{\partial v^1} (h_2 A_2) - \frac{\partial}{\partial v^2} (h_1 A_1) \right] \mathbf{u}_3,$$

$$\nabla \phi = \sum_{i=1}^{3} \frac{1}{h_i} \frac{\partial \phi}{\partial v^i} \mathbf{u}_i,$$

$$\nabla^2 \phi = \frac{1}{h_1 h_2 h_3} \left[\frac{\partial}{\partial v^1} \left(\frac{h_2 h_3}{h_1} \frac{\partial \phi}{\partial v^1} \right) + \frac{\partial}{\partial v^2} \left(\frac{h_3 h_1}{h_2} \frac{\partial \phi}{\partial v^2} \right) + \frac{\partial}{\partial v^3} \left(\frac{h_1 h_2}{h_3} \frac{\partial \phi}{\partial v^3} \right) \right].$$

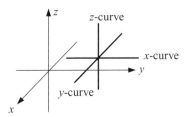

Figure B.4   Rectangular system

Example B.1 (Rectangular coordinate system): Let $v^1 = x, v^2 = y, v^3 = z$. Then $h_1 = h_2 = h_3 = 1$. The coordinate curves are shown in Figure B.4. The orthonormal basis vectors are $\mathbf{u}_i (i = x, y, z)$.   □

Example B.2 (Cylindrical coordinate system): Let $v^1 = \rho, v^2 = \varphi, v^3 = z$. (ρ, φ, z) are called cylindrical coordinates. They are related to the rectangular coordinates by the equations

$$x = \rho \cos \varphi, \ y = \rho \sin \varphi, \ z = z,$$
$$\rho > 0, 0 \leq \varphi < 2\pi.$$

The metrical coefficients are $h_1 = 1, h_2 = \rho, h_3 = 1$. The coordinate curves are shown in Figure B.5. The orthonormal basis vectors are

$$\mathbf{u}_\rho = \mathbf{u}_x \cos \varphi + \mathbf{u}_y \sin \varphi,$$
$$\mathbf{u}_\varphi = -\mathbf{u}_x \sin \varphi + \mathbf{u}_y \cos \varphi,$$
$$\mathbf{u}_z = \mathbf{u}_z.$$
                                                                                        □

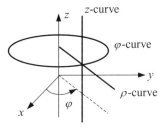

Figure B.5   Cylindrical coordinate system

Example B.3 (Spherical coordinate system): Let $v^1 = r, v^2 = \theta, v^3 = \varphi$. (r, θ, φ) are called spherical coordinates and are related to rectangular coordinates by

$$x = r \sin \theta \cos \varphi, \ y = r \sin \theta \sin \varphi, \ z = r \cos \theta,$$
$$r > 0, 0 \leq \theta \leq \pi, 0 \leq \varphi < 2\pi.$$

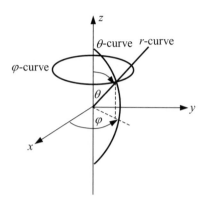

Figure B.6   Spherical coordinate system

The metrical coefficients are $h_1 = 1$, $h_2 = r$, $h_3 = r \sin \theta$. The coordinate curves are shown in Figure B.6. The orthonormal basis vectors are

$$\mathbf{u}_r = \mathbf{u}_x \sin \theta \cos \varphi + \mathbf{u}_y \sin \theta \sin \varphi + \mathbf{u}_z \cos \theta,$$
$$\mathbf{u}_\theta = \mathbf{u}_x \cos \theta \cos \varphi + \mathbf{u}_y \cos \theta \sin \varphi - \mathbf{u}_z \sin \theta,$$
$$\mathbf{u}_\varphi = -\mathbf{u}_x \sin \varphi + \mathbf{u}_y \cos \varphi.$$

□

Appendix C

Special Functions

Special functions refer to the mathematical functions that usually appear in the solutions of differential equations and have established names and notations.

C.1 Bessel Functions

The Bessel equation is

$$\frac{d^2 f}{dz^2} + \frac{1}{z}\frac{df}{dz} + \left(\mu^2 - \frac{p^2}{z^2}\right)f = 0.$$

One of the solutions is the Bessel function of the first kind

$$J_p(\mu z) = \sum_{m=0}^{\infty} \frac{(-1)^m}{\Gamma(m+1)\Gamma(p+m+1)} \left(\frac{\mu z}{2}\right)^{p+2m},$$

where the gamma function is defined by

$$\Gamma(\alpha) = \int_0^{\infty} x^{\alpha-1} e^{-x} dx, \alpha > 0.$$

If p is not an integer, a second independent solution is $J_{-p}(\mu z)$. If $p = n$ is an integer, $J_{-n}(\mu z)$ is related to $J_n(\mu z)$ by

$$J_{-n}(z) = (-1)^n J_n(z).$$

For this reason, we need to find other independent solutions. These include the Bessel function of the second kind defined by

$$N_p(\mu z) = \frac{\cos p\pi \, J_p(\mu z) - J_{-p}(\mu z)}{\sin p\pi},$$

Foundations of Applied Electrodynamics Geyi Wen
© 2010 John Wiley & Sons, Ltd

and the Bessel functions of the third (Hankel function of the first kind) and fourth kind (Hankel function of the second kind) defined by

$$H_p^{(1)}(\mu z) = J_p(\mu z) + j N_p(\mu z),$$
$$H_p^{(2)}(\mu z) = J_p(\mu z) - j N_p(\mu z).$$

Let $R_p(\mu z) = A J_p(\mu z) + B N_p(\mu z)$, where A and B are constant. We have the recurrence relations

$$\frac{2p}{\mu z} R_p(\mu z) = R_{p-1}(\mu z) + R_{p+1}(\mu z),$$

$$\frac{1}{\mu} \frac{d}{dz} R_p(\mu z) = \frac{1}{2} \left[R_{p-1}(\mu z) - R_{p+1}(\mu z) \right],$$

$$z \frac{d}{dz} R_p(\mu z) = p R_p(\mu z) - \mu z R_{p+1}(\mu z),$$

$$\frac{d}{dz} \left[z^p R_p(\mu z) \right] = \mu z^p R_{p-1}(\mu z),$$

$$\frac{d}{dz} \left[z^{-p} R_p(\mu z) \right] = -\mu z^{-p} R_{p+1}(\mu z).$$

C.2 Spherical Bessel Functions

The spherical Bessel equation is

$$\frac{d^2 z_n}{dz^2} + \frac{2}{z} \frac{dz_n}{dz} + \left[\mu^2 - \frac{n(n+1)}{z^2} \right] z_n = 0.$$

The spherical Bessel functions are the solution of this equation and are defined by

$$j_n(\mu z) = \sqrt{\frac{\pi}{2\mu z}} J_{n+\frac{1}{2}}(\mu z),\ n_n(\mu z) = \sqrt{\frac{\pi}{2\mu z}} N_{n+\frac{1}{2}}(\mu z),$$

$$h_n^{(1)}(\mu z) = \sqrt{\frac{\pi}{2\mu z}} H_{n+\frac{1}{2}}^{(1)}(\mu z),\ h_n^{(2)}(\mu z) = \sqrt{\frac{\pi}{2\mu z}} H_{n+\frac{1}{2}}^{(2)}(\mu z).$$

They are known as the spherical Bessel functions of the first, second, third and fourth kind respectively; $h_n^{(1)}$ and $h_n^{(2)}$ are also called the spherical Hankel functions of the first and second kind. Let $z_n(\mu z) = A j_n(\mu z) + B n_n(\mu z)$, where A and B are constant. We have the recurrence relations:

$$\frac{2n+1}{\mu z} z_n(\mu z) = z_{n-1}(\mu z) + z_{n+1}(\mu z),$$

$$\frac{2n+1}{\mu} \frac{d}{dz} z_n(\mu z) = n z_{n-1}(\mu z) - (n+1) z_{n+1}(\mu z),$$

$$\frac{d}{dz}\left[z^{n+1}z_n(\mu z)\right] = \mu z^{n+1}z_{n-1}(\mu z),$$

$$\frac{d}{dz}\left[z^{-n}z_n(\mu z)\right] = -\mu z^{-n}z_{n+1}(\mu z).$$

C.3   Legendre Functions and Associated Legendre Functions

The Legendre equation is

$$(1 - z^2)\frac{d^2 f}{dz^2} - 2z\frac{df}{dz} + n(n+1)f = 0. \tag{C.1}$$

It has two independent solutions. One is the Legendre function of the first kind, also called the Legendre polynomial denoted by $P_n(z)$

$$P_n(z) = \frac{1}{2^n n!}\frac{d^n}{dz^n}(z^2 - 1)^n,$$

where n is an integer. Another independent solution of (C.1) is the Legendre function of the second kind defined by

$$Q_n(z) = \frac{1}{2}P_n(z)\ln\frac{1+z}{1-z} - \sum_{r=1}^{n}\frac{1}{r}P_{r-1}(z)P_{n-r}(z).$$

The recurrence relations are

$$z\frac{d}{dz}P_n(z) - \frac{d}{dz}P_{n-1}(z) = nP_n(z),$$

$$(n+1)P_{n+1}(z) - (2n+1)zP_n(z) + nP_{n-1}(z) = 0,$$

$$(z^2 - 1)\frac{d}{dz}P_n(z) = nzP_n(z) - nP_{n-1}(z),$$

$$\frac{d}{dz}P_{n+1}(z) - \frac{d}{dz}P_{n-1}(z) = (2n+1)P_n(z).$$

The Legendre polynomials are orthogonal

$$\int_{-1}^{1} P_m(x)P_n(x)dx = \frac{2}{2n+1}\delta_{mn}.$$

The associated Legendre equation is

$$(1 - z^2)\frac{d^2 f}{dz^2} - 2z\frac{df}{dz} + \left[n(n+1) - \frac{m^2}{1-z^2}\right]f = 0.$$

The two linearly independent solutions are the associated Legendre function of the first and the second kind defined by

$$P_n^m(z) = \frac{(1-z^2)^{m/2}}{2^n n!} \frac{d^{m+n}}{dz^{m+n}} (z^2 - 1)^n,$$

and

$$Q_n^m(z) = (1-z^2)^{\frac{m}{2}} \frac{d^m}{dz^m} Q_n(z), \, m \le n,$$

respectively.

The following integrations are useful

$$\int_{-1}^{1} \frac{P_n^m(x) P_n^k(x)}{1 - x^2} dx = \frac{1}{m} \frac{(n+m)!}{(n-m)!} \delta_{mk},$$

$$\int_{-1}^{1} P_k^m(x) P_n^m(x) dx = \frac{2}{2k+1} \frac{(k+m)!}{(k-m)!} \delta_{kn},$$

$$\int_{0}^{\pi} \left[\frac{d P_n^m(\cos\theta)}{d\theta} \frac{d P_k^m(\cos\theta)}{d\theta} + \frac{m^2}{\sin^2\theta} P_n^m(\cos\theta) P_k^m(\cos\theta) \right] \sin\theta \, d\theta$$

$$= \frac{2}{2n+1} \frac{(n+m)!}{(n-m)!} n(n+1) \delta_{nk}.$$

Appendix D

SI Unit System

This book uses SI unit system in which mass M is measured in kilograms, length L in meters, time T in seconds, and charge Q in coulombs. Table (D.1) lists the electromagnetic quantities, their symbols, dimensions and SI unit.

Table D.1 Electromagnetic quantities, symbols, dimensions and SI unit.

Quantity	Symbol	SI unit	Dimensions
Charge	q	Coulomb	Q
Current	I	Ampere	Q/T
Resistance	R	Ohm	ML^2/TQ^2
Inductance	L	Henry	ML^2/Q^2
Capacitance	C	Farad	Q^2T^2/ML^2
Charge density	ρ	Coulomb/cubic meter	Q/L^3
Current density	$\mathbf{J}$	Ampere/square meter	Q/TL^2
Electric field intensity	$\mathbf{E}$	Volt/meter	ML/QT^2
Electric displacement	$\mathbf{D}$	Coulomb/square meter	Q/L^2
Electric dipole moment	$\mathbf{p}$	Coulomb-meter	QL
Polarization vector	$\mathbf{P}$	Coulomb/ square meter	Q/L^2
Magnetic field intensity	$\mathbf{H}$	Ampere/meter	Q/TL
Magnetic induction	$\mathbf{B}$	Weber/square meter	M/QT
Magnetic dipole moment	$\mathbf{m}$	Ampere-square meter	QL^2/T
Magnetization vector	$\mathbf{M}$	Ampere/meter	Q/TL
Vector potential	$\mathbf{A}$	Weber/Henry	ML/QT
Scalar potential	ϕ	Volt	ML^2/QT^2
Conductivity	σ	Mho/meter	Q^2T/ML^3
Permeability	μ	Henry/meter	ML/Q^2
Permittivity	ε	Farad/meter	Q^2T^2/ML^3
Frequency	f	Hertz	$1/T$
Force	$\mathbf{F}$	Newton	ML/T^2
Energy	W	Joule	ML^2/T^2
Power	P	Watt	ML^2/T^3
Poynting vector	$\mathbf{S}$	Watt/square meter	M/T^3

Table (D.2) shows the physical constants.

Table D.2 Physical constants.

Quantity	Symbol	Value
Speed of light	c	3.00×10^8 meter/second
Elementary charge	e	1.60×10^{-19} Coulomb
Electron mass	m_e	9.11×10^{-31} kilogram
Proton mass	m_p	1.67×10^{-27} kilogram
Permeability constant	μ_0	1.26×10^{-6} Henry/meter
Permittivity constant	ε_0	8.85×10^{-12} Farad/meter
Gravitational constant	G	6.67×10^{-11} Newton $\cdot$ square meter/square kilogram
Planck's constant	h	6.63×10^{-34} Joule $\cdot$ second
Boltzmann constant	k	1.3807×10^{-23} Joule/Kelvin

Bibliography

Abraham, R., J. E. Marsden and T. Ratiu, *Manifolds, Tensor Analysis, and Applications*, Springer-Verlag, 1988.

Aksoy, S. and O. A. Tretyakov, 'The evolution equations in study of the cavity oscillations excited by a digital signal', *IEEE Trans. Antennas and Propagation*, Vol. 52, No. 1, 263–270, 2004.

Adams, R. A., *Sobolev Spaces*, Academic Press, 1975.

Adler, R. B., L. J. Chu, and R. M. Fano, *Electromagnetic Energy Transmission and Radiation*, John Wiley & Sons, Ltd, Inc., 1960.

Ahner, J. F. and R. E. Kleimann, 'The exterior Neumann problem for the Helmholtz equation', *Arch. Rational Mech. Anal.*, Vol. 52, 26–43, 1973.

Albert, G. E. and J. L. Synge, 'The general problem of antenna radiation and the fundamental integral equation with application to an antenna of revolution-Part 1', *Quart. Appl. Math.*, Vol. 6, 117–131, April 1948.

Albertsen, N. C., J. E. Hansen, and N. E. Jensen, 'Computation of radiation from wire antennas on conducting bodies', *IEEE Trans. Antennas and Propagat.* Vol. AP-22, 200–206, No. 2, Mar 1974.

Alexandrov O. and G. Ciraolo, 'Wave propagation in a 3-D optical waveguide, Mathematical Models and Methods,' *Applied Sciences*, Vol. 14, No. 6, 819–852, 2004.

Ancona, C., 'On small antenna impedance in weakly dissipative media', *IEEE Trans. Antennas and Propagat.*, Vol. Ap-26, 341–343, Mar. 1978.

Angell, T. S. and A. Kirsh, *Optimization Method in Electromagnetic Radiation*, Springer, 2004.

Aydin, K. and A. Hizal, 'On the completeness of the spherical vector wave functions', *J. Math. Anal. & Appl.*, Vol. 117, 428–440, 1986.

Balanis, C. A., *Antenna Theory: Analysis and Design*, 2nd edn, John Wiley & Sons, Ltd, 1997.

Bamberger A. and A. S. Bonnet, 'Mathematical analysis of the guided modes of an optical fiber', *SIAM J. Math. Anal.*, Vol. 21, No. 6, 1487–1510, Nov. 1990.

Barut, A. O., *Electromagnetics and Classical Theory of Fields and Particles*, Macmillan, 1964.

Baum, C. E., 'Emerging technology for transient and broad-band analysis and synthesis of antennas and scatters', *Proc. IEEE*, Vol. 64, 1598–1616, 1976.

Baum, C. E., E. J. Rothwell, K. M. Chen and D. P. Nyquist, 'The singularity expansion method and its application to target identification', *Proc. IEEE*, Vol. 79, No. 10, 1481–1492, Oct. 1991.

Bladel, J. V., *Electromagnetic Fields*, McGraw-Hill, 1964.

Bladel, J. V., 'Low frequency asymptotic techniques', in *Modern Topics in Electromagnetics and Antennas* (Ed. R. Mittra), Peter Peregrines Ltd, 1977.

Bladel, J. V., *Relativity and Engineering*, Springer-Verlag, 1984.

Bluck, M. J., M. D. Pocock and S. P. Walker, 'An accurate method for the calculation of singular integrals arising in time-domain integral equation analysis of electromagnetic scattering', *IEEE Trans. Antennas and Propagat.* Vol. AP-45, No. 12, 1793–1798, Dec. 1997.

Bohm, D., *Quantum Theory*, Prentice-Hall, 1955.

Bondeson A., T. Rylander and P. Ingelström, *Computational Electromagnetics*, Springer, 2005.

Borgiotti, G. V., 'Maximum power transfer between two planar apertures in the Fresnel zone', *IEEE Trans. Antennas and Propagat.*, Vol. Ap-14, 158–163, Mar. 1966.

Borgiotti, G. V., 'On the reactive energy of an antenna', *IEEE Trans. Antennas and Propagat.* Vol. AP-15, 565–566, 1967.

Born, M. and E. Wolf, *Principles of Optics*, 6th edn., Pergamon Press, 1980.

Boyd, G. D. and H. Kogelnik, 'Generalized confocal resonator theory', *Bell. Sys. Tech. J.*, Vol. 41, 1347–1369, 1962.

Brau, C. A., *Modern Problems in Classical Electrodynamics*, Oxford University Press, 2004.

Brekhovskikh L. M., *Waves in Layered Media*, Academic Press, 1960.

Brezis, H. and F. Browder, 'Partial differential equations in the 20th century', *Advances in Mathematics*, Vol. 135, 76–144, 1998.

Brillouin, L., *Wave Propagation and Group Velocity*, Academic Press, 1960.

Brinkman, W. F. and D. V. Lang, 'Physics and the communication industry', *Reviews of Modern Physics*, Vol. 71, No. 2, 480–488, 1999.

Brown, W. C., 'The history of power transmission by radio waves', *IEEE Trans. Microwave Theory and Tech.*, Vol. MTT-32, 1230–1242, Sept. 1984.

Burton, A. J. and G. F. Miller, 'The application of integral equation methods to the numerical solution of some exterior boundary value problems', *Proc. Roy. Soc. Lond. A.*, Vol. 323, 201–210, 1971.

Byron, F. W. and R. W. Fuller, *Mathematics of Classical and Quantum Physics*, Addison-Wesley, 1969.

Calderon, A. P., 'The multiple expansion of radiation fields', *J. Rational Mech. Anal.*, Vol. 3, 523–537, 1954.

Carin, L. and L. B. Felsen, Eds., *Ultra-Wideband Short-Pulse Electromagnetics*, Plenum, 1995.

Carroll, S. M., *Spacetime and Geometry: An Introduction to General Relativity*, Benjamin Cummings, 2003.

Carson, J. R., 'A generalization of reciprocity theorem', *Bell Syst. Tech. J.*, Vol. 3, 393, 1924.

Carter, P. S., 'Circuit relations in radiating systems and applications to antenna problems', *Proc. IRE*, Vol. 20, No. 6, 1004–1041, June. 1932.

Chew, W. C., *Waves and Fields in Inhomogeneous Media*, Van Nostrand Reinhold, 1990.

Chew, W. C., *Fast and Efficient Algorithms in Computational Electromagnetics*, Artech House, 2001.

Choquet-Bruhat, Y., C. DeWitt-Morette, and M. Dillard Bleick, *Analysis, Manifolds, and Physics*, Vol. 1,2, North-Holland, 1988.

Chu, L. J., 'Physical limitations of omni-directional antennas', *J. Appl. Phys.*, Vol. 19, 1163–1175, 1948.

Chu, S. and S. Wong, 'Linear pulse propagation in an absorbing medium', *Phys. Rev. Lett.*, Vol. 48, No. 11, 738–741, 1982.

Cochran, J. A., *The Analysis of Linear Integral Equations*, McGraw-Hill, 1972.

Collardey, S., A. Sharaiha and K. Mahdjoubi, 'Evaluation of antenna radiation Q using FDTD method', *Electronics Letters*, Vol. 41, No. 12, 675–677, 9 June 2005.

Collardey, S., A. Sharaiha and K. Mahdjoubi, 'Calculation of small antennas quality factor using FDTD method', *IEEE Antennas and Wireless Propagation Letters*, No. 1, 191–194, 2006.

Collin, R. E., 'Stored energy Q and frequency sensitivity of planar aperture antennas', *IEEE Trans. Antennas and Propagat.* Vol. AP-15, 567–568, 1967.

Collin, R. E., *Antennas and Radio Wave Propagation*, McGraw-Hill, 1985.

Collin, R. E., *Field Theory of Guided Waves*, IEEE Press, 1991.

Collin, R. E., 'Minimum Q of small antennas', *Journal of Electromagnetic Waves and Applications*, Vol. 12, 1369–1393, 1998.

Collin, R. E., *Foundations for Microwave Engineering*, 2nd edn., IEEE Press, 2001.

Collin, R. E., and S. Rothschild, 'Evaluation of antenna Q', *IEEE Trans. Antennas and Propagat.* Vol. AP-12, 23–27, Jan. 1964.

Colton, D. and R. E. Kleimann, 'The direct and inverse scattering problems for an arbitrary cylinder: Dirichlet boundary conditions', *Proc. Royal. Soc.*, Edinburgh 86A, 29–42, 1980.

Colton, D. and R. Kress, *Integral Equation Methods in Scattering Theory*, John Wiley & Sons, Ltd, 1983.

Colton, D. and R. Kress, *Inverse Acoustic and Electromagnetic Scattering Theory*, Springer-Verlag, 1998.

Correia, L. M., 'A comparison of integral equations with unique solution in the resonance region for scattering by conduction bodies', *IEEE Trans. Antennas and Propagat.* Vol. AP-41, 52–58, Jan. 1993.

Costabel, M., 'A coercive bilinear form for Maxwell equations', *J. Math. Anal. & Appl.*, 157, 527–541, 1991.

Courant, R. and D. Hilbert, *Methods of Mathematical Physics*, Vol. 1–2, John Wiley & Sons, Ltd, 1953.

Craig, D. P and T. Thirunamachandran, *Molecular Quantum Electrodynamics: An Introduction to Radiation Molecule Interactions*, Academic Press, 1984.

Curtis, W. D. and F. R. Miller, *Differential Manifolds and Theoretical Physics*, Academic Press, 1985.

Deschamps, G. A., 'Impedance of an antenna in a conducting medium', *IRE Trans. Antennas Propagat.*, Vol. Ap-10, 648–649, Sept. 1962.

Diener, G., 'Superluminal group velocities and information transfer', *Physics Letters A*, Vol. 223, 327–331, Dec. 1996.

Diener, G., 'Energy transport in dispersive media and superluminal group velocities', *Physics Letters A*, Vol. 235, 118–124, Oct. 1998.

Dirac, P.A. M., *The Principles of Quantum Mechanics*, 4th edn., Oxford University Press, 1958.

Dolph, C. L. and S. K. Cho, 'On the relationship between the singularity expansion method and the mathematical theory of scattering', *IEEE Trans. Antennas and Propagat.* Vol. AP-28, No. 6, 888–897, Nov. 1980.

Dudley, D. G., *Mathematical Foundations for Electromagnetic Theory*, IEEE Press, 1994.

Duvaut, G. and J. L. Lions, *Inequalities in Mechanics and Physics*, Translated from French by C. W. John, Springer-Verlag, 1976.

Dyson, F., 'Feynman's proof of Maxwell equations', *Am. J. Phys.*, Vol. 58, 209–211, 1990.

Eisenhart, L. P., 'Separable systems of Stäckel', *Ann. Math.*, Vol. 35, No. 2, 284–305, 1934.

Elliott, R. S., *Antenna Theory and Design*, Prentice-Hall, 1981.

Elliott, R. S., *Electromagnetics: History, Theory and Applications*, IEEE Press, 1993a.

Elliott, R. S., *An Introduction to Guided Waves and Microwave Circuits*, Prentice-Hall, 1993b.

Erdelyi, A. (ed.), *Tables of Integral Transform, Bateman Manuscript Project*, Vol. 1, McGraw-Hill, 1954.

Fano, R. M., L. J. Chu and R. B. Adler, *Electromagnetic Fields, Energy, and Forces*, John Wiley & Sons, Ltd, and MIT Press, 1960.

Fante, R. L., 'Quality factor of general idea antennas', *IEEE Trans. Antennas and Propagat.* Vol. AP-17, 151–155, 1969.

Fante, R. L., 'Maximum possible gain for an arbitrary ideal antenna with specified quality factor', *IEEE Trans. Antennas and Propagat.* Vol. AP-40, 1586–1588, Dec. 1992.

Felsen, L. B. (ed.), *Transient Electromagnetic Fields*, Springer-Verlag, 1976.

Felsen, L. B. and N. Marcuwitz, *Radiation and Scattering of Electromagnetic Waves*, Prentice-Hall, 1973.

Franceschetti, G. and C. H. Papas, 'Pulsed antennas', *IEEE Trans. Antennas and Propagat.* Vol. AP-22, 651–661, Sept. 1974.

Frankl, D. R., *Electromagnetic Theory*, Prentice-Hall, 1986.

Friis, H. T., 'A note on a simple transmission formula', *Proc. IRE*, Vol. 34, 254–256, 1946.

Frisch, D. H. and L. Wilets, 'Development of the Maxwell-Lorentz equations from special theory of relativity and Gauss's law', *Am. J. Phys.*, Vol. 24, 574–579, 1956.

Fushchych, W., 'What is the velocity of the electromagnetic field?', *Journal of Nonlinear Mathematical Physics*, Vol. 5, No. 2, 159–161, 1998.

Galfand, I. and E. Shilov, *Generalized Functions*, Vols. 1–5, Academic Press, 1964.

Gandhi, O. P. *Microwave Engineering and Applications*, Pergamon Press, 1981.

Geyi, W., 'On the spurious solutions in boundary integral formulation for waveguide eigenvalue problems', *Proceedings of European Microwave Conference*, Vol. 2, 1311–1316, 1990.

Geyi, W., 'Neumann series solutions for low frequency electromagnetic scattering problems', *Chinese Journal of Electronics (English version)*, Vol. 4, No. 3, 1995, 89–92.

Geyi, W., 'Further research on the behavior of energy density of electromagnetic pulse', *Microwave and Optical Technology Letters*, Vol. 9, Aug. 20, 331–335, 1996.

Geyi, W., *Advances in Electromagnetic Theory*, National Defense Publishing House of China, 1999 (in Chinese).

Geyi, W., 'Physical limitations of antenna', *IEEE Trans., Antennas and Propagat.* Vol. AP-51, 2116–2123, Aug. 2003(a).

Geyi, W., 'A method for the evaluation of small antenna Q', *IEEE Trans. Antennas and Propagat.* Vol. AP-51, 2124–2129, Aug. 2003(b).

Geyi, W., 'A time-domain theory of waveguide', *Progress in Electromagnetics Research*, PIER 59, 267–297, 2006(a).

Geyi, W., 'New magnetic field integral equation for antenna system', *Progress in Electromagnetics Research*, PIER 63, 153–176, 2006(b).

Geyi, W., 'Reply to comments on 'The Foster reactance theorem for antennas and radiation Q'', *IEEE Trans. Antennas and Propagat.* Vol. AP-55, 1014–1016, 2007(a).

Geyi, W., 'Multi-antenna information theory', *Progress in Electromagnetics Research*, PIER 75, 11–50, 2007(b).

Geyi, W., 'Time-domain theory of metal cavity resonator', *Progress in Electromagnetics Research*, PIER 78, 219–253, 2008.

Geyi, W., Y. Chengli, L. Weigan, 'Unified theory of the backscattering of electromagnetic missiles by a perfectly conducting target', *J. Appl. Phys.*, Vol. 71, 3103–3106, Apr. 1992.

Geyi, W. and W. Hongshi, 'Solution of the resonant frequencies of cavity resonator by boundary element method', *IEE Proc., Microwaves, Antennas and Propagation*, Vol. 135, Pt. H, No. 6, 361–365, 1988(a).

Geyi, W. and W. Hongshi, 'Solution of the resonant frequencies of a microwave dielectric resonator using boundary element method', *IEE Proc., Microwaves, Antennas and Propagation*, Vol. 135, Pt. H, No. 5, 333–338, 1988(b).

Geyi, W., P. Jarmuszewski., Y. Qi, 'Foster reactance theorems for antennas and radiation Q', *IEEE Trans. Antennas and Propagat.*, Vol. AP-48, 401–408, Mar. 2000.

Geyi, W., Xueguan L., and Wanchun W., 'Solution of the characteristic impedance of an arbitrary shaped TEM transmission line using complex variable boundary element method', *IEE Proc., Microwaves, Antennas and Propagation*, Vol. 136, Pt. H, No. 1, 73–75, 1989.

Ginzburg, V. L., *The Propagation of Electromagnetic Waves in Plasmas*, Pergamon Press, 1964.

Glisson, A. W. and D. R. Wilton, 'Simple and efficient numerical methods for problems of electromagnetic radiation and scattering from surfaces', *IEEE Trans. Antennas and Propagat.* Vol. AP-28, 593–603, No. 5, Sept. 1980.

Good, R. H., 'Particle aspect of the electromagnetic field equations', *Phys. Rev.*, Vol. 105, No. 6, 1914–1919, 1957.

Gosling, W., *Radio Antennas and Propagation*, Newnes, 1998.

Goubau, G. (ed.), *Electromagnetic Waveguides and Cavities*, Pergamon, 1961.

Goubao G. and F. Schwinger, 'On the guided propagation of electromagnetic wave beams', *IRE Trans. Antennas and Propagat.*, Vol. AP-9, 248–256, May 1961.

Gradsheyn, L. S. and I. M. Ryzhik, *Tables of Integrals, Series, and Products*, Academic Press, 1994.

Graglia, R. D., 'On the numerical integration of the linear shape functions times the 3-D Green's function or its gradient on a plane triangle', *IEEE Trans. Antennas and Propagat.* Vol. AP-41, 1448–1455, Oct. 1993.

Greiner, W. and J. Reinhardt, *Field Quantization*, Springer-Verlag, 1996.

Griffiths, D. J., *Introduction to Electrodynamics*, Prentice-Hall, 1999.

Gustafson, K. E., *Partial Differential Equations and Hilbert Space Methods*, John Wiley & Sons, Ltd, 1987.

Hammond, P., *Energy Methods in Electromagnetism*, Clarendon Press, 1981.

Hansen, R. C., 'Fundamental limitations in antennas', *Proc. IEEE*, Vol. 69, 170–182, Feb. 1981.

Hansen, R.C., *Electrically Small, Superdirective, and Superconducting Antennas*, John Wiley & Sons, Ltd, 2006.

Hansen, W. W., 'A new type of expansion in radiation problems', *Phys. Rev.*, Vol. 47, 139–143, 1935.

Hanson, G. W. and A. B. Yakovlev, *Operator Theory for Electromagnetics: An Introduction*, Springer, 2002.

Harrington, R. F., 'On the gain and beamwidth of directional antennas', *IRE Trans. on Antennas and Propagat.* Vol. 6, 219–225, 1958.

Harrington, R. F., 'Effect of antenna size on gain, bandwidth, and efficiency', *J. Res. the National Bureau of Standards-D. Radio Propagation*, Vol. 64D, No. 1, Jan.–Feb. 1960.

Harrington, R. F., *Time-Harmonic Electromagnetic Fields*, McGraw-Hill Book Company, INC, 1961.

Harrington, R. F., *Field Computation by Moment Methods*, IEEE Press, 1993.

Harrington, R. F. and J. R. Mautz, 'A generalized formulation for aperture problems', *IEEE Trans. Antennas and Propagat.* Vol. AP-24, 870–873, Nov. 1976.

Harrington, R. F. and A. T. Villeneuve, 'Reciprocal relationships for gyrotropic media', *IRE Trans. Microwave Theory and Techniques*, Vol. MTT-6, 308–310, July 1958.

Hartemann, F. V., *High Field Electrodynamics*, CRC Press, 2002.

Hazard, C. and M. Lenoir, 'On the solution of time-harmonic scattering problems for Maxwell equations', *SIAM J. Math. Anal.*, Vol. 27, No. 6, 1597–1630, 1996.

Heras, J. A., 'How the potentials in different gauges yield the same retarded electric and magnetic field', *Am. J. Phys.*, Vol. 75, No. 2, 176–183, 2007.

Heurtley, J. C., ' Maximum power transfer between two finite antennas', *IEEE Trans. Antennas Propagat.*, Vol. Ap-15, 298–300, Mar. 1967.

Hooft, G., Nobel Lecture: 'A confrontation with infinity', *Reviews of Modern Physics*, Vol. 72, No. 2, 333–339, April 2000.

Hoop, T. A., *Handbook of Radiation and Scattering of Waves: Acoustic Waves in Fluids, Elastic Waves in Solids, Electromagnetic Waves*, Academic Press, 1995.

Horn, R. A. and C. R. Johnson, *Matrix Analysis*, Cambridge University Press, 1985.

Hsiao, G. C. and R. E. Kleinmann, 'Mathematical foundations for error estimation in numerical solutions of integral equations in electromagnetics', *IEEE Trans. Antennas and Propagat.* Vol. AP-45, 316–328, Mar. 1997.

Hu, M. K., 'Near zone power transmission formula', *IRE Nat'l Conv. Rec.*, Part 8, 128–135, 1958.

Huang, K., 'On the interaction between the radiation field and ionic crystals', *Proc. Roy. Soc. (London) A*, Vol. 208, 352–365, Sept. 1951.

Huygens, C., *Treatise on Light*, Dover Publications INC, 1962; first published in 1690.

Idemen, M., 'The Maxwell equations in the sense of distribution', *IEEE Trans. Antennas and Propagat.*, Vol. Ap-21, 736–738, Jul. 1973.

Inagaki N., 'Eigenfunctions of composite Hermitian operators with application to discrete and continuous radiating systems', *IEEE Trans. Antennas and Propagat.*, Vol. Ap-30, 571–575, Jul. 1982.

Ito, M., 'Dispersion of very short microwave pulses in waveguide', *IEEE Trans. Microwave Theory and Tech.*, Vol. 13, 357–364, May 1965.

Jackson, J. D., *Classical Electrodynamics*, 3rd edn., John Wiley & Sons, Ltd, 1999.

Järvenpää, S., M. Taskinen and P. Ylä-Oijala, 'Singularity extraction technique for integral equation methods with higher order basis functions on plane triangles and tetrahedral', *Int. J. Numer. Meth. Engng*, Vol. 58, 1149–1165, 2003.

Jones, D. S., *The Theory of Electromagnetism*, Pergamon Press, 1964.

Jones, D. S., *Methods in Electromagnetic Wave Propagation*, Clarendon Press, 1979.

Kahn, W. K. and H. Kurss, 'Minimum-scattering antennas', *IEEE Trans. Antennas and Propagat.*, Vol. Ap-13, 671–675, Sep. 1965.

Kalafus, R. M., 'On the evaluation of antenna quality factors', *IEEE Trans. Antennas and Propagat.* Vol. AP-17, 729–732, 1969.

Kartchevski E. M. *et al.*, 'Mathematical analysis of the generalized natural modes of an inhomogeneous optical fiber', *SIAM J. Appl. Math.*, Vol. 65, 2033–2048, 2005.

Kay, A. F., 'Near field gain of aperture antennas', *IRE Trans. Antennas and Propagat.*, Vol. Ap-8, 586–593, Nov. 1960.

Kellogg, O. D., *Foundations of Potential Theory*, Dover, 1953.

Kittel, C., *Introduction to Solid Physics*, 7th edn, John Wiley & Sons, Ltd, 1996.

Kleimann, R. E., 'Iterative solutions of boundary value problems', in *Function Theoretic Methods for Partial Differential Equations*, Springer, 1976.

Kleimann, R. E., 'Low frequency electromagnetic scattering', in *Electromagnetic Scattering*, (ed. P. L.E. Uslenghi), 1978.

Kleimann, R. E. and W. Wendland, 'On the Neumann's method for the exterior Neumann problem for the Helmholtz equation', *J. Math. Anal. & Appl.*, Vol. 57, 170–202, 1977.

Klein, C. and R. Mittra, 'Stability of matrix equations arising in electromagnetics', *IEEE Trans. Antennas and Propagat.* Vol. AP-21, No. 6, 902–905, Nov. 1973.

Kline, M. and I. W. Kay, *Electromagnetic Theory and Geometrical Optics*, Interscience, 1965.

Knepp, D. L. and J. Goldhirsh, 'Numerical analysis of electromagnetic radiation properties of smooth conducting bodies of arbitrary shape', *IEEE Trans. Antennas and Propagat.* Vol. AP-20, No. 3, 383–388, May 1972.

Kolner, B. H., 'Space-time duality and the theory of temporal imaging', *IEEE J. Quantum Electron.*, Vol. 30, 1951–1963, Aug. 1994.

Kong, J. A., *Electromagnetic Wave Theory*, Wiley-Interscience, 1990.

Kovetz, A., *Electromagnetic Theory*, Oxford University Press, 2000.

Krauss, J. D., *Electromagnetics*, McGraw-Hill, 1984.

Kraus, J. D. and D. A. Fleisch, *Electromagnetics with Applications*, McGraw-Hill, 1999.

Kreyszig, E., *Introductory Functional Analysis with Applications*, John Wiley & Sons, Ltd, 1978.

Kristensson, G., 'Transient electromagnetic wave propagation in waveguides', *J Electromagnetic Waves App*, Vol. 9, 645–671, 1995.

Kron, G., 'Equivalent circuits to represent the electromagnetic field equations', *Phys. Rev.*, Vol. 64, 126–128, 1943.

Kron, G., 'Equivalent circuit of the field equations of Maxwell-I', *Proc. IRE*, 289–299, May 1944.

Kron, G., 'Numerical solution of ordinary and partial differential equations by means of equivalent circuits', *J. Appl. Phys.*, Vol. 16, 172–186, 1945.

Kurokawa, K., *An Introduction to Microwave Circuits*, Academic Press, 1969.

Lamensdorf, D. and L. Susman, 'Baseband-pulse-antenna techniques', *IEEE Antennas and Propagation Magazine*, Vol. 36, No. 1, 20–30, 1994.

Landau, L. D., E. M. Lifshitz and L. P. Pitaevskii, *Electrodynamics of Continuous Media*, Pergamon, 1984.

Lee, K. F., *Principles of Antenna Theory*, John Wiley & Sons, Ltd, 1984.

Leis, R., *Initial Boundary Value Problems in Mathematical Physics*, John Wiley & Sons, Ltd, 1986.

Lewin, L., *Advanced Theory of Waveguides*, Illiffe and Sons, 1951.

Lo, Y. T. and S. W. Lee, *Antenna Handbook-Theory, Applications, and Design*, VNR, 1988.

Lovelock, D. and H. Rund, *Tensors, Differential Forms, and Variational Principles*, John Wiley & Sons, Ltd, 1975.

Ludvigsen, M., *General Relativity: A Geometric Approach*, Cambridge University Press, 1999.

MacMillan, W. D., *The Theory of the Potential*, Dover, 1958.

Marcuse, D., *Light Transmission Optics*, 2nd edn., Van Nostrand, 1982.

Marcuvitz, N., *Waveguide Handbook*, McGraw-Hill Book Company Inc., 1951.

Marion, J. B. and M. A. Heald, *Classical Electromagnetic Radiation*, 2nd edn, Academic Press, 1980.

Marin, L., 'Natural-mode representation of transient scattered fields', *IEEE Trans. Antennas and Propagat.* Vol. AP-21, No. 6, 809–818, Nov. 1973.

Marks, R. B., 'Application of the singular function expansion to an integral equation for scattering', *IEEE Trans. Antennas and Propagat.*, Vol. Ap-34, 725–728, May 1986.

Marks, R. B., 'The singular function expansion in time-dependent scattering', *IEEE Trans. Antennas Propagat.*, Vol. Ap-37, 1559–1565, Dec. 1989.

Mautz, J. R. and R.F. Harrington, 'Radiation and scattering from bodies of revolution', *Appl. Sce. Res.*, Vol. 20, 405–435, June 1969.

Mautz, J. R. and R. F. Harrington, 'H-field, E-field, and combined-field solutions for conducting bodies of revolution', *Arch. Elektron, Übertragungstech., Electron. Commun.*, Vol. 32, 19–164, 1978.

Maxwell, J. C., *A Treatise on Electricity and Magnetism*, 3rd edn., Vol. 1, Dover Publications, Inc., 1954; first published in 1891.

McGlinn, W. D., *Introduction to Relativity*, The Johns Hopkins University Press, 2003.

McIntosh, R. E. and J. E. Sarna, 'Bounds on the optimum performance of planar antennas for pulse radiation', *IEEE Trans. Antennas and Propagat.* Vol. AP-30, 381–389, July 1983.

McLean J. S., 'A re-examination of the fundamental limits on the radiation Q of electrically small antennas', *IEEE Trans. Antennas and Propagat.* Vol. AP-44, 672–676, 1996.

Mikhlin, S. G., *Linear Integral Equations*, Hindustan, 1960.

Mikhlin, S. G., *Variational Methods in Mathematical Physics*, Pergamon Press, 1964.

Miller, K. S., *Complex Stochastic Processes*, Addison-Wesley Publishing Company, 1974.

Miller, R. F., 'On the completeness of sets of solutions to the Helmholtz equation', *IMA J. Appl. Math.*, Vol. 30, 27–103, 1983.

Mittra, R., *Computer Techniques for Electromagnetics*, Pergamon Press, 1973.

Mittra, R. and W. W. Lee, *Analytical Techniques in the Theory of Guided Waves*, Macmillan, 1971.

Møller, C., *The Theory of Relativity*, Oxford University Press, 1952.

Monteath, G. D., *Applications of the Electromagnetic Reciprocity Principle*, Pergamon Press, 1973.

Montgomery, C. G., R. H. Dicke and E. M. Purcell, *Principles of Microwave Circuits*, McGraw-Hill, 1948.

Moon, P. and D. E. Spencer, *Field Theory Handbook*, Springer, 1988.

Morita, N., 'Surface integral representations for electromagnetic scattering from dielectric cylinders', *IEEE Trans. Antennas and Propagat.* Vol. AP-26, 261–266, No. 2, Mar. 1978.

Morita, N., 'Another method of extending the boundary condition for the problem of scattering by dielectric cylinders', *IEEE Trans. Antennas and Propagat.* Vol. AP-27, 97–99, No. 1, Jan. 1979a.

Morita, N., 'Resonant solutions involved in the integral equation approach to scattering from conducting and dielectric cylinders', *IEEE Trans. Antennas and Propagat.* Vol. AP-27, 869–871, No. 6, Nov. 1979b.

Morita, N., N. Kumagai and J. R. Mautz, *Integral Equation Methods for Electromagnetics*, Artech House, 1990.

Morse, P. M. and H. Feshbach, *Methods of Theoretical Physics*, McGraw-Hill, 1953.

Moses, H. E., 'Solutions of Maxwell equations in terms of a spinor notation: the direct and inverse problem', *Phys. Rev.*, Vol. 113, No. 6, 1670–1679, 1959.

Moses, H. E. and R. T. Prosser, 'Initial conditions sources, and currents for prescribed time-dependent acoustic and electromagnetic fields in three dimensions, Part 1: The inverse initial value problem, acoustic and electromagnetic bullets, expanding waves, and imploding waves', *IEEE Trans. Antennas and Propagat.* Vol. AP-34, 188–196, Feb. 1986.

Müller, C., 'Electromagnetic radiation patterns and sources', *IRE Trans. Antennas and Propagat.*, Vol. 4, 224–232, July 1956.

Müller, C., *Foundations of the Mathematical Theory of Electromagnetic Waves*, Springer, 1969.

Namiki, M. and K. Horiuchi, 'On the transient phenomena in the wave guide', *J. Phys. Soc.* Japan, Vol. 7, 190–193, 1952.

Nielsen M. A. and I. L. Chuang, *Quantum Computation and Quantum Information*, Cambridge University Press, 2000.

Oliner, A., 'Historical Perspectives on microwave field theory', *IEEE Trans. Microwave Theory and Tech.*, Vol. MTT-32, 1022–1045, Sept. 1984.

Packard, K. S., 'The origin of waveguides: a case of multiple rediscovery', *IEEE Trans. Microwave Theory and Tech.*, Vol. MTT-32, 961–969, Sept. 1984.

Panofsky, W. K. H. and M. Phillips, *Classical Electricity and Magnetism*, 2nd edn, Addison-Wesley, 1962.

Page, L., 'A derivation of the fundamental relations of electrodynamics from those of electrostatics', *Am. J. Sci.*, Vol. 44, 57–68, 1912.

Papas, C. H., *Theory of Electromagnetic Wave Propagation*, McGraw-Hill, 1965.

Pelzer, H., 'Energy density of monochromatic radiation in a dispersive medium', *Proc. Roy. Soc.* (London) A, Vol. 208, 365–366, Sept. 1951.

Peterson, A. F., 'The interior resonance problem associated with surface integral equations of electromagnetics: numerical consequences and a survey of remedies', *Electromagnetics*, Vol. 10, 293–312, July–Sept. 1990.

Peterson, A. F., S. L. Ray and R. Mittra, *Computational Methods for Electromagnetics*, Oxford University Press, 1998.

Pocock, M., M. J. Bluck and S. P. Walker, 'Electromagnetic scattering from 3-D curved dielectric bodies using time-domain integral equations', *IEEE Trans. Antennas and Propagat.* Vol. AP-46, No. 8, 1212–1219, Aug. 1998.

Poggio, A. J. and E. K. Miller, 'Integral equation solution of three dimensional scattering problems', in *Computer Techniques for Electromagnetics*, Pergamon Press, 1973.

Polyanin, A. P. and A. V. Manzhirov, *Handbook of Integral Equations*, CRC Press, 1988.

Popvić, B. D., 'Electromagnetic field theorems', *IEE Proc.* Vol. 128, Pt. A, 47–63, Jan. 1981.

Power, E. A., *Introductory Quantum Electrodynamics*, Longmans, 1964.

Pozar, D. M., 'Antenna synthesis and optimization using weighted Inagaki modes', *IEEE Trans. Antennas Propagat.*, Vol. Ap-32, 159–165, Feb. 1984.

Pozar, D. M., *Microwave Engineering*, John Wiley & Sons, Ltd, 1998.

Pozar, D. M. *et al.*, 'The optimum feed voltage for a dipole antenna for pulse radiation', *IEEE Trans. Antennas and Propagat.* Vol. AP-31, 563–569, July 1983.

Pozar, D. M. *et al.*, 'The optimum trasient radiation from an arbitrary antenna', *IEEE Trans. Antennas and Propagat.* Vol. AP-32, 633–640, June 1984.

Ramm, A. G., 'Non-selfadjoint operators in diffraction and scattering', *Math. Methods in Applied Science*, Vol. 2, 327–346, 1980a.

Ramm, A. G., 'Theoretical and practical aspects of the singularity and eigenmode expansion methods', *IEEE Trans. Antennas and Propagat.*, Vol. Ap-28, 897–901, Nov. 1980b.

Ramm, A. G., 'Mathematical foundations of the singularity and eigenmode expansion methods (SEM and EEM)', *J. Math. Anal. Appl.*, Vol. 86, 562–591, 1982.

Ramo, S. and J.R. Whinnery, *Fields and Waves in Modern Radio*, John and Wiley & Sons, Ltd, 1953.

Rao, S. M., D. R. Wilton and A. W. Glisson, 'Electromagnetic scattering by surfaces of arbitrary shape', *IEEE Trans. Antennas and Propagat.* Vol. AP-30, 409–417, No. 3, May 1982.

Rayleigh, L., *The Theory of Sound*, 2nd edn., Vol. 1, Dover Publications, Inc., 1945; first published in 1894.

Read, F. H., *Electromagnetic Radiation*, John Wiley & Sons, Ltd, 1980.

Reed, M. and B. Simon, *Methods of Modern Mathematical Physics*, Vol. 1-4, Academic Press, 1980.

Reddy, J. N., *Applied Functional Analysis and Variational Methods in Engineering*, McGraw-Hill, 1986.

Reitz, J. R., F. J. Milford and R. W. Christy, *Foundations of Electromagnetic Theory*, Addison-Wesley, 1979.

Rhodes, D. R., 'On the stored energy of planar apertures', *IEEE Trans. Antennas and Propagat.* Vol. AP-14, 676–683, 1966.

Rhodes, D. R., 'Observable stored energies of electromagnetic systems', *Journal of Franklin Institute*, Vol. 302, No. 3, 225–237, 1976.

Rhodes, D. R., 'A reactance theorem', *Proceedings of the Royal Society of London, Series A (Mathematical and Physical Science)*, Vol. 353, No. 1672, 1–10, 1977.

Richmond, J. H., 'A reaction theorem and its application to antenna impedance calculations', *IRE Trans. Antennas Propagat.*, Vol. Ap-9, 515–520, Nov. 1961.

Richtmyer R. D., 'Dielectric resonators', *J. Appl. Phys.*, Vol. 10, 391–398, 1939.

Rubinsten, I. and Rubinsten L., *Partial Differential Equations in Classical Mathematical Physics*, Cambridge University Press, 1998.

Rumsey, V. H., 'Reaction concept in electromagnetic theory', *Phys. Rev.*, Vol. 17, 952–956, 1954.

Rumsey, V. H., 'Some new forms of Huygens' principle', *IRE Trans. Antennas and Propagat.* Vol. AP-7, 103–116, Dec. 1959.

Rumsey, V. H., 'A new way of solving Maxwell equations', *IRE Trans. Antennas Propagat.*, Vol. Ap-9, 461–463, Sept. 1961.

Rumsey, V. H., 'A short way of solving advanced problems in electromagnetic fields and other linear systems', *IEEE Trans. Antennas and Propagat.*, Vol. Ap-11, 73–86, Jan. 1963.

Samaddar, S. N. and E. L. Mokole, 'Biconical antennas with unequal cone angles', *IEEE Trans. Antennas and Propagat.* Vol. AP-46, 181–193, 1998.

Schantz, H. G., 'Electromagnetic energy around Hertzian dipoles', *IEEE Antennas and Propagation Magazine*, Vol. 43, No. 2, 50–62, Apr. 2001.

Schechter, M., *Operator Methods in Quantum Mechanics*, Elsevier North Holland, 1981.

Schelkunoff, S.A., *Antennas: Theory and Practice*, John Wiley & Sons, Ltd, 1952.

Schiff, L., *Quantum Mechanics*, 3rd edn., McGraw-Hill, 1967.

Schulz-DuBois E.O., 'Sommerfeld pre- and postcursors in the context of waveguide transients', *IEEE Trans. Microwave Theory and Tech.*, Vol. MTT-18, 455–460, Aug. 1970.

Schutz, B. F., *A First Course in General Relativity*, Cambridge University Press, 1985.

Schwinger, J., L. L. DeRaad, Jr., K. A. Milton, W. Tsai, *Classical Electrodynamics*, Perseus Books, 1998.

Schwinger, J. and D.S. Saxon, *Discontinuities in Waveguides*, Gordon and Breach, 1968.

Sherman, J. W., 'Properties of focused apertures in the Fresnel region', *IRE Trans. Antennas Propagat.*, Vol. Ap-10, 399–408, July 1962.

Shlivinski, A. *et al.*, 'Antenna characterization in time domain', *IEEE Trans. Antennas and Propagat.* Vol. AP-45, 1140–1149, July 1997.

Shlivinski, A. and E. Heyman, 'Time domain near field analysis of short pulse antennas-Part 1 and Part 2', *IEEE Trans. Antennas and Propagat.* Vol. AP-47, 271–286, Feb. 1999.

Silver, S., *Microwave Antenna Theory and Design*, Dover Publications, 1949.

Slater, J. C., *Microwave Electronics*, Van Nostrand, 1950.

Slepian D. and H. O. Pollak, 'Prolate spherical wave functions, Fourier analysis and uncertainty-I, II', *Bell Sys. Tech. J.*, *Vol.* 40, 43–84, Jan. 1961.

Smith, G. S. and T. W. Hertel, 'On the transient radiation of energy from simple current distributions and linear antennas', *IEEE Antennas and Propagation Magazine*, Vol. 43, No. 3, 49–63, Jun. 2001.

Sobol, H., 'Microwave communications-A historical perspective', *IEEE Trans. Microwave Theory and Tech.*, Vol. MTT-32, 1170–1181, Sept. 1984.

Soejima, T., 'Fresnel gain of aperture aerials', *Proc. IEE (London)*, Vol. 110, 1021–1027, 1963.

Sommerfeld, A., *Electrodynamics*, Academic Press, 1949.

Spohn, H., *Dynamics of Charged Particles and Their Radiation Field*, Cambridge University Press, 2004.

Sten, J. C.-E. *et al.*, 'Quality factor of an electrically small antenna radiating close to a conducting plane', *IEEE Trans. Antennas and Propagat.* Vol. AP-49, 829–837, May 2001.

Stevenson, A. F., 'Solution of electromagnetic scattering problems as power series in the ratio (dimension of scatter/wavelength)', *J. Appl. Phys.*, Vol. 24, 1134–1142, 1953.

Stinson D. C., *Intermediate Mathematics of Electromagnetics*, Prentice-Hall, 1976.

Storer, J. E., 'Impedance of thin-wire loop antennas', *Trans. AIEEE (Communication and Electronics)*, Vol. 75, 606–619, 1956.

Stratton, J. A., *Electromagnetic Theory*, McGraw-Hill, 1941.

Stutzman, W. L. and G. A. Thiele, *Antenna Theory and Design*, John Wiley & Sons, Ltd, 1981.

Synge, J. L., *Relativity: The Special Theory*, North-Holland, 1965.

Tai, C-T., 'On the transposed radiating systems in an anisotropic medium', *IRE Trans. Antennas and Propagat.* Vol. AP-9, 502–503, Sept. 1961.

Takeshita S., 'Power transfer efficiency between focused circular antennas with Gaussian illumination in Fresnel region', *IEEE Trans. Antennas and Propagat.*, Vol. Ap-16, 305–309, May 1968.

Tonning, A., 'Energy density in continuous electromagnetic media', *IEEE Trans. Antennas and Propagat.* Vol. AP-8, 428–434, July 1960.

Tsang, L., J. A. Kong and K. Ding, *Scattering of Electromagnetic Waves*, John Wiley & Sons, Ltd, 2000.

Umashankar, K. and A. Taflove, *Computational Electromagnetics*, Artech House, 1993.

Vekua, I. N., 'About the completeness of the system of metaharmonic functions', *Dokl. Akad. Nauk.*, USSR 90, 715–718, 1953.

Vendelin, G. D., A. M. Pavio, U. L. Rhode, *Microwave Circuit Design Using Linear and Nonlinear Techniques*, 2nd edn, Wiley-Interscience, 2005.

Vichnevetsky, R., 'An analogy between dispersive wave propagation and special relativity' Dept. of Computer Science, Rutgers University, 1988.

Vilenkin N. Ya. *et al.*, *Functional Analysis*, Wolters-Noordhoff Publishing, 1972.

Volakis, J. L., *Antenna Engineering Handbook*, 4th edn, McGraw-Hill, 2007.

Wait, J. R., *Wave Propagation Theory*, Pergamon Press, 1981.

Walls, D. F. and G. J. Milburn, *Quantum Optics*, Springer-Verlag, 1995.

Wasylkiwskyj, W. and W. K. Kahn, 'Theory of mutual coupling among minimum-scattering antennas', *IEEE Trans. Antennas and Propagat.*, Vol. Ap-18, 204–216, Mar. 1970.

Waterman, P. C., 'Matrix formulation of electromagnetic scattering', *Proc. IEEE*, Vol. 53, 806–812, Aug. 1965.

Waterman, P. C., 'Symmetry, unitarity, and geometry in electromagnetic scattering', *Phys. Rev.*, Vol. D3, 825–839, 1971.

Weinberg, S., *Gravitation and Cosmology*, John Wiley & Sons, Ltd, 1972.

Weinberger, H. F., *Variational Methods for Eigenvalue Approximation*, Society for Industrial and Applied Mathematics, 1974.

Wheeler, H.A., 'Small antennas', *IEEE Trans. Antennas and Propagat.* Vol. AP-23, 462–469, No. 4, July 1975.

Whitham, G. B., *Linear and Nonlinear Waves*, John Wiley & Sons, Ltd, 1974.

Whittaker, E. T., *A History of the Theories of Aether and Electricity*, T. Nelson, 1951.

Wilcox, C. H., 'A generalization of theorems of Rellich and Atkinson', *Proc. Amer. Math. Soc.*, Vol. 7, 271–276, 1956a.

Wilcox, C. H., 'An expansion theorem for electromagnetic fields', *Comm. Pure Appl. Math.*, Vol. 9, 115–134, 1956b.

Wilcox, C. H., 'Debye potentials', *J. Math. Mech.*, Vol. 6, 167–202, 1957.

Wilton, D. R., S.M. Rao, A. W. Glisson, and D. H. Schaubert, 'Potential integrals for uniform and linear source distributions on polygonal polyhedral domains', *IEEE Trans. Antennas and Propagat.* Vol. AP-32, 276–281, No. 3, Mar. 1984.

Wolfson, R. and J. M. Pasachoff, *Physics*, HarperCollins Publishers, 1989.

Wong, L. J., A. Kuzmich and A. Dogariu, 'Gain-assisted superluminal light propagation', *Nature*, Vol. 406, 277, 2000.

Wu, T. T. and R. W. P. King, 'Transient response of linear antennas driven from a coaxial line', *IEEE Trans. Antenna and Propagat.* Vol. Ap-11, 17–23, Jan. 1963.

Yariv, A., *An Introduction to Theory and Applications of Quantum Mechanics*, John Wiley & Sons, Ltd, 1982.

Yee, K. S., Numerical solution of initial boundary value problems involving Maxwell's equations, *IEEE Trans. Antenna and Propagat.* Vol. Ap-14, 302–307, May 1966.

Yosida, K., *Functional Analysis*, 5th edn, Springer-Verlag, 1988.

Zeidler, E., *Applied Functional Analysis-Applications to Mathematical Physics*, Springer-Verlag, 1995.

Index